Probability and Probabilistic Reasoning for Electrical Engineering

Terrence L. Fine

School of Electrical & Computer Engineering
Cornell University

PEARSON
Prentice
Hall

Upper Saddle River, New Jersey 07458

Library of Congress Cataloging-in-Publication Data on file.

Vice President and Editorial Director, ECS: *Marcia J. Horton*
Associate Editor: *Alice Dworkin*
Executive Managing Editor: *Vince O'Brien*
Managing Editor: *David A. George*
Production Editor: *Daniel Sandin*
Director of Creative Services: *Paul Belfanti*
Creative Director: *Jayne Conte*
Cover Designer: *Bruce Kenselaar*
Art Editor: *Greg Dulles*
Manufacturing Buyer: *Lisa McDowell*
Senior Marketing Manager: *Holly Stark*

© 2006 by Pearson Education, Inc.
Pearson Prentice Hall
Pearson Education, Inc.
Upper Saddle River, New Jersey 07458

Pearson Prentice Hall™ is a trademark of Pearson Education, Inc.

Printed in the United States of America

10 9 8 7 6 5 4 3 2 1

ISBN 0-13-020591-5

Pearson Education Ltd., *London*
Pearson Education Australia Pty. Ltd., *Sydney*
Pearson Education Singapore, Pte. Ltd.
Pearson Education North Asia Ltd., *Hong Kong*
Pearson Education Canada Inc., *Toronto*
Pearson Educación de Mexico, S.A. de C.V.
Pearson Education—Japan, *Tokyo*
Pearson Education Malaysia, Pte. Ltd.
Pearson Education, Inc., *Upper Saddle River, New Jersey*

To my wife Dorian, without whom this would not have been finished;

To my children David and Jennifer;

To the many students in Cornell's EE/ECE310 whom I have had the privilege of teaching probabilistic reasoning;

and to the teaching assistants who were unfailingly well-informed, responsible, and caring of the needs of the students.

Contents

Part III   Conditional Probability and Its Applications   319

Chapter 11   Discrete Conditional Probability   321

Chapter 12   Mixed Conditional Probability   343

Chapter 13 MAP, MLE, and Neyman–Pearson Rules   **374**

Part IV   Characteristic Functions and Probability Bounds   457

Chapter 17 Characteristic Functions   459

Chapter 18 Probability Bounds and Sums   486

List of Notation

This listing is of roman and greek alphabetic characters and specially constructed mathematical symbols whose use in this text is of more than temporary value. The format for this listing is that each line starts with a notation symbol separated by a colon from its name, which is in turn separated by a colon from the section number in which this symbol is first used substantively. The arrangement of symbols is by first appearance in the text. Common mathematical symbols, as introduced through basic engineering calculus, are omitted.

$\mathbf{x}$:   vector, a column vector by default: 0.8

$\mathbf{x}^T$:   transpose of the vector: 0.8

$\cdot$:   scalar or dot product, $\mathbf{x} \cdot \mathbf{y}$: 0.8

$\mathbb{A}, \mathbb{B}$:   matrices: 0.8

$\mathcal{E}$:   random experiment: 1.4

Ω: sample space: 1.4

$\mathcal{S}$:   alternative sample space notation, also later as a system: 1.4, 10.2.1, 16.2

A, B, C: sets: 1.5, 1.9.1

I_A: indicator function of the event A: 1.5

$\mathcal{A}$:   event algebra or σ-algebra: 1.6, 3.5.1

$||A||$:   cardinality (size) of the set A: 1.6

2^{Ω}:   power set of Ω: 1.6

Π: partition: 1.6

$\hat{x}$:   sample mean of x: 1.7

$\hat{\sigma}$:   sample standard deviation: 1.7

$\in$:   element of: 1.9.1

$\notin$:   not element of: 1.9.1

N: the nonnegative integers: 1.9.1

N_n: the nonnegative integers no larger than n: 1.9.1

Z: all the integers, including the negative ones: 1.9.1

Z^+:   the positive, or natural, integers: 1.9.1

$\mathbb{R}$:   the real numbers: 1.9.1

$\emptyset$:   empty or null set: 1.9.1

$\subset$:   subset: 1.9.2

$\supset$:   superset: 1.9.2

$\iff$:   if and only if: 1.9.2

$\Rightarrow$:   implies: 1.9.2

A^c: complement of the set A: 1.9.2

A':   alternative notation for complement: 1.9.2

$\cup$:   cup or set union: 1.9.2

$\cap$:   cap or set intersection: 1.9.2

$-$:   set difference, $A - B$: 1.9.2

$\perp$: $A \perp B$, sets A and B are disjoint: 1.9.2

$\exists$:   universal quantifier for "there exists": 1.9.2

$\forall$:   universal quantifier "for all": 1.9.2

$\mathcal{I}$:   set that indexes a family or collection of sets: 1.9.2

(a, b):   ordered pair of elements a and b: 1.9.3

$A \times B$:   cartesian product of the sets A and B: 1.9.3

$\mathcal{D}$:   generic domain of a function f: 1.10

$\mathcal{R}$:   generic range of a function f: 1.10

$P(A)$:   probability of the event or set A: 2.2, 3.5.2

$\mu_{n,r}$:   number of ordered sequences of length r from alphabet size n: 2.3.1

$(n)_r$:   number of ordered sequences without replacement of length r: 2.3.3

$n!$: n factorial: 2.3.5

$\Gamma(x)$:   gamma function of x: 2.3.5

$\binom{n}{r}$:   binomial coefficient: 2.4

$H(p)$:   binary entropy function: 2.5

V: set of vertices or nodes of a graph: 2.6.2

E: set of edges or links of a graph: 2.6.2

G: graph (undirected, directed): 2.6.2, 2.6.4

G_n: graph with n labeled nodes: 2.6.2

Γ_n: number of undirected graphs on n labeled nodes: 2.6.2

$\Gamma_{n,m}$:   number of undirected graphs on n labeled nodes and m edges: 2.6.2

$\deg(v)$: degree of a node v: 2.6.3

$\overline{d}_{G_{n,m}}$: average node degree: 2.6.3

δ_G: diameter of a connected graph: 2.6.4

$\text{outdeg}(v)$: outdegree of node v in a directed graph: 2.6.5

$\text{indeg}(v)$: indegree of node v in a directed graph: 2.6.5

MB: Maxwell–Boltzmann statistics: 2.7.1

BE: Bose–Einstein statistics: 2.7.2

FD: Fermi–Dirac statistics: 2.7.3

$\binom{n}{n_1 \ n_2 \ ... \ n_r}$: multinomial coefficient: 2.8

$H(\mathbf{p})$: r-ary entropy function: 2.8

$P(A|B)$: conditional probability of A given B: 2.9.1, 11.2.1

$\perp\!\!\!\perp$: A and B are independent: 2.10, 14.2.1

$Av_n X$: time average of n outcomes of X: 3.4

$r_n(A)$: relative frequency of outcomes of event A: 3.4

P: probability measure: 3.5.2

p: probability mass function or pmf: 4.2

$\mathcal{R}_n$: random pmf: 4.4.1

$\mathcal{B}(n, p)$: binomial pmf: 4.4.2

$\mathcal{G}(\beta)$: geometric pmf: 4.4.3

$\mathcal{P}(\lambda)$: Poisson pmf: 4.4.4

$\mathcal{Z}(\alpha)$: Zeta or Zipf pmf: 4.4.5

$\mathbb{R}^n$: n-dimensional reals: 5.1

$\mathcal{B}, \mathcal{B}_X$: Borel σ-algebras: 5.1

$C_x, C_{\mathbf{x}}$: semi-infinite interval or orthant with vertex $\mathbf{x}$: 5.2, 7.2.1

F, F_X: univariate cumulative distribution function or cdf: 5.2

$U(x - x_0)$: unit step function: 5.2

P_X: probability measure on $\mathcal{B}_X$ determined by F_X: 5.2

$F(x^+)$: limit of F approaching x from the right: 5.3

$F(x^-)$: limit of F approaching x from the left: 5.3

F_d: purely discrete cdf: 5.4

F_{ac}: absolutely continuous cdf: 5.4

F_s: singular cdf: 5.4

$\hat{F}_n(x)$: empirical cdf: 5.5

f, f_X: probability density function or pdf: 6.2

$\delta(x)$:   unit impulse or delta function: 6.3

$\mathcal{U}(a, b)$:   uniform pdf: 6.5.1

$\beta(a, b)$:   Beta pdf: 6.5.2

$\mathcal{E}(\lambda)$:   exponential pdf: 6.6.1

$\Gamma(\alpha, \lambda)$:   gamma pdf: 6.6.2

$Par(\alpha, \tau)$:   Pareto pdf: 6.6.3

$\mathcal{L}(\alpha)$:   Laplacian pdf: 6.7.1

$\mathcal{N}(m, \sigma^2)$:   normal or Gaussian univariate pdf: 6.7.2

$F_{X_1,...,X_n}, F_{\mathbf{X}}$:   multivariate or joint cdf: 7.2.1

$f(\mathbf{x}), f_{\mathbf{X}}(\mathbf{x})$:   multivariate or joint pdf: 7.3

$\mathcal{D}(\mathbf{a})$:   Dirichlet multivariate pdf: 7.4.3

$\mathcal{N}(\mathbf{m}, \mathbb{C})$:   multivariate normal or Gaussian pdf: 7.4.4

$X, Y, Z, \mathbf{X}, \mathbf{Y}, \mathbf{Z}$:   random variables and random vectors: 8.2

$X^{-1}(B)$:   inverse image under X of the set B: 8.2.2

G^{-1}:   inverse of a cdf: 8.4.2

$\mathbb{J}$:   Jacobian matrix: 8.5

EX: expected value of the random variable X: 9.3

E: expectation operator: 9.5

$D(p||q)$:   Kullback–Leibler divergence: 9.9

$\mathcal{C}(a)$:   Cauchy pdf: 9.10.1

$\text{VAR}(X)$:   variance of X: 9.10.2

$\text{COV}(X, Y)$:   covariance of X and Y: 9.11

MSE: mean square error: 9.12.1

ρ_{XY}:   correlation coefficient of X and Y: 9.12.1

SNR: signal-to-noise ratio: 9.12.2

$\mathbb{R}_X$:   correlation matrix for the vector random variable $\mathbf{X}$: 9.14

$\mathbb{C}_X$:   covariance matrix for the vector random variable $\mathbf{X}$: 9.14

$\mathbb{R}_{X,Y}$:   cross-correlation matrix for the vector random variables $\mathbf{X}$ and $\mathbf{Y}$: 9.14

$\text{COV}(\mathbf{X}, \mathbf{Y})$:   cross-covariance matrix for the vector random variables $\mathbf{X}$ and $\mathbf{Y}$: 9.14

$\mathcal{S}$:   system, linear or otherwise, with inputs typically X_t and outputs Y_t: 10.2.1

$h(t, \tau)$:   impulse response or Green's function for a linear system: 10.2.2

Φ: state transition matrix for a linear system: 10.2.3

$R_X(t, s)$:   autocorrelation function of X_t: 10.3

$C_X(t, s)$:   autocovariance function of X_t: 10.3

$\mathbb{K}_k$:   Kalman filter gain matrix: 10.6.2

$\mathbb{P}_k$:   Kalman filter error covariance matrix: 10.6.3

$p_{Y|X}(y|x)$:   conditional pmf of discrete Y given X: 11.2.3

$F_{Y|X}(y|x)$:   conditional cdf: 11.2.3, 12.2

$f_{Y|X}(y|x)$:   conditional pdf: 12.3

x_m^n: the sequence $x_m, \ldots, x_n$: 12.4

$f_{X_{m+1}^n | X_1^m}$:   multivariate conditional pdf: 12.4

$P(X_{m+1}^n \in$ conditional 12.4

$A | X_1^m = x_1^m)$:   probability that X_{m+1}^n is in A given that X_1^m is x_1^m:

H_0, H_1:   states of the 13.2, world: 13.4.1

$\Lambda(x)$:   likelihood ratio $f_1(x)/f_0(x)$ when there are two pdfs f_1 and f_0: 13.3

$\phi(x)$:   decision rule when 13.3, observe x: 13.4.1

$P_D = \rho(P_{FA})$:   receiver operating characteristic (ROC): 13.4.2

ROC: receiver operating characteristic: 13.4.2

$\perp\!\!\!\perp \{A_1, \ldots, A_n\}$, mutually 14.3.1

$\perp\!\!\!\perp_{i=1}^n A_i$:   independent events:

$E(Y|\mathbf{X} = \mathbf{x})$:   conditional expectation of Y given $\mathbf{X} = \mathbf{x}$: 15.2

$Y^*(\mathbf{x}), \mathbf{Y}^*(\mathbf{x})$:   alternative notation for conditional expectation: 15.4.1

$\hat{Y}(\mathbf{x}), \hat{\mathbf{Y}}(\mathbf{x})$:   estimators of Y or $\mathbf{Y}$ given information $\mathbf{X} = \mathbf{x}$: 15.4.1, 15.4.2

$\mathcal{A} = \{a\}$:   also used as the set of actions in a decision problem: 16.2

$L(\theta, a)$:   loss function for pair of state of nature θ and action a: 16.2

$\mathcal{X}$:   space of observations: 16.2

$\mathcal{D} = \{d\}$:   set of possible decision functions: 16.2

$\pi(\theta)$:   pdf describing what is known about state θ prior to observation: 16.2

$\pi(\theta|x)$:   conditional pdf for state θ given observation x: 16.2

r^*:   Bayes risk: 16.3

d^*:   Bayes rule: 16.3

$\succ_\mathcal{D}, \succeq_\mathcal{D}, \approx_\mathcal{D}$:   preference relations between decision rules: 16.7.2

$\mathcal{C} = \{c\}$:   set of consequences: 16.7.3

$\mathcal{C}^* = \{c^*\}$:   set of pmfs over consequences: 16.7.3

$C(\theta, a)$:   consequence assignment function: 16.7.3

$p_d(c)$: probability that decision rule d will yield consequence c: 16.7.3

$\succeq_{c^*}$: preferences for gambles over consequences: 16.7.3

$v(c)$: value or utility function on consequences: 16.7.4

$\phi_X(u), \phi_{\mathbf{X}}(\mathbf{u})$: characteristic function of $u, \mathbf{u}$ for $X, \mathbf{X}$: 17.2

$G_X(s)$: generating function of s for X: 17.11

$T_X(s)$: tail probabilities generating function of s for X: 17.11

$\xrightarrow{ms}$: sequence $\{X_n\}$ converges in mean square to X: 19.3.1

$\rightarrow a.s.$: sequence $\{X_n\}$ converges almost surely (with probability 1) to X: 19.5.1

$\xrightarrow{r}$: sequence $\{X_n\}$ converges in rth mean to X: 19.5.1

$\xrightarrow{P}$: sequence $\{X_n\}$ converges in probability to X: 19.5.1

$\xrightarrow{D}$: sequence $\{X_n\}$ converges in distribution to X: 19.5.1

$\{X_t : t \in T\}$: random process of random variables X_t and t taking values in T: 20.2

T_τ: time shift operator, $T_\tau X_t = X_{t+\tau}$: 20.5.1

$\mathcal{I}$: invariant σ-algebra (not index set usage): 20.5.2

$\mathbb{P}$: one-step transition matrix for homogeneous Markov chain: 20.6.1

$\pi^{(t)}$: state probability distribution for Markov chain at time t: 20.6.2

$f_{i,j}^{(n)}$: first passage probability from state ξ_i to ξ_j in n steps for M.C.: 20.6.4

$w.s.s.$: wide-sense stationary: 20.10.1

$S_X(\omega)$: power spectral density function of radian frequency ω for $w.s.s.$ process X_t: 20.10.2

GRP: Gaussian random process: 20.11.1

Part I

Introduction to Random Phenomena and Probability

0

README: Learning and Teaching Probabilistic Reasoning

0.1 THE BIG PICTURE

Modern engineering, and particularly electrical engineering, deals with the design and construction of devices, varying in size from the nanoscale to that of buildings (e.g., a transistor in a computer chip to an electrical generator). The behavior of these devices exhibits significant random or unpredictable elements (e.g., self-generated noise, reliability). Electrical engineering and computer science are responsible for the design of often very large scale systems composed of such devices (e.g., a computer chip, a public switched telephone network, the Internet, a bulk electric power distribution network). Such systems then act upon, and act within, complex environments that themselves have significant elements of unpredictability. The engineering and physical science approach to mastering these unavoidable elements of randomness, chance, and uncertainty is to develop a mathematical theory that can describe them and then to use the understanding gained from this mathematical theory as the basis for optimal system design methodologies. The mathematical theory of random and uncertain phenomena is probability. Probabilistic reasoning is a useful term to indicate the combination of mathematical probability together with its associated methods for selecting appropriate probability models for given devices, systems, and environments and methods for defining our engineering objectives and characterizing optimal performance. The probability theory of random events, random variables, and random processes then manufactures our armory of possibly appropriate mathematical models. Statistics and statistical theories of decision making and estimation provide us with the means to select from our armory mathematical models appropriate for specific devices, systems, and environments. Statistical theories of decision making further enable us to formulate criteria of optimal performance and to design optimal systems in accordance with these criteria.

0.2 PURPOSE, BACKGROUND, AND ORGANIZATION

This text provides a comprehensive introduction to the mathematical theory of probability, its application to the modeling of random phenomena of the kinds encountered particularly in electrical and computer engineering, and its uses in making optimal decisions and inferences. It assumes the mathematical background commonly held by engineering students. The text focuses on those applications of probability to phenomena of random noise and system response that are typically encountered in electrical engineering. However, as no background is required in electrical engineering, readers with backgrounds and interests in other areas can also learn probabilistic reasoning from this text. An emphasis is placed on a motivated and orderly introduction of the many elements of probability that aims at a thorough discussion of concepts without assuming a more advanced mathematical background than that indicated.

Several questions face a reader of this text who is a first-time student of probability:

- Why is probability worthy of study?
- What are the considerations behind the organization of material, given that there is not just one way to develop the subject?
- What are some of the distinctive issues in learning probabilistic reasoning?
- What is included in, and omitted from, this text that is intended mainly, but not exclusively, for electrical engineering students?
- What are the prerequisites to learning from this text?

This chapter provides brief narrative responses to these questions. The ensuing chapters offer the complete response.

0.3 WHY STUDY PROBABILITY?

What will knowledge of probability enable you to understand and to do? Probability, and more generally probabilistic reasoning, sharpens our judgement concerning uncertainty and the likelihood of random events encountered in everyday life where we do not engage in explicit mathematical analyses. Probability enables us to make and productively use mathematical models of random phenomena, including the sorts of possibly random systems and random environments encountered in engineering and the physical sciences, and to systematically model, analyze, and design devices, processes, and systems whose complexity exceeds our intuitive grasp. With respect to design, it enables us to understand how to judge performance and quality in an uncertain world, how to formulate criteria of optimal performance for systems operating in a regime of random-ness, and how to design optimal systems to achieve desired engineering ends. Summarizing for future reference, motivation for the study of probabilistic reasoning is provided by the following three goals:

G1. Introduce and explore **mathematical models of random phenomena** (e.g., random devices, systems, signals) of importance in electrical engineering and allied disciplines.

G2. Begin to understand the **response of (electrical) systems to random excitations/signals** and to random changes in parameter values describing these systems.

G3. Learn the elements of making **inferences, estimates, and decisions** about signals and device/system parameters in the presence of chance and uncertainty.

Progress towards the first two goals enables us to formulate criteria of optimality when outcomes of actions cannot be known in advance and to make good decisions and inferences and reach well-grounded, properly guarded conclusions in the presence of chance and uncertainty. Processes for making such decisions and inferences become the basis of optimal systems.

We start with a few remarks on the prevalence of interesting and important random phenomena that will be amplified throughout this text. The need for mathematical models of such phenomena arises from our desire to understand them and to have the capacity to make rational plans and to construct rationally designed, possibly optimal systems to bend these random phenomena to our purposes. Encounters with uncertainty and chance and random phenomena have always been commonplace in everyday life (e.g., through our dependence on the weather, the occurrences and progressions of illnesses, our dependence on the actions of others). It is largely in the 19th and 20th centuries that there developed an appreciation of the unavoidable presence of these phenomena in the physical, social, and biological sciences. It is largely in the 20th century that the need to deal with these phenomena in a formal manner was recognized in engineering. More specifically, our interests are in the random phenomena that pervade electrical engineering, from the manufacturing processes (e.g., fluctuations in dopant implantation) used to create nanoscale devices and the (classical and quantum) noise generated by these devices, through rapidly evolving computing and worldwide communication systems that require resistance to random errors and designs robust with respect to a very wide range of randomly generated service demands, to measurements made to control and stabilize power systems spanning sizable portions of continents. We encounter random phenomena on both the smallest and the largest of spatial scales, in the behavior of both individual particles and the collective behavior of huge numbers of such particles, and over time scales ranging from nanoseconds to years. Probabilistic reasoning is the largely (but not completely!) mathematical discipline needed to understand and control such random phenomena, and it is now widely recognized as essential in the education of present-day (electrical) engineers. Indeed, an improved knowledge of probabilistic reasoning is of benefit to everyone.

We clarify our goals through the example of a basic system configuration that has an excitation or input random quantity (variable or signal) X that may be a transmitted message or other quantity whose value we need to know. X is then transformed by a possibly random (nondeterministic) system S that may be a communications channel into a response or output random variable $Y = S(X)$, known as the measurement, observation, or received message. Information theory postulates that there is no information transfer unless there is prior uncertainty at the receiving end about the X being transmitted. Not only do all channels distort the signals they transfer, either linearly through dispersion (e.g., in band-limited situations such as the use of twisted pair for telephony) or nonlinearly (e.g., through amplitude compression or through digital/analog conversion), but they also perturb the signal in a random fashion, typically by adding random noise (e.g., static) to the signal. Accordingly,

- G1 addresses the mathematical models essential for the descriptions of the components X, Y, and S.
- G2 enables us to compute the description of Y from those for X and S.
- G3 enables us to optimally infer from knowledge of Y and S to X, from received signal Y to transmitted signal X or from measured quantity Y to desired quantity X. The system S relates the two, either as a communications channel or a measurement system (e.g., radar, sonar, power system state estimator).

If the resulting optimal inference is insufficiently accurate for our purposes, then the models provided by G1 enable us to modify the communications system, say, through coding of X before transmission, so as to meet standards.

Finally, while not listed as an explicit goal, many of the mathematical techniques to be introduced are of independent interest and value for other engineering and scientific applications.

0.4 OUR ORGANIZATIONAL APPROACH AND TEACHING EXPERIENCE

Texts (and even their more liberated hypertext versions) are perforce written in a linear or sequential fashion with chapter following chapter and section following section. There is not a uniquely best way in which to organize this linear presentation that meets all the demands of a thorough understanding of probability. Choices must be made and compromises reached. The choices that we have made are not always the ones commonly found in other texts. However, our choices present material so that there is a sound and motivated connection to the material that preceded it. In particular, we have made an effort to motivate definitions of central concepts and, at times, to suggest alternatives, rather than just presenting a mathematical definition by fiat. Knowledge of alternatives not chosen is knowledge about what was chosen.

The mathematical essence of probability can be presented more compactly than we do. This essence is, of course, available by omitting some of the material contained here. However, it is important to realize that probabilistic reasoning poses a greater challenge to engineers than to mathematicians. Mathematicians are content to start with unambiguously stated assumptions and then rigorously derive their consequences and implications. This is an important process, but one that does not suffice for engineers. Engineers have to know the mathematical methods, not always in the detail of the mathematicians, but engineers also have to know of their applicability in the real world. Engineers cannot just start from mathematical "givens" that come from unknown sources. Engineers have to choose the "givens" and be prepared to defend their choices to others.

Our exposition addresses this often overlooked need, as well as the more widely understood one of providing the mathematical theory of probability.

You cannot easily understand a subject such as probability from a first exposure in a course in which you are hurried along to keep up with the often overwhelming linear flood of ideas, definitions, results, applications, problems, methods, and means of calculation—a flood that washes you onto the shoals of (weekly) problem sets and multiple examinations that are a necessary part of a course in probability. Understanding is gained from subsequent reflection and reconsideration of material already learned, when many connections are made between topics that were not adjacent in their presentation and when the linear ordering of lecture and text is replaced by a web of interconnected ideas. Such understanding grows from use of the material in subsequent courses or in engineering practice and from taking additional courses (e.g., ones in stochastic processes or time series models and in statistical methods) in which one sees probability used in a variety of contexts.

We feel strongly that carefully motivating the many elements of probability theory is essential if the student is to enjoy the benefits of an improved perspective on the broader issues of probabilistic reasoning. A sound perspective on probabilistic reasoning is of enduring value to all. Most students of probability who continue in the practice of engineering and the physical sciences will have a greater need for a conceptual understanding of probabilistic reasoning and for simple skills at computation than for proficiency with the mathematical and algorithmic details of probability theory. However, gaining a sound conceptual base does require an exposure to more than just this base; it needs examples of problem formulation and solution to ensure that you test your understanding in the precise context of a mathematical exercise. The many existing texts on probability (including both the most mathematically advanced ones and those with the most applied approach) rush past conceptual issues, such as motivation for definitions and implications of definitions and the dependencies among concepts, in order to enable you to prove theorems or solve probability exercises as quickly as possible. The result for many students is that they survive by learning in a stimulus–response fashion (a question of form A provokes an answer of form A') which produces learning of little value in the absence of the familiar stimuli that are not produced by the new contexts of a subsequent professional career—there is little carryover.

The student whose subsequent career will make significant technical use of probability will also require a sound conceptual introduction. However, this student will need subsequent courses that use probability techniques in order to gain the requisite proficiency in modeling and calculation. A few individuals will also need a treatment of probability from a mathematically more advanced standpoint than the one taken here. In sum, for all students of probability, a careful introduction to the conceptual structure of probabilistic reasoning, including the directions not taken, is of lasting benefit.

0.5 EXPOSITORY APPROACH

The formality of our treatment of applied probability may be uncongenial to you. It does have the advantage of a precise and organized development of theory and applications. However, what makes for an organized, careful presentation when viewed after you have learned the material is not always the easiest way to learn it in the first place. To help out, we supplement the mainstream text with less formal comments on what is to be, or has just been, developed and with many set-off examples and subsections that illustrate what has just been discussed.

We provide an undergraduate, calculus-level introduction to probability and its applications, especially to electrical engineering. Our expository pattern throughout this text is as follows:

M1. motivate a concept through qualitative discussion of its roles, applications, or intended uses;

M2. formalize the concept through a mathematical definition;

M3. draw out the elementary implications of the definition through lemmas and theorems;

M4. if appropriate, enrich the conceptual area by introducing associated definitions;

M5. illustrate the concept through worked-out examples and problems/exercises drawn primarily from electrical engineering.

Each of the 20 chapters begins with a section entitled, "Purpose, Background, and Organization" and ends with a section entitled "Summary" followed by appendices, if any, and "Exercises."

The material presented in these 20 chapters (see the table of contents and Section 0.9) is more than fits comfortably within a single semester. We emphasize the logical relationships between the concepts presented, feeling that too many treatments of probability follow the trails blazed by others without communicating the sense of direction possessed by the pioneers who first established these trails. A further consideration in our ordering of topics has been the desire to exhibit applications (goals G2, G3) of the concepts discussed as soon as possible. It is difficult to remain motivated when presented with a long series of abstract concepts and mathematical results whose bearing upon your primary interests seems remote. Frequent reference, even if brief, to important applications drawn from active areas of electrical engineering provides some of the motivation required to master applied probability.

There is no uniquely coherent way in which to develop the theory and application of probability in the linear fashion required by a text in which topics follow each other in serial fashion. We occasionally depart from serial presentation of concepts by providing (e.g., in Chapter 2) early, partial, and informal contact with these concepts before returning to careful study of them in later chapters. Thus, it may surprise those of you already knowledgeable about probability to find that we defer serious consideration of expectation to Chapter 9, of conditional probability to Chapters 11 and 12, and of stochastic independence to Chapter 14. We mitigate this by providing early first contact with the latter two issues in Sections 2.9 and 2.10, both in the simple context afforded by classical probability, and with expectation in Section 3.4. Curiously, there have been a few attempts to found all of probability upon conditional probability and other attempts to start from the notion of expectation or average outcome (e.g., the first published probability text written in 1657 by Huyghens [48]).

The subject of probabilistic reasoning, encompassing probability theory and the means by which it is applied to chance and uncertainty phenomena, is both an intellectually demanding one, requiring new abstract concepts and approaches to enable you to come to grips with the unfamiliar realm of nondeterministic phenomena, and a highly rewarding subject through the mastery gained over such prevalent phenomena. Expressing and operating with these new concepts requires facility with basic calculus, particularly with integration in one and two variables. Proofs and calculations establishing results are presented not so much to verify the results as to explain them and to provide you with the tools and methods needed to develop your own results, conclusions, and applications. Theorems are best understood in the light of their proofs, and the proof techniques used are valuable for establishing other results you will need. Similarly, the calculations provided illustrate techniques that will enable you to resolve the new problems you encounter.

This material is best learned when the text is accompanied by a course of lectures and frequent working of exercises. The many examples worked out in the text clarify the technical results that have just been presented and illustrate the solution methods for the exercises provided at the end of each chapter. While students working in groups has become something of a norm, reflective of the needs of industrial problem solving, and such collective work can be valuable, group work cannot substitute for your mastering exercises on your own. You learn by actively participating as well as by more passively listening and reading. The linear ordering of presentation in texts that was remarked on earlier also applies to lecture presentation of material, although in lectures there may be the opportunity to explore possible linkages through questions asked as they occur to you. Working with probability and taking additional courses in probability, even ones that are not substantially more advanced than one based on this text, leads you to think through the material and to find the cross-connections that are a large part of what we mean when we say we understand the material. An understanding based closely on the linear order of presentation usually suffices to do well in classes, but is insufficient in itself to provide a working comprehension.

0.6 WHY DEFINITIONS, THEOREMS, ETC.?

We make frequent (more than two hundred times!) use of the abstract mathematical format of proclaiming typographically set-off definitions, theorems, lemmas, and corollaries. We do so because these proclamations are precise, accurate, compact statements that are also easily referenced either through citations in the text or in the index. To make them easy to find, they are all given short titles indicative of their content. While these are the advantages of using these mathematical proclamations, there is also a significant disadvantage.

Like tersely written, efficient computer code without commentary, these mathematical proclamations are often hard to understand, and they rarely convey in themselves why they are of value to you. Once you have the idea of "what they are about," these compact accurate statements become of real value. Assistance in gaining this idea of "what they are about" is provided in comments that surround these proclamations and in the proofs that usually accompany the theorems, etc.

While you may have been taught (correctly) that proofs are provided to verify the correctness of theorems, lemmas, and corollaries, for our purposes proofs have the greater importance of teaching you both the meaning of these proclamations and how to establish new proclamations for yourself. The proof "unpacks" or expands on the meaning of the terse theorem statement by showing you how the theorem comes to be. You learn how to prove new results by coming to understand how old results were proven. New results may have grand implications to which your name will be attached throughout the decades to come or they may (more likely!) be small matters about which you seek clarity and certainty for yourself and others.

The formally presented definitions enable us to refer compactly by a short phrase to a longer statement. The role of a definition is clarified through its use, either in the examples or in the theorems. The very fact that we have taken the trouble to define a short "code word" for a longer expression signals that we are pointing to something of sufficient interest and utility that it will be referred to again.

We also provide commentary outside of this formal setting that is freer from the constraints of precision, accuracy, and compactness and that is intended to assist you in gaining an understanding, not only of what is being proclaimed, but of why this proclamation could be of interest to you. Many examples, set off typographically, are provided to help meet these goals.

0.7 MATHEMATICAL PREREQUISITES

Knowledge of electrical engineering is not required, although an elementary exposure of the kind given in introductory engineering and physics courses will help in appreciating some of the applications and in understanding a few of the exercises. What often distinguishes one probability text from another is how it motivates the student to learn the common mathematical theory. In our case, we use examples drawn from electrical engineering applications. However, just as generations of students have persevered with texts that discussed either games of chance or issues of manufacturing quality control, so students who do not share a particular interest in electrical engineering can also benefit from this text.

The mathematical prerequisites are nothing more than what is universally available from engineering calculus courses and, typically, no more than what is learned by the completion of a first full calculus course. However, our experience has been that when this material is evoked in the context of probability (a context in which it has not been learned), it does not readily come to mind. You need some facility with these basic calculus and precalculus ideas. The primary prerequisites are a college-level, first-year integral calculus course, together with its prerequisites of an understanding of sums and functions and familiarity with common functions (e.g., polynomial, exponential, trigonometric) and differential calculus (not differential equations); but you need to know these prerequisites well enough to use them comfortably and reliably. We review a few equations that will be of frequent use to us.

0.7.1 Sums

Consider the particular finite sum

$$S_n = \sum_{k=0}^{n} k.$$

The summation index k is a dummy variable in that it can be replaced by other symbols without changing the value S_n of the sum, thus:

$$S_n = \sum_{k=0}^{n} k = \sum_{j=0}^{n} j = \sum_{m=0}^{n} m.$$

We typically use the symbols i, j, k, l, m, and n to denote integer-valued summation indices, although we are logically free to use other symbols for the same purpose. The summation limit n

cannot be changed without consequence, since S_n is a function of the variable n (but for notational tradition, S_n would be better written as a function $S(n)$) and of no other variable. For each value of the variable n, generally assumed to be an integer, there is a corresponding value of the sum S_n. Hence, $S_0 = 0, S_1 = 1, S_2 = 3, \ldots$. The finite sum formula valid for any nonnegative integer n,

$$S_n = \sum_{k=0}^{n} k = \frac{n(n+1)}{2},$$

can be verified by finite induction on n. We carry out this proof because a number of subsequent proofs also rely on the method of **proof by finite induction**.

First, verify by direct calculation that the statement is true for an initial value of n. In this case, it is immediate that for $n = 0$, $S_0 = 0$ when evaluated either as the sum or as the proposed formula. Hence, the statement is verified for the initial case of $n = 0$. Second, make the inductive hypothesis that the statement is true for all $n \leq m$ and attempt to prove it true for the next case of $n = m + 1$. Observe that

$$S_{m+1} = \sum_{k=0}^{m+1} k = m + 1 + S_m.$$

By the inductive hypothesis

$$S_m = \frac{m(m+1)}{2}.$$

Hence,

$$S_{m+1} = m + 1 + \frac{m(m+1)}{2} = \frac{(m+1)(m+2)}{2},$$

and we have verified the statement for the next case of $n = m + 1$. The proof by induction is now complete, and the statement holds for all $n \geq 0$.

The sum of the geometric series,

$$\sum_{k=0}^{\infty} \beta^k = \frac{1}{1-\beta}, \quad \text{for } |\beta| < 1,$$

can be verified by assuming convergence and observing that

$$(1-\beta) \sum_{k=0}^{\infty} \beta^k = \sum_{k=0}^{\infty} (\beta^k - \beta^{k+1}) = \sum_{k=0}^{\infty} \beta^k - \sum_{k=1}^{\infty} \beta^k = 1.$$

Double sums are a frequent source of confusion. The double sum

$$\sum_{i=1}^{n} \sum_{j=1}^{m} a_{i,j}$$

can be expanded as a sum of nm terms in the form

$$a_{1,1} + a_{1,2} + \ldots + a_{1,m} +$$
$$a_{2,1} + \ldots + a_{2,m} +$$
$$\vdots$$
$$+ a_{n,1} + \ldots + a_{n,m}.$$

For example,

$$\sum_{i=1}^{2}\sum_{j=1}^{3} ij^2 = 1(1^2) + 1(2^2) + 1(3^2) + 2(1^2) + 2(2^2) + 2(3^2) = 1 + 4 + 9 + 2 + 8 + 18 = 42.$$

The square of a single sum is a double sum,

$$\left(\sum_{i=1}^{n} a_i\right)^2 = \sum_{i=1}^{n}\sum_{j=1}^{n} a_i a_j,$$

and in this case its expansion has n^2 terms. More generally, the product of two sums is a double sum,

$$\sum_{i=1}^{n} a_i \sum_{j=1}^{m} b_j = \sum_{i=1}^{n}\sum_{j=1}^{m} a_i b_j.$$

0.7.2 Vectors and Matrices

Vector and matrix notation is often helpful, particularly in discussing multivariate distributions and in studying linear systems. Our default assumption is that all vectors are column vectors unless otherwise specified. Throughout, we denote a vector-valued quantity by boldface, as in the example

$$\mathbf{x} = [x_i] = \begin{pmatrix} 1 \\ 4 \\ -2 \end{pmatrix}, \quad x_1 = 1, \; x_2 = 4, \; x_3 = -2.$$

The **dimension** is the number of components. In the example just given, the dimension is 3.

The **transpose** of a matrix is denoted by a superscript T:

$$\mathbf{x}^T = \begin{pmatrix} 1 & 4 & -2 \end{pmatrix}.$$

Given two column vectors $\mathbf{x}, \mathbf{y}$ of the same length n, we can define the **scalar product** (also known as the dot product):

$$\mathbf{x} \cdot \mathbf{y} = \mathbf{x}^T \mathbf{y} = \mathbf{y}^T \mathbf{x} = \sum_{i=1}^{n} x_i y_i.$$

If $\mathbf{x} \cdot \mathbf{y} = 0$, then $\mathbf{x}$ and $\mathbf{y}$ are said to be orthogonal.

Matrices are rectangular arrays of elements denoted by AMS blackboard font as in

$$\mathbb{A} = [A_{i,j}],$$

with $A_{i,j}$ being the value located in the ith row and jth column of the rectangular array that is the matrix $\mathbb{A}$. If the first subscript i ranges over the index set $\{1, \ldots, n\}$ and the second subscript j ranges over the index set $\{1, \ldots, m\}$, then $\mathbb{A}$ has n rows and m columns and is said to be an $n \times m$ matrix. A column vector is simply an $n \times 1$ matrix.

The product

$$\mathbf{y} = \mathbb{A}\mathbf{x}$$

of an $n \times m$ matrix $\mathbb{A}$ and an $m \times 1$ vector $\mathbf{x}$ yields an $n \times 1$ vector $\mathbf{y}$ whose ith entry is

$$y_i = \sum_{j=1}^{m} A_{i,j} x_j.$$

The product $\mathbb{A}\mathbb{B} = \mathbb{C}$ of a matrix $\mathbb{A}$ that is $n_1 \times m_1$ by a matrix $\mathbb{B}$ that is $n_2 \times m_2$ is defined and is an $n_1 \times m_2$ matrix $\mathbb{C}$ if and only if $m_1 = n_2$. In this case, the two matrices are said to conform, and their product can be expanded as

$$C_{i,j} = \sum_{k=1}^{m_1 = n_2} A_{i,k} B_{k,j}.$$

If $\mathbb{A}$ is $n \times m$, the matrix $\mathbb{A}^T$ is the $m \times n$ matrix with i, j element given by the j, i element of $\mathbb{A}$. Two important properties of the transpose are

$$\left(\mathbb{A}^T\right)^T = \mathbb{A} \text{ and } (\mathbb{A}\mathbb{B})^T = \mathbb{B}^T \mathbb{A}^T.$$

Note that the transpose of a product of matrices is the product of their transposes taken in the reverse order.

The scalar-valued quadratic form, requiring a square matrix $\mathbb{A}$ of dimension $n \times n$ and a conforming vector $\mathbf{x}$ of dimension n, is expressed as

$$\mathbf{x}^T \mathbb{A}\mathbf{x} = \sum_{i=1}^{n} \sum_{j=1}^{n} A_{i,j} x_i x_j.$$

We will need to be able to "complete the square" in that, for $\mathbb{A}$ a symmetric ($\mathbb{A}^T = \mathbb{A}$) matrix possessing an inverse $\mathbb{A}^{-1}$, we can write

$$\mathbf{x}^T \mathbb{A}\mathbf{x} + 2\mathbf{b}^T \mathbf{x} + c = (\mathbf{x} + \mathbb{A}^{-1}\mathbf{b})^T \mathbb{A}(\mathbf{x} + \mathbb{A}^{-1}\mathbf{b}) + c - \mathbf{b}^T \mathbb{A}^{-1}\mathbf{b}.$$

This expression can be verified by multiplying out the right-hand side of the identity.

0.7.3 Exponential Function

One needs to be at ease with the **exponential function** e^x. For real x, $e^x > 0$ and it increases rapidly with increasing x (e.g., doubling x to $2x$ yields e^{2x} that is the square $[e^x]^2$ of e^x). For any complex-valued x,

$$e^x e^y = e^{x+y}, \quad e^x = \sum_{k=0}^{\infty} \frac{x^k}{k!}, \quad \frac{de^x}{dx} = e^x.$$

It is important to distinguish between a product of two exponentials and raising one exponential e^x to a power y:

$$(e^x)^y = e^{xy}.$$

We easily verify the inequality

$$e^x \geq 1 + x,$$

for all real x, by noting that the left-hand and right-hand sides agree at $x = 0$, but for positive x the derivative of the left-hand side exceeds that of the right-hand side, whereas for negative x the derivative of the left-hand side is smaller than that of the right-hand side. For small values of x, we will often use the approximation

$$e^x \approx 1 + x$$

found by taking only the first two terms in the power series.

The symbol i will represent both a summation index that is a dummy integer-valued variable in a sum (e.g., $\sum_{i=1}^{5} i^2$) and the purely imaginary complex quantity $\sqrt{-1}$, with context specifying which meaning is in use. Hence, we have the **Euler formula** for exponentials, using $i = \sqrt{-1}$, namely,

$$e^{ix} = \cos(x) + i \sin(x),$$

which can be verified by power series expansions, and its immediate implications,

$$\cos(x) = \frac{e^{ix} + e^{-ix}}{2} \quad \text{and} \quad \sin(x) = \frac{e^{ix} - e^{-ix}}{2i}.$$

0.7.4 Calculus

Differentiation is usually well understood. A result that will be needed repeatedly is the derivative of a function $f(x)$ that is a composition of two functions $g(h(x))$. The value of the function f at the argument x is calculated as the value of the function g at the argument $h(x)$ that is in turn the value of the function h at its argument x. Let a superscript prime denote a derivative as in

$$\frac{df(x)}{dx} = f'(x).$$

Then the **chain rule of differentiation** informs us that

$$\frac{df(x)}{dx} = g'(h(x))h'(x).$$

Familiarity is expected with integration of functions of one and two variables, including the technique of changing variables in an integral and understanding the role of the **dummy variable** of integration. For example, given a function f of a real argument x,

$$\int_a^b f(x)\, dx = \int_a^b f(y)\, dy$$

is itself only a function of the limits a, b and not of the dummy variables x, y. A frequent source of error is losing track of the variables in a problem and producing as an answer an expression involving variables that can no longer be present. Hence, $\int_0^1 e^{x^2 y}\, dx$ is a function of y, but no longer a function of the dummy variable x, and we know this without having to be able to evaluate the integral.

We also require the ability to integrate functions of one and two variables (e.g., $f(x, y)$) over a specified one- or two-dimensional region A in the Cartesian or xy-plane by properly specifying the limits of integration and evaluating the result.

Familiarity with integration by parts is required:

$$\int_a^b u\, dv = uv|_a^b - \int_a^b v\, dv.$$

For example,

$$\int_a^b xe^{-x}\, dx = -xe^{-x}|_a^b + \int_a^b e^{-x}\, dx = (a+1)e^{-a} - (b+1)e^{-b}.$$

Students typically can perform required differentiations with the exception of the differentiation of an integral with respect to a parameter, a result known as **Leibniz's Rule** and given by

$$\frac{d}{dz} \int_{a(z)}^{b(z)} f(x, z)\, dx = b'(z) f(b(z), z) - a'(z) f(a(z), z) + \int_{a(z)}^{b(z)} \frac{\partial f(x, z)}{\partial z}\, dx.$$

Partial differentiation is needed for the integrand f when it is a function of more than one variable.

0.7.5 Sanity Checks

At various places in this text, we will point out "sanity checks." These will be simple, quick ways of checking the plausibility of an answer. In this era of reliance upon computers for most of our calculations, it is all too easy to make a programming or entry error and assume that the result must be correct because computers do not make mistakes. Similarly, it is very easy in carrying out calculations analytically to make errors, especially with the time pressure of an examination. You need to have these "sanity checks" to give you some notion of whether you are getting it right. In a number of cases, these checks will be as simple as recalling that the answer must be a nonnegative number (e.g., true for all probabilities and probability density functions) or what the value of the resulting computed function must be at certain special values of its argument (i.e.,

Does your program give you the right answer in the cases where you know in advance what the answer should be?) or that the variables in question are real valued and not integer. Keeping track of the variables that the answer is a function of is often another valuable "sanity check" (e.g., $\sum_{i=1}^{n} i^3$ does not depend on i).

0.8 BEYOND THE MATHEMATICAL PREREQUISITES: INTEGRATION

A correct, detailed treatment of a number of topics requires a mathematical background beyond that assumed of readers of this text. In particular, there are issues of the definition, existence, and properties of the integral that will be bypassed when they arise. Rather than caution you at each such occurrence, we use this introduction to indicate the nature of those issues whose resolution is left to more advanced texts.

There are several different ways to define a definite integral, and the names of some of the more common definitions are the Riemann, Riemann–Stieltjes, Lebesgue, Daniell, and Ito integrals. These integrals differ in the types of functions to which they apply and have somewhat different analytical properties. We assume calculus and therefore a knowledge of some integral, usually the Riemann integral, as it is the one most often taught to engineers. In the advanced theory of probability, the integral of choice is that of Lebesgue. However, even the Lebesgue integral does not always suffice for functions such as the sample functions or realizations of a stochastic or random process.

Throughout, we will freely use the term "integral" and the familiar symbol "$\int$" or "$\int_a^b f(x)\,dx$." We will assume that our integrals have the following basic properties:

Nonnegativity: If for all x, $f(x) \geq 0$ and $b \geq a$, then $\int_a^b f(x)\,dx \geq 0$. Nonnegative functions have nonnegative integrals.

Linearity: For any constants α and β and any integrable functions f and g, the function $h(x) = \alpha f(x) + \beta g(x)$ is integrable and

$$\int_a^b [\alpha f(x) + \beta g(x)]\,dx = \alpha \int_a^b f(x)\,dx + \beta \int_a^b g(x)\,dx.$$

The integral of a linear combination of two functions is the same linear combination of their individual integrals. This extends by induction to linear combinations of finitely many functions.

Dominated convergence: If we have a sequence of functions $\{f_n(x)\}$ that converges to a function $f(x)$ at each $a \leq x \leq b$,

$$\lim_{n \to \infty} f_n(x) = f(x),$$

and there is a function g such that

$$\int_a^b g(x)\,dx < \infty \text{ and } |f_n(x)| \leq g(x) \text{ for } a \leq x \leq b,$$

then

$$\lim_{n \to \infty} \int_a^b |f_n(x) - f(x)| \, dx = 0 \text{ and } \lim_{n \to \infty} \int_a^b f_n(x) \, dx = \int_a^b f(x) \, dx.$$

The point of this result is that it provides a condition under which we can interchange a limit on n with an integral. We will often make such an interchange without further justification.

0.9 OVERVIEW OF COVERAGE

This overview may be of more value to you subsequently as you proceed through the material and wish to see how the ideas presented fit together. The traditional single-semester course in probability is too short to cover all of the material in this text. Choices will need to be made of illustrative material, and some of the background to theoretical developments will have to be omitted.

We start in Chapter 1 with a wide-angle view of the manifold occurrences of random phenomena throughout history and with a narrower focus on the sorts of random phenomena of greater interest in physics and electrical engineering. Perhaps the strongest motivational point for the study of probability is the wealth of applications to diverse random phenomena. These phenomena are encountered in everyday life, natural phenomena of great complexity (e.g., the weather), social and economic phenomena that are of great complexity, physical phenomena derived from quantum and statistical mechanics processes, and engineering applications that are based on the interactions between the random behaviors of environmental and designed systems. One example drawn from the last category is a bridge or building subject to normal and designed-for random loads, as well as unexpected malicious behavior and extremes of weather and earthquakes. A second example is a communication system in which the system itself unavoidably generates noise and must then accurately recreate a signal about which little is known and which appears accompanied by noise generated in its transmission environment.

The formal starting point to reach Goal 1 (mathematical models for random phenomena) has two components. The first is the idea of a random experiment $\mathcal{E}$ that is given a mathematical formulation in terms of a set $\Omega = \{\omega\}$ of possible outcomes (e.g., possible audio/video signals observed over a given time interval and given bandwidth or which of the 52 cards in a deck of well-shuffled playing cards is uppermost in the deck). The second component is a family $\mathcal{A}$ of subsets of Ω, called events, that can be thought of as a list of yes/no questions of potential interest to you (e.g., was the word spoken in the half-second time interval "cat" or "bat"?) that are about the possible outcome ω of a random experiment $\mathcal{E}$.

In order to complete our description of $\mathcal{E}$, we need to assign probabilities that assess the tendencies for the questions posed in $\mathcal{A}$ to be answered affirmatively. Historically, the first formal theory of such probabilities originates in the latter half of the 17th century. This theory of probability, known as classical probability, is developed in Chapter 2 in the context of experiments having only finitely many possible outcomes that are all equally probable (e.g., familiar games of chance). The simple setting of classical probability enables us to present the basic ideas of probability, conditional probability, and independence, all of which are revisited in greater detail subsequently. The basis of probability calculations in classical probability is counting—the subject of combinatorics. An exposure to combinatorics and to graphs is of value in applied mathematics beyond the needs of probability.

A general theory of probability is presented in Chapter 3. We discuss at length the fundamental idea of probability P as a numerical assignment to a family of events $\mathcal{A}$ that satisfies a set of axioms. We then explore some of the commonly needed implications of these axioms. We also note the possibilities for nonnumerical notions of probability, such notions being common in everyday language, but as yet absent from scientific applications. In addition, we discuss a number of the meanings that have been proposed for probability. While we support a variety of meanings for the same mathematical concept, we encourage the reader to include the view that the probability of an event is a numerical measure of the tendency of that event to occur when $\mathcal{E}$ is performed. An unusual feature of this text is the introductory material of Chapter 3, which sketches alternative views of probability that are almost always omitted in probability texts.

We are primarily interested in random experiments that have associated numerical outcomes (e.g., number a deck of playing cards from 1 to 52, a voltage or current at a given time in an audio/video signal, a maximum stock price tomorrow, the remaining lifetime of a subsystem, the number of callers attempting to use a cellular mobile telephone system). Such numerically valued outcomes of random experiments are known as random variables, typically denoted by X, Y, Z, etc. In brief, we develop the probability description of finitely many random variables (e.g., voltages measured at several times, or at the same time but at different nodes in a circuit; or the lifetimes of several possibly related subsystems), define specific probability models of importance to electrical engineering, and address fundamental elements of the basic issues of system response and of estimation, inference, and decision making.

As in all of mathematics, pure and applied, having alternative representations of a concept is key to success in using it. Thus, vectors can be represented in alternative coordinate systems, and the proper choice of coordinate system can greatly simplify a problem. In this regard, Chapters 4, 5, 6, 7, and 17 are devoted to alternative representations of P for random variables in terms of probability mass functions, probability density functions, cumulative distribution functions, characteristic functions, and generating functions. The power of these different representations is demonstrated extensively throughout the text. We additionally emphasize about a dozen specific probability models, expressed through these representations, that are frequently needed in engineering practice. Relationships between these models that aid in understanding when they can be applied are discussed in a number of subsequent sections.

Probability is greatly enriched by introducing the associated concepts of expectation, conditioning, and independence. Chapter 9 introduces the expectation EX of a random variable X. Expectation has a number of meanings, but for now can be thought of either as a parameter partially specifying the probability model governing X and calculable from that probability model or as a typical or average value of a random variable. Expectation is essential in specifying design criteria for systems that have random responses (e.g., a communications receiver, electricity demand forecast, or an investment scheme) and in providing a single number to assess the size of a random variable that is the discrepancy between what is predicted and what occurs.

Probabilities are based on knowledge. When further knowledge is gained through observation or measurement, probabilities should be revised. (For example, I can now see from a distance that the top card in a well-shuffled deck is a red-faced card.) This process of revision, treated in Chapters 11 and 12, is known as conditioning, and it updates given probability specifications into ones compatible with the new information. Conditioning reflects how we learn from certain kinds of information. Once the notion of conditioning is in place, we can easily motivate and define the central notion of independence between random variables.

The notion of (stochastic, as distinct from logical) independence treated in Chapter 14 is one of the hallmark features of probability theory and one frequently assumed when making models of stochastic phenomena. Independence between two random variables X and Y is a statement that they are unlinked: One of them taking on a value does not physically cause the other to prefer certain values; the value taken by one of them is not informative about the value taken by the other. Thus, if I toss a coin and you then roll a die, the outcome of my coin toss tells you nothing about what will be the outcome of your roll of a die.

The limitation to finitely many random variables throughout the bulk of this text gives our treatment of probability a coherence and avoids certain mathematically sensitive areas. However, in the last two chapters we do introduce issues of limits of sequences of random variables and infinite collections of random variables. Chapter 19 provides a bridge to treating infinite collections of random variables, which is the province of a course on stochastic or random processes. Limits of sequences of random variables are defined and then used to establish the celebrated laws of large numbers that provide the conditions under which long-run time averages converge to statistical averages. The closing Chapter 20 provides an elementary introduction to a variety of stochastic processes that are commonly encountered in engineering practice.

While all of the preceding is clearly in support of reaching Goal 1, it also bears on Goal 2, describing the response Y of a system S driven by random input X. If X is random, then so will be Y. We use probability models to describe X and Y. When S is a deterministic system (repeatable response to each input), then we are considering S as a function of X with value Y. Techniques for describing Y are treated in Chapter 8. When the system S is itself random (e.g., a noisy communications channel), we need conditional probability, as presented in Chapters 11 and 12, to describe the system S. (The system output depends upon—is conditional on—the system input.) Results from conditional probability allow us to determine the probability description of the response Y from the probability description of the excitation X and the conditional probability description of S. We are also able to reverse the causal order of generating Y from X through S and to evaluate the conditional probabilities of X given or conditional on Y. This viewpoint is seen to be very useful once one thinks of Y as a noisy measurement produced by the measuring device S measuring X.

Goal 3 is about the use of probabilistic information to design systems optimal, or nearly so, for performing tasks such as signal detection in noise, and signal extraction from noise, and about learning probability models from prior assumptions (e.g., the model is Poisson, but the defining parameter λ is unknown) by estimating unknown parameters (e.g., λ) from data, in the form of measurements. As the practice of electrical engineering (and much more than electrical engineering) encounters a wide variety of such problems, we discuss (1) a number of approaches through linear (Chapters 9 and 10) and nonlinear least mean/expected squared error minimization (Chapter 15), (2) minimization of probability of error (Chapter 13), (3) minimization of more general error models (Chapter 16), (4) Neyman–Pearson hypothesis testing (in which detection probability is maximized, given a limit to allowable false-alarm probability), and (5) maximum likelihood estimation (both in Chapter 13).

Statistical techniques are covered, as noted previously, through an exposure to basic methods of estimation and decision making, including minimum error probability design, linear and non-linear least mean square estimation, maximum likelihood estimation, and Bayesian decision making. However, we do not cover statistical techniques used in data handling such as goodness-of-fit tests, treatment of outliers, design of experiments, etc. A course in statistical theory is a good supplement to one in probability. The reader needs to be careful in choosing such a course; many statistics courses merely supply the recipes by which practitioners of other disciplines can "cook" their data to extract support for particular models.

Extensive teaching experience suggests that the material presented in this text is too much to be either taught or learned well in one semester. Choices will have to be made and material will have to be omitted. A second course in random processes is desirable—one that can refresh the material learned in the first course and extend it to the study of random processes and their rich variety of applications.

While all of the omitted topics have an important place, we hope that the coherence of our approach is evident enough to avoid Harry Belafonte's memorable characterization of an explanation found in his "Man Piaba" calypso song: "It was clear as mud, but it covered the ground. . . ."

0.10 ELEMENTS OF A ONE-SEMESTER COURSE

There is intentionally more in this text than can be taught in one semester. It is my hope that the readers will find time to browse further, if not when they first study probability, then later. An emphasis of this text is an attempt to justify and carefully motivate the mathematical concepts and their uses, concepts that are too often given short shrift and adopted without much thought except for consistency with the traditions of teaching probability. Each instructor, however, will have his or her own emphases that will lead them to their choice of topics. Everyone is strongly encouraged to read all of Chapter 0 for its motivation to study probabilistic reasoning and for its explanation of our approach to this study. In addition, Sections 1.1 through 1.3 are essential introductions to the background and important uses of probabilistic reasoning in engineering and the physical sciences.

Those who wish to focus quickly on the mathematical theory, and then explain its relevance as they go, could start with Sections 3.5, 3.6, and 3.7. Chapters 4, 5, and 6 (except for 6.5.2) should be taught in their entirety. Sections 7.1 through 7.3 should be taught, as well as 7.4.1 and 7.4.4. Chapter 8 addresses the use of probability to calculate system response in the context of functions of random variables. Section 8.3 is essential and many will also want to treat Section 8.5 on multivariable functions. Expectation, a core concept, is treated in Chapter 9. At a minimum, Sections 9.3, 9.4, 9.8, and 9.10 through 9.14 should be covered. I view Chapter 10 on linear systems and linear least mean square estimation as providing important contact with the use of probability for system response and to make inferences. The material on Kalman filtering in Section 10.6, while popular, is complex and could probably best be omitted. Conditional probability is treated in Chapters 11 and 12. Chapter 11 should be covered fully. One could omit Section 12.2 and just start with conditional densities as a given. Chapter 13 contains important inference applications to minimum error probability design, to MAP and MLE rules, and to the

Neyman–Pearson approach to decision making (hypothesis testing) when prior probabilities are absent. The latter, in Section 13.4, could be omitted. Chapter 14 formally treats independence. It is possible to omit the initial discussion of independent events and start with Section 14.4 on independent random variables. One could then reconstruct the story of independent events by replacing them by their indicator function random variables. Chapter 15 on conditional expectation is essential, although Section 15.6 on Kalman filtering can be easily omitted as can Section 15.8. If time permits, elements of Chapter 16 on Bayesian inference are well worth presenting as important applications of conditional expectation. I believe that characteristic functions, as discussed in Chapter 17, provide an essential alternative description of probability that enables us to solve many problems that would otherwise be intractable. Chapter 18 on probability bounds has material of value in practice, but it can be omitted with the exception of Section 18.2 on Chebychev bounds. Chapters 19 and 20 really belong in a second course on random processes. They were included here to accommodate those who think otherwise.

EXERCISES

E0.1   Evaluate the following difference of two sums:

$$\delta = \sum_{k=0}^{100} k^2 e^{-k} - \sum_{m=1}^{100} m^2 e^{-m}.$$

E0.2   Prove by induction on n that

$$\sum_{k=0}^{n} k^2 = \frac{1}{6}(2n+1)(n+1)n.$$

E0.3   Evaluate the following infinite sums:

$$a = \sum_{n=0}^{\infty} 3^{-n};$$

$$b = \sum_{n=1}^{\infty} 2^{-n};$$

$$c = \sum_{n=0}^{\infty} \frac{2^n}{n!}.$$

E0.4   Evaluate the following finite double sums:

   a.

$$a = \sum_{n=1}^{4} \sum_{m=0}^{2} n 2^{m+n}.$$

b.

$$b = \sum_{n=0}^{5} \sum_{m=0}^{5} \frac{1}{(n+1)(m+1)}.$$

c.

$$c = \sum_{n=1}^{3} \sum_{m=0}^{2} \frac{1}{n+m}.$$

E0.5   Evaluate the dot product $\mathbf{x} \cdot \mathbf{y}$ for

$$\mathbf{x}^T = \begin{pmatrix} -1 & 0 & 1 \end{pmatrix}, \quad \mathbf{y}^T = \begin{pmatrix} 1 & 1 & -1 \end{pmatrix}.$$

Are $\mathbf{x}$ and $\mathbf{y}$ orthogonal?

E0.6   For each of the following equations, determine whether the calculation

$$\mathbf{y} = \mathbb{A}\mathbf{x} \text{ for } \mathbb{A} = \begin{pmatrix} 7 & 4 & -1 \\ 0 & 2 & 1 \\ -1 & -2 & 3 \end{pmatrix}$$

can be carried out, and if so, carry it out:

a.

$$\mathbf{x} = \begin{pmatrix} 1 & 1 & 6 \end{pmatrix}$$

b.

$$\mathbf{x} = \begin{pmatrix} 1 \\ 1 \\ 6 \end{pmatrix}$$

c.

$$\mathbf{x} = \begin{pmatrix} 1 \\ 1 \\ 0 \\ 6 \end{pmatrix}$$

E0.7   For the matrix $\mathbb{A}$ of the preceding problem, evaluate the matrix

$$\mathbb{B} = \mathbb{A}\mathbb{A}^T.$$

E0.8   Evaluate the quadratic form $q = \mathbf{x}^T \mathbb{A}\mathbf{x}$ for

$$\mathbf{x} = \begin{pmatrix} 0 \\ 2 \\ 1 \end{pmatrix}, \quad \mathbb{A} = \begin{pmatrix} 3 & 1 & 2 \\ 0 & 1 & 2 \\ 1 & 1 & 4 \end{pmatrix}.$$

E0.9   Complete the square in the following expression:

$$\mathbf{x}^T \begin{pmatrix} 3 & 1 & 2 \\ 1 & 1 & 3 \\ 2 & 3 & 4 \end{pmatrix} \mathbf{x} + \begin{pmatrix} -1 & 0 & 2 \end{pmatrix} \mathbf{x} + 5.$$

E0.10   If $x > 0$, how does

$$\frac{e^x}{1 + x + \frac{1}{2}x^2}$$

compare to 1?

E0.11 Can you find $x > 0$ such that

$$1 + x + x^2 > e^x?$$

E0.12 Without using a calculator, arrange $a, b,$ and c in nondecreasing order:

$$a = e^{33}e^{65}, \quad b = e^{20}e^{80}, \quad c = d^5 \text{ where } d = e^{22}.$$

E0.13 By using the power series for e^x, show that, for positive x,

$$e^x > \frac{e^x - 1}{x}.$$

E0.14 If the function $\Phi(x)$ has derivative

$$\frac{d\Phi(x)}{dx} = \frac{1}{\sqrt{2\pi}}e^{-\frac{x^2}{2}},$$

then evaluate the derivative f' of

$$f(x) = \Phi(\sqrt{x}).$$

E0.15 Evaluate dg/dx, given that

$$g(x) = \int_x^{x^2} [\log(x + y)]^2 \, dy.$$

E0.16 Show that the definite integral

$$\int_A f(x, y) \, dx \, dy$$

of the function

$$f(x, y) = e^{x+y}$$

over the region

$$A = \{(x, y) : 0 \le x \le y \le 1\}$$

is

$$\frac{1}{2}(e - 1)^2.$$

E0.17 Using integration by parts, show that

$$\int_a^b x^2 e^{-x} \, dx = (a^2 + 2a + 2)e^{-a} - (b^2 + 2b + 2)e^{-b}.$$

E0.18 Using integration by parts, show that if

$$I_k = \int_0^\infty x^k e^{-x} \, dx,$$

then

$$I_k = kI_{k-1}, \text{ for } k \ge 1.$$

Conclude from this that I_k is $k!$ (k factorial).

E0.19 Without carrying out the indicated calculations, provide reasons why the suggested answers are incorrect:

a.

$$\int_0^\pi \left(\cos(x)\right)^4 dx = \frac{1}{5}\left(\sin(x)\right)^5$$

b.

$$\sum_{n=0}^{100}\sum_{m=0}^{100}(-2)^{n+m} < 0$$

c.

$$\mathbb{A} = \begin{pmatrix} 2 & 4 & 3 & 5 & 1 \\ 3 & 5 & 2 & 1 & 4 \\ 5 & 1 & 4 & 2 & 3 \\ 1 & 4 & 5 & 3 & 2 \end{pmatrix}, \quad \mathbb{B} = \begin{pmatrix} 1 & 2 & 3 \\ 1 & 2 & 3 \\ 1 & 3 & 2 \\ 2 & 1 & 3 \\ 1 & 3 & 2 \end{pmatrix}, \quad \mathbb{AB} = \begin{pmatrix} 20 & 16 & 17 & 18 \\ 29 & 35 & 35 & 34 \\ 41 & 39 & 38 & 38 \end{pmatrix}$$

1

Background, Events, and Data

1.1 PURPOSE, BACKGROUND, AND ORGANIZATION

This chapter provides an introduction to probabilistic reasoning and its manifold applications, as it is seen from the multiple directions of its history, the wide variety of random phenomena that it serves to model, and the initial steps towards a mathematical theory based upon describing possible outcomes of random phenomena. An acquaintance with this background orients one to the domain inhabited by probability and probabilistic reasoning and suggests that the evolution of probabilistic reasoning is still continuing. The history of an explicit interest in random phenomena is millennia old. Dice, having a modern appearance with cubical faces marked by dots or pips, have been found prior to 1500 B.C.E. (The author actually saw such an ancient die at the Louvre in Paris). There are ancient interests in gaming and in divination and oracles (predicting the future) that are alive today. The first, nonquantitative steps towards a formal understanding were taken by the ancient Greeks. The modern period of quantitative probability begins in the mid-17th century.

Our present interest in probability stems from a profusion of random phenomena that are central to such subjects of engineering and societal importance as demographics, economics and the behavior of markets, reliability of complex systems, turbulence, communications, quantum phenomena in nanoscale devices, noise sources, and measurement errors. These phenomena are surveyed briefly. While each of these phenomena has been given multiple book-length treatments, our remarks should help justify to the reader that knowledge of probability theory is of value in a wide range of subjects, ranging from highly technical and specialized ones to phenomena of everyday life.

Mathematical modeling of random phenomena requires us to first model the outcomes or events of interest, and this is addressed in this chapter through sets, Boolean set operations, and collections of sets of interest. We close with a brief word on a few basic statistics (functions of data), like the sample mean and variance and histograms, that are used to transform data on the outcomes of many repetitions of a random phenomenon into the numerical beginnings or estimates of quantitative probabilities to be defined subsequently. While a thorough course in probability should give the student significant contact with data from actual random phenomena, such exposure is usually reserved to a first course in statistics.

1.2 BRIEF HISTORICAL BACKGROUND TO PROBABILITY

We take a quick tour through 3,500 years of the history of probability, with an emphasis on the last 300 years. Probability and probabilistic reasoning are not finished subjects, no matter how complex and well developed they are today. Some knowledge of their historical development can give you a perspective both on the great importance of probability in everyday life and science and technology and on its continuing conceptual development.

Probabilistic reasoning is a complex of approaches to navigating in the wide variety of realms of chance, random, uncertainty, and nondeterministic phenomena. This complex ranges from the informal, intuitive, qualitative reasoning of everyday life (e.g., "I'll probably see you tomorrow") to the formal, quantitative reasoning that underlies much of engineering applications (e.g., "The probability that the next transmitted bit will be in error is .0001"). Our focus will be on a formal presentation of quantitative probability, an element of probabilistic reasoning that has been vigorously explored in this past century. However, we start with a brief historical review of probabilistic reasoning.

An interest in gaming devices, such as dice, dates back to antiquity (David [22]). A three-thousand-year-old Egyptian die is shown in Figure 1.1, together with several gaming pieces.

Figure 1.1 Ancient Egyptian die and gaming elements.

The ancient Greeks struggled to develop a notion of *possibility* or *contingency*, as opposed to *necessity*. For example, a future event ("there will be a naval battle tomorrow") considered today need not necessarily happen or fail to happen. Lucretius in his *De Rerum Natura, II*, written prior to 55 B.C.E., introduced a notion of random motion in the concept of the *clinamen*, the uncaused swerving motions of atoms falling through space. However, they had no theory of probability, and it is plausible that they failed to develop one because they thought there were no regularities in random or chaotic behavior (Sambursky [79]). While there was much experience with gaming devices and such random phenomena as illnesses and the weather, there was no appreciable quantitative development of probabilistic notions until the Renaissance. The religious scholar Aquinas, writing in the 13th century, could only distinguish the four rough levels of probability corresponding to events occurring either always, frequently, rarely, or never. A survey of pre-17th-century thinking about probability can be found in Franklin [33].

The mathematical development of probability traditionally begins in the mid-17th century (Hacking [43], Hald [44], Maistrov [61], Stigler [84], and Todhunter [90]) with the 1654 correspondence between Pascal and Fermat on the "problem of points" suggested by the Chevalier de Mere. While Cardano [17] had written a book of advice on gaming, *Liber de Ludo Aleae*, in 1564, it remained unpublished until 1663. The first published textbook on probability, *De Ratiociniis in Ludo Aleae*, is by Huyghens [48] and was published in 1657.

Some of the highpoints of the 17th-century development were the estimates of probability related to data by Graunt and by the Dutch annuitants Hudde and de Witt (see Hacking [43]), the first use of the term *probability* in connection with a numerical value by Arnauld [3], and Leibniz's use of equally possible cases to formalize the calculation of probability. (See Chapter 2.) The crowning achievement of that century is surely Jacob Bernoulli's development of the first law of large numbers (see Chapter 19), linking probabilities to long-run relative frequencies. Bernoulli's *Ars Conjectandi, Pars Quarta* [9] was probably completed in 1692, but was published in 1713, after his death in 1705.

The 18th century (see Todhunter [90]) saw the birth of statistics, especially in the work of De Moivre (including the first central limit theorem), Bayes and the use of conditional probability to infer from effect to cause (so-called inverse probability), and Laplace [57], who contributed in several directions including the use of inverse probability and the mathematics of probability and who encouraged the everyday use of probability [57], saying

Probability theory is nothing but common sense reduced to calculation.

Laplace continued to contribute into the 19th century. Legendre (see Stigler [84]) developed a theory of errors that was based upon the notion of a least squares fit—a notion that continues to be in popular use to this day. Quetelet developed the study of social statistics and is credited with the notion of *the average man* (*l'homme moyenne*). John Venn pursued the frequentist, objective interpretation of probability [94]. The English statisticians Galton (see Section 5.2), Edgeworth, and Karl Pearson (in the 20th century, his son, Egon, contributed to the practice of hypothesis testing) developed statistical approaches to the biological sciences (especially heredity). (See Stigler [84].) Probability firmly entered the physical sciences in the 19th century with Boltzmann's and Gibbs's development of statistical mechanics. Boltzmann was the first to introduce probability into a fundamental law of physics, thereby breaking with the iron traditions of physics and philosophy that the fundamental laws must be deterministic.

At the start of the 20th century, Einstein [67] was much concerned with probabilistic issues in physics. In 1905, Einstein's *annus mirabilis* in which he published papers on special relativity and on the photoelectric effect (for which he won the Nobel prize), he also published a fundamental probabilistic paper on Brownian motion and showed how to determine Avogadro's number. Probability became central to 20th-century physics with Max Born's 1926 interpretation of the squared modulus $|\psi|^2$ of the Schrodinger wave function ψ as a probability density (Born [14], Pais [68]). Born remarks, "so only one interpretation is possible: Φ_{nm} [Φ is the wave function] represents [bestimmt die Wahrscheinlichkeit dafür] probability." In a footnote that follows immediately, Born amplifies this statement to assert that "probability is proportional to the square of the magnitude of Φ_{nm}." Born was awarded the Nobel prize in physics in 1954 with the citation, "for his fundamental research in quantum mechanics, especially for his statistical interpretation of the wavefunction." It is hard to overestimate the importance of 20th-century quantum mechanics and of its relation to probability. Probability firmly entered basic physics as associated with the intrinsic randomness of quantum phenomena (see Figure 1.8), and not due to "ignorance" that was potentially correctable. Probability appears in electrical engineering in the 1930s in the calculations by Thornton C. Fry [35] of the probability of a blocked call in telephony and its relation to the capacity of the telephone system.

The most widely accepted mathematical theory of probability was given its present form by Andrei N. Kolmogorov (Figures 1.2 and 1.3), with the first step in 1929 and the full presentation in Kolmogorov [54]. Alternative mathematical theories have been advanced to accommodate issues of modeling personal or subjective beliefs (see de Finetti [23, 24]; Walley [96]) and to quantum

Figure 1.2   A. N. Kolmogorov preparing a lecture (1973).

Figure 1.3   A. N. Kolmogorov with 1973 monograph given him by the author.

mechanical issues of variables that cannot be simultaneously measured (e.g., Heisenberg's uncertainty principle; see Jauch [49]). Richard von Mises [63, 64] and Bruno de Finetti [23, 24] contributed to the physical frequentist and subjective, personalistic conceptions of probability, respectively.

World War II encouraged the development of probabilistic approaches to problems of interest to electrical engineers particularly in the time series, control-oriented work of Kolmogorov and Norbert Wiener [99] and in the foundation of information theory by Claude Shannon [82]. Probability is now used throughout electrical engineering to describe random phenomena, whether in devices (e.g., thermal, flicker, and Barkhausen noises, quantum effects such as tunneling, $3°K$ background radiation) or in systems (e.g., control of processes operating in environments having random influences and inputs, as is the rule in communications, control, and computation). Probability enables us to characterize both physical and information (e.g., speech, text, video) sources of randomness.

1.3 IMPORTANT EXAMPLES OF RANDOM PHENOMENA

While we cannot devote much space to the multitude of particular applications that justify the importance of probabilistic reasoning, several of the phenomena to which we shall make reference throughout the text can be grouped as follows:

- Gaming Devices: vigorously tossed or spun coins; cards drawn from a well-shuffled deck; vigorously thrown dice; vigorously spun roulette wheels
- Lifetimes: people; light bulbs; household appliances; computer hard disks; transistors
- Waiting Times: times between successive arrivals of customers in a queue; client service requests from a computer file server; packets making up a single message on the Internet; keystrokes at a computer terminal; photons from a light source such as a laser or LED; electrons emitted by a cathode; α, β, or γ particles emitted by a radioactive source
- Background Noise: static; atmospherics; $3°K$ cosmic noise (see Figure 1.4)
- Communications Systems: communications channel additive noise (see Figure 1.4); received power fluctuations due to shadow fading and multipath in mobile cellular wireless systems; customer behavior in such systems; network packet statistics
- Complex Systems: bulk electric power distribution system; the Internet; the World Wide Web; computer chips and computing clusters
- Assessment: performance assessment and identification of bottlenecks in computing architectures
- Performance: artificial neural network performance and, more generally, machine learning in which the choice of statistical learning algorithms precedes the choice of a probability model for what is to be learned, but then there is a critical need to defend performance against a wide range of possible probability models
- Time Series: price fluctuations in the stock market on a time scale of hours and in the currency exchange market on a time scale of minutes (see Figures 1.5 and 1.6); EKG and EEG waveforms on a time scale of milliseconds; demand for electric power in a moderately

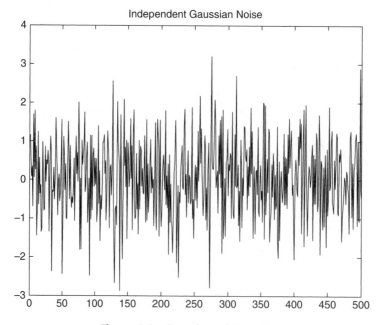

Figure 1.4   Gaussian white noise.

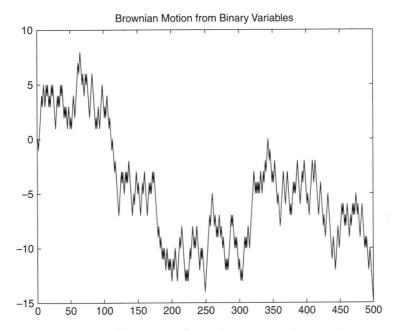

Figure 1.5   Brownian motion.

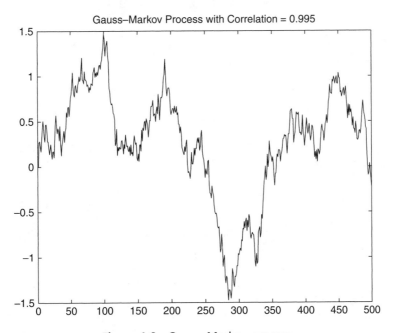

Figure 1.6   Gauss–Markov process.

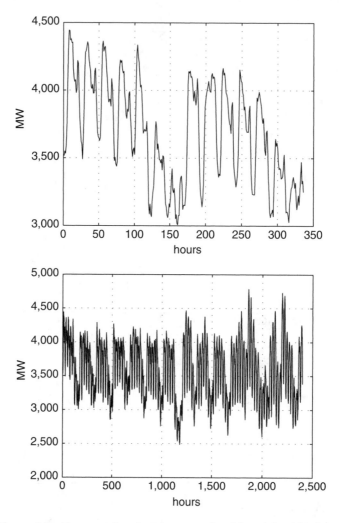

Figure 1.7   Demand for electric power for 14 and for 100 days.

sized geographical area (see Figure 1.7); numbers of customers held in a service queue observed over time

- Measurement Errors: errors made in measuring physical quantities such as voltage, time, mass, or length
- Information Sources: letters or words in an English text; phoneme strings or speech amplitude waveforms generated by natural language (English speech); images on the World Wide Web (e.g., JPEG files); video images
- Pseudorandom Sources: mathematically generated random numbers (see Section 1.11, Appendix 3) used in simulations, an increasingly important activity that complements theory and experiment, and in cryptography

- Devices: quantum effects in solid state devices; random emission of photons in optical sources; uneven distribution of dopants in semiconductors; thermal noise present in all dissipative physical systems such as electrical components; shot noise apparent at low currents when the "granularity" of the flow of small numbers of electrons becomes apparent as a noise; semiconductor/flicker/"$1/f$" noise, which has increasing power at low frequencies and limits low-frequency amplification and the stability of oscillators and clocks; Barkhausen noise generated by random movement of magnetic domains in magnetic materials; turbulence in plasma and fluid dynamics

Physical sources of chance can be grouped roughly as arising either from individual phenomena governed by quantum mechanics or collective or large-scale phenomena governed by statistical mechanics. Figure 1.8 shows the quantum interference between electrons arriving individually in a two-slit configuration. It is a demonstration of the statistical aspects of quantum phenomena. Successive frames show increasing numbers of electrons that passed through the configuration and the increasingly visible interference pattern they generate. Curiously, displays of this quality became available only in a 1989 article by Tonomura et al. [91], although the phenomenon had long been a talking point of quantum mechanics texts and there were precursor experiments.

Buildup of electron interference
pattern. Panels, from top:
10 electrons
100 electrons
3,000 electrons
20,000 electrons
70,000 electrons

Figure 1.8   A. Tonomura, et al., first single-electron interference demonstrations (from Tonomura, A., J. Endo, T. Matsuda, T. Kawasaki, and H. Ezawa. "Demonstration of single-electron buildup of an interference pattern." *American Journal of Physics*, vol. 57: 117–120, 1989).

1.4 MODELING RANDOM PHENOMENA I: RANDOM EXPERIMENT
AND SAMPLE SPACE

How do you recognize whether a phenomenon should be modeled as a random experiment? Guidelines, of which only one need apply, are provided by the following considerations encompassing the random phenomena noted in Section 1.3:

- repetition of the experiment with careful control of the initial conditions yields significantly different outcomes (chaos);
- the experiment is recognized as a transformation of another experiment previously determined to be a random experiment;
- palpable influence on the experimental outcome of irreducibly random quantum mechanical phenomena—as illustrated by electron diffraction, random electron and photon emission times, and quantum tunneling devices;
- enormous complexity of the experiment—as illustrated by the statistical mechanics of the huge number ($O(10^{23})$) of molecules in gases at macroscopic volumes and standard temperatures and pressures and by issues involving the behavior of large numbers of interacting individuals in a large community such as the Internet or an economy;
- enormous complexity of the outcomes of the experiment, with complexity understood as computational complexity and more specifically as a version of Kolmogorov complexity (see Li and Vitanyi [59])—implies inability to compress a data sequence beyond a fraction of its length;
- failure of careful study of data to identify an accurate deterministic prediction rule (function)—predicting future stock prices from present and past prices and outcomes of future coin tosses from past ones;
- lack of knowledge of what will be observed when the experiment is performed—as illustrated by the identity of the topmost card in a well-shuffled deck of cards or by the next utterance of a speaker.

In a *random experiment* $\mathcal{E}$, experimental conditions are established for its performance that produce the outcomes of interest. However, repeating the same experimental conditions may well result in a different experimental outcome (e.g., tossing a coin successive times, with due care being taken to toss it in the same manner each time). Putting aside the important specification of the experimental conditions defining the random experiment $\mathcal{E}$, the outcomes of $\mathcal{E}$ are characterized by the following three components:

RE1. its possible outcomes Ω
RE2. the collection of sets of outcomes of interest $\mathcal{A}$
RE3. a numerical assessment P of the likelihood or probability of occurrence of each of the outcomes of interest.

Turning first to the possible outcomes, we introduce the *sample space* denoted by Ω. (Others sometimes use the notation $\mathcal{S}$.) The formal introduction of a sample space was proposed first by Richard von Mises around 1919. There are four desiderata for a sample space:

SS1. list the possible outcomes of the experiment;

SS2. do so without duplication;

SS3. do so at a level of detail sufficient for our interests;

SS4. make this list complete in a practical sense, although usually not complete regarding all logically or physically possible outcomes.

This list or sample space Ω is modeled mathematically as a set. (See Section 1.9, Appendix 1.) A single toss of a coin might have the traditional sample space $\Omega = \{head, tail\}$, or we consider that the coin could physically balance on edge and have $\Omega = \{head, tail, edge\}$ (SS1). A less strained possibility is to take into account the (x, y) coordinates of the center of the coin when it lands after being tossed into the air or the angle θ the coin makes (referred, say, to the head being upright) with a chosen direction. Much more is known about the outcome of a coin toss than simply the traditional binary outcomes of *head* and *tail*. We ignore this additional information (SS3) on the unstated assumption that there is a bet with payoffs that depend only upon which side is uppermost and not upon the location or orientation of the coin (SS4). If we seek to model the number of packets received per second by a networked computer, then a possible Ω would be the set of nonnegative integers. However, there is an upper limit to the number set by the capacity of the communication link to the computer. Typically, we would ignore this upper bound, at least on a first modeling attempt. The measurement of the instantaneous voltage at a standard 120 VAC receptacle at a randomly chosen time could be modeled by a sample space of all the real numbers ranging between -170V and $+170$V. However, if we used a measuring instrument having finite precision, then this would be an idealization and only a discrete, finite set of numbers are possible measurements. Modeling speech utterances provides an application in which different choices are often made as to the level of detail provided by the sample space. We might model the sample space of speech utterances of a lecturer by the set of real-valued functions of time (speech amplitudes as recorded by a microphone) that are time limited to 50 minutes and band limited to 4 KHz. More crudely, we might model the utterance by the set of strings of phonemes of length no more than 30,000. Finally, we might simplify further and model the utterances by a text transcript of English words such as the one produced by a court stenographer.

1.5 MODELING RANDOM PHENOMENA II: EVENTS AND COMBINATIONS OF EVENTS

We employ an *event language* in which the statements about occurrences of combinations of events are represented by combinations of sets, constructed by using Boolean set operations of complementation, union, and intersection (see Section 1.9, Appendix 1), as in Definition 1.5.1.

Definition 1.5.1   (Event Language)

"A occurs" is represented by the set A;

"A does not occur" is represented by the set complement A^c;

"either A or B occurs" is represented by set union $A \cup B$;

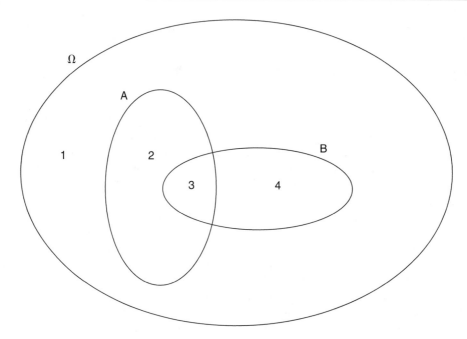

Figure 1.9   Venn diagram for two sets A, B in a sample space Ω.

"both A and B occur" is represented by set intersection $A \cap B$;
"all of $A_\alpha, \alpha \in \mathcal{I}$ occur" by intersection of multiple sets $\cap_{\alpha \in \mathcal{I}} A_\alpha$;
"at least one of $A_\alpha, \alpha \in \mathcal{I}$ occurs" by union of multiple sets $\cup_{\alpha \in \mathcal{I}} A_\alpha$.

Venn diagrams (see Figure 1.9) enable us to develop intuitive insights into these relationships, but the relationships they suggest between events may only be true for the particular two-dimensional picture you have drawn.

Example 1.1   Venn Diagram _____
In Figure 1.9, the sample space is the outer oval, and two sets, A and B, are denoted by inner overlapping ovals. In terms of the four numbered regions considered as sets,

$$\Omega = 1 \cup 2 \cup 3 \cup 4, \ A = 2 \cup 3, \ B = 3 \cup 4, \ A \cup B = 2 \cup 3 \cup 4, \ A \cap B = 3,$$

$$A^c = 1 \cup 4, \ B^c = 1 \cup 2, \ A - B = 2.$$

It is often convenient to represent a set A algebraically by a function I_A having domain (set of arguments of the function) Ω and binary-valued range (set of possible values of the function) $\{0, 1\}$.

Definition 1.5.2   (Indicator Function)   The indicator function I_A for an event or set A is given by

$$I_A : \Omega \to \{0, 1\}, \quad I_A(\omega) = \begin{cases} 1 & \text{if } \omega \in A \\ 0 & \text{otherwise} \end{cases} \quad A = \{\omega : I_A(\omega) = 1\}.$$

Example 1.2   Indicator Functions _____

If $I_A(\omega)$ is identically 1, then A is the sample space Ω.

If $I_A(\omega)$ is identically 0, then A is the empty set (set with no elements) $\emptyset$.

If $I_A(\omega)$ is 1 only for ω equal to a designated ω_0, then A is the set $\{\omega_0\}$ containing the single element ω_0.

We note that the *one-to-one relationship* (abbreviated as 1:1) between sets and their *indicator functions* is

$$A = B \iff (\forall \omega \in \Omega)\ I_A(\omega) = I_B(\omega),$$

where $\iff$ stands for *if and only if*. The fact that sets are equal if and only if their indicator functions are identical allows us to exploit the arithmetic of indicators:

$$I_{A^c} = 1 - I_A,$$

$$A \subset B \iff I_A \leq I_B,$$

$$I_{A \cap B} = \min(I_A, I_B) = I_A I_B,$$

$$I_{A \cup B} = \max(I_A, I_B) = I_A + I_B - I_{A \cap B},$$

to construct rigorous arguments regarding relationships between sets. In effect, we transform statements about sets into statements about functions and can then use our familiarity with algebra to resolve less familiar questions about sets. There are many references to set theory written at a variety of levels, with Halmos [45] and Suppes [87] being good sources.

1.6 MODELING RANDOM PHENOMENA III: EVENT COLLECTIONS AND ALGEBRAS

While one might think at first that, given the sample space Ω, we can be interested in all of its subsets, and this is occasionally true, we have three reasons for expecting that we may only be interested in certain of these subsets. First of all, the sample space may contain a level of detail beyond our present concerns. For example, it may represent a single toss of a die and have six elements, although the wager we are contemplating only depends upon whether the outcome is odd or even. Ω may contain not only which side of the coin landed uppermost, but also the (x, y) coordinates of the center of the coin, whereas the bet we have placed depends only upon which side is uppermost. Second, we intend to assign, to each event A, a numerical probability

$P(A)$. As such probabilities are based on some knowledge of the tendency of the event to occur or of the strength of our belief in its occurrence (see Section 3.3), our knowledge of P may not extend to all possible subsets of Ω. The third (and technical) reason for limiting the collection of events of interest is that the requirements on P imposed by the Kolmogorov axioms to be given in Section 3.5.2 may not allow P to be defined over all subsets of Ω when it is defined in reasonable ways over events (e.g., intervals when $\Omega = \mathbb{R}$) of great interest. The collection (set) $\mathcal{A}$ of subsets of the sample space Ω (be warned that $\mathcal{A}$ is a set whose elements are also sets!) that are the events of interest regarding the random experiment $\mathcal{E}$ and concerning which we have knowledge of their likelihood of occurrence is called an *event algebra*.

Definition 1.6.1   (Event Algebra)   An event algebra $\mathcal{A}$ is a collection of subsets of the sample space Ω such that it is

a. nonempty (this is equivalent to $\Omega \in \mathcal{A}$);
b. closed under complementation (if $A \in \mathcal{A}$ then $A^c \in \mathcal{A}$);
c. and closed under finite unions (if $A, B \in \mathcal{A}$ then $A \cup B \in \mathcal{A}$).

From the de Morgan laws (Section 1.9.2), we see that $\mathcal{A}$ is closed under finite intersection as well.

Example 1.3   Event Algebras _____

1. The smallest $\mathcal{A} = \{\phi, \Omega\}$, where ϕ denotes the *empty set*;
2. The largest $\mathcal{A}$ is the set of all subsets of Ω known as the *power set* and denoted by 2^Ω;
3. An intermediate example is

$$\Omega = \{1, 2, 3\}, \quad \mathcal{A} = \{\Omega, \phi, \{1\}, \{2, 3\}\}.$$

Theorem 1.6.1   (Intersections of Algebras)   Let $\mathcal{A}_1, \mathcal{A}_2$ be algebras of subsets of Ω and let $\mathcal{A} = \mathcal{A}_1 \cap \mathcal{A}_2$ be the collection of sets common to both algebras. Then $\mathcal{A}$ is an algebra.

Proof.   Since $\mathcal{A}_1, \mathcal{A}_2$ are algebras, they both contain Ω. Hence, Ω is in $\mathcal{A}$. If $A \in \mathcal{A}$, then A is in both $\mathcal{A}_1, \mathcal{A}_2$. Thus, A^c is in both $\mathcal{A}_1, \mathcal{A}_2$ and therefore in their intersection $\mathcal{A}$. If $A, B \in \mathcal{A}$, then they are in both $\mathcal{A}_1, \mathcal{A}_2$. Consequently, $A \cup B$ is in both $\mathcal{A}_1, \mathcal{A}_2$ and therefore in $\mathcal{A}$. We have now verified that the collection $\mathcal{A}$ satisfies the three conditions of Definition 1.6.1 and is therefore an algebra of events.   □

Corollary 1.6.1   There is a smallest (in the sense of inclusion) algebra containing any given family of subsets of Ω.

An example of such an algebra is the smallest algebra $\mathcal{A}$ of subsets of $\Omega = \mathbb{R}$ containing all of the four kinds of intervals. (See Section 1.9.1.) In this case, every set $A \in \mathcal{A}$ can be written as a finite union of intervals. To verify this, note that $\Omega = (-\infty, \infty) \in \mathcal{A}$, finite unions (and finite intersections) of finite unions of intervals are again finite unions of intervals, and the complement of a finite union of intervals is also a finite union of intervals.

Definition 1.6.2   (Cardinality)   The cardinality (or size) of a collection A, denoted $\|A\|$, is the number of elements of the collection. This number may be finite or infinite.

Examples can be found in Section 1.9.1.

Theorem 1.6.2   (Cardinality of Algebras)   An algebra of subsets of a finite set of n elements will always have a cardinality of the form 2^k, $k \leq n$; the cardinality of the power set is 2^n.

Example 1.4   Algebra Cardinality _____
For example, $\Omega = \{a, b, c, d\}$, $n = 4$, $\mathcal{A} = \{\emptyset, \Omega, \{a, c\}, \{b, d\}\}$ $\|\mathcal{A}\| = 2^2$.

An intermediate-sized event algebra can be constructed by first partitioning Ω into subsets and then forming the power set of these subsets, with an individual subset now playing the role of an individual outcome ω. Introduce the notion of an *index set* $\mathcal{I} = \{\alpha\}$ that is itself a set of labels or names for the subsets of Ω that are to be included in an event collection. Typically, if the collection is finite or countably infinite (as many members as there are integers), then we take $\mathcal{I}$ to be an appropriate subset of the integers. We first make precise the notion of a *partition* through Definition 1.6.3.

Definition 1.6.3   (Partitions)   A partition $\Pi = \{A_\alpha, \alpha \in \mathcal{I}\}$ of Ω is a collection of subsets of Ω (in this case, indexed or labeled by α taking values in an index or label set $\mathcal{I}$) satisfying

P1.   For all $\alpha \neq \beta$, $A_\alpha \cap A_\beta = \phi$, the empty set (see Section 1.9.1) (pairwise disjoint);
P2. $\cup_{\alpha \in \mathcal{I}} A_\alpha = \Omega$ (the collection covers Ω).

Hence, each element $\omega \in \Omega$ lies in one and only one of the sets $A_\alpha \in \Pi$.

Example 1.5   Algebra from Partitions _____
For example,

$$\Omega = \mathbb{R} = (-\infty, \infty), \quad \mathcal{I} = Z, \quad A_\alpha = (\alpha - 1, \alpha], \quad \Pi = \{A_\alpha, \alpha \in \mathcal{I}\}$$

is a partition of the real numbers into countably infinitely many intervals of unit length. If the partition Π is countable, then we can form an algebra $\mathcal{A}$ by including all unions of subcollections:

$$\mathcal{A} = \{\cup_{\alpha \in \mathcal{I}'} A_\alpha : \mathcal{I}' \subset \mathcal{I}\}.$$

1.7 DEALING WITH CHANCE DATA—STATISTICS

1.7.1 Statistics

While your study of probability must necessarily cover theory, it should also give you contact with the random phenomena, and the data they generate, that motivate and direct your study. It is these phenomena and data that we need to model and draw inferences from. Probabilities, the remaining element in our specification of a random experiment $\mathcal{E} = (\Omega, \mathcal{A}, P)$, are determined sometimes by physical theory, as in quantum mechanics and statistical mechanics, and sometimes by qualitative properties of $\mathcal{E}$ that are known to lead to certain probabilities, as in applications of the Central Limit Theorem to establish a normal distribution and in the derivations of the binomial, Poisson, and exponential probabilities given in the sequel. In engineering applications of probability, it is the rule that data play a large part in the selection of a unique probability model or assignment. Data do not speak for themselves—nor do facts. We require some background knowledge to understand the information conveyed by data. Nevertheless, data are essential in determining objective probabilities. We process data (e.g., repeated observations on some phenomena, long time records of some quantity) to extract its essentials in a model—to separate the wheat of forecastable regularity from the chaff of noise. Typically, if, after some effort expended on searching, we cannot identify a "simple" formula that will accurately predict, or enable us to reconstruct, the data, we then look for a probabilistic description. Ideally, we look for a few numerical characteristics of the data set, called *statistics*, such that we cannot effectively compress the observed data relative to all possible data sets sharing the same values of the statistics. Practically, we look for characteristics or functions of the data we can calculate that both inform us about the random experiment generating the data and that do not change their values significantly when we repeat the random experiment.

Statistics is the discipline concerned with chance/random data, with the modeling of such data by probabilities, and with the implications of probability models for observations from random phenomena. There is a very large literature on statistical theory and applications, written at a wide range of mathematical levels and from a variety of philosophical and applicational viewpoints. (See Blackwell [12] and Freedman, et al. [34] for an introduction.) We briefly digress into the realm of statistics to introduce several commonly employed tools for processing chance data to expose underlying significant relationships. These tools will be used to treat data sets introduced as examples of random phenomena. Subsequently, in pursuit of Goal 3, we will introduce formal elements of statistical estimation and decision making. For the present, we group our search for statistics into those providing stable information about the following topics:

- typical size of the random experimental outcome;
- degree of fluctuation in outcomes coming from repeated random experiments;
- dependencies between outcomes of repeated experiments or between experimental outcomes that are multidimensional (e.g., the height and weight of a given individual, voltages at two different nodes in a circuit).

1.7.2 Sample Mean, Median, Variance, and Histogram

Given a set of repeated, unlinked scalar (real-valued) measurements $x_1, \ldots, x_n$ of the same characteristic (e.g., number of photons emitted by a light source in consecutive time intervals

$[kT, (k + 1)T)$, grades of different students on the same examination), we often estimate the size of the typical measurement by the *sample mean*. (See expected value in Chapter 9.)

Definition 1.7.1   (Sample Mean)   The sample mean $\hat{x}$ of a given series $\mathbf{x} = (x_1, \ldots, x_n)$ is given by

$$\hat{x} = \frac{1}{n} \sum_{i=1}^{n} x_i.$$

Thus, if

$$x = 1, 2, 3, 4, 5, 6, 7, 8, 9, 10, 11, 48, \tag{1.7.1}$$

then $\hat{x} = 9.5$.

Example 1.6   Sample Mean _____

A more complex example is provided by the following matrix of data in which the 10 numbers in a column are the outcomes of a random experiment and the six rows correspond to repetitions of this experiment:

```
-1.1878    -1.1859     0.1286     0.8057    -0.3306    -0.1199
-2.2023    -1.0559     0.6565     0.2316    -0.8436    -0.0653
 0.9863     1.4725    -1.1678    -0.9898     0.4978     0.4853
-0.5186     0.0557    -0.4606     1.3396     1.4885    -0.5955
 0.3274    -1.2173    -0.2624     0.2895    -0.5465    -0.1497
 0.2341    -0.0412    -1.2132     1.4789    -0.8468    -0.4348
 0.0215    -1.1283    -1.3194     1.1380    -0.2463    -0.0793
-1.0039    -1.3493     0.9312    -0.6841     0.6630     1.5352
-0.9471    -0.2611     0.0112    -1.2919    -0.8542    -0.6065
-0.3744     0.9535    -0.6451    -0.0729    -1.2013    -1.3474
```

If we now compute the sample means of the individual columns, we obtain

```
-0.4665    -0.3757    -0.3341     0.2245    -0.2220    -0.1378
```

It is clear that the range of variation between the sample means (from $-.47$ to $.22$) is less than that found in the actual values (from -2.2 to 1.5). The example is more convincing when we lengthen the columns. Although it would take up too much space to do so, had we lengthened the columns, we would have found that the column means all started approaching the common value of zero.

A desirable characteristic of the sample mean is that linear scale changes of the data (e.g., shifting from the Fahrenheit to centigrade temperature scales or from miles to kilometers) induce the same linear scale change in the sample mean—the sample mean is invariant under linear scaling. A problem with the sample mean is that it is sensitive to extreme, but atypical, values in the data. Data whose values range over several orders of magnitude will have their sample mean determined by a few points. In the example of Equation 1.7.1, the sample mean of 9.5 becomes only 6 if we drop the single last measurement of 48.

This lack of *robustness* is corrected by an alternative common indicator of the size of the measurement given by the *sample median*. (See cumulative distribution functions in Section 5.2.) Roughly, $\hat{\mu}$ is a middle value such that half of the measurements are no larger than $\hat{\mu}$ and half are no smaller.

Definition 1.7.2 (Sample Median)   Given a series $\mathbf{x} = (x_1, \ldots, x_n)$, arrange it in nondecreasing order as $\mathbf{y} = (y_1, \ldots, y_n)$, with for all $1 \le i < n$, $y_i \le y_{i+1}$.

$$\hat{\mu} = \begin{cases} y_{(n+1)/2} & \text{if } n \text{ is an odd integer,} \\ 0.5y_{n/2} + 0.5y_{1+n/2} & \text{if } n \text{ is an even integer.} \end{cases}$$

For the 12 data points listed above in Equation 1.7.1, $\hat{\mu} = 6.5$; had we used only the first 11 of these points, then the median would have shifted only slightly, to 6. The sample median also enjoys the property of invariance under linear scale changes of the data.

In addition to measures of the size of a random quantity, we are also interested in measures of the randomness or fluctuation in the data. The most common such measure is the *sample standard deviation*. (See variance in Section 9.10.2.)

Definition 1.7.3 (Sample Standard Deviation)   Given a series $\mathbf{x} = (x_1, \ldots, x_n)$, we define the sample standard deviation

$$\hat{\sigma} = \sqrt{\frac{1}{n-1} \sum_{i=1}^{n} (x_i - \hat{x})^2},$$

which measures the departure of the data points from the sample mean.

The sample standard deviation of the 12 points listed above in Equation 1.7.1 is 3.61. A larger value of $\hat{\sigma}$ suggests greater variability or fluctuation in the data. The square of the sample standard deviation is known as the *sample variance*.

A more comprehensive measure of fluctuation and display of the data set $x_1, \ldots, x_n$ is provided by the *histogram*. (See the probability density function in Section 6.2 and the related empirical cdf in Section 5.5.) We assume that the order of presentation of the data set is irrelevant. The measurements $\{x_i\}$ could as well have been made in any other ordering. Thus, $\{x_i\}$ do not represent noisy measurements of the altitude of a vehicle moving along a parabolic trajectory, a history of stock prices, or a sequence of speech waveform amplitudes measured at equally spaced sampling times.

Definition 1.7.4 (Histogram)   Select an integer b to specify the number of bins in the histogram $H = \{H(1), \ldots H(b)\}$. Given a series $\mathbf{x} = (x_1, \ldots, x_n)$, let

$$y_1 = \min_i x_i \text{ and } y_n = \max_i x_i.$$

Define the number $H(j)$ of points of $\mathbf{x}$ lying in the jth bin as the number of points of $\mathbf{x}$ lying in the jth interval

$$\left[y_1 + \frac{(j-1)}{b}(y_n - y_1), y_1 + \frac{j}{b}(y_n - y_1) \right).$$

H is often presented as a bar plot with the jth bar of height $H(j)$.

There is a balance between the parameters b and n in that the larger b is, the more detailed is the histogram about the data, but the smaller $k = n/b$ is, the average number of data points in a bin, the less statistical stability and reliability there is in the histogram report. For the 12 data point set given in Equation 1.7.1, $\min_i x_i = 1$ and $\max_i x_i = 48$. If we choose $b = 4$, then the histogram intervals are $[1, 51/4), [51/4, 98/4), [98/4, 145/4) [145/4, 48)$ and $H(1) = 11$, $H(2) = H(3) = 0$, $H(4) = 1$. This is then plotted as a bar graph with $H(j)$ the height over the jth interval, to obtain the histogram display. Examples of histograms are provided in Figures 1.12 and 1.13 in Section 1.11, Appendix 3.

While we can always compute any of the aforementioned statistics, how are we to know that they reflect representative and stable characteristics of the process that generated the data? For example, if the amount of data were to double, would the newly calculated sample mean be close to the original sample mean? A useful statistic should exhibit some stability in that, as the amount of data increases, the fluctuations diminish in the sequence of recalculated statistics, and the sequence appears to converge or stabilize in a small range of values. We will examine the long-run behavior of statistics when we consider the laws of large numbers. The stability of the relative frequency of the number of occurrences of an event observed over a large number of repeated, unlinked performances of a random experiment will provide a statistical basis for probability and an interpretation of this concept, as discussed in Section 3.4.

1.7.3 Multivariable Data and Scatterplots

In many instances, the data set is multivariable or vector valued. For example, we might collect from each student in a class his or her height and weight $\mathbf{x}_i = (h_i, w_i)$; or we might have data on the past and present values of a time series (e.g, stock prices), together with a future value $(x_{t-k}, \dots, x_t, x_{t+1})$; or the vector might contain diagnostic measurements, as well as the true condition of the system. A simple way to examine the dependencies between these variables, and display such data, is through a *scatterplot* of two or three selected coordinates. For example, if we take the Gauss–Markov process data shown in Figure 1.6 and plot x_t vs. x_{t-1}, we obtain the scatterplot shown in Figure 1.10. Although it is not evident from Figure 1.6, there is a great deal of structure in the Gauss–Markov process that is exhibited by the tight linear dependence of x_t upon x_{t-1}, shown in Figure 1.10.

A real example is that of the dependence of demand for electric power, measured in megawatts (MW) in a moderately sized geographical region, on the average temperature, in degrees Fahrenheit, in that same region. Knowledge of the temperature, season, and day of the week can help the electric utilities responsible for providing power to that region plan ahead to meet the demand. Figure 1.11 plots the demand for power at 10 A.M. versus the 10 A.M. temperature. The dependence upon the day of the week is shown for the separation between weekdays (points shown as circles) and weekends (points shown as pluses). The total data set is for one year.

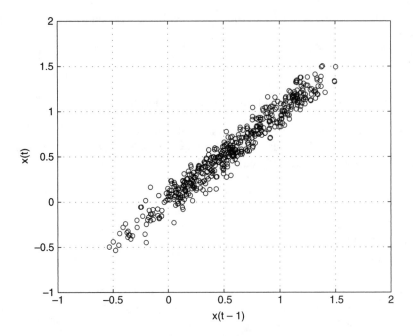

Figure 1.10 Scatterplot of Gauss–Markov Process $x(t)$ vs. $x(t-1)$.

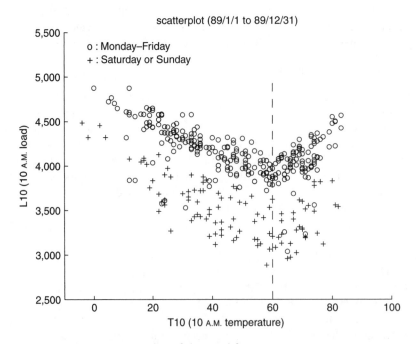

Figure 1.11 Scatterplot of demand for power vs. temperature.

There is clearly a nonlinear relationship between demand and temperature and a distinction between demand on weekdays and on weekends, although the latter data sets overlap. Furthermore, this relationship is stochastic in that, at any given approximate temperature, there is a range of demands for either type of day. How can we model this data, and, given the mathematical model, how do we make high-quality forecasts from temperature and day of the week to demand? While we will not return to answer this specific question, in the course of our study of probability we will provide approaches to both modeling and to optimal inference.

1.8 SUMMARY

Orientation and introduction is provided through a brief historical review of the development of ideas of possibility or contingency (lack of necessity), chance, numerical measures of chance, models of random phenomena, and applications of these models to everyday and scientific concerns through the development of statistics. The kinds of random phenomena of particular interest to present-day electrical engineering are illustrated and include the randomness inherent in electrical devices arising either from quantum mechanical processes that become evident in very small electrical structures; noise processes (especially thermal noise) and collective phenomena arising from statistical mechanical aspects of larger electrical structures; issues of reliability of components and systems; and understanding the performance of electrical systems having random elements that operate in random environments. The phenomena of interest to electrical engineering also include such nonphysical sources as information sources, examples of which are natural language either written as text or spoken as speech; data sources that have a formal structure and need to be stored, compressed, transmitted reliably, and interpreted; and the usual multimedia services and sources including text, sound, and image that drive today's growth in communications systems development and deployment.

We abstract random phenomena as arising from the performance of random experiments and denote a random experiment by $\mathcal{E}$. The first step in addressing G1 (mathematical models of random phenomena) is to formalize $\mathcal{E}$ through a mathematical model. This model has three components:

- a sample space Ω that is a set listing the possible outcomes of $\mathcal{E}$,
- an algebra $\mathcal{A}$ of subsets of Ω, called events (yes/no questions about the outcome of $\mathcal{E}$), that makes explicit our interests in modeling $\mathcal{E}$, and
- a probability measure P, assigning a numerical value $P(A)$ to an event A that is the probability of the occurrence of the event A when $\mathcal{E}$ is performed and represents the tendency for A to occur.

The sample space Ω and event algebra $\mathcal{A}$ are discussed and summaries of notation and basic results about sets and Boolean operations on sets are provided in Section 1.9, Appendix 1. The notions of set (e.g., A), complement (A^c), union ($A \cup B$), and intersection ($A \cap B$), and their properties and interrelationships will be needed in much of what is to follow. Occasional use will be made subsequently of the representation of sets by $\{0, 1\}$-valued indicator functions (I_A), enabling us to embed events into random variables introduced later. The probability measure P is introduced in a special case in Chapter 2 and in the general case in Chapter 3.

We close this chapter with a brief examination of basic statistical methods for summarizing the chance data generated by random experiments. These summary statistics include means and variances as well as the more informative histograms and scatter plots. The methods of statistics (e.g., see Blackwell [12]) are essential in linking the theory of probability to practice.

1.9 APPENDIX 1: SETS

1.9.1 Definitions of Sets and Examples

Definition 1.9.1   **(Set)**   A set (typically denoted by a capital Roman letter drawn from the initial part of the alphabet (e.g., $A, B, C, \ldots$), with the major exception of the sample space, is an unordered collection of distinct elements or members (typically denoted by lowercase Roman letters (e.g., $a, b, c, x, y, \ldots$), again with the sample space providing an exception).

See Halmos [45], Suppes [87], Leon-Garcia [58] for discussions of a set. A set may be specified by listing its elements within curly braces (often using ellipses "$\ldots$" for long lists whose rule for membership is implicit in the initial examples of members), as in

$$A = \{0, 1, 2, 3, 5, 8, 13\} \quad B = \{0, 1, 2, \ldots, 1000\}.$$

Alternatively, a set may be specified by stating a rule (proposition) for membership in the set, as in

$$C = \{x : \text{proposition about } x \text{ is true}\},$$

or

$$D = \{x : x \text{ positive, odd integer}\}.$$

A set is an unordered list; therefore,

$$\{1, 2, 3\} = \{2, 3, 1\}.$$

The relation of *set membership* or being an *element of the set* is undefined and represented by $\in$ as in $2 \in \{x : x \text{ a positive integer}\}$. Hence, we may write variously

$$\Omega = \{\omega : \omega \text{ positive integer less than } 7\} = \Omega = \{1, 2, 3, 4, 5, 6\}, \text{ and } 2 \in \Omega.$$

The relation of *not an element of* is denoted by $\notin$, as in

$$3 \notin \{1, 2, 4\}.$$

One needs to be careful to distinguish between an element such as 2 and the set $\{2\}$ containing that element. While $2 \in \{1, 2, 4\}, \{2\} \notin \{1, 2, 4\}$. When the context is understood, we may write simply $\Omega = \{\omega\}$ or $\omega \in \Omega$.

The size $\|A\|$ of a set is the number of its elements and is called its *cardinality*. Cardinalities can be *finite, countably infinite,* or *uncountably infinite.* A finite set can have its elements put into a one-to-one correspondence with the nonnegative integers less than a finite number—its elements can be counted. A countably infinite set has exactly as many members or elements as there are integers (i.e., the set can be put into a one-to-one correspondence with all of the integers). A set is countable if it is either finite or countably infinite. A set is uncountable if it is not countable.

Several important and familiar examples of sets are the following:

- finite and countably infinite sets of integers:

$$N_n = \{0, 1, 2, \ldots, n - 1\},$$

$$Z = \{x : x \text{ an integer}\} = \{\ldots, -3, -2, -1, 0, 1, 2, 3, \ldots\},$$

$$Z^+ = \{x : x \text{ a positive integer}\} = \{1, 2, 3, 4, \ldots\},$$

$$N = \{x : x \text{ a nonnegative integer}\} = \{0, 1, 2, 3, 4, \ldots\},$$

(The "natural" numbers do not include 0.)

- *Intervals* are uncountably infinite subsets of the set of real numbers or the line:

$$\mathbb{R} = \{x : x \text{ real number}\} = \{x : -\infty < x < \infty\},$$

$$(a, b) = \{x : a < x < b\},$$

$$(a, b] = \{x : a < x \leq b\},$$

$$[a, b) = \{x : a \leq x < b\},$$

$$[a, b] = \{x : a \leq x \leq b\}.$$

The first set N_n has exactly n consecutive integer elements. Clearly, $\|N_n\| = n$ and N_n is a finite set. The sets Z, Z^+, and N are all countably infinite, while the sets $\mathbb{R}, (a, b)$, for $a < b$, are uncountably infinite. The sets Z, Z^+, and N differ as sets of integers only in whether they include negative integers and 0. The sets $(a, b), (a, b], [a, b)$, and $[a, b]$ differ only in whether the endpoints of the intervals are included in the set. The special set with no elements, called the *empty or null set*, is denoted $\emptyset$. An example of it is

$$\emptyset = \{\} = \{x : x \in \mathbb{R} \text{ and } x < x\} \text{ or } \emptyset = (a, a).$$

The set $\emptyset$ has cardinality of 0 and is thus a finite set.

The set N_n provides a sample space for the number of *heads* in $n - 1$ tosses of a coin or the number of errors made in transmitting $n - 1$ bits. The set N of natural numbers provides a sample space for the number of photons emitted in a time T by a light source or the number of packets received by an Internet router in time T. The set $\mathbb{R}$ provides a sample space for measurements of voltage or current. The intervals $[-\pi, \pi]$ or $[0, 2\pi]$ are sample spaces for measurements of oscillator phase or the phase of a received communications (e.g., AM/FM) carrier.

1.9.2 Boolean Set Operations

Two sets A, B can be related through *inclusion* (denoted by $A \subset B$ and read "A is a *subset* of B" or "B *contains* A") when each element of A is also in B. We say that $A = B$ (read "A, B are equal") if and only if (iff, $\Longleftrightarrow$) $A \subset B$ and $B \subset A$. If $A \subset B$, then we can also say that $B \supset A$ (read "B is a *superset* of A"). An *event* A is always a subset of the sample space Ω. The notion of subset, as is true for the arithmetic notion of inequality $\leq$, has the properties of **reflexivity** ($A \subset A$) and **transitivity** ($A \subset B$, $B \subset C \Rightarrow A \subset C$). However, it does not share with $\leq$ the property of **completeness**: For any real numbers x, y, it must be that either $x \leq y$ or $y \leq x$, but it is not true for all sets A, B that either $A \subset B$ or $B \subset A$.

Identity or equality $A = B$ of two sets A, B means that they have precisely the same collection of elements. A basic method for proving $A = B$ is to first prove that $A \subset B$ (every element of A is in B) and then to prove that $B \subset A$ (every element of B is in A).

Sets can be transformed through the following *Boolean set operations*:

> **complementation** A^c or A', $A^c = \{\omega : \omega \text{ not in } A\} = \{\omega : \omega \notin A\}$,
> **union** $A \cup B = \{\omega : \omega \in A \text{ or } \omega \in B\}$,
> **intersection** $A \cap B = AB = \{\omega : \omega \in A \text{ and } \omega \in B\}$, and
> **difference** $A - B = A \cap B^c = \{\omega : \omega \in A \text{ and } \omega \notin B\}$.

These operations were formally introduced in the contexts of both deductive and probabilistic reasoning by Boole [13].

Example 1.7   Boolean Set Operations
As an illustration of these operations take

$$\Omega = N_8, \quad A = \{0, 1, 5\}, \quad B = \{1, 2, 3, 4\}.$$

It follows that

$$A^c = \{2, 3, 4, 6, 7\}, \quad A \cup B = \{0, 1, 2, 3, 4, 5\}, \quad A \cap B = \{1\}, \quad A - B = \{0, 5\}.$$

If $A \cap B = \emptyset$, then A, B have no elements in common, and we say that A, B are *disjoint* and use the (nonstandard) notation $A \perp B$.

Interrelationships and properties of these operations (letting $\circ$ denote either union or intersection, but fixed at either throughout an expression) include the following:

idempotence

$$(A^c)^c = A;$$

commutativity (symmetry)

$$A \circ B = B \circ A;$$

associativity

$$A \circ (B \circ C) = (A \circ B) \circ C;$$

distributivity

$$A \cap (B \cup C) = (A \cap B) \cup (A \cap C), \text{ and}$$

$$A \cup (B \cap C) = (A \cup B) \cap (A \cup C);$$

de Morgan laws

$$(A \cup B)^c = A^c \cap B^c \qquad (A \cap B)^c = A^c \cup B^c.$$

The de Morgan laws allow us to express union in terms of complements and intersection and intersection in terms of complements 2nd unions.

The notions of union and intersection extend to collections of arbitrarily many sets through the two *quantifiers*:

there exists $\exists$;
for all $\forall$.

If we have a collection $\{A_\alpha, \alpha \in \mathcal{I}\}$ of subsets of Ω indexed by an index set $\mathcal{I}$, then

$$\bigcup_{\alpha \in \mathcal{I}} A_\alpha = \{\omega : (\exists \alpha \in \mathcal{I}) \ \omega \in A_\alpha\}, \quad \bigcap_{\alpha \in \mathcal{I}} A_\alpha = \{\omega : (\forall \alpha \in \mathcal{I}) \ \omega \in A_\alpha\}.$$

For example, if $\Omega = N$, $\mathcal{I}$ is the set of positive integers divisible by 3 and $A_\alpha = N_\alpha$, then

$$\bigcup_{\alpha \in \mathcal{I}} N_\alpha = N, \quad \bigcap_{\alpha \in \mathcal{I}} N_\alpha = N_3.$$

1.9.3 Cartesian Products

A different construction of a new set C from two given sets A and B is that of the Cartesian product of sets. By an ordered pair (a, b) of elements $a \in A$ and $b \in B$, we mean an object such that

$$(a_1, b_1) = (a_2, b_2) \iff a_1 = a_2 \text{ and } b_1 = b_2.$$

Definition 1.9.2   (Cartesian Product)   The Cartesian product $A \times B$ of two given sets A and B is the set of all ordered pairs, the first element of which is from A and the second of which is from B:

$$A \times B = \{(a, b) : a \in A, \ b \in B\}.$$

For example, if $A = \{1, 2, 3\}$ and $B = \{c, d\}$, then

$$A \times B = \{(1, c), (1, d), (2, c), (2, d), (3, c), (3, d)\}.$$

We will make little explicit use of this construction, although it provides the means for constructing a new sample space for a joint experiment from the individual sample spaces.

1.10 APPENDIX 2: FUNCTIONS

Functions will be very familiar to any engineering student. Nevertheless, we review some important notation and points to keep in mind. We typically denote functions by lowercase or uppercase roman letters such as f and F. A function f, when supplied with an argument, often denoted by x, returns a value, often denoted by y. That a function f is a function of some argument is sometimes indicated by the notation $f(\cdot)$, where the center dot $\cdot$ is a placeholder for an argument of the function. When we need to be more explicit about the source of arguments and the possible values of the function f, we supply the *domain*, a set $\mathcal{D}$ of possible arguments of the function, and a *range*, a set $\mathcal{R}$ of possible values for the function. We list this information as

$$f : \mathcal{D} \to \mathcal{R},$$

which is read as "the function f maps from the domain set $\mathcal{D}$ to the range set $\mathcal{R}$." Thus, the common quadratic function $f(x) = x^2$ is typically understood as accepting any real-valued x and returning any nonnegative value x^2:

$$f : \mathbb{R} \to [0, \infty).$$

Somewhat less fussily, we might just say that

$$f : \mathbb{R} \to \mathbb{R}.$$

As another example, the cardinality $\|A\|$ of a subset A of a countable set Ω is a function on the power set 2^Ω (set of all of the subsets of Ω) to the nonnegative integers N:

$$f : 2^\Omega \to N.$$

How then are functions defined or specified? In the few instances in which the domain and range sets are (small) finite sets, we can specify a function by giving a list or table providing the element y of its range $\mathcal{R}$ corresponding to each element x of its domain $\mathcal{D}$. Thus,

$$\mathcal{D} = \mathcal{R} = \{0, 1\}, \quad f(0) = 1, \quad f(1) = 0,$$

is the function that complements a bit. In principle, this is the fundamental way to define a function. A function f is defined by providing a subset $\mathcal{G}$, called its *graph*, of the Cartesian product $\mathcal{D} \times \mathcal{R}$

of its domain and range. The graph $\mathcal{G}$ is a collection of ordered pairs $\{(x, y) : x \in \mathcal{D}, y \in \mathcal{R}\}$ defining f for each input or argument x through the second coordinate y. Although the functions we consider are all single valued (meaning that, for each $x \in \mathcal{D}$, there is a single $y \in \mathcal{R}$ such that the pair $(x, y) \in \mathcal{G}$), it is possible to define multivalued functions. In the common graph of the quadratic function, the set $\mathcal{D} \times \mathcal{R}$ is represented by all of the points in the plane (infinite sheet of paper), and the graph $\mathcal{G}$ is the set of "black" (pencil) points lying on the parabola. In practice, we often do not define functions in this abstract manner. Rather, we provide an algorithm or program for computing values of f given x.

It is common in basic mathematics courses to make little of the distinction between the arguments and values of a function, because most examples share the same set of real numbers. However, in applications of probability it becomes more important to understand this distinction. Just as keeping track of the units of physical quantities can eliminate possible formulas linking them (because the units produced by one formula may not agree with the units for the desired quantity; e.g., the units of $.5vt$, velocity times time, are those of distance and thus this cannot be an expression for energy, although $.5vt^2$ is an expression for kinetic energy), so can keeping track of the domain and range sets of a probability function guard us from misusing the function. For example, we will later encounter certain functions (probability mass functions) defined only for the domain N of integers and thereby know that they cannot apply to real-valued quantities such as voltages.

1.11 APPENDIX 3: PSEUDORANDOM NUMBERS GENERATED BY MATLAB

A brief discussion of a common arithmetical approach to generating pseudorandom numbers is provided in Section 8.4. The first set is generated by the Matlab command `randn`:

```
  1.1649535e+00    -3.6002963e-01     3.7504102e-01    -5.5709364e-01
  6.2683908e-01    -1.3557629e-01     1.1251618e+00    -3.3670570e-01
  7.5080155e-02    -1.3493385e+00     7.2864159e-01     4.1522746e-01
  3.5160690e-01    -1.2704499e+00    -2.3774543e+00     1.5578135e+00
 -6.9651254e-01     9.8457027e-01    -2.7378242e-01    -2.4442989e+00
  1.6961425e+00    -4.4880614e-02    -3.2293992e-01    -1.0981954e+00
  5.9059778e-02    -7.9894452e-01     3.1798792e-01     1.1226479e+00
  1.7970718e+00    -7.6517243e-01    -5.1117221e-01     5.8166726e-01
  2.6406853e-01     8.6173490e-01    -2.0413453e-03    -2.7135430e-01
  8.7167329e-01    -5.6225124e-02     1.6065110e+00     4.1419131e-01
 -1.4461715e+00     5.1347817e-01     8.4764863e-01    -9.7781423e-01
 -7.0116535e-01     3.9668087e-01     2.6810081e-01    -1.0214662e+00
  1.2459821e+00     7.5621897e-01    -9.2348909e-01     3.1768798e-01
 -6.3897700e-01     4.0048602e-01    -7.0499388e-02     1.5161078e+00
  5.7735022e-01    -1.3413807e+00     1.4789135e-01     7.4943245e-01
```

The second set is generated by the Matlab command `rand`:

```
2.1895919e-01    4.1748597e-01    3.2823423e-01    7.6649478e-01
4.7044616e-02    6.8677271e-01    6.3263857e-01    4.7773177e-01
6.7886472e-01    5.8897664e-01    7.5641049e-01    2.3777443e-01
6.7929641e-01    9.3043649e-01    9.9103739e-01    2.7490684e-01
9.3469290e-01    8.4616689e-01    3.6533867e-01    3.5926498e-01
3.8350208e-01    5.2692878e-01    2.4703889e-01    1.6650720e-01
5.1941637e-01    9.1964891e-02    9.8255029e-01    4.8651738e-01
8.3096535e-01    6.5391896e-01    7.2266040e-01    8.9765629e-01
3.4572111e-02    4.1599936e-01    7.5335583e-01    9.0920810e-01
5.3461635e-02    7.0119059e-01    6.5151857e-01    6.0564328e-02
5.2970019e-01    9.1032083e-01    7.2685883e-02    9.0465309e-01
6.7114938e-01    7.6219804e-01    6.3163472e-01    5.0452289e-01
7.6981862e-03    2.6245299e-01    8.8470713e-01    5.1629196e-01
3.8341565e-01    4.7464514e-02    2.7270997e-01    3.1903294e-01
6.6842238e-02    7.3608188e-01    4.3641141e-01    9.8664211e-01
```

Histograms for these two data sets are displayed in Figures 1.12 and 1.13.

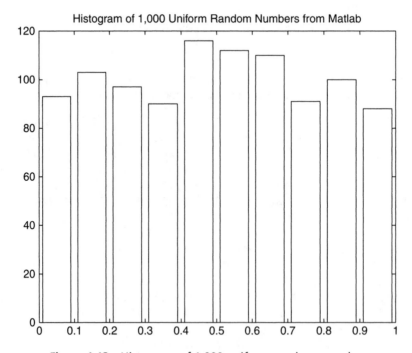

Figure 1.12   Histogram of 1,000 uniform random samples.

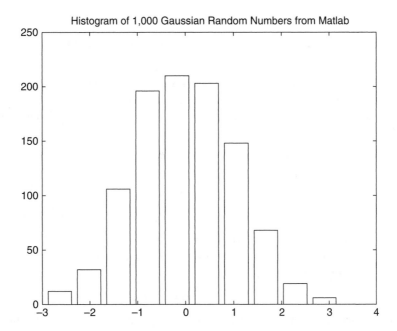

Figure 1.13 Histogram of 1,000 Gaussian random samples.

EXERCISES

E1.1 Specify the sample spaces appropriate for the following random experiments:

    a. The weight of a randomly selected student in your probability class.

    b. The lifetime (waiting time in days) to failure of a computer hard disk.

    c. Your lifetime (in years).

    d. Two consecutive characters appearing in a newspaper editorial.

    e. The instantaneous voltage at a 120-VAC electric receptacle, measured at a randomly chosen time.

    f. The position (to the nearest minute) of the minute hand on a clock face, observed at a randomly chosen time.

    g. The number of alpha particles emitted by a microgram of radium in 1 second.

    h. The number of photons emitted in 1 millisecond by an LED operating at 1 milliwatt.

    i. The number of green pixels on the first scan line of a computer monitor operating at 640×480 resolution.

E1.2 Let $\Omega = N_{10}$. Given the two subsets of Ω,

$$A = \{n : n \text{ is divisible by 3}\} \text{ and } B = \{n : n \text{ is divisible by 4}\},$$

specify, by listing the elements, the sets $A^c, A \cup B$, and $A \cap B$.

E1.3   Let $\Omega = \{(x, y) : x \in \mathbb{R}, y \in \mathbb{R}\} = \mathbb{R}^2$,

$$A = \{(x, y) : x^2 + y^2 \le 1\}, \text{ and } B = \{(x, y) : xy \ge 0\}.$$

Accurately sketch $A, B, C = A \cap B$ and $D = A \cup B$.

E1.4   A radar receiver computes the energies received by its antenna over a fixed time in both horizontal and vertical polarizations. Let the sample space Ω be the upper-right quadrant (first orthant) of the xy-plane—that is, $\Omega = \{(x, y) : x \ge 0, y \ge 0\}$. Given the two subsets of Ω,

$$A = \{(x, y) : x \ge 2y\} \text{ and } B = \{(x, y) : xy > 1\},$$

sketch the sets $A, B, A^c, B^c, A \cap B$, and $A \cup B$.

E1.5   If

$$\Omega = \mathbb{R}^2, \quad A = \{(x, y) : x + y \ge 1\}, \quad B = \{(x, y) : x \le 0\},$$

then accurately sketch $C = A \cup B$ and $D = A \cap B$.

E1.6   Evaluate the indicator function I_{A-B} in terms of I_A and I_B.

E1.7   Using indicator functions, verify that

$$A \subset B \iff B^c \subset A^c.$$

E1.8   Using the arithmetic of indicator functions, determine which of the following statements are correct and which are incorrect:

   a. $A \cup B = (A - B) \cup B$
   b. $(A \cup B) - C = A \cup (B - C)$
   c. $A \cup B \cup C = A \cup (B - A) \cup (C - (A \cup B))$

E1.9   Using indicator functions, prove the de Morgan law

$$A \cap B = (A^c \cup B^c)^c.$$

E1.10 Using the method of indicator functions, determine whether

$$(A - C) \cup (B - C) = AB^c C^c \cup A^c BC^c.$$

E1.11   a. Evaluate the indicator function I_D in terms of the indicator functions I_A, I_B, and I_C if D is the event that both A and B, but not C, occur.
   b. If $I_D = 1 - I_B$, then, using Boolean operations, explain the event D in terms of the events A, B, and C.
   c. If $I_D = I_A I_B I_C$, then, using Boolean operations, explain the event D in terms of the events A, B, and C.
   d. If $I_D = I_A + I_B$, can events A and B co-occur?

E1.12   a. Express the indicator function I_D in terms of the indicator functions I_A, I_B, and I_C when D is the event that exactly two of the events A, B, and C occur.
   b. Express the indicator function I_D in terms of the indicator functions I_A, I_B, and I_C when D is the event that exactly one of the events A, B, and C occurs.
   c. If $I_D = (1 - I_B)(1 - I_C)$, then use Boolean set operations to express D in terms of the sets A, B, and C.
   d. If $I_D = I_A - I_B$, what can you say about the relationship between sets A and B?

E1.13   a. Evaluate the indicator function I_D in terms of the indicator functions I_A, I_B, and I_C if D is the event that A or B, but not C, occurs.

b. If $I_D \leq I_A$, what can you say about the relation between A and D?

c. If $I_D = I_A^2$, what can you say about the relation between A and D?

d. If A, B, and C partition the sample space Ω, what can you conclude about $I_A + I_B + I_C$?

E1.14   The following parts are unrelated to each other:

a. If $I_A I_B$ is identically zero, what must be true of A and B? (Use our notation for set operations and relations.)

b. If $A \cap B^c = B \cap A^c$, what must be true of A and B?

c. If $I_A^2 + I_B^2 \equiv 1$, what can you conclude about A and B?

E1.15   The following parts are unrelated to each other:

a. If $I_A I_B$ is identically zero, what must be true of A and B? (Use our notation for set operations and relations.)

b. If $I_A - I_B \geq 0$, what must be true of A and B? (Use our notation for set operations and relations.)

c. If $I_A - I_B$ is identically 1, what must be true of A and B?

d. If $A \cap B = B \cup A$, what must be true of A and B?

e. If $A \cap B^c = B \cap A^c$, what must be true of A and B?

E1.16   If A, B, and C denote events, then provide expressions, using Boolean set operations, for the following events:

a. At least one of them occurs;   b. At most one of them occur;

c. Exactly one of them occurs;   d. None of them occur;

e. Only A occurs;   f. B occurs;

g. At least two occur;   h. Both A and B, but not C, occur;

i. At most two of them occur;   j. All of them occur.

E1.17   Simplify the following expressions:

a. $(A \cup B) \cup A$;   b. $A \cap (A \cap B)$;

c. $(A \cup B) \cap (A \cup C)$;   d. $(A \cap B) \cup (A \cap C)$.

E1.18   A computer file server has three client workstations: W_1, W_2, W_3. At a given time T, we are interested in which of these clients are requesting service.

a. Construct an appropriate sample space Ω containing eight points, where a given point corresponds to a possible configuration of requests.

b. Identify the events A_k, for $k = 0, 1, 2, 3$, that exactly k clients are requesting service.

c. Construct the smallest event algebra $\mathcal{A}$ containing each of these events A_k.

E1.19   If $\Omega = \{a, b, c, d, e, f\}$ and $A = \{a\}$, $B = \{b\}$, must $C = \{a, b\}$ or $D = \{d\}$ be in the smallest algebra $\mathcal{A}$ of subsets of Ω that contains both A and B?

E1.20   If $\Omega = \{a, b, c, d, e, f\}$ provide an example of an algebra $\mathcal{A}$ of subsets of Ω that contains $\{a, b\}$.

E1.21   a. Given a countably infinite sequence of sets $\{A_n\}$, provide an expression for the event F that only finitely many events of these events occur, also denoted by A_n $f.o.$

   b. Repeat part (a) for the event I that infinitely many of these events occur, also denoted by A_n $i.o.$ or by $\limsup A_n$.

E1.22 Select a sample of English text from newspaper articles.

   a. Record the binary sequence of vowels (a, e, i, o, u) and consonants in initial letters of 100 words; that is, replace each vowel occurrence by a 1 and each consonant by a 0.

   b. Informally, how predictable is the occurrence of a vowel V or consonant C from your data sequence? Can you think of a "simple" formula that predicts the location of the next vowel given the locations of the preceding vowels?

   c. Evaluate the sample mean, median, and standard deviation for the resulting binary-valued sequence.

   d. Plot the fraction of times vowels occur in the first n letters vs. n, for n a multiple of 5.

E1.23 We have data $\{x_i\} = \{4, 5, -2, 0, 1, 4, -2, 1\}$. Evaluate the sample mean $\bar{x}$ and the sample median $\bar{\mu}$.

E1.24 We have data $\{x_i\}$ consisting of the following six numbers:

$$0, \ -40, \ 2, \ -1, \ -5, \ 6.$$

Evaluate the sample mean $\bar{x}$ and the sample median $\bar{\mu}$.

E1.25 We have data $\{x_i\} = \{-2, 0, 1, 4, -2, -5\}$. Evaluate the sample mean $\bar{x}$ and the sample median $\bar{\mu}$.

E1.26 We have data $\{x_i\}$ consisting of the following four numbers:

$$-0.43, -1.66, 0.12, 0.29.$$

Evaluate the sample mean $\bar{x}$, the sample median $\bar{\mu}$, and the sample standard deviation $\bar{\sigma}$.

E1.27   a. Run the command *iostat -d 1* on a Unix-based client-server system and collect 100 seconds of data $\{p_1, \ldots, p_{100}\}$ on the number of packets sent by the server each second.

   b. Using a mathematics program such as Matlab, compute the sample mean and median and a measure of the fluctuation (sample standard deviation).

   c. Compute the histogram by using $b = 10$ bins.

   d. Examine a scatter plot of p_i vs. p_{i-1} to explore predictability. Can you think of a "simple" formula to predict p_i from $p_1, \ldots, p_{i-1}$?

E1.28 You can download several interesting data files from the website
http://lib.stat.cmu.edu/datasets.

   a. Select, say, the Dow Jones data and calculate the sample mean and median for 1980 and the sample standard deviation. Calculate the histogram of values, using $b = 20$.

   b. Explore the predictability of stock prices by examining a scatterplot of tomorrow's price vs. today's price.

2

Classical Probability

2.1 PURPOSE, BACKGROUND, AND ORGANIZATION

Classical probability, which is based upon the ratio of the number of outcomes favorable to the occurrence of the event of interest to the total number of possible outcomes, provided most of the probability models used prior to the 20th century. The limitations of this approach were recognized even in the late 17th century by those concerned with human mortality and by the pioneering master Jacob Bernoulli, who introduced the relative frequency interpretation, which is discussed in Chapters 3 and 19. Classical probability remains of importance today and provides the most accessible introduction to the more general theory of probability. As classical probability is based upon counting or enumerating outcomes satisfying the various conditions necessary for particular events to occur, it requires some familiarity with counting techniques that are the subject of *combinatorics*. The combinatorial methods introduced in this chapter are of independent value in much of applied mathematics and are well worth learning.

We also use the simple setting of classical probability to introduce such concepts as probability, conditional probability, inference, and independence. These concepts will be revisited in greater generality and depth in Chapters 3 through 7, and 11, 12, and 14. Thus, many texts introduce independence quite early, as it is indeed a core concept of applied probability. However, in

57

our view, independence is most clearly understood in the context of conditional probability, and conditional probability is best understood after gaining familiarity with unconditional probability. Hence, we delay significantly a full discussion of independence. Yet, independence is too important to be completely ignored for that long, and we introduce it in this chapter in the restricted setting of classical probability.

2.2 CHOOSING AT RANDOM

The classical account of probability developed from both simple games of chance (e.g., die tossing) that had evolved over time towards mechanical symmetries (from an irregular bone to a cube) implying equally likely outcomes and from notions of equipossibility discussed by Leibniz in the late 17th century. The bases for identifying "equipossibility" were often physical symmetry (e.g., a well-balanced die, made of homogeneous material in a cubical shape) or a balance of information or knowledge concerning the various possible outcomes. Equipossibility is meaningful only for finite sample spaces, and, in this case, the evaluation of probability is accomplished through the definition of classical probability.

Definition 2.2.1   (Classical Probability)   Given a finite sample space Ω, the *classical probability* of an event A is

$$P(A) = \frac{||A||}{||\Omega||}.$$

In traditional language, the probability of an event is the ratio of the number of cases favorable to the outcome of the event to the total number of possible cases. Today, we are more apt to refer to equipossible cases as ones *selected at random*. Probabilities can be evaluated for events whose elements are chosen at random by enumerating the number of elements in the event. Random choice was the main subject matter of probability texts through the early part of the 20th century.

Easily verified consequences of Definition 2.2.1, which hold even for general probability, are as follows:

$$P(A) \geq 0, \ P(\Omega) = 1, \ P(A^c) = 1 - P(A), \ P(\emptyset) = 0,$$

$$A \cap B = \emptyset \Rightarrow P(A \cup B) = P(A) + P(B).$$

It follows from

$$||A \cup B|| = ||A|| + ||B|| - ||A \cap B||$$

that

$$P(A \cup B) = P(A) + P(B) - P(A \cap B).$$

Furthermore, if $\Omega = \{\omega_1, \ldots, \omega_n\}$, then, letting

$$p_i \equiv p(\omega_i) \equiv P(\{\omega_i\}) = \frac{1}{n},$$

we see that, for any event,

$$P(A) = \sum_{\omega \in A} p(\omega).$$

The function p is called a probability mass function and will be discussed more fully in Chapter 4.

Example 2.1   Random Number _____
In a child's game, you are asked to pick a number (implied integer) at random from 1 to 10. In this case there are 10 possibilities, and the probability of selecting, say, a 7 is 1/10. In practice, 7 is actually chosen more often!

Example 2.2   Random Pixels _____
The 15-inch computer screen on which this book was written has a resolution $1,280 \times 854$ pixels or a pitch of 100 pixels/inch. Assume that it is displaying an image which is a green disk of diameter 300 pixels centered on a red background and that any pixel is equally likely to be selected. Hence, the sample space Ω of pixels has $||\Omega|| = 1,280(854) = 1,093,120$ elements. The probability of choosing any particular pixel—say, the one with coordinates (371,243)—is $1/1,093,120$. The probability of choosing a green pixel is the number of pixels in the green disk divided by 1,093,120. Ignoring round-off, there are approximately 70,680 green pixels, and the probability of randomly choosing a green pixel is $7,0680/1,093,120 = .065$.

2.3 ENUMERATION OF ORDERED SEQUENCES

2.3.1 Sampling with Replacement/Reuse

Evaluation of classical probability requires us to be able to count or enumerate finite sets to determine their cardinality. While small sets can be counted exhaustively (the brute-force approach), even sets of moderate size are difficult to count without use of mathematical techniques. *Combinatorics*, first developed by Lulle in the 13th century in connection with alchemy, is the branch of algebra that addresses such enumeration problems (e.g., Feller [27], Riordan [76], R. Stanley [83]).

Lemma 2.3.1   (Counting Ordered Sequences with Replacement)   Given a set (alphabet) of n distinct items (e.g., the first n letters of the alphabet or integers or the elements of N_{n-1}), the number $\mu_{n,r}$ of ways to select a distinct *ordered sequence* (word) of length r drawn from this set, with repeated selections of the same element being permitted (so-called *sampling with replacement*) is given by the recursion

$$\mu_{n,1} = n,$$

$$\mu_{n,r} = n\mu_{n,r-1}, \text{ for } r > 1,$$

based upon there being n choices for each of the r positions. The unique solution is

$$\mu_{n,r} = n^r.$$

Example 2.3 Sampling with Replacement _____

$$\Omega = \{a, b\}, \quad n = 2, r = 3, \quad \mu_{2,3} = 8 : aaa, aab, aba, baa, abb, bab, bba, bbb.$$

$$\Omega = \{a, b, c\}, \quad n = 3, r = 2, \quad \mu_{3,2} = 9 : aa, ab, ac, ba, bb, bc, ca, cb, cc.$$

This result applies to the number of possible outcomes of r tosses of a coin ($n = 2$) or die ($n = 6$) or the number of bytes ($r = 8, n = 2$).

Example 2.4 Number of Binary Strings or Subsets _____
The number of binary strings of length r is found to be 2^r by identifying $n = 2$. The number of subsets of a set $||\Omega|| = r$ can be determined by enumerating $\Omega = \{\omega_1, \ldots, \omega_r\}$ and describing each set A by a binary string $b_1, \ldots, b_r$ (values of the indicator function I_A), where $b_i = 1$ if and only if $\omega_i \in A$. As there are 2^r such binary strings, then there are 2^r subsets of a set of r elements.

Example 2.5 Counting Power Set _____

$$\Omega = \{a, b, c\}, \quad 2^\Omega = \{\emptyset, \Omega, \{a\}, \{b\}, \{c\}, \{a, b\}, \{a, c\}, \{b, c\}\}.$$

By this equation, if $||\Omega|| = r$, then $||2^\Omega|| = 2^r$, and this evaluation helps to explain the exponential notation for power sets.

Example 2.6 Probability of a Missing Symbol _____
Using an alphabet of 4 symbols $\{a, b, c, d\}$, a quaternary message source randomly generates a message of length 8. What is the probability of the event A that the symbol a will not appear in a message?

$$P(A) = \frac{||A||}{||\Omega||} = \frac{3^8}{4^8} = .75^8 \approx .10.$$

2.3.2 Chevalier de Mere's Scandal of Arithmetic

The Chevalier de Mere was a 17th-century nobleman and gambler who had a role in initiating the mathematical theory of probability by posing a problem that prompted the Pascal–Fermat correspondence in 1654. (See Todhunter [90].) The Chevalier also posed another problem he characterized as the "Scandal of Arithmetic":

> Which is more likely, obtaining at least one six in 4 tosses of a fair die (event A), or obtaining at least one double six in 24 tosses of a pair of dice (event B)?

He felt that these two cases should have equal probability, but knew from experience that the first case was more probable than the second case. Instead of evaluating $P(A)$, we evaluate

$P(A^c) = 1 - P(A)$. The number of ways to obtain no sixes in four tosses is the number of ordered sequences of length four from an alphabet of size $6 - 1 = 5$, with replacement, and is 5^4. The total number of possible outcomes (size $||\Omega||$ of Ω) is the number of ordered sequences of length four from an alphabet of size six and by Lemma 2.3.1 is 6^4. Hence,

$$P(A^c) = \frac{5^4}{6^4} = .482, \quad P(A) = .518.$$

Similarly, $P(B^c)$ is the ratio of the number of ways to obtain no double sixes in 24 tosses of a pair of dice to the total number of such outcomes. The number of outcomes of a single toss of a pair of dice is the number of ordered sequences of length 2 with an alphabet of size 6 and is $6^2 = 36$. Hence, the total number of possible outcomes is the number of ordered sequences of length 24 with an alphabet of size 36 and is 36^{24}. The number of such sequences in which there is no double six is the number of ordered sequences of length 24 with an alphabet size of 35 and is 35^{24}. Hence,

$$P(B^c) = \frac{35^{24}}{36^{24}} = .509, \quad P(B) = .491.$$

There is only a very small difference between the probabilities of the two events. While gamblers of this period did not have probability theory to rely upon, it is clear that their too-extensive empirical experience enabled them to correctly resolve very small differences in probability.

2.3.3 Sampling without Replacement

Lemma 2.3.2   (Counting Ordered Sequences without Replacement)   Given a set (alphabet) of n items, the number $(n)_r$ of ways to select a distinct ordered sequence (word) of length r drawn from this set, without repetitions being permitted (*sampling without replacement*), is given by the recursion

$$(n)_1 = n,$$

$$(n)_r = (n - (r - 1))(n)_{r-1}, \text{ for } r > 1,$$

with the unique solution

$$(n)_r = n(n - 1) \ldots (n - r + 1) = \prod_{i=0}^{r-1}(n - i).$$

Example 2.7   Sampling without Replacement _____

$$\Omega = \{a, b, c\}, \quad n = 3, r = 1, \ (3)_1 = 3 : a, b, c.$$

$$\Omega = \{a, b, c\}, \quad n = 3, r = 2, \ (3)_2 = 6 : ab, ac, ba, bc, ca, cb.$$

$$\Omega = \{a, b, c\}, \quad n = 3, r = 3, \ (3)_3 = 6 : abc, acb, bac, bca, cab, cba.$$

2.3.4 The Birthday Paradox

Lemma 2.3.2 may be used to solve the "birthday paradox"—the probability is about 0.5 that, in a group of 23 randomly selected people, at least two will share a birthday (assuming birthdays are equally likely to occur on any given day of the year). In fact, our experience with this in classrooms over the years shows that birthday sharing is more common than the probability calculation suggests. This must come from an uneven distribution of birthdays over the year. To find out the probability of the event M that there is at least one match in a group of size r when there are n 'birthdays,' we calculate

$$P(M^c) = \frac{(n)_r}{n^r} = \prod_{i=0}^{r-1}\left(1 - \frac{i}{n}\right).$$

If $n \gg r$, then we can use the approximation (see Section 0.7.3)

$$e^x \approx 1 + x,$$

which is valid for $|x| \ll 1$, based upon taking only the first two terms in the power series expansion for the exponential, to conclude that

$$\frac{(n)_r}{n^r} \approx e^{-\frac{1}{n}\sum_{i=0}^{r-1} i}.$$

The identity, verifiable by induction on r,

$$\sum_{i=0}^{r-1} i = \frac{r(r-1)}{2},$$

yields

$$P(M^c) \approx e^{-\frac{r(r-1)}{2n}}.$$

Hence,

$$P(M) \approx 1 - e^{-\frac{r(r-1)}{2n}},$$

which for $r = 23$, $n = 365$ is .500; the exact evaluation yields $P(M) = .507$.

As another example, if we attempt to assign three-digit identification numbers to members of a class of 150 students by letting them choose at random, then the probability that each student will have a unique identification number is approximately $\exp(-150(149)/2{,}000) = \exp(-11) \ll 1$. If instead we allow four-digit identification numbers in this class of 150 students, then the probability that individual random choices will yield unique identification numbers is approximately $\exp(-150(149)/20{,}000) \approx .327$.

2.3.5 Permutations and Factorial Function

An example of a *permutation* of distinct items is provided by the ordering of cards in a deck of 52 common playing cards. Each shuffle of the deck results in a new permutation or ordering or arrangement of the 52 cards. The number $n!$ of arrangements (permutations) of $n \geq 0$ distinct items is, from Lemma 2.3.2 and $r = n$,

$$n! = (n)_n,$$

where $n!$ denotes the *factorial function*. In terms of the factorial function, we may write

$$(n)_r = \frac{n!}{(n-r)!}.$$

Properties of the factorial function $n!$ include the following:

$$0! = 1! = 1;$$

$$n! = n(n-1)!;$$

$$n! = \int_0^\infty e^{-t} t^n \, dt.$$

The integral representation allows us to generalize the factorial function to arguments that need not be positive integers. This generalization is known as the Gamma function and is defined by

$$\Gamma(x) = \int_0^\infty e^{-t} t^{x-1} \, dt.$$

It is evident that $\Gamma(0) = \infty$ and $\Gamma(1) = 1$. Integration by parts yields the recursion

$$\Gamma(x) = (x-1)\Gamma(x-1).$$

Hence,

$$\Gamma(3) = 2 \cdot \Gamma(2) = 2 \cdot 1 \cdot \Gamma(1) = 2.$$

We conclude that

$$n! = \Gamma(n+1).$$

As we accept this generalization, we see that $0! = 1$ and that the factorial of a negative integer is plus infinity. Useful approximations to $n!$ that are increasingly accurate percentagewise as n increases are the following:

$$\sqrt{2\pi} \, n^{n+\frac{1}{2}} e^{-n} e^{\frac{1}{12n+1}} < n! < \sqrt{2\pi} \, n^{n+\frac{1}{2}} e^{-n} e^{\frac{1}{12n}}, \quad \text{(Robbins [77])};$$

$$\text{Stirling's formula:} \quad n! \approx s(n) = \sqrt{2\pi} \, n^{n+\frac{1}{2}} e^{-n}.$$

The accuracy of Stirling's formula is suggested by Table 2.1 and Figure 2.1.

Table 2.1   Accuracy of Stirling's Formula

n	1	2	3	4	5	6
$n!$	1	2	6	24	120	720
Stirling	.92	1.92	5.84	23.51	118.02	710.08
Perc. Err.	8	4	2.7	2.0	1.7	1.4

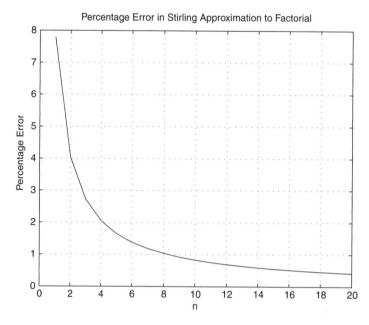

Figure 2.1   Percentage error of Stirling's approximation to factorial.

It follows from the upper and lower bounds provided immediately preceding Stirling's formula that, in the limit of increasing n, the fractional error between $n!$ and Stirling's formula $s(n)$ is bounded above by

$$0 < \frac{n! - s(n)}{n!} < \frac{e^{\frac{1}{12n}} - 1}{e^{\frac{1}{12n+1}}},$$

and vanishes with increasing n. As a consequence,

$$\lim_{n \to \infty} \frac{\sqrt{2\pi}\, n^{n+\frac{1}{2}} e^{-n}}{n!} = 1.$$

We can determine the growth rate of $n!$ by rewriting Stirling's formula as

$$n! \sim (\sqrt{2\pi e})\, e^{(n+\frac{1}{2})\log \frac{n}{e}}.$$

We see from this that $n!$ grows more rapidly than any exponential function of the form $e^{\alpha n}$.

2.4 ENUMERATION OF SETS: BINOMIAL COEFFICIENTS

Lemma 2.4.1   (Counting Sets)   The number of sets, or unordered collections, of size r drawn from an alphabet or sample space of size n, where, as appropriate for sets, we allow no duplication of elements (sampling without replacement), is given by the *binomial coefficient*

$$\binom{n}{r} = \frac{(n)_r}{r!} = \frac{n!}{(n-r)!\,r!}.$$

Proof. To verify this, note that the number of ordered such collections is

$$(n)_r = \frac{n!}{(n-r)!}.$$

As the elements in each such sequence of length r are distinct, the number of permutations (reshufflings) of each sequence is $r!$. Hence, the number of sequences of length r from an alphabet of size n when we sample without replacement and disregard the ordering of the elements in the sequence is

$$\frac{n!}{(n-r)!\,r!} = \binom{n}{r}. \qquad\qquad \square$$

The binomial coefficient has the properties

$$\text{(reflection property)}\quad \binom{n}{r} = \binom{n}{n-r},$$

$$\binom{n}{0} = 1,\quad \binom{n}{1} = n,\quad \binom{n}{r} = 0 \text{ if } n < r.$$

The binomial coefficient is also the number of subsets of size r that can be formed from a set of n elements.

Example 2.8   Classical Probability of a Binary Sequence _____

The number of 64-bit register sequences storing exactly 20 ones is $\binom{64}{20}$. If register sequences are selected at random, then the classical probability of seeing exactly 20 ones is

$$\frac{\binom{64}{20}}{2^{64}} = \frac{64!}{44!\,20!\,2^{64}},$$

and this will be approximated shortly with the use of the binary entropy function.

The behavior of the binomial coefficient as a function of r is displayed if we consider those r for which

$$\frac{\binom{n}{r}}{\binom{n}{r-1}} \geq 1.$$

This condition determines whether the binomial coefficient is nondecreasing as r increases, and simplifies to

$$r \leq \left\lfloor \frac{n+1}{2} \right\rfloor.$$

(The notation $\lfloor x \rfloor$ denotes the integer part of x as in $\lfloor 3.9 \rfloor = 3$.) Hence, so long as r is less than approximately $n/2$, the binomial coefficient is increasing. By the reflection property, the binomial coefficient is then decreasing in r greater than approximately $n/2$. Therefore,

$$\max_r \binom{n}{r} = \binom{n}{\lfloor \frac{n+1}{2} \rfloor}.$$

Induction on n can be used to verify the basic Binomial Theorem.

Theorem 2.4.1   (Binomial Theorem)

$$(x+y)^n = \sum_{r=0}^{n} \binom{n}{r} x^r y^{n-r}.$$

Hence,

$$\sum_{r=0}^{n} \binom{n}{r} = 2^n,$$

thereby confirming that the total number of subsets of a set of n points is 2^n.

2.5 APPLICATION TO ENTROPY AND DATA COMPRESSION

We introduce the *binary entropy function*

$$H(p) = -p \log(p) - (1-p) \log(1-p).$$

When logarithms are taken to base 2, we express the numerical value of $H(p)$ in *bits*. When we take logarithms to base e, we express H in *nats* (short for "natural" units). A concept of entropy was introduced by Boltzmann in thermodynamics in the 19th century and was used to measure the disorder of systems. On the basis of research he conducted during World War II, Claude Shannon published the foundational *A Mathematical Theory of Communication* [82] in 1948 and established entropy as a measure of uncertainty as the foundation of communication theory. It is most common in information and communication theory to use base 2 and express entropy in bits, and we do so in what follows.

When H is measured in bits, then Cover and Thomas [20], p. 284, show that

$$\frac{1}{n+1} 2^{nH\left(\frac{r}{n}\right)} \leq \binom{n}{r} \leq 2^{nH\left(\frac{r}{n}\right)}.$$

Applying the approximation just given to the binomial coefficient enables us to evaluate the probability of exactly 20 ones in a randomly selected 64-bit register sequence as

$$\frac{\binom{64}{20}}{2^{64}} \approx 2^{64(H(20/64)-1)} = 2^{64(-.379)} = 2 \times 10^{-7}.$$

A plot of $H(p)$ and its quadratic approximation,

$$H(p) \approx 4p(1-p),$$

is provided in Figure 2.2.

An introduction to data compression is to realize that message sources (e.g., a line of text in a given language or a frame of a video image) operate under constraints (e.g., grammar) that make many mathematically possible messages either highly unlikely or prohibited. The source is effectively prevented from generating all possible outputs. More accurately, we may be able to identify a set M of possible outputs that has probability close to 1 of being generated by the source and agree to ignore (permit very long representations of) those highly unlikely sequences lying in M^c. Thus, a message source may generate a sequence (word) of length n drawn from a symbol alphabet of size α. We take $\alpha = 2$ and consider messages to be binary strings

$$\mathbf{b} = (b_1, b_2 \ldots, b_n), \ b_i \in \{0, 1\}, \ \mathbf{b} \in \{0, 1\}^n,$$

as might be the case in digital communications and computer architectures.

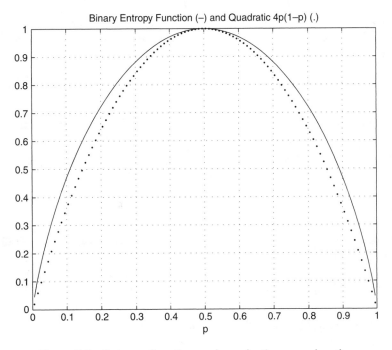

Figure 2.2   Entropy function and quadratic approximation.

We introduce a function w from all binary strings of length n to the integers

$$w : \{0, 1\}^n \to \{1, 2, \ldots, W\}$$

that serves to partition the set of all binary strings into W subsets

$$A_k = \{\mathbf{b} : w(\mathbf{b}) = k\} \subset \{0, 1\}^n.$$

For example, the *weight function*

$$w(\mathbf{b}) = \sum_{k=1}^{n} b_k$$

counts the number of ones in the binary sequence. Assume an enumeration of all of the binary sequences, say, by the integers for which they are the binary representation. We can then uniquely encode any given $\mathbf{b}$ by the pair of integers $(w(\mathbf{b}), \lambda(\mathbf{b}))$, where $\mathbf{b}$ is the λth sequence in the set $A_{w(\mathbf{b})}$. Given $\mathbf{b}$, we easily determine (w, λ), and given (w, λ), we easily determine $\mathbf{b}$. We then convert the pair (w, λ) into another binary string $\mathbf{c}$, that is the (compressed) code word for $\mathbf{b}$ by using approximately $\log_2 W$ bits to express $w(\mathbf{b})$ and follow this with approximately $\log_2 ||A_{w(\mathbf{b})}||$ bits to specify $\lambda(\mathbf{b})$. Compression will be achieved if

$$\log_2 W + \log_2 ||A_{w(\mathbf{b})}|| < n.$$

We simplify matters and assume that w is the weight function taking values only in $\{1, 2, \ldots, n\}$ and consider A_r to be the set of binary sequences having exactly r ones. We now know that

$$||A_r|| = \binom{n}{r} \leq 2^{nH\left(\frac{r}{n}\right)}.$$

Hence, we can uniquely label each message sequence of length n in A_r by a label that is a binary sequence of length $nH(r/n)$ (this is the job of an encoder); the inequality establishes that there are at least as many such shorter sequences (without constraints) as there are longer constrained (by their weight being r) sequences lying in A_r. We can now store or transmit the label of length

$$\log_2 n + nH\left(\frac{r}{n}\right)$$

and later uniquely reconstruct the original sequence of length n. (This is the job of the decoder.) Compression will be achieved if

$$\log_2 n + nH\left(\frac{r}{n}\right) < n \text{ or } 1 - H\left(\frac{r}{n}\right) > \frac{\log_2 n}{n}.$$

If r/n differs sufficiently from $1/2$, then, as can be seen from Figure 2.2, we can achieve a significant compression ratio of $H(r/n) < 1$. We refer to a source whose messages of length n form a set of size approximately 2^{nH} as having an *information rate* of H bits per symbol.

2.6 APPLICATION TO GRAPHS AS SYSTEM MODELS

2.6.1 Large Scale Systems as Networks

Systems of great engineering importance include the World Wide Web (WWW); Internet; electric power bulk distribution system; public-switched telephone network (PSTN, the familiar wired telephone network that also transports cellular phone calls once they reach a base station); mobile cellular telephone networks; computing local area networks having a common server; sensor networks; transportation systems such as air, interstate highway, and city streets; and such other systems as biological (e.g., main metabolic pathways between sequences of chemical processes), social (e.g., relationships of authority, collaboration, friendship, or kinship), and financial networks.

The Internet connects a variety of computers that run a common protocol TCP/IP and that can be represented by a set or list V of labeled nodes. These computers are physically connected by copper or fiber-optic cables and routing devices with the connections or links being elements in set E. The Internet supports the WWW. The WWW is largely a software system in which Web pages are the nodes or elements in the list V and connections between Web pages through hyperlinks (the familiar URLs) are the elements of a set E. A major study of the WWW was conducted by Broder et al. [15] on data collected during May 1999 and again in October 1999. They searched out (Web crawled) about 200 million Web pages and examined about 1.5 billion hyperlinks. They note that, "Much recent work has addressed the Web as a graph and applied algorithmic methods from graph theory in addressing a slew of search, retrieval, and [data] mining problems on the Web." We will return to some of the results of this study after we initiate our discussion of graphs as mathematical models for such complex systems.

The bulk electric power distribution system in the half of the United States known as the Eastern Interconnect has about 30,000 nodes (called "buses") that are points of connection between system elements. Figure 2.3 shows the New York state portion of the Eastern Interconnect. This portion has about 3,000 nodes, and the graph looks something like a map of New York. This somewhat complicated figure, best understood in terms of the colors in the original figure, indicates electric generators and substations (that step-down high voltage transmission to approximately 12kV for local distribution) by small blue bars. Transformers are labeled by pink dashed lines. The transmission lines connecting the generators and substations are at voltage levels such as 138 kV, 230 kV, 345 kV, 500 kV, and 765 kV, and different line colors and thicknesses indicate the appropriate voltage level.

So-called ad hoc sensor networks have sensor nodes (perhaps sensitive to acoustic, optical, magnetic, seismic, temperature, or chemical changes in their local environment) that report back to the network's client (e.g., a monitoring station for a chemical or biological attack–sensing network or a fire station for high temperatures in a building) through each other. These sensor nodes not only sense and report on aspects of their immediate environments, but also pass on reports from more distant nodes. The more reporting that they have to do, or the longer the distance over which they have to report, the more they consume electric power that is usually supplied by batteries with limited energy storage, and the shorter is their lifetime.

All of these systems have aspects (e.g., robustness to component failure) that can be understood in terms of models that emphasize their patterns of connectivity and deemphasize the individual differences between their components. (See Barabasi [6] and Strogatz [86].) Thus, if

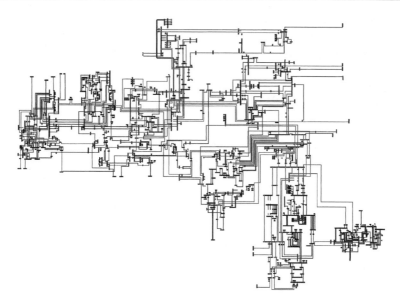

Figure 2.3   New York state electric power grid (map provided by J. S. Thorp and H. Wang).

a single electrical power transmission line fails, can all of the electrical generators still remain connected to all of the service loads? Can a new generator or load be accommodated by the existing electrical network or grid through a new set of transmission lines? Can the PSTN accept new telephone installations and the deletion of others? Clearly the exponential growth in numbers of users of the WWW and the Internet shows that these systems have a great capacity to accept new users that come and go. In computing and communication networks, we need to know the impact of the failure of certain links on maintaining access. In such networks, we are also interested in the number of links or connections that are needed to connect any two clients/elements of these networks. The larger the number of links, the greater the computational or communication delay and the greater the likelihood of degradation of service due to noise or interruptions on these connecting links. Finally, most of these networks evolve in time in a manner that is not centrally planned, with nodes and links being added or deleted by individual clients at their own local initiative. They also operate in a complex environment that exposes them to natural disturbances (e.g., tornadoes, winter storms, hurricanes, sunspot extremes) as well as to inadvertent or malicious human interference. All this is in addition to the traditional issues of the stochastic reliability of components and stochastic noise sources. The WWW is an outstanding example of very rapid evolution with nodes (Web pages) appearing and disappearing at the behest of individuals. The Web has grown by orders of magnitude since its inception as a means for communication between physicists at the high-energy physics research lab CERN, and it continues to grow at great rates. The ability of users or clients to find each other, an ability that is key to the popularity of the Web, depends critically on the structure of the Web. Paths between clients cannot be too long (too many "clicks" away), nor can such paths be hard to identify. How can we understand the ability of such systems to perform over time? What are the implications for the design of operational protocols?

2.6.2 Undirected Graphs

Mathematical models for connectivity in systems or networks are provided by graphs. These models enable us to study such questions as the connectedness of all of the elements of the network, the robustness of this connectedness to failure of connections between pairs of elements, the lengths of paths between pairs of elements, and the consequences for the networks of various patterns of evolution in the network through random models of the addition or deletion of network elements and of links between these elements. As knowledge of graphs is uncommon among engineering undergraduates, we make a digression into a few of their characteristics. Graph theory is a well-developed subject (e.g., see Bollobas [11], Tutte [93]) in which graphs have been categorized in a variety of ways and each category has then been studied in depth.

Definition 2.6.1   (Undirected Graph)   An undirected graph $G = (V, E)$ is defined by a set $V = \{v\}$ of labeled elements called nodes or vertices and a set $E \subset \{\{u, v\} : u, v \in V\} = V \times V$ of unordered pairs of nodes called links or edges. When needed for clarity, G_n denotes an undirected graph on n labeled nodes.

The notation $\{u, v\}$ is that of a set containing the two elements u and v, and, as a set, it is unordered in that $\{u, v\} = \{v, u\}$. The link $\{u, v\}$ is thought of as connecting the pair of nodes u and v, and these nodes are said to be *adjacent*. The special case of a link $\{u, u\}$ is called a *loop* or *self loop*. If, say, two telephones u and v have spoken to each other within a specified time period or two Websites bookmark each other, then a link is established between them and represented by the unordered pair or set of two elements $\{u, v\} \in E$. The graph is "undirected" in that if u is adjacent to v, then v is adjacent to u. Thus, if V is the set of people in a community, then we might have an edge $\{u, v\}$ between persons u and v if they have at least one parent in common. In a bulk electric power distribution system, we might have V be a set of generating stations and substations (these then connect to the loads or consumers of power) with an edge $\{u, v\}$ meaning that there is an electrical transmission line connecting u and v.

We can pictorially represent such a graph by a set of labeled points V in the plane and the edges/links shown as lines terminating at pairs of these points. We will assume that the nodes have been either enumerated and labeled by the integers, $V = \{1, 2, \ldots, n\}$, for a graph on n nodes, or labeled by $V = \{v_1, \ldots, v_n\}$. The role of the labels is to create a distinction between nodes. An example is provided in Figure 2.4.

In our brief examination of graphs, unless stated otherwise, we will assume the absence of loops. A loop in a communication system has the caller calling itself, and a loop in an electrical power distribution system short-circuits either a generator or a load.

Example 2.9   Number of Undirected Graphs on n Nodes _____
How many different undirected graphs Γ_n are there on a set V of n labeled nodes or vertices? Equivalently, how many sets of edges/links are there? The number of possible links is the number of ways to choose pairs of nodes from a set of size n. We learned in Section 2.3 that there are $\binom{n}{2}$ such subsets of unordered pairs drawn from a set of n elements. The number of different sets of links, and hence the number of different graphs, is then the number of subsets

of a set of size $\binom{n}{2}$. We learned in Section 2.3 that the number of subsets (size of the power set) of a set of $\binom{n}{2}$ items is

$$\Gamma_n = 2^{\binom{n}{2}} \approx 2^{\frac{1}{2}n^2}.$$

The number of graphs on n labeled nodes grows very rapidly with n.

How many different graphs $\Gamma_{n,m}$ are there on a set V of n labeled nodes that has $m = \|E\|$ links or edges? As there are $\binom{n}{2}$ possible links, we see that there are

$$\Gamma_{n,m} = \binom{\binom{n}{2}}{m}$$

subsets of size m from a set of size $\binom{n}{2}$, and hence, this many undirected graphs of n nodes and m links. Of course,

$$\Gamma_n = \sum_{m=0}^{\binom{n}{2}} \Gamma_{n,m}.$$

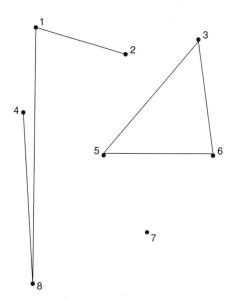

Figure 2.4 An undirected graph.

2.6.3 Node Degree

Some graphs, say, for air routes, have a few nodes (hub cities) with many connections to other cities, and many nodes (smaller cities) having few direct connections to other cities. Other graphs, such as city street intersections (nodes) connected by blocks (links), would rarely have intersections with more than five blocks arriving at an intersection. Similarly, some Websites (e.g., the search engines like Google) have connections to many (order of billions) other Websites, whereas the Websites of individuals usually only have hyperlinks (URLs) to perhaps tens of other Websites. This characteristic of the connectedness of nodes serves to differentiate among types of networks.

Definition 2.6.2   (Node Degree)   The degree $\deg(v)$ of a node v is the number of links in E connected to v.

Example 2.10   Node Degree _____

The maximum degree in a graph of n nodes without loops is $n - 1$. The minimum degree is 0 for an isolated node. The degrees of the nodes in the graph of Figure 2.4, given in the order of enumeration of the nodes 9 are: 2, 1, 2, 1, 2, 2, 0, 2.

We can also look at the set of degrees of all of the nodes in a graph.

Definition 2.6.3   (Degree Sequence)   The degree sequence is the set of node degrees for each of the nodes arranged in nonincreasing order.

Example 2.11   Degree Sequence _____

In Figure 2.4, the degree sequence is 2, 2, 2, 2, 2, 1, 1, 0. In Figure 2.5, treated as an undirected graph, the degree sequence is 7, 6, 5, 4, 4, 4, 4, 4, 3, 3, 3, 2, 2, 2, 1. The smallest degree sequence is the all zero sequence corresponding to a graph of isolated nodes. The largest degree sequence has all elements $n - 1$ for a completely connected graph on n nodes.

Not all degree sequences are possible. Note that an edge $\{u, v\}$ contributes a count of 1 to $\deg(u)$ and a count of 1 to $\deg(v)$. Hence, for any graph $G = (V, E)$ without loops,

$$\sum_{v \in V} \deg(v) = 2\|E\|,$$

and the sum of node degrees must be even. This result is borne out in our list of the degree sequence for the graph shown in Figure 2.4. If the graph is $G_{n,m}$ having n nodes and m links, then we can calculate the average node degree $\overline{d}_{G_{n,m}}$ from

$$\overline{d}_{G_{n,m}} = \frac{1}{n} \sum_{v \in V} \deg(v) = 2\frac{m}{n}.$$

2.6.4 Directed Graphs

While some connections are symmetric, others are not. For example, consider the social relation of u is a parent of v or that u is a work supervisor of v. Evidently, these are not symmetric relations. If the nodes are potential failure events in a power system, we may represent event u can cause event v by a link showing this. However, if u can cause v, then it need not follow that

v can cause u. On the Internet, in a certain period, the relation may be about which node requests service from another node. If I initiate a call to you in this period, you may or may not initiate a call to me. If one Web page has a hyperlink to another, then it does not follow that the second has a hyperlink to the first.

Definition 2.6.4   (Directed Graph)   A directed graph (also called a digraph) $G = (V, E)$ is a set $V = \{v\}$ of nodes or vertices and a set $E = \{(u, v) : u, v \in V\} \subset V \times V$ of *ordered* pairs (u, v) defining the directed links or edges connecting u to v and not necessarily the reverse. We distinguish (u, v) from (v, u) and from the unordered $\{u, v\}$.

Definition 2.6.5   (Directed Path and Walk)   A directed k-path is a subgraph of a directed graph G with $k + 1$ distinct nodes and k directed links such that the nodes can be enumerated as $v_0, \ldots, v_k$ in V and the directed links as $e_1, \ldots, e_k$ in E in such a way that $e_j = (v_{j-1}, v_j)$.

A directed walk of length k between nodes u and v exists if there are nodes $v_0, \ldots, v_k$, with $v_0 = u$ and $v_k = v$ (repetitions of nodes being allowed), and k directed links $e_1, \ldots, e_k$ in E such that $e_j = (v_{j-1}, v_j)$.

Definition 2.6.6   (Strongly and Weakly Connected)   A directed graph is *strongly connected* if between any two nodes there is a directed path linking them. A directed graph is *weakly connected* if it is connected when one considers the links as being undirected.

Example 2.12   Directed Graph _____

Thus, if $V = \{1, 2, 3\}$, then an example of a directed graph G has

$$E = \{(1, 2), (2, 1), (3, 2)\}.$$

In this example, node 1 has a link to node 2 and node 2 has a link to node 1. Node 3 has a link to node 2, but node 2 does not have a link to node 3. There are no other links in the graph. This graph is not strongly connected, since there are no directed paths from nodes 1 or 2 to node 3. However, it is weakly connected.

Example 2.13   World Wide Web _____

The study by Broder et al. [15], of about 200 million Web pages, reported that approximately 28% of the Web pages were strongly connected to each other and formed a strongly connected component (SCC). About 21%, called by them "IN," had links to the SCC, but there were no links from the SCC to them. Another 21%, called by them "OUT," had links from the SCC to them, but not from them to the SCC. An additional 22% of Web pages they called TENDRILS were connected either via outgoing links from IN or incoming links to OUT, but with no links to SCC. The total of IN, OUT, TENDRILS, and SCC formed a weakly connected component that amounted to 92% of the Web pages surveyed. This left about 8% of pages that were completely disconnected from the large weakly connected component.

While the number of nodes and the number of links provide some indication of the size of a graph, they do not tell us how many links separate given pairs of nodes.

Definition 2.6.7   (Diameter)   Let G be a connected graph, and let $d(u, v)$ be the length of a shortest path connecting nodes u and v. The diameter δ_G of G is

$$\delta_G = \max_{u,v \in V} d(u, v).$$

If the graph G is not connected, then δ_G is infinite.

Example 2.14   Connected Graph Diameter _____
In Figure 2.4, consider the connected component on the nodes 1, 2, 4, and 8. In this connected component,

$$d(1, 2) = d(1, 8) = d(4, 8) = 1, \ \ d(2, 8) = d(1, 4) = 2, \ \text{and } d(2, 4) = 3.$$

Hence, for that connected component, the diameter is 3. In any connected G_n, the diameter is at most $n - 1$.

Broder et al. [15] also studied the directed diameter, the diameter measured along walks that respect link directions. For the SCC of the WWW they estimate the diameter to be at least 28.

Example 2.15   Number of Directed Graphs _____
How many directed graphs without loops are there on n labeled nodes or vertices? There are $n(n - 1)$ ordered pairs of nodes in which the nodes in a pair are distinct. Hence, the number of possible sets of edges or links is the number of subsets of these ordered pairs and is given by

$$2^{n(n-1)}.$$

How many directed graphs without loops are there that have n labeled nodes and m links? Again, there are $n(n - 1)$ ordered pairs of distinct nodes and any m of them can be chosen for the set E. Hence, there are

$$\binom{n(n - 1)}{m}$$

such graphs.

2.6.5 Indegree and Outdegree

For directed graphs, we need to refine the notion of degree to distinguish between connections originating at a node and those terminating at that node.

Definition 2.6.8   (Indegree and Outdegree)   The outdegree of a node v, outdeg(v), is the number of nodes u for which there are links of the form (v, u).

The indegree of a node v, indeg(v), is the number of nodes u for which there are links of the form (u, v).

As each link, say the one from u to v, contributes 1 to the outdegree of u and 1 to the indegree of v, it follows that

$$\sum_{v \in V} \text{indeg}(v) = \sum_{v \in V} \text{outdeg}(v),$$

although the indegree and outdegree sequences may differ. The $\text{outdeg}(v)$ is the number of nodes that can be accessed from v by a path of length 1, and the $\text{indeg}(v)$ is the number of nodes that can access v by a path of length 1.

Example 2.16   Indegree and Outdegree _____
In Example 2.13, the indegrees of nodes 1, 2, 3 are 1, 2, 0, and the outdegrees are 1, 1, 1.

Example 2.17   Number of Graphs of Given Outdegree _____
How many directed graphs on n nodes are there in which each node has outdegree d? For each node, we have $\binom{n-1}{d}$ choices for our d outgoing links. These choices can be made independently for each of the n nodes. Hence, the number of such graphs is

$$\binom{n-1}{d}^n.$$

Example 2.18   Indegree and Outdegree Distributions _____
The WWW study by Broder et al. [15] discussed in Example 2.14 also had results on indegrees and outdegrees of nodes in their sample of the Web. Their analytical fits to the data on indegree and outdegree are expressions for the fraction $p_I(d)$ of Web pages having indegree d and the fraction $p_O(d)$ of Web pages having outdegree d that are given by

$$p_I(d) \propto \frac{1}{d^{2.1}}, \; p_O(d) \propto \frac{1}{d^{2.72}}.$$

This behavior is an example of a *power law* in which the fraction decays with increasing degree as an algebraic power of the degree. Indegree and outdegree distributions provide examples of the Zeta or Zipf probability mass function to be discussed in Section 4.4.5.

2.6.6 Random Graphs

The organization of such complex systems as the Internet, the WWW, and ad hoc sensor networks has not been centrally planned. Indeed, the full organization of the WWW is not known. These systems evolve subject to the wishes of a very large number of individuals and sometimes subject to a randomly fluctuating environment. To some extent, this is also true of the bulk electric power distribution system. Furthermore, these systems are so large that an intelligible model is not one that fully specifies them, even if it could be constructed. The random evolution of these systems and the need to intelligibly model a system with a huge number of similar elements suggests the use of probability methods. At this point in our development of such methods, we are only able to describe random selections in which all possibilities have the same probability. Starting with the next chapter, this limitation will be removed. As an introduction to probabilistic system modeling for very complex systems, we introduce randomness by focusing on a family of graphs and assuming that all members of this family are equally likely to be chosen. Many results on the mathematics of *random graphs* can be found in Bollobas [11] and Palmer [69].

Example 2.19   Probability of a Graph with Given Number of Links _____

What is the probability that a randomly chosen undirected graph G_n on n nodes will have m links? The set of graphs on n nodes was shown in Section 2.6.2 to have $\Gamma_n = 2^{\binom{n}{2}}$ members. Of these, the number $\Gamma_{n,m}$ that have m edges is

$$\binom{\binom{n}{2}}{m}.$$

Hence, for a given n,

$$P(m \text{ links}) = \frac{\Gamma_{n,m}}{\Gamma_n} = \frac{\binom{\binom{n}{2}}{m}}{2^{\binom{n}{2}}}.$$

Similarly, the probability of choosing a directed graph on n nodes with m links given that you have a directed graph on n nodes is

$$P(m \text{ links}) = \frac{\binom{n(n-1)}{m}}{2^{n(n-1)}}.$$

Example 2.20   Probability of Given Outdegree _____

What is the probability that in a randomly chosen directed graph on n nodes each node will have outdegree d? From the preceding subsection, we know that there are

$$2^{n(n-1)}$$

equally probable directed graphs and

$$\binom{n-1}{d}^n$$

of them have each node with outdegree d. Hence,

$$P((\forall v \in V) \text{ outdeg}(v) = d) = \frac{\binom{n-1}{d}^n}{2^{n(n-1)}} = \left[\frac{\binom{n-1}{d}}{2^{n-1}} \right]^n.$$

2.7 APPLICATION TO STATISTICAL MECHANICS

Statistical mechanics studies the behavior of systems of large numbers of interacting particles. In contrast to thermodynamics, which is also concerned with such systems but studies them at a macroscopic level (e.g., by using such macroscopic variables as temperature and pressure in studying gases), statistical mechanics studies these systems at a microscopic level. At this microscopic level quantum mechanical effects are often important. Statistical mechanics makes use of three distributions of particles in cells of *phase space* that are referred to as the particle statistics. (See Feller [27].) Phase space refers to the coordinate, or state, space describing a particular physical system. In electrical engineering, particularly in circuit and control theory, phase space is referred to as state space. The dynamics of a circuit composed of resistors, inductors, and capacitors can be described by the state variables consisting of the individual inductor currents and capacitor voltages. For example, n classical molecules in a gas have their motion described by the 6^n-dimensional space of 3 position and 3 momentum coordinates for each of the molecules. Quantum mechanics introduces quantum numbers, such as spin, to describe the state of a particle. The term *cells* refers to the elements of a particular partition of the phase space. Throughout, cells are assumed to be distinguishable. Particles are assumed to be distinguishable when we look at matters classically. However, they are indistinguishable when matters are considered quantum mechanically.

2.7.1 Maxwell–Boltzmann Statistics

The classical *Maxwell–Boltzmann statistics* (MB) of particles in phase space cells is based on the number of distinguishable arrangements of r distinguishable particles in n cells, and, from Lemma 2.3.1, that number is n^r, where the n cells are reusable as we construct a word of length r. If a particle is equally probable to be in any of the cells, then classical probability can be used to calculate the probabilities of any arrangements of interest (e.g., all of the gas molecules occupying only one-half of the possible cells, pressure and temperature fluctuations, etc.). MB statistics are used to model classical games of chance and molecules in the thermodynamics of gases. An argument persisting through the 16th century concerned the correct way to count the outcome of, say, several dice and was eventually settled (e.g., by Galileo writing in 1606) in conformity with what we now call MB statistics and is based upon the distinguishability of individual dice.

2.7.2 Bose–Einstein Statistics

However, if the particles are indistinguishable, then there are $\binom{n-1+r}{r}$ distinguishable arrangements. This is the basis for *Bose–Einstein statistics* (BE), and it applies to photons and other bosons. Bosons are particles with integral values of quantum mechanical spin and provide the carriers of force in various fields (e.g, photons for the electromagnetic field). To verify the Bose–Einstein enumeration, consider, as an example,

$$* * * * \mid * * \mid \mid \mid \mid * \mid$$

in which the stars represent the r indistinguishable particles and the vertical bars represent the $n - 1$ cell walls required to establish the n cells. Every such picture represents a possible arrangement of r indistinguishable particles among the n cells. The number of such arrangements is the number of ways of choosing the r locations for the particles from among the $n - 1 + r$ locations for either bars or stars. (Ever since Feller published this account in 1950, the argument is referred to as "bars and stars.") Such a count of arrangements also applies to the number of ways to write a positive integer n as a sum of r nonnegative integers with the order of integers in the sum being relevant (e.g., $n = 10$, $r = 4$, $10 = 5 + 1 + 0 + 4 = 1 + 5 + 4 + 0 = 2 + 2 + 2 + 4$, etc.).

2.7.3 Fermi–Dirac Statistics

Finally, if no fully defined (e.g., through appropriate quantum numbers) phase space cell can contain more than one particle (Pauli exclusion principle in quantum mechanics), we have $\binom{n}{r}$ such arrangements. This latter case is referred to as *Fermi–Dirac statistics* (FD) and applies to fermions. Fermions are particles with half-integral quantum mechanical spins and include electrons, protons, and neutrons. If the phase space cells are large enough (not tightly defined), then multiple particles may occupy the same cell. For example, in an atom, we may have multiple electrons in the same ring. Example 2.25 will be one of particle-type detection; the goal is to discriminate between bosons and fermions only through knowledge of their occupancy numbers in cells of phase space.

2.8 MULTINOMIAL COUNTING

If we have r types of symbols and n_i indistinguishable copies/tokens of a type i symbol (e.g., the word `probable` is composed of two uses of the letter b and one use each of the letters `a,e,l,o,p,r`), then we can arrange the $n = \sum_1^r n_i$ tokens to form

$$\binom{n}{n_1}\binom{n - n_1}{n_2}\binom{n - n_1 - n_2}{n_3} \cdots 1 = \frac{n!}{\prod_1^r n_i!}$$

distinct ordered sequences or words of length n. This count is known as the *multinomial coefficient*, denoted by

$$\binom{n}{n_1 \; n_2 \; \ldots \; n_r} \quad \text{where} \quad \sum_{i=1}^{r} n_i = n.$$

To verify this enumeration, note that, of the n positions in the word of length n, we can choose n_1 positions for the indistinguishable symbols of type 1 in $\binom{n}{n_1}$ ways. Of the remaining $n - n_1$ symbols in the word, we can choose positions for the n_2 indistinguishable symbols of type 2 in exactly $\binom{n - n_1}{n_2}$ ways. Finally, we are left with only n_r positions in which to place the last n_r symbols of type r, and this can be done in only one way. The total number of possible words is the product of the numbers of ways of placing each of the letter types. The multinomial coefficient also calculates the number of ways in which we can partition n distinguishable objects into r subsets with given sizes $n_1, \ldots, n_r$. Multinomial enumeration applies to the number of texts of length n using r symbols when we use exactly n_i tokens of the ith symbol.

Example 2.21   Multinomial Counting _____

A computer monitor screen having $n = 1,280 \times 854$ pixel resolution, with $r = 3$ (green, blue, red) possible colors for each pixel, can display

$$\binom{n}{i_1\ i_2\ i_3}$$

patterns having i_1 green, i_2 blue, and $i_3 = n - i_1 - i_2$ red pixels. The total number of such arrangements, of varying compositions $i_1, \ldots, i_r$, $\sum_{k=1}^{r} i_k = n$, is r^n, and it follows by specializing $x_i = 1$ in the Multinomial Theorem, presented shortly.

Example 2.22   Cellular Communications and Multinomial Counting _____

One form of cellular mobile wireless communications uses frequency division multiple access (FDMA). In FDMA, a service provider purchases a fixed bandwidth that translates into a maximum number (say, f) of available channels or calling frequencies. We assume that, at a given time, there are n callers distributed at random over r cells. In order to avoid crosstalk, all callers within a cell and the immediately adjacent cells must be assigned unique channels or frequencies. Frequencies can be reused only for callers in cells that are not adjacent. Hence, if $\{n_i\}$ are the cell occupancy numbers, and if i and j are adjacent cells, then, for proper operation, $n_i + n_j \leq f$. Thus, we are interested in the cell occupancy numbers. For purposes of channel assignment, the ordering of users within a cell is irrelevant. Consequently, we are partitioning the n users into r subsets of sizes $n_1, n_2, \ldots, n_r$. Hence, the number of arrangements of n users over r cells with specified occupancy numbers is given by the multinomial coefficient. As the total number of arrangements of n distinguishable users into r cells is r^n, the probability of a particular set of occupancy numbers is

$$P(n_1, \ldots, n_r) = \frac{\binom{n}{n_1\ n_2\ \ldots\ n_r}}{r^n}.$$

This probability can then be used to determine the probability that f channels will suffice.

Theorem 2.8.1   (Multinomial Theorem)

$$(x_1 + \cdots + x_r)^n = \sum_{i_1=0}^{n} \sum_{i_2=0}^{n-i_1} \cdots \sum_{i_{r-1}=0}^{n-\sum_{j<r-1} i_j} \binom{n}{i_1 i_2 \ldots i_r} \prod_{k=1}^{r} x_k^{i_k},$$

where $i_r = n - \sum_{j<r} i_j$.

Note that the summands are monomials in which the sum of the r integer exponents is n.

The multinomial coefficient can also be approximated by use of the r-ary entropy function (expressed as before in bits)

$$H(\mathbf{p}) = -\sum_{i=1}^{r} p_i \log_2 p_i, \text{ for } p_i \geq 0, \sum_{i=1}^{r} p_i = 1,$$

which is a direct generalization of the binary entropy function. Letting $p_i = n_i/n$, we note that if $\sum_{i=1}^{r} n_i = n$, then

$$\binom{n}{n_1 \dots n_r} = \frac{n!}{\prod_1^r n_i!} \approx 2^{nH(\mathbf{p})}.$$

2.9 CONDITIONAL CLASSICAL PROBABILITY

2.9.1 Definition of Conditional Classical Probability

We provide a first introduction to the important concept of conditional probability in the relatively simple setting of classical probability in the hope that the concepts can be communicated more clearly. This topic will be revisited in more generality in Chapters 11 and 12. We inquire into updating classical probability when we are given additional information of the particular kind that the outcome of $\mathcal{E}$ is now known to lie in a particular $B \subset \Omega$ and we have no further knowledge of its location. We learn that event B occurred, and we wish to take this into account. For example, a playing card has been chosen at some distance from us, and we can tell only that it is a red face card. We can now eliminate elements of $\Omega - B$ from consideration. Our new sample space becomes the set B, and the information is postulated to provide no reason to change our mind about the relative probabilities of the subsets of B.

Definition 2.9.1   (Conditional Classical Probability)   The conditional classical probability $P(A|B)$ of event A, given that event $B \neq \emptyset$ occurred, is given by

$$P(A|B) = \frac{\|A \cap B\|}{\|B\|}.$$

The event B replaces the sample space Ω, and we take into account only those elements of A that are also in B by replacing the event A by $A \cap B$.

Example 2.23   Conditional Classical Probability _____
If all 26 letters of the alphabet are equally likely to be chosen as the first letter of a word, then the probability that the first letter is e is 1/26 and the probability that the first letter is b is also 1/26. However, if we are informed that the first letter is an unknown vowel (a,e,i,o,u), then the conditional probability that the first letter is e is now 1/5, whereas the conditional probability that it is b is 0.

2.9.2 Properties of Conditional Classical Probability

Easily verified consequences of the definition of conditional classical probability are given in the next theorem.

Theorem 2.9.1   (Properties of Conditional Classical Probability)

$$P(A|B) = \frac{P(A \cap B)}{P(B)},$$

$$P(A|B) \geq 0, \; P(\Omega|B) = 1, \; P(B|B) = 1, \; P(A|B) = P(A \cap B|B),$$

$$A \supset B \Rightarrow P(A|B) = 1.$$

If $A \perp C$ (A and C are disjoint)

$$P(A \cup C|B) = \frac{P((A \cap B) \cup (C \cap B))}{P(B)} = \frac{P(A \cap B) + P(C \cap B)}{P(B)}$$

$$= P(A|B) + P(C|B).$$

Note that the first result enables us to express the conditional probability in terms of the unconditional probability. This result will become the basis of the general definition of discrete conditional probability in Chapter 11. So long as $P(B) > 0$, Theorem 2.9.1 will be seen to be valid for general probabilities and not just for classical probabilities.

Example 2.24   Bytes with Errors

Consider a message source that produces bytes (8 bits) with the error detection characteristic that each byte has only an even number of ones. Hence, the size $||\Omega||$ of the sample space is

$$\binom{8}{0} + \binom{8}{2} + \binom{8}{4} + \binom{8}{6} + \binom{8}{8} = 2\left[\binom{8}{0} + \binom{8}{2}\right] + \binom{8}{4} = 2(1 + 28) + 70 = 128.$$

The sample space Ω contains half of the possible $2^8 = 256$ bytes. Let 1_j be the event that the jth digit is a one. This event occurs if the remaining seven elements of the byte have an odd sum. Thus,

$$P(1_j) = \frac{\binom{7}{1} + \binom{7}{3} + \binom{7}{5} + \binom{7}{7}}{128} = \frac{7 + 35 + 21 + 1}{128} = \frac{1}{2}.$$

Therefore, any bit in the byte is equally probable to be either a one or a zero. We are now told that the sum of the bits in the byte is not just even, but has the value k (event S_k). We wish to see how that changes the probability of the last bit

$$P(1_8|S_k) = \frac{||1_8 \cap S_k||}{||S_k||}.$$

If $k = 0$, then $||1_8 \cap S_0|| = 0$ and $P(1_8|S_0) = 0$. Next, assume that $k > 0$. In this case,

$$||S_k|| = \binom{8}{k}, \quad ||1_8 \cap S_k|| = \binom{7}{k-1}.$$

Properties of the binomial coefficient and of factorial, presented earlier, enable us to conclude that

$$P(1_8|S_k) = \frac{\binom{7}{k-1}}{\binom{8}{k}} = \frac{k}{8}.$$

Thus, it is only in the case of the event S_4 that the conditional probability of 1_8 remains equal to the unconditional probability of $1/2$.

Example 2.25 Bosons or Fermions

A second example is that of discriminating between bosons and fermions. We assume that, in advance of observation, we have f possible sources of fermions—say, electrons—and b possible sources of bosons—say, photons. A source S is selected at random. Letting F be the case that the source is one of fermions and B that it is one of bosons, we see that by random selection

$$P(S = F) = \frac{f}{b+f}, \quad P(S = B) = \frac{b}{b+f}.$$

In either case, the source emits exactly r particles. The observation or measurement A that is made analyzes the incoming particles by determining which of the n cells of quantum phase space they are observed to occupy. The n-dimensional vector $\mathbf{a}$ of possible occupancy numbers for bosons satisfies

$$\mathbf{a} = \{a_i\}, \ a_i \in \{0, 1, 2, 3, \ldots, r\}, \ \sum_{i=1}^{n} a_i = r,$$

and the set of such possibilities will be denoted A_B. From Section 2.7, for the fermions, we have the additional restriction that $a_i \in \{0, 1\}$, at most one fermion in a cell, and the set of such possibilities will be denoted by A_F. Note that $A_B \supset A_F$. From Section 2.7, we have

$$||A_F|| = \begin{cases} 0 & \text{if } r > n \\ \binom{n}{r} & \text{otherwise} \end{cases} \quad ||A_B|| = \binom{r+n-1}{r}.$$

The results of Section 2.7 enable us to write the conditional probabilities for a measurement given a source type as

$$P(A = \mathbf{a}|S = F) = \frac{I_{A_F}(\mathbf{a})}{\binom{n}{r}} = \begin{cases} \dfrac{1}{\binom{n}{r}} & \text{if } \mathbf{a} \in A_F \\ 0 & \text{otherwise} \end{cases},$$

$$P(A = \mathbf{a}|S = B) = \frac{I_{A_B}(\mathbf{a})}{\binom{r+n-1}{r}} = \begin{cases} \dfrac{1}{\binom{r+n-1}{r}} & \text{if } \mathbf{a} \in A_B \\ 0 & \text{otherwise} \end{cases}.$$

The conditional probabilities in the preceding example describe the measurement process in that they specify the probabilities of a given observation (effect) **a** given the possible true type of particle (cause). How, though, are we to determine the unconditional (absolute) probability $P(A = \mathbf{a})$ of a particular measurement value, and how are we to interpret a given measurement **a** in terms of the probability of a particle type being present? We answer these two questions in the next few sections.

2.9.3 Effects from Causes: Total Probability

Recall that our first goal is to provide mathematical models for random phenomena, and our introduction of classical probability and classical conditional probability are first steps in this direction. If we use the language of "cause and effect," then we can restate this goal as providing mathematical models for the production of random causes. Our second goal, of calculating system response, can be restated as calculating probabilities of effects given causes when causes do not deterministically select effects. Let the partition $\{B_i\}$ of Ω correspond to the identification of mutually exclusive and exhaustive causes—one and only one cause can operate. Let the event A correspond to a particular effect or result of the operation of a cause. The stochastic relation between cause and effect is specified by a conditional probability $P(A|B_i)$ that states the probability that a cause B_i will produce an effect A. The probability of cause B_i operating is $P(B_i)$. We wish to determine $P(A)$. Dropping the language of cause and effect, a method for calculating unconditional probability from conditional probability is given by the total classical probability theorem.

Theorem 2.9.2 **(Total Classical Probability)** If $\{B_1, \ldots, B_n\}$ is a partition of Ω, then

$$P(A) = \sum_{i=1}^{n} P(A|B_i)P(B_i).$$

Proof. From the definition of conditional probability,

$$P(A|B_i)P(B_i) = P(A \cap B_i).$$

From Section 2.2, for disjoint sets $\{A \cap B_i\}$ (disjointness following from that postulated for $\{B_i\}$),

$$P\left(\bigcup_{i=1}^{n}(A \cap B_i)\right) = \sum_{i=1}^{n} P(A \cap B_i).$$

It remains only to recognize that

$$\bigcup_{i=1}^{n}(A \cap B_i) = A \cap \left(\bigcup_{i=1}^{n} B_i\right) = A \cap \Omega = A. \qquad \square$$

An illustration of the use of the Total Probability Theorem is provided by the notion of a *two-stage experiment*, also called a compound experiment. In a two-stage experiment, we first perform a random experiment—say, $\mathcal{E}_0$—in which the sample space $\Omega_0 = \{\omega_1, \ldots, \omega_T\}$ contains the labels for further random experiments $\mathcal{E}_1, \ldots, \mathcal{E}_T$. If we observe ω_i, then we next perform the unrelated experiment $\mathcal{E}_i$ and obtain outcome event—say, A. To be more concrete, consider a basic communication system in which $\mathcal{E}_0$ selects a message m to be transmitted. The message m is then transmitted through a noisy channel whose characteristics are described by the conditional probability $P(A|m)$ for observing event A at the channel output when message m was transmitted. We wish to determine the probability $P(A)$ without regard to which message is selected. A more detailed example is that of an urn model and is described in this context in Section 11.5.

Example 2.26   Classical Total Probability ────────────────────────────────

A specific example of such a two-stage experiment can be identified in Example 2.25 of particle-type detection. The measurement A results from first randomly selecting a source type (boson or fermion) and then measuring the occupancy vector $\mathbf{a}$ for the r received particles. The calculations at the end of Section 2.7, when combined with the Total Probability Theorem of this section, yield

$$P(A) = \frac{f}{b+f} \frac{I_{A_F}(\mathbf{a})}{\binom{n}{r}} + \frac{b}{b+f} \frac{I_{A_B}(\mathbf{a})}{\binom{r+n-1}{r}}.$$

2.9.4 Causes from Effects: Bayes' Theorem

Our third goal is to understand how to make inferences, estimates, and decisions from our probability characterizations of random phenomena. In this case, we can consider the inference from an observed effect—say, event A—to one of a set of possible causes $\{B_i\}$. This is the problem of diagnosis in medicine, where A are the symptoms and $\{B_i\}$ are the possible diseases that could cause the symptoms. In communications, it is the problem of inferring from the received signal A to which of the possible messages $\{B_i\}$ was transmitted. A solution to the problem is provided by the celebrated Bayes' classical probability theorem.

Theorem 2.9.3   (Bayes' Theorem for Classical Probability)   If $\{B_i\}$ partition Ω, $B_k \neq \emptyset$, then

$$P(B_k|A) = \frac{P(A|B_k)P(B_k)}{\sum_i P(A|B_i)P(B_i)}.$$

Proof. Theorem 2.9.2 reduces the denominator of the right-hand side of the equation to $P(A)$. Recourse to the elementary properties of conditional probability given in Theorem 2.9.1 suffices to complete the proof.   □

Example 2.27   Classical Bayes' Theorem ──────────────────────────────────

Consider an image formed of $n \times m$ pixels with the kth row or scan line containing $d_k(\leq m)$ defective pixels. In the first stage experiment, a row is chosen at random, with the choice unknown to us. We then examine a randomly selected pixel in that row and find that the event D is a defective pixel. Let $R = k$ be the event that this pixel was in the kth row of the image. Bayes' theorem allows us to determine, from

$$P(R = k) = \frac{1}{n} \text{ and } P(D|R = k) = \frac{d_k}{m},$$

that

$$P(R = k|D) = \frac{\frac{1}{n}\frac{d_k}{m}}{\sum_{i=1}^{n}\frac{1}{n}\frac{d_i}{m}} = \frac{d_k}{\sum_{i=1}^{n} d_i}.$$

Hence, even though the row was initially selected at random, given the event of randomly finding a defective pixel in that row, it is now more probable that it is a row containing a larger number d_k of defective pixels.

Example 2.28 Bayesian Determination of Particle Type _____

Recall our recent example of the detection of particle type. Bayes' theorem and our calculations at the close of the preceding subsections yield

$$P(S = F|A = \mathbf{a}) = 1 - P(S = B|A = \mathbf{a}) = \frac{\frac{f}{b+f}\frac{I_{A_F}(\mathbf{a})}{\binom{n}{r}}}{\frac{f}{b+f}\frac{I_{A_F}(\mathbf{a})}{\binom{n}{r}} + \frac{b}{b+f}\frac{I_{A_B}(\mathbf{a})}{\binom{r+n-1}{r}}}.$$

This expression can be simplified by dividing the numerator and denominator by the numerator and expanding the binomial coefficients to arrive at

$$P(S = F|A = \mathbf{a}) = \left[1 + \frac{bn!(n-1)!I_{A_B}(\mathbf{a})}{f(n-r)!(r+n-1)!I_{A_F}(\mathbf{a})}\right]^{-1}.$$

Given the particular measurement $\mathbf{a} \in A_F$, if

$$bn!(n-1)! < f(n-r)!(r+n-1)!,$$

then the particle type is more probably F than it is B.

2.10 INDEPENDENCE IN CLASSICAL PROBABILITY

We introduce the key concept of *independence*, also known as *stochastic independence*. We will treat this concept in greater detail in Chapter 14. Event A is independent of event B if knowledge of the occurrence of B does not lead us to revise the probability of occurrence of A. B is uninformative about A and they are *unlinked*, perhaps because there is no physical link of causation between them or because, given our knowledge base, they are inferentially unrelated.

Thus, if I have drawn a card from a deck of cards and you then toss a die, there is no causal effect of my draw upon the outcome of your toss. Given the concept of classical conditional probability, we can restate this more formally as A is independent of B if

$$P(A|B) = P(A).$$

Of course, not all events are independent of each other. Knowing that an English word has the first two letters th is informative about the third letter being more likely an e than a b. Knowing the signal obtained by a communications receiver constrains the messages that could have been transmitted. From Theorem 2.9.1, we see that $P(A|B) = P(A)$ can be restated as

$$P(A \cap B) = P(A)P(B).$$

This condition is symmetric between A and B (i.e., A is independent of B if and only if B is independent of A). Hence, we can introduce the definition of two independent events.

Definition 2.10.1   (Two Independent Events) $A \perp\!\!\!\perp B$, read as "A and B are independent,"

$$A \perp\!\!\!\perp B \iff P(A \cap B) = P(A)P(B).$$

In terms of classical probability,

$$A \perp\!\!\!\perp B \iff ||A \cap B|| \cdot ||\Omega|| = ||A|| \cdot ||B||.$$

Lemma 2.10.1   (Implications of Two Independent Events)   Implications of this definition of independence are

$$A \perp\!\!\!\perp B \Rightarrow B \perp\!\!\!\perp A$$

and

$$A \perp\!\!\!\perp B \Rightarrow A \perp\!\!\!\perp B^c.$$

Proof. The first implication is immediate from the observation that

$$P(A \cap B) = P(A)P(B) \iff P(B \cap A) = P(B)P(A).$$

To verify the second implication, consider

$$A = (A \cap B) \cup (A \cap B^c) \Rightarrow P(A) = P(A \cap B) + P(A \cap B^c).$$

Hence, from $A \perp\!\!\!\perp B$, we conclude that

$$P(A \cap B^c) = P(A)(1 - P(B)) = P(A)P(B^c),$$

and therefore $A \perp\!\!\!\perp B^c$. $\qquad\qquad\qquad\qquad\qquad\qquad\qquad\qquad\qquad\qquad\qquad\square$

Example 2.29   Independence in Classical Probability _____

Let us reconsider the example in the last section of a message source producing bytes constrained only to have an even number of ones. From the evaluation of conditional probability, we see that $P(1_8|S_k) = P(1_8)$ only in the case of $k = 4$. Hence, the events 1_8 and S_4 are independent, but the events 1_8 and S_k are not independent for any choice of $k = 2, 6, 8$.

Example 2.30   Independence in Classical Probability (Continued) _____

Consider two quantized voltages X and Y that can take on only the five discrete levels $\{-2, -1, 0, 1, 2\}$. Assume that all of the $5 \times 5 = 25$ possible pairs of values $X = x_i, Y = y_j$ are equally probable. We first ask whether event A, that $X = x_i$, is independent of event B, that $Y = y_j$? Clearly, $||A \cap B|| = 1$, $||\Omega|| = 25$, $||A|| = ||B|| = 5$. Hence, the events are independent, and the value taken on by the quantized voltage X is independent of the value taken on by the quantized voltage Y.

Now consider the sum, $Z = X + Y$, of the quantized voltages. Clearly Z can take on all integer values from -4 to $+4$. Is $Z = 2$ independent of $X = 1$? The event A that $Z = 2$ is given in terms of the values (x_i, y_j) by the set $\{(0, 2), (1, 1), (2, 0)\}$. Hence, $||A|| = 3$. The event

$$B = \{(1, -2), (1, -1), (1, 0), (1, 1), (1, 2)\}$$

that $X = 1$ has cardinality 5. The event $A \cap B = (1, 1)$ has cardinality 1. However, $1 \cdot 25 = 25 \neq 3 \cdot 5 = 15$, and the two events are not independent.

2.11 SUMMARY

We complete the mathematical model for $\mathcal{E}$ for the case of Ω a finite set. It is assumed that all outcomes listed in Ω are equally probable, likely, or possible; albeit, this does not address the question of the meaning of such an assertion. The probability $P(A) = k/n$ of an event $A = \{\omega_1, \ldots, \omega_k\}$ depends only on the number k of elementary outcomes or points ω_i that make up the set A and the cardinality (size) n of the sample space Ω. Computations in classical probability require us to have facility with enumerating sets having prescribed characteristics, and Sections 2.3, 2.4, and 2.8 introduce us to the important and independently valuable subject of combinatorics.

Historically, classical probability grew out of games of chance, and games of chance evolved so that their outcomes were equally probable (e.g., later dice were more symmetrical than earlier ones). Symmetry is the basis of most gambling games, including some dating to antiquity. Section 2.5 touches on modern applications to entropy and data compression, Section 2.6 introduces us to graphs as models for important large-scale systems, and Section 2.7 makes contact with particle statistical mechanics.

The simple case of classical probability enables us to introduce all of the elements of a probability model without mathematical complications. The properties observed to hold for classical probability will also be required in the more general theory developed in the next chapter. We then drew out the elementary consequences of this model and enriched the theory by introducing definitions of additional concepts of conditional probability and independence. Conditional probability $P(A|B)$ is the updated probability of the event A given that we now know that B has occurred. We present a more general notion of conditional probability in Chapters 11 and 12. The notion of conditional probability then provides access to a formalization, $P(A|B) = P(A)$, of unlinkedness or uninformativeness between events that is known as (stochastic) independence and is discussed more fully in Chapter 14.

EXERCISES

Many good problems can be found in the classic text by Feller [27], although they are not in the language of electrical engineering.

E2.1   (Scrabble is not random.) In what follows, we refer to the usual 26 alphabetic characters of English and restrict vowels to a, e, i, o, u.

   a.   How many words are there of length 2 without regard to the constraints of English?

   b.   In Scrabble (as of March 2, 1998), there are 96 acceptable two-letter words. What is the probability that a randomly selected pair of letters produces a Scrabble word?

   c.   In Scrabble, there are 37 three-letter words that begin with the letter c. What is the probability that a randomly selected two-letter continuation of c will produce a Scrabble word?

   d.   In Scrabble, there are 62 acceptable four-letter words containing exactly three vowels. What is the conditional probability that a randomly selected four-letter word will be a Scrabble word given that it has exactly three vowels?

E2.2   How many distinct words (not necessarily meaningful ones) can you form from the following, using all of the given letters:

   a.   The author's first name TERRENCE?

   b.   The author's last name FINE?

   c.   Your first name?

   d.   Your last name?

   e.   The word PROBABILITY?

E2.3   Message source M produces a word of length eight characters, with the characters drawn from the ternary alphabet $\{0, 1, 2\}$, and all such words are equally probable. What is the probability that M produces a word that looks like a byte (i.e., no appearance of '2')?

E2.4   A ternary communications system sends packets composed of characters drawn from the alphabet $\{0, 1, 2\}$.

   a.   If a packet has nine characters, how many possible packets are there?

   b.   If all possible packets are equally probable, then what is the probability of a packet that does not contain the character '2'?

c. What is the probability of a packet containing equal numbers of each of the characters?

E2.5   A signal transmission path has five links that are serially/cascade connected, each of which can be in any one of 10 distinct states.

   a. How many distinct transmission paths can there be (assuming that the sequence/order of the links matters)?
   b. What is the probability of such a randomly chosen path having the second link in state "1"?
   c. How many paths can there be if we use three special links in place of the first three of the original links and these special links can be in any one of 20 states?

E2.6   Consider the following simplified packet routing problem: A packet is sent to a first router and this router uses 2 bytes to address a second router. The selected second router then uses a 1-byte address to send the packet to the end user.

   a. How many different second routers can the first router reach?
   b. How many end users can this system send a packet to?

E2.7   In QPSK, there are four possible signals $\{q_0, q_1, q_2, q_3\}$. What is the minimum length n of messages (sequence of QPSK signals) needed to ensure that there are at least 1,000 different messages?

E2.8   Given the integers $1, \ldots, n$, how many subsets of size $k \leq n$ have their largest member the integer $i \leq n$?

E2.9   What is the probability of the event A that when a k-tuple of integers is drawn at random with replacement from $1, \ldots, n$, the largest integer chosen is i?

E2.10 Verify the Fermat combinatorial identity

$$\binom{n}{k} = \sum_{i=k}^{n} \binom{i-1}{k-1},$$

for $n \geq k$.

E2.11   a. How many k-tuples of positive integers $x_1, \ldots, x_k$ satisfy

$$\sum_{i=1}^{k} i = k + 1?$$

   b. Verify that there are $\binom{n-1}{n-k}$ k-tuples of positive integers $x_1, \ldots, x_k$ satisfying

$$\sum_{i=1}^{k} x_i = n.$$

E2.12 What is the probability that a randomly selected 64-bit register sequence will start with 00 and have exactly four zeros?

E2.13   a. Let f_n denote the number of $\{0, 1\}$-valued binary sequences of length n in which there are no adjacent ones. (Codes for magnetic recording have constraints of this type.) Evaluate f_0, f_1, and f_2.

b. Verify the Fibonacci recursion

$$f_n = f_{n-1} + f_{n-2}$$

for $n \geq 2$.

c. If all binary sequences $x_1, \ldots, x_n$ of length n have equal probability, evaluate the probability p_n that we observe a sequence having no adjacent ones.

E2.14 A message source generates equally probable messages of length 10 from a symbol alphabet $\{-1, 0, 1\}$, satisfying the constraint that no two consecutive symbols are the same in any message (e.g., cannot have $\cdots 00 \cdots$).

a. How many possible messages can this source generate that satisfy the constraint?

b. How many possible sequences of length 10 are eliminated by the constraint?

c. What is the probability of a message beginning with the sequence of symbols $-1, 1$?

d. What is the probability of a message having -1 in the fourth position?

E2.15 A random byte is received consisting of eight bits $b_1, \ldots, b_8$, $b_i \in \{0, 1\}$.

a. If all bit patterns are equally probable, what is the probability of receiving 10101010?

b. What is the probability that $b_1 = 1$?

c. What is the probability that all of the bits have value 1?

d. What is the probability that at least one bit has the value 1?

e. What is the probability that $b_1 + b_2 = 1$?

E2.16 How many binary strings of length $2n$ have r_1 ones in the first half and r_2 ones in the second half?

E2.17 Consider the 30 leading digits in the first two columns of Gaussian random number data listed in Section 1.11. We hypothesize that these digits are drawn at random from $-9, -8, \ldots, 0, 1, \ldots, 9$. It is clear from the presence of duplications that these draws must be with replacement.

a. If this is the case, then what is the probability of observing as many of -1 as we did?

b. Is the hypothesis we made of random draws supported by the probability of this event?

E2.18 a. What is the probability of drawing five cards, all different, from the usual well-shuffled deck of 52 cards?

b. Repeat the calculation if the five cards are now drawn with replacement.

c. What is the probability of drawing five cards (without replacement) having the same suit?

E2.19 At an IEEE Banquet there were 13 faculty and 65 student attendees. Seven tee shirts were awarded by random drawing without replacement.

a. What is the probability (evaluated by classical probability) that the first two individuals selected were faculty?

b. Provide an expression for the probability (evaluated by classical probability) that the students receive no more than three tee shirts (i.e., no more than three students among the first seven individuals selected).

E2.20   a. If a language has 200 syllables, how many words can you form having no more than three syllables?

   b. How many binary sequences are there of length 100 with exactly 40% ones?

E2.21   A message source generates binary messages of length 1,000. Each message has exactly 300 ones, and all such messages are equally probable.

   a. What is the probability of a message with all ones in the initial 300 positions?

   b. What is the probability of a message with all ones in the first 600 positions?

   c. If all possible messages are to be compressed to the same length m, how small can m be?

   d. What is the bit-per-symbol entropy of this message source?

E2.22   a. How many undirected labeled graphs are there on 10 nodes?

   b. How many undirected labeled graphs on 10 nodes have five links?

   c. For what number m^* of links do we have the largest number of undirected graphs on n nodes?

E2.23   If possible, sketch a graph on six nodes having the degree sequence 3, 2, 2, 2, 1, 0.

E2.24   What is the probability that a randomly selected undirected G_4 will be connected?

E2.25   Is it more probable that a randomly selected undirected G_6 will be connected than that a randomly selected undirected G_5 will be connected?

E2.26   a. How probable is it that a randomly selected undirected graph on six nodes will have exactly 10 links?

   b. How probable is it that a randomly selected directed graph on six nodes will have exactly 10 links?

E2.27   A communications network or bulk power transfer grid is described by the graph of Figure 2.5 in which an edge or straight-line segment denotes a communications link or power transmission line and a numbered box is a node or vertex and denotes a switch or receiver of communications or a load or switching substation. The links will be assumed to be directed in that a message or power can only pass from a lower numbered node to a higher numbered node.

   a. Verify that there are exactly three directed paths from node 1 to node 6.

   b. How many possible directed paths are there from node 1 to node 13?

   c. How many possible directed paths are there from node 1 to node 14 if the path must contain node 5?

E2.28   For the directed graph depicted in Figure 2.5 and the conventions about links directed from lower numbered nodes to higher numbered nodes, evaluate the indegree and outdegree sequences.

E2.29   For the directed graph of Figure 20.3, list the sequences of outdegrees and of indegrees.

E2.30   a. For the WWW, how many more pages are there with 10 links pointing to them than with 100?

   b. How many more Web pages are there that point to 10 other pages than that point to 100?

E2.31   A message source M generates equally probable messages of length 10, from a quaternary symbol alphabet $\{e^{i\frac{\pi}{2}k}, k = 0, 1, 2, 3\}$, that must satisfy the constraint that no two imaginary symbols can be adjacent.

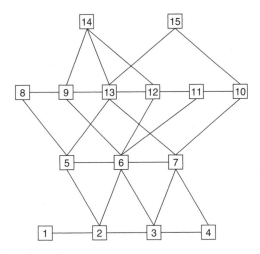

Figure 2.5   Communications or Power Grid Network Graph.

    a. How many possible message sequences can M generate satisfying the constraint?

    b. How many messages are eliminated by the constraint that no two imaginary symbols can be adjacent?

    c. What is the probability of a message beginning with the symbols $i \, -i$?

    d. What is the probability of a message having -1 in the fifth position?

E2.32 Consider a message source S that uses a ternary alphabet $\{a, b, c\}$ to generate a message of $n > 2$ symbols. S satisfies the additional constraint that the first and last symbols in a message must be the same. All messages satisfying this constraint are equally probable. What is the probability that a message contains exactly two occurrences of the symbol a?

E2.33 A message source S, using the quaternary alphabet $\{a, b, c, d\}$, produces equally probable messages of length 8 that satisfy the constraint that each message must contain exactly two uses of the symbol a.

    a. How many different possible messages m can S emit?

    b. What is the probability of event A of a message having its first two symbols be aa?

    c. Given that the message starts with aa, what is the probability of event B that the last symbol is a b?

    d. Given event C that the message starts with ab, what is the probability of event D that the last symbol is a?

E2.34 Professor Phynne believes that $100!$ is larger than 100^{75}. Use Stirling's formula to determine whether he is correct. (Hint: $e^{2.3} \approx 10$.)

E2.35 What is the probability that there will be exactly one matching pair of birthdays in a group of $r \geq 2$ people when the number of birthdays $n = 365$ and we assume that all birthdays are equally probable?

E2.36 In mobile cellular communications systems, a slotted Aloha protocol is used to resolve collisions between callers seeking to place a call in the same time slot. If two callers "collide" (attempt to call in the same time slot), then a simplified form of the protocol is that caller c_i attempts to place her call at a randomly selected time slot $t_i \in N_{10}$ and that all pairs of time slots t_1, t_2 are equally probable.

    a. Given that c_1 and c_2 have collided, what is the probability that caller c_1 will attempt to call at time slot $t_1 = 2$ and caller c_2 attempt to call at time slot $t_2 = 9$?

    b. What is the probability that two callers will succeed in placing their calls ($t_1 \neq t_2$)?

E2.37 In a mobile cellular telephone system, the service region is divided into c distinct cells, each serviced by a base station. Assume that, at a given time, there are $u > c$ distinct clients and that all allocations of clients to cells are equally probable. A base station in a cell can handle at most f clients, where $f < u < cf$.

    a. How many distinct allocations of clients are there?

    b. What is the probability that Cell 1 cannot serve all of its clients?

    c. Provide an expression for the probability that all clients can be provided with service; that is, no cell has more than f clients.

E2.38   a. How many texts of length 52 are there that use each of the 26 letters of the alphabet exactly twice?

    b. What is the probability that, in a text generated as in (a), you will observe the consecutive occurrence dd?

E2.39 A radar partitions its surveillance range set $\mathcal{R}$ into nonoverlapping range gate sets $\{G_1, \ldots, G_g\}$. At a particular time, there are exactly a aircraft in $\mathcal{R}$. Assume that all arrangements of aircraft in range gate sets are equally probable.

    a. What is the probability P_a that no two aircraft share the same gate set?

    b. What is the probability P_b that at least two aircraft share the same gate set?

    c. What is the probability $P_c(k)$ of exactly k aircraft in gate set G_1?

    d. Show, using the Binomial Theorem, that $P_c(k)$ sums over k to unity, as it should.

E2.40 A device has $E \geq 1$ energy levels and $N \geq E$ free electrons, with any number of electrons allowed to occupy each energy level.

    a. How many distinguishable energy level occupancy arrangements are there?

    b. If all such arrangements are equally probable, what is the probability P_b that all of the energy levels will be filled by at least one electron?

    c. Verify that $0 \leq P_b \leq 1$.

E2.41 k photons are known to have arrived in the time interval $[0, N)$, which we take to be partitioned into unit-length intervals, $[0, N) = \cup_{i=1}^{N}[i - 1, i)$. All arrangements, of the k indistinguishable photons in the $N (> k)$ given intervals of unit length, are equally probable.

    a. What is the probability P_a that no two photons occupy any of the unit length intervals of the given partition?

    b. What is the probability P_b that at least two photons share some unit length interval?

    c. What is the probability P_c of exactly two photons in $[0, 1)$?

E2.42 In a given experiment, we observe three photons occupying some of 10 cells $\{q_1, \ldots, q_{10}\}$ in quantum phase space.

a. How many physically distinguishable arrangements m are there for placing the photons in the cells?

b. If all of these arrangements are equally probable, what is the probability of event A that all three photons are in cell q_1?

c. What is the probability of event B that none of the photons are in cell q_1?

E2.43 In a given experiment, we observe three electrons satisfying Fermi–Dirac statistics occupying some of 10 cells $\{q_1, \ldots, q_{10}\}$ in quantum phase space.

a. How many physically distinguishable arrangements m are there for placing the electrons in the cells?

b. If all of these arrangements are equally probable, what is the probability of event A that there is an electron in each of cells q_1, q_4, q_9?

c. What is the probability of event B that none of the electrons are in cell q_1?

E2.44 A message source S, using the quaternary alphabet $\{a, b, c, d\}$, produces equally probable messages of length 8 that satisfy the constraint that each message must contain exactly three uses of the symbol a.

a. How many different possible messages m can S emit?

b. What is the probability of event A of a message having its first two symbols be aa?

c. How many messages n are there in which a and b appear exactly once?

E2.45 An image is composed of a 20×20 array of pixels that are either black or white.

a. How many possible images are there?

b. If an image is chosen at random, what is the probability of its having exactly 100 black pixels?

c. What is the most probable number of white pixels?

E2.46 a. How many ways are there to arrange r students in a lecture room with n seats?

b. If $r = 130$ and $n = 200$, evaluate your answer, using Stirling's formula.

E2.47 A 10×10 pixel image is composed of 20 red pixels, 30 green pixels, and the remainder blue pixels.

a. How many such images can you form?

b. In such a randomly generated image, what is the probability that the first pixel is red?

c. If an image is selected at random, what is the probability that the first row will contain no blue pixels?

d. What is the probability that the first pixel is red and the last pixel is green?

E2.48 An image of a character is represented by a 5×7 pixel array, with each pixel taking on one of four possible gray-levels $g_1, \ldots, g_4$.

a. How many possible image representations are there?

b. What is the probability that a randomly selected representation will have no pixels containing level g_4?

c. What is the probability that a randomly selected representation will have a first row of pixels all at level g_1?

E2.49   In a given classical probability problem it is determined that the events $\{B_1, B_2, B_3\}$ are pairwise disjoint and that their union is the sample space. These events have the following probabilities: $P(B_1) = .2, P(B_2) = .3$. There is another event A about which it is known that $P(A|B_1) = .3, P(A|B_2) = .4, P(A|B_3) = .1$.

   a. Evaluate $P(A)$.
   b. Evaluate $P(B_2|A)$.

E2.50   Is it always true that, for $P(B) > 0$,

$$P(A|B) \leq \frac{P(A)}{P(B)} \ ?$$

You must give reasons for your answer.

E2.51   In a given communication system, all bytes are equally probable. Define the weight w of a byte **b** to be

$$\mathbf{b} = [b_1, b_2, \ldots, b_8], \ b_i \in \{0, 1\}, \ w = \sum_1^8 b_i.$$

Define the following events:

$$A = \{\mathbf{b} : b_1 = b_2 = 1\}, \ B = \{\mathbf{b} : w \text{ is odd}\}.$$

Evaluate $P(A), P(B), P(B|A)$, and $P(A|B)$.

E2.52   a. If we toss two well-balanced dice and observe the event A that the sum of spots showing uppermost is 9, then what is the conditional probability of the event B that the first die is showing 4 spots?
   b. Are the events A and B independent?

E2.53   a. If all bytes are equally probable, then what is the probability of observing one containing exactly 3 ones?
   b. If we observe exactly 3 ones, what is the probability that the first bit is a one?
   c. Are these two events independent?

E2.54   In information theory, we are interested in so-called typical sequences: sequences having a specified composition. In this problem, we consider sequences of length 30, with elements drawn from $\{0, 1, 2\}$. All such sequences are equally probable.

   a. What is the probability that a sequence will be composed of elements that are just ones and twos?
   b. What is the probability that a sequence will have 5 zeros, 10 ones, and 15 twos?
   c. What is the probability that a sequence will have zeros in its first 5 positions, given that it has the composition specified in part (b)?

E2.55   We observe an $X = S + N$ that is the sum of a signal S taking values in $\{0, 1\}$ and a noise N taking values in $\{-1, 0, 1\}$. Assume that all possible pairs (s_i, n_j) of values s_i for S and n_j for N are equally probable.

   a. What is the probability $p_k = P(X = k)$ for each value k taken on by X?
   b. Evaluate $P(S = 0)$.

    c. Evaluate $P(S = 0|X = k)$ for each value k taken on by X.

    d. Are $S = 0$ and $X = -1$ independent?

E2.56 We observe $Z = XY$, where $X \in \{0, 1\}$ and $Y \in \{-1, 0, 1\}$. Assume that all possible pairs (x_i, y_j) of values x_i for X and y_j for Y are equally probable.

    a. What is the probability $p_k = P(Z = k)$ for all possible values k taken by Z?

    b. Evaluate $P(Y = 0)$.

    c. Evaluate $P(Y = 0|Z = k)$ for all possible values k for Z.

E2.57 A woman and a man (unrelated to each other) each have two children. Assume that all four possible arrangements of "boy" and "girl" for the pair of younger and older child, bb, bg, gb, gg, are equally probable. At least one of the woman's children is a boy, and the man's older child is a boy. (Hint: use Bayes.)

    a. Show that the probability that the man has two boys, conditional upon the given information, is 1/2.

    b. Show that the probability that the woman has two boys, conditional upon the given information, is 1/3.

E2.58 We have a population of $N = n_0 + n_1$ items of the two types 0 and 1 in which the n_i items of type i are indistinguishable from each other, but distinguishable from the n_{1-i} items of type $1 - i$. We draw a sample of size m at random and without replacement from this population. The probability $P(M_0 = m_0)$ that we will have selected m_0 members of population 0 is given by

$$P(M_0 = m_0) = \frac{\binom{n_0}{m_0}\binom{n_1}{m-m_0}}{\binom{N}{m}}$$

and is known as the *hypergeometric pmf*. The number of different subsets (the items of each type are indistinguishable) of size m from a set of size N is $\binom{N}{m}$. All of these subsets are equally probable. There are $\binom{n_0}{m_0}$ ways to choose the m_0 objects of type 0 from their population of size n_0, and, similarly, there are $\binom{n_1}{m-m_0}$ ways to choose the $m - m_0$ objects of type 1 from their population of size n_1.

    a. If a binary sequence of length 64 has 16 ones, what is the probability that if we randomly select 10 terms in this sequence, without replacement, 3 of them will be ones?

    b. For given n_0, n_1 and m, with $n_0 + n_1 \geq m$, what is the most probable value of m_0?

E2.59 A computer monitor with p pixels has f fixed-location defective pixels. What is the probability that a random image of size $q < p$ pixels will contain $q_0 < q$ defective pixels?

Part II

Unconditional Probability and Its Applications

3

Probability Foundations

3.1 PURPOSE, BACKGROUND, AND ORGANIZATION

It is fair to say that the foundations of probability are still under construction, with one mathematical approach far better developed than all others and three interpretations (classical of Chapter 2, frequentist, and subjective) dominant. As noted in Chapters 0 and 1, probabilistic reasoning comes in many forms and addresses a wide variety of chance, random, uncertainty and nondeterministic phenomena. In everyday life, probabilistic reasoning is commonplace and probabilistic judgements are expressed in ordinary language. "I will probably see you tomorrow." "It is likely that tomorrow will be warmer than today." "It is unlikely that the failure of the system was caused by the failure of this component." We express probability judgements both by our statements and by our actions. Passing a car on a two-lane road with oncoming cars visible in the distance implies that we have assessed the risks, judged the speeds and distances, and are aware of the serious consequences of errors in our judgement, but deem them sufficiently unlikely. Everyday probability judgements and expressions run the gamut from those based on weakly held beliefs and virtually no expressible empirical evidence, to those (e.g., prices to be paid for insurance policies) based on extensive analyses of millions of records, and to those based upon the firmest of convictions (e.g., belief in the testimony of one's senses). Probability reasoning in everyday life, while not mathematically developed along conventional lines, needs to be taken very seriously. It is more our guide to life than the formal theory we will present.

In the physical sciences, we expect probabilities to be determined by some underlying theory (e.g., statistical mechanics, quantum mechanics) and to be expressed by number. Engineering treats a wide range of phenomena, including human behavior, natural phenomena, and phenomena for which there is no detailed physical understanding. Thus, there needs to be room to incorporate probabilistic knowledge that is both qualitative and expressed linguistically, as well as the quantitative knowledge that is expressed numerically. Before settling on one account of probability, we provide a glimpse into the space of alternatives.

We can classify the forms of probabilistic reasoning along the following dimensions:

- degree of precision—the structural concept
- meaning of, or interpretation to be attached to, probability
- formal mathematical structure of probability as provided by a set of axioms.

The structural concept determines the precision with which we can expect probabilities to represent random phenomena. The interpretation provides the basis on which probability is to be assessed or determined and indicates what we can expect to learn from it—what a probability statement means. The interpretation also governs the choice of application for probability. The structural concept and interpretation then guide the axiomatization. The axiom set, however, can only capture a portion of what is implied or understood in the interpretation.

We open our discussion with a sketch of a variety of structural concepts and interpretations for probability. We then settle upon the traditional numerical probability satisfying the celebrated Kolmogorov axioms and upon an interpretation related to frequencies of occurrence and their propensity to occur. The mathematical theory is then developed by presentation of key implications of the Kolmogorov axioms and examples of their use.

3.2 HIERARCHY OF PROBABILITY STRUCTURAL CONCEPTS

The following are examples of a variety of different structural concepts of probability:

Possibly: "Possibly A" is the most rudimentary, least precise concept, and the one examined by the classical Greeks in attempting to distinguish between the necessary and the contingent. There are a number of modal concepts of *possibility*, including the following:

logical possibility, in the sense of freedom from logical self-contradiction;

epistemic possibility, in which the occurrence of A does not contradict our knowledge, which includes, but also extends beyond, mere logic;

physical possibility, the occurrence of A is compatible with physical laws although it might be exceedingly unlikely—the example of a tossed coin landing and balancing on edge on a hard surface;

practical possibility, the everyday notion or ordinary language concept wherein A is practically possible if it has at least some "not too small" likelihood of occurring.

Probably: "Probably A" is a strengthening of possibility to mean more likely than not. While it might correspond to the numerical probability of A being at least $1/2$, this concept does not require any commitment to numerical probability or the precise state of knowledge that numerical probability assumes. It is capable of refinement in everyday use as in the expression "it is highly probable/likely that." In legal usage, there are the standards of "preponderance of the evidence," used in civil trials, "clear and convincing," and "beyond a reasonable doubt," used in criminal trials.

Comparative Probability: "A is at least as probable as B." The comparative probability relation includes "probably A" through "A is at least as probable as A^c." It can be related to numerical probability through $P(A) \geq P(B)$; although, as in the previous two examples, comparative probability does not require a commitment to numerical probability. Insofar as one can make many comparisons between events, one may be in the position of making more precise comparative statements. Thus, "rain tomorrow is less probable than drawing any card from 2 through 9 in a well-shuffled deck of cards and more probable than drawing any card from 2 through 8."

Interval-Valued Probability: "A has interval-valued, or lower and upper, probability $(\underline{P}(A), \overline{P}(A))$." This allows for a variable degree of indeterminacy expressed through a range of probability values without commitment to the "truth" of any particular value in the range. For example, $\underline{P}(A)$ might correspond to the inferior (lower) limit (liminf) of relative frequency (see Section 3.4) and $\overline{P}(A)$ to the superior (upper) limit (limsup) of the relative frequency, and there need not be any limiting relative frequency.

Set-Valued Probability: Related to interval-valued probability wherein the interval for the probability of an event A is generated by taking minima and maxima of a specified set $\mathcal{M} = \{P_\alpha\}$ of probability assignments P_α. This structure for probability has been closely studied in Walley [96] and related by him to subjective probability noted in the next section.

Numerical Probability: "The probability of A is the real number $P(A)$." This is the usual notion that will occupy us beyond these preliminary comments. While this concept has absorbed almost all of the attention of those concerned with uncertainty and chance phenomena and has proven fruitful in engineering and scientific practice, it is not the only concept used in ordinary discourse and everyday reasoning. It is doubtful that numerical probability is adequate for the full range of uses to which it is put, and it is likely that it has inhibited the development of mathematical theories appropriate for other random phenomena.

A discussion of this hierarchy is available in Walley and Fine [95], Walley [96], and Fine [29, 30]. In conformity with overwhelming practice, henceforth we concentrate on the final, most precise structural concept of numerical probability.

3.3 INTERPRETATIONS OF PROBABILITY

An overview of this complex and controversial area is available from Fine [29, 30] and the references cited therein. Probability does not appear to be reducible to other concepts; it is a notion unto itself. The best that can be done is to relate probability to other concepts through an interpretation. The five most common groups of interpretations are the following:

1. **Logical:** degree of confirmation of a hypothesis proposition that "*A* occurs" by an evidence proposition "*B* occurred" (Carnap [18]). This is a concept attached to a formal logical system and not, say, to the physical world. It is meant to render inductive reasoning quantitative. When the evidence or premises are insufficient to deductively lead to the hypothesis or conclusion, we may still be able quantitatively to measure the degree of support of hypothesis by evidence through logical probability. A (distant) goal might be to take the official transcript of a criminal trial and, from it, mathematically deduce the degree to which the transcript supports the hypothesis of guilt. Complexity theory as developed initially by Chaitin, Kolmogorov, and Solomonoff (see Li and Vitanyi [59]) provides a reworking of this concept in the framework of computability theory and, in principle, provides a quantitative measure of the extent to which the evidence explains a hypothesis.

2. **Subjective:** refers to your degree of personal belief in the occurrence of the event *A* as made operational or publicly visible through the behavioral interpretation of your willingness to bet (de Finetti [23, 24], Savage [80], Walley [96]) or act. This interpretation reflects the importance of our beliefs and judgements in deriving probability statements that can guide behavior and provide it with some coherence (a term with a technical meaning in this interpretation) that ensures some consistency among our various beliefs without having to assert their truth. Unfortunately, even naive reflection and introspection suggests that our probabilistic beliefs are rarely sharp enough to be expressed numerically.

3. **Frequentist:** limiting relative frequency of occurrence of event *A* in unlinked repeated performances of the random experiment $\mathcal{E}$ (Venn [94] and von Mises [63]). This is the most common interpretation offered in the physical sciences. The concept has been reworked through Kolmogorov complexity theory as discussed in Li and Vitanyi [59]. As limits are beyond the ken of experiment, limiting relative frequencies are a problematic idealization.

4. **Propensity:** tendency, propensity, or disposition for an event *A* to occur (Suppes [88]).

5. **Classical:** based on an enumeration of *equally likely cases* or equally possible cases as discussed in Chapter 2 (attributed by Hacking [43] to Leibniz in 1678).

While there is substantial controversy over the meaning of probability, for much of engineering, we offer the following *propensity view of probability*:

> *The probability $P(A)$ of the event A is a precise numerical measure of the propensity or tendency of the event A to occur in a (not necessarily repeatable) performance of a random experiment $\mathcal{E}$.*

> *This probabilistic propensity can have links through theory to other physical concepts (e.g., electric potential V through Schrodinger's equation in quantum mechanics or mechanical symmetry through statistical mechanics).*

> *For indefinitely repeatable experiments probabilistic propensity is displayed in the long-run relative frequency of the occurrence of A in n repeated, unlinked performances $\mathcal{E}_1, \ldots, \mathcal{E}_n$ of the random experiment $\mathcal{E}$. The laws of large numbers establish the consistency between probability and its display by long-run relative frequencies.*

The notion of "propensity" is debated within the philosophy of science community. As Humphreys [47] has observed, there are difficulties with the breadth of applicability of the propensity interpretation when we consider conditional probabilities. The notion of "unlinked performances" is formalized as independence in Chapter 14 and the laws of large numbers are discussed in Chapter 19.

3.4 LONG-RUN TIME AVERAGES AND RELATIVE FREQUENCIES

It remains for us to discuss the third element of modeling, the assignment of a numerical measure to events that represents their likelihood of occurrence. The properties of this assignment are largely motivated by those of long-run averages of outcomes. Consider a collection of random experiments $\{\mathcal{E}_i\}$ sharing the same event algebra $\mathcal{A}$ and having individual outcomes $\{\omega_i\}$ that are not necessarily numerical. Let $X(\omega)$ be a fixed real-valued function of outcomes, with $X_i = X(\omega_i)$ being the numerical value associated with the ith experimental outcome ω_i. For example, let $\mathcal{E}_i$ be the experiment of the selection of the ith word in some long text, ω_i be the actual ith word, and $X(\omega_i) = X_i$ the length of this word. Denote by

$$Av_n X = \frac{1}{n} \sum_{i=1}^{n} X_i$$

the *time average* of the outcomes $\{X_i\}$ of the first n random experiments. For mathematical simplicity, let us assume that the function X is drawn from a family $\mathcal{F}$ of functions that can take on only finitely many distinct numerical values (as is the case for word length in a natural language). Fixing a given sequence of outcomes $\{\omega_i\}$, it is easy to verify the following properties of Av_n:

Av0. $Av_n : \mathcal{F} \to \mathbb{R}$.
Av1. If for all ω, $X(\omega) \geq 0$, then $Av_n X \geq 0$.
Av2. If X is the constant function (takes on only one value), then $Av_n X = X$.
Av3. For all $X, Y \in \mathcal{F}$, for all $\alpha, \beta \in \mathbb{R}$,

$$Av_n(\alpha X + \beta Y) = \alpha Av_n X + \beta Av_n Y.$$

In particular, if we are interested in an event A and take $X(\omega) = I_A(\omega)$, a binary-valued function, then the time average is known as the relative frequency of A. The set $\mathcal{F}$ of functions taking only finitely many values is now restricted further to the set of binary-valued functions indexed by the events in $\mathcal{A}$. Correspondingly restricting the notion of Av_n yields the definition of relative frequency.

Definition 3.4.1   (Relative Frequency)   The *relative frequency* of an event A, determined from the outcomes $\{\omega_1, \ldots, \omega_n\}$ of n random experiments, is

$$r_n(A) = \frac{1}{n} \sum_{1}^{n} I_A(\omega_i) = \frac{N_n(A)}{n}.$$

In Figure 3.3, we plot $r_n(H)$ for the occurrence of "heads" in repeated, unlinked tosses of a coin that is biased towards "heads." Key properties of relative frequency are

RF0. $r_n : \mathcal{A} \to \mathbb{R}$.
RF1. $r_n(A) \geq 0$.
RF2. $r_n(\Omega) = 1$.
RF3. If $A \perp B$ then $r_n(A \cup B) = r_n(A) + r_n(B)$.

While relative frequency has other properties (e.g., it is a rational number), many of them follow from RF0–RF3.

Although r_n is a restriction to a smaller class of functions than is Av_n, we can express Av_n in terms of r_n. Given a function X taking on values in the finite set (range of X) $\{x_1, \ldots, x_k\}$, consider the k events $\{A_i = \{\omega : X(\omega) = x_i\}, i = 1, \ldots, k\}$. We can now regroup the summands in Av_nX to write it as

$$Av_nX = \sum_{i=1}^{k} x_i r_n(A_i).$$

This is written more informatively as

$$Av_nX = \sum_{i=1}^{k} x_i r_n(X = x_i).$$

Thus, for any n and X, we can compute the time average Av_nX from the range set of X and the relative frequencies $r_n(X = x_i)$. In particular, if, for each i, we have convergence of the sequence $r_1(X = x_i), \ldots, r_n(X = x_i), \ldots$ to a limit p_i, then we will also have convergence of the long-run time average Av_nX,

$$\lim_{n \to \infty} Av_nX = \sum_{i=1}^{k} x_i p_i.$$

Figures 3.1 and 3.2 show plots of the averages of word length for text taken from the first page of Kolmogorov's foundational monograph [54]. Figure 3.1 is for the first 50 words and Figure 3.2, for the first 329 words. These figures suggest an increasing stabilization of the average word length as the number of words increases. The limiting long-run value of $\{Av_nX\}$, when it exists, is an interpretive basis for the essential concept of expectation EX or mean of a random numerically valued quantity X. The preceding argument leads us to a definition of expected value, at least in this simple case, wherein X takes on only finitely many values, as

$$EX = \sum_{i} x_i p_i.$$

We discuss expectation more fully in Chapter 9.

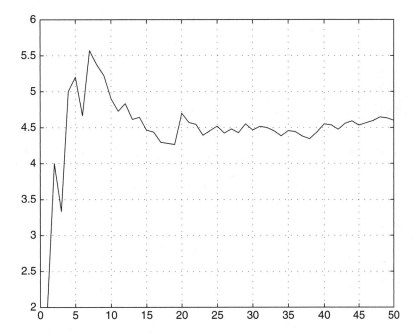

Figure 3.1 Averages of first 50 word lengths from Kolmogorov.

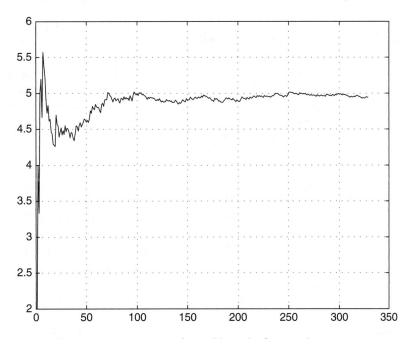

Figure 3.2 Averages of word lengths from Kolmogorov.

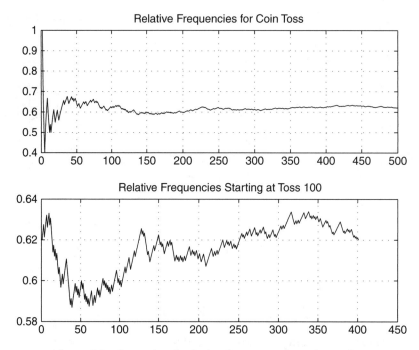

Figure 3.3   Example of relative frequency for coin tossing.

Our interest in the long-run averages introduced previously for random phenomena is based upon the underlying belief that for many repeatable random phenomena, while there is unpredictability for individual repetitions, there are predictable aspects when we average over many repetitions. It appears from Figures 3.3 and 3.4 that this is the case in some sense for relative frequencies, but that the notion of "convergence" of this fluctuating sequence needs to be explicated. Figure 3.3 plots relative frequencies for "heads" in 500 simulated tosses of a coin and provides more detail about the last 400 tosses in a second plot. Figure 3.4 plots relative frequencies for the occurrences of consonants as initial letters of the words that appear on the first page of the English translation of Kolmogorov [54]. The eventual probability-theoretic link between relative frequency and probability will be through laws of large numbers. (See Chapter 19.) However, the meaning of this link is still the subject of philosophical debate. On our propensity account, relative frequency is just one of several ways in which the probabilistic propensity can display itself.

With this background, we proceed as if there is some (empirical or metaphysical) grounds for asserting that

$$r_n(A) \to P(A),$$

although the sense of convergence as n grows will not be specified until Chapter 19. If this is the case, then P should inherit the properties of limits of sequences of r_n.

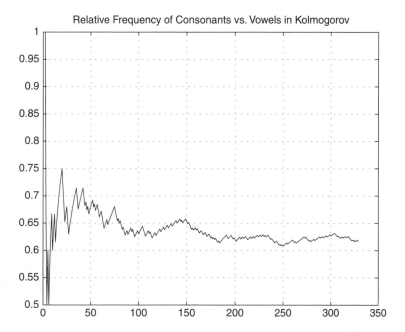

Figure 3.4   Example of relative frequency for consonants vs. vowels.

3.5 MODELING RANDOM PHENOMENA IV: KOLMOGOROV AXIOMS

3.5.1 Event Algebras

An advantage of an axiomatization of a probability structure is the clarity it affords when engaged in probabilistic reasoning; this may become clearer in Section 11.7, when we resolve three paradoxical problems. Moving probabilistic reasoning into the mathematical arena through an axiomatization, which will be subsequently enriched through definitions of associated concepts of expectation, conditional probability, independence, etc., enables us to proceed surefootedly even if sometimes rather slowly. Of course, adopting an axiomatization also constrains us, and an inadequate set of axioms can lead us to a poor exchange of mathematical certainty for faithfulness to reality. This is always a concern with mathematical approaches to engineering and scientific problems.

The theory of probability focuses on *collections of events*, called event algebras (see Section 1.6) and typically denoted $\mathcal{A}$, that contain all the events of interest to us, and are such that we have knowledge of their likelihood of occurrence.

We assume that if we are interested in certain events, then we are also interested in the new events that can be made from them by application of the Boolean operations discussed in Section 1.9.2. This assumption is not as innocuous as most consider it to be. Quantum mechanics provides a clear instance of an important application of probability theory in which we may be interested in both an experimentally observable event A concerning, say, the position of an electron and an experimentally observable event B concerning the momentum of the electron.

However, the Heisenberg uncertainty principle tells us that we cannot be interested in $A \cap B$; the determination of the occurrence of $A \cap B$ does not correspond to a physical experiment. Nonetheless, we proceed in the usual fashion and accept an amended Definition 1.6.1 for the algebra of events with which we will be concerned. The further technical restriction, needed for the formal theory of probability, is given in the following:

Definition 3.5.1 (σ-algebra) A σ-algebra $\mathcal{A}$ is an event algebra that is also closed under countable unions,

$$(\forall i \in Z) \, A_i \in \mathcal{A} \Rightarrow \cup_{i \in Z} A_i \in \mathcal{A}.$$

The *Borel algebra* $\mathcal{B}$ of subsets of the reals is the smallest σ-algebra containing the intervals and is the usual algebra when we deal with real- or vector-valued quantities. The choice of algebra involves technical issues, however, that are beyond our scope. Our needs will not require us to delve into these issues beyond being assured that events we discuss are constructed out of intervals and repeated set operations on these intervals and these constructions will not lead us out of $\mathcal{B}$. In particular, countable unions of intervals (e.g., the set of rational numbers), their complements (e.g., the set of irrational numbers), and much, much more are in $\mathcal{B}$.

3.5.2 Kolmogorov's Axioms for Probability

The axioms to be given for probability are not meant to uniquely describe a probability. Probabilities are expected to vary with the application, and this is even true of the dependence of classical probability upon the size of the sample space. Rather, the axioms circumscribe a family of probabilities models. The nontrivial specific choice of a probability satisfying the axioms is then left to the engineer or modeler/statistician familiar with the random phenomena being described, and it is discussed in Chapters 4, 5, 6, and 7. With the axioms in place, you can use mathematical methods to uncover properties that will be true for any application-specific probability. Motivated by the connection to relative frequencies maintained by the propensity interpretation, and by the properties of classical probability introduced in Chapter 2, we impose the properties RF0–RF3 of relative frequencies and of classical probability (Section 2.2) upon probability P through the first four Kolmogorov axioms (Kolmogorov [54]).

K0. **Setup:** The random experiment $\mathcal{E}$ is described by a *probability space* $(\mathcal{A}, P)$ consisting of an event σ-algebra $\mathcal{A}$ and a real-valued function

$$P : \mathcal{A} \to \mathbb{R}.$$

K1. **Nonnegativity:** $\forall A \in \mathcal{A}, \quad P(A) \geq 0.$
K2. **Unit normalization:** $P(\Omega) = 1.$
K3. **Finite additivity:** If A, B are disjoint, then

$$P(A \cup B) = P(A) + P(B).$$

A fifth axiom, while neither a property of limits of sequences of relative frequencies nor meaningful in the finite sample space context of classical probability, is offered by Kolmogorov to ensure a degree of mathematical closure under limiting operations.

K4. Monotone continuity: If $(\forall i > 1) A_{i+1} \subset A_i$ and $\bigcap_i A_i = \phi$ (a nested series of sets shrinking to the empty set), then

$$\lim_{i \to \infty} P(A_i) = 0,$$

K4 is an idealization that is rejected by some accounts (usually subjectivist) of probability. Kolmogorov ([54], p. 15) comments on this axiom that,

> it is almost impossible to elucidate its empirical meaning, ... This limitation [the axiom] has been found expedient in researches of the most diverse sort.

An equivalent form of K4 is as follows:

K4′. Countable or σ-additivity: If $\{A_i\}$ is a countable collection of pairwise disjoint (no overlap) events, then

$$P\left(\bigcup_{i=1}^{\infty} A_i\right) = \sum_{i=1}^{\infty} P(A_i).$$

Theorem 3.5.1   (K4, K4′ Equivalence)   If P satisfies K0−K3, then it satisfies K4′ if and only if it satisfies K4.

Proof reference. See the appendix.   □

Definition 3.5.2   (Probability Measure)   A function satisfying K0−K4 is called a *probability measure*.

Example 3.1   Uniform Discrete Probability _____

If Ω is a finite set, then classical probability provides an example of a measure

$$P(A) = \frac{||A||}{||\Omega||}$$

defined for any subset A of Ω. The basic facts about cardinality, that $0 \leq ||A|| \leq ||\Omega||$ and

$$||A \cup B|| = ||A|| + ||B|| - ||A \cap B||,$$

enable you to verify that P satisfies the Kolmogorov axioms.

Example 3.2   Weighted Discrete Probability ——————————————————————

A second example of a probability measure P takes a sample space $\Omega = \{00, 01, 10, 11\}$ that represents two bits of information or two successive tosses of a coin. (Say, "head" is represented by "1".) The algebra $\mathcal{A}$ is taken to be the power set of all $2^4 = 16$ subsets of Ω. Let

$$p_{ij} \geq 0, \quad \sum_{i=0}^{1} \sum_{j=0}^{1} p_{ij} = 1.$$

We can then construct P satisfying the Kolmogorov axioms through the following assignments of probability to events:

$$P(\emptyset) = 0;$$

$$P(\{00\}) = p_{00}; \ P(\{01\}) = p_{01}; \ P(\{10\}) = p_{10}; \ P(\{11\}) = p_{11};$$

$$P(\{00, 01\}) = p_{00} + p_{01}; \ P(\{00, 11\}) = p_{00} + p_{11}; \ P(\{00, 10\}) = p_{00} + p_{10};$$

$$P(\{10, 01\}) = p_{10} + p_{01}; \ P(\{11, 01\}) = p_{11} + p_{01}; \ P(\{10, 11\}) = p_{10} + p_{11};$$

$$P(\{00, 01, 10\}) = p_{00} + p_{01} + p_{10}; \ P(\{00, 01, 11\}) = p_{00} + p_{01} + p_{11};$$

$$P(\{00, 10, 11\}) = p_{00} + p_{10} + p_{11}; \ P(\{10, 01, 11\}) = p_{10} + p_{01} + p_{11};$$

$$P(\Omega) = 1.$$

Example 3.3   Uniform Continuous Probability ——————————————————————

A third example takes a sample space $\Omega = [0, 1]$ of the unit interval and corresponds to a random experiment in which a number is drawn at random from the unit interval. Perhaps the experiment consists of examining the fractional part of a second displayed on a highly precise clock when it is viewed at a random time; this is done in some computer routines for generating random numbers. In this case, we take the Borel event algebra (the algebra built up from the intervals) and assign

$$P(A) = \int_0^1 I_A(x)\,dx = \int_A dx.$$

$P(A)$ is the length of the set A, and it is known as the uniform distribution. (See Section 6.5.1.)

An additional example is given in the false-alarm probability calculation performed in Section 3.10. Many additional examples of probability measures are provided in succeeding chapters. Some examples that are not probability measures and illustrate the failure of each of the axioms K1, K2, and K3 are as follows:

Example 3.4   Failure of Axiom K1

Take $\Omega = \{a, b, c\}$ with

$$P(\{a\}) = .6, \quad P(\{b\}) = .5, \quad P(\{c\}) = -.1, \text{ and } P(A) = \sum_{\omega \in A} P(\{\omega\}).$$

Thus, K1 fails, since $P(\{c\}) = -.1 < 0$. However,

$$P(\Omega) = P(\{a, b, c\}) = P(\{a\}) + P(\{b\}) + P(\{c\}) = .6 + .5 - .1 = 1,$$

and K2 is satisfied. Since we assign

$$P(A) = \sum_{\omega \in A} P(\{\omega\}),$$

it follows from the properties of sums that K3 is satisfied.

Example 3.5   Failure of Axiom K2

Simply take

$$\Omega = \{a, b\}, P(\{a\}) = \frac{2}{3} = P(\{b\}), \text{ and } P(A) = \sum_{\omega \in A} P(\{\omega\}).$$

It is immediate that

$$P(\Omega) = P(\{a, b\}) = P(\{a\}) + P(\{b\}) = \frac{4}{3} > 1.$$

It is also immediate that $P(A) \geq 0$, since it is a sum of positive terms, or it is 0 for the empty set $\emptyset$. The additivity of P follows as before.

Example 3.6   Failure of Axiom K3

Take the P in the example showing the failure of K1, and consider

$$Q(A) = [P(A)]^2 = \left(\sum_{\omega \in A} P(\{\omega\}) \right)^2.$$

Since P satisfies K2,

$$Q(\Omega) = [P(\Omega)]^2 = 1^2 = 1,$$

and Q also satisfies K2. To show that Q does not satisfy K3, consider

$$Q(\{a, b\}) = [P(\{a, b\})]^2 = [.6 + .5]^2 = 1.21,$$

whereas

$$Q(\{a\}) + Q(\{b\}) = [.6]^2 + [.5]^2 = .61 \neq 1.21.$$

While the Kolmogorov axioms are used nearly universally and reflect properties of classical probability and relative frequencies, critical comments and a different perspective on them can be found in Chapter 3 of Fine [29]. One can ask, for example, whether the axioms can be further supplemented to narrow the class of probability measures. There is an interest in this in the context of defining random processes, uncountable collections of random variables, introduced in Chapter 20. We will shortly turn to examine more narrowly defined classes of probability measures that prove useful for certain classes of random phenomena. One can also ask about the relevance to engineering and science of other structural concepts (e.g., interval-valued probability) satisfying quite different axiom systems. This, however, is still a relatively unexplored area.

3.6 ELEMENTARY CONSEQUENCES OF THE AXIOMS K0–K4

3.6.1 Consequences for a Single Event

Consequences involving only a single event are as follows:

Lemma 3.6.1   **(Basic Properties of P)**

$$P(A^c) = 1 - P(A);$$

$$P(\emptyset) = 0;$$

$$P(A) \leq 1.$$

The first result follows from

$$A \cup A^c = \Omega, A \perp A^c.$$

Thus, by K3,

$$P(\Omega) = 1 = P(A \cup A^c) = P(A) + P(A^c).$$

Hence, $P(A)$ directly determines $P(A^c)$. It is sometimes easier to calculate $P(A)$ for some event of interest by first calculating $P(A^c)$. An example of this was provided in our discussion of the "birthday paradox" in Section 2.3.4. While we were interested in the event A that at least two people shared the same birthday, it was easier to consider the event A^c that no two people shared the same birthday. The evaluation of $P(\emptyset)$ follows from the complementation result and K2. That probability is always less than or equal to 1 follows from complementation and K1. Hence, probabilities of events are always numbers lying in the unit interval [0, 1]:

Probabilities cannot be negative, nor can they exceed 1.

3.6.2 Consequences for Two Events

We now consider consequences of the axioms involving two events. If $A \supset B$, then the occurrence of B implies the occurrence of A. For example, the event A might be the occurrence of either of the three letters c, e, and t in a randomly chosen word from some English text, and the event B might be the similar occurrence of either of the two letters e and t. Thus, if P is a measure of the tendency or propensity of an event to occur, then we expect that $P(A) \geq P(B)$, with the inequality usually being strict. The expected *monotonicity of probability* with respect to set inclusion

$$A \supseteq B \Rightarrow P(A) \geq P(B)$$

follows from

$$A = B \cup (A - B), B \perp (A - B),$$

and the use of K3 to establish

$$P(A) = P(B) + P(A - B).$$

Invoke K1 to recall that $P(A - B) \geq 0$, and the conclusion of monotonicity follows. Note that if A, B are in the event algebra, then so is the event $A - B$, and therefore $P(A - B)$ is defined. Lemmas 3.6.2 and 3.6.3 are an immediate consequence of $A \cup B \supset A \supset A \cap B$.

Lemma 3.6.2   (Probability Inequalities)

$$P(A \cup B) \geq \max(P(A), P(B)) \geq \min(P(A), P(B)) \geq P(A \cap B).$$

Lemma 3.6.3   (Probability of Union)   An exact expression for the probability of a nondisjoint union is given by

$$P(A \cup B) = P(A) + P(B) - P(A \cap B).$$

Proof. This result follows from K3 and

$$A \cup B = A \cup (B - A), A \perp (B - A) \Rightarrow P(A \cup B) = P(A) + P(B - A),$$

$$B = (A \cap B) \cup (B - A), A \cap B \perp (B - A) \Rightarrow P(B - A) = P(B) - P(A \cap B). \qquad \square$$

When $A \perp B$, it follows that $A \cap B = \emptyset$, and from the preceding, $P(\emptyset) = 0$. In this case, the result specializes to K3.

For example, using the events $A = \{c, e, t\}, B = \{e, o, t\}$ for letters of the alphabet as they might be found in a randomly chosen word,

$$P(A \cup B) = P(\{c, e, o, t\}) = P(\{c, e, t\}) + P(\{e, o, t\}) - P(\{e, t\}),$$

and further evaluation requires knowing three of the four probabilities found in this equation. (See Section 4.3 for actual English letter probabilities of occurrence.)

3.6.3 Consequences for More than Two Events

The extension of K3 from two sets to finitely many pairwise disjoint sets is provided by Lemma 3.6.4.

Lemma 3.6.4 (Finite Disjoint Unions) If K0–K3, then

$$(\forall i \neq j)\ A_i \perp A_j \Rightarrow P\left(\bigcup_{i=1}^{n} A_i\right) = \sum_{i=1}^{n} P(A_i).$$

Proof. While this lemma is an immediate consequence of K4', it in fact does not require K4'. Finite induction on n establishes the lemma based solely on K0–K3. It is true for $n = 2$ by K3. Assume that it is true for all $j < n$. Write

$$\bigcup_{i=1}^{n} A_i = A_n \cup \left(\bigcup_{i=1}^{n-1} A_i\right).$$

Since $A_n \perp \cup_{i=1}^{n-1} A_i$, it follows from K3 that

$$P\left(\bigcup_{i=1}^{n} A_i\right) = P(A_n) + P\left(\bigcup_{i=1}^{n-1} A_i\right).$$

By the induction hypothesis,

$$P\left(\bigcup_{i=1}^{n-1} A_i\right) = \sum_{i=1}^{n-1} P(A_i),$$

and the lemma follows. $\qquad\square$

For example, Section 4.3 informs us that English text letter frequencies include the following probabilities:

$$P(\{c\}) = .032,\ P(\{e\}) = .120,\ P(\{o\}) = .067,\ P(\{t\}) = .085.$$

Hence, the events A, B just described have probabilities

$$P(A) = P(\{c\}) + P(\{e\}) + P(\{t\}) = .032 + .120 + .085 = .237,$$

and

$$P(B) = P(\{e\}) + P(\{o\}) + P(\{t\}) = .120 + .067 + .085 = .272,$$

respectively. More generally, if $A = \{\omega_1, \ldots, \omega_k\}$ is a set made of k elements of the sample space Ω, then

$$P(A) = \sum_{i=1}^{k} P(\{\omega_i\});$$

the probability of an event is the sum of the probabilities of its constituent elements.

In classical probability discussed in Chapter 2, all outcomes ω_i have equal probability, say, p. Hence,

$$P(\Omega) = \sum_{\omega \in \Omega} p = ||\Omega|| p,$$

and by the unit normalization axiom, we conclude that $p = 1/||\Omega||$. Given the event A described previously, a classical probability evaluation would yield

$$P(A) = \frac{k}{||\Omega||} = \frac{||A||}{||\Omega||},$$

as was assumed in classical probability.

Lemma 3.6.5   (Countable Disjoint Unions)   The combination of K3, K4, or just K4′, yields an extension to countably infinite disjoint unions

$$(\forall i \neq j)\, A_i \perp A_j \Rightarrow P\left(\bigcup_{i=1}^{\infty} A_i\right) = \sum_{i=1}^{\infty} P(A_i).$$

Example 3.7   Probability of a Countable Union _____

Let $\Omega = N$, the set of nonnegative integers, and take $\mathcal{A} = 2^{\Omega}$, the power set of Ω. Let A_n be the event that the outcome is exactly n, and assign

$$P(A_n) = \frac{1}{2^{n+1}}.$$

From the summation formula for a geometric series,

$$\sum_{n=0}^{\infty} P(A_n) = \frac{1}{2} \sum_{n=0}^{\infty} \frac{1}{2^n} = \frac{1}{2} \frac{1}{1 - \frac{1}{2}} = 1,$$

as required for P to be a probability measure on Ω. Let B be the event that the outcome is an odd integer. We write

$$B = \bigcup_{k=0}^{\infty} A_{2k+1}.$$

Hence, by the lemma or K4′,

$$P(B) = P\left(\bigcup_{k=0}^{\infty} A_{2k+1}\right) = \sum_{k=0}^{\infty} P(A_{2k+1})$$

$$= \sum_{k=0}^{\infty} \frac{1}{2^{2k+2}} = \frac{1}{4} \sum_{k=0}^{\infty} \frac{1}{4^k} = \frac{1}{4} \frac{1}{1 - \frac{1}{4}} = \frac{1}{3}.$$

Lemma 3.6.5 yields

Lemma 3.6.6   (Probability from Partitions)   If $\Pi = \{A_i\}$ is a countable partition of Ω made of sets in $\mathcal{A}$, then for any $B \in \mathcal{A}$

$$P(B) = \sum_i P(B \cap A_i).$$

Proof. Since Π is a partition (see Definition 1.6.3), it follows that

$$B = B \cap \Omega = B \cap (\cup_i A_i) = \cup_i (B \cap A_i),$$

and the sets $C_i = B \cap A_i$ are pairwise disjoint. Hence, by Lemma 3.6.5 or by K4′,

$$P(B) = P(\cup_i (B \cap A_i)) = \sum_i P(B \cap A_i). \qquad \Box$$

Example 3.8   Probability from Partitions _____
Let $\Omega = [0, \infty)$ and define

$$P(A) = \int_0^\infty e^{-x} I_A(x)\, dx = \int_A e^{-x}\, dx.$$

Section 6.6.1 establishes that this definition of P satisfies the Kolmogorov axioms. We wish to evaluate the probability of the event $A = \{x : x \in \Omega,\ \sin(2\pi x) \geq 0\}$. Select the partition $\{B_i\}$ given by $B_i = [i, i+1)$ for nonnegative integer i. Taking into account that on B_i, $\sin(2\pi x) \geq 0$ only over the subinterval $[i, i+\frac{1}{2}]$, we have that

$$P(A \cap B_i) = \int_i^{i+\frac{1}{2}} e^{-x}\, dx = (1 - e^{-\frac{1}{2}})e^{-i}.$$

Recourse to Lemma 3.6.6, and the summation formula for a geometric series, yields

$$P(A) = \sum_{i=0}^\infty (1 - e^{-\frac{1}{2}})e^{-i} = \frac{1 - e^{-\frac{1}{2}}}{1 - e^{-1}}.$$

3.7 BOOLE'S INEQUALITY

We can see from the result for the exact probability of a union of two, possibly overlapping, events that this probability requires us to know more than just the probabilities of the individual events. However, from the probabilities of the individual events, we can upper-bound the probability of their union through what is known as the *union bound* or Boole's Inequality.

Lemma 3.7.1   (Boole's Inequality, Union Bound)   For n arbitrary events $\{A_1, \ldots, A_n\}$, Boole's Inequality is

$$P\left(\bigcup_{i=1}^{n} A_i\right) \le \sum_{i=1}^{n} P(A_i).$$

Remark. Of course, this upper bound to a probability is of no interest if it exceeds 1. Furthermore, Lemma 3.6.5 establishes that Boole's Inequality is achieved with equality when the events are pairwise disjoint.

Proof. We proceed by induction on n. The inequality is trivially true for $n = 1$ and true for $n = 2$ from the previously cited result on the probability of an arbitrary union of two sets. Hence, assume that this inequality holds for any $j < n$ and prove that it holds for n. To do so, write

$$\bigcup_{i=1}^{n} A_i = A_n \cup \bigcup_{i=1}^{n-1} A_i.$$

By the inequality for $n = 2$,

$$P\left(\bigcup_{i=1}^{n} A_i\right) \le P(A_n) + P\left(\bigcup_{i=1}^{n-1} A_i\right).$$

By the induction hypothesis for $j = n - 1 < n$,

$$P\left(\bigcup_{i=1}^{n-1} A_i\right) \le \sum_{i=1}^{n-1} P(A_i),$$

and Boole's Inequality follows.   □

Use of axiom K4$'$ allows us to show that this inequality also holds for any countably infinite union of events,

$$P\left(\bigcup_{i=1}^{\infty} A_i\right) \le \sum_{i=1}^{\infty} P(A_i).$$

The value of Boole's Inequality is that we can estimate a probability of a union without knowing more than the individual probabilities of the sets in the union.

Example 3.9   Boole's Inequality and Letter Probabilities _____

Let $A_{t,\lambda}$ denote the event that the letter t occurs in a word of length λ randomly chosen from some large text. The probability that any given letter will be a t can be taken to be .085, although this actually varies with the position in the word. As there are λ positions in the word at which a t might be found, the Boole's Inequality yields

$$P(A_{t,\lambda}) \le .085\lambda,$$

and is of interest only for $\lambda \leq 11$. Note that as there are words containing multiple occurrences of t, the occurrences within a word are not disjoint events. Thus, this upper bound is strict.

Example 3.10   Boole's Inequality and Permutations

There are n distinct objects and n distinct labels for these objects. However, the labels have been applied at random to the objects. There are $n!$ equally probable permutations for the assignment of labels to objects. We first ask for the probability of the event A_i that the ith object is correctly labeled. Given that the ith object is known to be correctly labeled, there are $(n-1)!$ permutations of the remaining labels that are consistent with the correct placement of label i. From classical probability,

$$P(A_i) = \frac{||A_i||}{||\Omega||} = \frac{(n-1)!}{n!} = \frac{1}{n}.$$

We now ask for the probability of the event B_m that at least one label is correctly placed on the first m objects. From Boole's Inequality,

$$P(B_m) = P\left(\bigcup_{i=1}^{m} A_i\right) \leq \sum_{i=1}^{m} P(A_i) = \frac{m}{n}.$$

This result is a strict upper bound for $m > 1$. The events A_i and A_j are not pairwise disjoint. Indeed, for $i \neq j$,

$$||A_i \cap A_j|| = (n-2)!$$

and

$$P(A_i \cap A_j) = \frac{(n-2)!}{n!} = \frac{1}{n(n-1)} > 0.$$

Example 3.11   Probability of "Occurring Finitely Often"

While our concerns prior to Chapters 19 and 20 are with finitely many random variables, we can use Boole's Inequality to consider whether a countably infinite series of events $\{A_n\}$ has events that occur only finitely often. The event

$$F = \bigcup_{n=1}^{\infty} \bigcap_{i \geq n} A_i^c,$$

is the event that there exists an n such that for all $i \geq n$ the A_i do not occur. Hence, F is the event that $\{A_n\}$ occur only finitely often. To evaluate $P(F)$, consider

$$P(F) = P\left(\bigcup_{n=1}^{\infty} \bigcap_{i \geq n} A_i^c\right) \leq \sum_{n=1}^{\infty} P\left(\bigcap_{i \geq n} A_i^c\right),$$

where we have used Boole's Inequality to obtain the preceding upper bound to $P(F)$. The obvious generalization to many events of Lemma 3.6.2 yields

$$P(\cap_{i \geq n} A_i^c) \leq \min_{i \geq n} P(A_i^c).$$

If we now assume that

$$\lim_{i \to \infty} P(A_i) = 1,$$

then

$$\lim_{i \to \infty} P(A_i^c) = 0,$$

and it follows that

$$(\forall n) \; P(\cap_{i \geq n} A_i^c) = 0.$$

We then have that

$$P(F) \leq \sum_{n=1}^{\infty} 0 = 0,$$

and in this case the events $\{A_i\}$ occur finitely often with probability zero or infinitely often (the complement) with probability one.

Another example of the use of Boole's Inequality, for estimating false-alarm probability with multiple detectors, is given in the third paragraph of Section 3.10. Boole's Inequality will also be used in Section 20.3 to prove the Borel–Cantelli Lemma. As this lemma is concerned only with properties of infinite collections of events, we defer its consideration until our discussions of random processes (infinite collections of random variables) in Chapter 20.

A corollary of Boole's Inequality is

Corollary 3.7.1   For n arbitrary events $\{A_1, \ldots, A_n\}$,

$$P\left(\bigcap_{i=1}^{n} A_i\right) \geq \sum_{i=1}^{n} P(A_i) - (n - 1).$$

Remark. Of course, this lower bound to a probability is of no interest when it becomes negative.

Proof. Use de Morgan's laws to rewrite the Boole's Inequality for events $\{B_1, \ldots, B_n\}$ as

$$P\left(\bigcup_{i=1}^{n} B_i\right) = 1 - P\left(\bigcap_{i=1}^{n} B_i^c\right) \leq \sum_{i=1}^{n} (1 - P(B_i^c)).$$

A little arithmetic rearrangement and replacement of the arbitrary B_i^c by A_i yields the desired conclusion. $\square$

From the proof just given and the remark following Boole's Inequality, we see that the preceding inequality is achieved with equality for $\{A_i\}$ when the events $\{A_i^c\}$ are pairwise disjoint. Using de Morgan's laws allows us to restate this conclusion as equality is achieved when for each pair of events $A_i \cup A_j = \Omega$ or for each pair A_i, A_j, at least one of these events must occur. The value of the inequality provided by this corollary is that it enables us to lower-bound the probability that each one of a collection of events will occur.

3.8 INCLUSION–EXCLUSION PRINCIPLE

An exact evaluation of $P(\bigcup_{i=1}^n A_i)$ that extends the result given earlier for $n = 2$ is provided by Lemma 3.8.1.

Lemma 3.8.1   (Inclusion–Exclusion Principle)   Let I denote a generic set of indices that is an arbitrary nonempty subset of $\{1, \ldots, n\}$. For arbitrary events $\{A_1, \ldots, A_n\}$,

$$P\left(\bigcup_{i=1}^n A_i\right) = \sum_{\phi \neq I \subset \{1,\ldots,n\}} (-1)^{\|I\|+1} P\left(\bigcap_{i \in I} A_i\right),$$

where the summation is over all of the $2^n - 1$ such index sets excluding the empty set.

Example 3.12   Probability of a Nondisjoint Union _____
When $n = 3$,

$$P(A_1 \cup A_2 \cup A_3)$$
$$= P(A_1) + P(A_2) + P(A_3)$$
$$- P(A_1 \cap A_2) - P(A_1 \cap A_3) - P(A_2 \cap A_3)$$
$$+ P(A_1 \cap A_2 \cap A_3).$$

Proof. The proof is by induction on n. The result is trivially true for $n = 1$ and has been established earlier for $n = 2$. As in the proof of Boole's Inequality, we write

$$\bigcup_{i=1}^n A_i = A_n \cup \bigcup_{i=1}^{n-1} A_i,$$

and apply the $n = 2$ result to conclude that

$$P\left(\bigcup_{i=1}^n A_i\right) = P(A_n) + P\left(\bigcup_{i=1}^{n-1} A_i\right) - P\left(A_n \cap \bigcup_{i=1}^{n-1} A_i\right).$$

Rewriting the last term as $P(\bigcup_{i=1}^{n-1}(A_n \cap A_i))$ yields an expression that involves a union of exactly $n-1$ sets. Hence, we can invoke the induction hypothesis for the two rightmost terms to find

$$
P\left(\bigcup_{i=1}^{n} A_i\right) = P(A_n) + \sum_{\phi \neq I \subset \{1,\dots,n-1\}} (-1)^{\|I\|+1} P\left(\bigcap_{i \in I} A_i\right)
$$
$$
- \sum_{\phi \neq I \subset \{1,\dots,n-1\}} (-1)^{\|I\|+1} P\left(\bigcap_{i \in I}(A_n \cap A_i)\right).
$$

Rearrangement of terms yields the desired conclusion.   □

An example of the use of this result is provided by evaluating the false-alarm probability for multiple detectors that is provided in the fourth paragraph of Section 3.10. The inclusion–exclusion result is also a basis for a number of upper and lower bounds that are known as *Bonferroni inequalities* [36].

3.9 CONVEX COMBINATIONS OF PROBABILITY MEASURES

Consider a compound experiment in which we have available a variety of (often different) random experiments $\mathcal{E}_1, \dots, \mathcal{E}_m$. Suppose we choose to perform $\mathcal{E}_i$ on the basis of another random experiment $\mathcal{E}_0$, having m possible outcomes with probability λ_i for the ith such outcome. Suppose also that these outcomes are unlinked to those of the experiments $\{\mathcal{E}_i, i > 0\}$. For example, we may have $m = 6$ urns, each filled with colored balls of the same size, weight, and texture, and experiment $\mathcal{E}_0$ has us choose one of the six urns by tossing a fair die. We then sample from the chosen urn to achieve an end result (e.g., color of ball chosen). The probability measure P for this compound experiment is defined in terms of the ones describing the component experiments through a process of *convex combination*.

It is easiest to first think of convex combinations geometrically as applied to vectors. Given the two vectors $\mathbf{x}, \mathbf{y}$, then any vector $\mathbf{z}$ with endpoint on the line segment joining the endpoints of these two vectors can be expressed in the form $\mathbf{z} = \lambda \mathbf{x} + (1 - \lambda)\mathbf{y}$. Clearly, if $\lambda = 0$, then $\mathbf{z} = \mathbf{y}$; and if $\lambda = 1$, then $\mathbf{z} = \mathbf{x}$. If $0 < \lambda < 1$, then $\mathbf{z}$ lies strictly between the endpoints $\mathbf{x}, \mathbf{y}$ or on the line segment joining them. This notion of convex combination is readily extended to countably many vectors.

Definition 3.9.1   (Convex Combinations of Vectors)   A *convex combination* of vectors $\mathbf{x}_1, \dots, \mathbf{x}_n$ is a vector $\mathbf{z}$ determined by the nonnegative numbers $\lambda_1, \dots, \lambda_n$, $\sum_i \lambda_i = 1$ through

$$
\mathbf{z} = \sum_{i=1}^{n} \lambda_i \mathbf{x}_i.
$$

These ideas extend immediately to functions.

Definition 3.9.2   (**Convex Combinations of Functions**)   A *convex combination* of real- or vector-valued functions $f_1, \ldots, f_n$ is a real- or vector-valued function g determined by the nonnegative numbers

$$\lambda_1, \ldots, \lambda_n, \quad \sum_i \lambda_i = 1,$$

through

$$g(x) = \sum_{i=1}^{n} \lambda_i f_i(x).$$

Given a common event algebra $\mathcal{A}$, probability measures $P_1, \ldots, P_m$, and numbers

$$\lambda_1, \ldots, \lambda_m, \quad \lambda_i \geq 0, \quad \sum_{1}^{m} \lambda_i = 1,$$

we can define a new probability measure

$$P(A) = \sum_{1}^{m} \lambda_i P_i(A).$$

It is easy to verify that P satisfies the Kolmogorov axioms K0–K4 when each of the P_i do.

Theorem 3.9.1   (**Closure under Convex Combinations**)   A convex combination P of probability measures $\{P_i\}$ is a probability measure.

Proof remark.   To see that convexity is essential, note that if $\sum_i \lambda_i = \theta \neq 1$, then $P(\Omega) = \theta \neq 1$, in violation of K2. If one of the $\lambda_i < 0$, then it is possible that for some event A, $P(A) < 0$, in violation of K1. $\square$

For example, speakers $S_1, \ldots, S_m$ compete for access to the same communications channel, with probability λ_i that speaker S_i is the one with current access. The speakers generate symbols (e.g., phonemes) from a common alphabet Ω. However, they select symbols with speaker-dependent probabilities; speaker S_i is described by probability measure P_i. The channel then sees a composite source S described by probability measure $P = \sum_i \lambda_i P_i$.

3.10 APPLICATION TO TARGET DETECTION

The subject of *target detection* is addressed from a more developed standpoint in Chapter 13. We introduce it here as a means of exhibiting the value of several of the ideas introduced in this basic chapter. We are interested in detecting the presence or absence of a target vehicle. Detection is based upon sensor (e.g., radar or sonar, and more recently a host of modalities including acoustic, infrared, magnetic, and seismic) responses and probabilistically guided processing of the outputs of such sensors so as to make rational decisions. A *false alarm* occurs when the decision is made that the target is present and it is in fact absent. Such false alarms occur because of inevitable noise introduced by the sensors themselves (e.g., thermal noise) and because of environmental or background noises (e.g., static or other random fluctuations in our surveillance environment). A sensor makes an observation whose particular value lies in a known set Ω of possible observations. We will choose a sample space that might correspond to energy measurements made in two time slots or in two frequency bands:

$$\Omega = [0, \infty)^2 = \{(x, y) : x \geq 0, y \geq 0\}.$$

If no target vehicle is in fact present, then some probability measure P_0 describes the background noise and interference and thus the tendency of subsets of Ω to occur. However, if a target vehicle is present, then some other measure P_1 is needed to account for, say, the target reflecting radar or sonar energy back to our detector. Typically, P_1 places a higher probability on larger energies (x, y) than does P_0. On the basis of knowledge of P_0 and P_1 (see Chapter 13), the sample space Ω is then partitioned into two decision sets D_0 and D_1, $\Omega = D_0 \cup D_1$, $D_0 = D_1^c$, with the observation $(x, y) \in D_0$ meaning that our *decision* is that the target is absent, and $(x, y) \in D_1$ meaning that our decision is that the target is present. To be more specific about P_0 and P_1, we assume (see Chapter 6) that they can both be written in terms of nonnegative functions f_0 and f_1, whose individual double integrals over all of Ω are each unity, as follows:

$$P_0(A) = \int_0^\infty \int_0^\infty I_A(x, y) f_0(x, y) \, dx \, dy = \int_A f_0(x, y) \, dx \, dy$$

$$\text{and} \quad P_1(A) = \int_A f_1(x, y) \, dx \, dy.$$

The problem we are addressing is known as the Neyman–Pearson hypothesis testing problem and is treated in Section 13.4. The answer to finding the optimal decision region, say, D_1, that a target is present, is given by

$$D_1 = \{(x, y) : \text{ decide target present }\} = \left\{ (x, y) : \frac{f_1(x, y)}{f_0(x, y)} \geq \tau \right\}.$$

The parameter τ is your choice to achieve a desired balance between a low false-alarm probability $P_0(D_1)$, which is the probability of the event D_1 evaluated by the probability measure P_0 describing target absent, and a high probability of detection $P_1(D_1)$, which is the probability of the same event evaluated according to a different probability measure.

Example 3.13   False-Alarm Probability Calculation

For specificity we assume $\gamma > 1$ and

$$f_0(x, y) = \begin{cases} \gamma^2 e^{-\gamma x - \gamma y} & \text{if } x \geq 0, y \geq 0 \\ 0 & \text{if otherwise,} \end{cases}$$

$$f_1(x, y) = \begin{cases} e^{-x - y} & \text{if } x \geq 0, y \geq 0 \\ 0 & \text{if otherwise.} \end{cases}$$

Note that both f_0 and f_1 are nonnegative and

$$P(\Omega) = \int_{-\infty}^{\infty} \int_{-\infty}^{\infty} \gamma^2 e^{-\gamma x - \gamma y} \, dx \, dy = \left(\int_{-\infty}^{\infty} \gamma e^{-\gamma x} \, dx \right)^2 = 1.$$

Hence, P_0 and P_1 are indeed probability measures satisfying the Kolmogorov axioms.
  We now employ the prescription of Section 13.4 to identify

$$D_1 = \{(x, y) : \text{ decide target present }\} = \left\{ (x, y) : \frac{f_1(x, y)}{f_0(x, y)} \geq \tau \right\}$$

$$= \left\{ (x, y) : \frac{e^{-x-y}}{\gamma^2 e^{-\gamma x - \gamma y}} \geq \tau \right\} = \{(x, y) : x + y \geq \frac{1}{\gamma - 1} \log(\gamma^2 \tau) = c\}.$$

The constant c is ours to determine to achieve, say, a desired false-alarm probability. It enters
into the false-alarm probability $P_{FA} = P_0(D_1)$ through

$$P_{FA} = \int_{\{(x,y):x+y \geq c, x \geq 0, y \geq 0\}} \gamma^2 e^{-\gamma x - \gamma y} \, dx \, dy$$

$$= \int_0^c dx \int_{c-x}^{\infty} \gamma^2 e^{-\gamma x - \gamma y} \, dy + \int_c^{\infty} dx \int_0^{\infty} \gamma^2 e^{-\gamma x - \gamma y} \, dy$$

$$= \int_0^c dx \int_{c-x}^{\infty} \gamma^2 e^{-\gamma x - \gamma y} \, dy + e^{-\gamma c}$$

$$= \int_0^c dx \gamma e^{-\gamma x} \int_{c-x}^{\infty} \gamma e^{-\gamma y} \, dy + e^{-\gamma c} = \int_0^c \gamma e^{-\gamma x} e^{-\gamma(c-x)} \, dx + e^{-\gamma c} = (\gamma c + 1) e^{-\gamma c}.$$

Note that when $c = 0$, then $D_1 = \Omega$, and we decide a target is present no matter what observation (x, y) we make on received energies. In this case, $P_{FA} = 1$, as it should. This is not
a useful state of affairs. Correspondingly, when $c = \infty$, then $D_1 = \emptyset$ and $D_0 = D_1^c = \Omega$, and
we decide target absent no matter what observation we make. In this case $P_{FA} = 0$. In either
case, there was no point in building the system! If we arbitrarily take $\gamma c = 5$, then

$$D_1 = \{(x, y) : x + y \geq 1\},$$

and we decide a target is present if and only if the sum of the received energies x and y are
at least 1. In this case $P_{FA} = 6e^{-5} = .040$, a plausible setting. Approximately once every 25
measurements, when the target is absent, we can expect a false alarm.

Example 3.14   Detection Probability _____
From the work above, we are also in a position to evaluate the detection probability $P_D =
P_1(D_1)$ through

$$P_D = P_1(D_1) = P_1(\{(x, y) : x + y \geq c\}) = \int_{\{(x,y):x+y \geq c, x \geq 0, y \geq 0\}} e^{-x-y} \, dx \, dy,$$

where we have used f_1 in place of f_0 in the similar equation given in the preceding example. The end result, paralleling the preceding example (calculated by setting $\gamma = 1$), is

$$P_D = (c + 1)e^{-c}.$$

When $c = 0$, then $D_1 = \Omega$ and $P_D = 1$. When $c = \infty$, then $D_1 = \emptyset$ and $P_D = 0$. Neither choice requires the detection system to be built. If we choose $c = 1$, then $P_D = 2e^{-1} = .736$. About three-quarters of the time that a target is present, we will detect it.

We now enrich this illustration of the use of probability ideas by introducing different versions of P_1, denoted by $\{P_{1,t_k}\}$, that correspond to different target types $\{t_k\}$ that might be expected within the surveillance region. If we are examining the same surveillance region with the same system, then the probability measure P_0 for target absent remains unchanged. However, in this situation we have several different optimal partitions of Ω depending upon which version of P_1 we assume to hold.

Example 3.15   Multiple Detectors _____

Assume, as above, that P_{1,t_k} is described by

$$P_{1,t_k}(A) = \int_A f_{1,t_k}(x, y)\, dx\, dy,$$

where, once again, each f_{1,t_k} is a nonnegative function that integrates to 1 over Ω. Thus, the optimal decision region for target type t_k is as in the preceding discussion (see Section 13.4)

$$D_{1,t_k} = \{(x, y): \text{ decide target type } t_k \text{ present }\} = \left\{(x, y): \frac{f_{1,t_k}(x, y)}{f_0(x, y)} \geq \tau\right\}.$$

The corresponding false-alarm probabilities $\{P_{FA,k}\}$ are given as above by

$$P_{FA,t_k} = P_0(D_{1,t_k}) = \int_{D_{1,t_k}} f_0(x, y)\, dx\, dy.$$

The corresponding detection probabilities are

$$P_{D,t_k} = P_{1,t_k}(D_{1,t_k}) = \int_{D_{1,t_k}} f_{1,t_k}(x, y)\, dx\, dy.$$

As we are running these different detectors simultaneously, we will have a false alarm if any one of them responds with a false alarm. Of course, all of them will respond, some of them with correct detections and others with false alarms. Hence, a false alarm will occur on the event

$$D_1 = \bigcup_k D_{1,t_k}$$

with probability

$$P_{FA} = P_0 \left(\bigcup_k D_{1,t_k} \right).$$

Observe that this union is of sets that likely overlap. While we can, in principle, carry out an exact evaluation of the false-alarm probability, Boole's Inequality of Lemma 3.7.1 immediately informs us that

$$P_{FA} \le \sum_k P_0(D_{1,t_i}).$$

This upper bound is only of potential value if it is less than 1. As false-alarm probabilities tend to be set to small values, there is a good chance that this will be the case. Carrying out the same upper bound for detection probabilities is unlikely to yield a useful result. Detection probabilities are usually set to larger values, and it is unlikely that the upper bound will be less than 1. An exact result is also available from the Inclusion-Exclusion Principle of Lemma 3.8.1, although its evaluation is going to be tedious if there are many target types.

If we had the additional knowledge that when a target is present it is a single target type t_k with probability λ_k, then rather than having different versions of P_1 we would, by the properties of convex combination, have a composite target source described by the new measure

$$P_1 = \sum_k \lambda_i P_{1,t_k},$$

and we would use it to design a single target detection set D_1.

3.11 SUMMARY

In this chapter, we motivated and established the foundations of probability. Motivation is important so that engineers and others who seek to apply probability can understand the assumptions they are making in using a particular mathematical approach. We opened with a survey of alternative formal concepts of probability that ranged from mere possibility to precise numerical probability. We then noted five interpretations of the meaning of probability $P(A)$ and settled upon one based upon chance as a propensity or tendency for an event A to occur in the performance of a random experiment $\mathcal{E}$, with this propensity having a display through relative frequencies of occurrence in unlinked repetitions of this random experiment. The meaning or interpretation of probability is still a controversial and unresolved issue. Probability, as we use it in daily life and in technical applications, has more than one meaning.

With this background, we presented the celebrated axioms of probability proposed by Andrei N. Kolmogorov in 1933. These axioms, with the exception of the final one of monotone continuity, model properties that are readily observed to be true for classical probability and for relative frequencies. The final axiom of Kolmogorov is one of closure, valuable from a mathematical viewpoint, but not one well rooted in our understanding of probability.

Following our expository pattern, we explored the consequences of these axioms and thereby introduced the basic properties of probability that will be used repeatedly in subsequent chapters. The consequences were organized by whether they involved a single event, two events, or more than two events. In particular, we provided an exact evaluation of the probability of a union of n events through the Inclusion-Exclusion Principle, and an upper bound to this probability through Boole's Inequality. In many instances, we know only the probabilities of individual events and can only use Boole's Inequality.

We closed with a connection to Goal 3 through an extended application to target detection. We will revisit target detection more thoroughly in Chapter 13.

3.12 APPENDIX: PROOF OF EQUIVALENCE THEOREM

Theorem 3.12.1   If P satisfies K0–K3, then it satisfies K4$'$ if and only if it satisfies K4.

Proof. We first show that K0–K4 imply the countable additivity axiom K4$'$. Let $\{A_i\}$ be any countably infinite collection of pairwise disjoint events, and define for any n

$$B_n = \cup_{i>n} A_i,$$

$$\cup_{i=1}^{\infty} A_i = B_n \cup (\cup_{i=1}^{n} A_i).$$

Clearly, for all $i \leq n$ we have that $A_i \perp B_n$. By K3 (extended in Lemma 3.6.4 to cover finite unions)

$$P(\cup_{i=1}^{\infty} A_i) = P(B_n) + \sum_{i=1}^{n} P(A_i).$$

The limit

$$\lim_{n \to \infty} \sum_{i=1}^{n} P(A_i) = \sum_{i=1}^{\infty} P(A_i),$$

exists because $\sum_1^n P(A_i) \leq 1$ and is nondecreasing. K4$'$ will follow if we can also guarantee that

$$\lim_{n \to \infty} P(B_n) = 0.$$

Note that $B_{n+1} \subset B_n$ and that $\cap_1^{\infty} B_n = \emptyset$. Immediate application of K4 assures us that the last limit is zero and K4$'$ holds.

We now show that K0–K3, K4$'$ imply the monotone continuity axiom K4. Let $\{B_n\}$ be any countably infinite event collection satisfying the hypotheses of K4,

$$B_{n+1} \subset B_n \quad \text{and} \quad \cap_{n=1}^{\infty} B_n = \emptyset.$$

Define

$$A_n = B_n - B_{n+1},$$

and observe that $\{A_n\}$ is a countably infinite collection of pairwise disjoint events (because the $\{B_n\}$ are nested). Furthermore,

$$B_n = (\cup_{j \geq n} A_j) \cup (\cap_{k=1}^{\infty} B_k) = \cup_{j \geq n} A_j.$$

From K4$'$, we have that

$$P(B_n) = \sum_{j \geq n} P(A_j).$$

Because the infinite sum

$$\sum_1^{\infty} P(A_j) \leq 1,$$

we conclude that

$$\lim_{n \to \infty} P(B_n) = \lim_{n \to \infty} \sum_{j \geq n} P(A_j) = 0,$$

as desired to verify K4.   □

EXERCISES

E3.1   (Subjective Probability) If you are currently taking a course in probability, we are interested in how precisely you can estimate your subjective probability of the event G that your final grade in this course will exceed your current GPA. In order to do so, compare the likelihood of G to the likelihoods of events concerning the identity of the topmost card in a well-shuffled deck of playing cards. For example, is G less probable than the topmost card being either a heart, diamond, or club? Is G more probable than the topmost card being a nine? By considering events that are more probable than G and ones that are less probable, determine the narrowest interval of probability in which you have sufficient confidence to bet accordingly with small sums.

E3.2   (Relative Frequency) Select the Dow Jones data set described in Problem E1.28. Turn it into a binary-valued data set by assigning a 1 if $d_j \geq d_{j-1}$ and a 0 otherwise (i.e., a 1 if the price did not decrease from that of the previous trading day). Choosing a range of years such as 1980–1989, does the relative frequency of ones appear to converge?

E3.3   a. Toss a penny 100 times and record the sequence of heads and tails. How often did heads occur? Plot the fraction of times heads occur in the first n tosses vs. n, for n a multiple of 5.

                b. Repeat this experiment only now spinning the penny on a tabletop, rather than tossing it. Do the results on occurrences of heads seem to agree?

E3.4   Select a thumbtack and toss it 100 times (be careful!), recording the sequence of outcomes as to whether the tack lands on its back/head (B) (assign this case a 1) or rests on its edge and point (E) (assign this case a 0).

a. Informally, how predictable is the occurrence of the outcome B from your data sequence? Can you think of a "simple" formula that predicts the location of the next occurrence of B given the locations of the previous occurrences?

b. Evaluate the sample mean, median, and standard deviation for the resulting binary-valued sequence.

c. Plot the fraction of times B occurs in the first n tosses vs. n, for n a multiple of 5.

E3.5   a. Prove that the relative frequency $r_n(A)$ is the following convex combination:

$$r_n(A) = \left(1 - \frac{1}{n}\right) r_{n-1}(A) + \frac{1}{n} I_A(\omega_n).$$

b. Prove that

$$|r_n(A) - r_{n-1}(A)| \le \frac{1}{n}.$$

c. Does the fact that successive terms in the sequence $\{r_n(A)\}$ necessarily converge to each other,

$$\lim_{n \to \infty} |r_n(A) - r_{n-1}(A)| = 0,$$

imply that the sequence of relative frequencies itself must converge?

E3.6   Let $\Omega = [0, 1]$ be the unit interval. Define an event collection $\mathcal{A}$ as the set of all events that are finite unions of intervals. (The intervals can be of any of the four kinds.)

a. Prove that $\mathcal{A}$ is an algebra of events.

b. Does $\mathcal{A}$ include the set R of rational numbers in the unit interval?

c. Prove that $\mathcal{A}$ is not a σ-algebra.

E3.7   Let $\Omega = [0, 1]^2$ be the unit square. Define an event collection $\mathcal{A}$ as the set of all subsets of the unit square that are finite unions of rectangles.

a. Prove that $\mathcal{A}$ is an algebra of events.

b. Is the interior T of the triangle with vertices $(0, 0)$, $(1/2, 1)$, $(1, 0)$ a set in $\mathcal{A}$?

c. Prove that $\mathcal{A}$ is not a σ-algebra.

E3.8   Prof. Phynne has attempted to calculate $p = P(A)$ and has determined that it is a root of the following polynomial of degree five:

$$(p - 2)(p - 2\sqrt{-1})(p + 2\sqrt{-1})(p + .5)(p - .5) = 0.$$

Can you help him choose an answer?

E3.9   If $\Omega = \{a, b, c\}$, the algebra $\mathcal{A}$ is the set of all subsets of Ω, and the probability measure P is partially defined by

$$P(\{a, b\}) = .5, \quad P(\{b, c\}) = .8, \quad P(\{a, c\}) = .7,$$

then complete the specification of P for all remaining events.

E3.10 Consider the event $A = \{\omega_1, \ldots, \omega_m\}$, where it is known that $0 < a \le P(\{\omega_k\}) = p_k$ for $k = 1, 2, \ldots, m$, and that $P(A) \le b$.

Show that $b/a \ge m$.

E3.11   If $\{A_i\}$ is a countably infinite partition of Ω and $P(A_i) = ab^i, i \geq 0$, then what are the constraints on a and b for P to be a probability measure?

E3.12   Show that (a) and (b) are incompatible with the Kolmogorov axioms. Construct a classical probability example on a sample space of 10 points to show that (c) is compatible with the axioms.

     a. $P(A - B) = .7, P(B) = .4$;

     b. $P(A) = .6, P(B) = .5, P(A \cap B) = .05$;

     c. $P(A) = .6, P(B) = .5, P(A \cap B) = .2, P(A \cup B) = .9$.

E3.13   a.   If $P(A) = .7$, evaluate $P(A^c)$.

     b.   If $P(A) = .1$ and $P(B) = .2$, what can you say about $P(A \cup B) + P(A \cap B)$?

     c.   If $P(A) = .2$ and $P(B) = .4$, what can you say about $P(A \cup B)$?

E3.14   a.   If $P(A \cap B) = .1$, and $P(A \cup B) \leq .5$, how large can $P(B)$ be?

     b.   If $P(A \cap B) = .2$, what is the largest possible value for the product $P(A)P(B)$? Provide reasons for your answer.

E3.15   a.   Verify the following by appeal to the de Morgan laws and Kolmogorov axioms:

$$P(A) + P(B) + P(A^c \cap B^c) = 1 + P(A \cap B).$$

     b.   Is it always true that

$$P(A \cap B) \geq P(A) + P(B) - 1?$$

E3.16   Prove that

$$P(A) = P(B) = 1 \Rightarrow P(A \cap B) = 1$$

$$\text{and } A \perp B \Rightarrow P(A) \leq P(B^c).$$

E3.17   a.   If $P(A) = .4$, $P(B) = .5$, $P(A \cap B) = .3$, evaluate $P(A \cup B)$.

     b.   If $P(A \cap B) = 1$, what can you say about the following: $P(A)$, $\min(P(A), P(B))$?

     c.   If $P(A \cup B) = 1$, what can you say about the following: $P(A)$, $\max(P(A), P(B))$, and $\min(P(A), P(B))$?

E3.18   a.   If $P(A) = .2$, $P(B - A) = .3$, $P(C \cap A^c \cap B^c) = .1$, evaluate $P(A \cup B \cup C)$.

     b.   If $P(A) = .2, P(B) = .3, P(C) = .1, AB = AC = BC = \emptyset$, evaluate $P(A \cup B \cup C)$.

E3.19   Is there a probability measure satisfying

$$P(A) = .1, \quad P(B) = .2, \quad P(C) = .3, \quad P(A \cup B \cup C) = .7?$$

E3.20 $\Omega = \{0, 1, \ldots, 9\}$, $A = \{\omega : \omega \text{ even}\}$, $B = \{\omega : \omega \text{ odd}\}$, and $C = \{1, 2, 3\}$.

     a.   Can $P(C) > \max(P(A), P(B))$?

     b.   Can $P(C) < \min(P(A), P(B))$?

(If it is possible, provide a probability assignment to the elements of Ω for which it is true. If it is not possible, prove your claim.)

E3.21   In a given random experiment $\mathcal{E}$, it is known that

$$P(\{0, 1, 2, 3\}) = .5, \quad P(\{0, 1\}) = .2.$$

What can you conclude about $P(\{0, 1, 2\})$, and $P(\{3\})$?

E3.22 If we know that

$$P(\{0\}) = .2, P(\{1\}) = .15, P(\{2\}) = .25, P(\{3\}) = .01,$$

evaluate the probability that at least one of numbers 0,1,2,3 is the outcome of the experiment.

E3.23 Given three events defined as the sets

$$A_1 = \{0, 2\}, \quad A_2 = [1, 4], \quad A_3 = \{3, 5\},$$

and having probabilities $P(A_1) = .3$, $P(A_2) = .2$, $P(A_3) = .4$, what can you conclude about each of the following probabilities?

    a. $P(A_1^c)$, $P(A_1 \cup A_2)$, $P(A_1 \cup A_3)$.
    b. $P(\{2\})$, $P(A_1 \cap A_3)$.
    c. $P(\cup_{i=1}^{3} A_i)$.

E3.24 In Problem E1.18, the event A_k is that k clients are requesting service and we know that $k \leq 3$. We are given the probability assignment:

$$P(A_1) = .243, \quad P(A_2) = .027, \quad P(A_3) = .001.$$

    a. What is the probability $P(A_0)$ that no client is requesting service?
    b. What is the probability that at least two clients are requesting service?

E3.25 If $P(A) = .3, P(B) = .5, P(A \cup B) = .7$, what can you conclude about $P(A \cap B)$?

E3.26 If $\Omega = \{0, 1, 2, \ldots, 9\}$, $A = \{0, 1, 4\}, B = \{3, 5\}, P(A) = .8, P(B) = .1$, what can you conclude about the probability of $C = \{0, 1, 2, 3, 4\}$?

E3.27 Repeat Example 3.7 by determining the probability $P(C)$ that the outcome ω is a multiple of 3.

E3.28 Given three events defined as the intervals

$$A_1 = [0, 2], \quad A_2 = [1, 4], \quad A_3 = [3, 5],$$

and having probabilities $P(A_1) = .1$, $P(A_2) = .2$, $P(A_3) = .3$, what can you conclude about each of the following probabilities?

    a. $P(A_1^c)$, $P(A_1 \cup A_2)$, $P(A_1 \cup A_3)$.
    b. $P(A_1 \cap A_2)$, $P(A_1 \cap A_3)$.
    c. $P(\cup_{i=1}^{3} A_i)$.

E3.29 Evaluate $P(A \cup B \cup C)$ when you know that

$$P(A) = .3, P(B) = .4, P(C) = .5, P(A \cap B) = .2, P(A \cap C) = .1,$$

$$P(B \cap C) = .15, P(A \cap B \cap C) = .1.$$

E3.30 Two fair dice are tossed independently so that all ordered pairs of outcomes are equally probable. (See Section 2.10.) What is the probability that the sum of the outcomes (number of dots that are uppermost) is 11?

E3.31 If for each i, $P(A_i) = .1$, what can you conclude about $P(\cup_{i=1}^{5} A_i)$?

E3.32 In a given message source that generates bytes, it is known that the probability of seeing a consecutive "01" is .05 no matter which of the positions $1, 2, \ldots, 7$ we examine for the leading "0." What can you say about the probability of at least one occurrence of "01" in a byte?

E3.33   Take $\Omega = [0, 1]$, the unit interval, the event algebra $\mathcal{A} = \mathcal{B}$ as the usual Boolean algebra generated by including all intervals in Ω, and select the so-called uniform distribution

$$P(A) = \int_0^1 I_A(x)\, dx.$$

Define the countable collection of disjoint events $\{A_i\}$, as nonoverlapping intervals of decreasing length, through

$$A_i = [a_i, b_i], a_1 = \frac{1}{2}, b_1 = 1, a_{i+1} = \frac{a_i}{3}, b_{i+1} = \frac{b_i}{3}.$$

From this specification, we observe that

$$a_i = \frac{1}{2} 3^{1-i}, b_i = 3^{1-i}, l_i = b_i - a_i = a_i, b_{i+1} < a_i.$$

   a. Evaluate the probability $P(A_i)$.
   b. Evaluate the probability that at least one of the events $\{A_i\}$ will occur.

E3.34   In a photon-counting experiment, it is known that the probability of observing exactly k photons is $e^{-1}/k!$ (where $k!$ is factorial k) and this holds for all nonnegative integer values k. Evaluate the probability of observing an even number of photons.

   (Hint: Consider the power series expansion for $e^x + e^{-x}$.)

E3.35   There are n letters addressed to n distinct individuals and n appropriately uniquely addressed envelopes. The letters are randomly inserted in the envelopes. Let B_n be the event that at least one letter is in a correct envelope.

   a. Exactly evaluate $P(B_n)$ by using the Inclusion–Exclusion Principle.
   b. Examine $P(B_n)$ as n diverges.

E3.36   Let $\Omega = \{\omega_1, \ldots, \omega_n\}$ and take $\mathcal{A}$ to be the power set of Ω.

   a. Verify that for any $\omega^* \in \Omega$

$$P(A) = I_A(\omega^*)$$

   satisfies the Kolmogorov axioms.

   b. Can any probability measure P on $\mathcal{A}$ be written as a convex combination of the measures $\{P_1, \ldots, P_n\}$, where

$$P_i(A) = I_A(\omega_i)?$$

E3.37   Consider the target detection example of Section 3.10 with the sole revision that

$$P_0(A) = \int_0^\infty dx \int_0^\infty dy I_A(x, y) \gamma^4 xy e^{-\gamma x - \gamma y}.$$

   a. Evaluate the false-alarm probability when there is a single detector and the upper-corner region $D_1 = C_{a,b}^u$ for $a > 0, b > 0$.
   b. Use Boole's Inequality to upper-bound the false-alarm probability P_{FA} when there are three detectors being used simultaneously and we claim detection if any of the detectors do.

    c. Use the Inclusion–Exclusion Principle to evaluate P_{FA} exactly in the case described in (b).

E3.38 Let $\Omega = N = \{0, 1, \ldots\}$ and

$$P(\{\omega : \omega \leq k\}) = 1 - \left(\frac{1}{2}\right)^k,$$

for integer $k \geq 0$.

    a. Evaluate $P(A_n)$ for $A_n = \{\omega : \omega > n\}$.
    b. Evaluate $P(\cup_{i \geq n} A_i)$.
    c. Do the events $\{A_n\}$ occur infinitely often with probability 1?

4

Describing Probability I: Countable Ω

4.1 PURPOSE, BACKGROUND, AND ORGANIZATION

In Chapter 3, we provided the mathematical foundations of numerical probability. However, we provided neither the means by which probability models are constructed in practice nor the handful of models that are in constant use. In Chapters 4, 5, 6, and 7, we present the methods that are used in practice for convenient and tractable specification of probability measures (functions satisfying the Kolmogorov axioms) and present a number of examples of the most commonly encountered probability models. We organize our presentation by the type of sample space and start in this chapter with sample spaces that have either finitely or countably infinitely many elements, the so-called countable sample spaces.

4.2 PROBABILITY MASS FUNCTIONS

The common approach to simplifying the description of a probability measure P is to specify a restriction P_S of P to a much smaller subset S of the event algebra $\mathcal{A}$. The choice of S can be made so that there are much simpler constraints placed on P_S than are placed on P by the Kolmogorov axioms. Furthermore, it will turn out that, by knowing P_S, we can uniquely reconstruct all of P on $\mathcal{A}$. A different method of describing probability measures is deferred to Chapter 17 where we discuss characteristic functions.

A sample space $\Omega = \{\omega_i\}$ is countable if it is either finite or countably infinite. It is countably infinite if it has as many elements as there are integers. In either case, the elements of Ω can be enumerated as, say, $\omega_1, \omega_2, \ldots$. If the event algebra $\mathcal{A}$ contains each singleton set $\{\omega_i\}$ (from which it follows that $\mathcal{A}$ is the power set of Ω), then we specify probabilities satisfying the Kolmogorov axioms through a restriction P_S to the set $\mathcal{S} = \{\{\omega_i\}\}$ of singleton events.

Definition 4.2.1   (Probability Mass Function)   A *probability mass function (pmf)* p is a function on Ω that satisfies the following two conditions:

PMF1.

$$p : \Omega \to [0, 1];$$

PMF2.

$$\sum_{\omega \in \Omega} p(\omega) = 1.$$

When the elements of Ω are enumerated, then it is common to abbreviate $p(\omega_i) = p_i$. The next theorem asserts that pmfs suffice to describe all, and only, probability measures on countable sample spaces.

Theorem 4.2.1   (PMFs Completely Describe P)   There is a one-to-one correspondence between probability measures P satisfying K0–K4 on the power set of a countable sample space Ω and pmfs p, and it is given by

$$(a) \; p(\omega) = P(\{\omega\}),$$

$$(b) \; P(A) = \sum_{\omega \in A} p(\omega).$$

Proof. When P is given satisfying K0–K4 and p is defined by (a), it is immediate from K1 and the elementary consequences that PMF1 holds. That PMF2 holds follows from K4' (σ-additivity) and K2 (unit normalization). When p is given satisfying PMF1, PMF2, it is immediate that P satisfies K0–K2. That P also satisfies K3, K4' follows directly from (b) and the properties of sums. $\square$

Hence, every pmf p defines a probability measure P satisfying K0–K4 and conversely. The convenience of a specification by pmf becomes clear when Ω is a finite set of, say, n elements. The algebra $\mathcal{A}$ then contains 2^n events. Specifying P requires specifying 2^n values, one for each event, and doing so in a manner that is consistent with the Kolmogorov axioms. However, specifying p requires only providing n values, one for each element of Ω, satisfying the simple constraints of nonnegativity and addition to 1; specifying the pmf is far less complex than specifying the probability measure and automatically ensures satisfaction of the Kolmogorov axioms.

A final property of pmfs is that a convex combination p of pmfs $\{p^{(i)}\}$ is readily verified to be a pmf, and it is the pmf corresponding to the convex combination P of measures $\{P_i\}$:

$$P(A) = \sum_i \lambda_i P_i(A) \text{ iff } p(\omega) = \sum_i \lambda_i p^{(i)}(\omega).$$

Convex combinations can be interpreted through a first random selection with probability λ_i of a source described by measure P_i followed by an independent draw according to the measure P_i. (See Section 3.9.)

4.3 APPLICATION TO MODELING ENGLISH TEXT LETTER FREQUENCIES

As an example of a pmf, consider the probability of occurrence of a letter from the Roman alphabet in an English text. While there is some question about the stability of the frequencies of occurrence of letters across different bodies of natural language text, Welsh [98] provides a table of letter frequencies that form a probability mass function for the occurrence of a letter in English text. (We have modified his table for Example 4.1 by reducing the probability for s by .001 so that the total probability sums to 1.) An early use of estimates of letter frequencies was to the layout of the typewriter keyboard designed by the portrait painter and telegraphy inventor Samuel Morse. (The most frequently occurring letters were placed under the left hand to slow down the typing rate of the more common right handers so as to reduce the occurrence of jammed keys on early mechanical typewriters). Another early use of letter frequencies was in the cryptanalysis of simple substitution ciphers (a one-to-one correspondence between letters of the plain text and the cipher text).*

Example 4.1   PMF for English Letters
We list the probabilities in descending order.

ω	e	t	a	i	n	o	s
$p(\omega)$	.120	.085	.077	.076	.067	.067	.066
ω	r	h	d	l	u	c	f
$p(\omega)$	.059	.050	.042	.042	.037	.032	.024
ω	m	w	y	p	b	g	v
$p(\omega)$	.024	.022	.022	.020	.017	.017	.012
ω	k	q	j	x	z		
$p(\omega)$	.007	.005	.004	.004	.002		

Example 4.2   Probability of a Vowel
Thus, we can determine the probability of a vowel as

$$P(\text{vowel}) = P(a,e,i,o,u) = .077 + .120 + .076 + .067 + .037 = .377,$$

*If, say, in the lengthy encrypted text the most commonly occurring letter is q and the second most commonly occurring letter is r, then q is the plaintext e and r is the plaintext t.

or $P(a,b) = p(a) + p(b) = .077 + .017 = .094$. If we did not have reliable information about, say, the 10 least commonly occurring letters, then all we would know is that $P(y,p,b,g,v,k,q,j,x,z) = .110$, with this being calculated as the probability of the complement of the 16 letters whose probabilities we do know. In this case, we could only bound $P(a,b) = .077 + p(b)$ by bounding $0 < p(b) < .11$ (we use strict inequality to incorporate our knowledge that all of the letters have a positive probability of occurring) and concluding that $.077 < P(a,b) < .187$.

4.4 COMMONLY ENCOUNTERED PMFs

Important, frequently employed families of pmfs are presented in the next subsections. Our expository pattern will be to start with a definition of a family of pmfs, discuss the constraints on the parameters associated with this family, verify that each pmf in the family satisfies the two properties of nonnegativity and summation to 1, provide a plot of the pmf for a specific choice of parameter, discuss its applications, and, if needed, expose the analytical properties of the pmf.

4.4.1 Random $\mathcal{R}(n)$

Definition 4.4.1   (Random PMF $\mathcal{R}(n)$)   If the sample space $\Omega = \{\omega_1, \ldots, \omega_n\}$, then the $\mathcal{R}(n)$ family of pmfs specifies

$$P(\{\omega_i\}) = p_i = \begin{cases} \dfrac{1}{n} & \text{if } 0 \le i \le n-1 \\ 0 & \text{otherwise} \end{cases}.$$

It is immediate that this pmf is nonnegative (taking only the values 0 and $1/n$) and that it sums to 1. Hence, it is a pmf.

The random $\mathcal{R}(n)$ pmf models fair gaming devices (e.g., well-balanced coins and dice, well-shuffled decks of cards) and high-rate encoded digital data having n distinct, equally probable outcomes. A plot of the $\mathcal{R}(8)$ pmf for $\Omega = \{1, 2, \ldots, 8\}$ is given in Figure 4.1.

All of our discussion of classical probability in Chapter 2 illustrates the use of this model of "choosing at random."

4.4.2 Binomial $\mathcal{B}(n, p)$

Definition 4.4.2   (Binomial PMF $\mathcal{B}(n, p)$)   The Binomial $\mathcal{B}(n, p)$ pmf for $0 \le p \le 1$ and n a positive integer, is specified by

$$p_k = \begin{cases} \binom{n}{k} p^k (1-p)^{n-k} & \text{for } 0 \le k \le n \\ 0 & \text{otherwise} \end{cases}.$$

A common alternative notation for the Binomial p_k is $b(k, n, p)$.

If one accepts the convention that the factorial of a negative integer is infinite, then we do not need to spell out the case of k outside of $0, 1, \ldots, n$. This convention for factorial is a fact

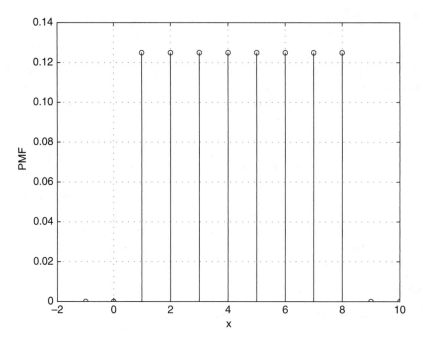

Figure 4.1   Random pmf on 8 outcomes.

for the Gamma function that generalizes the factorial function. If $p < 0$, then $p_1 < 0$ in violation of the properties of a pmf. Similarly, if $p > 1$, then either p_0 or p_1 is negative. Hence, we can only have $0 \le p \le 1$.

The Binomial Theorem (Section 2.4) is used to verify the second condition on a pmf that

$$\sum_{i=0}^{n} p_i = \sum_{i=0}^{n} \binom{n}{i} p^i (1-p)^{n-i} = (p+1-p)^n = 1.$$

A plot of $\mathcal{B}(8, .3)$ is provided in Figure 4.2.

The $\mathcal{B}(n, p)$ is found in the context of unlinked (independent) repetitions $\mathcal{E}_1, \ldots, \mathcal{E}_n$ of a random experiment $\mathcal{E} = (\Omega, \mathcal{A}, P)$. We are interested in a particular event $A \in \mathcal{A}$ having $P(A) = p$. Let K denote the random number of times that A is observed to occur among the n outcomes of the repeated experiments. Then $P(K = k)$ is given by $\mathcal{B}(n, p)$. In many instances, if the outcomes of the repeated experiments are the individual symbols in a transmitted message of length n drawn from the alphabet Ω and A is the event that a symbol is transmitted in error, then $\mathcal{B}(n, p)$ models the number of unlinked (independent) errors made in n symbols of text when p is the probability of an error in a single symbol of text and also the number of "heads" in n unlinked, repeated tosses of a coin with probability p for a "head" on a single toss. When $p = 1/2$, we have a model for the number of ones in a randomly chosen binary sequence of length n or the number of "heads" in n tosses of a fair (well-balanced, symmetrical) coin. The binomial probability law will be derived in Section 14.5.2.

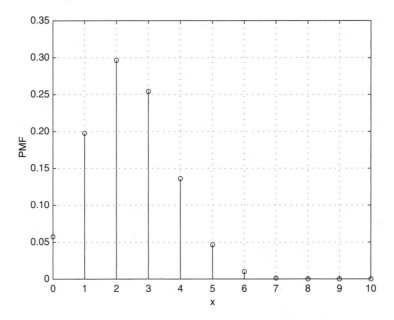

Figure 4.2 Binomial pmf for $n = 8$ and $p = .3$.

We study the binomial pmf analytically by examining the ratio of successive terms

$$\frac{p_{i+1}}{p_i} = \frac{n-i}{i+1}\frac{p}{1-p},$$

and observing that it is strictly decreasing in i. If

$$\frac{p_1}{p_0} = \frac{np}{1-p} > 1,$$

then the sequence $\{p_i\}$ starts by strictly increasing. If $np/(1-p) < 1$, then the sequence just decreases throughout. If

$$\frac{p_n}{p_{n-1}} = \frac{p}{n(1-p)} < 1,$$

then the sequence terminates by strictly decreasing. If $p/n(1-p) > 1$, then the sequence is increasing throughout. If

$$\frac{1}{n} < \frac{p}{1-p} < n,$$

then the sequence $\{p_i\}$ is an unimodal function—one that is increasing to the left of its maximum and decreasing to the right. The maximum or most probable value occurs when this ratio is approximately 1. Taking into account the fact that the ratio will generally not be exactly 1 for any integer i, we find a maximum probability value to be $\lfloor (n+1)p \rfloor$, or approximately np (which may not be an integer).

Example 4.3   Binomial Symbol Error Model _____

The per symbol error probability p is .001 in a message transmission system. If 2,000 symbols were sent, then the probability of k errors is

$$P(K = k) = \binom{2,000}{k} .001^k (1 - .001 = .999)^{2,000-k}.$$

The most probable number of errors is $\lfloor (2,001).001 \rfloor = 2$, and it has probability

$$P(K = 2) = \binom{2,000}{2} .001^2 (1 - .001)^{2,000-2} = 1,000(1,999)10^{-6}(1 - .001)^{1,998} \approx 2e^{-2},$$

where we have used the approximation

$$1 - x \approx e^{-x}$$

valid for small x and approximated 1,998 and 1,999 by 2,000. The probability of an error-free message is

$$P(K = 0) = \binom{2,000}{0} .001^0 (1 - .001)^{2,000} = (1 - .001)^{2,000} \approx e^{-2}.$$

Example 4.4   Binomial Symbol Error Model (Continued) _____

If, in the preceding example, we were not given p, but were told that

$$P(K = 2) = 2P(K = 0),$$

then we can solve for p through

$$\binom{2,000}{2} p^2 (1 - p)^{2,000-2} = 2 \binom{2,000}{0} p^0 (1 - p)^{2,000} \iff 1,000(1,999)p^2 = 2(1 - p)^2$$

$$\iff 1,000(\sqrt{1.999/2})p = 1 - p.$$

Hence, to a good approximation, $p = .001$.

4.4.3 Geometric $\mathcal{G}(\beta)$

Definition 4.4.3   (Geometric PMF $\mathcal{G}(\beta)$)   The Geometric $\mathcal{G}(\beta)$ pmf for Ω the nonnegative integers and $0 \le \beta < 1$ is specified by

$$P(\omega = k) = p_k = (1 - \beta)\beta^k, \quad k = 0, 1, \ldots.$$

A variant of $\mathcal{G}(\beta)$ for Ω the strictly positive integers is specified by

$$p_k = (1 - \beta)\beta^{k-1}, \quad k = 1, 2, \ldots.$$

If $\beta = 0$, then, in the first version, $P(\omega = 0) = 1$, and in the second version, $P(\omega = 1) = 1$. If $\beta = 1$, then, for all k, $P(\omega = k) = 0$, this assignment sums to 0 and not 1, and contradicts this possibility. If $\beta < 0$, then, say, in the first version, $P(\omega = 1) < 0$, contradicting this possibility. If $\beta > 1$, then, for all k, $P(\omega = k) < 0$, contradicting this possibility. Hence, we can only have $0 \le \beta < 1$. In order to verify that the pmf sums to 1, as required, we use the sum of a geometric series to establish that, say, in the first version,

$$\sum_{k=0}^{\infty} p_k = (1 - \beta) \sum_{k=0}^{\infty} \beta^k = (1 - \beta)\frac{1}{1 - \beta} = 1.$$

We conclude that the Geometric $\mathcal{G}$ is indeed a pmf. A plot of $\mathcal{G}(.5)$, starting at $k = 0$, is shown in Figure 4.3.

$\mathcal{G}$ models the following phenomena:

- the lifetimes of components, measured in integer time units starting from 0, when they fail catastrophically, that is without degradation due to aging (we will subsequently use a continuous time model for this given by the exponential in Section 6.6.1)
- waiting times measured in integer units (starting from 1) until the arrival of the next customer in a queue or the next radioactive disintegration or photon emission (also subsequently replaced by the continuous time exponential)

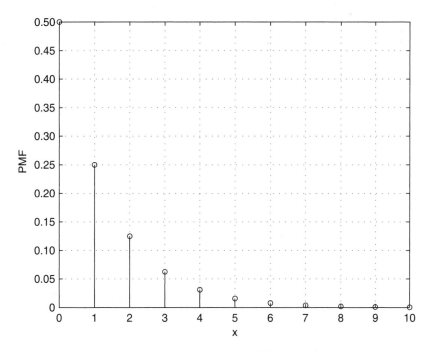

Figure 4.3 Geometric pmf for $\beta = .5$.

- the delay or number (starting from 0) of repeated, unlinked/independent random experiments that must be performed prior to the first occurrence of a given event A (e.g., number of coin tosses prior to the first appearance of a "head" or the number of transmitted symbols before the first error)
- trial number (starting from 1) of the first appearance of A in repeated, unlinked/independent random experiments

To understand the origins of the Geometric, consider the classical or random case of binary sequences of length n, all of the 2^n such sequences being equally probable. Let T_1 be the index of the first term in the sequence that is a 1; $T_1 = 1$ if the sequence starts with a 1, is 2 if the sequence starts with 01, etc., and is infinite if the event does not occur. Hence, $T_1 = k < n$ if and only if the sequence starts with a run of $k - 1$ zeros, followed by a 1 in the kth position. There are 2^{n-k} binary sequences in which $T_1 = k$; the first k terms are fixed and the remaining $n - k$ terms are arbitrary. Therefore, by classical probability, $P(T_1 = k) = 2^{-k}$ for $k \leq n$. Of course, with probability 2^{-n}, we may have a sequence of all zeros and $T_1 = \infty$. If we let $n \to \infty$, then T_k has a nonzero probability for all positive values of k, $P(T_1 = k) = 2^{-k}$, $k \geq 1$, and the waiting time T_1 to a first occurrence of a 1 has a Geometric model, albeit the variant that starts at $k = 1$ rather than at $k = 0$.

We generalize this argument for the case of an unlimited number of unlinked repetitions $\mathcal{E}_1, \ldots, \mathcal{E}_n, \ldots$ of a random experiment $\mathcal{E} = (\Omega, \mathcal{A}, P)$. Select an event $A \in \mathcal{A}$ with $P(A) = 1 - \beta$ or $\beta = P(A^c)$, and define K to be the random *delay* until the first appearance of A (i.e., $K = k \geq 0$ if A appears as an outcome of $\mathcal{E}_{k+1}$ and is not an outcome of $\mathcal{E}_j$ for any $j \leq k$. (We could also have chosen to define the *waiting time* as $K = k$ if A is first observed on $\mathcal{E}_k$. In this case, $\mathcal{G}(\beta)$ is adjusted so that $P(k = k) = (1 - \beta)\beta^{k-1}$ for $k = 1, 2, \ldots$). This construction will be the basis for a derivation of $\mathcal{G}$ in Section 14.5.1.

For the case when we start counting at 0, the parameter β relates to the mean m or average waiting time through

$$\beta = \frac{m}{m + 1}.$$

When we start counting at 1, m is replaced by $m - 1$ to yield

$$\beta = \frac{m - 1}{m}.$$

These results can be established once we have defined means in Section 9.3. At present, they provide a second interpretation of the parameter β that can help us in modeling applications of the kind described previously.

Example 4.5   Geometric Waiting Time
We are interested in the first occurrence of a "6" in repeated tosses of an unsymmetrical die. We are told that the average delay ($T = 0$ if this happens on the first trial) $m = 6.5$ to this first occurrence. What is the probability $P(T = 0)$ that the delay T is 0 or that "6" is observed on

the first toss? For $\mathcal{G}(\beta)$,

$$\beta = \frac{m}{1+m} = \frac{6.5}{1+6.5} = \frac{13}{15}.$$

Hence,

$$P(T = k) = (1 - \beta)\beta^k,$$

and

$$P(T = 0) = 1 - \beta = \frac{2}{15}.$$

The most probable value of T is always at 0 for the Geometric.

Example 4.6 Geometric Waiting Time (Continued) _____

If, in the preceding example, we were asked for the smallest value of k such that it is more likely that $T \leq k$ than that $T > k$, then

$$P(T \leq k) \geq P(T > k) = 1 - P(T \leq k) \iff P(T \leq k) \geq \frac{1}{2},$$

and we seek the smallest such k satisfying

$$\sum_{j=0}^{k}(1 - \beta)\beta^j \geq \frac{1}{2}.$$

Recall that for the geometric series

$$\sum_{j=0}^{k}\beta^j = \frac{1 - \beta^{k+1}}{1 - \beta}.$$

Hence, we seek the smallest k such that

$$1 - \beta^{k+1} = 1 - \left(\frac{13}{15}\right)^{k+1} \geq \frac{1}{2}.$$

Calculation reveals that this occurs for $k = 4$. For this example, it is more probable than not that the delay will be no more than 4 or that a "6" will appear no later than the fifth toss.

4.4.4 Poisson $\mathcal{P}(\lambda)$

Definition 4.4.4 (Poisson PMF $\mathcal{P}(\lambda)$)   The Poisson pmf $\mathcal{P}(\lambda)$ is specified over all of the nonnegative integers, in terms of a parameter $\lambda \geq 0$, through

$$p_k = e^{-\lambda}\frac{\lambda^k}{k!}, \quad k = 0, 1, \ldots.$$

If $\lambda < 0$, then $p_1 < 0$, thereby contradicting this as a pmf. Hence, λ must be nonnegative. If $\lambda = 0$, then $P(\omega = 0) = p_0 = 1$. To verify that $\mathcal{P}$ sums to 1, we recall the power series for the exponential

$$e^x = \sum_{k=0}^{\infty} \frac{x^k}{k!}.$$

Hence,

$$\sum_{k=0}^{\infty} p_k = \sum_{k=0}^{\infty} e^{-\lambda} \frac{\lambda^k}{k!} = e^{-\lambda} \left[\sum_{k=0}^{\infty} \frac{\lambda^k}{k!} = e^\lambda \right] = 1,$$

as required for a pmf.

The Poisson is about counting the number of occurrences of random events in a given time T. $\mathcal{P}(\lambda)$ models the following phenomena:

- the number of photons emitted by a light source of intensity I photons/sec in time T seconds ($\lambda = IT$);
- the number of atoms of radioactive material of half-life L_h and mass m undergoing decay in time T ($\lambda \propto mT/L_h$) (the mass m determines the total number of atoms through the atomic weight of the material and Avogadro's number);
- the unit of radiation given by the *curie* corresponds to the radiation given off by one gram of radium and equals 3.7×10^{10} emissions/second; if the time interval is 1 nanosecond, then $\lambda = 37$;
- number of dopant atoms deposited to make a nanosized device such as an FET;
- number of customers arriving in a queue or workstations requesting service from a file server in time T ($\lambda \propto T$);
- number of occurrences of rare events in time T ($\lambda \propto T$).

The parameter λ interprets as the mean or average number of counts and can be considered as a product IT of a particle source intensity or current I and an observation time T.

Examining the ratio of successive values of the pmf

$$\frac{p_{i+1}}{p_i} = \frac{\lambda}{i+1}$$

shows it to be strictly decreasing in i. Thus, $\{p_i\}$ is decreasing throughout if $\lambda < 1$, decreasing after $p_0 = p_1$ for $\lambda = 1$, and initially increasing if $\lambda > 1$ and eventually decreasing no matter what the value of λ. A most probable value (there may be equal probabilities for immediately adjacent values) is k^* if

$$p_{k^*+1} \leq p_{k^*} \text{ and } p_{k^*-1} \leq p_{k^*}.$$

Equivalently,

$$k^* \leq \lambda \leq k^* + 1$$

or

$$\lambda - 1 \leq k^* \leq \lambda.$$

We take $k^* = \lfloor \lambda \rfloor$ for definiteness. (Note that $\lfloor \lambda \rfloor$ is the integer part of λ; for example, $\lfloor 3.9 \rfloor = 3$.) The corresponding peak value of the pmf, which may be shared between two terms, is

$$p_{\lfloor \lambda \rfloor} = e^{-\lambda} \frac{\lambda^{\lfloor \lambda \rfloor}}{\lfloor \lambda \rfloor!} \approx \frac{1}{\sqrt{2\pi\lambda}},$$

where the last term follows by assuming $\lambda \gg 1$ and by using Stirling's formula for factorial. (See Section 2.3.5.) The Poisson model will be derived from the Binomial in Section 4.5.

Example 4.7 Poisson Current Model _____

The random current I crossing a junction has an average of $\overline{I}$ electrons per unit time. What is the probability that $E = k$ electrons cross the junction in time τ? The Poisson is the model with $\lambda = \overline{I}\tau$. Hence,

$$P(E = k) = e^{-\overline{I}\tau} \frac{(\overline{I}\tau)^k}{k!}.$$

The most probable number of electrons to cross in time τ is $\lfloor \overline{I}\tau \rfloor$ or close to the average value.

Example 4.8 Poisson Current Model (Continued) _____

If $\overline{I}$ in the previous example is unknown, but we know that it is as probable that no electrons will pass in time τ as that one electron will pass, then

$$P(E = 0) = e^{-\lambda} = P(E = 1) = \lambda e^{-\lambda}.$$

Hence, we find that $\lambda = 1$ and

$$P(E = k) = e^{-1} \frac{1}{k!}.$$

4.4.5 Zeta or Zipf $\mathcal{Z}(\alpha)$

Definition 4.4.5 (Zeta or Zipf PMF $\mathcal{Z}(\alpha)$)   The Zeta or Zipf $\mathcal{Z}(\alpha)$, for $\alpha > 1$, is specified for the positive integers by

$$p_k = \frac{k^{-\alpha}}{\zeta(\alpha)}, \quad k = 1, 2, \ldots.$$

$\zeta(\alpha)$ is the Riemann zeta function.

The unit normalization is ensured by division by

$$\zeta(\alpha) = \sum_{k=1}^{\infty} k^{-\alpha},$$

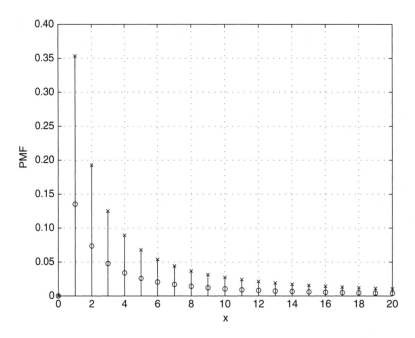

Figure 4.4   Zeta or Zipf pmf for $\alpha = 3/2$ (circle) and $\alpha = 4$ (x).

a well-studied function known as the Riemann zeta function that has a closed form expression only in certain cases. The sum in the definition of the zeta function converges if and only if $\alpha > 1$. For this choice of parameter α, it is immediate that $\mathcal{Z}(\alpha)$ is positive and sums to 1, as is necessary and sufficient for a pmf. It is evident that the pmf sequence $\{p_k\}$ is strictly decreasing for any $\alpha > 1$. A plot of $\mathcal{Z}(1.5)$ and $\mathcal{Z}(4)$ is shown in Figure 4.4. The $\alpha = 4$ pmf decreases more rapidly than the $\alpha = 3/2$ pmf and for large enough k will become smaller.

The most probable value is easily seen to be the initial value of 1.

The Zeta or Zipf is also called a *heavy-tailed*, *power law*, or *scale-free* probability distribution of pmf. "Heavy tailed" and "power law" refer to its rate of decrease with increasing k being only algebraic and hence much slower than exponential. "Scale free" refers to the absence of a scale parameter such as λ in the Poisson. This heavy-tailed model is a simple example of a class of probability laws whose importance has increased dramatically since the mid-1990s (e.g., see Axtell [5] for references and distribution of firm sizes and Adler et al. [2] for communications and financial applications). Variables encountered in modern computing and communication systems and in a host of other phenomena (e.g., those arising from self-organized criticality) have required such heavy-tailed models. These applications of the Zeta include the following:

- number of customers affected by a power system blackout;
- the distribution of Unix file sizes stored on file servers (see [2]);
- sizes of files requested in Web transfers ([2]);
- randomly selected Web pages on the WWW (see Section 2.6.5) have indegree and outdegree that are described by the Zeta pmf (see Section 2.6.6);

- packet delays on the Internet ([2]);
- running times for NP-hard problems as a function of certain problem complexity parameters reveals a heavy-tailed distribution for such running times.

The continuous Pareto distribution discussed in Section 6.6.3, introduced by Pareto to describe the distribution of wealth, is the heavy-tailed model that has been in longest use. Recent applications to foreign currency exchange rates can be found in [2]. While it is doubtful that any of the aforementioned heavy-tailed phenomena are exactly describable by a Zeta or Zipf distribution, these distributions provide a good approximation. We will use $\mathcal{Z}$ for a discrete version of the Pareto whenever we are modeling a counting variable that is heavy tailed.

Example 4.9   Zeta/Zipf Graph Degree Distribution _____

We observed in Section 2.6.5 that a randomly selected Web page will have an indegree I of size d that can be idealized as having probability

$$P(I = d) \propto \frac{1}{d^{2.1}}.$$

Recognizing that indegree takes on the positive integer values and that a pmf sums to 1, we see that

$$P(I = d) = \frac{d^{-2.1}}{\zeta(2.1)}.$$

Thus,

$$\frac{P(I = 1)}{P(I = 10)} = 10^{2.1} \approx 126.$$

4.5 POISSON AS A RARE EVENTS LIMIT OF THE BINOMIAL

The Poisson can be derived from the Binomial through the so-called *rare events limit*. Consider the binomial $\mathcal{B}(n, p_n)$, where, for an experiment with outcomes $\{0, 1, \ldots, n\}$, we let the probability p_n depend on n. If we choose this dependence such that

$$\lim_{n \to \infty} n p_n = \lambda > 0,$$

then the sequence of binomials $\mathcal{B}(n, p_n)$ converges to $\mathcal{P}(\lambda)$ in the sense that the pmf $p_{k,n}$ of the $\mathcal{B}(n, p_n)$ converges to $e^{-\lambda} \frac{\lambda^k}{k!}$. Basically, n has to be very large and p has to be appropriately small. This is called the "rare events" limit because it can be interpreted as applying to the number of occurrences of a rare (p small) event in a very large population n of individuals or trials.

Theorem 4.5.1   (Rare Events Limit)   If $\lim_{n \to \infty} n p_n = \lambda > 0$, then

$$\lim_{n \to \infty} \binom{n}{k} p_n^k (1 - p_n)^{n-k} = e^{-\lambda} \frac{\lambda^k}{k!}.$$

Proof. For n large and k fixed, we have

$$\frac{(n-k)^k}{k!} < \binom{n}{k} = \frac{n!}{k!(n-k)!} < \frac{n^k}{k!},$$

$$\lim_{n\to\infty} \left(\frac{n-k}{n}\right)^k = 1.$$

Hence,

$$\binom{n}{k} p_n^k (1-p_n)^{n-k} \approx \frac{(np_n)^k}{k!}(1-p_n)^{n-k},$$

with the approximation "$\approx$" understood as the ratio of the two sides tends to unity as n tends to infinity. By hypothesis, $\lim_{n\to\infty} np_n = \lambda > 0$. Thus, $p_n \to 0$, and we can approximate[†]

$$(1-p_n)^{n-k} \approx e^{-(n-k)p_n}.$$

Taking limits yields

$$\lim_{n\to\infty} \binom{n}{k} p_n^k (1-p_n)^{n-k} = e^{-\lambda} \frac{\lambda^k}{k!},$$

which is the Poisson pmf. $\square$

   This limit also provides us with a basis for recognizing when the Poisson will be applicable. For example, in the case of radioactive decay, we are dealing with a very large number of atoms in even a small sample of radioactive material, with atoms decaying independently and with a very small probability of any particular atom decaying in a given moderate time. A similar argument applies to photons emitted by a large population of potential emitters/atoms, each of which is unlikely to emit a photon in the time interval over which we are counting/observing. The first use of the Poisson model is said to have been by a Prussian physician, von Bortkiewicz, who found that the annual number of late-19th-century Prussian soldiers kicked to death by horses followed a Poisson distribution—a large number of soldiers interacting with horses and a low incidence of fatal interactions.
   There is yet another way to derive the Poisson probability law. The Poisson describes the number of events that occur in a given time T if we assume that the waiting times between successive events (e.g., customer arrival) are independent (unlinked, see Chapter 14) and that they can be described by an exponential density. (See Section 6.6.1.) This property is examined in Section 20.7.1.

[†]For $x \geq 0$, $1 - x \leq \exp(-x)$ and, hence, $(1-x)^n \leq \exp(-nx)$. For $0 \leq x \leq 1/2$ (the value of $1/2$ is one for which the result is easily shown to be correct, although a larger value of about .68 will also do), $-x - x^2 \leq \log(1-x)$, implying $(1-x)^n \geq \exp(-nx - nx^2)$. Under our assumption, $np_n \to \lambda > 0$, it follows that $np_n^2 \to 0$, and our approximation is justified by these converging upper and lower bounds.

4.6 SUMMARY

Alternative means of describing the same object are of great value in gaining an understanding of the object and are often essential to success in working or computing with the object. The applications of probability that motivate our study of this subject find their successes largely through convenient means of specifying probability measures and reliance upon a handful of specific families of probability measures that occur frequently in practice. This chapter treated the specification of a probability measure P, when the sample space Ω is either finite or countably infinite, through a probability mass function (pmf) p, a function that specifies the probabilities of each of the individual elements of Ω. Given the pmf, we can construct $P(A)$ by summation of the pmf over the elements in the set A

$$P(A) = \sum_{\omega \in A} p(\omega).$$

Given the measure P, we obtain the pmf through

$$p(\omega) = P(\{\omega\}).$$

Hence, there is a one-to-one relationship between probability mass functions and probability measures on the power set (set of all subsets) of countable Ω. Probability mass functions that are particularly important in practice are the random, binomial, geometric, Poisson, and Zeta/Zipf; and these are defined and their applications and basic properties established.

As we will do subsequently for other probability models, we show that one model (Poisson) can arise from another model (binomial). These demonstrations enrich our understanding of the meaning and conditions for applicability of the individual models.

EXERCISES

E4.1   If $\Omega = \{\omega_1, \omega_2, \omega_3\}$, $\mathcal{A} = 2^{\Omega}$, and

$$P(\{\omega_1, \omega_2\}) = .65, P(\{\omega_2, \omega_3\}) = .85,$$

then determine the pmf $\{p_i\}$ and $P(\{\omega_1, \omega_3\})$.

E4.2   In a certain communication system a byte (8 bits) is transmitted with a bit error probability of .1. If the system can correct at most one error made in each byte, what is the probability of a message being received correctly? What is the most probable number of errors in a byte?

E4.3   If the most probable number of errors in a byte is 2, what can you say about the probability of no errors?

E4.4   Use the binomial pmf to evaluate the probability that, in a message of length 5, there will be fewer than two errors when the probability of an error on a single symbol is 0.1.

E4.5   If a byte is as likely to contain exactly three errors as it is to contain exactly four errors, what is the probability of its containing exactly three errors?

E4.6   If all bytes are equally likely to be received, evaluate the probability of receiving a byte having at least four more ones than zeros.

E4.7   In many samples of text of length 8 characters, it is observed that the most probable number of errors is 2.

    a. What probability model do you suggest for the number X of errors?

    b. Is a per symbol error probability of 1/3 compatible with the observed most probable number of errors being two?

    c. If the per symbol error probability is 1/3, then what is the probability that the number of errors is two?

E4.8   In a certain bit transmission system in which either a "0" or a "1" is sent, it is observed that the probability of making two errors in 10 bits is .0045. What is the probability of making an error on the first bit?

E4.9   In the transmission of a packet of 100 bits that is received with possible errors, the probability of there being exactly one error in the packet is twice the probability of there being exactly two errors. Selecting an appropriate probability model and making reasonable approximations, what is the average number of errors?

E4.10 A packet of length 100 bits is transmitted over a noisy channel with error correction that can correct no more than two errors. The probability that the ith bit will be in error is .01.

    a. What is the probability p_c that the packet, after possible error correction, will be received correctly?

    b. What is the most probable number of errors?

E4.11 A communications channel transmits a block of S symbols in time T, with the probability p of a given symbol being in error. An error correction system allows the receiver to correct up to k errors, and if more than k errors are made over the channel, then the block is incorrectly decoded by the receiver. What is the probability that the block will be correctly decoded if $S = 5$, $k = 1$, $p = .01$?

E4.12 A sequence of 10 unlinked (independent) bytes has been received. It is known that the probability is .3 that a first symbol in a byte is a 0. Let K be the number of received bytes having a 0 as first symbol. What is $P(K = 2)$?

E4.13 In a certain communication system, the probability is .1 that a packet will arrive that is in error.

    a. What is the probability of no errors in the next 10 packets received?

    b. What is the most probable number of errors made in the 10 packets and how probable is it?

E4.14 In a long sequence of n calls to computer memory, we find that calls to cache and calls to main memory are made so that the probability of any particular sequence of these n calls involving exactly c calls to the cache is given by $p^c(1 - p)^{n-c}$, for known $0 < p < 1$.

    a. What is the probability of there being exactly c calls to the cache without regard to the sequencing of the calls?

    b. What is the most probable number of calls to cache?

    c. Show that if $c_1 \leq n_1 < n$, then the probability of c_1 calls to cache in the first n_1
       calls to memory is $\binom{n_1}{c_1}p^{c_1}(1-p)^{n_1-c_1}$.

    d. What is the probability that the first t calls are only to the cache?

    e. What is the probability of the tth call being the first call to main memory?

E4.15 A component has a random lifetime L, measured in units of days, such that the probability
$P(L = n)$ of failure on day n is given by the geometric distribution $\mathcal{G}(\beta)$,

$$P(L = n) = (1 - \beta)\beta^n, n \in \mathbb{N}.$$

If $\beta = \frac{1}{2}$, then calculate the probability of the event F_i of no failure before day i,

$$F_i = \{i, i + 1, i + 2, \ldots\} = \{n : n \geq i\}.$$

(Recall that for $0 < b < 1$, $\sum_{k=n}^{\infty} b^k = b^n/(1-b)$.)

E4.16 If the mean or average lifetime L of a component is 10 days, then what is an appropriate
probability model for L?

E4.17 A device that fails without aging effects is known to have a mean lifetime of 3 hours.
What is the probability that it will last either less than 1 or more than 5 hours?

E4.18 What is the pmf for the trial number T to the first occurrence of a "3" in repeated tosses
of a fair die?

E4.19 Let $T = 1, 2, \ldots$ be the trial number on which a 5 or a 6 first appears in successive
tosses of a fair die. What is $P(T = k)$?

E4.20 In a particular implantation process the average number of ions that are implanted in
a small surface is 100. The device works successfully if the number of ions actually
implanted is in the range $[80, 120]$. Provide an expression for the probability $P(S)$ that
a sample device will work successfully.

E4.21   a. What is the probability that a laser emitting an average of γ photons/sec will be
       undetectable because it emits no photons in time T?

    b. What is the most probable number of photons emitted by this source in time T and
       how probable is it?

E4.22 If the average number of clients requesting service from a file server in one minute is 3,
what is the probability of no clients requesting service in one minute?

E4.23 An optical communications receiver is confronted with a light source such that, over
a 1-μsec interval, we are twice as likely to observe 3 photons as we are to observe 4
photons.

    a. What is the most likely number of photons to be observed in 1-μsec?

    b. What is the probability that we will observe any photons in a 1-μsec interval (detect
       the presence of the light source)?

E4.24 If the probability of 0 photons being emitted in time T is .1, then what is the probability
of at least 2 photons being emitted in T?

E4.25 The probability of observing 2 photons in a given time is twice the probability of observing
3 photons. What is the most probable number of photons to be observed?

E4.26 If a Geiger counter measures an average rate of $2\frac{1}{2}$ decays/second, then what is the
probability of its measuring no decays in a given time interval of length 1 second? What
is the most probable number of particles to be observed?

E4.27 The number K_T of ions emitted from a source in time T satisfies $P(K_T = 1) = .1$.

   a. What is the mean or average number of ions emitted in T?
   b. What is the probability $P(K_{2T} = 2)$ of 2 ions arriving in time $2T$?

E4.28 In a randomly generated binary sequence of length m, the probability of a run of zeros of exact length $r(< m - 1)$, starting at position k, is $p^r(1 - p)^2$ if $1 < k < m - r$, the probability is $p^r(1 - p)$ if $k = 1, m - r$ and is zero if $k > m - r$. Use Boole's Inequality to upper-bound the probability that the binary sequence will contain a run of exact length r.

E4.29 The number N of voice communication packets arriving in time T is such that $P(N = 0) = P(N = 1)$. What is the probability description of N?

E4.30 If the probability is .9 of observing at least one emitted photon in 1 second, what is the probability of observing exactly one photon in 1 second? What is the most probable number of photons to be observed in 1 second?

E4.31 We know that the probability is e^{-10} of no packets being received by a client computer in 1 second.

   a. What is the probability law for the random variable N of number of packets received in 1 second?
   b. What is the probability of no more than two packets being received in 1 second?
   c. What is the most probable number n^* of packets to be received in 2 seconds and how probable is it?

E4.32 The number X of ions implanted in 1 second in a small area a of a device is such that $P(X = 2) = 3P(X = 3)$.

   a. Fully specify the probability mass function p_X.
   b. What is the probability of at least three ions being implanted?

E4.33 Specify the pmf needed to model the following random sources:

   a. number N of clicks in a Geiger counter in 5 seconds when the average number of clicks in 1 second is 3;
   b. number N of errors made in transmitting a text of 100 characters when the average number of errors in such a text is 1;

E4.34 The probability of five customers arriving in a queue in 2 minutes is twice as great as the probability of two customers arriving in 2 minutes. What is the probability of no customers arriving in 5 minutes?

E4.35 What is the probability that at least two packets will be received in 3 milliseconds when the average number of packets received in 1 millisecond is two?

E4.36 a. Numerically evaluate p_1 for $\mathcal{Z}(3/2)$.
   b. What is p_2?

E4.37 The sizes of files stored on a large Unix file system follows a $\mathcal{Z}(\alpha)$ pmf, when file sizes are measured in units of kilobytes.

   a. If file sizes of 1 KB are 10,000 more probable than file sizes of 1 MB, then what is the parameter α?
   b. How much more probable are file sizes of 1 MB than those of 1 GB?

E4.38 If the probability p_1 that a file requested on the Web will have size 1 KB is 10 times as large as the probability p_2 that the file size is 2 KB, then compare p_1 with the probability p_8 that the file size is 8 KB.

E4.39 The random number N of repeated, independent experiments $\mathcal{E}_1, \ldots, \mathcal{E}_N$ that must be performed until we first observe the rth occurrence of a specified event A of probability $P(A) = p$ is described by the *negative binomial pmf* given by

$$P(N = n) = p_n^{(r)} = \binom{n-1}{r-1} p^r (1-p)^{n-r}, \ n = r, r+1, \ldots.$$

This pmf can be understood by realizing that any specific sequence of r occurrences of A and $n - r$ occurrences of A^c has probability $p^r (1-p)^{n-r}$. However, if n is to be the time of occurrence of the rth appearance of A, then there had to be $r - 1$ occurrences in the preceding $n - 1$ trials and the rth occurrence of A precisely on the nth experimental outcome. There are $\binom{n-1}{r-1}$ sequences with this property.

  a. If $r = 1$, relate this to the Geometric pmf $\mathcal{G}(\beta)$.
  b. For a given p and r, what is the most probable value of N?

E4.40 a. What is the probability that a fourth "head" will appear on the eighth toss of a fair coin?
  b. What is the probability if the coin is biased with probability 1/3 for "head"?

5

Describing Probability II: Uncountable Ω and Distributions

5.1 PURPOSE, BACKGROUND, AND ORGANIZATION

If Ω is not countable, then it is uncountable. We extend the discussion begun in the previous chapter, for countable Ω, to include sample spaces that are uncountably infinite. A basic example of an uncountable set is the set of real numbers that is a nonempty interval $[a, b] = \{x : a \leq x \leq b\}$. More complex instances of uncountable sample spaces are encountered in the study of random or stochastic processes (see Chapter 20) where the sample space is a space of functions. The only examples of uncountable sample spaces that we assume up to Chapter 19 are either the real numbers $\mathbb{R} = (-\infty, \infty)$ or n-tuples (vectors $\mathbf{x}$) of reals

$$\mathbb{R}^n = (-\infty, \infty)^n = \{\mathbf{x} = (x_1, \ldots, x_n) : (\forall i \leq n) \, x_i \in \mathbb{R}\}.$$

In many applications, we will only be interested in sample spaces that are finite subintervals of the reals or in the nonnegative reals $[0, \infty)$. However, the techniques we introduce allow us to consider these uncountable subsets of the reals as subsets of the larger set $\mathbb{R}$.

The usual event algebra $\mathcal{A}$ in this case is called the *Borel algebra* $\mathcal{B}$ and is the smallest σ-algebra (algebra closed under formation of sets through complements, countably many unions, and intersections) containing all of the finite intervals in $\mathbb{R}$. We will not concern ourselves further

156

with $\mathcal{A}$ beyond noting that it contains all the sets that all but the mathematically trained or gifted can think of.

Our engineering interest in probability for uncountable sample spaces is in modeling random experiments having real-valued outcomes. More will be said about the terminology *random variable* in Section 8.2. For our present purposes, it simply refers to a numerical (integer or real-valued) quantity such as an instantaneous thermal noise voltage V at a node in a circuit, the magnitude of electric field strength E (volts/meter) at an antenna, the weight W of an individual, the number N of photons emitted by a source in a given time period, or the waiting time T to the occurrence of an event such as the next Internet packet receipt, electron emission, or system failure. We commonly denote random variables by uppercase roman letters drawn from the end of the alphabet (e.g., $X, Y, Z, X_1, \ldots, X_n$). A random experiment $\mathcal{E}_X = (\mathbb{R}, \mathcal{B}_X, P_X)$ for a numerical random variable X is described by the sample space $\mathbb{R}$ (used even when the outcomes might be only integers), the Borel algebra $\mathcal{B}_X$ of events (appropriate subsets of $\mathbb{R}$), and by a probability measure P_X defined on $\mathcal{B}_X$ that satisfies the Kolmogorov axioms. We will often delete the identifying subscript X when there is no confusion about the random variables being described.

Probability measures defined on the huge collection of events in $\mathcal{B}$, however, are quite complicated and difficult to specify. A key to the successful use of probability in engineering and the sciences has been the ability to conveniently specify probability measures through much simpler alternative representations. This is the subject of Chapters 4, 5, 6, 7, and 15. The approach we take was demonstrated first in Chapter 4, where we introduced the pmf p defined on Ω, rather than on the power set $\mathcal{A} = 2^\Omega$, and were then able to recover the full $P(A) = \sum_{\omega \in A} p(\omega)$. We will focus on describing the probability measure P_X only those events in $\mathcal{B}_X$ determined by the outcomes of the random variable X being less than or equal to a prescribed value x.

For random variables, a basic and simple way to describe their governing probability measure P is through a cumulative distribution function. Derivatives of cumulative distribution functions provide access in Chapter 6 to probability density functions that are the analogs of the probability mass functions of Chapter 4. Cumulative distribution functions and probability density functions provide the working methods for dealing with probability measures describing random experiments. As is our expository pattern, we define cumulative distribution functions, provide examples to clarify the definition, extract the elementary implications of the definition, and then provide some deeper insight into the nature and role of cumulative distribution functions. We close with the example of a cumulative distribution function that is based on data and used to estimate an unknown cumulative distribution function.

5.2 CUMULATIVE DISTRIBUTION FUNCTIONS

Introduce the *corner* set (this name is better justified by the generalization to several dimensions given in Section 7.2)

$$C_x = \{X : X \le x\},$$

often abbreviated to $X \in (-\infty, x]$ or just $(-\infty, x]$. The first alternative specification for a random variable X is provided by the important concept of a *cumulative distribution function (cdf)*, also called a *distribution function*, and is a restriction of the probability measure P to just those events of the form C_x.

Definition 5.2.1 **(CDF)**   The *univariate cumulative distribution function*

$$F_X : \mathbb{R} \to [0, 1],$$

$$F_X(x) = P(X \in (-\infty, x]) = P(\{X : X \leq x\}) = P(X \leq x).$$

Throughout this text, we will make use of the *unit step function* denoted by

$$U(x) = \begin{cases} 0 & \text{if } x < 0 \\ 1 & \text{if } x \geq 0 \end{cases}.$$

Unlike the many other appearances of the unit step in electrical engineering, defining $U(0) = 1$ is essential to us.

Example 5.1   CDF of a Constant

The first example corresponds to a random variable X that is constant at the value x_0, $P(X = x_0) = 1$,

$$F_X(x) = P(X \leq x) = \begin{cases} 0 & \text{if } x < x_0 \\ 1 & \text{if } x \geq x_0 \end{cases} = U(x - x_0).$$

Example 5.2   CDF of the Binomial

A less trivial example is provided by the binomial $\mathcal{B}(2, 1/5)$ probability model,

$$P(X = 0) = \frac{16}{25}, \ P(X = 1) = \frac{8}{25}, \ P(X = 2) = \frac{1}{25},$$

$$F_X(x) = P(X \leq x) = \begin{cases} 0 & \text{if } x < 0 \\ \dfrac{16}{25} & \text{if } 0 \leq x < 1 \\ \dfrac{24}{25} & \text{if } 1 \leq x < 2 \\ 1 & \text{if } 2 \leq x. \end{cases} = \frac{16}{25}U(x) + \frac{8}{25}U(x-1) + \frac{1}{25}U(x-2).$$

The cdf F_X is plotted in Figure 5.1, with circles showing the values of the function at the points of discontinuity of the cdf.

Example 5.3   CDF of the Uniform and Exponential

A probability measure P describing the random choice of a number from the unit interval $[0, 1]$, as described in Section 3.5.2, has a cdf given by (see Figure 5.3)

$$F_X(x) = \int_0^1 I_{(-\infty, x]}(y) \, dy = \begin{cases} 0 & \text{if } x < 0 \\ x & \text{if } 0 \leq x < 1 \\ 1 & \text{if } 1 \leq x. \end{cases}$$

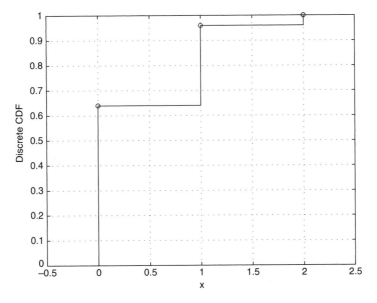

Figure 5.1   Discrete cdf.

If

$$P(A) = \int_A e^{-x} U(x)\, dx,$$

then the cdf for the exponential is given by

$$F_X(x) = \int_0^\infty I_{[0,x]}(y) e^{-y}\, dy = \begin{cases} 0 & \text{if } x < 0 \\ 1 - e^{-x} & \text{if } x \geq 0 \end{cases} = (1 - e^{-x}) U(x).$$

The restriction of P to semi-infinite interval events of the form $X \leq x$ yields the cdf F_X. Surprisingly, while F_X only directly determines the probability measure P on the semi-infinite intervals, this suffices to determine P for all events involving the random variable X. The next theorem restates this conclusion. For our purposes, we can replace the term "Lebesgue–Stieltjes" by the more familiar "Riemann."

Theorem 5.2.1   (CDF Determines Measure)   We can uniquely recover the probability measure P_X defined on the Borel algebra $\mathcal{B}_X$ from its restriction to a cdf F_X through the (Lebesgue–Stieltjes) integral

$$P_X(A) = P_X(X \in A) = \int_{-\infty}^\infty I_A(x)\, dF_X(x).$$

Proof remark. It is beyond the scope of this text to prove this result. However, we will motivate it in the next section.   □

If the cdf $F_X(x)$ has, for example, a piecewise continuous derivative $f_X(x)$ of which it is the integral, and the set A is a finite union of intervals, then we can replace the Lebesgue–Stieltjes integral by the common Riemann integral

$$P(A) = \int_{-\infty}^{\infty} I_A(x) f_X(x)\, dx = \int_A f_X(x)\, dx.$$

Hence, the cdf F_X contains a full probabilistic description of the random variable X. However, the scalar function of a scalar variable, $F_X(x)$, is a much simpler and more familiar entity than the set function $P_X(A)$. The cdf is defined for all real numbers $x \in \mathbb{R}$ whereas the probability measure is defined for all sets A in the far more complex entity $\mathcal{B}_X$. The cdf in turn gives rise to pmfs and to probability density functions (pdfs) (introduced in the next chapter). It is the pmf and the pdf that provide the working basis for engineering applications of probability.

5.3 PROPERTIES OF UNIVARIATE CDFs

We establish the elementary properties of a cdf F that follow from its relationship, given in Definition 5.2.1, to a probability measure P satisfying axioms K0–K4. Theorem 5.2.1 guarantees that the cdf F_X defined by a given probability measure P_X for the random variable X in turn defines precisely the same measure P_X. Theorem 5.3.1 will establish the different result that any function satisfying certain of the properties of a cdf is itself a cdf and therefore defines a probability P_X satisfying the Kolmogorov axioms. For notational simplicity, in what follows we drop the subscript X as being understood.

Because $F(x)$ is a probability, it follows that

$$1 \geq F(a) \geq 0.$$

For $a < b$, it follows from finite additivity K3, via

$$(-\infty, b] = (-\infty, a] \cup (a, b] \Rightarrow F(b) = F(a) + P((a, b]),$$

that

$$P((a, b]) = F(b) - F(a). \tag{5.1}$$

Equation (5.1) and nonnegativity (K2) of $P((y, x])$ yield

$$x > y \Rightarrow F(x) \geq F(y). \tag{5.2}$$

This establishes that F is nondecreasing.

The cdf is a nonnegative, bounded by 0 and 1, nondecreasing function F that can have only positive jump discontinuities of finite height. It follows that it must have limits from both the right,

$$F(x^+) = \lim_{n \to \infty} F\left(x + \frac{1}{n}\right), \tag{5.3a}$$

and from the left,

$$F(x^-) = \lim_{n \to \infty} F\left(x - \frac{1}{n}\right). \tag{5.3b}$$

A consequence of Eq. (5.3a) is that

$$n < m \Rightarrow A_m = \left\{\omega : \omega \le x + \frac{1}{m}\right\} \subset A_n, \ \cap_n A_n = (-\infty, x],$$

and K4 is

$$\text{Right Continuity:} \quad F(x^+) = F(x). \tag{5.4}$$

The cdfs described in Examples 5.1 and 5.2 display the property of right continuity. The cdf of Example 5.3 is continuous and therefore satisfies right continuity (as well as left continuity).

The probabilities of each of the four kinds of intervals are given by

$$P((a, b]) = F(b) - F(a), \ P((a, b)) = F(b^-) - F(a), \tag{5.5a}$$

$$P([a, b)) = F(b^-) - F(a^-), \ P([a, b]) = F(b) - F(a^-). \tag{5.5b}$$

We observe from Eq. (5.5b) that, if $F(x)$ has a jump discontinuity at $x = a$, then

$$P(X = a) = P([a, a]) = F(a) - F(a^-);$$

the jump height at a *is the probability with which* a *is taken on exactly by* X.

If $F(x)$ is continuous at a, then the jump height is 0 and $P(X = a) = 0$.

Letting $A_n = \{X : X \le -n\}$ and observing that $A_{n+1} \subset A_n$, $\cap_n^\infty A_n = \emptyset$, enables us to invoke K4 to establish that

$$\lim_{n \to \infty} F(-n) = \lim_{n \to \infty} P(A_n) = 0.$$

Because $F(x)$ is nonnegative and nondecreasing, we can conclude that

$$\lim_{x \to -\infty} F(x) = 0. \tag{5.6a}$$

Similarly, one shows that

$$\lim_{x \to \infty} F(x) = 1. \tag{5.6b}$$

Thus, not only does any cdf F lie in the unit interval [0, 1], but the values of 0 and 1 are taken on in the appropriate limit (and sometimes for finite values of x).

The next theorem establishes when a function F is a cdf, and therefore determines a probability measure P.

Theorem 5.3.1   (Characterization of a CDF) If $\Omega = \mathbb{R}$, $\mathcal{B}$ is the Borel algebra of subsets of $\mathbb{R}$, and a function F satisfies Eqs. (5.2), (5.4), and (5.6), then F is a cdf for some probability measure P on $\mathcal{B}$.

Remark. While we do not offer a proof of this theorem, the necessity for a cdf to satisfy properties (5.2), (5.4), and (5.6) has been established in the observations preceding the statement of the theorem. The sufficiency of these conditions (i.e., given a function F satisfying the conditions, there exists a random variable X having F for its cdf) will follow from the ability to be developed in Chapter 8 to construct such a random variable starting from one having a uniform distribution.

Example 5.4   CDF Characterization _____

An example of the use of Theorem 5.3.1 is that it assures us that the function

$$F(x) = \frac{1}{1 + e^{-x}},$$

which satisfies Eqs. (5.2), (5.4), and (5.6), is a cdf and defines a probability measure P even when we did not start out knowing P.

We can construct a new cdf F from given cdfs $\{F_i\}$ through a process of convex combination that, as explained in Section 3.9, corresponds to a compound experiment.

Theorem 5.3.2   (CDF Closure under Convex Combinations)   A convex combination F of cdfs $\{F_i\}$ is a cdf.

Proof. We need to verify that

$$F(x) = \sum_{i=1}^{n} \lambda_i F_i(x)$$

is a cdf by proving that F has the properties given by Eqs. (5.2), (5.4), and (5.6). Because each $F_i(x)$ is nondecreasing in x and the λ_i are nonnegative, it is immediate that $F(x)$ is nondecreasing in x and Eq. (5.2) holds. Similarly, because each $F_i(x)$ is continuous from the right, it follows that a weighted finite sum (the weights need not be nonnegative) is also continuous from the right; we need only interchange limits and finite sums. Finally, Eq. (5.6) follows by interchanging limits, as x goes either to plus or minus infinity, with the finite sum. Clearly, when each summand converges to zero then so does the sum and Eq. (5.6a) is verified. When each $F_i(x)$ converges to one then so does the sum if we invoke the condition $\sum_i \lambda_i = 1$. $\square$

Example 5.5   Empirical CDF _____

The function

$$\hat{F}_n(x) = \frac{1}{n} \sum_{i=0}^{n-1} U(x - i)$$

is easily seen to be a convex combination of the function $\{U(x - i)\}$ with equal weights of $\lambda = 1/n$. As we know, from our first example in this chapter, that $U(x - i)$ is the cdf for the degenerate probability measure $P(X = i) = 1$, we can conclude from this theorem that F_n is also a cdf. The cdf F_n is piecewise constant with all of its points of increase being jumps of height $1/n$ at the integer values $0, 1, \ldots, n-1$. We could also have used Theorem 5.3.1 to prove that F_n is a cdf. Of course, F_n can be also recognized as the cdf corresponding to the probability mass function $\mathcal{R}_n$ of Section 4.4.

5.4 GENERAL REPRESENTATION AND DECOMPOSITION OF UNIVARIATE CDFs

A cdf can always be represented as a convex combination of three basic types of cdfs known as the discrete cdf (denoted by F_d), the absolutely continuous cdf (denoted by F_{ac}), and the singular cdf (denoted by F_s). We first introduce these three basic types and then cite the representation theorem. It will turn out that we generally do not encounter the singular cdf and we will subsequently ignore this mathematical possibility.

Definition 5.4.1   (Discrete CDF)   A *discrete cdf* F_d can be written in the form

$$F_d(x) = \sum_i p_i U(x - x_i),$$

where U is the unit step function, $\{x_i\}$ is an arbitrary countable set of real numbers, and $\{p_i\}$ is a countable set of positive numbers that sum to 1.

Two examples of a discrete cdf are given by Examples 5.1 and 5.2. For F_d, we have that

$$P(X = x_i) = F_d(x_i) - F_d(x_i^-) = p_i,$$

and the random variable X described by F_d takes on only countably many values with positive probability. The collection of probabilities $\{p_i\}$ is a probability mass function and has been discussed in Chapter 4.

Definition 5.4.2   (Absolutely Continuous CDF)   An *absolutely continuous cdf* F_{ac} can be written in the form

$$F_{ac}(x) = \int_{-\infty}^{x} f(z)\, dz,$$

where the integrand,

$$f(x) = \frac{dF_{ac}(x)}{dx},$$

is defined everywhere, with the possible exception of a set of points of zero length (technically, Lebesgue measure), and is a nonnegative, integrable function (possibly having discontinuities) satisfying

$$\int_{-\infty}^{\infty} f(z)\, dz = 1.$$

Strictly speaking, the integral in the definition of F_{ac} is the Lebesgue integral. However, for our purposes we will take it to be the more familiar Riemann integral of engineering calculus. The cdf of Example 5.3 is an example of this type of cdf. In these cases, no value of the random variable X is taken on with positive probability,

$$(\forall x) \; P(X = x) = 0.$$

The integrand f is known as a *probability density function (pdf)* and will be discussed further in Chapter 6.

Definition 5.4.3   (Singular CDF)   A *singular cdf* F_s is a continuous cdf (satisfies Eqs. (5.2), (5.4), and (5.6) and has no jumps) all of whose points of increase form a set I of length (Lebesgue measure) 0.

The most puzzling type of cdf is surely the singular one. If you have not seen a construction of such a function then you are unlikely to be able to think of one. While singular cdfs are not as bizarre as they first appear, they are rarely encountered in engineering.

We now present the representation theorem for arbitrary cdfs.

Theorem 5.4.1   (Decomposition of CDF)   Any univariate cdf F can be represented as a convex combination of the three basic types of cdf through

$$F(x) = \lambda_d F_d(x) + \lambda_{ac} F_{ac}(x) + \lambda_s F_s(x),$$

where the three coefficients are nonnegative and sum to 1.

Proof remark.   See Loeve [60] or Ash [4].   □

Hence, neglecting the singular part ($\lambda_s = 0$), we can, in practice, represent any univariate cdf F as a weighted sum of step functions added to an integral of a nonnegative function. We sum up the situation for nonsingular cdfs in Theorem 5.4.2.

Theorem 5.4.2   (CDF Representations)   When there is no singular part, then there are only the following three cases:

1. If $\Omega = \{x_i\} \subset \mathbb{R}$ is countable and real valued with probability measure P described by pmf $p(x)$, then

$$F(x) = \sum_{\{i : x_i \leq x\}} p(x_i) = \sum_i p(x_i) U(x - x_i),$$

and it is purely discrete. Furthermore,

$$P(A) = \sum_{i : x_i \in A} p(x_i).$$

2. If $\Omega = \mathbb{R}$ with P described by an absolutely continuous cdf F, then there exists a pdf f and

$$F(x) = \int_{-\infty}^{x} f(y)\, dy.$$

In this case, it is immediate that

$$f(x) = \frac{dF(x)}{dx}.$$

Furthermore, for arbitrary $A \in \mathcal{B}$,

$$P(A) = \int_{A} f(x)\, dx.$$

3. If $\Omega = \mathbb{R}$ and the cdf F is a mixture of discrete and absolutely continuous types, then there is a $0 < \lambda < 1$ and

$$F(x) = \lambda \sum_{\{i : x_i \leq x\}} p(d_i) + (1 - \lambda) \int_{-\infty}^{x} f(y)\, dy,$$

and

$$P(A) = \lambda \sum_{i : x_i \in A} p(d_i) + (1 - \lambda) \int_{A} f(x)\, dx.$$

In case (3), we can uniquely decompose the cdf $F(x)$ by first identifying its at most countably many points of discontinuity $\{d_i\}$ and the corresponding discontinuous changes or jumps

$$j_i = F(d_i) - F(d_i^-).$$

We then know that the discrete part

$$\lambda F_d(x) = \sum_i j_i U(x - d_i),$$

where

$$\lambda = \sum_i j_i$$

follows from $\lim_{x \to \infty} F_d(x) = 1$. Similarly, we can identify the absolutely continuous part from

$$(1 - \lambda) F_{ac}(x) = F(x) - \lambda F_d(x).$$

Example 5.6   Rayleigh Noise _____

The noise random variable X at the output of AM envelope detector when no signal is present is *Rayleigh* if it has the cdf

$$F_X(x) = (1 - e^{-\alpha x^2}) U(x) \quad \text{for } \alpha > 0. \tag{5.7}$$

Taking the derivative yields the pdf

$$f_X(x) = 2\alpha x e^{-\alpha x^2} U(x),$$

and

$$P(X > t) = 1 - F_X(t) = e^{-\alpha t^2}$$

for $t \geq 0$ and is 1 for $t < 0$.

Example 5.7   Discrete and Continuous CDF

An example of a mixed cdf is provided by the lifetime X of, say, a lightbulb. Imperfections in manufacturing and mechanical damage in shipment mean that there is a positive probability p_0

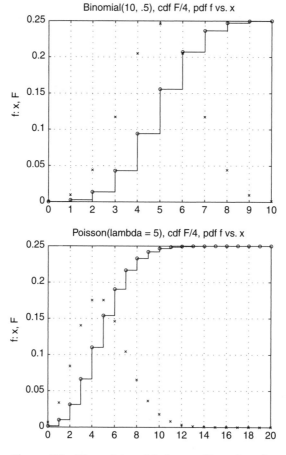

Figure 5.2   Binomial and Poisson cdfs and pmfs.

that the lightbulb will fail to light at the first moment it is installed. Hence, $P(X = 0) = p_0 > 0$. However, if it does light initially, then its lifetime will have a continuous cdf with no particular lifetime x_0 having positive probability. In this case, we might have

$$F_X(x) = p_0 U(x) + (1 - p_0)(1 - e^{-\alpha x})U(x).$$

Illustrations of cdfs and companion pmfs and pdfs are provided in Figures 5.2 through 5.4. In Figure 5.2 the cdf has been scaled down by a factor of 4 to make it comparable to the pmf/pdf whose values at the integers are marked by "x". The actual cdf is piecewise constant (flat). In Figures 5.3 and 5.4 the cdf is shown with a solid line, and in Figure 5.4 the cdf has been reduced

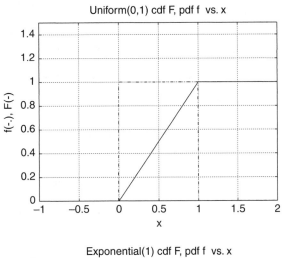

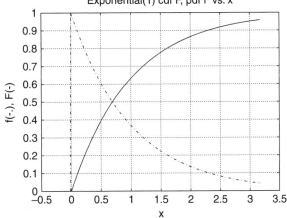

Figure 5.3   Uniform and exponential cdfs.

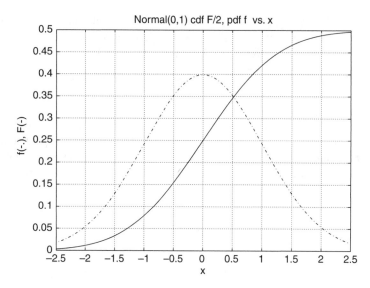

Figure 5.4   Normal/Gaussian cdf and its derivative.

in magnitude by a factor of 2. Additional examples of important pdfs and plots of cdfs and pdfs will be given in Chapter 6.

5.5 THE EMPIRICAL CDF

Probability models for random experiments should be grounded in data collected from repeated performances of unlinked/independent experiments. In some cases physical arguments can be advanced to establish the functional form of the probability model to within a few real parameters. In the absence of a well-grounded physical or mathematical model for the cdf F_X for a random experiment $\mathcal{E}$, we can estimate the cdf from data on the actual outcomes $x_1, \ldots, x_n$ of n unlinked repetitions of $\mathcal{E}$ through the *empirical distribution function*

$$\hat{F}_n(x) = \frac{1}{n} \sum_{i=1}^{n} U(x - x_i).$$

In Example 5.5, we verified that the piecewise constant function $\hat{F}_n$ has the properties of a cdf. Figure 5.5 shows an empirical cdf for the grades on a class exam. The histogram introduced in Section 1.7 can be calculated from the empirical cdf with the number of observations lying in the interval $(a, b]$ (the height of the histogram) being $n(\hat{F}_n(b) - \hat{F}_n(a))$.

The Glivenko–Cantelli theorem (e.g., see Gnedenko [39], Loeve [60], I, p. 20) establishes conditions under which $\sup_x |\hat{F}_n(x) - F_X(x)|$ converges to 0 and the Kolmogorov–Smirnov theorem (e.g., see Gnedenko [39] and Wilks [100]) provides information on the asymptotic distribution of $\sqrt{n} \sup_x |\hat{F}_n(x) - F_X(x)|$. These strong results establish that the empirical cdf converges,

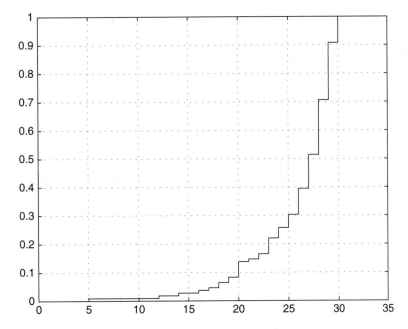

Figure 5.5 Empirical cdf of exam grades.

simultaneously at all arguments $x \in \mathbb{R}$, to the true cdf and furthermore the rate of convergence is $O(1/\sqrt{n})$. A more complete discussion requires knowledge of the various forms of convergence of random variables, and this is treated in Chapter 19.

5.6 SUMMARY

The specification of a probability measure on an uncountably infinite sample space depends upon the particulars of the space. This chapter deals with the simplest, but still important, sample spaces Ω that are intervals $[a, b] \subset \mathbb{R}$ of the real numbers. For such spaces the usual event algebra is the Borel algebra $\mathcal{B}$ generated as the smallest σ-algebra containing the intervals. The essential concept of a random variable, a real-valued function on Ω that models real-valued random outcomes, is introduced. Random variables are denoted typically by uppercase roman letters from the last part of the alphabet. The probability of intervals is described by the cumulative distribution function

$$F_X(x) = P(\{X : X \leq x\}), \text{ with } P(\{X : \alpha < X \leq \beta\}) = F_X(\alpha) - F_X(\beta).$$

It is then true, but beyond the scope of this text, that from the relatively simple F_X (a nondecreasing $[0, 1]$-valued function of a real argument), we can reconstruct the full measure P on $\mathcal{B}$. The cdf F_X is far easier to specify consistently than is the probability measure P to which it is equivalent for a random variable X. We study the properties of such cdfs and present a representation for any cdf F as a convex combination of three types of cdf (discrete, absolutely continuous, and singular) that will provide the basis for introducing pdfs in Chapter 6.

We closed with an example of a discrete cdf, the empirical cdf, $\hat{F}_n$, that is often used in statistics to estimate (Goal 3) the cdf when we have data but little theoretical insight into the true cdf.

EXERCISES

E5.1   By considering the elementary properties given by Eqs. (5.2), (5.4), and (5.6) that characterize a cdf, determine which of the following are cdfs for $\Omega = \mathbb{R}$ (specify the property that fails when the function is not a cdf):

a.
$$e^x/(1+e^x)$$

b.
$$U(x) + [1 - U(x)](1 + e^x)/2$$

c.
$$e^{-|x|}$$

d.
$$0 \text{ if } x < 0 \text{ and } 1 \text{ if } x \geq 0$$

e.
$$0 \text{ if } x \leq 0 \text{ and } 1 \text{ if } x \geq 0.$$

E5.2   The number K_T of ions emitted from a source in time T has a cdf F_K such that $F_K(1) - F_K(1/2) = .1$.

a. What is $P(K_T = 1)$?
b. What is $F_K(1)$?

E5.3   Evaluate and sketch the cdf $F(x)$ for the exponential pdf given by
$$f(x) = \alpha e^{-\alpha x} U(x),$$
when $\Omega = \mathbb{R}$.

E5.4   Use the exponential to evaluate the probabilities of the following events:

a. $\{\omega : \omega \leq x/\alpha\}$;
b. $\bigcup_{k=0}^{\infty} \{\omega : 2k/\alpha < \omega \leq (2k+1)/\alpha\}$.

E5.5   Evaluate and sketch the cdf $F(x)$ for the Laplacian pdf given by
$$f(x) = \frac{\alpha}{2} e^{-\alpha|x|},$$
when $\Omega = \mathbb{R}$.

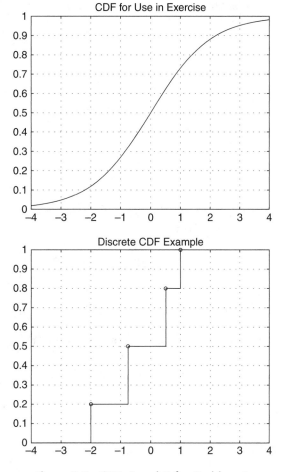

Figure 5.6 CDFs F and G for Problem 4.

E5.6 Use the Laplacian to evaluate the probabilities of the following events:

a. $\{\omega : \omega \le x/\alpha\}$;

b. $\bigcup_{k=0}^{\infty} \{\omega : 2k/\alpha < \omega \le (2k+1)/\alpha\}$.

E5.7 If the cdf

$$F_X(x) = \frac{1}{1 + e^{-x}},$$

evaluate $P(0 \le X \le 1)$.

E5.8 Consider the two different cdfs F and G shown in Figure 5.6 for a random variable X. For each of them, evaluate the following probabilities:

$$P(X < -3), \ P(X \le -1), \ P(X = 0), \ P(X > 2).$$

E5.9   a. Determine and accurately sketch the cdf F for the $\mathcal{R}_4$ pmf when $\Omega = \mathbb{R}$.
       b. Using F, evaluate $P(\{\omega : -.5 < \omega < .5\})$.

E5.10   Determine and accurately sketch the cdf F for the binomial $\mathcal{B}(3, \frac{1}{3})$.

E5.11   If

$$F_X(x) = \begin{cases} 1 & \text{if } x \geq 2 \\ \dfrac{x^2}{4} & \text{if } 0 \leq x < 2 \, , \\ 0 & \text{otherwise} \end{cases}$$

    then determine $P(|X| < 1)$.

E5.12   a. If the cdf

$$F_Y(y) = \begin{cases} 0 & \text{if } y < -2 \\ .1(y + 2) & \text{if } -2 \leq y < 1 \\ .5 & \text{if } 1 \leq y < 2 \quad , \\ .25y & \text{if } 2 \leq y < 4 \\ 1 & \text{if } 4 \leq y \end{cases}$$

    evaluate $P(Y \leq -1), P(Y = 1), P(Y = 2)$. (It may help to sketch F_Y.)
    b. Decompose F_Y into a discrete cdf F_d and a continuous cdf F_{ac}, and determine the coefficient λ for F_d.

E5.13   Determine and accurately sketch, for $x \leq 3$, the cdf F for the Poisson with $\lambda = 1$.

E5.14   A sequence of 10 unlinked (independent) bytes has been received. It is known that the probability is .3 that a first symbol in a byte is a 0. Let K be the number of received bytes having a 0 as first symbol.

    a. What is $P(K = 2)$?
    b. What is $F_K(1)$ for the cdf F_K?

E5.15   The cdf (it may help to sketch it) of interest is

$$F_X(x) = \begin{cases} 0 & \text{if } x < 0 \\ \dfrac{2}{3}x^2 & \text{if } 0 \leq x < 1 \, . \\ 1 & \text{otherwise} \end{cases}$$

  a. Determine

$$P(X \leq -1), P(X \leq 1/2), P(|X| > 1/2), P(X = 1).$$

    b. Decompose the cdf F_X into a discrete cdf F_d and a continuous cdf F_{ac} and determine the coefficient λ for F_d.

E5.16   The cdf of interest is

$$F_X(x) = \begin{cases} 0 & \text{if } x < -2 \\ .3 & \text{if } -2 \leq x < 0 \, . \\ 1 - .5e^{-x} & \text{if } 0 \leq x \end{cases}$$

    a. Evaluate $P(|X| < 1)$.

    b. Decompose F_X into its discrete and continuous parts, F_d and F_{ac}, and determine the coefficient λ for F_d.

E5.17 If $f_X(x) = x^{-2}U(x - 1)$, determine and sketch the cdf F_X.

E5.18 We are given that

$$F_Y(y) = \begin{cases} e^y & \text{if } y < -1 \\ \dfrac{1}{2} & \text{if } -1 \le y < 1 \\ 1 - e^{-y} & \text{if } 1 \le y \end{cases}.$$

    a. Evaluate $P(Y = 1)$, $P(Y \le \frac{1}{2})$, and $P(Y = 0)$.

    b. Decompose F_Y into its discrete and continuous part F_d and F_{ac}, and determine the coefficient λ for F_d.

    c. Evaluate the density f for F_{ac}.

E5.19 The cdf

$$F_X(x) = \begin{cases} e^{x-3} & \text{if } x < 1 \\ \dfrac{1}{2} & \text{if } 1 \le x < 2 \\ 1 & \text{if } 2 \le x \end{cases}.$$

    a. Evaluate $P(X > -1)$, $P(X = 1)$.

    b. Decompose F_X into its discrete and continuous parts F_d and F_{ac}, and determine the coefficient λ for F_d.

    c. Evaluate the density f for F_{ac}.

E5.20 Consider the cdf

$$F_X(x) = \begin{cases} 0 & \text{if } x < 0 \\ 1 - \dfrac{1}{2}e^{-x} & \text{if } x \ge 0 \end{cases}.$$

    a. Evaluate the following:

$$P(X \le 0), \ P(X > 1).$$

    b. Decompose F_X into its discrete and continuous parts F_d and F_{ac}, and determine the coefficient λ for F_d.

E5.21   The cdf (it will help to sketch it)

$$
F_X(x) = \begin{cases}
\dfrac{1}{3}e^x & \text{if } x < 0 \\[2ex]
\dfrac{1}{3} & \text{if } 0 \le x < 1 \\[2ex]
\dfrac{2}{3} & \text{if } 1 \le x < 2 \\[2ex]
1 - \dfrac{1}{4}e^{2-x} & \text{if } 2 \le x
\end{cases}.
$$

a. Evaluate $P(X \le 1)$ and $P(X \ge 2)$.
b. Evaluate $P(\frac{1}{2} < X \le \frac{3}{4})$.
c. Evaluate $P(X = 0)$ and $P(X = 1)$.
d. Decompose the cdf F_X into its discrete and continuous parts F_d and F_{ac}, and specify the coefficient λ of F_d.

6

Describing Probability III: Uncountable Ω and Densities

6.1 PURPOSE, BACKGROUND, AND ORGANIZATION

We are now prepared by the previous chapter for the introduction of the important working concept of a probability density function as a means of defining a probability measure. We relate the probability density function to the probability mass function and see that expressions from Chapter 5 involving the mass function can often be converted into ones involving the density function by replacing sums by integrals. Furthermore, we will supply means, through use of the Dirac delta function, to combine the discrete and continuous parts of the cdf into a single pdf. Finally, we discuss a number of specific families of probability density functions, defined to within either one or two numerical parameters, that are called upon frequently to model random experiments of engineering and scientific importance.

6.2 PROBABILITY DENSITY FUNCTIONS

We assume the same background to the sample space of $\mathbb{R}$ and event algebra $\mathcal{B}$ that was discussed in Section 5.1. In these cases, we can describe the probability measure P through

Definition 6.2.1   (Probability Density Function)   A function f is a *pdf* (commonly called a *density*) if and only if it satisfies the following two conditions:

PDF1. $f(x) \geq 0$;
PDF2. $\int_{\mathbb{R}} f(x)\, dx = 1$.

Example 6.1   Uniform PDF _____

A first example of a pdf is the function $f(x) = U(x)U(1-x)$ that is 1 over the unit interval $[0, 1]$ and zero elsewhere. A second example is $f(x) = x^{-2}U(x-1)$ that is zero for any $x < 1$. A third example is $f(x) = \frac{1}{2}e^{-|x|}$ that is nonzero everywhere. It is immediately apparent in all three examples that f is nonnegative and satisfies PDF1. Verifying PDF2 requires elementary integration:

$$\int_{-\infty}^{\infty} U(x)U(1-x)\, dx = \int_{0}^{1} 1\, dx = 1;$$

$$\int_{-\infty}^{\infty} x^{-2}U(x-1)\, dx = \int_{1}^{\infty} x^{-2}\, dx = \frac{-1}{x}\Big|_{1}^{\infty} = 1;$$

$$\int_{-\infty}^{\infty} \frac{1}{2}e^{-|x|}\, dx = \frac{1}{2}\int_{-\infty}^{0} e^{x}\, dx + \frac{1}{2}\int_{0}^{\infty} e^{-x}\, dx = \frac{1}{2}(1+1) = 1.$$

The role of a pdf f is to describe a probability measure P through evaluating $P(A)$ by integrating the pdf f over the set A,

$$P(A) = \int_{A} f(x)\, dx = \int_{-\infty}^{\infty} I_A(x)f(x)\, dx,$$

where $A \subseteq \mathbb{R}$ or $A \subseteq \mathbb{R}^n$.

Example 6.2   Pareto PDF _____

If $f(x) = x^{-2}U(x-1)$ and $A = \{x : 2 \geq x \geq 0\}$, then

$$P(A) = \int_{A} f(x)\, dx = \int_{0}^{2} x^{-2}U(x-1)\, dx = \int_{1}^{2} x^{-2}\, dx = 1 - \frac{1}{2} = \frac{1}{2}.$$

The necessity for PDF1 follows from K1 requiring that $P(A) \geq 0$. Hence, if the pdf f took on negative values over a set of nonzero volume (nonzero length, if $\Omega = \mathbb{R}$), we would have the contradiction that, for arbitrarily small $\epsilon > 0$,

$$P(\{x : f(x) < -\epsilon\}) = \int_{\{x : f(x) < -\epsilon\}} f(x)\, dx <$$

$$-\epsilon \text{ volume}(\{x : f(x) < -\epsilon\}) < 0.$$

PDF2 follows from K2 requiring that $P(\Omega) = 1$; hence, it is necessary that

$$\int_\Omega f(x)\, dx = 1.$$

Theorem 6.2.1   (Probability Density Function)   If f is a pdf and

$$P(A) = \int_A f(x)\, dx,$$

then P satisfies K0–K4.

Proof sketch. That P satisfies K0–K2 follows immediately from PDF1, PDF2. That P satisfies K3, K4, or K4′ follows from the additivity property

$$\int_A (f(x) + g(x))\, dx = \int_A f(x)\, dx + \int_A g(x)\, dx$$

and the continuity property of the integral.   □

Unlike the pmf p and cdf F, a pdf f is not a probability.

Remark (Interpretation of PDF)   When $\Omega \subseteq \mathbb{R}$, we interpret the pdf f by observing that, for small δ,

$$P\left(\left\{x : x_0 - \frac{\delta}{2} \le x \le x_0 + \frac{\delta}{2}\right\}\right) = P\left(\left[x_0 - \frac{\delta}{2}, x_0 + \frac{\delta}{2}\right]\right)$$

$$= \int_{x_0 - \frac{\delta}{2}}^{x_0 + \frac{\delta}{2}} f(x)\, dx \approx f(x_0)\delta,$$

provided that f is slowly varying over the interval in question (thus giving meaning to δ being "small"). Hence, while the pdf $f(x_0)$ is not itself a probability (while nonnegative, it can exceed unity in value, as in Figure 6.1), it is closely related to the probability that the value x_0 is nearly (to within $\delta/2$) observed.

Remark (Equivalence of P, F, f)   The probability measure P induces a cdf F through $F(x) = P(\{X : X \le x\})$, and the cdf F induces a pdf f through $f(x) = dF/dx$. From the pdf f, we can recover the measure P through $P(A) = \int_A f(x)\, dx$. Hence, for real-valued outcomes and the case of probability measures whose cdf does not have a singular part, there is a one-to-one correspondence between the measure, cdf, and pdf. Either function can be used to define the other two. It is a matter of convenience which representation we choose, and the pdf is usually the simplest.

6.3 REPRESENTING PMFs BY PDFs

When $\Omega \subseteq \mathbb{R}$, it is often convenient to embed a pmf p in a pdf f through the device of a *unit-impulse* or *Dirac delta function* denoted $\delta(x)$. Informally, a delta function can be thought of as an extremely narrow and symmetric pulse function $p(x)$ such that the area under the pulse is 1.

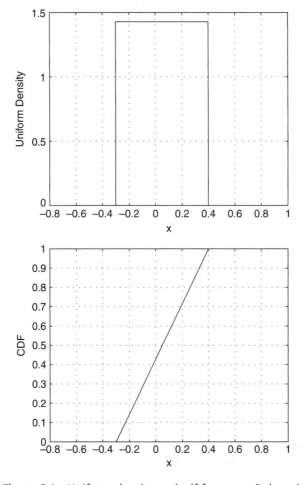

Figure 6.1   Uniform density and cdf for $a = -.3$, $b = .4$.

An example of such a pulse is $\alpha e^{-\alpha|x|}$, where α is very large (nearly infinite). More formally, we assume symmetry,

$$\delta(x) = \delta(-x),$$

and an integration property given in the next lemma.

Lemma 6.3.1   (Sifting Property)   A *delta function* δ has the property that for any function g continuous in an interval (a, b) about 0, with $a < 0 \le b$,

$$g(0) = \int_a^b \delta(x) g(x)\, dx.$$

The integral is zero if 0 is not in $[a, b]$.

Some care is needed in the use of delta functions in order to maintain consistency with the cdf being continuous from the right. Thus, we observe the convention that

$$\int_{-\epsilon}^{0} g(x)\delta(x)\,dx = g(0).$$

The delta function, also known as the unit impulse function, can be thought of as the "derivative"

$$\delta(x) = \frac{dU(x)}{dx},$$

where U is the *unit step function* ($U(x) = 1$ if $x \geq 0$ and $U(x) = 0$ otherwise).

We can now transform a pmf for $\Omega \subseteq \mathbb{R}$ (real-valued outcomes),

$$p(x_i) = p_i = P(\{\omega = x_i\}),$$

into a pdf through

$$f(\omega) = \sum_i p_i \delta(\omega - x_i).$$

The meaning of this embedding of a pmf in a pdf is as follows: Given any countable set A of points $\{x_i\}$, we create a set $\hat{A}$ by replacing the points in A by short intervals $[x_i - \epsilon, x_i + \epsilon]$ centered about each of the x_i in A, with $0 < \epsilon < \min_{i \neq j} |x_i - x_j|$ chosen so that the intervals do not overlap. We can now evaluate

$$P(A) = \sum_{\{i : x_i \in A\}} p_i = \int_{\hat{A}} f(\omega)\,d\omega = P(\hat{A}).$$

In practice, we omit explicit introduction of $\hat{A}$ and proceed with the given set A as if it were $\hat{A}$.

Example 6.3   Impulse Representation of Binomial _____

If X is distributed as $\mathcal{B}(n, p)$, then (subscripting the density f by X to clarify what it is a density for)

$$f_X(x) = \sum_{k=0}^{n} \binom{n}{k} p^k (1-p)^{n-k} \delta(x - k).$$

Take $\epsilon < 1$ to evaluate

$$P(X = j) = \int_{j-\epsilon}^{j+\epsilon} f_X(x)\,dx = \sum_{k=0}^{n} \binom{n}{k} p^k (1-p)^{n-k} \int_{k-\epsilon}^{k+\epsilon} \delta(x - k)\,dx$$

$$= \binom{n}{j} p^j (1-p)^{n-j}.$$

6.4 CLOSURE UNDER CONVEX COMBINATIONS

Theorem 6.4.1   (PDF Closure under Convex Combinations)   If the probability measure P on $\Omega \subseteq \mathbb{R}$ arises from a convex combination of measures $\{P_i\}$ on Ω,

$$P(A) = \sum_i \lambda_i P_i(A),$$

and the individual P_i are described by pdfs f_i, then the pdf f describing P is given by

$$f(x) = \sum_i \lambda_i f_i(x).$$

Proof. It is immediate from the cdf F_i as a restriction of P_i to semi-infinite intervals, that the cdf F corresponding to P is given by

$$F(x) = \sum_i \lambda_i F_i(x).$$

It follows, by differentiation with respect to x, that a convex combination of pdfs yields a pdf

$$f(x) = \sum_i \lambda_i f_i(x)$$

that satisfies the two defining properties of a pdf and describes P. $\square$

Example 6.4   Reliability and Convex Combination of PDFs _____
Consider the application of convex combinations to *reliability assessment*. For many components that do not degrade significantly due to aging, it is known that, due to manufacturing errors or shipping shocks, they may fail on first use with small probability λ, but if they do not fail immediately, then their lifetime L follows the exponential law $L \sim \mathcal{E}(\alpha)$ treated subsequently. In this case,

$$f_L(x) = \lambda \delta(x) + (1 - \lambda)\alpha e^{-\alpha x} U(x),$$

where $\delta(x)$ is the Dirac delta function or unit impulse. A more realistic model would incorporate degradation due to aging.

Example 6.5   Aircraft Maneuvering and Convex Combination of PDFs _____
An aircraft has several modes of maneuvering (e.g., level straight flight, climbing, rolling). It is in mode i with probability λ_i. A radar response to an aircraft traveling in mode i can be described by the pdf f_i. When we do not know which mode the aircraft is operating in, then

we model the radar response by the convex mixture pdf

$$f(x) = \sum_i \lambda_i f_i(x).$$

Such a model of aircraft flight is used in algorithms for tracking aircraft. Estimates are made of which mode the aircraft is in and then a tracking algorithm tuned to that mode is used to track the aircraft until a mode change is detected.

6.5 PDFS OVER FINITE INTERVALS

We organize our presentation, of important examples of pdfs that recur constantly in practice, by whether they are nonzero only on a finite interval, on a semi-infinite interval, or on all of $\mathbb{R}$ itself.

6.5.1 Uniform Density $\mathcal{U}(a, b)$

Definition 6.5.1   (Uniform Density)   The uniform density, denoted $\mathcal{U}(a, b)$, is specified by the pdf

$$f(x) = \frac{1}{b-a} U(x-a) U(b-x) \text{ for } a < b, \ (a,b) \subset \Omega = \mathbb{R}.$$

If $b \le a$, then f is either negative or undefined, and in either case it is not a pdf. Note that if $b - a < 1$, then the pdf will exceed 1. A plot of this density and cdf, for $a = -.3$ and $b = .4$, is provided in Figure 6.1.

The corresponding cdf is

$$F(x) = P(X \le x) = \frac{1}{b-a} \int_a^b I_{y \le x}(y) \, dy = \begin{cases} 0 & \text{if } x < a \\ \dfrac{x-a}{b-a} & \text{if } a \le x \le b \\ 1 & \text{if } x \ge b \end{cases}.$$

$\mathcal{U}(a, b)$ models a continuous version of $\mathcal{R}_n$, and it is often improperly used to represent "complete ignorance" about the values of a random parameter that is only known to lie in the finite interval $[a, b]$. It applies to the phase of oscillators (e.g., $a = -\pi, b = \pi$) and phase of received signals in incoherent communications. Multipath is a critical phenomenon in mobile cellular communications in urban or semi-urban areas. The multipath signals are received over different paths due to reflections from buildings, bridges, vehicles, etc. When they arrive at a receiver, they add together either constructively or destructively, depending upon the relative phases of the component signals. These phases are modeled as unlinked (independent) and uniformly distributed.

We shall sometimes use the $\sim$ notation "$X \sim \mathcal{U}(a, b)$," read as "X is distributed as $\mathcal{U}(a, b)$." As we introduce other probability models, the notational use of "$\sim$" will expand to accommodate to them.

Example 6.6   Uniform PDF continued

It is known that a client is equally likely to request service anywhere in the service availability interval $[t_0, t_1]$. If the time to complete the requested service is $\tau < t_1 - t_0$, what is the probability π that the service will be completed before the end of service availability interval? If A denotes the client request arrival time, then

$$f_A(x) = \frac{1}{t_1 - t_0} U(x - t_0) U(t_1 - x) \text{ and } F_A(x) = \frac{x - t_0}{t_1 - t_0} U(x - t_0) \text{ for } x \leq t_1.$$

It follows that

$$\pi = P(A \leq t_1 - \tau) = F_A(t_1 - \tau) = \frac{t_1 - \tau - t_0}{t_1 - t_0} = 1 - \frac{\tau}{t_1 - t_0}.$$

6.5.2 Beta Density $\beta(a, b)$

Definition 6.5.2   (Beta Density)   The Beta density, denoted $\beta(a, b)$, has a pdf given by

$$f(x) = \frac{\Gamma(a + b)}{\Gamma(a)\Gamma(b)} x^{a-1}(1 - x)^{b-1} U(x) U(1 - x), \ a > 0, b > 0.$$

The Beta pdf is zero outside of the unit interval $[0, 1]$. If either a or b is less than or equal to 0, then the density f will have a nonintegrable singularity either at 0 (in the case of $a \leq 0$) or at 1 (in the case of $b \leq 0$). Hence, we require that this not happen.

The Gamma function was introduced in Section 2.3.5, where it was shown that it satisfied the recursion

$$\Gamma(1) = 1, \ \Gamma(x) = (x - 1)\Gamma(x - 1).$$

If x is an integer, then

$$\Gamma(x) = (x - 1)!.$$

If $a = m$ and $b = n$ are integers, then

$$\frac{\Gamma(m + n)}{\Gamma(m)\Gamma(n)} = \frac{(m + n - 1)!}{(m - 1)!(n - 1)!}.$$

From the nonnegativity of the Gamma function and the restriction $0 \leq x \leq 1$, we see that the Beta density is nonnegative, as required. Appendix 6.9 verifies that the Beta density integrates to one and completes the verification that the Beta is indeed a density or pdf.

If $a = b = 1$, then the Beta reduces to the $\mathcal{U}(0, 1)$. If $a = b$, then the Beta is symmetric about $1/2$ in that $f(x) = f(1 - x)$. If $b < a$ then $P(X > 1/2) > P(X < 1/2)$. A plot of this density for $a = 2$ and $b = 3$ is provided in Figure 6.2.

The Beta density is often used to represent prior knowledge about a single probability $P(A)$ for some event A (e.g., $P(A)$ is the probability p of a "head" in the binomial pdf $\mathcal{B}(n, p)$). One can interpret the parameters $a = m$ and $b = n$ as relating to the number of times in which A was

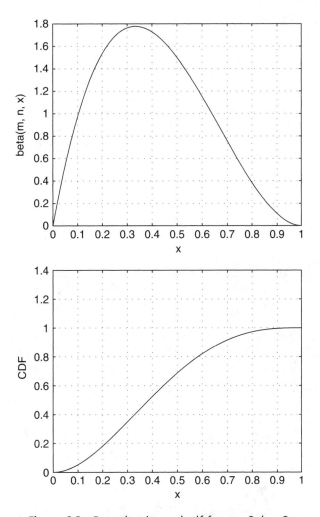

Figure 6.2 Beta density and cdf for $a = 2, b = 3$.

observed to have occurred in repeated, unlinked random experiments. While this matter will be discussed in Chapter 12, we assert here that if p is chosen with prior pdf f and p is the success probability for $\mathcal{B}(n, p)$, then

$$P(K = k) = \int_0^1 \left(\binom{n}{k} p^k (1 - p)^{n-k} \right) f(p) \, dp.$$

Hence, if the prior pdf f is the $\beta(a, b)$, then

$$P(K = k) = \binom{n}{k} \frac{\Gamma(a + b)}{\Gamma(a)\Gamma(b)} \int_0^1 p^{k+a-1} (1 - p)^{n-k+b-1} \, dp.$$

Noting, from the unit normalization of the Beta itself, that

$$\int_0^1 p^{k+a-1}(1-p)^{n-k+b-1}\,dp = \frac{\Gamma(k+a)\Gamma(n-k+b)}{\Gamma(n+a+b)},$$

we find that

$$P(K=k) = \binom{n}{k}\frac{\Gamma(a+b)\Gamma(k+a)\Gamma(n-k+b)}{\Gamma(a)\Gamma(b)\Gamma(n+a+b)}.$$

When a and b are both integers, we can replace the Γ function by a factorial to find that

$$P(K=k) = \binom{n}{k}\frac{(a+b-1)!(k+a-1)!(n-k+b-1)!}{(a-1)!(b-1)!(n+a+b-1)!}.$$

A generalization of the Beta to model prior knowledge about several probabilities, as in the multinomial pmf, is given by the Dirichlet distribution discussed in Section 7.4.3.

Example 6.7   Beta PDF

We have a poorly understood system that will either fail or succeed in operation over a given period of time. We believe that the probability p of success is likely to be greater than the probability of failure and that neither is close to 0 or to 1. Hence, a uniform prior $\mathcal{U}(0,1)$ for p would not reflect the vague knowledge we have. We might, somewhat arbitrarily, adopt the asymmetric $(a > b)$ $\beta(3,2)$ prior reflecting our belief that the system is more likely to succeed than to fail but not emphatically so. Proceeding as done earlier for the $\mathcal{B}(n,p)$ case, and recognizing that we have this case with $n = k = 1$, we can write

$$P(S) = P(K=1) = \binom{1}{1}\frac{(3+2-1)!(1+3-1)!(1-1+2-1)!}{(3-1)!(2-1)!(1+3+2-1)!} = \frac{3}{5}.$$

The cdf

$$F(x) = \begin{cases} 0 & \text{if } x \le 0 \\ \dfrac{\Gamma(a+b)}{\Gamma(a)\Gamma(b)}\displaystyle\int_0^x y^{a-1}(1-y)^{b-1}\,dy & \text{if } 0 < x < 1 \\ 1 & \text{if } x \ge 1 \end{cases}$$

does not have a convenient expression. If we take $b = n$ as an integer, then we can use the binomial theorem expansion to write

$$y^{a-1}(1-y)^{n-1} = \sum_{k=0}^{n-1}\binom{n-1}{k}(-1)^{n-1}y^{a-1+k}.$$

Introducing this expression in the integrand of $F(x)$ enables us to evaluate it for $0 < x < 1$ as

$$F(x) = \frac{\Gamma(a+n)}{\Gamma(a)\Gamma(n)}x^a\sum_{k=0}^{n-1}\binom{n-1}{k}\frac{(-x)^k}{k+a}.$$

Thus, we have the cdf as a polynomial in x that is easily calculated if n is not too large. F is shown in Figure 6.2 for the parameters $a = 2$ and $b = 3$ that were used in plotting the pdf.

6.6 PDFS OVER SEMI-INFINITE INTERVALS

6.6.1 Exponential Density $\mathcal{E}(\lambda)$

Definition 6.6.1   (Exponential Density)   The exponential density is nonzero only for nonnegative arguments, parameterized by a positive parameter λ, and is given by

$$f(x) = \lambda e^{-\lambda x} U(x).$$

$\mathcal{E}(\lambda)$ will have a nonnegative density if and only if $\lambda \geq 0$. It will integrate to 1 if and only if $\lambda > 0$,

$$\int_{-\infty}^{\infty} \lambda e^{-\lambda x} U(x)\, dx = \int_{0}^{\infty} \lambda e^{-\lambda x}\, dx = \int_{0}^{\infty} e^{-y}\, dy = 1,$$

where we made the substitution $y = \lambda x$.

The corresponding cdf is

$$F(x) = P(X \leq x) = \int_{-\infty}^{x} \lambda e^{-\lambda y} U(y)\, dy = (1 - e^{-\lambda x}) U(x).$$

The exponential density and cdf are shown, for $\lambda = 1$, in Figure 6.3.

$\mathcal{E}(\lambda)$ is a continuous time version of the Geometric $\mathcal{G}$ that models the following phenomena:

- lifetimes of components that fail without aging;
- waiting times between successive photon arrivals, electron emissions from a cathode, radioactive decays, or customer arrivals;
- duration of telephone or wireless calls (not `ftp` sessions!).

The exponential will be derived in Section 11.2.2 by using conditional probabilities to rigorously define the "memorylessness" or lack of aging property. The parameter λ is the reciprocal of the mean/average (see Section 9.3) lifetime or waiting time. If there are unlinked exponentially distributed waiting times between the occurrences of successive events, then it will be shown in Section 20.7.1 that the random number of occurrences of these events in a given observation time T has a Poisson distribution.

Example 6.8   Exponential and Aging _____

People change with age. (I know, you do not need probability theory to tell you this!) If there were no aging, then the pdf for our age at death would be exponential. To see that we change with age, let μ be the median lifetime, $P(L > \mu) = .5$. For the exponential pdf, then,

$$P(L > 2\tau) = 1 - F_L(2\tau) = e^{-2\lambda\tau} = (P(L > \tau))^2.$$

Hence, $P(L > 2\mu) = .25$. The state of Connecticut reported that in 1998, the median age at death in that state was 79. It follows that if people failed without aging, we could expect that 1/4 of them would survive past 158 years!

6.6.2 Gamma Density $\Gamma(\alpha, \lambda)$

Definition 6.6.2   (Gamma Density)   The $\Gamma(\alpha, \lambda)$ density is specified for positive α and λ by

$$f(x) = \frac{\lambda}{\Gamma(\alpha)} (\lambda x)^{\alpha-1} e^{-\lambda x} U(x).$$

If $\alpha \le 0$, then f will have a nonintegrable singularity at 0. If $\lambda \le 0$, then f will either be identically 0 or be nonintegrable at infinity. Hence, we must choose α and λ positive if f is to

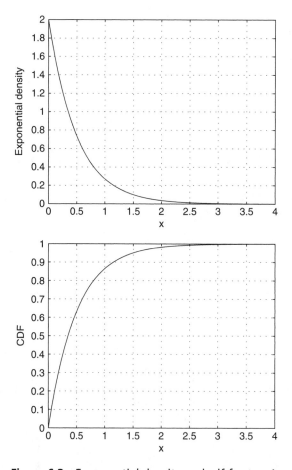

Figure 6.3   Exponential density and cdf for $\lambda = 1$.

be a pdf. That f integrates to 1, and therefore is a pdf, follows from the definition of the gamma function $\Gamma(\alpha)$ that is used to normalize f. The Gamma density and cdf are shown, for $\alpha = 3$ and $\lambda = 2$, in Figure 6.4.

The Gamma cdf for $x \geq 0$ is

$$F_{\alpha,\lambda}(x) = \frac{\lambda}{\Gamma(\alpha)} \int_0^x (\lambda y)^{\alpha-1} e^{-\lambda y} \, dy.$$

Letting $z = \lambda y$ yields

$$F_{\alpha,\lambda}(x) = \frac{1}{\Gamma(\alpha)} \int_0^{\lambda x} z^{\alpha-1} e^{-z} \, dz.$$

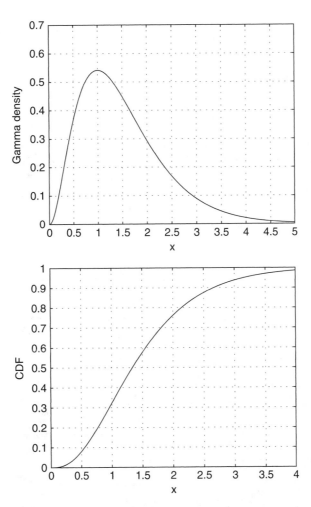

Figure 6.4   Gamma density and cdf for $\alpha = 3$, $\lambda = 2$.

Letting

$$G_\alpha(y) = \Gamma(\alpha)F_{\alpha,1} = \int_0^y z^{\alpha-1}e^{-z}\, dz,$$

and integrating by parts with $u = z^{\alpha-1}$ and $dv = e^{-z}dz$, yields the recursion

$$G_\alpha(y) = -y^{\alpha-1}e^{-y} + (\alpha - 1)G_{\alpha-1}(y).$$

We easily see that, for $y \geq 0$,

$$G_1(y) = 1 - e^{-y}.$$

Hence,

$$G_2(y) = -ye^{-y} + 1 - e^{-y} = 1 - (1 + y)e^{-y},$$

and

$$F_{2,\lambda}(x) = \frac{1}{\Gamma(2)}G_2(\lambda x) = 1 - (1 + \lambda x)e^{-\lambda x}.$$

This recursion can be used to obtain a closed form expression for $F_{\alpha,\lambda}$ for any integer value of α.

Regarding applications of the Gamma, for $\alpha = 1$, $\Gamma(1, \lambda)$ is just the exponential $\mathcal{E}(\lambda)$. For positive integer α the Γ can be shown to be the distribution of a sum of α random variables that are independent (unlinked, see Chapter 14) and distributed as $\mathcal{E}(\lambda)$. Let W_i be the waiting time between the arrivals of the $i - 1$st and ith particle or customer, and let $W_i \sim \mathcal{E}(\lambda)$. Then the random waiting time T_α to the arrival of the αth customer or particle at a detector is

$$T_\alpha = \sum_{i=1}^{\alpha} W_i \sim \Gamma(\alpha, \lambda).$$

The parameter α, however, need not be an integer. We will see in Chapter 8 that the square of a mean zero, unit variance normal random variable (see Section 6.7.2) is a gamma-distributed random variable with $\alpha = 1/2$.

Example 6.9   Gamma and Waiting Times _____

Let T_2 be the waiting time to the emission of the second electron from a cathode. If the mean current is I electrons/second, then the mean waiting time m between successive electrons is $1/I$ seconds and the exponential distribution for the waiting time between successive electrons is given by $\mathcal{E}(I)$. Hence, $T_2 \sim \Gamma(2, I)$ with the cdf $F_{2,\lambda}$ just given. The median waiting time μ_2 for the second electron is defined by

$$P(T_2 > \mu_2) = \frac{1}{2} = (1 + I\mu_2)e^{-I\mu_2}.$$

Determining μ_2 requires us to solve the transcendental equation

$$(1 + \theta)e^{-\theta} = \frac{1}{2},$$

where $\theta = I \mu_2$. Using Matlab or other computational resources reveals that $\theta = 1.68$ and that the median waiting time to the emission of the second electron is

$$\mu_2 = \frac{1.68}{I}.$$

For positive integer k and $\alpha = k/2$ and $\lambda = 1/2$, the Gamma is known as the chi-squared distribution, χ_k^2, of k degrees of freedom. χ_k^2 is the distribution of the sum of the squares of k $\mathcal{N}(0, 1)$ (see subsequently) random variables that are independent. χ_k^2 is encountered in statistical tests of significance. A key property of the Gamma is that if $X_i \sim \Gamma(\alpha_1, \lambda)$ and X_1 and X_2 are unlinked or independent (see Section 14.4), then their sum $Y = X_1 + X_2 \sim \Gamma(\alpha_1 + \alpha_2, \lambda)$; a sum of several independent Gamma-distributed random variables, sharing the same value of λ, is also Gamma distributed.

6.6.3 Pareto Density $\mathcal{P}ar(\alpha, \tau)$

Definition 6.6.3   (Pareto Density)   Given $\alpha > 0$ and $\tau > 0$, the Pareto density is specified by

$$f(x) = \alpha \tau^\alpha x^{-\alpha - 1} U(x - \tau).$$

If $\alpha \le 0$, then the function f is either identically 0 or negative and cannot be a pdf. If $\tau \le 0$, then the function f is either identically 0 or can take on negative or complex values, depending upon α, and f cannot be a pdf. However, if α and τ are both greater than 0, then the function f is nonnegative and integrable. That f integrates to 1 follows from the expression for the cdf

$$F(x) = \left(1 - \left(\frac{x}{\tau}\right)^{-\alpha}\right) U(x - \tau).$$

Observe in Figure 6.5 that the cdf is approaching its asymptotic value of 1 rather slowly. While the median $\mu = \sqrt{2}$, the value of $F(5)$ is only .96. The Pareto is the basic continuous example of a so-called heavy-tailed density. For large argument x, the density $f(x)$ and cdf $F(x)$ converge to 0 and 1, respectively, algebraically (as a power of x) rather than exponentially or more rapidly. While such distributions had until recently found only a few applications to such areas as economics (the distribution of wealth) and hydrology (flood heights of the Nile river), Mandelbrot [62] had long argued for their importance in communications as well. With the advent of widely used Internet-based communications, we find such heavy-tailed models essential for a host of new communications variables. Examples (see Adler et al. [2] and also Section 4.4.5) of such variables are the waiting times between successive keystrokes at computer terminals, delays in packet transmission, and the duration of Internet sessions (e.g., the length of time taken to `ftp` a message). Recent studies of computational running times for NP-hard problems

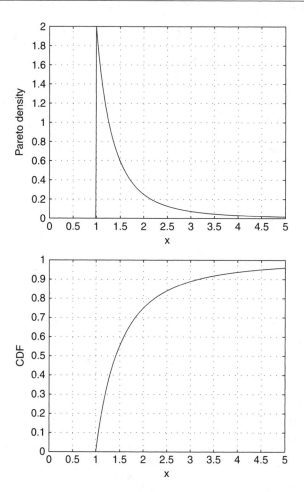

Figure 6.5   Pareto density and cdf $\tau = 1$ and $\alpha = 2$.

as a function of certain parameters reveals a heavy-tailed distribution for such running times. While it is not necessarily the case that the precise heavy-tailed model needed for these various random phenomena is the Pareto, the Pareto model is the simplest and has essential features of other heavy-tailed models. For our purposes, we assume that the Pareto model applies to the phenomena noted earlier. We will see in Section 8.3.4 that a Pareto random variable Y can be generated by exponentiating an exponentially distributed X through $Y = \tau e^X$.

Example 6.10   Pareto and FTP Session _____

If the probability of the duration D of an `ftp` session exceeding d milliseconds is given by

$$P(D > d) = \left(\frac{d}{10}\right)^{-3} = 1 - F_D(d),$$

then the median duration μ_D being defined by

$$P(D > \mu_D) = \left(\frac{\mu_D}{10}\right)^{-3} = \frac{1}{2},$$

yields $\mu_D = 2,000^{1/3} \approx 12.6$ ms. The probability that the duration of a session will exceed twice that of the median is

$$P(D > 2\mu_D) = \left(\frac{2\mu_D}{10}\right)^{-3} = 2^{-3} P(D > \mu_D) = 2^{-3} 2^{-1} = \frac{1}{16}.$$

6.7 PDFS OVER ALL OF $\mathbb{R}$

6.7.1 Laplacian Density $\mathcal{L}(\alpha)$

Definition 6.7.1   (Laplacian Density)   The Laplacian density is a family of two-sided exponentials parameterized by a single positive constant α and specified by

$$f(x) = \frac{\alpha}{2} e^{-\alpha|x|}.$$

If $\alpha \leq 0$ then either f is identically 0 or it is nonintegrable and negative. That f is a pdf follows from its integrating to 1 as established by its cdf

$$F(x) = \int_{-\infty}^{x} \frac{\alpha}{2} e^{-\alpha|y|} \, dy = \int_{-\infty}^{\alpha x} \frac{1}{2} e^{-|z|} \, dz = \begin{cases} \int_{-\infty}^{\alpha x} \frac{1}{2} e^{z} \, dz & \text{if } x \leq 0 \\ \int_{-\infty}^{0} \frac{1}{2} e^{z} \, dz + \int_{0}^{\alpha x} \frac{1}{2} e^{-z} \, dz & \text{if } x > 0 \end{cases}$$

$$= \begin{cases} \frac{1}{2} e^{\alpha x} & \text{if } x \leq 0 \\ \frac{1}{2} + \frac{1}{2}(1 - e^{-\alpha x}) & \text{if } x > 0 \end{cases} = \frac{1}{2} e^{\alpha x}(1 - U(x)) + \left(1 - \frac{1}{2} e^{-\alpha x}\right) U(x).$$

Figure 6.6 shows the Laplacian density and cdf for $\alpha = 1$.

The Laplacian models amplitudes of speech signals and amplitudes of differences of intensities between adjacent pixels in an image. The Laplacian can be derived as the difference between two independent, identically distributed exponentials. (See the exercises in Chapter 17.)

Example 6.11   Laplacian and Speech Amplitude _____

If the probability that a speech signal amplitude A exceeds 1 volt is .1, then what is the probability that it is less than -1 volt? Given the pdf,

$$P(A > 1) = \int_{1}^{\infty} \frac{\alpha}{2} e^{-\alpha|x|} \, dx = \int_{1}^{\infty} \frac{\alpha}{2} e^{-\alpha x} \, dx = .1.$$

Similarly,

$$P(A \leq -1) = \int_{-\infty}^{-1} \frac{\alpha}{2} e^{-\alpha|x|} \, dx = \int_{-\infty}^{-1} \frac{\alpha}{2} e^{\alpha x} \, dx = \int_{1}^{\infty} \frac{\alpha}{2} e^{-\alpha y} \, dy = P(A \geq 1),$$

when we substitute $y = -x$. Hence, $P(A \le 1) = P(A > 1) = .1$. Since the random variable A has a continuous cdf F_A, $P(A = x) = 0$ and $P(A < -1) = P(A \le -1)$. We observe from this calculation that, for the symmetric density of the Laplacian, it is always the case that

$$P(A \le -x < 0) = P(A \ge x).$$

The information $P(A > 1) = .1$ also allows us to solve for the parameter α from

$$P(A > 1) = \int_1^\infty \frac{\alpha}{2} e^{-\alpha x} \, dx = .1 = \int_\alpha^\infty \frac{1}{2} e^{-y} \, dy = \frac{1}{2} e^{-\alpha},$$

where we made the substitution $y = \alpha x$. Hence, $\alpha = \log_e 5$.

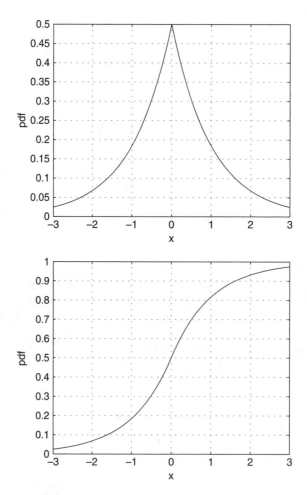

Figure 6.6   Laplacian density and cdf for $\alpha = 1$.

6.7.2 Normal or Gaussian Density $\mathcal{N}(m, \sigma^2)$

Definition 6.7.2 (Normal or Gaussian Density) The normal or Gaussian density is parameterized by a mean m and variance σ^2 and is specified by

$$f(x) = \frac{1}{\sigma \sqrt{2\pi}} e^{\frac{-(x-m)^2}{2\sigma^2}},$$

for $\Omega = \mathbb{R}$.

The mean (see Section 9.3) parameter m can be any real number. By convention σ, known as the standard deviation (see Section 9.10.2), is taken to be nonnegative. Figure 6.7 shows the Gaussian density and the corresponding cdf for $m = 0$ and σ^2 either 1 (solid lines) or 2 (dotted lines).

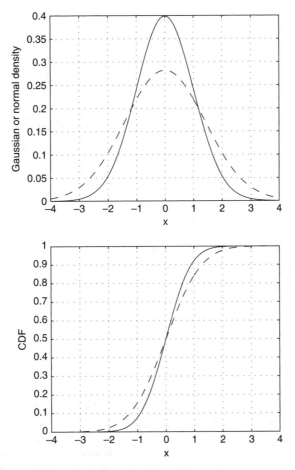

Figure 6.7 Gaussian density and cdf for $m = 0$, $\sigma^2 = 1$ (solid), $\sigma^2 = 2$ (dotted).

The corresponding cdf cannot be expressed in closed form in terms of finitely many of the common elementary functions.

The parameter m is called the *mean*, the parameter σ^2 is the **variance**, and $\sigma > 0$ is known as the *standard deviation* and is nonnegative. (See Section 1.7 for the data-based counterparts.) This probability law can be derived through the Central Limit Theorem (Section 17.9).

Historically, this distribution was called the "normal" because it was so widely applicable to biological and social phenomena that it was the anticipated or normal distribution. The numerous applications of this model include thermal noise in resistors and in all physical systems that have a dissipative component; 3 °K universal background radiation; 290 °K radiation from the Earth as seen from space (e.g., see Pisacane and Moore [71]); 6,000 °K radiation from our Sun's surface at optical frequencies, but 1,000,000 °K radiation from our Sun's corona at radio frequencies; shot noise produced by the random arrivals of individual photons or electrons (in shot noise, the granularity of current flow becomes evident at very low currents); low-frequency noise ($1/f$, flicker, semiconductor, excess noises) as found in low-frequency amplifiers and in variations in quartz crystal oscillator frequency; and variabilities in parameters of manufactured components and of biological organisms (e.g., height, weight, intelligence).

Many of the noise phenomena just noted can be schematized as observations of certain characteristics of large-scale or macroscopic systems formed out of many loosely interacting small-scale or microscopic components wherein the macroscopic observable is a sum of the microscopic quantities. This description will be formalized when we discuss the central limit theorem in Section 17.9.

The normal is also used to describe many kinds of measurement errors. The measured value Y is often related to the true quantity of interest X through an unlinked noise N in an additive signal-plus-noise model

$$Y = X + N.$$

It is frequently the case that the noise $N \sim \mathcal{N}(m, \sigma^2)$. If $m = 0$, then the measurement Y is said to be *unbiased* for the true quantity X.

Thermal noise (see also Section 20.11) is produced by all dissipative (i.e., nonsupercon-ducting) physical systems operating above absolute zero temperature and is as common as friction in mechanical systems. When we measure or observe, over a single-sided (only positive, physical frequencies) bandwidth B Hz, the voltage V across a resistor of R ohms that is at an absolute temperature of T °K, we find that V is distributed as (we will use $\sim$ to stand for the phrase "is distributed as") $\mathcal{N}(0, \sigma^2)$ with

$$\sigma^2 = 4kTRB,$$

and $k = 1.38 \times 10^{-23}$ is Boltzmann's constant in units of watts per Hz per °K; in physics, the units are commonly given as joules per degree Kelvin. Thus, the voltage V across a resistor of 1 megohm at approximate room temperature T of 300 °K that is observed over a 1-MHz bandwidth is described by $\mathcal{N}(0, 16.5 \times 10^{-9})$.

We can make an equivalent circuit for a (source) resistor R_S at temperature T by replacing this ideal resistor of R_S ohms by a series connection of an ideal resistance of this value and a voltage source V. The voltage source produces a noise voltage that has no DC component (there is no physical asymmetry to choose one direction of current flow over the opposite one) and that has a variance (noise power) σ^2 just given. If we connect a load resistance R_L across the terminals of this equivalent circuit, then the random noise power delivered to the load is

$$V^2 \frac{R_L}{(R_S + R_L)^2},$$

and the average (the concept of average or expected value is discussed fully in Chapter 9) noise power delivered is

$$P_L = \sigma^2 \frac{R_L}{(R_S + R_L)^2}.$$

The maximum power that can be transferred when the load resistance R_L is properly matched to the source resistance R_S is

$$P_L = kTB$$

and is achieved by setting $R_L = R_S$. It is sometimes more insightful to consider a noise source at absolute temperature T observed over a bandwidth B Hz as potentially delivering average noise power of kTB watts. When, as is common in applications, a device (e.g., the Earth viewed by a microwave communications antenna in space) is said to have a noise temperature of $T \, ^\circ\mathrm{K} = 290 \, ^\circ\mathrm{K}$, this is to be interpreted as specifying the average power $P = kTB$ that would be delivered to a matched resistive load R_L. For an antenna, this power delivery is achieved when the antenna impedance is matched to its transmission line impedance (say, 50 ohms) and thence to a resistive load with the same impedance (say, 50 ohms).

An amusing instance of a plot of the Gaussian is due to the statistician W. J. Youden (cited in Tufte [92]—where it makes a better visual impression in this unusual treatise) who composed the display of Figure 6.8.

THE

NORMAL

LAW OF ERROR

STANDS OUT IN THE

EXPERIENCE OF MANKIND

AS ONE OF THE BROADEST

GENERALIZATIONS OF NATURAL

PHILOSOPHY. IT SERVES AS THE

GUIDING INSTRUMENT IN RESEARCHES

IN THE PHYSICAL AND SOCIAL SCIENCES AND

IN MEDICINE AGRICULTURE AND ENGINEERING

IT IS AN INDISPENSABLE TOOL FOR THE ANALYSIS AND THE

INTERPRETATION OF THE BASIC DATA OBTAINED BY OBSERVATION AND EXPERIMENT

Figure 6.8   The Normal in words by Youden.

6.7.3 Analytical Properties of the Normal Distribution

While the normal density has many desirable properties, it cannot be integrated in familiar terms. If we wish to compute, say,

$$P(a < X \le b) = \int_a^b f_X(x)\,dx = \int_a^b \frac{1}{\sigma\sqrt{2\pi}} e^{-\frac{(x-m)^2}{2\sigma^2}}\,dx,$$

then the result is not expressible as a finite composition of the usual elementary functions (e.g., algebraic, exponential, trigonometric, logarithmic). A first step in dealing with such an integral is to bring the integrand to a standard form through the simple substitution

$$y = \frac{x - m}{\sigma}.$$

This substitution reduces $X \sim \mathcal{N}(m, \sigma^2)$ to $Y \sim \mathcal{N}(0, 1)$. We now introduce four new functions that are closely related:

- The standard normal cdf

$$\Phi(z) = \int_{-\infty}^z \frac{1}{\sqrt{2\pi}} e^{-\frac{x^2}{2}}\,dx,$$

  takes values in $[0, 1]$. Typical values are

$$\Phi(-\infty) = 0,\ \ \Phi(0) = .5,\ \ \Phi(1) = .841,\ \ \Phi(2) = .977,\ \ \Phi(3) = .9987,\ \Phi(\infty) = 1.$$

  $\frac{1}{2} - \Phi(z)$ is easily seen to be symmetric about $z = 0$ in that for $z \ge 0$,

$$\frac{1}{2} - \Phi(-z) = \Phi(z) - \frac{1}{2}.$$

- The complementary standard normal cdf

$$\overline{\Phi}(z) = 1 - \Phi(z) = \int_z^\infty \frac{1}{\sqrt{2\pi}} e^{-\frac{x^2}{2}}\,dx.$$

- The error function

$$\operatorname{erf}(z) = \frac{2}{\sqrt{\pi}} \int_0^z e^{-x^2}\,dx,$$

  is an odd function in that $\operatorname{erf}(-z) = -\operatorname{erf}(z)$. Clearly, $\operatorname{erf}(0) = 0$ and $\operatorname{erf}(\infty) = 1$. erf takes values in $[-1, 1]$, but is positive for positive argument. erf is readily shown to be expressible in terms of the standard normal Φ through

$$\operatorname{erf}(z) = 2\left(\Phi(z) - \frac{1}{2}\right).$$

- The complementary error function is given by

$$\mathrm{cerf}(z) = 1 - \mathrm{erf}(z) = \frac{2}{\sqrt{\pi}} \int_z^\infty e^{-x^2}\, dx.$$

These functions have been tabulated. While we consider Φ, the standard normal, to be the basic quantity of the four given, it is erf that is found commonly on calculators and in Matlab. In terms of these functions, we can express

$$P(a < X \le b) = \Phi\left(\frac{b - m}{\sigma}\right) - \Phi\left(\frac{a - m}{\sigma}\right).$$

Note that, as the normal has a continuous cdf, the preceding probability is unchanged if we replace $<$ with $\le$ and vice versa. For $z \gg 1$, we have the useful asymptotic approximation (derivable from integration by parts) that

$$\overline{\Phi}(z) \sim \frac{e^{-\frac{z^2}{2}}}{z\sqrt{2\pi}}.$$

The asymptotic notation $f(x) \sim g(x)$ means that

$$\lim_{x \to \infty} [f(x)/g(x)] = 1.$$

Example 6.12 Gaussian CDF

Consider the two random variables

$$X \sim \mathcal{N}(1, 3) \text{ and } Y \sim \mathcal{N}(2, 4).$$

We ask, Which of them has a higher probability of being negative?

$$P(X < 0) = P(X \le 0) = \Phi\left(\frac{0 - 1}{\sqrt{3}}\right) \text{ and } P(Y < 0) = \Phi\left(\frac{0 - 2}{2}\right).$$

As $-1 < -1/\sqrt{3}$ and $\Phi(z)$ is strictly increasing in z, it is immediate that X has a higher probability than Y of being negative.

Example 6.13 Gaussian and Thermal Noise

A two-terminal device with an equivalent resistance of 1 megohm operates at a room temperature of about 300 °K. The thermal voltage V that it generates is observed in the band from 1.5 GHz to 2.5 GHz. What is the probability that V is greater than 8 millivolts? From the information given earlier, we know that $V \sim \mathcal{N}(0, \sigma^2)$ with $\sigma^2 = 4kTRB$. Hence,

$$\sigma^2 = 4(1.38 \times 10^{-23})(300)10^6(10^9) = 16.5 \times 10^{-6}.$$

Thus, $\sigma \approx .004$, and

$$P(V > .008) = 1 - \Phi\left(\frac{.008 - 0}{.004}\right) = \overline{\Phi}(2) = 1 - .977 = .023.$$

If, instead, we asked for the probability of V exceeding 20 millivolts, then

$$P(V > .02) = \overline{\Phi}(5) \sim \frac{\exp(-\frac{5^2}{2})}{5\sqrt{2\pi}} \approx \frac{3.7 \times 10^{-6}}{12.5} = .3 \times 10^{-6}.$$

If we asked for the probability that the magnitude of V exceeded 8 millivolts, then

$$P(|V| > .008) = P(V > .008) + P(V < -.008).$$

However, for $\mathcal{N}(0, \sigma^2)$, the pdf $f_X(x) = f_X(-x)$ is symmetric about 0 and

$$P(X > x) = \int_x^\infty f_X(x)\,dx = \int_{-\infty}^{-x} f_X(x)\,dx = P(X < -x).$$

Hence, $P(|V| > .008) = 2P(V > .008) = .046$. Finally, if we asked for the probability that V was between 8 and 20 millivolts, then

$$P(.008 < V < .02) = \Phi(5) - \Phi(2) = 1 - \overline{\Phi}(5) - .977 \approx .023 - .3 \times 10^{-6} \approx .023.$$

6.8 SUMMARY

We continued the discussion begun in Chapter 5 on specifying a probability measure P for a sample space that is uncountably infinite and, more specifically, an interval of the real numbers. In Chapter 5, we learned that we could make the specification through a simpler object: the cumulative distribution function (cdf). We also learned of a characterization of such cdfs as a convex combination of three types of cdfs. When the cdf F_X is absolutely continuous, then its derivative f_X exists and F_X is the integral of its derivative f_X. The derivative is known as a probability density function (pdf), or for short, a density. Any function f_X having the following two properties is a pdf:

$$f_X(x) \geq 0, \quad \int_{-\infty}^\infty f_X(x)\,dx = 1.$$

Densities, being easily recognized from the two preceding properties, provide the most common method for specifying probability measures. The measure P and cdf F_X are determined from the density by integration:

$$P(A) = \int_A f_X(x)\,dx = \int_{-\infty}^\infty I_A(x)f_X(x)\,dx,$$

$$F_X(x) = \int_{-\infty}^x f_X(y)\,dy.$$

We then extended our ability to specify probability measures via pdfs by including the case of a discrete component cdf through the agency of allowing Dirac delta functions (or unit impulse functions) in the density f_X. This device enabled us to incorporate probability mass functions as a special case of the probability density function. We closed our treatment of pdfs by introducing the families of densities (uniform, exponential, Beta, Pareto, Gamma, Laplacian, and Gaussian) that are commonly encountered in practice, indicating their application areas, and discussing their analytical properties. The pdf provides what is usually the most convenient way to specify a probability measure when the sample space is a subset of all of the real numbers. In the sequel, we will make frequent use of these probability models; they provide the foundation for the successes of applied probability.

6.9 APPENDIX: BETA DENSITY UNIT NORMALIZATION

It remains to be verified that β integrates to 1 to show that it is a density. We do so for $a = m$ and $b = n$ integers through

$$I_{m,n} = \int_0^1 x^{m-1}(1-x)^{n-1}\,dx = \frac{\Gamma(m)\Gamma(n)}{\Gamma(m+n)}.$$

By the change of variables $y = 1 - x$, we see that

$$I_{m,n} = I_{n,m}.$$

Use integration by parts with $u = (1-x)^{n-1}$ and $dv = x^{m-1}\,dx$ to establish that

$$I_{m,n} = \frac{1}{m}x^m(1-x)^{n-1}\Big|_0^1 + \frac{n-1}{m}\int_0^1 x^m(1-x)^{n-2}\,dv.$$

Yielding,

$$I_{m,n} = \begin{cases} \dfrac{n-1}{m}I_{m+1,n-1} & \text{if } m \geq 2 \text{ and } n \geq 2 \\[2mm] \dfrac{1}{m} & \text{if } n = 1 \\[2mm] \dfrac{1}{n} & \text{if } m = 1 \end{cases} \quad .$$

Assuming that $m \geq 2$ and $n \geq 2$, we have

$$I_{m,n} = \frac{n-1}{m}I_{m+1,n-1}.$$

Repeating this iteration to see the pattern yields

$$I_{m,n} = \frac{n-1}{m}\left(\frac{n-2}{m+1}I_{m+2,n-2}\right) = \frac{(n-1)(n-2)}{m(m+1)}I_{m+2,n-2}.$$

Iteration gives

$$I_{m,n} = \frac{(n-1)(n-2)\cdots 2 \cdot 1}{m(m+1)\cdots(m+n-2)} I_{m+n-1,1}.$$

Referring to the $n = 1$ case yields

$$I_{m,n} = \frac{(n-1)!}{m(m+1)\cdots(m+n-1)} = \frac{(n-1)!(m-1)!}{(m+n-1)!} = \frac{\Gamma(n)\Gamma(m)}{\Gamma(m+n)}.$$

We have completed the verification that the $\beta(m,n)$ integrates to 1 and is a legitimate pdf. This result holds as well for $\beta(a,b)$ where a and b are not integers.

EXERCISES

E6.1   If the cdf

$$F(x) = \begin{cases} e^{x-2} & \text{if } x < 1 \\ \dfrac{1}{2} & \text{if } 1 \le x < 2, \\ 1 & \text{if } 2 \le x \end{cases}$$

       evaluate the pdf $f(x)$.

E6.2   If

$$F_X(x) = \begin{cases} 0 & \text{if } x < -1 \\ \dfrac{x+1}{2} & \text{if } -1 \le x < 0 \\ \dfrac{x+6}{8} & \text{if } 0 \le x < 2 \\ 1 & \text{if } x \ge 2 \end{cases},$$

       then evaluate the pdf $f_X(x)$.

E6.3   What is the probability that the phase Φ of the AM carrier received on my car radio at a given time is within 30° of zero phase?

E6.4   If X is random over the interval $[-1, 2]$, then evaluate $P(0 \le X \le 1)$.

E6.5   Show that the location of the maximum of the $\beta(a,b)$ density, for a and b greater than 1, is at $(a-1)/(a+b-2)$. Is this maximum unique?

E6.6   If the probability of heads p for a coin is chosen according to $\beta(2,2)$, what is the probability of a head?

E6.7   What is the probability of 2 successes in 4 unlinked repeated trials when the probability p for a success in a single trial is chosen according to the $\beta(3,3)$ prior?

E6.8   It is observed that, for a particular kind of chip, it is as likely to fail before 5,000 hours as it is to last longer than 5,000 hours.

       a. Can you determine the mean chip life?

b. What is the probability that a chip will last either less than 1,000 or more than 10,000 hours?

E6.9 If a component that fails without aging has a mean or average lifetime of 5 years, what is the probability that it will fail before 10 years?

E6.10 The probability is .5 that a phone call will exceed 1 minute in length. What is the probability that its duration will exceed .5 minute?

E6.11 The waiting time T between successive arrivals of dopant atoms in an implant process is such that $P(T > 1) = .1$. What is the probability description of T?

E6.12 If the probability of a waiting time T between the arrival of successive dopant atoms exceeding 5 units is .1, what is the probability of $T > 10$?

E6.13 Making reasonable assumptions, if the mean time to failure of a power transistor is 10,000 hours, then what is the probability of failure in the first 1,000 hours of operation?

E6.14 If the waiting time T to the next photon emission is such that $P(T > 2) = .09$, what is the probability that if we observe this photon source for time τ, we will see at least one photon?

E6.15 The waiting time between successive emissions of an electron from a cathode is a random variable T whose average value is 1 nanosecond (ns). A counter for these electrons will count correctly provided that no two electrons are emitted within less than .05 ns. What is the probability that the third emitted electron will not be confused with the second emitted electron (that their arrivals are at least .05 ns apart)?

E6.16 a. If the mean waiting time between the arrivals of successive photons at a detector is 1 ns, what is the pdf for the waiting time T_2 between the arrivals of the first and third photons?

b. What is the cdf for T_2?

E6.17 Photons arrive at a detector so that the mean waiting time between arrivals of successive photons is 2.

a. What is the pdf of the waiting time T_3 to the arrival of the third photon?

b. Evaluate the cdf F_{T_3} for T_3.

c. Using your computational resources, determine the median value of T_3?

E6.18 For a particular class of computationally difficult problems, it is known that if the algorithm is initialized randomly, then the running time R is at least one time unit and the probability is .5 that the running time $R > 10$ time units. What is the probability that the running time will exceed 1,000 time units?

E6.19 A computer center providing Internet access observes that half of all sessions last for at least 20 minutes. If the minimum length of a session is one minute, what is the probability of a session length exceeding one hour?

E6.20 An `ftp` session of duration L has a minimum time (in appropriate units) of τ_0. We have determined that $P(L \geq 2\tau_0) = 4P(L \geq 4\tau_0)$.

a. Evaluate the probability that the session length will exceed $10\tau_0$.

b. Sketch the cdf $F_L(x)$.

E6.21 For a given site, it is known that the probability is $1/2$ that an internet session will last longer than τ_1 seconds. It is also known that the shortest possible such session is 1 second. Specify and sketch the cdf F_T for the duration T in seconds of such sessions.

E6.22   The instantaneous amplitude A of a speech signal is a real-valued random variable that is described by the Laplacian density

$$f_A(x) = \beta e^{-\alpha|x|}, \quad \Omega = \mathbb{R}.$$

      a. Express β in terms of α.

      b. What is the probability of an amplitude exceeding $\frac{1}{\alpha}$?

E6.23   If a speech amplitude A is such that $P(A > 1) = .3$, then what is the $P(A > 2)$?

E6.24   A speech amplitude S is such that $P(|S| > 1) = P(S > 0)$. What is the probability description of S?

E6.25   Let X denote the difference in intensities of a randomly selected pair of successive pixels in a scanned video image.

      a. What probability model do you suggest for X?

      b. If it is known that $P(X \le -5) = .1$, then what is the probability that $0 \le X \le 5$?

E6.26   A speech signal digitizer has quantization levels spanning the range $[-5, +5]$ units. If the probability of a signal amplitude X exceeding this range is e^{-10}, what is the probability model f_X for X?

E6.27   In designing an A/D quantizer for speech recording, we need to know the probabilities of large amplitudes so as to set the dynamic range of the quantizer correctly.

      a. Specify the pdf f_A for the random speech amplitude A at a given time.

      b. If we know that $P(|A| > 1) = e^{-1}$, for what value τ is $P(|A| > \tau) = e^{-6} \approx 1/400$?

E6.28   A noisy voltage V in a circuit is as likely to be negative as it is to be positive. If $P(V > 1) = \overline{\Phi}(2)$, then evaluate $P(V < \frac{1}{2})$.

E6.29   If a thermal noise voltage V is such that $P(-.001 < V < .001) = .006$, then estimate $P(1 - .001 < V < 1 + .001)$. (Make reasonable approximations.)

E6.30   If a noisy current I has mean m and variance σ^2, then evaluate $P(|I| \le 2)$ as fully as you can.

E6.31   If the probability of a thermal noise voltage V exceeding 1 μv is .1, what can you say about the pdf f_V?

E6.32   A thermal noise voltage V observed over a bandwidth of 1 GHz across a resistor of 1 megohm is such that the probability is .001 that its magnitude $|V|$ is less than 2 microvolts (2×10^{-6} v).

      a. What is the temperature T of the resistor?

      b. What can you say about the probability that $|V|$ will be less than 1 millivolt (10^{-3} v)?

E6.33   If the probability of a thermal noise voltage V exceeding 1 μv is .49, what can you say about the pdf f_V?

E6.34   a. A thermal noise source is observed to transfer 1 nanowatt of average noise power to a load resistance of 1,000 ohms. Noting that the variance σ^2 of the voltage is the average squared voltage, describe the voltage V across this load resistance due to the thermal noise source.

      b. Approximately evaluate $P(|V| \le 10^{-6})$.

E6.35 A satellite antenna, matched to a 50-ohm transmission line, is pointed at the Earth and a voltage V is measured across a 50-ohm load in a 1-GHz bandwidth. Neglect other sources (e.g., microwaves, etc.) of noise.

    a. Evaluate the average thermal noise power P dissipated in the load.

    b. Provide an expression for the probability that the magnitude of the thermal noise voltage V appearing across the load will be less than 2 μV.

E6.36 The random voltage generated by an antenna is modeled by $\Omega = \mathbb{R}$ and the $\mathcal{N}(0, \sigma^2)$ density or probability law. We are interested in the events

$$A_k = \{\omega : |\omega| > k\sigma\}, \quad B = \{\omega : \omega > 2\sigma\}.$$

Evaluate $P(A_k)$ for $k = 1, 2, 3, 4$, and $P(B)$.

E6.37 In a standardized test like the SAT, where the outcome is described by $\mathcal{N}(500, 10,000)$, how likely is it that a student's score will exceed 700?

E6.38 Specify the pdf needed to model the following random sources:

    a. speech amplitude A, when $P(|A| \leq 1) = .5$;

    b. lifetime L of a lightbulb having an average lifetime of 1,000 hours;

    c. voltage V recorded across a resistor R at temperature T when the voltage is measured across a bandwidth W Hz;

    d. duration of an `ftp` session when it is four times as probable that it will last as long as 5 time units as that it will last as long as 10 time units.

E6.39 An AM receiver that is not receiving a signal will have a random output $Z(t)$ from its envelope detector that is due to channel and front-end noise. It is often the case that Z at a fixed time has what is called a *Rayleigh distribution* and it can be described as follows: $\Omega = [0, \infty)$,

$$P(A) = \int_A \alpha^2 z \epsilon^{-(\alpha z)^2/2} \, dz.$$

    a. Verify that this specification satisfies the basic probability axioms K1, K2, and K3.

    b. Evaluate the probabilities of the following events:

$$A_0 = \{z\}, \quad A_1 = [0, z), \quad A_2 = (z/\alpha, \infty).$$

E6.40 Provide an expression for the probability, when no signal is present, that the output of an AM envelope detector will exceed a value of v.

E6.41 In cellular communications, it is common to model the received power P due to fading from a base station by the *lognormal distribution*,

$$f_P(x) = \frac{1}{x\sigma\sqrt{2\pi}} e^{-\frac{\log^2 x}{2\sigma^2}} U(x).$$

Verify that this is a pdf.

E6.42 Assume that the maximum power P (in gigawatts) demanded daily from a large power plant is a random quantity, fluctuating from day to day, with probability $P(A)$ of lying in the set A given by

$$P(A) = \int_{A \cap [0,\infty)} cp^2 e^{-2p} \, dp.$$

a. Find the value of c needed to ensure that P satisfies the Kolmogorov axioms. Recall that

$$\int_0^\infty x^k e^{-x} \, dx = k!.$$

b. If the plant has a maximum capacity of 3 gigawatts, then what is the probability that on a given day demand will exceed capacity?

7

Multivariate Distribution Functions and Densities

7.1 PURPOSE, BACKGROUND, AND ORGANIZATION

You are already familiar with the generalization from scalar-valued functions to such vector-valued functions as the force applied to an object, the resulting acceleration of the object, and the electric and magnetic fields described by Maxwell's equations. To accommodate such vector-valued quantities, we need to extend our descriptions of probability measures from a sample space of real numbers to a sample space

$$\Omega = \mathbb{R}^n = \{(x_1, \ldots, x_n) : (\forall i) \; -\infty < x_i < \infty\}$$

that contains the n-tuples of real numbers. In effect, we extend our notion of a random variable X taking real values to a random vector $\mathbf{X} = (X_1, \ldots, X_n)$ taking values that are n-tuples of reals. We easily generalize the notion of a cdf, introduced in Chapter 5, to a cdf for random vectors. Most of the cdfs encountered in practice can then be represented as the integral of a probability density function of n variables, thereby generalizing the probability density function introduced in Chapter 6. The resulting concept of a multivariate pdf is illustrated through several constructions of such functions and through an application to the lifetime of a system composed of several subsystems. With this chapter we will have come a long way towards reaching our Goal 1 of developing mathematical models for random phenomena.

7.2 MULTIVARIATE OR JOINT CDFs

7.2.1 Definition of Multivariate CDF

The easiest way to describe vector-valued random variables is through the multivariate or joint density function. (See Definition 7.2.1.) However, we start by paralleling the univariate case with a more general approach in terms of cdfs. We extend our previous definition of a cdf (see Chapter 5) to describe probability measures when $\Omega \subset \mathbb{R}^n$. We now have a random vector variable $\mathbf{X}$ (with components $X_1, \ldots, X_n$) taking on values $\mathbf{x} \in \mathbb{R}^n$. For example, $\mathbf{X}$ could represent various node voltages in a circuit, the state variables in a dynamical system, the recent history of stock prices, or time-sampled speech amplitudes. For $n > 1$, such cdfs are referred to either as "multivariate" or as "joint." In order to define a multivariate cdf that will be the restriction of the probability measure to certain events involving $\mathbf{X}$, we first introduce a family of sets we call *corners*, indexed by $\mathbf{x} = (x_1, \ldots, x_n)$ with generic member

$$C_{\mathbf{x}} = \{(X_1, \ldots, X_n) : (\forall i \leq n)X_i \leq x_i\}.$$

The set $C_{\mathbf{x}}$ is an orthantlike or semi-infinite corner with "northeast" vertex (vertex in the direction of the first orthant) specified by the point $\mathbf{x}$.

Definition 7.2.1 (Multivariate or Joint CDF)   If $\Omega \subset \mathbb{R}^n$, $\mathcal{B}$ the corresponding Borel algebra, then the *multivariate cdf* is

$$F_{X_1, \ldots, X_n}(x_1, \ldots, x_n) = F_{\mathbf{X}}(\mathbf{x}) = P(C_{\mathbf{x}}).$$

Example 7.1   Bivariate CDF _____

A simple example of a multivariate cdf for $n = 2$ is defined in terms of an arbitrarily selected point $\mathbf{a} = (a_1, a_2)$ by

$$F_{X_1, X_2}(x_1, x_2) = U(x_1 - a_1)U(x_2 - a_2) = \begin{cases} 1 & \text{if } x_1 \geq a_1, x_2 \geq a_2 \\ 0 & \text{otherwise} \end{cases}.$$

This corresponds to $P(\mathbf{X} = \mathbf{a})$.

7.2.2 Basic Properties of the CDF

We explore the elementary consequences of this definition by first verifying that $F_{X_1, \ldots, X_n}(x_1, \ldots, x_n)$ is nondecreasing in each variable. Let $\mathbf{x} \leq \mathbf{y}$ mean that, for each $i \leq n$, the ith component x_i of $\mathbf{x}$ is less than or equal to the ith component y_i of $\mathbf{y}$; that is, $(\forall i \leq n)y_i \geq x_i$. It then follows that the corner $C_{\mathbf{x}} \subset C_{\mathbf{y}}$. Hence, by the monotonicity of P with respect to set inclusion,

$$(\mathbf{x} \leq \mathbf{y}) \Rightarrow F_{\mathbf{X}}(\mathbf{x}) = P(C_{\mathbf{x}}) \leq P(C_{\mathbf{y}}) = F_{\mathbf{X}}(\mathbf{y}). \tag{7.1}$$

We see that for any $1 \le i \le n$, if $x_i \to -\infty$, then $C_{\mathbf{x}}$ shrinks (monotonically decreases) to the empty set $\emptyset$. Hence, from K4,

$$\lim_{x_i \to -\infty} F_{X_1,\dots,X_n}(x_1,\dots,x_n) = \lim_{x_i \to -\infty} P(C_{\mathbf{x}}) = 0. \tag{7.2}$$

In order to examine the consequences of letting $x_i \to +\infty$, observe that as x_i increases, the probability $P(\mathbf{X} \in C_{\mathbf{x}})$ increases to that of $P(X_1 \le x_1, \dots, X_{i-1} \le x_{i-1}, X_{i+1} \le x_{i+1}, \dots, X_n \le x_n)$—the constraint on X_i has been removed. Hence, expressing this in terms of cdfs yields

$$\lim_{x_i \to \infty} F_{X_1,\dots,X_n}(x_1,\dots,x_n) = F_{X_1,\dots,X_{i-1},X_{i+1},\dots,X_n}(x_1,\dots,x_{i-1},x_{i+1},\dots,x_n). \tag{7.3}$$

Thus, the cdf for $X_1, \dots, X_{n-1}$ is easily determined from the cdf for $X_1, \dots, X_n$ by taking the limit as $x_n \to \infty$.

Higher order (dimension) cdfs determine the lower order ones, but not conversely.

In particular, it follows that

$$\lim_{x_1,\dots,x_n \to \infty} F_{X_1,\dots,X_n}(x_1,\dots,x_n) = 1. \tag{7.4}$$

Example 7.2   Marginal CDF from Bivariate

Consider the bivariate cdf

$$F_{X_1,X_2}(x_1,x_2) = \frac{1}{1 + e^{-2x} + 3e^{-3x}}.$$

The marginal cdf F_{X_1} for X_1 alone is

$$P(X_1 \le x_1) = F_{X_1}(x_1) = \lim_{x_2 \to \infty} F_{X_1,X_2}(x_1,x_2)$$

$$= \lim_{x_2 \to \infty} \frac{1}{1 + e^{-2x} + 3e^{-3x}} = \frac{1}{1 + e^{-2x}}.$$

Recalling the notational convention that $P(A, B) = P(A \cap B)$, for $n = 2$ we can verify that, for $a_1 \le b_1$ and $a_2 \le b_2$,

$$P(a_1 < X_1 \le b_1, a_2 < X_2 \le b_2)$$

$$= F_{\mathbf{X}}(b_1,b_2) - F_{\mathbf{X}}(a_1,b_2) - F_{\mathbf{X}}(b_1,a_2) + F_{\mathbf{X}}(a_1,a_2). \tag{7.5}$$

This result is an immediate consequence of the set identity for the union of disjoint sets given by

$$C_{b_1,b_2} = \{a_1 < X_1 \le b_1, a_2 < X_2 \le b_2\} \cup (C_{a_1,b_2} - C_{a_1,a_2}) \cup C_{b_1,a_2}.$$

A sketch of the sets in question will clarify this identity.

Example 7.3   Probability of Rectangles _____

Introduce the cdf

$$F_{X_1,X_2} = (1 - e^{-2x_1})U(x_1)\left(1 - \frac{3}{x_2}\right)U(x_2 - 3).$$

$$P(0 < X_1 \le 1, 2 < X_2 \le 4) = F_{X_1,X_2}(1, 4) - F_{X_1,X_2}(0, 4) - F_{X_1,X_2}(1, 2) + F_{X_1,X_2}(0, 2)$$

$$= (1 - e^{-2})\left(1 - \frac{3}{4}\right) - (1 - e^0)\left(1 - \frac{3}{4}\right) - (1 - e^{-2})\left(1 - \frac{3}{2}\right)U(2 - 3)$$

$$+ (1 - e^0)\left(1 - \frac{3}{2}\right)U(2 - 3) = \frac{1}{4}(1 - e^{-2}) - 0 - 0 + 0 = \frac{1}{4}(1 - e^{-2}).$$

In this case, it happens that $P(0 < X_1 \le 1, 2 < X_2 \le 4) = F_{X_1,X_2}(1, 4)$.

7.3 MULTIVARIATE PDF REPRESENTATION OF MULTIVARIATE CDF

Definition 7.3.1   (Multivariate PDF)   A function f is a *multivariate or joint pdf* (commonly called a *density*) if and only if it satisfies the following two conditions:

PDF1. $(\forall \mathbf{x} \in \mathbb{R}^n)\, f(\mathbf{x}) \ge 0$;
PDF2. $\int_{\mathbb{R}^n} f(\mathbf{x})d\mathbf{x} = 1$.

Example 7.4   Bivariate PDF _____

Take $n = 2$ and let f, g, and h be univariate ($n = 1$) pdfs of the kind discussed in Chapter 6. We can form

$$f_{X_1,X_2}(x_1, x_2) = f(x_1)g(x_2)$$

or, as a second example,

$$f_{X_1,X_2}(x_1, x_2) = h(x_2 - x_1)f(x_1).$$

It is immediate from the nonnegativity of f and g that the bivariate pdf is also nonnegative. Satisfying PDF2 is readily verified. In the first example, the double integral factors as a product of integrals, each of which has unit value:

$$\int_{-\infty}^{\infty} \int_{-\infty}^{\infty} f_{X_1,X_2}(x_1, x_2)\, dx_1\, dx_2 = \int_{-\infty}^{\infty} f(x_1)\, dx_1 \int_{-\infty}^{\infty} g(x_2)\, dx_2 = 1 \cdot 1 = 1.$$

In the second example, we carry out the double integration by integrating first on x_2 and changing variables to $z = x_2 - x_1$ in that integral:

$$\int_{-\infty}^{\infty} \int_{-\infty}^{\infty} f_{X_1,X_2}(x_1, x_2)\, dx_1\, dx_2 = \int_{-\infty}^{\infty} f(x_1)\, dx_1 \int_{-\infty}^{\infty} h(x_2 - x_1)\, dx_2$$

$$= \int_{-\infty}^{\infty} f(x_1) \int_{-\infty}^{\infty} h(z)\, dz\, dx_1 = \int_{-\infty}^{\infty} f(x_1) \cdot 1\, dx_1 = 1.$$

The essential role of the multivariate pdf f_X is that it can determine the probability measure P through integration.

Theorem 7.3.1   (Measure Representation by PDF)   If f_X is a multivariate pdf, then, for all $A \in \mathcal{B}$,

$$P(A) = \int_A f_X(\mathbf{x}) d\mathbf{x} = \int_{-\infty}^{\infty} \cdots \int_{-\infty}^{\infty} I_A(x_1, \ldots, x_n) f_{X_1, \ldots, X_n}(x_1, \ldots, x_n) \, dx_1 \ldots dx_n$$

determines a probability measure on the Borel subsets of $\mathbb{R}^n$. Furthermore, P induces a multivariate cdf

$$F_X(\mathbf{x}) = P(C_\mathbf{x}) = \int_{-\infty}^{x_1} \cdots \int_{-\infty}^{x_n} f_X(y_1, \ldots, y_n) \, dy_1 \ldots dy_n.$$

Example 7.5   Probability of the Unit Disk _____

Introduce the bivariate pdf

$$f_{X_1, X_2}(x_1, x_2) = \frac{1}{2\pi} e^{-x_1} U(x_1) U(\pi - |x_2|)$$

as a product of $\mathcal{E}(1)$ and $\mathcal{U}(-\pi, \pi)$ univariate pdfs. We ask for the probability that (X_1, X_2) lies inside the unit disk, or

$$P(X_1^2 + X_2^2 \leq 1) = \int_0^1 e^{-x_1} \left(\int_{-\sqrt{1-x_1^2}}^{\sqrt{1-x_1^2}} \frac{1}{2\pi} \, dx_2 \right) dx_1 = \frac{1}{\pi} \int_0^1 e^{-x_1} \sqrt{1 - x_1^2} \, dx_1.$$

The limits of integration were set up to cover the unit disk, taking into account the fact that the $\mathcal{E}(1)$ pdf is zero for negative x_1. Had we integrated first on x_1 and then on x_2, we would have found that

$$P(X_1^2 + X_2^2 \leq 1) = \frac{1}{\pi} \int_0^1 \left(1 - e^{-\sqrt{1-x_2^2}} \right) dx_2.$$

Numerical integration yields $P(X_1^2 + X_2^2 \leq 1) \approx .169$.

The probability measure P determines the cdf F_X, which (ignoring technicalities of absolute continuity) determines the pdf f_X as a nonnegative function through

$$f_{X_1, \ldots, X_n}(x_1, \ldots, x_n) = \frac{\partial^n F_{X_1, \ldots, X_n}(x_1, \ldots, x_n)}{\partial x_1 \ldots \partial x_n}.$$

The pdf in turn determines the probability measure P, with $P(A)$ being evaluated by integrating $f_{X_1, \ldots, X_n}$ over the multidimensional set A. Hence, from either of these three descriptions, we can reconstruct the other two. Recall that, in the event of discontinuities in the cdf, we accommodate by allowing the pdf to contain unit impulse functions. A unit impulse in $\mathbb{R}^n$ concentrated at the point $\mathbf{x}^0$ is a product $\prod_1^n \delta(x_i - x_i^0)$ of unit impulse functions in each dimension.

The integrability of the pdf f implies that

$$(\forall i) \lim_{x_i \to \pm\infty} f(x_1, \ldots, x_n) = 0.$$

The conclusion about higher order cdfs specifying the lower order ones carries over as well to the multivariate densities. In this case, a simple limiting operation for the cdf is replaced by an integration over unwanted variables in the pdf.

Lemma 7.3.1 (Marginal PDFs) Given a pdf $f_{X_1,\ldots,X_n}(x_1, \ldots, x_n)$ defined for n random variables $X_1, \ldots, X_n$, we can determine the pdf for any $n - 1$ of them by integrating out on the other one. In particular,

$$f_{X_1,\ldots,X_{n-1}}(x_1, \ldots, x_{n-1}) = \int_{-\infty}^{\infty} f_{X_1,\ldots,X_n}(x_1, \ldots, x_n)\, dx_n.$$

Proof sketch. The lemma follows immediately from the property of joint cdfs that

$$F_{X_1,\ldots,X_{n-1}}(x_1, \ldots, x_{n-1}) = \lim_{x_n \to \infty} F_{X_1,\ldots,X_n}(x_1, \ldots, x_n). \qquad \square$$

Example 7.6   Univariate PDF from Bivariate PDF _____

Consider the bivariate pdf of the kind given in the second example of Example 7.4, namely,

$$f_{X_1,X_2}(x_1, x_2) = h(x_2 - x_1)f(x_1),$$

and take

$$h(z) = \frac{1}{\sqrt{2\pi}} e^{-\frac{(z-1)^2}{2}}, \quad f(x) = 2e^{-2x} U(x)$$

to yield

$$f_{X_1,X_2}(x_1, x_2) = \sqrt{\frac{2}{\pi}} e^{-\frac{(x_2-x_1-1)^2}{2}} e^{-2x_1} U(x_1).$$

We determine the univariate $f_{X_2}(x_2)$ by integrating over x_1 in $f_{X_1,X_2}(x_1, x_2)$:

$$f_{X_2}(x_2) = \int_{-\infty}^{\infty} \sqrt{\frac{2}{\pi}} e^{-\frac{(x_2-x_1-1)^2}{2}} e^{-2x_1} U(x_1)\, dx_1 = \int_0^{\infty} \sqrt{\frac{2}{\pi}} e^{-\frac{(x_2-x_1-1)^2}{2}} e^{-2x_1}\, dx_1.$$

This integral can be expressed in terms of exponentials and the error functions discussed in Section 6.7.2, but there is little point in doing so at present.

7.4 EXAMPLES OF MULTIVARIATE PDFs

7.4.1 Product or Independent PDF

An important class of multivariate or joint densities is based on a *product or independence construction*. Given univariate densities $f_{X_1}(x_1), \ldots, f_{X_n}(x_n)$, we can define a joint pdf

$$f_{\mathbf{X}}(x_1, \ldots, x_n) = \prod_{1}^{n} f_{X_i}(x_i). \tag{7.6}$$

It is straightforward to verify that this results in a nonnegative function whose integral over $\Omega = \mathbb{R}^n$ is 1. Hence, we have defined a pdf in n variables built up from pdfs in a single variable. For example,

$$f_{\mathbf{X}}(x_i, \ldots, x_n) = \left(\prod_{1}^{n} \alpha_i \right) e^{-\sum_{1}^{n} \alpha_i x_i} \prod_{1}^{n} U(x_i)$$

defines a multivariate density built from a collection of exponential densities. The first example in Example 7.4 is a bivariate pdf of the product type. In the case of a multivariate density of the product or independence type, the corresponding multivariate cdf is easily seen to be

$$F_{\mathbf{X}}(\mathbf{x}) = \prod_{1}^{n} F_{X_i}(x_i),$$

where F_{X_i} is the univariate cdf corresponding to f_{X_i}. The special case of all random variables $\{X_i\}$ sharing the same individual pdf or cdf is referred to by saying that the random variables are *independent and identically distributed (i.i.d.)*.

While we will be discussing the important concept of independence in Chapter 14, we note that this construction makes sense only when we believe that the component outcomes $\{X_i\}$ are unlinked physically or are information theoretically unrelated to each other. Knowledge of the value of, say, $X_1 = x_1$ tells us nothing about the values of the remaining variables. Thus, the height H_1 of individual 1 and the weight W_2 of an unrelated individual 2 are independent, although H_1 and W_1 would be dependent.

7.4.2 Innovations Model

An important class of multivariate pdfs in which there is a linkage between successive random variables in the enumeration $X_1, X_2, \ldots, X_n$ is given by the following case of an *innovations model*: Let $f_1, \ldots, f_n$ be univariate pdfs of the kind discussed in Chapter 6, and let

$$g_1(x_1), \; g_2(x_1, x_2), \ldots, \; g_{n-1}(x_1, \ldots, x_{n-1})$$

be arbitrary functions, with g_j being a function of the first j variables. The idea is that g_{j-1} $(X_1, \ldots, X_{j-1})$ summarizes the part of the "past" that is relevant to the prediction of the next term X_j. The discrepancy or innovation

$$Y_j = X_j - g_{j-1}(X_1, \ldots, X_{j-1})$$

is the novelty in the next term X_j. The innovation Y_j, described by the pdf f_j, is assumed here to be unlinked with (independent from) the past $X_1, \ldots, X_{j-1}$. For example, if we take $g_j(X_1, \ldots, X_j) = X_j$, so that

$$X_j = Y_j + X_{j-1},$$

then we have what is known as an independent increments process. Such a process might represent our fortune after j stages of gambling, with Y_j our net winnings at the jth stage, based upon our gambling strategy given by g_j and X_j our cumulative winnings. We can now introduce the multivariate innovations pdf

$$f_{X_1, \ldots, X_n}(x_1, \ldots, x_n) = f_{X_1}(x_1) \prod_{j=2}^{n} f_j\left(x_j - g_{j-1}(x_1, \ldots, x_{j-1})\right). \tag{7.7}$$

The nonnegativity of the proposed multivariate pdf is immediate from the nonnegativity of each of the univariate pdfs appearing in the product. That the multivariate pdf integrates to 1 is verified by organizing the multiple integral to integrate in reverse order, starting with x_n and finishing with x_1. In integration over x_j, we make the change of variables $y_j = x_j - g_{j-1}(x_1, \ldots, x_{j-1})$.

Example 7.7 n-variate PDF

$$f_{X_1, \ldots, X_n}(x_1, \ldots, x_n) = \delta(\mathbf{x}_1) \prod_{j=2}^{n} \frac{1}{\sigma\sqrt{2\pi}} e^{-\frac{(x_j - x_{j-1})^2}{2\sigma^2}}$$

defines a discrete-time Brownian motion process (see Section 20.7.2) that starts at time 1 at amplitude 0 and then has unlinked Gaussian increments $\{Y_j\}$ of mean zero and variance σ^2:

$$X_j = Y_j + X_{j-1}.$$

Another class of dependent multivariate pdfs is provided by the concept of Markov dependence discussed in Section 12.6.2 and in Section 20.6. In Markov dependence, successive variables are linked to each other. In effect, variable X_i has a causative relation to X_{i+1}, and given $X_1, \ldots, X_i$, we need only know X_i to specify X_{i+1}. The example just given is indeed one of Markov dependence as well as being an innovations model. However, the definition of Markov dependence is postponed, as it requires the notion of conditional density.

7.4.3 Dirichlet Distribution $\mathcal{D}$

The Dirichlet is an example of a family of multivariate densities on a subset of $\mathbb{R}^n$ in which the component outcomes are dependent. The conditions specified next in its definition make sense if one thinks of the Dirichlet as a pdf over possible pmfs on $n + 1$ outcomes, with the probability p_{n+1} for the $(n + 1)$st outcome being suppressed, as it is uniquely determined by $1 - \sum_1^n p_k$.

Definition 7.4.1   (Dirichlet Density)   For any $n \geq 1$, the Dirichlet $\mathcal{D}(\mathbf{a})$ is a family of pdfs indexed by an $(n+1)$-dimensional parameter vector $\mathbf{a}$ of positive components and defined on an n-dimensional argument $\mathbf{x} = (x_1, \ldots, x_n)$. The pdf $f_{\mathbf{a}}(\mathbf{x})$ is specified by

$$f_{\mathbf{a}}(x_1, \ldots, x_n) = f_{\mathbf{a}}(\mathbf{x}) = \frac{\Gamma(\sum_{k=1}^{n+1} \alpha_k)}{\prod_{k=1}^{n+1} \Gamma(\alpha_k)} \left(1 - \sum_{k=1}^{n} x_k\right)^{\alpha_{n+1}-1} \prod_{k=1}^{n} x_k^{\alpha_k-1}$$

for $\mathbf{x}$ satisfying the pmf conditions for n of $n+1$ terms, namely,

$$(\forall k \leq n)\ x_k \geq 0,$$

$$\sum_{k=1}^{n} x_k \leq 1.$$

$f_{\mathbf{a}}(\mathbf{x})$ is zero if the n components of $\mathbf{x}$ do not satisfy the conditions on a pmf for $n+1$ outcomes. (The $(n+1)$st term $x_{n+1} = 1 - \sum_{k=1}^{n}$ is suppressed.)

It is immediate that the proposed density is nonnegative. It is not immediate that it integrates to 1. Indeed, if we allowed any component of $\mathbf{a}$ to be less than or equal to 0, then we would have a nonintegrable singularity at the origin, and the value of the integral would be infinite. Therefore, we have required that each component of $\mathbf{a}$ be positive to ensure a finite value of the integral of the density.

Example 7.8   Dirichlet PDF ⎯⎯
It will be shown subsequently that, for $n = 1$, the Dirichlet reduces to $\beta(\alpha_1, \alpha_2)$.
    For $n = 2$ and for the components of $\mathbf{a}$ integers, the Dirichlet can be written as

$$f_{\mathbf{a}}(x_1, x_2) = \frac{(\alpha_1 + \alpha_2 + \alpha_3 - 1)!}{(\alpha_1 - 1)!(\alpha_2 - 1)!(\alpha_3 - 1)!} (1 - x_1 - x_2)^{\alpha_3 - 1} x_1^{\alpha_1 - 1} x_2^{\alpha_2 - 1},$$

for $x_1 \geq 0$, $x_2 \geq 0$, and $x_1 + x_2 \leq 1$. A plot of this pdf for $\mathbf{a} = (2, 4, 6)$ is shown in Figure 7.1. The contour plot on the right side of the figure makes it evident that the Dirichlet density is 0 when $x_1 + x_2$ exceeds 1 and also makes evident the asymmetry in the Dirichlet when the components of $\mathbf{a}$ are unequal.

We will be able to verify that the Dirichlet pdf does integrate to 1 once we establish the interesting property of a Dirichlet that its lower dimensional marginals are also Dirichlets.

Lemma 7.4.1   (Dirichlet Marginals)   Given a Dirichlet pdf $f_{\mathbf{a}}(x_1, \ldots, x_n)$ with $(n+1)$-dimensional parameter vector $\mathbf{a}$, the marginal pdf for the variables $x_1, \ldots, x_{n-1}$ is also a Dirichlet pdf $\mathcal{D}(\mathbf{b})$ with n-dimensional parameter vector $\mathbf{b}$ given by

$$(\forall k < n)\ \beta_k = \alpha_k, \quad \beta_n = \alpha_n + \alpha_{n+1}.$$

See Section 7.7 for a proof.

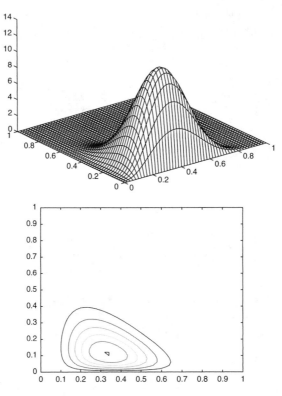

Figure 7.1   Dirichlet density for $a_1 = 2, a_2 = 4, a_3 = 6$.

With this lemma, we now see that any $\mathcal{D}(\mathbf{a})$ for n variables $x_1, \ldots, x_n$ can be reduced through successive integrations to a Dirichlet of a single variable x_1. Specifically, when integrated over $x_2, \ldots, x_n$, the Dirichlet $\mathcal{D}(\mathbf{a})$ becomes

$$f(x_1) = \frac{\Gamma(\sum_1^{n+1} \alpha_k)}{\Gamma(\alpha_1)\Gamma(\sum_2^{n+1} \alpha_k)} x_1^{\alpha_1 - 1} (1 - x_1)^{(\sum_2^{n+1} \alpha_k) - 1}.$$

This result is the Dirichlet for $n = 1$, and it is recognized as the $\beta(\alpha_1, \sum_2^{n+1} \alpha_k)$ of Section 6.5.2. Hence, integration over the last remaining variable x_1 amounts to integration of the Beta, and this is 1 for any density of a single variable. We conclude that $\mathcal{D}(\mathbf{a})$ is a proper pdf.

As noted before, the Dirichlet can be thought of as a distribution over pmfs for $n + 1$ possible outcomes given by $\mathbf{p} = (p_1, \ldots, p_n)$, together with $p_{n+1} = 1 - \sum_k p_k$. If the components of $\mathbf{a}$ are all integers, then we can replace the Gamma functions by factorials ($\Gamma(n + 1) = n!$) to get

$$f_{\mathbf{a}}(\mathbf{x}) = \frac{(\sum_{k=1}^{n+1} a_k - 1)!}{\prod_{k=1}^{n+1}(a_k - 1)!} \left(1 - \sum_{k=1}^{n} x_k\right) \prod_{k=1}^{n} x_k^{a_k - 1}.$$

The pdf has a mathematical form similar to that of the multinomial pmf of Section 2.8; we replace $a_k - 1$ by n_k and understand p_k to be the probability of drawing a symbol of the kth type. If we do so, then only the denominator of the multiplicative constant is the same, but the numerator is different. A more significant difference between the multinomial distribution and the Dirichlet is that the Dirichlet is a pdf for the probabilities themselves considered as random quantities, with the a_k considered fixed and known. Hence, the Dirichlet pdf integrates to 1 when integrated over the elements of $\mathbf{p}$. The multinomial, on the other hand, provides the probabilities for the n_k as random outcomes of draws with fixed and known probabilities p_k. Thus, the multinomial sums to 1 when summed over the $\{n_k\}$.

$\mathcal{D}(\mathbf{a})$ finds many applications in statistics in connection with the multinomial distribution of Section 2.8. If we have $n + 1$ distinct types of objects—say, letters of the alphabet—and we draw a random sample with replacement of total size s having s_i repetitions of the ith letter, $s = \sum_1^{n+1} s_i$, then, conditional on knowing that p_i is the probability of drawing the ith letter, $\sum_1^{n+1} p_i = 1$, the multinomial distribution tells us that the probability of this observation is

$$P(s_1, \ldots, s_{n+1} | p_1, \ldots, p_{n+1}) = \frac{s!}{\prod_1^{n+1} s_i!} \prod_1^{n+1} p_i^{s_i}.$$

If we now believe that the probabilities themselves came from a $\mathcal{D}(\mathbf{a})$, then we can ask for the distribution of the observations. The justification for the calculation we are about to make can be found in the Total Probability Theorem developed in Sections 11.5 and 12.7. Invoking this theorem for our current purposes of demonstrating the roles for the Dirichlet, we find that

$$P(s_1, \ldots, s_{n+1}) = \int_0^1 \cdots \int_0^1 P(s_1, \ldots, s_{n+1} | p_1, \ldots, p_{n+1}) f_\mathbf{a}(p_1, \ldots, p_n) \, dp_1 \ldots dp_n$$

$$= \int_0^1 \cdots \int_0^1 \frac{s!}{\prod_1^{n+1} s_i!} \prod_1^{n+1} p_i^{s_i} \frac{\Gamma(\sum_{k=1}^{n+1} \alpha_k)}{\prod_{k=1}^{n+1} \Gamma(\alpha_k)} \left(1 - \sum_{k=1}^n p_k\right)^{\alpha_{n+1}-1} \left(\prod_{k=1}^n p_k^{\alpha_k - 1}\right) dp_1 \ldots dp_n.$$

The integrand consists of a multiplicative constant term

$$I = \frac{s!}{\prod_1^{n+1} s_i!} \frac{\Gamma(\sum_{k=1}^{n+1} \alpha_k)}{\prod_{k=1}^{n+1} \Gamma(\alpha_k)}$$

and a term

$$II = \int_0^1 \cdots \int_0^1 \left(1 - \sum_{k=1}^n p_k\right)^{s_{n+1}+\alpha_{n+1}-1} \left(\prod_{k=1}^n p_k^{s_k + \alpha_k - 1}\right) dp_1 \ldots dp_n$$

involving the variables of integration. We immediately recognize that II, complicated as it may appear, is just the integral of the $\mathcal{D}$ integrand for a parameter vector that is $\mathbf{s} + \mathbf{a}$! Hence, from our knowledge of this integral, we see that

$$II = \frac{\prod_{k=1}^{n+1} \Gamma(s_k + \alpha_k)}{\Gamma(s + \sum_{k=1}^{n+1} \alpha_k)}.$$

We have arrived at a closed-form solution

$$P(s_1, \ldots, s_{n+1}) = (I)(II) = \frac{s!}{\prod_1^{n+1} s_i!} \frac{\Gamma(\sum_{k=1}^{n+1} \alpha_k) \prod_{k=1}^{n+1} \Gamma(s_k + \alpha_k)}{\prod_{k=1}^{n+1} \Gamma(\alpha_k) \; \Gamma(s + \sum_{k=1}^{n+1} \alpha_k)}$$

for the probability of observing s_k appearances of letter k in s draws with replacement when the probability p_k of drawing letter k was itself selected randomly in accordance with $\mathcal{D}(\mathbf{a})$. The same tools just employed can also be used to update our knowledge of the probabilities $p_1, \ldots, p_{n+1}$, given that we have observed the counts $s_1, \ldots, s_{n+1}$. However, such calculations are better reserved until we have discussed conditional probability in Chapters 11 and 12. Additional properties of the Dirichlet that are of value in statistical applications can be found in Section 7.7 of Wilks [100].

7.4.4 Multivariate Normal PDF

Another example of a multivariate construction that does not impose independence is the *multivariate normal density* specified by a vector $\mathbf{m}$, known as its mean, and a symmetric, nonnegative definite matrix $\mathbb{C}$ known as a covariance matrix. (See Section 9.14.)

Definition 7.4.2 (Multivariate Normal PDF) The n-dimensional column vector of random variables $\mathbf{X}$ is multivariate normal or Gaussian if there is an arbitrary n-vector $\mathbf{m}$ and a covariance matrix $\mathbb{C}$ such that

$$f_{\mathbf{X}}(\mathbf{x}) = (2\pi)^{-\frac{n}{2}} \det(\mathbb{C})^{-\frac{1}{2}} e^{-\frac{1}{2}(\mathbf{x}-\mathbf{m})^T \mathbb{C}^{-1}(\mathbf{x}-\mathbf{m})}. \tag{7.8}$$

This condition can also be stated as $\mathbf{X} \sim \mathcal{N}(\mathbf{m}, \mathbb{C})$.

Note that

$$(2\pi)^{-\frac{n}{2}} \det(\mathbb{C})^{-\frac{1}{2}} = \det(2\pi \mathbb{C})^{-\frac{1}{2}}.$$

Example 7.9   Bivariate Gaussian/Normal _____

The bivariate Gaussian pdf with parameters

$$\mathbf{m} = \begin{pmatrix} 0 \\ 0 \end{pmatrix}, \text{ and } \mathbb{C} = \begin{pmatrix} 3 & -2 \\ -2 & 3 \end{pmatrix},$$

has the inverse covariance matrix

$$\mathbb{C}^{-1} = \begin{pmatrix} .6 & .4 \\ .4 & .6 \end{pmatrix},$$

and the corresponding pdf is given by

$$f_{X_1, X_2}(x_1, x_2) = \frac{1}{2\pi} \frac{1}{\sqrt{5}} \exp\left(-\frac{1}{2}(.6x_1^2 + .8x_1 x_2 + .6x_2^2)\right).$$

A plot of this pdf is shown in Figure 7.2. The arguments x_1 and x_2 each take values in the interval $[-5, 5]$.

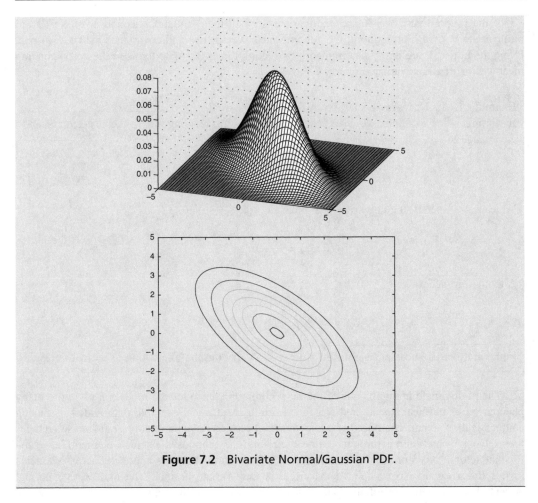

Figure 7.2   Bivariate Normal/Gaussian PDF.

We will provide an alternative specification of the multivariate normal density in Section 17.6 based upon the tool of characteristic functions introduced in Chapter 17. Specifying the multivariate normal by characteristic functions will enable us to easily establish the following lemma:

Lemma 7.4.2   (Marginals of Multivariate Normals)   Let $\mathbf{X}$ be an n-dimensional multivariate normal $\mathcal{N}(\mathbf{m}, \mathbb{C})$ and $\mathbf{Y}$ a k-dimensional vector, $k < n$, formed by selecting the components $X_{i_1}, \ldots, X_{i_k}$ of $\mathbf{X}$. Define

$$\mathbf{m}_Y = (m_{i_1}, \ldots, m_{i_k})^T$$

and $\mathbb{C}_Y$ as the $(k \times k)$-dimensional matrix formed by selecting those elements of $\mathbb{C}$ lying in the rows and columns indexed by $i_1, \ldots, i_k$. Then

$$\mathbf{Y} \sim \mathcal{N}(\mathbf{m}_Y, \mathbb{C}_Y).$$

If the random variables $\mathbf{X}$ are multivariate normal, then any subcollection $\mathbf{Y}$ of $\mathbf{X}$ is itself multivariate normal. Furthermore, the defining mean vector $\mathbf{m}_Y$ and covariance matrix $\mathbb{C}_Y$ specifying the pdf of $\mathbf{Y}$ are easily determined by selecting the appropriate terms in the corresponding mean vector $\mathbf{m}$ and covariance matrix $\mathbb{C}$ for $\mathbf{X}$.

Example 7.10   Univariate PDF from Bivariate Normal _____

In the preceding example, we specified a bivariate normal having a zero mean vector $\mathbf{m}$ and covariance matrix

$$\mathbb{C} = \begin{pmatrix} 3 & -2 \\ -2 & 3 \end{pmatrix}.$$

We can now specify the univariate marginal through $X_1 \sim \mathcal{N}(0, 3)$, or

$$f_{X_1}(x_1) = \frac{1}{\sqrt{2\pi \cdot 3}} e^{-\frac{x_1^2}{6}}.$$

Although, in general,

$$f_{X_1}(x_1) = \int_{-\infty}^{\infty} f_{X_1, X_2}(x_1, x_2)\, dx_2,$$

in the multivariate normal case we do not need to carry out the integration to know the result.

The multivariate normal is one of the most commonly encountered models applying to many phenomena of physical, social, and biological origin. A physical example is provided by analog communication signals composed of a known signal plus additive thermal noise (or noise voltages themselves) being filtered through a linear circuit (one composed of such elements as resistors, inductors, capacitors, and linear amplifiers). The various voltages appearing across elements in this linear circuit have a joint density that is multivariate normal. The multivariate normal is central to the definition of the Gaussian random process (GRP) discussed in Section 20.11. Examples of a GRP are the thermal, flicker, shot, and Barkhausen noises that were noted in Section 1.3.

7.5 APPLICATION TO RELIABILITY

Reliability theory is concerned with determining the distribution (cdf) F_L for the lifetime L of a system $\mathcal{S}$ that is composed of subsystems, $\mathcal{S} = \mathcal{S}(\mathcal{S}_1, \ldots, \mathcal{S}_n)$, with an individual subsystem $\mathcal{S}_i$ having lifetime L_i described by the cdf F_{L_i}. There are two issues. The first is the dependence of the overall system lifetime $L = L(L_1, \ldots, L_n)$ on the individual lifetimes. The second issue is the dependencies between the subsystem lifetimes. In the well-known incident involving the Brown's Ferry nuclear reactor, three safety subsystems were all dependent upon a common DC power supply, and failure of this power supply meant unanticipated simultaneous failure of all three safety systems. The challenge of reliability modeling for the kinds of highly complex systems on which our society has become dependent (e.g., electric power) is one of understanding the often hidden dependencies between subsystems. However, in our illustrative application to reliability, we assume that $n = 2$ and that the two subsystem lifetimes L_1, L_2 are described by a joint pdf

$$f_{L_1,L_2}(x_1, x_2) = f_{L_1}(x_1)f_{L_2}(x_2)$$

that is of the product or independence type; this assumption implies that failures of the two subsystems are unrelated to each other. It follows from the product construction that

$$F_{L_1,L_2}(x_1, x_2) = F_{L_1}(x_1)F_{L_2}(x_2).$$

More specifically, we might assume that $L_i \sim \mathcal{E}(\alpha_i)$ or that

$$f_{L_1,L_2}(x_1, x_2) = \alpha_1 \alpha_2 e^{-\alpha_1 x_1 - \alpha_2 x_2} U(x_1)U(x_2),$$

$$F_{L_1,L_2}(x_1, x_2) = (1 - e^{-\alpha_1 x_1})(1 - e^{-\alpha_2 x_2})U(x_1)U(x_2).$$

We examine three possible relationships between the overall system lifetime L and the component subsystem lifetimes L_1, L_2:

- In the parallel mode of operation, the system $\mathcal{S}$ continues to operate so long as a single subsystem is operating; the two subsystems are redundant. Hence, the system lifetime L_p is that of the longest-lived subsystem. It follows that $L_p = \max(L_1, L_2)$.
- In the serial mode of operation (think of a chain carrying a load that drops the load if any of its links break), both subsystems are essential to the operation of the overall system and the first subsystem to fail causes the failure of $\mathcal{S}$. In this case, $L_s = \min(L_1, L_2)$.
- In the spare parts mode of operation, both subsystems are redundant. Subsystem $\mathcal{S}_1$ is deployed first, and when it fails, subsystem $\mathcal{S}_2$ is deployed. In this case, $L_{sp} = L_1 + L_2$.

The calculations of F_L that follow for each of the preceding three modes of operation serve as a bridge to the discussion of functions of random variables that is the subject of the next chapter.

Example 7.11   Lifetimes in Parallel Mode _____

In the parallel mode,

$$F_{L_p}(x) = P(L_p \leq x) = P(\max(L_1, L_2) \leq x) = P((L_1 \leq x) \cap (L_2 \leq x))$$

$$= P(L_1 \leq x, L_2 \leq x) = F_{L_1,L_2}(x, x) = F_{L_1}(x)F_{L_2}(x).$$

The corresponding pdf, derived by differentiation with respect to x, is

$$f_{L_p}(x) = f_{L_1}(x)F_{L_2}(x) + F_{L_1}(x)f_{L_2}(x).$$

Using our specific hypotheses, we find that

$$F_{L_p}(x) = (1 - e^{-\alpha_1 x})(1 - e^{-\alpha_2 x})U(x).$$

Example 7.12   Lifetimes in Serial Mode

In the serial mode,

$$F_{L_s}(x) = P(L_s \leq x) = P(\min(L_1, L_2) \leq x) = P((L_1 \leq x) \cup (L_2 \leq x)).$$

Recall that $P(A \cup B) = P(A) + P(B) - P(A \cap B)$ to write

$$P((L_1 \leq x) \cup (L_2 \leq x)) = P(L_1 \leq x) + P(L_2 \leq x)$$

$$-P((L_1 \leq x) \cap (L_2 \leq x)).$$

Hence,

$$F_{L_s}(x) = F_{L_1}(x) + F_{L_2}(x) - F_{L_1, L_2}(x, x)$$

$$= (1 - e^{-\alpha_1 x})U(x) + (1 - e^{-\alpha_2 x})U(x) - (1 - e^{-\alpha_1 x})(1 - e^{-\alpha_2 x})U(x).$$

The corresponding pdf in the general case is

$$f_{L_s}(x) = f_{L_1}(x) + f_{L_2}(x) - f_{L_1}(x)F_{L_2}(x) - F_{L_1}(x)f_{L_2}(x)$$

$$= (1 - F_{L_2}(x))f_{L_1}(x) + (1 - F_{L_1}(x))f_{L_2}(x),$$

where we have organized the terms to make it clear that f_{L_s} is nonnegative, as required.

Example 7.13   Lifetimes in Spare-Parts Mode

In the spare-parts mode,

$$F_{L_{sp}}(x) = P(L \leq x) = P(L_1 + L_2 \leq x) = \int_{\{(x_1, x_2) : x_1 + x_2 \leq x\}} f_{L_1, L_2}(x_1, x_2) \, dx_1 \, dx_2.$$

Hence, we must integrate over the set of points (x_1, x_2) lying below the line $x_1 + x_2 = x$,

$$F_{L_{sp}}(x) = \int_{-\infty}^{\infty} dx_1 \int_{-\infty}^{x - x_1} dx_2 f_{L_1}(x_1) f_{L_2}(x_2)$$

$$= \int_{-\infty}^{\infty} f_{L_1}(x_1) \, dx_1 \int_{-\infty}^{x - x_1} f_{L_2}(x_2) \, dx_2 = \int_{-\infty}^{\infty} f_{L_1}(x_1) F_{L_2}(x - x_1) \, dx_1.$$

Differentiation with respect to x yields the pdf

$$f_{L_{sp}}(x) = \int_{-\infty}^{\infty} f_{L_1}(x_1) f_{L_2}(x - x_1) \, dx_1,$$

as a convolution of f_{L_1} with f_{L_2}. It is reasonable to assume in the spare-parts case that the two subsystems have identically distributed lifetimes ($\alpha_1 = \alpha_2 = \alpha$). In this case, we find that, for

$x \geq 0$ (if $x < 0$, then $F_{L_{sp}}(x) = 0$),

$$F_{L_{sp}}(x) = \int_{-\infty}^{\infty} \alpha e^{-\alpha x_1} U(x_1)(1 - e^{-\alpha(x-x_1)}) U(x - x_1)\, dx_1$$

$$= \int_{0}^{x} \alpha e^{-\alpha x_1}(1 - e^{-\alpha(x-x_1)})\, dx_1 = (1 - e^{-\alpha x} - \alpha x e^{-\alpha x}) U(x).$$

The corresponding pdf

$$f_{L_{sp}}(x) = \frac{dF_{L_{sp}}(x)}{dx} = \alpha^2 x e^{-\alpha x} U(x)$$

is of the Rayleigh type.

Note that $L_s < L_p < L_{sp}$ is an immediate consequence of

$$\min(L_1, L_2) \leq \max(L_1, L_2) < L_1 + L_2.$$

Hence, it follows that

$$(\forall x)\ F_{L_{sp}}(x) \leq F_{L_p}(x) \leq F_{L_s}(x).$$

7.6 SUMMARY

We extend the discussion in Chapters 5 and 6 to cover uncountably infinite sample spaces that are now subsets of the sample space $\mathbb{R}^n$ of n-tuples of real numbers and that allow us to model vector-valued random variables. The specification of a probability measure P is now through multivariate or joint cdfs (Section 7.2) and pdfs (Section 7.3), with a multivariate pdf $f_{\mathbf{X}}(\mathbf{x})$ needing to satisfy only the two conditions

$$(\forall \mathbf{x} \in \mathbb{R}^n) f_{\mathbf{X}}(\mathbf{x}) \geq 0, \qquad \int_{\mathbb{R}^n} f_{\mathbf{X}}(x_1, \ldots, x_n)\, dx_1 \ldots dx_n = 1.$$

This extension introduces little conceptual novelty, but does increase the complexity of the mathematical manipulations. Important families of multivariate pdfs are the product or independence, innovations, and multivariate normal pdfs. The product or independence cdf or pdf is a product of univariate cdfs or pdfs. We discuss the meaning of this construction in Chapter 14 and find that it is the correct model when the random variables $X_1, \ldots, X_n$ are unlinked to each other. The innovations pdf applies when there is sequential linkage in an appropriate enumeration of the random variables, when, say, X_j depends upon the "past" variables $X_1, \ldots, X_{j-1}$ only through, at most, some function g_{j-1} of these past variables. In the case of the innovations model, an increment Y_j is added to g_{j-1} that is independent of g_{j-1} and has pdf f_{Y_j}. We note that there is an important class of models known as Markov processes whose discussion is deferred to Sections 12.6.2 and 20.6, as it requires the concept of conditional probability. Finally, we introduce the multivariate normal, or Gaussian, pdf as a generalization of the univariate normal introduced in Section 6.7.2. The multivariate normal will be given an easier characterization in Section 17.6. At this point, we have introduced all of the basic methods for representing probability and describing random variables, with the exception of the characteristic and generating functions of Chapter 17. We have also presented the most commonly encountered specific families of probability models, although we will have more to say about their interrelationships.

The chapter closes with an extended application of the independence pdf to assessing system lifetime, a topic in reliability theory.

7.7 APPENDIX: PROOF OF LEMMA 7.4.1

Proof of Lemma 7.4.1 From Lemma 7.3.1, the marginal pdf

$$f(x_1, \ldots, x_{n-1}) = \int_{-\infty}^{\infty} f_{\mathbf{a}}(x_1, \ldots, x_n)\, dx_n.$$

Note that the integration is over the variable x_n that appears in $f_{\mathbf{a}}$ only in the term

$$x_n^{\alpha_n - 1} \left(1 - \sum_1^{n-1} x_k - x_n\right)^{\alpha_{n+1} - 1}.$$

Taking into account the limits of integration, we have to evaluate

$$I = \int_0^{1 - \sum_1^{n-1} x_k} x_n^{\alpha_n - 1} \left(1 - \sum_1^{n-1} x_k - x_n\right)^{\alpha_{n+1} - 1} dx_n.$$

Changing the integration variable to $y = x_n/(1 - \sum_1^{n-1} x_k)$ yields

$$I = \left(1 - \sum_1^{n-1} x_k - x_n\right)^{\alpha_n + \alpha_{n+1} - 1} \int_0^1 y^{\alpha_n - 1}(1 - y)^{\alpha_{n+1} - 1}\, dy.$$

Recalling the beta density discussed in Section 6.5.2, we see that

$$\int_0^1 y^{\alpha_n - 1}(1 - y)^{\alpha_{n+1} - 1}\, dy = \frac{\Gamma(\alpha_n)\Gamma(\alpha_{n+1})}{\Gamma(\alpha_n + \alpha_{n+1})}.$$

Restoring the terms in $f_{\mathbf{a}}$ that depended upon $x_1, \ldots, x_{n-1}$, we find that

$$f(x_1, \ldots, x_{n-1}) = \frac{\Gamma(\sum_{k=1}^{n+1} \alpha_k)}{\Gamma(\alpha_n + \alpha_{n+1}) \prod_{k=1}^{n-1} \Gamma(\alpha_k)} \left(1 - \sum_{k=1}^{n-2} x_k\right)^{\alpha_n + \alpha_{n+1} - 1} \prod_{k=1}^{n-1} x_k^{\alpha_k - 1}.$$

Thus, $f(x_1, \ldots, x_{n-1})$ is precisely of the form of a Dirichlet $\mathcal{D}(\mathbf{b})$ when we identify

$$(\forall k \leq n - 1) \ \beta_k = \alpha_k, \ \beta_n = \alpha_n + \alpha_{n+1},$$

as was to be shown. $\square$

EXERCISES

E7.1 Let each F_i be a univariate cdf and define the bivariate function

$$F(x, y) = 0.5 F_1(x) F_2(y) + 0.5 F_3(x) F_4(y).$$

   a. Evaluate $\lim_{x \to \infty, y \to \infty} F(x, y)$.
   b. Is $\lim_{x \to \infty} F(x, y)$ a univariate cdf?
   c. Evaluate

$$f(x, y) = \frac{\partial^2 F(x, y)}{\partial x \partial y}.$$

   d. Is f a bivariate pdf?
   e. Is F a bivariate cdf?

E7.2 Let each F_i be a univariate cdf and define the trivariate function

$$F(x, y, z) = \alpha F_1(x) F_2(y) F_3(z) + \beta F_4(x) F_5(y) F_6(z).$$

   a. What are necessary constraints on α and β for F to be a cdf?
   b. Assuming that these constraints are satisfied, evaluate the bivariate cdf $F(x, z)$ that is determined by $F(x, y, z)$.

E7.3 If the bivariate cdf

$$F_{X,Y}(x, y) = 0.3\Phi(x - 1)U(y + 1) + .7U(x)(1 - e^{-3y})U(y)$$

   (Φ is the $\mathcal{N}(0, 1)$ cdf), then evaluate the joint probability that both $-2 < X \leq -1$ and $Y > -0.5$.

E7.4 The joint pdf

$$f_{X,Y}(x, y) = e^{-x} U(x) U(y) U(1 - y).$$

   Evaluate the probabilities of A, A^c, and B for

$$A = \{(X, Y) : X \geq Y\} \text{ and } B = \{(X, Y) : X = Y\}.$$

E7.5 A joint or bivariate pdf

$$f_{X,Y}(x, y) = \begin{cases} 3x^2 & \text{if } 0 \leq x \leq 1, 0 \leq y \leq 1 \\ 0 & \text{if otherwise} \end{cases}.$$

    a. Evaluate the pdf f_Y.
    b. Evaluate the cdf F_Y.
    c. Evaluate $P(X \geq Y)$.
    d. Is this joint density of the product or innovation type?

E7.6   a. If the density $f_{X,Y}(x, y) = e^{-2y} U(y)U(x)U(2 - x)$, then evaluate the density f_X and the cdf F_X.
    b. Is this bivariate density of the product or innovation type?
    c. Evaluate the probability that X exceeds Y, $P(X > Y)$.

E7.7 In a particle accelerator, a pointlike particle strikes a circular target of radius 1. The impact point is equally likely to fall anywhere within the target area.

    a. What is the density $f_{X,Y}(x, y)$ when X, Y are Cartesian coordinates for the plane of the target with the origin at the center of the target?
    b. Is this joint density of the product or innovation type?
    c. What is the probability that the particle will strike within a distance r of the center of the target?
    d. Evaluate the cdf $F_X(x)$ (for the x-coordinate for $x < 0$).

E7.8 There are two components C_1, C_2, having random lifetimes L_1, L_2, respectively, that have the density

$$f_{L_1,L_2}(x, y) = \alpha\beta e^{-\alpha x - \beta y} U(x)U(y),$$

where $U(z)$ is the unit step function and α, β are positive.

    a. Is this joint density of the product or innovation type?
    b. Evaluate the cdf F_{L_1,L_2}.
    c. Evaluate the cdf F_{L_1} of L_1.
    d. Evaluate the probabilities of the following three events:

$$L_1 = t_1; \quad L_1 > t_1; \quad \text{both components fail prior to time } t.$$

E7.9 If the cdf $F_{X,Y}$ for $\Omega = \mathbb{R}^2$ is

$$F_{X,Y}(x, y) = (1 - e^{-\alpha x})(1 - e^{-\alpha y})U(x)U(y),$$

then evaluate the joint density $f_{X,Y}$, the marginal cdf

$$F_X(x) = P(\{(X, Y) : X \leq x\}),$$

and its density f_X.

E7.10   a. If

$$F_{X,Y}(x,y) = \begin{cases} 0 & \text{if } x < 0 \text{ or } y < 0 \\ 1 & \text{if } x > 1, y > 1 \\ xy & \text{if } 0 \le x \le 1, 0 \le y \le 1 \\ x & \text{if } 0 \le x \le 1, y > 1 \\ y & \text{if } 0 \le y \le 1, x > 1 \end{cases},$$

then evaluate F_X.

   b. Evaluate the pdf $f_{X,Y}$.

E7.11 We are given that

$$f_{X,Y}(x,y) = \frac{1}{2}\delta(x)\delta(y) + \frac{1}{2\pi}U(1 - x^2 - y^2).$$

   a. Verify that $f_{X,Y}$ is a pdf.
   b. Evaluate and sketch f_X.
   c. Evaluate and sketch the cdf F_X.
   d. Evaluate $P(X^2 + Y^2 \le \frac{1}{2})$.

E7.12 If the joint pdf

$$f_{X,Y}(x,y) = \begin{cases} x + y & \text{if } 0 \le x \le 1, 0 \le y \le 1 \\ 0 & \text{if otherwise} \end{cases},$$

then evaluate $P(Y \ge X + \frac{1}{2})$.

E7.13 The joint pdf, for waiting times T_1, T_2 to the first and second particle emissions, is given by

$$f_{T_1,T_2}(t_1, t_2) = \alpha^2 e^{-\alpha t_2} U(t_2 - t_1)U(t_1) = \begin{cases} \alpha^2 e^{-\alpha t_2} & \text{if } t_2 \ge t_1 \ge 0 \\ 0 & \text{if otherwise} \end{cases}.$$

   a. Is this joint density of the innovations type?
   b. Evaluate the pdf $f_{T_1}(t_1)$.
   c. Evaluate the pdf $f_{T_2}(t_2)$.

E7.14 In many circumstances, we measure a signal X that is a sum $S + N$ of a quantity S that we are interested in and a measurement noise N. When S and N are unlinked, it can be shown (see Section 13.5) that in this case the joint density

$$f_{X,S}(x,s) = f_S(s)f_N(x - s).$$

   a. Is $f_{X,S}$ of the product or innovation type?
   b. If $N \sim \mathcal{N}(0, \sigma_n^2)$ (a frequent assumption about measurement noise) and $S \sim \mathcal{N}(m_s, \sigma_s^2)$, evaluate $f_{X,S}$.
   c. Is $f_{X,S}$ of the bivariate normal type?

E7.15   If the joint cdf

$$F_{X,Y,Z}(x,y,z) = \frac{1}{1 + e^{-x} + e^{-y} + e^{-z}},$$

then evaluate $P(X \geq 1)$. (This does not require a lot of calculation.)

E7.16   a. Show that, for any $n \geq 1$ and positive $\alpha_1, \ldots, \alpha_n$ and positive $\beta_1, \ldots, \beta_n$, the function

$$G(x_1, \ldots, x_n) = [1 + \sum_{i=1}^{n} \beta_i e^{-\alpha_i x_i}]^{-1}$$

is a joint or multivariate cdf. (Hint: Show that the corresponding density is nonnegative, and use the properties of G to show that this density integrates to 1.)

   b. Is G of the product or innovation type?

E7.17   Let $f_{\mathbf{a}}(x_1, x_2)$ be a Dirichlet pdf with $\alpha_1 = \alpha_2$.

   a. Is $f_{\mathbf{a}}$ a symmetric function in that, for any x_1 and x_2, $f_{\mathbf{a}}(x_1, x_2) = f_{\mathbf{a}}(x_2, x_1)$?
   b. If $\alpha_1 = \alpha_2 = 2$ and $\alpha_3 = 3$, what is the pdf for X_1?

E7.18   Given that $X \sim \beta(\alpha_1, \alpha_2)$ and $Y \sim \beta(\alpha_3, \alpha_4)$, what are the constraints on the $\{\alpha_i\}$ so that X and Y are bivariate Dirichlet?

E7.19   A message source has an alphabet of the three symbols a, b, c, and it generates a message of length 100. The symbols are chosen with respective probabilities p_a, p_b, p_c, and successive symbols are independent of each other or unlinked with each other. The symbol probabilities are unknown to us, but we know that they themselves were chosen according to a Dirichlet, with parameters $\alpha_a = 1, \alpha_b = 2, \alpha_c = 3$ for the respective probabilities.

   a. What is the probability that the message will contain 30 appearances of the symbol a and 25 appearances of the symbol b?
   b. What is the probability that the same source will generate a sequence having 50 appearances of a?

E7.20   A trivariate normal random vector $\mathbf{X}$ has mean vector

$$\mathbf{m} = \begin{pmatrix} 0 \\ -1 \\ 2 \end{pmatrix} \text{ and covariance matrix } \mathbb{C} = \begin{pmatrix} 6 & 7 & 2 \\ 7 & 10 & 3 \\ 2 & 3 & 4 \end{pmatrix}.$$

   a. Write out the pdf $f_{\mathbf{X}}(x_1, x_2, x_3)$.
   b. Write out the pdf $f_{X_1, X_3}(x_1, x_3)$.
   c. Write out the pdf $f_{X_2}(x_2)$.
   d. Which of X_2 or X_3 has the greater probability of being positive?

E7.21   Let g_1, g_2, and g_3 be univariate pdfs, and define

$$f_{X,Y,Z}(x,y,z) = g_1(z)g_2(y - z^2)g_3(x - yz).$$

   a. Verify that $f_{X,Y,Z}$ is a trivariate pdf and determine its type.
   b. Determine $f_{Y,Z}$.
   c. Provide an expression for f_X.

E7.22 a. If a system with lifetime L operates three subsystems in parallel mode, and the subsystem lifetimes L_1, L_2, L_3 are described by

$$F_{L_1,L_2,L_3}(x_1, x_2, x_3) = \prod_{i=1}^{3}(1 - e^{-\alpha_i x_i})U(x_i),$$

then evaluate the overall system lifetime cdf $F_L(x)$ and pdf $f_L(x)$.

b. Repeat this analysis for serial mode.

c. Repeat this analysis for spare parts mode.

E7.23 The bivariate Cauchy for $\alpha > 0$ is given by the bivariate pdf

$$f_{X,Y}(x, y) = \frac{\alpha}{2\pi}(\alpha^2 + x^2 + y^2)^{-\frac{3}{2}}.$$

a. Verify that this yields the univariate Cauchy

$$f_X(x) = \frac{1}{\pi}\frac{\alpha}{\alpha^2 + x^2}.$$

b. Hence, verify that $f_{X,Y}$ is indeed a bivariate pdf.

8

Functions of Random Variables

8.1 PURPOSE, BACKGROUND, AND ORGANIZATION

Hitherto, we have been concerned almost exclusively with reaching Goal 1—with developing probability models for random phenomena. In this chapter, we focus on Goal 2—on beginning to understand how to determine the response of a system to random phenomena. We open by discussing in greater detail the concept of random variables and random vectors. This level of detail is not essential to our present purposes, but is essential as an introduction to random variables as they are used in mathematically more advanced treatments of probability. In brief, a random variable is a function which transforms outcomes that may themselves be other random variables or the outcomes in the sample space itself. Moreover, random variables are restricted to functions such that, given a probability measure describing their arguments or inputs, one can calculate a new probability measure that describes their values or outputs. We then proceed to study how to carry out this calculation in general and in detail for specific families of functions. In the case of vector-valued random variables, we study in detail the basic case of linear transformations between vectors specified by matrices, more general one-to-one differentiable transformations, and the problem of quadrature modulation in communications.

    An important application is made to creating random variables having arbitrary desired cdfs. Such random variables are required in creating simulations, an area of activity that has grown greatly in importance with the growth of computing power and the desire to engineer ever more complex systems.

8.2 RANDOM VARIABLES

8.2.1 Background to Random Variables

In pursuit of our second goal of understanding the response of systems having random excitations, we consider the simplest case of memoryless systems or functions. In general, we have an input vector $\mathbf{X} = \{X_1, \ldots, X_n\}$ that gives rise to an output vector $\mathbf{Y} = \{Y_1, \ldots, Y_m\}$ through the relationship

$$\mathbf{Y} = \mathbf{g}(\mathbf{X}).$$

For example, X might be the random maximum Fahrenheit temperature reached tomorrow and Y its centigrade version, $Y = 5/9(X - 32)$. A multivariate example is provided in Section 7.5, where we computed the overall system lifetime L, given as a function of the lifetimes L_1, L_2 of its two subsystems. Unlike the case of deterministic systems, treated in the usual physics and electrical systems courses that have been taken by most readers of this text, our objective is not to determine the numerical values $\mathbf{Y} = \mathbf{y}$ given the function $\mathbf{g}$ and the argument $\mathbf{X} = \mathbf{x}$. What we require now of a description of the output $\mathbf{Y}$ is its probability description or law, a more difficult item. Given a probabilistic description of the inputs, say, through the multivariate or joint density function $f_{\mathbf{X}}$, we wish to determine the corresponding output description $f_{\mathbf{Y}}$. In order to do so, we need to define the impact of mappings on probability, and we do so by first considering more carefully the concept of a *random variable*.

The term "random variable" is a somewhat confusing label used primarily to denote a numerical quantity X (e.g., the height of the next individual to enter the room, the number of Geiger count clicks observed in the next second, the waiting time to the arrival of the next packet, the voltage measured across an antenna at a given time) that somewhat unpredictably takes on a specific numerical value x when the appropriate random experiment or measurement is performed. We also use the term to describe vector-valued random quantities (e.g., the height, weight, and blood pressure of the next person to enter the room, the state vector of capacitor voltages and inductor currents in a circuit, the sequence of waiting times between the arrivals of the first four photons, the sequence of discrete amplitudes recorded on a music CD) and will emphasize the vector nature of the random variable by the notation $\mathbf{X}$.

8.2.2 Functions and Inverse Images

As used in the mathematical theory of probability, the term "random variable" refers synonymously to a function, mapping, or transformation from a given initial probability space triple $\Omega, \mathcal{A}, P$ to a final probability space triple $\Omega_X, \mathcal{B}, P_X$:

$$X : \Omega \mapsto \Omega_X, \quad X(\omega) = x \in \Omega_X.$$

In some cases, this view of a random variable as a function is somewhat strained, and the initial and final probability spaces can be taken to be the same. That is, if X denotes the number of spots showing uppermost in one toss of a die, then we might take the initial sample space $\Omega = \{1 \, dot, 2 \, dots, 3 \, dots, \ldots, 6 \, dots\}$ to have the six elements that are the patterns of dots to be found on the six distinct faces of a die. However, we may also take Ω to be simply $\{1, 2, 3, 4, 5, 6\}$.

In the latter case, $\Omega = \{1, 2, 3, 4, 5, 6\} = \Omega_X$, $\mathcal{A} = \mathcal{B} = 2^\Omega$. Functions from one probability space to another, as indicated at the outset of this section, are of great interest to us, for they represent systems transforming or processing random quantities taking values in Ω into ones taking values in Ω_X. Thus, from probabilistic knowledge of the thermal noise voltage V across a resistor R, we may wish to gain probabilistic knowledge of the random thermal noise power $P = V^2/R$. Or, from probabilistic knowledge of a measured value Y, we may wish to infer to probabilistic knowledge of the true quantity X that was measured.

While the notion of a function X mapping from one set into another one is nothing new to you mathematically, what is new is the need to determine how the original probabilities $P(A)$, determined for events $A \in \mathcal{A}$, transform into the final probabilities $P_X(B), B \in \mathcal{B}$. The basic assumption—and it is an assumption that does not follow from the Kolmogorov axioms—is that a set $B \in \mathcal{B}$ "inherits" its probability from the particular set of all of the possible inputs that could have been transformed by X into B. Such a set is known as the *inverse image* of B under X and is denoted

$$X^{-1}(B) = \{\omega : X(\omega) \in B\}, \tag{8.1}$$

where $X^{-1}(B) \in \mathcal{A}$ and $B \in \mathcal{B}$ are subsets, respectively, of the domain Ω and range Ω_X of the function X.

Example 8.1   Inverse Image _____

$$\Omega = \mathbb{R}, \ \Omega_X = \mathbb{R}, \ X(\omega) = e^\omega, \ B = [-a, a], a > 0,$$

$$X^{-1}(B) = (-\infty, \log a).$$

A more complex example is

$$\Omega = [0, \infty), \Omega_X = \mathbb{R}, X(\omega) = \omega^2 - \omega = \omega(\omega - 1), B = [-a, a], a \geq 0,$$

$$X^{-1}(B) = \begin{cases} \left[0, \frac{1}{2} + \sqrt{a + \frac{1}{4}} \right] & \text{if } a \geq \frac{1}{4} \\[4mm] \left[0, \frac{1}{2} - \sqrt{-a + \frac{1}{4}} \right] \cup \left[\frac{1}{2} + \sqrt{-a + \frac{1}{4}}, \frac{1}{2} + \sqrt{a + \frac{1}{4}} \right] & \text{if } a < \frac{1}{4} \end{cases}.$$

(A sketch of the quadratic function $\omega^2 - \omega$, with zeros at 0 and 1, will help.)

The points in $X^{-1}(B) \subset \Omega$ are the ones that could have given rise to an observation of the event $B \subset \Omega_X$. Thus, we are encouraged to assign $P_X(B) = P(X^{-1}(B))$. However, this will be possible only if $X^{-1}(B) \in \mathcal{A}$. For otherwise, $P(X^{-1}(B))$ will not be defined. In the second example, if we take $\mathcal{A}$ to be the usual Borel algebra of all subsets that can be generated from the intervals in $\Omega = [0, \infty)$, then, for every $B \subset \mathbb{R}$, it will follow that $X^{-1}(B) \in \mathcal{A}$. However, if we restrict $\mathcal{A}$ to be generated only from those intervals in $[0, \infty)$ that have integer endpoints (e.g.,

[2, 5)), then $X^{-1}([-1, 1]) \notin \mathcal{A}$, since $\frac{1}{2} + \sqrt{1 + \frac{1}{4}}$ is irrational and cannot be the endpoint of an interval in $\mathcal{A}$.

8.2.3 Measurability

Thus, we are led to impose a condition on the random variable X to ensure that $P(X^{-1}(B))$ will always be defined, and this technical condition is known as *measurability*.

Definition 8.2.1   (Measurability)   The function X is measurable with respect to the algebras $\mathcal{A}$ and $\mathcal{B}$ if and only if

$$(\forall B \in \mathcal{B}) \ X^{-1}(B) = \{\omega : X(\omega) \in B\} \in \mathcal{A}.$$

Random variables are henceforth assumed to be measurable with respect to the relevant σ-algebras.

We are now in a position to consider how probability is inherited under a random variable mapping. We incorporate the universally accepted solution to this question into the definition of a random variable.

Definition 8.2.2   (Random Variable)   An initial probability space $\Omega, \mathcal{A}, P$ and a final probability space $\Omega_X, \mathcal{B}, P_X$ are linked by a random variable X whenever

$$X : \Omega \to \Omega_X$$

is measurable with respect to $\mathcal{A}$ and $\mathcal{B}$ and

$$(\forall B \in \mathcal{B}) P_X(B) = P(X^{-1}(B)).$$

8.2.4 Functions of Random Variables

Hence, returning to our situation of $\mathbf{Y} = \mathbf{g}(\mathbf{X})$, $\Omega = \mathbb{R}^n$, $\Omega_{\mathbf{Y}} = \mathbb{R}^m$, and assuming the usual Borel σ-algebras of subsets (the smallest σ-algebra containing all of the multidimensional rectangles), we adopt the relationship

$$P_{\mathbf{Y}}(\mathbf{Y} \in A \subset \mathbb{R}^m) = P_{\mathbf{X}}(\{\mathbf{X} : g(\mathbf{X}) \in A\}) = \int_{\{\mathbf{x}:g(\mathbf{x})\in A\}} f_{\mathbf{X}}(\mathbf{x})d\mathbf{x}.$$

If we specialize this result to the semi-infinite corner (see Section 7.2)

$$C_{\mathbf{y}} = \{(Y_1, \ldots, Y_m) : Y_1 \leq y_1, \ldots, Y_m \leq y_m\},$$

then we have the multivariate cdf for the output:

$$F_{\mathbf{Y}}(\mathbf{y}) = P_{\mathbf{Y}}(\mathbf{Y} \in C_{\mathbf{y}}) = P_{\mathbf{X}}(\{\mathbf{X} : g(\mathbf{X}) \in C_{\mathbf{y}}\}).$$

The multivariate output density $f_{\mathbf{Y}}$ can then be calculated by differentiating the multivariate cdf. We turn now to carry out this process in detail for several common cases. Throughout, it will be important to remember that while we start with $F_{\mathbf{X}}(\mathbf{x})$ or $f_{\mathbf{X}}(\mathbf{x})$, functions of $\mathbf{x}$, we end with $F_{\mathbf{Y}}(\mathbf{y})$ or $f_{\mathbf{Y}}(\mathbf{y})$, functions of $\mathbf{y}$.

8.3 SINGLE-INPUT–SINGLE-OUTPUT (SISO) FUNCTIONS

8.3.1 Approach to SISO Functions

The typical problem has as its given information a random variable X described by a probability law that is usually specified as a pmf p_X, a pdf f_X, or a cdf F_X. We are given a function g that generates a new random variable $Y = g(X)$. We then need to determine a probability description of Y, usually through F_Y or f_Y. We start by determining

$$F_Y(y) = P(Y \le y) = P(\{X : g(X) \le y\}) = P(g^{-1}((-\infty, y])).$$

Sketch $y = g(x)$, and combine it with a horizontal line crossing the Y-axis at level y, to better identify the inverse image sets $\{x : g(x) \le y\}$. Calculate

$$F_Y(y) = \int_{\{x:g(x)\le y\}} f_X(x)\, dx.$$

Now calculate

$$f_Y(y) = \frac{dF_Y(y)}{dy}.$$

It is sometimes easier to calculate the pdf f_Y directly by first differentiating the integral expression for $F_Y(y)$ with respect to y. From f_Y, we can calculate

$$P_Y(Y \in A) = \int_A f_Y(y)\, dy$$

and thereby gain a complete probabilistic description of the new random variable Y. In the subsections that follow, we illustrate this approach by solving a number of sample problems for different SISO functions.

8.3.2 Linear Function $Y = aX + b$

Let

$$F_Y(y) = P(Y \le y) = P(aX + b \le y) = P(aX \le y - b).$$

We wish to divide through by a, but must take care as to its sign. Dividing both sides of an inequality by a negative number reverses the sign of the inequality.

Case 1, $a > 0$:

$$F_Y(y) = P\left(X \le \frac{y - b}{a}\right) = F_X\left(\frac{y - b}{a}\right), \quad f_Y(y) = \frac{dF_Y(y)}{dy} = \frac{1}{a} f_X\left(\frac{y - b}{a}\right).$$

Case 2, $a < 0$:

$$F_Y(y) = P\left(X \ge \frac{y - b}{a}\right) = 1 - F_X\left(\left(\frac{y - b}{a}\right)^-\right).$$

Here, $F(z^-)$ is the value of the cdf F as z is approached from the left (the lower cdf value if there is a jump discontinuity at z; otherwise $F(z^-) = F(z)$). If we assume, for simplicity, that F_X is continuous and differentiable, then

$$F_Y(y) = 1 - F_X\left(\frac{y-b}{a}\right), \quad f_Y(y) = -\frac{1}{a}f_X\left(\frac{y-b}{a}\right).$$

Putting the two cases together results in

$$f_Y(y) = \frac{1}{|a|}f_X\left(\frac{y-b}{a}\right). \tag{8.2}$$

If, for example, $X \sim \mathcal{N}(m, \sigma^2)$, then we find that

$$Y = aX + b \Rightarrow Y \sim \mathcal{N}(am + b, a^2\sigma^2).$$

Thus, a linearly transformed normal random variable is again normal. This property of being preserved under an arbitrary linear transformation, while true for the multivariate normal, is generally false for other probability models.

8.3.3 Power Law Function $Y = X^n$

Again, we distinguish two cases and treat first

Case 1: n odd.

A simple case is that of $Y = aX$, as just discussed. A cubic relationship obtains for the volume of a three-dimensional object having fixed proportions in its dependence upon a characteristic length. (This is true, say, for three-dimensional ellipsoids having a fixed ratio between the lengths of their major and minor axes, with the volume then proportional to the cube of the length of the major axis. It is approximately true for, say, human males, with the characteristic length being their height). In this case,

$$F_Y(y) = P(X^n \le y) = P(X \le y^{\frac{1}{n}}) = F_X(y^{\frac{1}{n}}).$$

$$f_Y(y) = \frac{1}{n}y^{\frac{1}{n}-1}f_X(y^{\frac{1}{n}}).$$

This result is generalized in the next subsection.

Case 2: n even.

In this case, if $y < 0$, then $\{X : x^n \le y < 0\} = \emptyset$, and $F_Y(y) = P(X^n \le y) = 0$. Assuming $y \ge 0$, we then have

$$F_Y(y) = P(X^n \le y) = P(-y^{\frac{1}{n}} \le X \le y^{\frac{1}{n}}) = F_X(y^{\frac{1}{n}}) - F_X((-y^{\frac{1}{n}})^-).$$

Assuming that F_X is continuous and differentiable yields

$$f_Y(y) = \frac{1}{n}y^{\frac{1}{n}-1}\left[f_X(y^{\frac{1}{n}}) + f_X(-y^{\frac{1}{n}})\right]. \tag{8.3}$$

For example, we can inquire into the power dissipated over a bandwidth B by an R-ohm resistor at absolute temperature T due to its internally generated thermal noise voltage V. We know that $V \sim \mathcal{N}(0, \sigma^2)$, where $\sigma^2 = 4kTRB$. The power

$$P = \frac{V^2}{R}.$$

Clearly, dissipated power P is nonnegative; for $p < 0$, we have $F_P(p) = 0$. For $p \geq 0$,

$$F_P(p) = P(P \leq p) = P\left(\frac{V^2}{R} \leq p\right) = P\left(-\sqrt{Rp} \leq V \leq \sqrt{Rp}\right)$$

$$= F_V\left(\sqrt{Rp}\right) - F_V\left((-\sqrt{Rp})^-\right),$$

from which we conclude (note that F_V is differentiable) that

$$f_P(p) = \frac{\sqrt{R}}{2\sqrt{p}}\left[f_V(\sqrt{Rp}) + f_V(-\sqrt{Rp})\right] U(p) = \sqrt{\frac{R}{p}} \frac{1}{\sigma\sqrt{2\pi}} e^{-\frac{Rp}{2\sigma^2}} U(p)$$

$$= \frac{1}{\sqrt{8\pi kTBp}} e^{-\frac{p}{8kTB}} U(p).$$

This density is in the gamma family of Section 6.6.2.

Note that the density for the dissipated thermal noise power P is independent of the resistance R and depends only upon the temperature T of the resistor and the bandwidth B over which we are making our measurement. This partially explains why you may encounter specifications of thermal noise sources (e.g., the planet Earth as seen by a satellite antenna, or, more generally, black body radiators) in which it suffices for received noise power calculations to specify a *noise temperature* for the thermal noise source.

8.3.4 Monotone and Continuous Function g

Now assume that $Y = g(X)$, where g is continuous and nondecreasing (e.g., power law for odd n, but not for even n). Our basic approach in Section 8.3.1 requires us to calculate the probability of the set $\{X : Y = g(X) \leq y\}$, and this is easy for nondecreasing g's that have an inverse g^{-1} such that

$$g(x) \leq y \iff x \leq g^{-1}(y).$$

In this case,

$$F_Y(y) = P(\{Y : Y \leq y\}) = P(\{X : g(X) \leq Y\})$$

$$= P(\{X : X \leq g^{-1}(y)\}) = F_X(g^{-1}(y)).$$

For example, this condition holds if $g(x) = e^x$ and we take $g^{-1}(y) = \log(y)$. We introduce a version of the inverse (see Equation 8.1) by noting that, since g is nondecreasing and continuous, we can define

$$g^{-1}(y) = \underline{x}(y) = \min\{x : g(x) \geq y\}.$$

The postulated continuity of g is needed to ensure that, when the minimum is finite, it is achieved. The set $\{x : g(x) \leq y\}$ is a semi-infinite interval with upper endpoint (possibly infinite), say, $\underline{x}(y)$. The continuity of g assures us that this upper endpoint (if finite) is attached to the interval

$$\{x : g(x) \leq y\} = (-\infty, \underline{x}(y)].$$

It is easily verified that, for g continuous and nondecreasing,

$$g(x) \leq y \iff x \leq g^{-1}(y),$$

as desired. Hence,

$$F_Y(y) = F_X(g^{-1}(y)).$$

If g has an interval where it is constant at, say, the value g_0 (the graph of g is flat over this interval), then F_Y will have a jump discontinuity of height $P_X(\{X : g(X) = g_0\})$ that is usually positive, the continuity of g and even of F_X notwithstanding.

If we now assume that g is continuous and strictly increasing and that F_X is differentiable, then we obtain

$$f_Y(y) = \frac{dg^{-1}(y)}{dy} f_X(g^{-1}(y)).$$

A similar result (with the sign change needed to ensure that the resulting density is nonnegative) obtains if g is strictly decreasing:

$$f_Y(y) = -\frac{dg^{-1}(y)}{dy} f_X(g^{-1}(y)).$$

Putting the two cases together yields

$$f_Y(y) = \left| \frac{dg^{-1}(y)}{dy} \right| f_X(g^{-1}(y)). \tag{8.4}$$

for differentiable and monotone g.

Example 8.2   Derivation of Pareto _____

As an illustration, we show that a Pareto random variable $Y \sim Par(\alpha, \tau)$ can be constructed by an exponential transformation $Y = \tau e^X$, $\tau > 0$, on an exponentially distributed random variable $X \sim \mathcal{E}(\alpha)$. We have

$$g^{-1}(y) = \log(y/\tau), \quad \frac{dg^{-1}(y)}{dy} = \frac{1}{y},$$

$$f_Y(y) = \frac{\alpha}{y} e^{-\alpha \log(y/\tau)} U(y - \tau) = \alpha y^{-1-\alpha} \tau^\alpha U(y - \tau),$$

where the condition $y \geq \tau$ comes from $y = \tau e^x$ for nonnegative x.

8.4 SIMULATION AND THE PROBABILITY INTEGRAL TRANSFORMATION

Simulation of complex systems has become an increasingly important activity in engineering (e.g., the design of neural networks, nanoscale devices, protocols for wireless mobile cellular communications systems, studies of nuclear weapons designs in an era where testing is banned) and in the sciences (e.g., the needs of high-energy physics provided much of the impetus behind the supercomputing centers of the late 1980s and early 1990s). As the costs of prototyping systems becomes great or the number of systems one wishes to explore becomes very large, there is a great incentive to simulate such systems by running a computational model. Simulation has joined experiment and theory as a means of understanding and designing physical and mathematical systems. Simulation, which can be thought of as computational experimentation with theory, requires us to convert a mathematical model of a system or experiment into a computational model and then to run this model repeatedly, varying its parameters and inputs. Running the model typically requires rapidly generating very many appropriately random inputs—a form of simulation known as Monte Carlo. Hence, we need to be able to use the computer to efficiently create large numbers of random variables having appropriate distributions. Where can we obtain these random variables?

8.4.1 Generating a Uniform Random Variable

There is an option of random numbers generated by appropriate physical sources of randomness, such as the exponentially distributed waiting times between successive radioactive or photonic emissions from a source or the Gaussianly distributed amplitudes of many noise sources, especially filtered thermal noise. In practice, the physical sources have generally been limited in their rate of production of such numbers, particularly if several significant digits are needed, and their experimental realizations have had dependencies between successively generated numbers and nonstationarities that are not present in the ideal physical phenomenon being exploited. Furthermore, the sequence of outcomes was unrepeatable, a defect if we wish to compare the performances of several variants of a system excited by the same inputs. While, ideally, these physical sources are preferable to deterministic algorithmic sources and may still be preferable for random sources that must meet stringent statistical standards, as in cryptographic applications, they are not widely used in engineering simulations.

In the mid-20th century, use was made of a book-length table of a million Gaussian random numbers [74]! Interestingly, the book was published with errata. However, simulation often requires very large numbers of random variables, needs them produced at a rate that can keep up with high-speed computation, and cannot rely on such off-line production.

As we will explain following this subsection, random variables having a given distribution can be constructed by appropriate deterministic transformations of $\mathcal{U}(0, 1)$ random variables. Thus, if we can find a reliable, high-speed, repeatable (if needed) source of $\mathcal{U}(0, 1)$ random variables, then we can largely meet the needs of today's simulations. Indeed, all computing systems provide arithmetically based methods for generating (calculating) long sequences of numbers whose distribution is $\mathcal{U}(0, 1)$ (e.g., we can use the Matlab command `rand(n,m)` to generate an $n \times m$ matrix of unlinked $\mathcal{U}(0, 1)$ random variables). Strictly speaking, these numbers are *pseudorandom*, not random, and are calculated deterministically from an initially chosen "seed number." Well-designed arithmetic algorithms produce sequences of numbers that pass many statistical tests for randomness, and this justifies our treating them as if they were truly random. Still, one needs to be alert to the possibility that some simulations may inadvertently exploit the hidden lack of true randomness. Thus, uses of algorithmically generated random numbers in cryptographic applications have to pass much more stringent tests than when the same algorithms are used to simulate, say, customers arriving in a queue.

An arithmetical method that has long been used is that of a *linear congruential generator*, or *LCG* (see Dagpunar [21] and citations on the Web), that iteratively generates a sequence $\{L_n\}$ of real numbers through the recursion

$$L_n = (aL_{n-1} + c) \bmod m.$$

The parameter m is a positive integer known as the "modulus," a is a positive integer known as the "multiplier," and c is a nonnegative integer known as the increment. The initial value L_0 is known as the "seed." L_0, a, and c all take values in $[0, m-1]$. The operation $a \bmod b$ returns the remainder of a after division by b. Thus, (7 mod 5) is 2 and (23 mod 4) is 3. It is immediate from the properties of mod that L_n, a remainder after division by the modulus m, can take only integer values in $[0, m-1]$. The desired simulated sequence of $\mathcal{U}(0, 1)$ random variables

$$X_n = \frac{L_n}{m}$$

is now confined to the unit interval.

The period p of this algorithm is the smallest positive integer such that $X_p = X_0$. Clearly, once the same seed value is reached, the LCG will repeat the same sequence of values that preceded this return. Hence, the LCG output is useful only for generating $p - 1$ terms. As there are just m possible values for X_n, the LCG period p cannot exceed m: There must be a repetition in the output value if $n > m$. We are required to take a very large value for m if we are to be able to generate long random sequences.

Design conditions for the choice of the parameters of the LCG algorithm that will ensure a large value of the period include the following:

1. m should be as large as possible;
2. c is relatively prime to m (no common divisors);
3. $(a - 1)$ is a multiple of each of the prime number factors of m;
4. $(a - 1)$ is a multiple of 4 if m is.

The selection of a random-number-generating algorithm and of its parameters is an art form, with regular patterns appearing in the resulting sequence of pseudorandom numbers if the wrong choices are made. Fundamentally, X_n is a deterministic function of X_{n-1}. It is a linear function over segments of the sequence in which $b = aX_{n-1} + c < m$. For, in this case, the $b \bmod m$ function returns its argument b unchanged. Random-number-generating algorithms are studied to see that they produce sequences that can pass a variety of statistical tests for independent and uniformly distributed random variables. Such tests would include histograms and scatterplots of successive terms in the output sequence. For our purposes, we will assume that you have available a computing system with a well-studied, sufficiently reliable generator of independent $\mathcal{U}(0, 1)$ random variables.

8.4.2 Generating an Arbitrary Random Variable

We wish to choose a function g to convert a given random variable $U \sim \mathcal{U}(0, 1)$ to a random variable $Y = g(U)$ having a desired cdf G. The cdf G is the goal or target and the cdf F_Y is what has been achieved in transforming U into Y via the function g. For example, if we desire $Y \sim \mathcal{E}(\alpha)$, then $F_Y(y) = G(y) = 1 - e^{-\alpha y}$. As we have already done for g nondecreasing, we introduce the inverse function G^{-1} for the cdf G. For our exponential cdf example, $G^{-1}(x) = \frac{1}{\alpha} \log \frac{1}{1-x}$ and satisfies $G(G^{-1}(x)) = x$. If $X \sim \mathcal{U}(0, 1)$, then

$$F_Y(y) = P(\{X : G^{-1}(X) \leq y\}) = F_X(G(y)) = \begin{cases} 1 & \text{if } G(y) > 1 \\ G(y) & \text{if } 0 \leq G(y) \leq 1 \\ 0 & \text{if otherwise} \end{cases}.$$

Inverses of the most general kinds of cdfs (e.g., they may have discontinuities and flat regions) are obtained through the following definition:

Definition 8.4.1 **(Inverse CDF)**   The inverse G^{-1} to a cdf G is

$$G^{-1}(u) = \min\{y : G(y) \geq u\}. \tag{8.5}$$

The continuity from the right of cdfs guarantees that the minimum invoked in the definition is achieved and need not be replaced by an infimum, and the definition is well posed. Observe that both G and G^{-1} are nondecreasing functions. Hence, it follows from the definition that

$$G(G^{-1}(u)) \geq u \text{ and } G^{-1}(G(y)) \leq y. \tag{8.6}$$

Given $U \sim \mathcal{U}(0, 1)$ and a desired arbitrary cdf G, we define the new random variable (see Figure 8.1)

$$Y = G^{-1}(U).$$

Informally, we use $U \sim \mathcal{U}(0, 1)$ with observed $U = u$ to select a random value of the desired cdf $u = F_Y(y)$. (Note that $U \in [0, 1]$ takes on only the values that can be taken on by a cdf.) We then determine the desired $Y = y$ from the inverse $y = F_Y^{-1}(u)$. If we have graphed the desired cdf $F_Y(y)$, then we follow the horizontal line with the F_Y-axis value of $U = u$ until it intercepts the cdf and then drop down vertically to determine the corresponding argument $Y = y$.

To verify that this approach works, we calculate

$$F_Y(y) = P(Y = G^{-1}(U) \leq y) = P(\{U : G^{-1}(U) \leq y\}).$$

Let $A = \{U : G^{-1}(U) \leq y\}$ denote the event such that $F_Y(y) = P(A)$. We claim that A can be reexpressed as

$$B = \{U : U \leq G(y)\}.$$

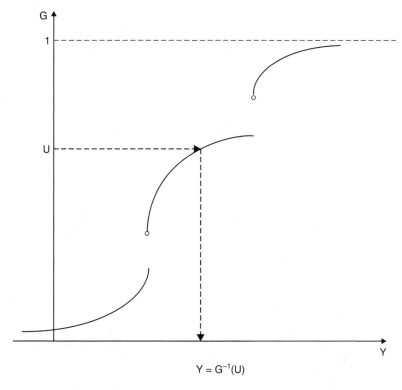

$$Y = G^{-1}(U)$$

Figure 8.1   Using U to find Y.

To verify that $A = B$, we have to show that $u \in A$ implies $u \in B$ and, conversely, $u \in B$ implies $u \in A$. Assume that $u \in A$, so that $G^{-1}(u) \leq y$. Because G is nondecreasing, it follows that $G(G^{-1}(u)) \leq G(y)$. However, from Eq. (8.6), we have

$$u \leq G(G^{-1}(u)) \leq G(y),$$

and $u \in B$. Now assume that $u \in B$, so that $u \leq G(y)$. From the fact that G^{-1} is also a nondecreasing function, we see that $G^{-1}(u) \leq G^{-1}(G(y))$. Invoking Equation 8.6 yields

$$G^{-1}(u) \leq G^{-1}(G(y)) \leq y,$$

and it follows that $u \in A$. Therefore, $A = B$ and

$$F_Y(y) = P(A) = P(B) = P(\{U : U \leq G(y)\}).$$

Recalling that $U \sim \mathcal{U}(0, 1)$ yields the desired conclusion, we have

$$F_Y(y) = P(\{U : U \leq G(y)\}) = G(y).$$

Since G is a cdf, it automatically takes values only in the interval [0,1].

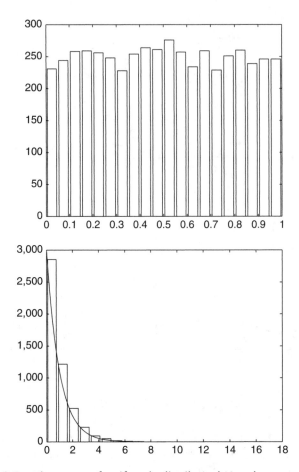

Figure 8.2 Histogram of uniformly distributed X and exponential Y.

As an illustration, the first half of Figure 8.2 shows the histogram (see Section 1.7; the histogram is a data-based estimator of the pdf) of a sample of 5,000 $\mathcal{U}(0, 1)$-distributed random variables created by the Matlab command `rand(5000,1)`. These variables were then transformed via $Y = -\log(1 - X)$ and the histogram of Y plotted in the second half of the figure. The overlapping plot of a true exponential density strongly suggests that Y is exponentially distributed, as desired.

Example 8.3   Simulating a Binary-Valued Variable _____

A simple discrete example is the generation of $P(Y = 1) = p, P(Y = 0) = 1 - p$ from $X \sim \mathcal{U}(0, 1)$. The solution is

$$G^{-1}(X) = \begin{cases} Y = 0 & \text{if } 0 \le X \le 1 - p \\ Y = 1 & \text{otherwise} \end{cases}.$$

An inverse to the previous example is to turn a given random variable Y having a strictly increasing continuous cdf F into a random variable $X \sim \mathcal{U}(0, 1)$ through the monotone transformation

$$X = F(Y)$$

known as the *probability integral transformation*. Note that F^{-1} is continuous and strictly increasing and that $X \in [0, 1]$. Hence,

$$
\begin{aligned}
F_X(x) = P(X \leq x) &= P(\{Y : F(Y) \leq x\}) \\
&= P(\{Y : F^{-1}(F(Y)) \leq F^{-1}(x)\}) = P(\{Y : Y \leq F^{-1}(x)\}) \\
&= F(F^{-1}(x)) = x,
\end{aligned}
$$

and $X \sim \mathcal{U}(0, 1)$ as claimed.

8.5 MULTIPLE-INPUT–MULTIPLE-OUTPUT (MIMO): $N = M$

We first consider the MIMO case where the dimension of the output equals that of the input. In Section 8.7, we examine the remaining case of unequal dimensions. In what follows, we treat only the case of a one-to-one nonlinear transformation $\mathbf{g}$ from $\mathbf{X}$ to $\mathbf{Y}$, $\mathbf{Y} = \mathbf{g}(\mathbf{X})$, with a unique differentiable inverse $\mathbf{g}^{-1}$, $\mathbf{X} = \mathbf{g}^{-1}(\mathbf{Y})$. The general solution is given in terms of a *Jacobian matrix*

$$
\mathbb{J} = [J_{i,j}], \ J_{i,j} = \frac{\partial x_i}{\partial y_j} = \frac{\partial g_i^{-1}(\mathbf{y})}{\partial y_j}
$$

(first derivatives are taken of the original or input variables with respect to the new or output variables) as

$$
f_{\mathbf{Y}}(\mathbf{y}) = f_{\mathbf{X}}\left(g^{-1}(\mathbf{y})\right) |\det(\mathbb{J})|. \tag{8.7}
$$

Note the presence of the absolute value of the determinant of the Jacobian; failure to do so can yield negative pdfs!

Rapid informal access to this result is obtained by considering Figure 8.3 and assuming that the lengths of the sides of the rectangle R are sufficiently small that

$$
P(\mathbf{Y} \in R) \approx f_{\mathbf{Y}}(\mathbf{y})\mathrm{vol}(R).
$$

An alternative evaluation is

$$
P(\mathbf{Y} \in R) = P(\mathbf{X} \in R^{-1}) \approx f_{\mathbf{X}}(\mathbf{x})\mathrm{vol}(R^{-1}),
$$

where $\mathbf{x} = \mathbf{g}^{-1}(\mathbf{y})$. Equating results yields

$$
f_{\mathbf{Y}}(\mathbf{y}) \approx f_{\mathbf{X}}\left(\mathbf{g}^{-1}(\mathbf{y})\right) \frac{\mathrm{vol}(R^{-1})}{\mathrm{vol}(R)}.
$$

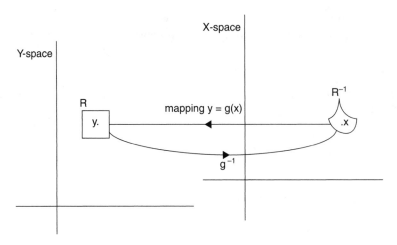

Figure 8.3   Nonlinear transformation.

To make the approximation exact, we take the limit as the maximum dimension of R shrinks to zero. The limiting ratio of the volume of R^{-1} to that of R is given by the magnitude of the determinant of the *Jacobian* (see R. Strichartz [85])

$$\mathbb{J} = \frac{\partial(x_1, \ldots, x_n)}{\partial(y_1, \ldots, y_n)} = \begin{pmatrix} \dfrac{\partial x_1}{\partial y_1} & \cdots & \dfrac{\partial x_1}{\partial y_n} \\ \vdots & & \vdots \\ \dfrac{\partial x_n}{\partial y_1} & \cdots & \dfrac{\partial x_n}{\partial y_n} \end{pmatrix},$$

$$\lim_{\mathrm{maxdim}(R) \to 0} \frac{\mathrm{vol}(R^{-1})}{\mathrm{vol}(R)} = |\det(\mathbb{J})|,$$

a result from transformations of coordinate systems in the calculus of multivariable functions.

It is sometimes computationally convenient to evaluate $\det(\mathbb{J})$ as

$$\det \begin{pmatrix} \dfrac{\partial g_1^{-1}(y_1, \ldots, y_n)}{\partial y_1} & \cdots & \dfrac{\partial g_1^{-1}(y_1, \ldots, y_n)}{\partial y_n} \\ \vdots & & \vdots \\ \dfrac{\partial g_n^{-1}(y_1, \ldots, y_n)}{\partial y_1} & \cdots & \dfrac{\partial g_n^{-1}(y_1, \ldots, y_n)}{\partial y_n} \end{pmatrix} = \frac{1}{\det \begin{pmatrix} \dfrac{\partial g_1(\mathbf{x})}{\partial x_1} & \cdots & \dfrac{\partial g_1(\mathbf{x})}{\partial x_n} \\ \vdots & & \vdots \\ \dfrac{\partial g_n(\mathbf{x})}{\partial x_1} & \cdots & \dfrac{\partial g_n(\mathbf{x})}{\partial x_n} \end{pmatrix}},$$

recognizing that we will need to replace $\mathbf{x}$ in the result by $\mathbf{y} = \mathbf{g}(\mathbf{x})$ to obtain $f_\mathbf{Y}$ as a function of $\mathbf{y}$. This alternative result looks more familiar when thought of as

$$\frac{dx}{dy} = \frac{1}{\frac{dy}{dx}}.$$

8.6 MIMO APPLICATIONS

8.6.1 MIMO Linear Transformations

A first application is to nonsingular linear transformations

$$\mathbf{Y} = \mathbb{A}\mathbf{X} + \mathbf{b},$$

where the square matrix $\mathbb{A}$ has a nonzero determinant and, therefore, an inverse $\mathbb{A}^{-1} = \mathbb{B}$, so that

$$\mathbf{X} = \mathbf{g}^{-1}(\mathbf{y}) = \mathbb{B}(\mathbf{Y} - \mathbf{b}) = \mathbb{B}\mathbf{Y} + \mathbf{c}.$$

Such a transformation arises when $\mathbf{X}$ are the inputs to a finite-memory discrete-time linear system with impulse response coefficients (see Section 10.2) $h_{i,j} = A_{i,j}$. The elements of the Jacobian matrix are given by

$$J_{i,j} = \frac{\partial x_i}{\partial y_j} = B_{i,j}.$$

Hence, we conclude that

$$f_{\mathbf{Y}}(\mathbf{y}) = f_{\mathbf{X}}\left(\mathbb{B}(\mathbf{y} - \mathbf{b})\right) |\det(\mathbb{B})|. \tag{8.8}$$

It is the case that

$$\det(\mathbb{B}) = 1/\det(\mathbb{A}).$$

Theorem 8.6.1   (Preservation of Normality)   If real-valued $\mathbf{X}$ is distributed as $\mathcal{N}(\mathbf{m}_X, \mathbb{C}_X)$ and $\mathbf{Y} = \mathbb{A}\mathbf{X} + \mathbf{b}$, for real-valued, nonrandom $\mathbb{A}, \mathbf{b}$, then $\mathbf{Y}$ is distributed as $\mathcal{N}(\mathbb{A}\mathbf{m}_X + \mathbf{b}, \mathbb{A}\mathbb{C}_X\mathbb{A}^T)$. Hence, an affine transformation of a Gaussian random vector $\mathbf{X}$ results in a new Gaussian random vector $\mathbf{Y}$.

Proof. Let $\propto$ denote proportionality to within factors that do not involve the variables $\mathbf{x}$ or $\mathbf{y}$, and define

$$S_X = (\mathbf{x} - \mathbf{m}_X)^T \mathbb{C}_X^{-1} (\mathbf{x} - \mathbf{m}_X)$$

in order to write

$$f_{\mathbf{X}}(\mathbf{x}) \propto \exp\left(-\frac{1}{2}S_X\right).$$

From Eq. (8.8), the corresponding

$$f_{\mathbf{Y}}(\mathbf{y}) \propto f_{\mathbf{X}}\left(\mathbb{B}(\mathbf{y} - \mathbf{b})\right) \propto \exp\left(-\frac{1}{2}S_Y\right),$$

with

$$S_Y = \left(\mathbb{B}(\mathbf{y} - \mathbf{b}) - \mathbf{m}_X\right)^T \mathbb{C}_X^{-1} \left(\mathbb{B}(\mathbf{y} - \mathbf{b}) - \mathbf{m}_X\right).$$

Using $\mathbb{A}$ as the inverse of the matrix $\mathbb{B}$, rewrite

$$\mathbb{B}(\mathbf{y} - \mathbf{b}) - \mathbf{m}_X = \mathbb{B}(\mathbf{y} - \mathbf{b} - \mathbb{A}\mathbf{m}_X),$$

to obtain

$$S_Y = (\mathbb{B}(\mathbf{y} - \mathbf{b} - \mathbb{A}\mathbf{m}_X))^T \, C_X^{-1} \, (\mathbb{B}(\mathbf{y} - \mathbf{b} - \mathbb{A}\mathbf{m}_X))$$
$$= (\mathbf{y} - \mathbf{b} - \mathbb{A}\mathbf{m}_X)^T \mathbb{B}^T C_X^{-1} \mathbb{B}(\mathbf{y} - \mathbf{b} - \mathbb{A}\mathbf{m}_X).$$

Direct multiplication of the two terms verifies that the inverse of $\mathbb{B}^T C_X^{-1}\mathbb{B}$ is $\mathbb{A}C_X\mathbb{A}^T$. Comparison of the forms of S_Y and S_X reveals the desired conclusion: $\mathbf{Y} \sim \mathcal{N}(\mathbf{b} + \mathbb{A}\mathbf{m}_X, \mathbb{A}C_X\mathbb{A}^T)$. □

An even easier derivation of this result is given in Section 17.6.

8.6.2 Quadrature Modulation

A second application of the results on MIMO transformations can be made to represent a quadrature-modulated communications signal

$$S(t) = X(t)\cos(\omega_c t) - Y(t)\sin(\omega_c t),$$

having narrowband (relative to the carrier frequency) baseband signal components $X(t)$ and $Y(t)$ transmitted at a carrier frequency of $f_c = \omega_c/2\pi$. An alternative representation of $S(t)$ can be given in terms of the signal amplitude known as the *envelope* $R(t)$, satisfying $R(t) \geq 0$, and time-varying *phase* $\Theta(t)$, satisfying $|\Theta(t)| \leq \pi$:

$$S(t) = R(t)\cos(\omega_c t + \Theta(t)).$$

The transformation from the random variables $X(t)$ and $Y(t)$ to the random variables $R(t)$ and $\Theta(t)$ is that of the transformation from Cartesian coordinates (x, y) to polar coordinates (r, θ) through

$$x = r\cos\theta, \; y = r\sin\theta \quad r = \sqrt{x^2 + y^2}, \; \theta = \tan^{-1}\left(\frac{y}{x}\right).$$

In this case, we find that $|\det(\mathbb{J})| = r$, or the well-known

$$dxdy = rdrd\theta.$$

Thus,

$$f_{R,\Theta}(r, \theta) = f_{X,Y}(x = r\cos\theta, y = r\sin\theta)r.$$

Quadrature modulation in the context of narrowband random processes is discussed in Section 20.11.3.

8.6.3 Transforming Uniforms into Normals

The procedure used in practice to transform uniformly distributed random variables into normally distributed ones is not the one described earlier in the SISO section, due largely to the awkwardness of determining the inverse of the error function. Rather, it is common to transform pairs of uniforms into pairs of normals as follows: Let $\mathbf{X} = (X_1, X_2)$ be $i.i.d.$ $\mathcal{U}(0, 1)$ (where $i.i.d.$ is discussed in Chapter 14 and means that we use the product bivariate density described in Section 7.4.1) random variables having joint pdf

$$f_{\mathbf{X}}(\mathbf{x}) = f_{X_1}(x_1)f_{X_2}(x_2) = U(x_1)U(1 - x_1)U(x_2)U(1 - x_2).$$

Consider the transformation to $\mathbf{Y} = (Y_1, Y_2)$ given by

$$y_1 = \sqrt{-2\sigma^2 \log x_1} \cos(2\pi x_2), \quad y_2 = \sqrt{-2\sigma^2 \log x_1} \sin(2\pi x_2).$$

Applying the preceding pdf yields

$$f_{\mathbf{Y}}(\mathbf{y}) = f_{\mathbf{X}}\left(x_1 = e^{-\frac{y_1^2+y_2^2}{2\sigma^2}}, x_2 = \frac{1}{2\pi}\tan^{-1}\left(\frac{y_2}{y_1}\right)\right)\frac{1}{2\pi\sigma^2}e^{-\frac{y_1^2+y_2^2}{2\sigma^2}}$$

$$= \frac{1}{2\pi\sigma^2}e^{-\frac{y_1^2+y_2^2}{2\sigma^2}},$$

where we have used the fact that $f_{\mathbf{X}} = 1$ for arguments in the interval [0,1]. Hence, we see from $f_{\mathbf{Y}}$ that Y_1 and Y_2 are bivariate $i.i.d.$ $\mathcal{N}(0, \sigma^2)$ (i.e., $f_{Y_1,Y_2}(y_1, y_2) = f_{Y_1}(y_1)f_{Y_2}(y_2)$, $Y_i \sim \mathcal{N}(0, \sigma^2)$), as desired.

8.7 MIMO TRANSFORMATION: $M ¡ N$

We now consider the case of a many-to-one mapping. The basic approach when $m < n$ is to reduce this case to the previous one by *augmenting* $\mathbf{Y}$ to $\mathbf{Y}'$, $\dim(\mathbf{Y}') = \dim(\mathbf{X})$, through the addition of $n - m$ variables to $\mathbf{Y}$ so that the new overall transformation is one-to-one and has a unique differentiable inverse. We illustrate this approach in the next two examples.

Example 8.4   Augmentation

For a first example, consider $Y = X_1 + X_2$. We augment Y to (Y, Z) with $Z = X_1$. There are infinitely many choices for Z that will work, requiring only that they yield a Jacobian with nonzero determinant. However, some choices lead to easier calculations than others, and we take a simple one. Clearly, the inverse function is given by

$$X_1 = Z, \ X_2 = Y - Z.$$

The Jacobian matrix is

$$\mathbb{J} = \begin{pmatrix} \dfrac{\partial x_1}{\partial y} & \dfrac{\partial x_1}{\partial z} \\ \dfrac{\partial x_2}{\partial y} & \dfrac{\partial x_2}{\partial z} \end{pmatrix} = \begin{pmatrix} 0 & 1 \\ 1 & -1 \end{pmatrix}.$$

It is immediate that $|\det(\mathbb{J})| = 1$. Thus,

$$f_{Y,Z}(y, z) = f_{X_1,X_2}(z, y - z).$$

Eliminating Z with the use of Lemma 7.3.1, we find that

$$f_Y(y) = \int_{-\infty}^{\infty} f_{X_1,X_2}(z, y - z)\, dz.$$

Proceeding further would require knowing more about f_{X_1,X_2}.

Example 8.5   **Augmentation (Continued)** _____

As a second example, assume that X_1, X_2 are described by the pdf $f_{X_1,X_2}(x_1, x_2)$ and that we are interested in $Y_1 = X_1^2 + X_2$. Because this two-to-one transformation does not have an inverse, we augment it by introducing a new variable. A simple choice is $Y_2 = X_1$, and this ensures that the mapping $\mathbf{Y} = \mathbf{g}(\mathbf{X})$ has a unique differentiable inverse

$$x_1 = g_1^{-1}(y_1, y_2) = y_2, \quad x_2 = g_2^{-1}(y_1, y_2) = y_1 - y_2^2.$$

We can now evaluate

$$\det \begin{pmatrix} \dfrac{\partial x_1}{\partial y_1} & \cdots & \dfrac{\partial x_1}{\partial y_n} \\ \vdots & & \vdots \\ \dfrac{\partial x_n}{\partial y_1} & \cdots & \dfrac{\partial x_n}{\partial y_n} \end{pmatrix} = \det \begin{pmatrix} 0 & 1 \\ 1 & -2y_2 \end{pmatrix} = -1.$$

Hence, from the preceding analysis, we see that

$$f_{Y_1,Y_2}(y_1, y_2) = f_{\mathbf{X}}\big(x_1 = y_2, x_2 = y_1 - y_2^2\big).$$

To find the desired f_{Y_1} from f_{Y_1,Y_2} we need to integrate over the variable Y_2 that we introduced:

$$f_{Y_1}(y_1) = \int_{-\infty}^{\infty} f_{Y_1,Y_2}(y_1, y_2)\, dy_2.$$

If, for example,

$$f_{X_1,X_2}(x_1, x_2) = U(x_1)U(1 - x_1)U(x_2)U(1 - x_2),$$

corresponding to X_1 and X_2 being *i.i.d.* (see Chapter 14) and uniform, then

$$f_{Y_1,Y_2}(y_1, y_2) = U(y_2)U(1 - y_2)U(y_1 - y_2^2)U(1 - y_1 + y_2^2).$$

To evaluate $f_{Y_1}(y_1)$, the first case for a nonzero density is that $0 \le y_1 \le 1$, and then

$$f_{Y_1}(y_1) = \int_0^{\sqrt{y_1}} 1\, dy_2 = \sqrt{y_1}.$$

The second case is that $1 < y_1 \leq 2$, so that

$$f_{Y_1}(y_1) = \int_{\sqrt{y_1-1}}^{1} 1 \, dy_2 = 1 - \sqrt{y_1 - 1}.$$

The density f_{Y_1} is 0 for $y_1 < 0$ and for $y_1 > 2$.

The remaining case of $\dim(\mathbf{Y}) > \dim(\mathbf{X})$ is of less interest, for it implies a degenerate relationship between the components of $\mathbf{Y}$. This degeneracy, due to a deterministic relationship between the many components of $\mathbf{Y}$, would exhibit itself through the appearance of delta functions (unit impulses) in the resulting $f_{\mathbf{Y}}$.

8.8 SUMMARY

We exercise what has been learned about probability models by turning to consider G2 and examining a class of deterministic systems that have finite memory. Such a system S is most familiarly thought of simply as a real- or vector-valued function of real- or vector-valued variables; input and output dimensions need not agree. We then show how the probability description of the input X to a general deterministic system S transforms into a probability description for the output $Y = S(X)$. Note that

> *System response now does not mean determining Y for a given X, but rather determining the probability model for Y from that of X.*

We are considering the concept of system response at a higher level of abstraction than was customary in the usual prior engineering courses on deterministic signals and systems.

The general method is then applied to a variety of special cases. For real-valued inputs and outputs, we study linear, power law, and monotone transformations. For vector-valued inputs and outputs, we study linear and one-to-one transformations and then the case of inputs of higher dimension than outputs. The calculus technique of Jacobians is central to the one-to-one case.

An important application of these results is to shaping a random variable Y to have a desired probability description (e.g., cdf F_Y), given a random variable X having a different (and usually simpler) probability description F_X. Typically, X is assumed to have a uniform distribution $\mathcal{U}(0, 1)$, since algorithms for generating such distributions are commonly available. An ability to do this is essential for simulation, an area of endeavor that now ranks with theory and experiment in importance and in effort expended.

EXERCISES

E8.1   a. If X is measured in degrees centigrade and we wish to convert to Y measured in degrees Fahrenheit ($Y = 32 + (9/5)X$), then determine f_Y in terms of f_X.

   b. If $X \sim \mathcal{N}(m, \sigma^2)$, what is Y?

E8.2   a. If

$$Y = 1 - X^2, \quad f_X(x) = \frac{x+1}{2} U(x+1)U(1-x),$$

then evaluate $f_Y(y)$. (Hint: Start from the basic method.)

b. Repeat for $X \sim \mathcal{U}(-1, 1)$.

E8.3   If $X \sim \mathcal{B}(4, \frac{1}{2})$ and $Y = (X - 2)^2$, then evaluate and sketch the cdf $F_Y(y)$.

E8.4   a. If $X \sim Par(\alpha, \tau)$ and $Y = X^2$, verify that Y is also Pareto $Par(\beta, \tau')$ for some β and τ'.

b. Does this conclusion generalize for $Y = X^n$?

E8.5   a. Let $Y = e^X$. Find F_Y, f_Y in terms of F_X, f_X.

b. Specialize this result for f_Y to the case of $X \sim \mathcal{N}$. (Y is said to have the lognormal density, and this is used as a model for received power at the handset of a mobile cellular wireless system due to fading.)

E8.6   a. If

$$F_X(x) = \frac{1 - e^{-x}}{1 + e^{-x}} U(x), \text{ and } Y = e^{-X^2},$$

then evaluate the cdf F_Y.

b. If we quantize X distributed as in (a) as

$$Q = \begin{cases} 3 & \text{if } 2 < X \\ 1 & \text{if } 0 < X \le 2 \\ -1 & \text{if } X \le 0 \end{cases},$$

then evaluate and sketch the cdf $F_Q(q)$.

E8.7   If $X \sim \mathcal{U}(-\frac{\pi}{2}, \frac{\pi}{2})$, then show that $Y = a\tan(X)$ has a Cauchy density

$$f_Y(y) = \frac{1}{\pi} \frac{a}{a^2 + y^2}.$$

E8.8   Time-sampled speech with amplitude X, having $f_X(x) = \frac{1}{2}e^{-|x|}$, is passed through a 2-bit quantizer

$$q(x) = \begin{cases} q_2 & \text{if } x > \tau \\ q_1 & \text{if } 0 < x \le \tau \\ -q_1 & \text{if } -\tau \le x \le 0 \\ -q_2 & \text{if } x < -\tau \end{cases},$$

with $q(-x) = -q(x)$.

a. Determine the pmf for the quantizer output $Y = q(X)$.

b. Design the threshold τ so that all four quantization levels occur with equal probability.

E8.9   We are given that $X \sim \mathcal{U}(-\frac{1}{3}, \frac{2}{3})$ and $Y = X^2 U(X)$.

a. Determine $P(Y = 0)$.

b. Determine $F_Y(y)$ for $y \le 0$.

      c. Determine $F_Y(y)$ for $y > 0$.

      d. Determine $f_Y(y)$.

E8.10 If the voltage $V \sim \mathcal{N}(0, \sigma^2)$ is full-wave rectified to $Y = |V|$, determine f_Y.

E8.11 If the voltage $V \sim \mathcal{N}(m, \sigma^2)$ is half-wave rectified, $Y = VU(V)$, determine f_Y.

E8.12 If $Y = X^2 U(X)$, $X \sim \mathcal{N}(0, \sigma^2)$, then determine $f_Y(y)$. (Hint: $F_Y(0) \neq 0$.)

E8.13   a. If $Y = g(X)$, where

$$g(x) = \begin{cases} -a & \text{if } x < -a \\ x & \text{if } |x| \le a \\ a & \text{if } x > a \end{cases},$$

      then determine F_Y, f_y in terms of F_X, f_X.

      b. Specialize this result to the case where X is Laplacian, so that $f_X(x) = \frac{1}{2} e^{-|x|}$.

E8.14 If

$$Y = \begin{cases} X + 1 & \text{if } X \le 0 \\ X - 1 & \text{if } X > 0 \end{cases}$$

and $f_X(x) = \frac{1}{2} e^{-|x|}$, then evaluate f_Y for $|y| < 1$.

E8.15 Measuring the power in decibels of a signal X is done through

$$Y = 10 \log_{10}(X^2).$$

If $X \sim \mathcal{N}(0, 4)$, determine the cdf $F_Y(y)$.

E8.16 The thermal noise voltage in a resistor of $1\ \Omega$ at temperature T, when measured in a bandwidth W, has variance σ^2. We wish to represent the thermal noise power $P = V^2$ in decibels through the formula

$$\rho = 10 \log_{10}(P).$$

Evaluate the density $f_\rho(x)$.

E8.17 A so-called artificial neural network is composed of interconnected nonlinear nodes of the form

$$Y = \frac{1}{1 + e^{-X}}.$$

In order to understand the behavior of such a node, calculate f_Y, given that $X \sim \mathcal{N}(m, \sigma^2)$.

E8.18 In a wireless communication system, the power P received by a mobile communication device at a given distance from a base station is a random variable that is modeled as

$$\log\left(\frac{P}{p_\mu}\right) = X, \quad X \sim \mathcal{N}(0, \sigma^2),$$

where p_μ is a given constant known as the median power.

      a. What is the density $f_{\log(P)}(z)$ of $\log(P)$?

      b. What is the density $f_P(p)$ of the power P?

E8.19 In measuring speech signal power in db, we calculate $Z = c \log(A^2)$, where A is the amplitude, log is the natural logarithm, and $c = 10/\log(10)$.

      a. What is the probability model for A?

      b. Evaluate the cdf $F_Z(z)$.

      c. What is $P(Z > 2c)$?

E8.20 Show how to generate $Y \sim \mathcal{B}(3, 1/2)$ from $X \sim \mathcal{U}(0, 1)$.

E8.21 We are given $X \sim \mathcal{U}(0, 1)$ and wish to simulate Y with

$$P(Y = 1/2) = .3, \;\; P(Y = 2) = .25, \;\; P(Y = 3) = .45.$$

Explain in detail how to do so.

E8.22 Given $X \sim \mathcal{U}(0, 1)$ find a function g, $Y = g(X)$, to use so that

$$F_Y(y) = \begin{cases} 0 & \text{if } y < 0 \\ y^2 & \text{if } 0 \le y < 1 \\ 1 & \text{if } y \ge 1 \end{cases}.$$

E8.23 Show how to generate Y with cdf $F_Y(y) = (1 - e^{-y^2})U(y)$ from $X \sim \mathcal{U}(0, 1)$.

E8.24 You are given $X \sim \mathcal{U}(0, 1)$ and desire a random variable Y with cdf

$$F_Y(y) = \frac{1}{1 + e^{-y}}.$$

How can you create Y?

E8.25 We are given a random variable $X \sim \mathcal{U}(0, 1)$ and wish to simulate a random variable Y with pdf $f_Y(y) = \frac{3}{2}y^2 U(1 - |y|)$. Explain how to do so.

E8.26 For a simulation study of WWW file sizes, we need to create Y with Pareto pdf $\mathcal{P}ar(\alpha, 1)$. Show how to do this, starting from $X \sim \mathcal{U}(0, 1)$.

E8.27 Given $X \sim \mathcal{U}(0, 1)$, show how to create Y with

$$F_Y(y) = \left(1 - \frac{1}{2}e^{-y}\right)U(y).$$

E8.28 Determine the function g, so that if $X \sim \mathcal{U}(0, 1)$, then $Y = g(X)$ has cdf

$$F_Y(y) = (1 - e^{-y^4})U(y).$$

E8.29 We are given $X \sim \mathcal{U}(0, 1)$. Explain how to simulate Y with

$$f_Y(y) = \frac{1}{2}U(y)U(1 - y) + \frac{1}{2}\delta(y).$$

(Hint: Sketch the cdf F_Y.)

E8.30 Explain in detail how to convert $X \sim \mathcal{U}(0, 1)$ into Y with

$$f_Y(y) = 4y^3 U(y)U(1 - y).$$

E8.31   a. Given $X \sim \mathcal{U}(0, 1)$, construct Y with pdf

$$f_Y(y) = \begin{cases} \dfrac{1}{2}\cos(y) & \text{if } |y| \le \dfrac{\pi}{2} \\ 0 & \text{otherwise} \end{cases}.$$

      b. Construct Z from X with $P(Z = 1) = .6$ and $P(Z = 0) = .4$.

E8.32 a. Given $X \sim \mathcal{U}(0, 1)$, construct Y having density function

$$f_Y(y) = 2e^{-2y} U(y).$$

b. Given X as in (a), construct $Z \sim \mathcal{B}(2, 1/2)$.

E8.33 If $f_X(x) = 2x U(x) U(1 - x)$, $Y = g(X)$, then find g so that $Y \sim \mathcal{E}(1)$.

E8.34 a. If

$$f_{X_1, X_2}(x_1, x_2) = e^{-x_1 - 2|x_2|} U(x_1),$$

$$Y_1 = X_1^2 + X_2, \ Y_2 = 3X_1^2 - X_2,$$

then determine f_{Y_1, Y_2}.

b. Evaluate $P(Y_1 + Y_2 \geq 0)$.

E8.35 a. Calculate $f_{X,Y}(x, y)$ for

$$F_{X,Y}(x, y) = \frac{1}{1 + e^{-2x} + e^{-y}}.$$

b. If

$$W = X + Y, \ V = X^2 - Y^2,$$

solve for X, Y as functions of W, V.

c. Calculate $f_{W,V}(w, v)$.

d. Provide an expression for $f_V(v)$.

E8.36 A random power $P = VI$ is produced by voltage V and current I described by $f_{V,I}(v, y) = e^{-v-y} U(v) U(y)$.

a. Provide an expression for the cdf $F_P(z)$ by using our basic method. (A sketch will help.)

b. Evaluate $P(P < 0)$.

c. Provide an expression for the pdf f_P.

E8.37 Provide an expression for the pdf $f_P(x)$ of the power $P = I^2 R$ dissipated by a thermal noise current I of variance σ^2 passing through an independent randomly selected resistance $R \sim \mathcal{U}(r_0 - 1, r_0 + 1)$ for $r_0 > 1$. (Hint: Calculate F_P from the basic method.)

E8.38 You are given

$$f_{X_1, X_2}(x_1, x_2) = e^{-x_2} U(x_2) U(x_1) U(1 - x_1), \ Y_1 = X_1 X_2, \ Y_2 = \frac{X_1}{X_2}.$$

a. Evaluate f_{Y_1, Y_2}.

b. Evaluate $P(Y_1 Y_2 > 1)$.

E8.39 An artificial neural network uses nodes of the form

$$Y = \tanh(\mathbf{w}^T \mathbf{X}), \quad \tanh(z) = \frac{e^z - e^{-z}}{e^z + e^{-z}} = 1 - \frac{2}{1 + e^{2z}},$$

for a given vector $\mathbf{w}$. Observe that $\tanh(z)$ is increasing in z.

Calculate f_Y, given that $\mathbf{X} \sim \mathcal{N}(\underline{0}, \mathbb{C})$. (It is easier to first calculate $f_Z, Z = \mathbf{w}^T \mathbf{X}$.)

E8.40   An oscillator

$$X(t) = A \cos(\Omega t + \Theta),$$

has a fixed amplitude A and phase $\Theta = 0$, but an unstable random frequency $\Omega \sim \mathcal{N}(\omega_0, \sigma^2)$.

Determine $f_{X(t)}(x)$. (Hint: Go back to basics and start with a sketch.)

E8.41   The output Y of an AM radio envelope detector is $\sqrt{X_1^2 + X_2^2}$, where X_1, X_2 are $i.i.d.$ $\mathcal{N}(0, \sigma^2)$ (i.e., $f_{X_1, X_2}(x_1, x_2) = f_{X_1}(x_1) f_{X_2}(x_2)$, $X_i \sim \mathcal{N}(0, \sigma^2)$). If $Z = \tan^{-1}(X_1/X_2)$, then evaluate $f_{Y,Z}$. (Hint: Cartesian (X_1, X_2) to polar (Y, Z).)

E8.42   Let X_1, X_2 be $i.i.d.$ $\mathcal{E}(\alpha)$ (product factorization of the joint density as in the preceding problem, but with $X_i \sim \mathcal{E}(\alpha)$), and $Y_1 = X_1 + X_2$, $Y_2 = X_1/Y_1$. Calculate the joint density f_{Y_1, Y_2} and verify that it is also of the product form $f_{Y_1}(y_1) f_{Y_2}(y_2)$.

E8.43   We know that

$$f_{X_1, X_2}(x_1, x_2) = \frac{1}{\pi} e^{-(x_1^2 + x_2^2)}, \quad Y = 3X_1 + X_2^2.$$

a.   Augment Y by a properly selected random variable Z so that you can fairly easily determine $f_{Y,Z}(y, z)$.

b.   Provide an expression for the pdf $f_Y(y)$.

E8.44   a.   If $Y = X_1 X_2$, then use augmentation to derive an expression for f_Y as an integral when we assume that f_{X_1, X_2} is known.

b.   Evaluate $f_Y(y)$ for $0 < y < 1$ when X_1 and X_2 are both $\mathcal{U}(0, 1)$.

E8.45   If $Y = X_1^2 + 2X_2$ with X_1, X_2, $i.i.d.$ $\mathcal{N}(0, 1)$ (see the preceding problems for $i.i.d.$), then evaluate f_Y.

E8.46   If $X_t = A \cos(\omega t + \Theta)$, where ω is a fixed constant, A and Θ are independent random variables (meaning that $f_{A,\Theta}(a, \theta) = f_A(a) f_\Theta(\theta)$) with A being Rayleigh distributed, $f_A(a) = \frac{a}{\alpha} e^{-\frac{a^2}{2\alpha}} U(a)$, and $\Theta \sim \mathcal{U}(-\pi, \pi)$, then evaluate the density of X_0. (Hint: Augment this problem and use Jacobians.)

This problem arises when one attempts to understand narrowband Gaussian noise, a process commonly encountered in communications systems. The density would be the same for X_t.

E8.47   In communications, we encounter a signal S represented in terms of amplitude A and phase Θ through $S = A \cos(\Theta)$. The amplitude and phase are such that

$$f_{A,\Theta}(a, \theta) = f_A(a) f_\Theta(\theta) \quad \text{(independent)}$$

$$f_A(a) = a e^{-\frac{a^2}{2}} U(a), \quad f_\Theta(\theta) = \begin{cases} \dfrac{1}{2\pi} & \text{if } |\theta| \leq \pi \\ 0 & \text{otherwise} \end{cases}.$$

Select an appropriate augmentation random variable Z and determine the joint pdf $f_{S,Z}(s, z)$, being careful to specify the ranges of validity.

E8.48   a. The random power P supplied is the product of voltage V and current I ($P = VI$), with

$$f_{V,I}(v, u) = \begin{cases} v + u & \text{if } 0 \leq v \leq 1, \ 0 \leq u \leq 1 \\ 0 & \text{otherwise} \end{cases}.$$

Select an augmentation random variable Z and evaluate the joint pdf $f_{P,Z}(p, z)$.

   b. Determine $f_P(p)$.

9

Expectation and Moments

9.1 PURPOSE, BACKGROUND, AND ORGANIZATION

Expectation, as used in the 17th century, and still so today, had mostly to do with fair prices for gambles or for random variables and assigned a single real value to a random variable. We provide four interpretations of expectation that are in current use and a mathematical definition that suffices for our purposes and much of practice, but lacks an element of abstract generality which can be supplied only by more advanced notions of integration that are beyond our scope. The basic statistical quantities of mean, moment, variance, correlation, and covariance are then introduced and explored for both scalar and vector random variables. With respect to Goal 1, these quantities provide only partial descriptions of random variables. However, these descriptions have historically proven very useful in practice. We will also address Goal 2 with respect to linear transformations by examining how expectations of inputs transform into expectations of outputs. Expectation plays a major role in setting system performance specifications that are needed to define Goal 3. We take a step in the direction of reaching Goal 3 by studying a simple case of optimal linear least mean square estimation. Such optimal linear estimation is revisited in detail in Chapter 10.

9.2 THE CONCEPT OF EXPECTATION

The concept of *expectation*, or the *expected value EX of a random variable X*, or the "average" is about as old as that of probability itself. The first published text on probability by C. Huyghens [48] in 1657 based probability on expectation. (See Hacking [43].) This is still a viable direction from which to develop probability theory, although it is uncommon.

We identify four distinct interpretations or roles for expectation:

1. parameter m of a probability measure, distribution function F_X, or density f_X, often called a *mean*;
2. linear functional E (*definite integral*) on a set $\mathcal{X} = \{X\}$ of random variables (functions on the sample space Ω) that yields a typical value of the random variable;
3. *long-run average outcome* for repeated, unlinked experiments;
4. *fair price* for a gamble with payoffs X.

Although Interpretation 4 came first historically, the easiest access to a discussion of expectation is through Interpretation 1.

9.3 INTERPRETATION 1: MEAN

We motivate the definition of the mean by considering the calculation of, say, the average grade received by all students in a college at the end of a given semester. This is an instance in which there would typically be tens of thousands of grades.

Example 9.1 Average Grade of All Students _____
Assume that grades are converted to a numerical scale of $A+ = 4.3$, $A = 4.0$, $A- = 3.7$, $B+ = 3.3, \ldots, D- = .7, F = 0$. One way to calculate the average or mean grade would be to list all of the grades in whatever order (perhaps alphabetically by student), add them all up, and then divide by the total number of grades. Another way would be to first group the grades by those sharing the same numerical value and determine the fraction $p(G)$ of all of the grades that have the value G. We could then calculate the same average grade by the weighted summation

$$4.3p(A+) + 4.0p(A) + 3.7p(A-) + \ldots + .7p(D-) + 0P(F).$$

The fraction $p(A+)$ of all grades assigned that semester which were $A+$ will be treated by us as the probability of the random variable grade being assigned $A+$.

The next definition of the mean generalizes this approach to take into account continuous as well as discrete random variables, and it is best suited to calculation.

Definition 9.3.1 (Expectation) The expectation of a random variable X having a cdf

$$F_X(x) = \lambda \sum_i p_i U(x - x_i) + (1 - \lambda) \int_{-\infty}^{x} f_X(y) \, dy$$

is given by the formula

(E0)
$$EX = \lambda \left[\sum_{\{i:x_i < 0\}} x_i p_i + \sum_{\{i:x_i \geq 0\}} x_i p_i \right] + (1 - \lambda) \left[\int_{-\infty}^{0} x f_X(x) \, dx + \int_{0}^{\infty} x f_X(x) \, dx \right],$$

provided that there is no divergence of both a positive and a negative term. In the event of divergence of both, we say that the expectation does not exist.

Note that this interpretation defines *EX* as a single number derived uniquely from a given cdf, pmf, or pdf. While it might seem to be just a fussy detail that we do not attempt to define expectation when both integrals diverge, we shall see in Section 19.8 that doing otherwise would lead to a definition of expectation that is not consistent with long-run averages of outcomes of unlinked, repeated random experiments. Our definition maintains consistency between Interpretations 1 and 3.

Remark. By allowing the derivatives of the unit step function (unit impulses or Dirac delta functions) in the pdf f_X, we can abbreviate the previous result to

$$EX = \int_{-\infty}^{0} x f_X(x) \, dx + \int_{0}^{\infty} x f_X(x) \, dx,$$

provided that at least one of the two integrals is finite.

Remark. When the random variable is discrete valued, we can express expectation in terms of a pmf $p_i = P(X = x_i)$ through

$$EX = \sum_{\{i:x_i < 0\}} x_i P(X = x_i) + \sum_{\{i:x_i \geq 0\}} x_i P(X = x_i) = \sum_{\{i:x_i < 0\}} x_i p_i + \sum_{\{i:x_i \geq 0\}} x_i p_i,$$

provided that at least one of these sums is finite.

Example 9.2 Discrete Probability Case _____

A first example of expectation for a discrete probability is

$$P(X = -1) = .25, P(X = 0) = .5, P(X = 2) = .25 \Rightarrow$$

$$EX = -1(.25) + 0(.5) + 2(.25) = .25.$$

A second example is

$$P(X = -a) = P(X = a) = \frac{1}{2} \Rightarrow EX = \frac{1}{2}(-a + a) = 0.$$

Hence, we see that many different distributions (e.g., vary a in the previous example) can yield the same expected value.

Example 9.3   Continuous Probability Case _____

$$f_X(x) = \frac{1}{2}(x + 1)U(1 - |x|) \Rightarrow$$

$$EX = \int_{-1}^{0} \frac{1}{2}(x + 1)x \, dx + \int_{0}^{1} \frac{1}{2}(x + 1)x \, dx = \int_{-1}^{1} \frac{1}{2}(x + 1)x \, dx = \frac{1}{3}.$$

Example 9.4   Cauchy Density _____

$$f_X(x) = \frac{1}{\pi} \frac{1}{1 + x^2} \Rightarrow$$

$$EX = \int_{-\infty}^{0} \frac{1}{\pi} \frac{x}{1 + x^2} \, dx + \int_{0}^{\infty} \frac{1}{\pi} \frac{x}{1 + x^2} \, dx = -\infty + \infty.$$

Hence, EX does not exist in this case.

Verification Table 9.1 is a useful exercise.

Example 9.5   Expectation as "Size" _____
One view of the mean EX is that it is a measure of the "size" of a random variable X. No one number can capture the intuitive notion of "size" in all circumstances. However, taking this viewpoint for the time being, we can ask, For what values of n is $K \sim \mathcal{B}(n, \frac{1}{2})$ larger than

Table 9.1   Expected Values for Various Distributions

Model	$\mathcal{R}(n)$	$\mathcal{B}(n, p)$	$\mathcal{G}(\beta)$	$\mathcal{P}(\lambda)$	
EX	$\dfrac{n + 1}{2}$	np	$\dfrac{\beta}{1 - \beta}$	λ	

Model	$\mathcal{U}(a, b)$	$\mathcal{E}(\alpha)$	$\mathcal{L}(\alpha)$	$\mathcal{N}(m, \sigma^2)$	$Par(\alpha, \tau)$
EX	$\dfrac{a + b}{2}$	$\dfrac{1}{\alpha}$	0	m	$\dfrac{\alpha}{\alpha - 1}\tau$

$X \sim \mathcal{E}(\alpha)$? From the preceding discussion, we know that $EK = 0.5n$ and $EX = 1/\alpha$. Hence, we might think that the number K of "heads" in n tosses of a fair coin will be in some sense larger than the component lifetime X if $n > 2/\alpha$.

An alternative method of evaluating the mean that is sometimes more convenient is provided by the next lemma.

Lemma 9.3.1   (Mean from CDF)   If

$$\lim_{x \to -\infty} xF_X(x) = \lim_{x \to \infty} x(1 - F_X(x)) = 0,$$

then

(E1)
$$EX = -\int_{-\infty}^{0} F_X(x)\, dx + \int_{0}^{\infty} (1 - F_X(x))\, dx,$$

provided that at least one of these two integrals is finite.

Proof. Property E1 is derived from E0 by integrating by parts in the original definition. We have

$$EX = \int_{-\infty}^{0} x f_X(x)\, dx + \int_{0}^{\infty} x f_X(x)\, dx,$$

$$dv = f_X\, dx, u = x.$$

For the $(-\infty, 0]$ case, we take $v = F_X(x)$; for the $[0, \infty)$ case, we take $v = F_X(x) - 1$. For example, we find that

$$\int_{-\infty}^{0} x f_X(x)\, dx = x F_X(x)\Big|_{-\infty}^{0} - \int_{-\infty}^{0} F_X(x)\, dx.$$

The hypothesis of the lemma assures us that the first term is zero and that

$$\int_{-\infty}^{0} x f_X(x)\, dx = -\int_{-\infty}^{0} F_X(x)\, dx.$$

A similar analysis of the second term yields the desired result.   □

Example 9.6   Expectation from the CDF
We have

$$F_X(x) = \begin{cases} \dfrac{1}{2}(1 + x) & \text{if } -1 < x < 0 \\[2mm] 1 - \dfrac{1}{2}e^{-\frac{x^2}{2}} & \text{if } x \geq 0 \end{cases}$$

$$\Rightarrow EX = -\int_{-1}^{0} \frac{1}{2}(1 + x)\, dx + \int_{0}^{\infty} \frac{1}{2}e^{-\frac{x^2}{2}}\, dx = -\frac{1}{4} + \frac{\sqrt{\pi}}{2\sqrt{2}},$$

where the evaluation of the second integral comes from the known integral of the normal density $\mathcal{N}(0, 1)$.

9.4 ELEMENTARY PROPERTIES OF EXPECTATION

The expectation properties E0 and E1 of Section 9.3, properties E2 through E7 developed in this section, and property E8 of Section 9.8 are summarized in Appendix 1, Section 9.17.

Properties E2 and E3 are immediate from the definition of expectation:

(E2) $$P(X = c) = 1 \Rightarrow EX = c,$$

(E3) $$X \geq 0) = 1 \Rightarrow EX \geq 0.$$

Thus, the average of outcomes, all of which take on the same value c, should be c. Similarly, the average of outcomes, all of which are nonnegative, should be nonnegative.

The property

(E4) $$E(aX) = aE(X)$$

follows easily from our derivation (Section 8.3.2) of the density f_Y of $Y = aX$,

$$f_Y(y) = \frac{1}{|a|} f_X\left(\frac{y}{a}\right),$$

and its use in $EY = \int y f_Y \, dy$. Thus, if X is a distance measured in microns with $EX = m$ and we convert X to Y measured in nanometers, then $EY = 1{,}000m$. The average value of a set of numbers should transform properly when we change the measurement unit of these numbers.

If we have two different sets of numbers—say, our profits over time from investments X and Y—then our average profits on both should be the sum of our average profits on them individually, or

(E5) $$E(X + Y) = EX + EY \quad \text{when finite.}$$

This formula follows from a calculation of $Z = X + Y, f_Z(z) = \int f_{X,Y}(x, z - x) \, dx$. Hence,

$$EZ = E(X + Y) = \int_{-\infty}^{\infty} \int_{-\infty}^{\infty} z f_{X,Y}(x, z - x) \, dx \, dz.$$

Making the 1:1 change of variables $y = z - x, x = x$ yields

$$EZ = E(X + Y) = \int_{-\infty}^{\infty} \int_{-\infty}^{\infty} (y + x) f_{X,Y}(x, y) \, dx \, dy.$$

Use

$$f_Y(y) = \int_{-\infty}^{\infty} f_{X,Y}(x, y) \, dx$$

to conclude that

$$\int_{-\infty}^{\infty} \int_{-\infty}^{\infty} y f_{X,Y}(x,y)\, dx\, dy = \int_{-\infty}^{\infty} y f_Y(y)\, dy = EY,$$

and similarly,

$$\int_{-\infty}^{\infty} \int_{-\infty}^{\infty} x f_{X,Y}(x,y)\, dy\, dx = \int_{-\infty}^{\infty} x f_X(x)\, dx = EX.$$

Thus,

$$EZ = E(X + Y) = EX + EY.$$

An immediate corollary of E5 is its extension to sums of many random variables. This is proven by finite induction on the number of summands. By E5, it is true for a sum of two random variables. Assume that it is true for all $k < n$. Then, by E5,

$$E\left(\sum_1^n X_i\right) = E\left(X_n + \sum_1^{n-1} X_i\right) = EX_n + E\left(\sum_1^{n-1} X_i\right).$$

However, by the induction hypothesis,

$$E\left(\sum_1^{n-1} X_i\right) = \sum_1^{n-1} EX_i,$$

and the desired conclusion follows.

The expectation of a sum of random variables is the sum of expected values.

Combining properties E2, E4, and E5 then yields

(E6) $$\qquad\qquad E(aX + bY + c) = aEX + bEY + c.$$

If a random variable X is always at least as large as a random variable Y, then a proper definition of an average should yield $EX \geq EY$. This statement is asserted in

(E7) $$\qquad\qquad P(X \geq Y) = 1 \Rightarrow EX \geq EY,$$

for at least one of EX, EY finite. Property E7 follows from properties E3, E4, and E5 via the formula

$$P(X \geq Y) = P(X - Y \geq 0).$$

Note that, from property E3, $E(X - Y) \geq 0$, and from property E5, $E(X - Y) = EX + E(-Y)$. From property E4, it follows that $E(X - Y) = EX - EY$. Hence, we conclude that $EX - EY \geq 0$.

9.5 INTERPRETATION 2: DEFINITE INTEGRAL

Under Interpretation 2, the expectation EX of a random variable X is thought of as a definite integral of the function $X(\omega)$, $\omega \in \Omega$, weighted by the probability measure P. Schematically (although this can be made precise),

$$EX = \int_{\Omega} X(\omega)P(d\omega).$$

Properties E2 through E6 enable us to follow Interpretation 2 and interpret E as a definite integral (a positive linear functional assigning numbers to the functions that are random variables) on the set $\mathcal{X}$ of, say, bounded (to ensure finiteness of all terms, more generally we would use integrable) random variables:

$$\mathcal{X} = \{X : (\exists B)P(|X| < B) = 1\}, \quad E : \mathcal{X} \to \mathbb{R}.$$

The linear functional E assigns a finite real number to each bounded random variable $X \in \mathcal{X}$ much as the definite integral $\int_{-\infty}^{\infty} f \, dx$ assigns a real number to any integrable function f. The resulting integral, average, or mean is thought of as a "typical value" of the random variable, with the caveat that a single number cannot fully describe a random variable. More precisely, the expectation is often thought of as a *measure of location* of the random variable.

9.6 INTERPRETATION 3: LONG-RUN AVERAGE OUTCOME

Interpretation 3 was introduced in Section 9.2. Mathematically, it is the subject of the *laws of large numbers*, as discussed in Chapter 19. If X is a random variable associated with a random experiment $\mathcal{E}$ (e.g., $\mathcal{E}$ is a toss of a coin with $X = 1$ for heads and $X = 0$ for tails) and $\{X_i\}$ are the versions associated with the unlinked repetitions $\{\mathcal{E}_i\}$ of $\mathcal{E}$, then there are several senses discussed in Chapter 19 in which the average outcome of the first n repetitions, $(1/n)\sum_{1}^{n} X_i$, converges to EX. These laws of large numbers underlie the understanding that EX is approximately what we can expect to see if we conduct a large number of identically prepared and unlinked random experiments and average all of the outcomes. A special case of this conclusion is that the relative frequency $r_n(A)$ for an event A will, for large enough n, closely approximate $P(A) = EI_A(X)$.

9.7 INTERPRETATION 4: FAIR PRICE

An early use of expectation, in accordance with Interpretation 4, resulted in the *St. Petersburg Paradox*. You are offered a gamble in which a fair coin is tossed repeatedly until the first occurrence of a "head." If the first "head" occurs on trial $n \geq 1$, then you are paid $X = \$2^n$. How much would you pay to be offered this gamble? Would you be willing to pay, say, $10,000 to have the gamble? As the probability of a first "head" on trial n is 2^{-n}, it is easy to see that

$$EX = \sum_{1}^{\infty} 2^{-n} 2^n = \sum_{1}^{\infty} 1 = \infty.$$

The paradox is our unwillingness to pay much for a gamble evaluated at being worth infinitely much. The eventual resolution of this paradox was to introduce the concept of a *utility function u*, as was first done by Daniel Bernoulli [8] in the 18th century. The role of the utility function is to numerically rescale the payoffs of a gamble to better accord with your relative strengths of preferences for these payoffs. Contemplating preferences for gambles became the tool for numerically measuring these relative strengths. The notion of utility is of importance in present-day economics and rational decision making. (See Section 16.7.) D. Bernoulli suggested taking

$$u(p) = \log p,$$

where the choice of base for the logarithm is arbitrary and will be taken to be 2. This choice was made without regard to the individual making the assessment, in disagreement with the current understanding that utility varies with the individual and his or her current situation. Note that, with this choice, the utility $u(10^7) = 7 \log_2 10$ of, say, \$10^7 is much less than 10 times that of $u(10^6) = 6 \log_2 10$. The expected utility of the St. Petersburg gamble is now

$$Eu = \sum_1^\infty 2^{-n} u(2^n) = \sum_1^\infty n 2^{-n} < \infty.$$

If we use the identity,

$$\sum_0^\infty n \beta^{-n} = \frac{\beta}{(\beta - 1)^2},$$

derivable from the derivative of the geometric sum, then we find that, for the St. Petersburg gamble, $Eu = 2$. Since $u(\$4) = 2$, this suggests that \$4 is a fair price.

9.8 EXPECTATION OF A FUNCTION OF A RANDOM VARIABLE

Property E8, given next, enables us to compute $Eg(X)$ directly rather than by first finding $f_Z(z)$ for $Z = g(X)$ and then evaluating $\int z f_Z(z) \, dz$.

Lemma 9.8.1   (Expectation of a Function of a Random Variable)

(E8) $$Eg(X) = \int_{-\infty}^{\infty} g(x) f_X(x) \, dx.$$

Proof. A derivation of property E8 proceeds as follows:

$$Eg(X) = EZ = \int_{-\infty}^{\infty} z f_Z(z) \, dz = \sum_{k=-\infty}^{\infty} \int_{\frac{k}{n}}^{\frac{k+1}{n}} z f_Z(z) \, dz = \lim_{n \to \infty} \sum_{k=-\infty}^{\infty} \int_{\frac{k}{n}}^{\frac{k+1}{n}} z f_Z(z) \, dz$$

$$= \lim_{n \to \infty} \sum_{k=-\infty}^{\infty} \frac{k}{n} \int_{\frac{k}{n}}^{\frac{k+1}{n}} f_Z(z) \, dz = \lim_{n \to \infty} \sum_{k=-\infty}^{\infty} \frac{k}{n} P\left(\frac{k}{n} < Z = g(X) \le \frac{k+1}{n} \right)$$

$$= \lim_{n \to \infty} \sum_{k=-\infty}^{\infty} \frac{k}{n} \int_{\{x : \frac{k}{n} < g(x) \le \frac{k+1}{n}\}} f_X(x)\, dx$$

$$= \lim_{n \to \infty} \sum_{k=-\infty}^{\infty} \int_{\{x : \frac{k}{n} < g(x) \le \frac{k+1}{n}\}} g(x) f_X(x)\, dx$$

$$= \lim_{n \to \infty} \int_{-\infty}^{\infty} g(x) f_X(x)\, dx = \int_{-\infty}^{\infty} g(x) f_X(x)\, dx. \qquad \Box$$

Example 9.7   Probability and Expectation of Indicators

As a first example of the use of E8, consider the formula

$$g(x) = U(t - x), \quad Eg(X) = EU(t - X) = \int_{-\infty}^{t} f_X(x)\, dx = F_X(t).$$

Hence, knowing E on $\mathcal{X}$, we can recover the cdf F_X and thence the pdf f_X and probability measure P. It is this observation that would allow us to base a theory of probability upon expectation. A generalization is that, for a random variable X and event $A \subset \mathbb{R}$,

$$EI_A(X) = P(A).$$

Example 9.8   Shannon's Entropy

For a second example, we generalize the Shannon entropy function defined in Section 2.5 for a binary random variable. When X is discrete valued and described by a pmf, we adopt $-\log P(X = x)$ as the information provided by a particular realization of the random variable $X = x$. The average value $H(X)$ of this information is then defined to be

$$H(X) = -E[\log P(X)]$$

and is the appropriate extension of the entropy function introduced in Section 2.5. If X is a discrete-valued random variable taking on n values and having pdf $\{p_i\}$, then

$$H(X) = -\sum_{i=1}^{n} p_i \log p_i.$$

If X is a random variable described by a pdf f_X, then we adopt $-\log f_X(x)$ as the information provided by the realization $X = x$, and the entropy becomes

$$H(X) = -E[\log f_X(X)] = -\int_{-\infty}^{\infty} f_X(x) \log(f_X(x))\, dx.$$

The value of this definition of "information" is supplied by its central role in the coding theorems of information and communication theory (e.g., see Cover and Thomas [20]). Further insight into the properties of the entropy function is provided in the next section.

9.9 CONVEX FUNCTIONS AND JENSEN'S INEQUALITY

The results of this section may be omitted as being of specialized interest.

Convex combinations of vectors and of probability measures were introduced in Section 3.9. We now consider the associated notion of a convex function κ of a vector $\mathbf{x}$. Our first approach is through the next definition.

Definition 9.9.1   (Convex Function)   A real-valued function κ of a vector argument in $\mathbb{R}^n$ is a *convex function* if, for all $\mathbf{x}, \mathbf{y} \in \mathbb{R}^n$ and for all $0 \le \lambda \le 1$,

$$\kappa(\lambda \mathbf{x} + (1 - \lambda)\mathbf{y}) \ge \lambda \kappa(\mathbf{x}) + (1 - \lambda)\kappa(\mathbf{y}).$$

Hence, a convex function κ of a convex combination of two vectors is greater than or equal to the same convex combination of the values $\kappa(\mathbf{x}_i)$. This definition easily implies an extension to convex combinations of multiple vectors. It can be shown that, when κ has real-valued arguments and is twice continuously differentiable, κ is convex if and only if it has a nonnegative second derivative. Thus, $\kappa(x) = x^2$ is convex because $\kappa''(x) = 2 \ge 0$. Similarly, $\kappa(x) = e^{\alpha x}$ is convex because $\kappa''(x) = (\alpha)^2 e^{\alpha x} \ge 0$. A *concave function* is defined to be the negative of a convex function. Thus, $\log x$ has a negative second derivative and is therefore a concave function. A characterization of a convex function of a vector-valued argument is given in the next lemma.

Lemma 9.9.1   (Convex Function)   The function κ is convex if and only if, at each vector point $\mathbf{y}$, there is a supporting hyperplane to the graph of the function κ at the point $\mathbf{y}$ of the form $\kappa(\mathbf{y}) + \mathbf{w} \cdot (\mathbf{x} - \mathbf{y})$. Restated, κ is a convex function if and only if

$$(\forall \mathbf{y})(\exists \mathbf{w})(\forall \mathbf{x}) \quad \kappa(\mathbf{x}) \ge \kappa(\mathbf{y}) + \mathbf{w} \cdot (\mathbf{x} - \mathbf{y}).$$

The characterization of a convex function provided by the lemma allows immediate access to the frequently useful Jensen's inequality.

Theorem 9.9.1   (Jensen's Inequality)   If κ is a convex function of a random variable X, then

$$E\kappa(X) \ge \kappa(EX).$$

Proof.   Apply the preceding lemma in the scalar case, taking $\mathbf{y} = EX$ and $\mathbf{x} = X$, and take expectations of both sides of the inequality. Then

$$E\kappa(X) \ge \kappa(EX) + \mathbf{w}(EX - EX) = \kappa(EX). \qquad \square$$

Example 9.9   Use of Jensen's Inequality _____
We can conclude from the convexity of x^2 that

$$EX^2 \ge (EX)^2,$$

and the second moment is always at least as large as the square of the first moment.

Other inequalities follow from the convexity of the exponential,

$$Ee^{\alpha X} \geq e^{\alpha EX},$$

and the concavity of the logarithm,

$$E(\log X) \leq \log(EX).$$

An important quantity in information theory and statistics (e.g., see Cover and Thomas [20], Section 2.3) is the *Kullback–Leibler (KL) divergence* or distance $D(p||q)$. This is a measure of the separation between two pdfs p and q given by

$$D(p||q) = E_p\left[\log\frac{p(X)}{q(X)}\right],$$

where E_p denotes that the expectation over the random variable X is taken according to the pdf p. Note that, in general, $D(p||q) \neq D(q||p)$. The nonnegativity of KL divergence then follows by the use of Jensen's inequality through the following relationships:

$$-D(p||q) = E_p\left[\log\left(\frac{q(X)}{p(X)}\right)\right] \leq \log\left(E_p\left[\frac{q(X)}{p(X)}\right]\right) = \log(E_q 1) = \log(1) = 0.$$

Example 9.10   Kullback–Leibler Divergence _____

If we are observing, say, the number K of errors made in a message of n symbols, but know only that the probability of an error on a single symbol is either p_0 or p_1, then there are statistical tests to decide which of these two cases is more likely to be correct, given the actual value of K. The performance of such tests turns out to depend upon the Kullback–Leibler divergence as a measure of the distance between the two binomial distributions $\mathcal{B}(n, p_0)$ and $\mathcal{B}(n, p_1)$. We have

$$D(p_1||p_0) = \sum_{k=0}^{n}\binom{n}{k}p_1^k(1-p_1)^{n-k}\log\left(\frac{\binom{n}{k}p_1^k(1-p_1)^{n-k}}{\binom{n}{k}p_0^k(1-p_0)^{n-k}}\right)$$

$$= \sum_{k=0}^{n}\binom{n}{k}p_1^k(1-p_1)^{n-k}\left[k\log\frac{p_1}{p_0} + (n-k)\log\frac{1-p_1}{1-p_0}\right].$$

Recall that, for $K \sim \mathcal{B}(n, p_1)$,

$$EK = \sum_{k=0}^{n}\binom{n}{k}p_1^k(1-p_1)^{n-k}k = np_1.$$

Hence,

$$D(p_1||p_0) = n\left[p_1\log\frac{p_1}{p_0} + (1-p_1)\log\frac{1-p_1}{1-p_0}\right]$$

$$= -n\left(H(p_1) + p_1\log p_0 + (1-p_1)\log(1-p_0)\right).$$

If we take logarithms to the base 2 (units of "bits") and choose $p_1 = 0.5$, then

$$D(p_1 \| p_0) = -n \left(1 + \frac{1}{2} \log_2 p_0 (1 - p_0) \right).$$

9.10 MOMENTS, ESPECIALLY VARIANCE

9.10.1 Moments

Moments provide partial information about the probability measure P, the cdf F_X, or the pdf f_X. Sometimes this is the only information readily available. Moments of X are means of powers of X.

Definition 9.10.1   (nth Moment)   From E8, for nonnegative integer n, the nth moment of a random variable X is

$$EX^n = \int_{-\infty}^{\infty} x^n f_X(x)\, dx$$

if this integral exists.

Moments of all orders need not exist.

Example 9.11   Existence of Higher Order Moments _____
Consider the two-sided Pareto-like density

$$f_X(x) = \frac{\alpha}{2|x|^{\alpha+1}} U(|x| - 1), \quad \alpha > 0.$$

This is a density, since it is nonnegative and integrates to 1. Furthermore, we see that

$$EX^n = \int_{-\infty}^{-1} \frac{\alpha}{2} x^n |x|^{-\alpha-1}\, dx + \int_1^{\infty} \frac{\alpha}{2} x^{n-\alpha-1}\, dx$$

$$= \int_1^{\infty} (-1)^n \frac{\alpha}{2} x^{n-\alpha-1}\, dx + \int_1^{\infty} \frac{\alpha}{2} x^{n-\alpha-1}\, dx.$$

If $n \geq \alpha$, then the first integral is $-\infty$ for n odd and ∞ for n even. The second integral is ∞ for any $n \geq \alpha$. Both integrals are finite for $n < \alpha$. We conclude that X only has finite moments of order n strictly less than α, infinite moments for $n \geq \alpha$ and even, and undefined moments for $n \geq \alpha$ and odd.

Example 9.12   Moments of the Cauchy Density _____
The Cauchy density $\mathcal{C}(a)$,

$$f_X(x) = \frac{a}{\pi(a^2 + x^2)},$$

is an even function that has no finite moments at all. It has infinite nth moments for n even and no moments for n odd.

As a final example, recall that we say that a function $g(x)$ is even if, for all x, $g(x) = g(-x)$ (g has a graph that is symmetrical about the origin), and g is odd if, for all x, $g(-x) = -g(x)$. Note that if g is an odd function, then

$$\int_{-a}^{a} g(x)\,dx = 0$$

whenever the integral from 0 to a is finite. Hence, if the pdf $f_X(x)$ is an even function (e.g., $X \sim \mathcal{N}(0, 1)$), then all odd moments are zero if they are finite.

A useful fact about the finiteness of moments is given in the next theorem.

Theorem 9.10.1   (Finiteness of Moments)   If $j < k$, then

$$|EX^k| < \infty \Rightarrow |EX^j| < \infty.$$

Proof.   By hypothesis, there exist finite constants

$$c_1 > \int_{0}^{\infty} x^k f_X\,dx \geq 0, \quad c_2 > \left| \int_{-\infty}^{0} x^k f_X\,dx \right|.$$

Hence, for $j < k$,

$$c_1 > \int_{0}^{\infty} x^k f_X\,dx \geq \int_{1}^{\infty} x^k f_X\,dx \geq \int_{1}^{\infty} x^j f_X\,dx \geq$$

$$\int_{0}^{\infty} x^j f_X\,dx - \int_{0}^{1} f_X\,dx \geq \int_{0}^{\infty} x^j f_X\,dx - 1.$$

Therefore,

$$0 \leq \int_{0}^{\infty} x^j f_X\,dx \leq c_1 + 1 < \infty.$$

A similar calculation reveals that $\int_{-\infty}^{0} x^j f_X\,dx$ is finite. $\square$

The existence of finite higher order moments implies the existence of finite lower order moments.

9.10.2 Central Moments

Definition 9.10.2   (Central Moment)   The nth central moment is

$$E(X - EX)^n = \int_{-\infty}^{\infty} (x - EX)^n f_X(x)\,dx$$

if this integral exists.

The first central moment is zero. The second central moment is known as the *variance* and is denoted by $\mathrm{VAR}(X)$. From the binomial theorem, we see the relationships

$$E(X - EX)^n = \sum_{k=0}^{n} \binom{n}{k} (-EX)^{n-k} EX^k$$

and

$$EX^n = \sum_{k=0}^{n} \binom{n}{k} (EX)^{n-k} E(X - EX)^k.$$

As a corollary, we see that the nth moment exists and is finite if and only if the nth central moment exists and is finite. In general,

$$E(X - EX) = 0, \quad \mathrm{VAR}(X) = E(X^2) - (EX)^2.$$

Example 9.13   Moments of the Normal _____

For example, if $X \sim \mathcal{N}(m, \sigma^2)$, then

$$EX = m, \quad \mathrm{VAR}(X) = \sigma^2.$$

Furthermore, if $Y = (X - m)/\sigma \sim \mathcal{N}(0, 1)$, then

$$EX = m, \quad \mathrm{VAR}(X) = \sigma^2, \quad E(X - EX)^k = \sigma^k EY^k.$$

The odd moments of Y for the even density f_Y are 0. The even moments of Y can be derived in a number of ways, including integration by parts. An easier method will become available in Section 17.4 when we can use characteristic functions. The results are

$$EY^k = \frac{1}{\sqrt{2\pi}} \int_{-\infty}^{\infty} y^k e^{-\frac{y^2}{2}} \, dy = \begin{cases} 0 & \text{if } k \text{ odd} \\ (k-1)(k-3)\cdots 1 & \text{otherwise} \end{cases}.$$

Thus, if $Y \sim \mathcal{N}(0, 1)$, then

$$EY = 0, EY^2 = 1, EY^3 = 0, EY^4 = 3, EY^6 = 15, \text{ etc.}$$

Example 9.14   All Pairs of Mean and Variance Are Possible _____

$$P(X = m - a) = P(X = m + a) = \frac{1}{2} \Rightarrow EX^k = \frac{1}{2}[(m-a)^k + (m+a)^k].$$

$$EX = m, \quad EX^2 = \frac{1}{2}[2m^2 + 2a^2] = m^2 + a^2, \quad \mathrm{VAR}(X) = EX^2 - (EX)^2 = a^2.$$

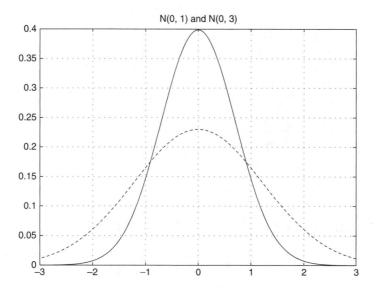

Figure 9.1   Normals of variances 1 and 3.

This example shows that we can find a very simple random variable having any prescribed mean and variance. Combined with the normal distribution, the example shows that different distributions can share the same mean and variance.

Variance is a measure of *fluctuation* or *dispersion*—a measure of how variable a random variable is. In general, the larger the variance, the larger is the variability or spread. This property is well illustrated by the two normal densities shown in Figure 9.1, with the broader dashed curve having variance 3 and the solid curve having variance 1.

Basic properties of variance are listed in the next lemma.

Lemma 9.10.1   (Variance Properties)

$$\text{VAR}(X) \geq 0, \ \text{VAR}(c) = 0,$$

$$\text{VAR}(X + c) = \text{VAR}(X), \ \text{VAR}(aX) = a^2 \, \text{VAR}(X).$$

Hence, the variance of a constant is as small as it can be, and the variance of a random variable X is unchanged by its being shifted (changed in "location"). Variance can be infinite (e.g., for the Cauchy density), in which case it is no longer a good measure of dispersion. Among random variables constrained by $P(a \leq X \leq b) = 1$, the one having the largest possible variance is $P(X = a) = P(X = b) = \frac{1}{2}$ (see the Exercises); this clearly has the largest possible intuitive variability or dispersion.

Example 9.15   Variance as a Measure of "Fluctuation" _____

Returning to an earlier example in which we used the mean to compare the "size" of $K \sim \mathcal{B}(n, \frac{1}{2})$ with $X \sim \mathcal{E}(\alpha)$, we can now compare K and X as to their variability. From Table 9.2

Table 9.2   Variances of Common Distributions

Model	$\mathcal{R}(n)$	$\mathcal{B}(n,p)$	$\mathcal{G}(\beta)$	$\mathcal{P}(\lambda)$	
VAR(X)	$\dfrac{n^2-1}{12}$	$np(1-p)$	$\dfrac{\beta}{(1-\beta)^2}$	λ	
Model	$\mathcal{U}(a,b)$	$\mathcal{E}(\alpha)$	$\mathcal{L}(\alpha)$	$\mathcal{N}(m,\sigma^2)$	$\mathcal{P}ar(\alpha,\tau)$
VAR(X)	$\dfrac{(b-a)^2}{12}$	$\dfrac{1}{\alpha^2}$	$\dfrac{2}{\alpha^2}$	σ^2	$\dfrac{\alpha\tau^2}{(\alpha-1)^2(\alpha-2)}$

we see that $\text{VAR}(K) = n/4$ and $\text{VAR}(X) = 1/\alpha^2$. Hence, K is more variable than X (in this particular sense) if $n > 4/\alpha^2$.

9.11 CORRELATION, COVARIANCE, AND THE SCHWARZ INEQUALITY

Correlation and covariance are momentlike quantities that provide simple measures of the strength of *linear dependence* (see Section 9.12) between two or more variables.

Definition 9.11.1 **(Correlation and Covariance)** The correlation between two random variables X and Y is

$$E(XY) = \int_{-\infty}^{\infty} \int_{-\infty}^{\infty} xy f_{X,Y}(x,y)\, dx\, dy,$$

and their covariance is

$$\text{COV}(X,Y) = E[(X - EX)(Y - EY)] = E(XY) - (EX)(EY).$$

Observe that

$$\text{VAR}(X) = \text{COV}(X,X).$$

Example 9.16   Correlation and Covariance _____
Consider the joint pdf

$$f_{X,Y}(x,y) = \begin{cases} 2 & \text{if } 0 \le x \le y \le 1 \\ 0 & \text{otherwise} \end{cases},$$

with positive value of 2 over a triangular region and 0 elsewhere. In order to evaluate the moments, we first determine the marginal densities:

$$f_X(x) = \int_0^1 f_{X,Y}(x,y)\, dy = \int_x^1 2\, dy = \begin{cases} 2(1-x) & \text{if } 0 \le x \le 1 \\ 0 & \text{otherwise} \end{cases},$$

$$f_Y(y) = \int_0^1 f_{X,Y}(x,y)\, dx = \int_0^y 2\, dx = \begin{cases} 2y & \text{if } 0 \le y \le 1 \\ 0 & \text{otherwise} \end{cases}.$$

We now evaluate the moments of X and Y from

$$EX^n = \int_0^1 x^n 2(1-x)\,dx = \frac{2}{(n+1)(n+2)},$$

$$EY^n = \int_0^1 y^n 2y\,dy = \frac{2}{n+2}.$$

Hence,

$$EX = \frac{1}{3}, \quad EX^2 = \frac{1}{6}, \quad \text{VAR}(X) = EX^2 - (EX)^2 = \frac{1}{18},$$

$$EY = \frac{2}{3}, \quad EY^2 = \frac{1}{2}, \quad \text{VAR}(Y) = EY^2 - (EY)^2 = \frac{1}{18}.$$

The correlation is given by

$$E(XY) = \int_0^1 dx \int_x^1 dy\, 2xy = \int_0^1 x\,dx \left(\int_x^1 2y\,dy = 1 - x^2 \right) = \frac{1}{4}.$$

The covariance is given by

$$\text{COV}(X, Y) = E(XY) - (EX)(EY) = \frac{1}{36}.$$

Definition 9.11.2   (Uncorrelated) X and Y are uncorrelated if

$$E(XY) = (EX)(EY) \text{ or } \text{COV}(X, Y) = 0.$$

Clearly, the definitions of correlation and covariance are symmetrical in X and Y. A related concept is that of orthogonality.

Definition 9.11.3   (Orthogonality) X and Y are orthogonal if

$$E(XY) = 0.$$

It is evident that X and Y as defined in Example 9.16 are neither uncorrelated nor orthogonal. If one of X or Y has zero mean, then orthogonality is equivalent to uncorrelatedness. Orthogonality is an analog of geometrical orthogonality, or perpendicularity between vectors $(\mathbf{X} \cdot \mathbf{Y} = 0)$.

A relation between correlation or covariance and moments is given by the celebrated Schwarz inequality.

Theorem 9.11.1   (Schwarz Inequality)

$$[E(XY)]^2 \le EX^2 EY^2.$$

$$[\text{COV}(X, Y)]^2 \le \text{VAR}(X)\, \text{VAR}(Y).$$

Proof.

$$(aX + Y)^2 \geq 0 \Rightarrow E(aX + Y)^2 \geq 0 \Rightarrow (EX^2)a^2 + 2(EXY)a + EY^2 \geq 0.$$

Hence, this quadratic in a cannot have two unequal real roots; if it did, then it would become negative for some real values of a. Thus, from the discriminant arising in the solution of quadratic equations, we see that

$$[E(XY)]^2 - EX^2EY^2 \leq 0,$$

and the first inequality follows. The second inequality follows from the first by replacing X by $X - EX$ and Y by $Y - EY$. □

The condition for equality in this inequality is also seen to be precisely that

$$(\exists a^*) \; E(Y - a^*X)^2 = 0,$$

or that $Y = a^*X$ with probability 1. As they must, the random variables X and Y defined in Example 9.16 satisfy the Schwarz inequality:

$$[E(XY)]^2 = \frac{1}{16} \leq EX^2EY^2 = \frac{1}{12}.$$

Correlation and covariance appear in calculations of second moments of sums of random variables. In these evaluations, we make use of the fact that the square of a sum is a double sum and that the expected value of a sum or double sum is the sum or double sum of expected values. We have

$$E\left(\sum_1^n X_i\right)^2 = E\sum_{i=1}^n\sum_{j=1}^n X_iX_j = \sum_{i=1}^n\sum_{j=1}^n E(X_iX_j),$$

$$\mathrm{VAR}\left(\sum_1^n X_i\right) = E\left(\sum_1^n (X_i - EX_i)\right)^2 = \sum_{i=1}^n\sum_{j=1}^n \mathrm{COV}(X_i, X_j).$$

Hence, the expected value of the square of a sum of random variables is the double sum of the correlations, and the variance of a sum is a double sum of the covariances.

9.12 LINEAR DEPENDENCE AND LEAST MEAN SQUARE ESTIMATION

9.12.1 Basics

To understand the issue of linear dependence, consider making a linear (technically, affine, since we have allowed an additive constant) approximation $\hat{Y} = aX + b$ to Y; perhaps we observe X and would like to make a simple inference to Y by using an *estimator* (a function of the observed data and not of the quantity to be estimated or inferred) $\hat{Y}(X)$, based on the data X. If we are to compare different estimator functions to determine a good or an optimal one, then we typically rely upon a numerical assessment of quality and rank the estimators on the basis of the usual ordering of real numbers. One commonly used measure of the size of the error, $Y - \hat{Y}$, is the mean square error :

$$\text{mean square error (MSE)} = E(Y - \hat{Y}(X))^2.$$

By squaring the error $Y - \hat{Y}$ we are assuming that negative errors are equivalent to positive errors of the same magnitude, that "small" errors are relatively negligible (since squaring them makes them even smaller), and that "large" errors are highly significant (since squaring them makes them even larger). There are, of course, many other functions of the error that we can use, but squaring has advantages of computational ease and a long tradition (dating back to Gauss) of use. Just squaring the error would yield a nonnegative random variable. Taking the expectation of this term gives us a single typical numerical value of "size" that can then be used to compare the performances of different estimators.

We first provide a general condition on estimators such that the mean of the minimum MSE estimator equals the mean of the quantity being estimated. Consider the rather general estimator $\hat{Y}(X) = f(X) + b$, where f is an arbitrary function having a finite second moment. We wish to select the constant b so as to minimize the estimation error $E(Y - \hat{Y})^2 = E(Y - f(X) - b)^2$. The first and second derivatives of the error with respect to b are

$$\frac{dE(Y - f(X) - b)^2}{db} = -2E(Y - f(X) - b), \quad \frac{d^2E(Y - f(X) - b)^2}{db^2} = 2.$$

From the positivity of the second derivative, we see that setting the first derivative equal to zero yields a minimum, as desired. Setting the first derivative equal to zero yields

$$b + Ef(X) = E\hat{Y} = EY.$$

Lemma 9.12.1 (Estimator Mean) If an estimator $\hat{Y}(X) = f(X) + b$ and Ef^2 is finite, then the mean square error $E(Y - \hat{Y})^2$ is minimized by choosing the constant term b so that $E\hat{Y} = EY$.

We will also establish this general conclusion in the specific case of a linear estimator without recourse to calculus.

Returning to our linear estimator, we have

$$E(Y - aX - b)^2 = E((Y - EY) - a(X - EX) - c)^2,$$

where $c = b - EY + aEX$. Proceeding further, we find that

$$E(Y - aX - b)^2 = \text{VAR}(Y) + a^2 \text{VAR}(X) + c^2 - 2a \text{COV}(X, Y).$$

If we select the values a^*, c^* that yield a minimum mean square error (*MMSE*), then

$$c^* = 0, \quad a^* = \frac{\text{COV}(X, Y)}{\text{VAR}(X)}, \quad E(Y - a^*X - b^*)^2 = \text{VAR}(Y) - \frac{\text{COV}(X, Y)^2}{\text{VAR}(X)}.$$

Observe that $c^* = 0$ is better understood as the condition that $E\hat{Y} = EY$, with the estimator $\hat{Y}$ chosen to have the same mean as that of the quantity Y being estimated. If we introduce the *correlation coefficient*,

$$\rho_{X,Y} = \frac{\text{COV}(X,Y)}{\sqrt{\text{VAR}(X)\,\text{VAR}(Y)}},$$

a normalized form of covariance, then

$$E(Y - a^*X - b^*)^2 = \text{VAR}(Y)(1 - \rho_{X,Y}^2).$$

Hence, the closer the magnitude of the correlation coefficient is to unity, the better is the linear approximation. Note that if X and Y are uncorrelated, then $\text{COV}(X,Y) = 0$ and $a^* = 0$. Hence, the best linear approximation of Y by X ignores X when they are uncorrelated.

The next inequality is a corollary of the Schwarz inequality.

Corollary 9.12.1 $|\rho_{X,Y}| \leq 1$.

The correlation coefficient $\rho_{X,Y}$ can also be understood in terms of scatterplots. If you look back to the scatterplot of $x(t)$ vs. $x(t-1)$ provided in Figure 1.9, you will see that $|\rho_{x(t-1),x(t)}|$ is a measure of the thickness of the cloud of points about the regression line. The closer the magnitude of the correlation coefficient is to unity, the more tightly the points cluster about the regression line.

While a large value of $\rho_{X,Y}$ indicates a high degree of linear dependence between X and Y, a small value of this index does not mean that they are only weakly dependent. There can be a high degree of nonlinear dependence that is not detected by the covariance.

Example 9.17   Completely Dependent Variables Can Be Uncorrelated _____
For example, if $EX = EX^3 = 0$, as might arise if X was symmetrically distributed about 0 (i.e., f_X even) and $Y = X^2$, then Y is completely determined by X. However,

$$E(XY) = EX^3 = 0 = EXEY, \ \ \text{COV}(X,Y) = 0 = \rho_{X,Y}.$$

9.12.2 Application to Signal-Plus-Noise Measurements

In measurement, we have an unknown quantity or a signal amplitude S whose value we wish to know. We assume that, even before making a measurement, we do know that $ES = m_S$ and $\text{VAR}(S) = \sigma_S^2$. We have a measuring instrument that returns an observation X of S. However, X is only noisily related to S. The specific model we make here is that of *signal plus noise* (see Section 13.5), or

$$X = S + N,$$

with N the additive noise introduced by the inaccuracies in the measurement instrument. The instrument is said to be *unbiased* if $EX = ES$, or, in our case, if $EN = 0$. The accuracy of the measurement process is often represented by the noise or error variance $\text{VAR}(N) = \sigma_N^2$. (Recall that one interpretation of variance is as a measure of fluctuation or variability.) Finally, we assume that the true quantity S and the measurement noise N are uncorrelated, or $\text{COV}(S, N) = 0$.

The naive estimator $\hat{S}_0$ of S is to take $\hat{S}_0 = X$. The measurement is itself the estimate of the quantity being measured, particularly since, by $EX = ES$, it is correct, on average. The mean square error performance of this naive estimator is then

$$E(S - \hat{S}_0)^2 = E(S - X)^2 = EN^2 = \sigma_N^2.$$

We now examine whether we can improve our estimation by taking a linear estimator

$$\hat{S} = aX + b.$$

From the results of the preceding subsection, we see that, for a minimum mean square estimator, we must have

$$E\hat{S} = ES \Rightarrow a(ES + EN) + b = ES \Rightarrow b = (1 - a)m_S.$$

Furthermore, from the preceding subsection,

$$a = \frac{\text{COV}(S, X)}{\text{VAR}(X)}.$$

Straightforward evaluations yield

$$\text{COV}(S, X) = E\left[(S - m_S)(S - m_S + N)\right] = \text{VAR}(S) + \text{COV}(S, N) = \sigma_S^2,$$

$$\text{VAR}(X) = E(S - m_S + N)^2 = \text{VAR}(S) + \text{VAR}(N) + 2\,\text{COV}(S, N) = \sigma_S^2 + \sigma_N^2.$$

Define the *signal-to-noise ratio (SNR)* (the ratio of signal power as measured by its variance to the noise power) by

$$\text{SNR} = \frac{\sigma_S^2}{\sigma_N^2}.$$

Combining these results yields

$$a = \frac{\text{SNR}}{1 + \text{SNR}}, \quad \hat{S}(X) = \frac{\text{SNR}}{1 + \text{SNR}}(X - m_S) + m_S.$$

Using the relationships $\text{VAR}(aZ) = a^2\,\text{VAR}(Z)$ and $\text{COV}(aZ, Y) = a\,\text{COV}(Z, Y)$, we determine that the performance of this estimator is

$$E(S - \hat{S})^2 = \text{VAR}(S) + \text{VAR}(X)\left(\frac{\text{SNR}}{1 + \text{SNR}}\right)^2 - 2\frac{\text{SNR}}{1 + \text{SNR}}\,\text{COV}(S, X)$$

$$= \sigma_S^2 + (\sigma_S^2 + \sigma_N^2)\left(\frac{\text{SNR}}{1 + \text{SNR}}\right)^2 - 2\sigma_S^2\frac{\text{SNR}}{1 + \text{SNR}} = \frac{\sigma_S^2}{1 + \text{SNR}}.$$

This conclusion can be rewritten, using the definition of SNR, as

$$E(S - \hat{S})^2 = \sigma_N^2 \frac{\text{SNR}}{1 + \text{SNR}}.$$

Observe that, because $\text{SNR} \geq 0$, and therefore that

$$\frac{\text{SNR}}{1 + \text{SNR}} \leq 1,$$

we see that the performance $E(S - \hat{S})^2$ is better than that σ_N^2 of the naive estimator $\hat{S}_0$.

If we rewrite the performance of our linear estimator as

$$E(S - \hat{S})^2 = \frac{\sigma_S^2}{1 + \text{SNR}},$$

then we see that, for $\text{SNR} \ll 1$, the estimation error is nearly σ_S^2 and corresponds to estimating S by its mean m_S; this is reasonable, since $\sigma_S^2 \ll \sigma_N^2$ and the measurement X is too noisy to be useful. However, when $\text{SNR} \gg 1$, we have a good measurement and the mean square error performance of our linear estimator is nearly zero.

9.13 EXTENSIONS OF EXPECTATION TO VECTOR, MATRIX, AND COMPLEX-VALUED VARIABLES

The extension of the definition of expectation from scalar random variables to vector-valued or matrix-valued random variables is easily accomplished through the next definition.

Definition 9.13.1   (Matrix-Valued Expectation)   If $\mathbb{M} = [M_{i,j}]$ is a matrix or $\mathbf{X} = [X_i]$ is a vector with random variable elements, then

$$E\mathbb{M} = [EM_{i,j}], \quad E\mathbf{X} = [EX_i].$$

In other words, the expectation of a matrix (vector) is just the matrix (vector) of expectations of the individual elements.

The extension to complex-valued random variables is given by the next definition.

Definition 9.13.2   (Complex-Valued Expectation)   If Z is a complex-valued random variable having real part X and imaginary part Y, then

$$Z = X + iY \Rightarrow EZ = EX + iEY.$$

Hence, we just combine the usual expectation of the real and imaginary parts taken separately.

The basic properties of expectation still hold in these extensions. For example, the extension of property E6 is as follows: If $\mathbb{A}, \mathbb{B}, \mathbf{c}$ are nonrandom, then

$$E(\mathbb{A}\mathbf{X} + \mathbb{B}\mathbf{Y} + \mathbf{c}) = \mathbb{A}E\mathbf{X} + \mathbb{B}E\mathbf{Y} + \mathbf{c}.$$

We verify this equation by noting that if

$$\mathbf{Z} = \mathbb{A}\mathbf{X} + \mathbb{B}\mathbf{Y} + \mathbf{c},$$

then its ith component

$$Z_i = \sum_{j=1}^{n} A_{i,j} X_j + \sum_{j=1}^{m} B_{i,j} Y_j + c_i.$$

Applying E6 to EZ_i yields

$$EZ_i = \sum_{j=1}^{n} A_{i,j} EX_j + \sum_{j=1}^{m} B_{i,j} EY_j + c_i.$$

Rewriting this formula in matrix–vector form, we obtain

$$E\mathbf{Z} = [EZ_i] = \mathbb{A}E\mathbf{X} + \mathbb{B}E\mathbf{Y} + \mathbf{c},$$

as expected. Thus, in a linear network, the voltages $\mathbf{V}$ and currents $\mathbf{I}$ are related through an impedance matrix $\mathbb{Z}$, $\mathbf{V} = \mathbb{Z}\mathbf{I}$. Hence, if the impedance is nonrandom, then

$$E\mathbf{V} = \mathbb{Z}E\mathbf{I}.$$

9.14 CORRELATION AND COVARIANCE FOR REAL RANDOM VECTORS

The extensions of the ideas of correlation and covariance to possibly complex-valued vectors are important in practice. However, the sequence of definitions and characterizations is unfortunately dry and may not be familiar. We repeat the discussion presented here for real random vectors in an appendix in Section 9.18 adapted to complex random vectors. In the current section, we will learn the definitions of correlation and covariance matrices for a real random vector $\mathbf{X}$. The essential characteristic is that the (i, j)th element of the matrix be the corresponding correlation or covariance of the pair X_i and X_j. Defining a covariance matrix then enables us to better explain the multivariate normal distribution first introduced in Section 7.4.4. Finally, we provide necessary and sufficient conditions for a matrix to be a correlation or covariance matrix.

Definition 9.14.1   (Correlation Matrix)   The column vector $\mathbf{X}$ has correlation matrix $\mathbb{R}_X = E\mathbf{X}\mathbf{X}^T$.

Note that $\mathbb{R}_X$ is symmetric in that it equals its own transpose; that is,

$$\mathbb{R}_X^T = E(\mathbf{X}\mathbf{X}^T)^T = E(\mathbf{X}\mathbf{X}^T),$$

where we use $(\mathbb{A}\mathbb{B})^T = \mathbb{B}^T\mathbb{A}^T$.

The extension of covariance to vectors is given in the next definition.

Definition 9.14.2   (Covariance Matrix)   The covariance matrix $\mathbb{C}_{\mathbf{X}}$ for the column vector $\mathbf{X}$ is

$$\mathbb{C}_{\mathbf{X}} = E\left[(\mathbf{X} - E(\mathbf{X}))(\mathbf{X} - E(\mathbf{X}))^T\right] = \mathbb{C}_{\mathbf{X}}^T,$$

$$C_{i,j} = E\left[(X_i - EX_i)(X_j - EX_j)\right].$$

The covariance matrix has the same mathematical properties as the correlation matrix.

Example 9.18   Mean and Covariance in Multivariate Normal _____

We can now assert that the multivariate normal law $\mathcal{N}(\mathbf{m}, \mathbb{C})$ is parameterized by

$$\mathbf{m} = E\mathbf{X}, \quad \mathbb{C} = \text{COV}(\mathbf{X}, \mathbf{X}) = E\left[(\mathbf{X} - E(\mathbf{X}))(\mathbf{X} - E(\mathbf{X}))^T\right].$$

Given random vectors $\mathbf{X}, \mathbf{Y}$, we easily extend the preceding definitions for a single vector and introduce two more definitions.

Definition 9.14.3   (Cross-Correlation Matrix)   The cross-correlation matrix between vectors $\mathbf{X}, \mathbf{Y}$ is $\mathbb{R}_{X,Y} = E\mathbf{X}\mathbf{Y}^T$.

Note that the cross-correlation is not symmetric, in general, between $\mathbf{X}$ and $\mathbf{Y}$. The cross-correlation between $\mathbf{X}$ and itself is simply the correlation matrix $\mathbb{R}_X$.

Definition 9.14.4   (Cross-Covariance Matrix)   The cross-covariance matrix between vectors $\mathbf{X}, \mathbf{Y}$ is

$$\text{COV}(\mathbf{X}, \mathbf{Y}) = E\left[(\mathbf{X} - E\mathbf{X})(\mathbf{Y} - E\mathbf{Y})^T\right].$$

A mathematical characterization of the class of correlation and covariance matrices is based on the notion of a nonnegative definite matrix. (See Horn and Johnson [46] for details we omit.)

Definition 9.14.5   (Nonnegative Definite Matrix)   An $n \times n$ matrix $\mathbb{A}$ is nonnegative definite (nnd) if, for any column vector $\mathbf{a}$,

$$\mathbf{a}^T \mathbb{A} \mathbf{a} = \sum_i \sum_j A_{i,j} a_i a_j \geq 0.$$

Lemma 9.14.1   A symmetric real matrix $\mathbb{A}$ is an nnd matrix if and only if all of its eigenvalues are nonnegative.

Proof citation.   See Horn and Johnson [46], Theorem 7.2.1. $\square$

Theorem 9.14.1   (Correlation Matrix)   A real matrix $\mathbb{A}$ is a correlation or a covariance matrix if and only if it is symmetric and nonnegative definite.

Proof citation. See Section 9.18, Appendix 2 for a proof of necessity in the complex case. $\square$

An important special case of a nonnegative definite matrix is a *positive definite (pd)* matrix that satisfies the strict inequality

$$\mathbf{a} \neq \underline{0} \Rightarrow \mathbf{a}^T \mathbb{A}\mathbf{a} > 0.$$

Lemma 9.14.2   A real symmetric matrix $\mathbb{A}$ is positive definite if and only if all of its eigenvalues are positive. In this case $\mathbb{A}$ has a positive determinant, given by the product of eigenvalues, and therefore has an inverse.

Proof citation. See Horn and Johnson [46], Theorem 7.2.1. $\square$

We can easily construct examples of, say, positive definite $n \times n$ covariance matrices by selecting any n positive numbers $\lambda_1, \ldots, \lambda_n$ and any n column vectors $\mathbf{e}_1, \ldots, \mathbf{e}_n$ of dimension n that are *orthonormal* in that the collection of vectors satisfies the following two properties:

$$\text{unit normalization} \quad (\forall i) \ \mathbf{e}_i^T \mathbf{e}_i = 1;$$

$$\text{orthogonality} \quad (\forall i \neq j) \ \mathbf{e}_i^T \mathbf{e}_j = 0.$$

The covariance matrix $\mathbb{C}$ is then formed as

$$\mathbb{C} = \sum_{i=1}^{n} \lambda_i \mathbf{e}_i \mathbf{e}_i^T.$$

Note that $\mathbf{e}_i \mathbf{e}_i^T$ is an $n \times n$ matrix.

Example 9.19   Positive Definite Matrix _____

As a specific example, take $n = 2$, $\lambda_1 = 1$, $\lambda_2 = 4$, and choose

$$\mathbf{e}_1 = \frac{1}{\sqrt{2}} \begin{pmatrix} 1 \\ -1 \end{pmatrix}, \ \mathbf{e}_2 = \frac{1}{\sqrt{2}} \begin{pmatrix} 1 \\ 1 \end{pmatrix}.$$

These choices yield the positive definite matrix

$$\mathbb{C} = \frac{1}{2} \begin{pmatrix} 5 & 3 \\ 3 & 5 \end{pmatrix}.$$

9.15 LINEAR TRANSFORMATION AND SYNTHESIS OF (GAUSSIAN) RANDOM VECTORS

9.15.1 Propagation of Correlation Matrices

The effects on correlation and covariance matrices of a nonlinear transformation from $\mathbf{X}$ to $\mathbf{Y}$ are difficult to discern, as they depend upon both the transformation and the original density $f_{\mathbf{X}}$.

However, the effect of an affine or linear transformation is readily determined from the linearity of expectation. If

$$\mathbf{Y} = \mathbb{A}\mathbf{X} + \mathbf{b}, \ \mathbb{R}_\mathbf{X} = E\mathbf{X}\mathbf{X}^T, \ E\mathbf{X} = \mathbf{m},$$

then

$$E\mathbf{Y} = \mathbb{A}\mathbf{m} + \mathbf{b}, \tag{9.1}$$

$$\mathbb{R}_\mathbf{Y} = E(\mathbb{A}\mathbf{X} + \mathbf{b})(\mathbb{A}\mathbf{X} + \mathbf{b})^T$$

$$= E(\mathbb{A}\mathbf{X})(\mathbf{X}^\dagger \mathbb{A}^T) + \mathbf{b}\mathbf{b}^T + \mathbf{b}E\mathbf{X}^T\mathbb{A}^T + \mathbb{A}E\mathbf{X}\mathbf{b}^T$$

$$= \mathbb{A}\mathbb{R}_\mathbf{X}\mathbb{A}^T + \mathbf{b}\mathbf{b}^T + \mathbf{b}\mathbf{m}^T\mathbb{A}^T + \mathbb{A}\mathbf{m}\mathbf{b}^T.$$

The corresponding result for the covariance matrix follows from the result for the correlation matrix by setting means and additive constants to zero:

$$\mathbb{C}_\mathbf{Y} = \mathbb{A}\mathbb{C}_\mathbf{X}\mathbb{A}^T. \tag{9.2}$$

9.15.2 Synthesis of Covariance Matrices

We show here how to synthesize a random vector $\mathbf{Y}$ having a desired positive definite covariance matrix $\mathbb{C}_\mathbf{Y} = \mathbb{C}$ from a random vector $\mathbf{X}$ having equal variance, uncorrelated components and, thus, covariance matrix

$$\mathbb{C}_\mathbf{X} = \sigma^2 \mathbb{I}.$$

From the preceding treatment,

$$\mathbf{Y} = \mathbb{A}\mathbf{X} \Rightarrow \mathbb{C}_\mathbf{Y} = \mathbb{A}\mathbb{C}_\mathbf{X}\mathbb{A}^T.$$

Hence, if $\mathbf{X}$ is as assumed before, then

$$\mathbb{C}_\mathbf{Y} = \sigma^2 \mathbb{A}\mathbb{A}^T.$$

Given a desired covariance matrix $\mathbb{C}_\mathbf{Y}$, we need to factor it into the form just shown. Such a factorization yields a matrix $\mathbb{A}$ that is known as the square root of $\mathbb{C}_\mathbf{Y}$. To proceed further, we state the Cholesky decomposition.

    Lemma 9.15.1   **(Cholesky Decomposition)**   A matrix $\mathbb{C}$ is positive definite if and only if there is a (nonsingular lower triangular) matrix $\mathbb{L}$ such that

$$\mathbb{C} = \mathbb{L}\mathbb{L}^T.$$

    Proof citation. See Horn and Johnson [46], Corollary 7.2.9.   □

    Given a positive definite matrix $\mathbb{A}$, the Matlab command `chol(A)` will return $\mathbb{L}^T$.

Example 9.20   Cholesky Decomposition _____

Take

$$\mathbb{A} = \begin{pmatrix} 2 & -1 \\ -1 & 3 \end{pmatrix} \Rightarrow \mathbb{L} = (chol(\mathbb{A}))^T = \begin{pmatrix} 1.4142 & 0 \\ -0.7071 & 1.5811 \end{pmatrix}.$$

Thus, if **Y** has a positive definite covariance matrix, then this matrix can be represented as

$$\mathbb{C} = \mathbb{L}\mathbb{L}^T,$$

and it follows from Eq. (9.2) that, for **X** a vector of equal variance, uncorrelated components,

$$\mathbf{Y} = \sigma^{-1}\mathbb{L}\mathbf{X}$$

will yield a random vector with the desired covariance matrix $\mathbb{C}$ for **Y**. Continuing with the previous example, if we start with

$$\mathbb{C}_\mathbf{X} = \begin{pmatrix} 1 & 0 \\ 0 & 1 \end{pmatrix}$$

and desire **Y** with covariance matrix

$$\mathbb{A} = \begin{pmatrix} 2 & -1 \\ -1 & 3 \end{pmatrix},$$

then

$$\mathbf{Y} = \begin{pmatrix} 1.4142 & 0 \\ -0.7071 & 1.5811 \end{pmatrix}\mathbf{X}$$

succeeds.

A converse to the foregoing is to take a random vector **Y** and convert it to a random vector **X** that has equal variance, uncorrelated components. This process is referred to as *whitening*, a term best understood in the context of power spectral ideas that lie outside our scope. Given a Cholesky decomposition of

$$\mathbb{C}_\mathbf{Y} = \mathbb{L}\mathbb{L}^\dagger,$$

take

$$\mathbf{X} = \mathbb{L}^{-1}\mathbf{Y}$$

and invoke Eq. (9.2) to determine that

$$\mathbb{C}_\mathbf{X} = \mathbb{L}^{-1}\mathbb{C}_\mathbf{Y}\mathbb{L}^{-\dagger} = \mathbb{L}^{-1}\mathbb{L}\mathbb{L}^\dagger\mathbb{L}^{-\dagger} = \mathbb{I},$$

as desired.

We are now in a position to synthesize Gaussian random vectors from simpler ones. From Section 8.6.1, we know that if real-valued $\mathbf{X} \sim \mathcal{N}(\mathbf{m}_X, \mathbb{C}_X)$ and $\mathbf{Y} = \mathbb{A}\mathbf{X} + \mathbf{b}$, for real-valued, nonrandom $\mathbb{A}, \mathbf{b}$, then $\mathbf{Y} \sim \mathcal{N}(\mathbb{A}\mathbf{m}_X + \mathbf{b}, \mathbb{A}\mathbb{C}_X\mathbb{A}^T)$. From Section 8.6.3, we learned how to generate unlinked Gaussian random variables with zero mean and unit variance from a process that can give us unlinked $\mathcal{U}(0, 1)$ random variables. In Matlab, the command rand(n,1) yields a column vector of size n such uniformly distributed random variables. Of course, in Matlab, the command `randn(n,1)` directly yields a column vector $\mathbf{X} \sim \mathcal{N}(\mathbf{0}, \mathbb{I})$. Hence, we can safely assume the availability of the simple Gaussian random vector $\mathbf{X} \sim \mathcal{N}(\mathbf{0}, \mathbb{I})$ having zero mean and uncorrelated components of unit variance. In order to obtain $\mathbf{Y} \sim \mathcal{N}(\mathbf{m}_Y, \mathbb{C}_Y)$, with $\mathbb{C}_Y$ positive definite, we choose $\mathbf{Y} = \mathbb{A}\mathbf{X} + \mathbf{b}$, where

$$\mathbf{b} = \mathbf{m}_Y$$

and $\mathbb{A}$ is a Cholesky decomposition factor of $\mathbb{C}_Y$, or

$$\mathbb{C}_Y = \mathbb{A}\mathbb{A}^T.$$

Correlation and covariance and their application to linear estimation is discussed further in the next chapter.

9.16 SUMMARY

We return to G1 and enrich probability theory by introducing the fundamental concept of expectation. The expected value EX of a random variable X can be thought of in four different ways: EX as a parameter called the mean of F_X or of f_X; EX as a definite integral of the function of ω that is the random variable $X(\omega)$; in accordance with the frequentist interpretation of probability, EX as the long-run time average of the values of an infinite sequence of repeated unlinked copies $\{X_i(\omega)\}$ of X corresponding to unlinked repetitions of the random experiment $\mathcal{E}$; and, finally, in accordance with the subjective interpretation of probability as a degree of belief that is evidenced by a willingness to bet, EX as the fair price for a gamble paying off $X(\omega)$ when $\omega \in \Omega$ is chosen. Curiously, it is possible to reverse our (very traditional) order of presentation and build a theory of probability by starting with expectation and then deriving probability from $P(A) = EI_A(X)$, a course taken by Huyghens in his—the first published—text on probability. This is also the approach taken today in a promising theory of interval-valued probability. (See Walley [96].)

In accordance with our expository practice, we formally defined expectation following the view of a parameter, elucidated its mathematical properties, and examined its application in the definition of the statistically important notions of moments (especially the mean and variance) and correlations. Our reason for deferring treatment of expectation until we had treated functions of a random variable in Chapter 8 was so that we could make good sense of these uses of expectation so important to statistical reasoning about random phenomena. The role of these moments and correlation is illustrated by a G3 application to finding a linear function $\hat{Y}$ of data $\mathbf{X}$ that best estimates another random variable Y in the sense of minimizing the expected squared error $E(Y - \hat{Y})^2$.

The concept of expectation was extended to include complex-valued random variables and random vectors and matrices. We closed the chapter with the synthesis of random vectors having arbitrary covariance matrices by means of linear transformations of random vectors having simple (diagonal) covariance matrices. In particular, we became able to synthesize arbitrary Gaussian random vectors, a skill having many uses, including supporting simulations.

9.17 APPENDIX 1: SUMMARY OF BASIC PROPERTIES OF EXPECTATION

(E0)
$$EX = \lambda \left[\sum_{\{i : x_i < 0\}} x_i p_i + \sum_{\{i : x_i \geq 0\}} x_i p_i \right]$$
$$+ (1 - \lambda) \left[\int_{-\infty}^{0} x f_X(x)\, dx + \int_{0}^{\infty} x f_X(x)\, dx \right].$$

(E2)
$$P(X = c) = 1 \Rightarrow EX = c.$$

(E3)
$$P(X \geq 0) = 1 \Rightarrow EX \geq 0.$$

(E4)
$$E(aX) = aE(X).$$

(E5)
$$E(X + Y) = EX + EY \text{ when finite.}$$
$$E\left(\sum_{1}^{n-1} X_i \right) = \sum_{1}^{n-1} EX_i.$$

(E6)
$$E(aX + bY + c) = aEX + bEY + c.$$

(E7)
$$P(X \geq Y) = 1 \Rightarrow EX \geq EY.$$

(E8)
$$Eg(X) = \int_{-\infty}^{\infty} g(x) f_X(x)\, dx.$$

9.18 APPENDIX 2: CORRELATION FOR COMPLEX RANDOM VECTORS

The following generalizations are not used in this text.

We now rewrite the treatment of Section 9.14 to account for complex-valued random vectors. A symmetric matrix is replaced by a Hermitian matrix that accounts not only for the transpose, but for complex conjugation, whereas, for any real random vector

$$E(\mathbf{X}\mathbf{X}^T) \geq 0,$$

this is no longer true if $\mathbf{X}$ is complex valued. The necessary modification to transpose is given by the *Hermitian adjoint operator*.

Definition 9.18.1 (Hermitian Adjoint Operator)   The Hermitian adjoint operator $\dagger$ applied to a matrix $\mathbb{M}$ yields a matrix $\mathbb{M}^{\dagger}$, called the Hermitian adjoint of the matrix $\mathbb{M}$, that is the transpose of the matrix $\mathbb{M}$ with all elements then replaced by their complex conjugate (*).

Note that if $\mathbb{M}$ is a real-valued matrix, then $\mathbb{M}^{\dagger} = \mathbb{M}^T$, the usual transpose. Note further that $(\mathbb{M}^{\dagger})^{\dagger} = \mathbb{M}$; that is, the Hermitian adjoint operator is idempotent.

Definition 9.18.2 (Hermitian Matrix)   A matrix $\mathbb{M}$ is Hermitian if $\mathbb{M} = \mathbb{M}^{\dagger}$.

Observe that a Hermitian real-valued matrix is simply a symmetric matrix.

Definition 9.18.3 (Correlation Matrix) The column vector $\mathbf{X}$ has correlation matrix $\mathbb{R}_X = E\mathbf{X}\mathbf{X}^\dagger$.

Note that $\mathbb{R}_X$ is Hermitian; that is,

$$\mathbb{R}_X^\dagger = E(\mathbf{X}\mathbf{X}^\dagger)^\dagger = E\mathbf{X}\mathbf{X}^\dagger.$$

If the components of $\mathbf{X}$ are samples from a random signal, with component X_i equal to signal amplitude X_{t_i}, then the element $R_{i,j}$ of the correlation matrix $\mathbb{R}_X$ is related to the random signal autocorrelation function $R(t, s)$ through $R_{i,j} = R(t_i - t_j)$. The extension of covariance to vectors is given in the next definition.

Definition 9.18.4 (Covariance Matrix) The covariance matrix for the column vector $\mathbf{X}$ is

$$\mathbb{C} = E(\mathbf{X} - E(\mathbf{X}))(\mathbf{X} - E(\mathbf{X}))^\dagger = \mathbb{C}^\dagger, \quad C_{i,j} = E(X_i - EX_i)(X_j - EX_j)^*.$$

The covariance matrix has the same mathematical properties as the correlation matrix. Given random vectors $\mathbf{X}, \mathbf{Y}$, we introduce the next definition.

Definition 9.18.5 (Cross-Correlation Matrix) The cross-correlation matrix between vectors $\mathbf{X}, \mathbf{Y}$ is $\mathbb{R}_{X,Y} = E\mathbf{X}\mathbf{Y}^\dagger$.

Note that the cross-correlation is not symmetric, in general, between $\mathbf{X}$ and $\mathbf{Y}$. The cross-correlation between $\mathbf{X}$ and itself is simply the correlation matrix $\mathbb{R}_X$.

Definition 9.18.6 (Cross-Covariance Matrix) The cross-covariance matrix between vectors $\mathbf{X}, \mathbf{Y}$ is

$$\text{COV}(\mathbf{X}, \mathbf{Y}) = E(\mathbf{X} - E\mathbf{X})(\mathbf{Y} - E\mathbf{Y})^\dagger.$$

Generally, we will be using correlation and covariance only for real-valued random vectors and can replace the operator $\dagger$ by the transpose T.

We can now assert that the multivariate normal law $\mathcal{N}(\mathbf{m}, \mathbb{C})$ is parameterized by

$$\mathbf{m} = E\mathbf{X}, \quad \mathbb{C} = \text{COV}(\mathbf{X}, \mathbf{X}) = E(\mathbf{X} - E(\mathbf{X}))(\mathbf{X} - E(\mathbf{X}))^\dagger.$$

A mathematical characterization of the class of correlation and covariance matrices is based on the notion of a nonnegative definite matrix. (See Horn and Johnson [46] for details we omit.)

Definition 9.18.7 (Nonnegative Definite Matrix) An $n \times n$ matrix $\mathbb{A}$ is nonnegative definite (nnd) if, for any column vector $\mathbf{a}$,

$$\mathbf{a}^\dagger \mathbb{A} \mathbf{a} \geq 0.$$

Lemma 9.18.1 A Hermitian matrix $\mathbb{A}$ is an nnd matrix if and only if all of its eigenvalues are nonnegative.

Proof citation. See Horn and Johnson [46], Theorem 7.2.1. □

Theorem 9.18.1   (Characterization of a Correlation Matrix)   A matrix $\mathbb{A}$ is a correlation or a covariance matrix if and only if it is Hermitian and nonnegative definite.

Proof. We establish only the necessity of this condition. Say,

$$\mathbb{A} = E\mathbf{X}\mathbf{X}^\dagger.$$

Then

$$\mathbf{a}^\dagger E\mathbf{X}\mathbf{X}^\dagger\mathbf{a} = E(\mathbf{a}^\dagger\mathbf{X})(\mathbf{X}^\dagger\mathbf{a}) = E(\mathbf{a}^\dagger\mathbf{X})(\mathbf{a}^\dagger\mathbf{X})^\dagger = EZZ^*,$$

where $Z = \mathbf{a}^\dagger\mathbf{X}$ is a scalar random variable. However, $EZZ^* = E|Z|^2 \geq 0$. Thus, $\mathbb{A}$ is nnd if it is a correlation (or covariance) matrix. The Hermitian property follows from

$$\mathbb{A} = E\mathbf{X}\mathbf{X}^\dagger \Rightarrow \mathbb{A}^\dagger = E(\mathbf{X}\mathbf{X}^\dagger)^\dagger = E\mathbf{X}\mathbf{X}^\dagger,$$

where we used the idempotent property $(\mathbf{X}^\dagger)^\dagger = \mathbf{X}$. $\square$

An important special case of a nonnegative definite matrix is a *positive definite (pd)* matrix that satisfies the strict inequality

$$\mathbf{a} \neq \underline{0} \Rightarrow \mathbf{a}^\dagger\mathbb{A}\mathbf{a} > 0.$$

Lemma 9.18.2   A Hermitian matrix $\mathbb{A}$ is positive definite if and only if all of its eigenvalues are positive. In this case, $\mathbb{A}$ has a positive determinant and therefore has an inverse.

Proof citation. See Horn and Johnson [46], Theorem 7.2.1. $\square$

EXERCISES

E9.1   If

$$f_X(x) = 2xU(x)U(1-x),$$

then evaluate EX and $\text{VAR}(X)$.

E9.2   If $f_X(x) = \frac{2}{x^3}U(x-1)$, evaluate EX, $\text{VAR}(X)$.

E9.3   If $P(X=0) = .2, P(X=1) = .5, P(X=2) = .3$, determine EX, $\text{VAR}(X)$.

E9.4   If $P(X=-1) = .2, P(X=0) = .3, P(X=3) = .5$, evaluate EX, $\text{VAR}(X)$.

E9.5   Verify the table of expected values in Section 9.3.

E9.6   Verify the table of variances in Section 9.10.

E9.7   Evaluate EX, EX^2, and $\text{VAR}(X)$ for X distributed as $\mathcal{E}(\alpha)$ and also as $Par(\alpha, \tau)$.
Repeat for $X \sim \mathcal{P}(\lambda)$ by first evaluating EX and then $EX(X-1)$.

E9.8   If $P(a \leq X \leq b) = 1$, then what is the largest possible EX, and what is the distribution that achieves it? Also, what is the largest possible $\text{VAR}(X)$ and the distribution that achieves it? (Hint: Consider $Y = X - (b+a)/2$, the relation between $\text{VAR}(Y)$ and $\text{VAR}(X)$, the maximum value of EY^2 when $|Y| \leq (b-a)/2$, and the inequality $EY^2 \geq \text{VAR}(Y)$.)

E9.9   a. If T is a random temperature, will its mean ET be higher when it is reported in centigrade rather than in Fahrenheit degrees?

         b. Will its variance be higher when T is reported in centigrade rather than in Fahrenheit degrees?

E9.10   If $X \sim \mathcal{U}(a, b)$ and $Y \sim \mathcal{B}(n, p)$, then evaluate EX^k, EY^k for $k = 0, 1, 2, 3$.

E9.11   If $X \sim \mathcal{U}(0, 1)$ and $Y = e^X$, then evaluate EY firstly by use of property E8 of Section 9.8 and secondly by computing the density f_Y and using the basic definition of EY from Section 9.3.

E9.12   Evaluate the mean and variance for $X \sim \mathcal{U}(a, b)$.

E9.13   If $X \sim \mathcal{U}(0, 1)$ and $Y \sim \mathcal{N}(2, 3)$ then evaluate $E(\frac{1}{2}X + Y)$.

E9.14   If

$$
F_X(x) = \begin{cases} \dfrac{1}{2}e^x & \text{if } x < 0 \\[2mm] 1 - \dfrac{1}{2}e^{-2x} & \text{if } x \geq 0 \end{cases},
$$

then evaluate VAR(X).

E9.15   If the joint probability mass function

$$p_{X,Y}(0, 0) = .1, \ p_{X,Y}(0, 1) = .3, \ p_{X,Y}(1, 0) = .2, \ p_{X,Y}(1, 1) = .4,$$

evaluate the pmf $p_X(x)$ and $E(XY)$.

E9.16   A component that fails without aging has a lifetime T such that its median lifetime μ is 1,000 hours. What is its expected lifetime in hours, and what is the variance of that lifetime?

E9.17   Find EP for the power P dissipated in a 4-Ω resistor across which is a voltage $V \sim \mathcal{N}(1, 3)$.

E9.18   a. A speech signal has Laplacian-distributed amplitude X such that VAR$(X) = 2$. What is f_X?

         b. If the waiting time T to the next photon emission has $ET = 3$, what is f_T?

E9.19   a. If the mean square lifetime $EL^2 = 4$ for a component that fails without aging, what is $P(L > 3)$?

         b. If V is a thermal noise voltage with $EV^2 = 2$, what is the pdf $f_V(v)$?

         c. If T is a waiting time between successive keystrokes such that $ET^2 = \infty$, $ET = 2$ and $\tau_0 = 1$ is the largest value such that $P(T \geq \tau_0) = 1$, what is the pdf $f_T(t)$?

E9.20   a. We observe that $X \sim Par(4, 1)$. Evaluate EX, EX^2, VAR(X).

         b. We observe that $Y \sim \mathcal{U}(0, 1)$. Evaluate EX, EX^2, VAR(X).

         c. We are informed that COV$(X, Y) = 1/8$. Evaluate $E(XY)$.

E9.21   Show by evaluating the left-hand side that

$$\text{COV}(aX + b, cY + d) = ac\text{COV}(X, Y).$$

E9.22   The noise variance for the thermal noise voltage V produced by a resistor of R ohms at an absolute temperature T, when measured over a bandwidth W, is given in terms of the Boltzmann constant k by $\sigma^2 = 4kTRW$.

         a. Evaluate the expected power P dissipated in R under these conditions when it is short-circuited.

b. If, under the given circumstances, we have two resistors of R_1, R_2 ohms each, and they generate uncorrelated thermal noises V_1, V_2, then evaluate $E(V_1 + V_2)^2$.

E9.23   a. If S is the amplitude of a speech signal with variance 4, then evaluate $P(S > 1)$.

b. If $EX = 1$, $VAR(X) = 1$, $EY = 1.5$, what can you conclude about the relative sizes of EX^2, EY^2?

E9.24   a. We observe $X \sim \mathcal{N}(2, 1)$. Evaluate $EX, VAR(X), EX^2$.

b. We wish to infer to $Y \sim \mathcal{E}(1)$. Evaluate $EY, EY^2, VAR(Y)$.

c. We are informed that $COV(X, Y) = 1/2$. What is the correlation $E(XY)$?

d. Which of the two estimators of Y given by $Y_1 = 0.5X$ and $Y_2 = X - 1$ has the lower mean square error?

E9.25   a. We observe that $X \sim \mathcal{Par}(4, 1)$. Evaluate $EX, EX^2, VAR(X)$.

b. We observe that $Y \sim \mathcal{U}(0, 1)$. Evaluate $EX, EX^2, VAR(X)$.

c. We are informed that $COV(X, Y) = 1/8$. Evaluate $E(XY)$.

d. Find the linear least mean square estimator $\hat{Y} = aX + b$.

E9.26   a. We measure X having $EX = -1, EX^2 = 2$ and wish to estimate Y having $EY = 1, EY^2 = 3$. We know that $E(XY) = 0$. If we choose an estimator $\hat{Y} = aX + b$, evaluate a and b to achieve the minimum mean squared error $E(\hat{Y} - Y)^2$.

b. If, for the information given in (a), we decided to use $a = b = 1$ as an estimator, evaluate its performance $E(\hat{Y} - Y)^2$.

E9.27   (1-Bit Quantization) Let

$$X \sim \mathcal{E}(1), \quad Y = q(X) = \begin{cases} q_1 & \text{if } X > \tau \\ q_0 & \text{otherwise} \end{cases}.$$

The random variable Y is a 1-bit quantized (A/D converted) version of X with quantizer threshold τ and reconstruction levels $0 < q_0 < q_1$.

a. Find F_Y and the mean and variance of Y.

b. Evaluate the mean square approximation error $E(Y - X)^2$.

c. We wish to design a 1-bit quantizer, by the choice of τ, q_0, q_1, to minimize the mean squared error. Show that, for given q_0, q_1, the best choice is

$$\tau = \frac{q_0 + q_1}{2}.$$

E9.28   You are given that

$$EY = 3, EX = 4, VAR(Y) = 1, VAR(X) = 2, COV(X, Y) = -0.5.$$

a. If we decide to estimate Y by $\hat{Y} = 2X + b$, choose b to minimize the mean square error.

b. If we decide to estimate Y by $\hat{Y} = \frac{1}{2}X + 1$, what is the mean square error of this estimator?

c. Using the information given at the outset, evaluate the correlation $E(XY)$ and EX^2.

E9.29   We are given

$$f_{X,Y}(x, y) = \begin{cases} 3x^2 & \text{if } 0 \le x \le 1, 0 \le y \le 1 \\ 0 & \text{otherwise} \end{cases}.$$

    a. Evaluate EX^k and EY^k.

    b. Evaluate VAR(X), VAR(Y).

    c. Evaluate $E(XY)$.

    d. Evaluate COV(X, Y).

    e. For given a, evaluate $E(a(X - EX) - (Y - EY))^2$.

    f. Find the MMSE affine estimator $\hat{Y}(X)$ of Y based on X.

E9.30   a. If an oscillator

$$X(t) = A\cos(\omega_0 t + \Theta), \quad \Theta \sim \mathcal{U}(-\pi, \pi), \quad A \sim \mathcal{E}(\alpha),$$

$$f_{A,\Theta}(a, \theta) = f_A(a)f_\Theta(\theta),$$

    then evaluate $EX(t)$, COV($X(t), X(s)$), VAR($X(t)$).

    b. Design the linear, least mean square predictor $\hat{X}(t)$ of $X(t)$ based upon observing $X(s)$ for some $s < t$.

E9.31 We know that $T \sim \mathcal{E}(1)$, and we wish to approximate to T^2 by a linear function $aT + b$.

    a. Find the least mean square coefficients a^*, b^* by carrying out the minimization calculation.

    b. If $a = 1/2, b = 0$, then evaluate the mean square approximation error.

E9.32 We wish to infer $Z = Y^3$ from observed X and know that the joint density $f_{X,Y}(x, y)$ is equal to $1/2$ over the rectangle $0 \le x \le 2, 0 \le y \le 1$ and is 0 elsewhere.

    a. Evaluate EZ, VAR(Z).

    b. Evaluate COV(X, Z).

    c. Evaluate $E[(Z - EZ) - a(X - EX)]^2$ as a function of a.

    d. Find the linear minimum mean square estimator $\hat{Z} = cX + d$ of Z.

E9.33

$$f_{X,Y}(x, y) = \begin{cases} x + y & \text{if } 0 \le x \le 1, \ 0 \le y \le 1 \\ 0 & \text{otherwise} \end{cases}.$$

    a. Evaluate EX^k and EY^k.

    b. Evaluate VAR(X), VAR(Y).

    c. Evaluate $E(XY)$.

    d. Evaluate COV(X, Y).

    e. For given a, evaluate $E(a(X - EX) - (Y - EY))^2$.

    f. Find the MMSE affine estimator $\hat{Y}(X) = aX + b$ of Y based on X.

E9.34 We observe that $X = S + N$ with

$$S \sim \mathcal{U}(0, 1), \quad N \sim \mathcal{N}(0, \sigma^2), \quad \text{COV}(S, N) = 0.$$

    a. Determine the linear, least mean square estimator $\hat{S}(X)$ for S.

    b. Determine the performance of $\hat{S}$.

E9.35 A voltage $V \sim \mathcal{E}(1)$ is measured by an instrument having response $X = V + N$ that contains an uncorrelated normally distributed error N of zero mean and variance σ^2.

We attempt to infer V from $\hat{V}(X) = aX + b$. What are the choices of a, b that yield a minimum mean square estimate?

E9.36 The two-dimensional random vector $\mathbf{X}$ has covariance matrix

$$\mathbb{C}_\mathbf{X} = \begin{pmatrix} \sigma_1^2 & \rho\sigma_1\sigma_2 \\ \rho\sigma_1\sigma_2 & \sigma_2^2 \end{pmatrix}.$$

The random vector $\mathbf{Y} = \mathbb{U}\mathbf{X}$, where $\mathbb{U}$ is the linear transformation of rotation through an angle θ, or

$$\mathbb{U} = \begin{pmatrix} \cos\theta & -\sin\theta \\ \sin\theta & \cos\theta \end{pmatrix}.$$

Evaluate the covariance matrix $\mathbb{C}_\mathbf{Y}$.

E9.37 In a certain linear system the inputs $\mathbf{X}$ are related to the outputs $\mathbf{Y}$ through the formula $\mathbf{Y} = \mathbb{A}\mathbf{X}$, where

$$\mathbb{A} = \begin{pmatrix} 1 & 1 \\ -1 & 1 \end{pmatrix}.$$

  a. If

$$\mathbb{R}_X = \begin{pmatrix} 2 & 1 \\ 1 & 3 \end{pmatrix},$$

find the correlation matrix $\mathbb{R}_Y$ for the outputs.

  b. Can you evaluate the covariance matrix for $\mathbf{Y}$?

E9.38 If

$$\mathbf{X} \sim \mathcal{N}\left(\begin{bmatrix} 1 \\ -2 \\ 0 \end{bmatrix}, \begin{bmatrix} 5 & 2 & -3 \\ 2 & 10 & -6 \\ -3 & -6 & 13 \end{bmatrix} \right),$$

evaluate EX_2 and $COV(X_1, X_3)$.

E9.39 In a given circuit, the currents $\mathbf{I}$ are related to the voltages $\mathbf{V}$ through the formula

$$\mathbf{I} = \begin{pmatrix} 3 & -2 \\ -2 & 5 \end{pmatrix}\mathbf{V} + \begin{pmatrix} 1 \\ 0 \end{pmatrix}.$$

It is known that

$$E\mathbf{V} = \begin{pmatrix} 1 \\ 2 \end{pmatrix}, \quad \mathbb{C}_V = COV(\mathbf{V}, \mathbf{V}) = \begin{pmatrix} 4 & 2 \\ 2 & 5 \end{pmatrix}.$$

  a. Evaluate $E\mathbf{I}$.

  b. Evaluate $COV(\mathbf{I}, \mathbf{I}) = \mathbb{C}_I$.

E9.40 In a two-mesh resistive network with two noisy resistors, the relationship between the two mesh currents $\mathbf{I}$ and the two equivalent noise voltages $\mathbf{V}$ is given by the formula

$$\begin{pmatrix} 5 & -3 \\ -3 & 7 \end{pmatrix}\mathbf{I} = \begin{pmatrix} 1 & 0 \\ 0 & 1 \end{pmatrix}\mathbf{V}.$$

We know that

$$f_{\mathbf{V}}(v_1, v_2) = \prod_{i=1}^{2} \frac{1}{\sigma_i \sqrt{2\pi}} e^{-\frac{v_i^2}{2\sigma_i^2}}.$$

Evaluate $f_{\mathbf{I}}(i_1, i_2)$.

E9.41   a. If

$$\mathbf{Z} = \begin{pmatrix} X \\ Y \end{pmatrix} \sim \mathcal{N}(\mathbf{m}, \mathbb{C}), \quad \mathbf{m} = \begin{pmatrix} 1 \\ 0 \end{pmatrix},$$

$$\mathbb{C} = E(\mathbf{Z} - E\mathbf{Z})(\mathbf{Z} - E\mathbf{Z})^T = \begin{pmatrix} 1 & \frac{1}{2} \\ \frac{1}{2} & 2 \end{pmatrix},$$

evaluate EX, EY, $\mathrm{COV}(X, Y)$, $\mathrm{VAR}(X)$, and $\mathrm{VAR}(Y)$.

b. Evaluate the mean square error performance of

$$\hat{Y} = \frac{1}{2}(X - 1)$$

as an estimator of Y.

E9.42   For $i = 1, 2$, we observe that $X_i = S + N_i$, where $S \sim \mathcal{N}(m, 1)$, $N_i \sim \mathcal{N}(0, \sigma_i^2)$, and $\mathrm{COV}(S, N_i) = \mathrm{COV}(N_1, N_2) = 0$.

a. Evaluate $\mathrm{COV}(S, X_i)$.

b. Evaluate the covariance matrix $\mathbb{C}_X = \mathrm{COV}(\mathbf{X}, \mathbf{X})$.

E9.43   The random vector

$$\mathbf{X} = \begin{pmatrix} X_1 \\ X_2 \end{pmatrix} \sim \mathcal{N}(\mathbf{m}, \mathbb{C}), \quad \mathbf{m} = \begin{pmatrix} 0 \\ 1 \end{pmatrix}, \quad \mathbb{C} = \begin{pmatrix} 2 & -1 \\ -1 & 3 \end{pmatrix}.$$

a. Evaluate $E\mathbf{X}$, $\mathrm{VAR}(X_1)$, $\mathrm{VAR}(X_2)$, $\mathrm{COV}(X_1, X_2)$.

b. Let $\hat{X}_2(X_1) = aX_1 + b$ be a linear estimator of X_2 based on X_1. Select a and b to minimize $E(\hat{X}_2 - X_2)^2$.

c. Evaluate the performance of this optimum linear estimator.

d. If

$$\mathbf{Y} = \mathbb{A}\mathbf{X}, \quad \mathbb{A} = \begin{pmatrix} 1 & 2 \\ 3 & -1 \\ -2 & 2 \end{pmatrix},$$

evaluate $E\mathbf{Y}$ and $\mathrm{COV}(\mathbf{Y}, \mathbf{Y})$.

E9.44   Show, by applying the definition of positive definiteness in terms of quadratic forms, that if we select an arbitrary $n \times n$ real-valued matrix $\mathbb{A}$ and any $\epsilon > 0$, then

$$\mathbb{H} = \mathbb{A}^T \mathbb{A} + \epsilon \mathbb{I}$$

is an $n \times n$ positive definite matrix.

E9.45 Given $\mathbf{X}$ with 3×3 covariance matrix $\mathbb{I}$, show how to synthesize $\mathbf{Y}$ having covariance matrix

$$\mathbb{C}_\mathbf{Y} = \begin{pmatrix} 2 & 0 & 0 \\ 0 & 1 & 0 \\ 0 & 0 & 3 \end{pmatrix}.$$

E9.46 If you start with $\mathbf{X}$ having covariance matrix

$$\mathbb{C}_\mathbf{X} = \begin{pmatrix} 2 & 0 \\ 0 & 1 \end{pmatrix},$$

then show how to synthesize $\mathbf{Y}$ with covariance matrix

$$\mathbb{C}_\mathbf{Y} = \begin{pmatrix} 2 & -1 \\ -1 & 3 \end{pmatrix}.$$

E9.47 Starting from the Matlab command `randn(3,1)`, generate a Gaussian random vector $\mathbf{Y}$ with $E\mathbf{Y} = \begin{pmatrix} 0 & -1 & 1 \end{pmatrix}^T$ and covariance matrix

$$\mathbb{C}_Y = \begin{pmatrix} 1.6840 & 0.3459 & 0.6776 \\ 0.3459 & 4.0900 & -1.5864 \\ 0.6776 & -1.5864 & 1.5411 \end{pmatrix}.$$

10

Linear Systems, Signals, and Filtering

10.1 PURPOSE, BACKGROUND, AND ORGANIZATION
10.2 BACKGROUND TO LINEAR SYSTEMS
10.3 RANDOM SIGNALS AND AUTOCORRELATION FUNCTIONS
10.4 MOMENTS OF OUTPUTS OF LINEAR SYSTEMS
10.5 APPLICATION TO WIENER FILTERING
10.6 APPLICATION TO KALMAN FILTERING
10.7 SUMMARY
    EXERCISES

10.1 PURPOSE, BACKGROUND, AND ORGANIZATION

Historically, much of electrical engineering systems theory and design was limited to analog circuits that were linear systems built out of relatively few (on the order of 100) components. This, of course, changed radically with the advent of very large scale integrated circuits, computers, digital signal processing, and computer-aided design (CAD). However, we still have a meaningful interest in such linear systems for purposes of filtering (e.g., band-passing a signal when you tune a receiver, or amplification as in audio equipment). In this chapter, we review basic concepts of linear systems, starting with that of linearity and moving to representations of linear systems either through impulse response (Green's functions to physicists) or through a state transition matrix. A brief introduction is then provided to modeling random signals that are the inputs and outputs of such linear systems, through the mean and correlation. This subject will be revisited in Chapter 20.

In pursuit of Goal 2, we discuss how to compute the response of linear systems to random inputs with respect to such basic probability quantities or statistics as the mean, variance, correlation, and covariance. An important and unusual characteristic of linear systems is that all we need to know to compute these statistics at the output of the system is the same kind of statistics at the input. Linear systems preserve our ability to calculate a description in terms of means, correlations, and covariances for their output when we have this description for their input. Finally, we apply these ideas to achieving Goal 3: using probability to form optimal estimates and inferences and to make decisions. We discuss Wiener filtering and Kalman filtering, with Wiener filtering relating to impulse response representation of linear systems and Kalman filtering relating to state transition matrix representation. The important ideas of least mean square estimation are developed further from their introduction in Section 9.12.

10.2 BACKGROUND TO LINEAR SYSTEMS

10.2.1 Linearity

A *discrete time system* $\mathcal{S}$ is a function with argument an infinite (possibly complex-valued) sequence $\mathcal{X} = (\ldots, x_{-2}, x_{-1}, x_0, x_1, \ldots)$, called its input, and value an infinite (possibly complex-valued) sequence $\mathcal{Y} = (\ldots, y_{-2}, y_{-1}, y_0, y_1, \ldots)$, called its output. The term x_k is the input at time $t_k < t_{k+1}$, and the term y_k is the corresponding output at time t_k. An example of a (nonlinear, time-varying) system is

$$y_k = k x_{k-1} x_{k+1}^2.$$

A system $\mathcal{S}$ is *linear* if, for any (complex-valued) α,

$$\mathcal{Z} = \alpha \mathcal{X} = (\ldots, \alpha x_{-2}, \ldots, \alpha x_k, \ldots) \Rightarrow \mathcal{S}(\mathcal{Z}) = \mathcal{W} = \alpha \mathcal{Y} = (\ldots, \alpha y_{-2}, \ldots, \alpha y_k, \ldots).$$

Hence, in a linear system, if we amplify all inputs by a uniform factor of α, then we correspondingly amplify all outputs by the same factor. An example of a linear system is

$$y_k = \sum_{j=k-l}^{k} (k - j) x_j.$$

Much more information about linear systems is available from a variety of standard texts, including Oppenheim, Willsky, and Nawab [66].

10.2.2 Characterization by Impulse Response

A discrete-time linear system with input X_t and real-valued output Y_t can be characterized by an *impulse response* h (known in physics as a *Green's function*) through

$$Y_t = \sum_{\tau=-\infty}^{\infty} h(t, \tau) X_\tau,$$

where we have assumed that inputs occur at equally spaced unit times. If the system is *causal*, then an output Y_t at time t cannot depend upon future inputs $X_\tau, \tau > t$. Hence, for a causal linear system,

$$h(t, \tau) = 0, \text{ for } \tau > t,$$

$$Y_t = \sum_{\tau=-\infty}^{t} h(t, \tau) X_\tau.$$

A linear system is *time invariant* if time-shifting the inputs results in the same time shift of the output—that is, if

$$Z_t = X_{t+s}, \quad V_t = \sum_\tau h(t, \tau) Z_\tau = Y_{t+s}.$$

The condition of time invariance is equivalent to requiring that

$$(\exists g)(\forall t, \tau) \; h(t, \tau) = g(t - \tau).$$

We follow the usual abuse of notation in this case and continue to refer to the impulse response by $h(t - \tau)$, so that

$$Y_t = \sum_{\tau=-\infty}^{\infty} h(t - \tau) X_\tau.$$

A linear system is *stable* if "bounded" inputs produce "bounded" outputs. There are a variety of ways to make the concept of stability more precise. We will assume that a linear system is stable if

$$(\exists B < \infty)(\forall t) \sum_{\tau=-\infty}^{\infty} h^2(t, \tau) < B.$$

This choice of stability has the advantage that inputs of finite mean square will yield outputs that are also of finite mean square. Thus, random inputs X_t to a linear system result in random outputs Y_t that are weighted sums of the inputs, and this observation focuses our attention on sums of random variables.

10.2.3 Characterization by State Transition Matrix

The material in this subsection is used only in connection with Kalman filtering as discussed in Sections 10.6 and 15.6. It may be omitted otherwise.

   Many, but not all, physical systems admit of a linear representation in terms of a finite-dimensional state vector $\mathbf{x}$. That is, knowledge of the *state vector* $\mathbf{x}_t$ at the present time t is all we need to know about the past and present of the system for us to be able to predict the future response of the system. The state vector at a given time summarizes the past history of the system. For example, in a mechanical system composed of particles (e.g., a gas-filled region), the current positions and momenta of all particles, together with whatever forcing may be present, determine the future evolution of the system. In an electrical system or circuit composed of inductors, capacitors, and resistors, the inductor currents and capacitor voltages, together with the sources, determine the future system response. More generally, a physical system whose response $y(t)$ can be described by a linear differential equation of order n, with possibly time varying coefficients, or

$$D^k y = \frac{d^k y(t)}{dt^k}, \quad D^n y + \sum_{k=0}^{n-1} a_k(t) D^k y = f(t),$$

is a system in which there is an n-dimensional system state composed of y and its first $n - 1$ derivatives. To see this, we introduce the state vector

$$x_k(t) = D^k y(t), \quad \mathbf{x} = \begin{pmatrix} y & Dy & \dots & D^{n-1}y \end{pmatrix}^T$$

and rewrite the system differential equation as

$$Dx_n(t) = -\sum_{k=0}^{n-1} a_k(t)x_k(t) + f(t), \ Dx_k(t) = x_{k+1}(t) \text{ for } k < n.$$

We use Φ for the state transition matrix, as is traditional, although it does not have a blackboard font. Finally, written in matrix–vector form, we have

$$\mathbb{D}\mathbf{x} = \Phi\mathbf{x} + \mathbb{B}f(t), \quad \mathbb{B} = \begin{pmatrix} 0 \\ \vdots \\ 0 \\ 1 \end{pmatrix},$$

$$\Phi = \begin{pmatrix} 0 & 1 & 0 & \dots & 0 & 0 \\ 0 & 0 & 1 & 0 & \dots & 0 \\ \vdots & \vdots & \ddots & \vdots & \vdots & \vdots \\ 0 & 0 & \dots & 0 & 0 & 1 \\ -a_0 & -a_1 & -a_2 & \dots & -a_{n-2} & -a_{n-1} \end{pmatrix}.$$

In discrete time, which is our concern, the preceding are all examples of systems in which a response y_{k+1} at a future time t_{k+1} is determinable from the present value of the state vector $\mathbf{x}_k$ and knowledge of the present forcing (input) function $\mathbf{u}_k$ through

$$y_{k+1} = f(\mathbf{x}_k, \mathbf{u}_k).$$

Furthermore, this dependence of the future state only upon the present state is also true of the state vector itself; the state vector $\mathbf{x}_{k+1} = \mathbf{x}(t_{k+1})$ at the next time t_{k+1} is determinable from the state vector $\mathbf{x}_k = \mathbf{x}(t_k)$ at the present time t_k, together with the current excitation $\mathbf{u}_k$ to the system. In a linear, possibly time-varying, system, $\mathbf{x}_{k+1}$ is a linear function of $\mathbf{x}_k$ and $\mathbf{u}_k$ and can therefore be represented as

$$\mathbf{x}_{k+1} = \Phi_k \mathbf{x}_k + \mathbb{B}_k \mathbf{u}_k.$$

The matrix Φ_k is known as the *one-step state transition matrix*. We can combine $\mathbb{B}_k\mathbf{u}_k = \mathbf{w}_k$ to write the representation

$$\mathbf{x}_{k+1} = \Phi_k \mathbf{x}_k + \mathbf{w}_k.$$

We need to provide an initial condition, $\mathbf{x}_0$ at time t_0, to fully specify the evolution of this system after time t_0. We will return to this state vector representation when we discuss Kalman filtering in Section 10.6.

10.3 RANDOM SIGNALS AND AUTOCORRELATION FUNCTIONS

It is common in electrical engineering applications to deal with an (assumed here to be real-valued) *random signal* X_t, also denoted $X(t)$, where X_t is the random amplitude of the random signal measured at time t. The time t is drawn from a generally infinite time set T that is usually an interval $[\tau_0, \tau_1]$ with possibly infinite endpoints. The discussion we give here departs (as do Chapters 19 and 20) from our context of finitely many random variables, perhaps assembled as components of a finite-dimensional vector, by introducing a model for correlation that is most often used in the context of a random signal described by infinitely many time samples of amplitude. For example, X_t might be the amplitude of a speech signal at time t, a component of an electromagnetic signal measured at an antenna at time t, or the thermal noise signal generated at time t by any dissipative electrical component. Another example of X_t is what is called a *point process*: X_t is binary valued and has the value 1 if an event—say, the arrival of a photon at a detector or a packet at a receiver—occurs at time t and has the value 0 if otherwise. When time is discrete, we have a *time series* that is a (generally infinite) sequence $\{X_{t_i} = X_i\}$ of random variables. A full probabilistic description of a random signal requires that, for any n and any $t_1, \ldots, t_n$ drawn from the time set T, we can specify the multivariate pdf

$$(\forall x_1, \ldots, x_n) \, f_{X_{t_1}, \ldots, X_{t_n}}(x_1, \ldots, x_n).$$

This is clearly a demanding requirement and one frequently satisfied in practice by more or less warranted assumptions that the signal in question can be described by some familiar model (e.g., multivariate Gaussian). A selection of such models of random processes is provided in Chapter 20.

A partial description of a random signal can be given in terms of its first two moments. We introduce the *mean function*

$$m_X(t) = EX_t,$$

which is a function of the time parameter $t \in T$, and the *autocorrelation function*

$$R_X(t, s) = E(X_t X_s),$$

which is a function of a pair (t, s) of time variables. We can then express the *autocovariance function*

$$C_X(t, s) = R_X(t, s) - m_X(t) m_X(s).$$

The relative ease of estimation of the mean function and the autocorrelation and their relative simplicity compared with the full description through a multivariate pdf has encouraged engineering practice to assume that knowledge of the mean and autocorrelation functions is available. As we shall see in this chapter, methods of signal processing have been developed that rely only on this characterization of random signals.

The mean function m_t, corresponding to the average or expected signal, can be any function of time. However, the autocorrelation function R_X is restricted. Clearly, from its definition,

$$R_X(t, s) = R_X(s, t),$$

and $R_X(t, s)$ is a symmetric function of its two arguments. From the fact that

$$0 \leq E\left(\sum_1^n a_i X_{t_i}\right)^2 = \sum_{i=1}^n \sum_{j=1}^n a_i a_j R_X(t_i, t_j),$$

we see that, for any choice of constants $\{a_i\}$ and any choice of times $t_1, \ldots, t_n$, we must have

$$\sum_{i=1}^n \sum_{j=1}^n a_i a_j R_X(t_i, t_j) \geq 0.$$

Functions satisfying this condition are said to be *nonnegative definite functions*. An important example of such a function is

$$R_X(t, s) = \sigma^2 e^{-\alpha|t-s|}, \quad \alpha > 0.$$

Lemma 10.3.1   (Autocorrelation Function)   A function R_X is an autocorrelation function if and only if it is a nonnegative definite function.

Proof sketch. Our remarks preceding the lemma prove the "only if" part. A proof of the "if" part can be based upon the construction of a multivariate normal distribution having a correlation function that is any prescribed nonnegative definite function.   □

The notion of nonnegative definiteness for correlation matrices was discussed in Section 9.14 in the setting of vector-valued random variables.

10.4 MOMENTS OF OUTPUTS OF LINEAR SYSTEMS

10.4.1 Impulse Function Representation

We summarize the material on the response of linear systems and consider

$$S_n = \sum_{k=-\infty}^{\infty} a_k X_k = \mathbf{a}^T \mathbf{X}.$$

We organize our development of moments of S_n by increasing specialization of the random variables: general $\{X_k\}$, uncorrelated $\{X_k\}$, independent $\{X_k\}$, and $i.i.d.$ $\{X_k\}$. From the linearity of expectation stated in property E6,

$$ES_n = \sum_{k=-\infty}^{\infty} a_k EX_k; \tag{10.1}$$

for simplicity, we let $\{a_k\}$ represent the impulse response for a fixed value of n. Rigorously, we would also need to assume that only finitely many of $\{a_k\}$ are nonzero, although conditions can be provided to deal with the case of infinitely many summands. In what follows, we restrict ourselves explicitly to the finite case. In particular, the expectation of the average ($a_k = 1/n$) is given by

$$ES_n = \frac{1}{n} \sum_{k=1}^{n} EX_k.$$

Hence, when there is a common expected value m,

$$ES_n = m.$$

We use the fact that the square of a sum is a double sum, or

$$\left(\sum_{1}^{n} b_i \right)^2 = \sum_{i=1}^{n} \sum_{j=1}^{n} b_i b_j,$$

to calculate the mean square response:

$$ES_n^2 = E\left(\sum_{1}^{n} a_k EX_k \right)^2 = E\left(\sum_{i=1}^{n} \sum_{j=1}^{n} a_i X_k a_j X_j \right) = \sum_{i=1}^{n} \sum_{j=1}^{n} a_i a_j E(X_i X_j). \tag{10.2}$$

Repeating this calculation yields

$$\text{VAR}(S_n) = E(S_n - ES_n)^2 = E\left(\sum_{1}^{n} a_i (X_i - EX_i) \right)^2 = \sum_{i=1}^{n} \sum_{j=1}^{n} a_i a_j \, \text{COV}(X_i, X_j). \tag{10.3}$$

If the summands are uncorrelated, then

$$\text{VAR}(S_n) = \sum_{1}^{n} a_i^2 \, \text{VAR}(X_i). \tag{10.4}$$

If the uncorrelated summands have common variance σ^2, then

$$\text{VAR}(S_n) = \sigma^2 \sum_{1}^{n} a_i^2,$$

and the variance of the average is

$$\text{VAR}(S_n) = \frac{1}{n} \sigma^2.$$

Therefore an average of $i.i.d.$ random variables has an expectation equal to the common mean m and a variance that approaches 0:

If EX^2 is finite, then there is less and less fluctuation about the mean as we increase the number of $i.i.d.$ terms being averaged.

This result can be used to establish a weak law of large numbers concerning the limiting behavior (known as convergence in mean square) of $\{S_n\}$ as n approaches infinity. (See Section 19.3.)

While moments are relatively easy to calculate and provide useful information about a random variable, finitely many of them cannot determine the probability law of the random variable. If we wish to gain a precise grip on the distribution of S_n, then we can write out the cdf F_{S_n} as a multiple integral. Letting $Y_k = a_k X_k$ and $S_n = \sum_1^n Y_k$, we have

$$F_{S_n}(s) = \int_{-\infty}^{+\infty} dy_n \int_{-\infty}^{s-y_n} dy_{n-1} \int_{-\infty}^{s-y_n-y_{n-1}} dy_{n-2} \cdots$$
$$\int_{-\infty}^{s-\sum_{j>1} y_j} dy_1 \, f_{Y_1,\ldots,Y_n}(y_1, \ldots, y_n).$$

From the results in Section 8.6.1 on linear transformations, we see that

$$F_{S_n}(s) = \frac{1}{|\prod_1^n a_i|} \int_{-\infty}^{+\infty} dy_n \int_{-\infty}^{s-y_n} dy_{n-1} \int_{-\infty}^{s-y_n-y_{n-1}} dy_{n-2} \cdots$$
$$\int_{-\infty}^{s-\sum_{j>1} y_j} dy_1 \, f_{X_1,\ldots,X_n}\left(\frac{y_1}{a_1}, \ldots, \frac{y_n}{a_n}\right).$$

While this result can then be differentiated with respect to s to provide an expression for the density function, the result provides little insight into the behavior of sums of random variables. A better tool is that of characteristic functions treated in Chapter 17 and applied to sums of independent random variables in Section 17.8.

Finally, we can determine the correlation between sums:

$$ES_n S_m = \sum_{i=1}^n \sum_{j=1}^m a_i a_j E(X_i X_j).$$

The corresponding covariance is

$$\mathrm{COV}(S_n, S_m) = \sum_{i=1}^n \sum_{j=1}^m a_i a_j \, \mathrm{COV}(X_i, X_j).$$

If we assume that the summands $\{X_i\}$ are uncorrelated and, for convenience, that $n > m$, then

$$\mathrm{COV}(X_i, X_j) = \begin{cases} \mathrm{VAR}(X_i) & \text{if } i = j \\ 0 & \text{if } i \neq j \end{cases},$$

$$\mathrm{COV}(S_n, S_m) = \sum_{j=1}^m a_j^2 \, \mathrm{VAR}(X_j).$$

10.4.2 Correlation under Linear Transformation

The equations of Section 9.15 are readily applied to compute means and covariance matrices for a random state vector $\mathbf{X}_k$ in the state vector representation

$$\mathbf{X}_{k+1} = \Phi_k \mathbf{X}_k + \mathbf{W}_k,$$

with nonrandom one-step transition matrix Φ_k, inputs $\mathbf{W}_k$ that are random vectors, and a possibly random initial condition $\mathbf{X}_0$. Taking expectations yields a recursion

$$E\mathbf{X}_{k+1} = \Phi_k E\mathbf{X}_k + E\mathbf{W}_k$$

for the expected state vector at t_{k+1} in terms of its expectation at time t_k. If the external forcing is such that $E\mathbf{W}_j = \mathbf{0}$, then

$$E\mathbf{X}_{k+1} = \left(\prod_{j=0}^{k} \Phi_j \right) E\mathbf{X}_0.$$

Now let $E\mathbf{W}_k \mathbf{W}_K^T = \mathbb{R}_k$ (our usage in the remainder of this chapter of $\mathbb{R}$ as a correlation matrix is not to be confused with our usual usage of it as representing the real numbers), and assume that the current forcing $\mathbf{W}_k$ is orthogonal to the current state $\mathbf{X}_k$, $E\mathbf{W}_k \mathbf{X}_k^T = \mathbb{O}$, where the matrix $\mathbb{O}$ is the matrix with all zero entries. We obtain a recursion for the state covariance matrices

$$E\mathbf{X}_{k+1}\mathbf{X}_{k+1}^T = \Phi_k E\mathbf{X}_k \mathbf{X}_k^T \Phi_k^T + \mathbb{R}_k,$$

and can solve for $E\mathbf{X}_{k+1}\mathbf{X}_{k+1}^T$ if, say, we know $E\mathbf{X}_0\mathbf{X}_0^T$.

10.5 APPLICATION TO WIENER FILTERING

10.5.1 Setup

In support of our Goal 3, in this case optimal signal processing, we examine how to extract information about a signal or random variable Y (sometimes we will use the notation S for "signal" or "parameter" for this quantity) from several observations forming a random vector $\mathbf{X}$, thereby generalizing the treatment given in Section 9.12. For example, Y could be a voltage and X_i the ith noisy measurement of this voltage in a series of repeated measurements, possibly made with different instruments; Y could also be a broadcast value and $\mathbf{X}$ samples of values received over a noisy channel; Y could be a latitude–longitude vector and $\mathbf{X}$ GPS receiver measurements from several satellites. The *estimator* $\hat{Y}$ of Y is a function of the observations $\mathbf{X}$, $\hat{Y} = \hat{Y}(\mathbf{X})$, and not, of course, a function of the unknown Y. We would like to choose an estimator $\hat{Y}$ that lies close to Y. As both Y and $\hat{Y}$ are random variables, we need a statistical measure of closeness.

The most commonly used such measure of the closeness between two real-valued random variables is the *mean squared error (MSE)* given by $E(\hat{Y} - Y)^2$, with the expectation of $(\hat{Y} - Y)^2$

yielding a single numerical measure of its size. As first discussed in Section 9.12, this measure treats a positive error $e = \hat{Y} - Y$ identically to a negative error, minimizes the effects of small errors (squaring a small number makes it even smaller), and pays attention to large errors (squaring a large number makes it even larger). If, for example, the quantity Y is the altitude above an airport runway being measured by an altimeter on an aircraft, then we might want to treat positive and negative errors differently; in the one case we might have a bumpy landing when we think we are touching down but find ourselves 2 meters above the runway, while in the other case we might crash thinking we were still above the runway when we struck it forcibly. While other error measures are sometimes used and will be treated when we discuss Bayesian estimation and decision making in Chapter 16, the MSE is often the most easily applied. This tractability has encouraged its widespread (over)use.

We would like to choose the function $\hat{Y}(\mathbf{X})$ that minimizes $E(\hat{Y} - Y)^2$. However, to do so in full generality requires knowledge of conditional expectation (see Chapter 15) and rather extensive knowledge of the joint distribution of $Y, \mathbf{X}$. This latter knowledge is often unavailable. The approach of Wiener filtering, pioneered by Norbert Wiener [99] (and also by Andrei Kolmogorov), was developed in World War II to improve the accuracy of antiaircraft fire. It then found a host of postwar applications that continue to this day. A computationally effective form of this filtering based upon a recursive formulation was proposed by Kalman [51] and is the basis of a large industry in Kalman filtering. (See Section 10.6.) In (nonrecursive) Wiener filtering, we restrict $\hat{Y}$ to be an affine function of $\mathbf{X}$—that is,

$$\hat{Y}(\mathbf{X}) = \mathbf{a}^T \mathbf{X} + b, \tag{10.5}$$

and choose $\mathbf{a}, b$ so as to minimize the MSE.

10.5.2 Calculation of the Wiener Filter Coefficients

Note that the true Y does not depend upon our choices for $\mathbf{a}, b$. Consider first

$$\frac{\partial E(\hat{Y} - Y)^2}{\partial b} = 2E(\hat{Y} - Y)\frac{\partial \hat{Y}}{\partial b}.$$

Of course,

$$\frac{\partial \hat{Y}}{\partial b} = 1.$$

Hence,

$$\frac{\partial E(\hat{Y} - Y)^2}{\partial b} = 2E(\hat{Y} - Y).$$

To find the optimal value of b, set the derivative equal to zero to conclude that, for the optimal $b = b^*$,

$$E\hat{Y} = \mathbf{a}^T E\mathbf{X} + b^* = EY, \ \ b^* = EY - \mathbf{a}^T E\mathbf{X}. \tag{10.6}$$

As was shown generally in Lemma 9.12.1, the optimal affine estimator $\hat{Y}$ chooses b to match its expectation to that of the unknown quantity Y, or

$$E\hat{Y} = EY.$$

We now substitute this conclusion into the MSE:

$$E(\hat{Y} - Y)^2 = E\left[\mathbf{a}^T(\mathbf{X} - E\mathbf{X}) - (Y - EY)\right]^2.$$

It remains for us to determine the optimal $\mathbf{a} = \mathbf{a}^*$. Clearly,

$$\frac{\partial E\left[\mathbf{a}^T(\mathbf{X} - E\mathbf{X}) - (Y - EY)\right]^2}{\partial a_i} = 2E\left[(\hat{Y} - Y)\frac{\partial \mathbf{a}^T(\mathbf{X} - E\mathbf{X})}{\partial a_i}\right]$$

and

$$\frac{\partial \mathbf{a}^T(\mathbf{X} - E\mathbf{X})}{\partial a_i} = \frac{\partial \sum_j a_j(X_j - EX_j)}{\partial a_i} = X_i - EX_i.$$

For optimality, we set these partial derivatives with respect to the a_i to zero. The resulting condition is known as the *orthogonality condition*, or *orthogonality principle*,

$$(\forall i)\, E\left[(\hat{Y} - Y)(X_i - EX_i)\right] = 0.$$

In words, the error $\hat{Y} - Y$ of the optimal affine estimator should be orthogonal to each mean-corrected component X_i of the measurement $\mathbf{X}$. Substituting for the estimator yields

$$E\left[\left((\mathbf{a}^{*T}(\mathbf{X} - E\mathbf{X}) - (Y - EY)\right)(\mathbf{X} - E\mathbf{X})\right] = \mathbf{0}.$$

Using the fact that some of the terms are scalars and therefore equal to their own transposes and using the notion of covariance yields,

$$E\left[\mathbf{a}^{*T}(\mathbf{X} - E\mathbf{X})(\mathbf{X} - E\mathbf{X})\right] = E\left[(\mathbf{X} - E\mathbf{X})(\mathbf{X} - E\mathbf{X})^T\mathbf{a}^*\right] = \mathrm{COV}(\mathbf{X}, \mathbf{X})\mathbf{a}^* = \mathbb{C}_{\mathbf{X}}\mathbf{a}^*,$$

and

$$E\left[(Y - EY)(\mathbf{X} - E\mathbf{X})\right] = E\left[(\mathbf{X} - E\mathbf{X})(Y - EY)^T\right] = \mathrm{COV}(\mathbf{X}, Y).$$

Hence, the design condition for $\mathbf{a}^*$ becomes

$$\mathbb{C}_{\mathbf{X}}\mathbf{a}^* = \mathrm{COV}(\mathbf{X}, Y) \Rightarrow \mathbf{a}^* = \mathbb{C}_{\mathbf{X}}^{-1}\,\mathrm{COV}(\mathbf{X}, Y). \tag{10.7}$$

We have solved for the optimal MSE affine estimator

$$\hat{Y} = \mathbf{a}^{*T}\mathbf{X} + \mathbf{b} = \mathbf{a}^{*T}(\mathbf{X} - E\mathbf{X}) + EY$$

$$= \left[\mathbb{C}_{\mathbf{X}}^{-1}\,\mathrm{COV}(\mathbf{X}, Y)\right]^T(\mathbf{X} - E\mathbf{X}) + EY$$

$$= \mathrm{COV}(Y, \mathbf{X})\mathbb{C}_{\mathbf{X}}^{-1}(\mathbf{X} - E\mathbf{X}) + EY \tag{10.8}$$

of Y, provided that we know $EY, E\mathbf{X}, \mathbb{C}_{\mathbf{X}}$, and $\mathrm{COV}(\mathbf{X}, Y)$.

We also need to know the performance of the optimal estimator given by

$$E(\hat{Y} - Y)^2 = E\left[\mathbf{a}^{*T}(\mathbf{X} - E\mathbf{X}) - (Y - EY)\right]^2$$

$$= E(\mathbf{a}^{*T}(\mathbf{X} - E\mathbf{X}))^2 + \text{VAR}(Y) - 2\mathbf{a}^{*T}\text{COV}(\mathbf{X}, Y).$$

Taking the first and third terms individually yields

$$E(\mathbf{a}^{*T}(\mathbf{X} - E\mathbf{X}))^2 = E\left[\mathbf{a}^{*T}(\mathbf{X} - E\mathbf{X})(\mathbf{X} - E\mathbf{X})^T\mathbf{a}^*\right] = \mathbf{a}^{*T}\mathbb{C}_{\mathbf{X}}\mathbf{a}^*$$

$$= \text{COV}(Y, \mathbf{X})\mathbb{C}_{\mathbf{X}}^{-1}\mathbb{C}_{\mathbf{X}}\mathbb{C}_{\mathbf{X}}^{-1}\text{COV}(\mathbf{X}, Y) = \text{COV}(Y, \mathbf{X})\mathbb{C}_{\mathbf{X}}^{-1}\text{COV}(\mathbf{X}, Y);$$

$$\mathbf{a}^{*T}\text{COV}(\mathbf{X}, Y) = \text{COV}(Y, \mathbf{X})\mathbb{C}_{\mathbf{X}}^{-1}\text{COV}(\mathbf{X}, Y).$$

Hence, the minimum mean square error is

$$E(\hat{Y} - Y)^2 = \text{VAR}(Y) - \text{COV}(Y, \mathbf{X})\mathbb{C}_{\mathbf{X}}^{-1}\text{COV}(\mathbf{X}, Y), \tag{10.9}$$

when we use an estimator of the form $\hat{Y} = \mathbf{a}^T\mathbf{X} + b$. If the resulting mean square error is too large for our purposes, then we either need to consider a nonlinear function $\hat{Y}(\mathbf{X})$ or obtain better data than $\mathbf{X}$.

10.5.3 Example of Wiener Filtering: Signal-Plus-Noise

Our model is that, at time t, the measured value X_t is the sum of a time-varying signal S_t and an additive noise N_t:

$$X_t = S_t + N_t.$$

For simplicity, we assume that we measure only at the times $t_1 < t_2$ and wish to infer the signal S_{t_0} at the time t_0. Wiener, in his classic work [99], notes the three problems of *smoothing*, which we interpret as $t_0 < t_1$; *interpolation*, where $t_1 \leq t_0 \leq t_2$; and *extrapolation (prediction)* for $t_2 < t_0$. We assume that, at all pairs of times (s, t), the signal S_t and noise N_s are uncorrelated, or

$$\text{COV}(S_t, N_s) = 0.$$

Assume further that the noise process N_t satisfies

$$EN_t = 0, \quad \text{COV}(N_s, N_t) = \sigma_N^2 e^{-\alpha|t-s|}, \quad \alpha > 0.$$

We assume that the signal has

$$ES_t = m_S(t), \quad \text{COV}(S_s, S_t) = \sigma_S^2 e^{-\beta|t-s|}, \quad \beta > 0.$$

Hence,

$$EX_t = ES_t + EN_t = m_S(t),$$

and, using the fact that N_t and S_s are uncorrelated, we obtain

$$\text{COV}(X_s, S_t) = E\left[(S_s - m_S(s) + N_s)(S_t - m_S(t))\right] = \text{COV}(S_s, S_t) = \sigma_S^2 e^{-\beta|t-s|},$$

$$\text{COV}(X_s, N_t) = E\left[(S_s - m_S(s) + N_s)N_t\right] = \text{COV}(N_s, N_t) = \sigma_N^2 e^{-\alpha|t-s|},$$

$$\text{COV}(X_s, X_t) = \text{COV}(X_s, S_t) + \text{COV}(X_s, N_t) = \sigma_N^2 e^{-\alpha|t-s|} + \sigma_S^2 e^{-\beta|t-s|}.$$

Let $\mathbf{X}$ be the column vector $(X_{t_1}, X_{t_2})^T$ and

$$\mathbf{m}_S = E\mathbf{X} = \begin{pmatrix} m_S(t_1) \\ m_S(t_2) \end{pmatrix}.$$

The covariance matrix is

$$\mathbb{C}_X = \text{COV}(\mathbf{X}, \mathbf{X}) = \begin{pmatrix} \sigma_N^2 + \sigma_S^2 & \sigma_N^2 e^{-\alpha(t_2-t_1)} + \sigma_S^2 e^{-\beta(t_2-t_1)} \\ \sigma_N^2 e^{-\alpha(t_2-t_1)} + \sigma_S^2 e^{-\beta(t_2-t_1)} & \sigma_N^2 + \sigma_S^2 \end{pmatrix},$$

and the covariance vector is

$$\text{COV}(\mathbf{X}, S_{t_0}) = \begin{pmatrix} \sigma_S^2 e^{-\beta|t_1-t_0|} \\ \sigma_S^2 e^{-\beta|t_2-t_0|} \end{pmatrix}.$$

Note that we cannot remove the absolute-value signs appearing in the covariance vector without knowing whether we are dealing with smoothing, interpolation, or extrapolation. From Eq. (10.8), we can now express

$$\hat{S}_{t_0} = \text{COV}(S_{t_0}, \mathbf{X})\mathbb{C}_{\mathbf{X}}^{-1}(\mathbf{X} - \mathbf{m}_S) + \mathbf{m}_S.$$

From Eq. (10.9), the performance of this estimator is found to be

$$E(\hat{S}_{t_0} - S_{t_0})^2 = \sigma_S^2 - \text{COV}(S_{t_0}, \mathbf{X})\mathbb{C}_{\mathbf{X}}^{-1}\text{COV}(\mathbf{X}, S_{t_0}).$$

Substituting the detailed expressions for the various covariance matrices would yield a cumbersome expression that is omitted.

10.5.4 Extension to Estimating a Vector-Valued Y

We can extend the preceding discussion to construct a linear estimator $\hat{\mathbf{Y}}(\mathbf{X})$ of a random vector $\mathbf{Y}$ of dimension m from observation of a random vector $\mathbf{X}$ of dimension n. The form of the estimator is

$$\hat{\mathbf{Y}} = \mathbb{A}\mathbf{X} + \mathbf{b} \Rightarrow \hat{Y}_i = \sum_j A_{ij}X_j + b_i,$$

where the $m \times n$ matrix $\mathbb{A}$ and $m \times 1$ vector $\mathbf{b}$ are nonrandom. Note that there is no constraint between the solution for, say, $\hat{Y}_j$ and that for $\hat{Y}_k$: They can be chosen independently, since $\mathbb{A}$ and $\mathbf{b}$ are arbitrary.

The MSE criterion in the vector case is taken as $E[(\hat{\mathbf{Y}} - \mathbf{Y})^T\mathbb{M}(\hat{\mathbf{Y}} - \mathbf{Y})]$. The case of any positive definite weighting matrix $\mathbb{M}$ is treated in Section 15.4.2. The optimal estimator turns out not to depend upon the choice of positive definite $\mathbb{M}$. With this in mind, we simplify matters and choose $\mathbb{M} = \mathbb{I}$, in which case

$$E[(\hat{\mathbf{Y}} - \mathbf{Y})^T \mathbb{M}(\hat{\mathbf{Y}} - \mathbf{Y})] = \sum_{i=1}^{m} E(\hat{Y}_i - Y_i)^2.$$

We can minimize this expression by minimizing each of the m terms in the sum. The flexibility offered by $\mathbb{A}$ and $\mathbf{b}$ allows us to minimize any term without limiting our ability to minimize any other term. In effect, we now have m individual estimation problems of the type already discussed. From the solution given to the scalar $m = 1$ case in Eq. (10.8), we see that $\mathbf{b}$ is chosen so that

$$E\mathbf{Y} = E\hat{\mathbf{Y}} \Rightarrow E\mathbf{Y} = \mathbb{A}E\mathbf{X} + \mathbf{b} \Rightarrow \mathbf{b} = E\mathbf{Y} - \mathbb{A}E\mathbf{X},$$

and the matrix $\mathbb{A}$ has ith row $\mathbf{r}_i$ that is a solution vector:

$$\mathbf{r}_i = [\mathbb{C}_{\mathbf{X}}^{-1} \, \mathrm{COV}(\mathbf{X}, Y_i)]^T.$$

Hence,

$$\mathbb{A} = [\mathbb{C}_{\mathbf{X}}^{-1} \, \mathrm{COV}(\mathbf{X}, \mathbf{Y})]^T = \mathrm{COV}(\mathbf{Y}, \mathbf{X})\mathbb{C}_{\mathbf{X}}^{-1}.$$

The optimum estimator in the vector case is then

$$\hat{\mathbf{Y}} = \mathrm{COV}(\mathbf{Y}, \mathbf{X})\mathbb{C}_{\mathbf{X}}^{-1}(\mathbf{X} - E\mathbf{X}) + E\mathbf{Y}. \tag{10.10}$$

Summing the results from Eq. (10.9) for the performance of the individual optimal scalar estimator, the performance of this optimal vector estimator

$$E(\hat{\mathbf{Y}} - \mathbf{Y})^T(\hat{\mathbf{Y}} - \mathbf{Y}) = \sum_{i=1}^{m} E(\hat{Y}_i - Y_i)^2$$

$$= \sum_{i=1}^{m} \mathrm{VAR}(Y_i) - \sum_{i=1}^{m} \mathrm{COV}(Y_i, \mathbf{X})\mathbb{C}_{\mathbf{X}}^{-1} \, \mathrm{COV}(\mathbf{X}, Y_i).$$

Rewriting the equation using the matrix trace operation (sum of the diagonal elements of a matrix), we have

$$E(\hat{\mathbf{Y}} - \mathbf{Y})^T(\hat{\mathbf{Y}} - \mathbf{Y}) = \mathrm{Trace}(\mathbb{C}_{\mathbf{Y}}) - \mathrm{Trace}\left(\mathrm{COV}(\mathbf{Y}, \mathbf{X})\mathbb{C}_{\mathbf{X}}^{-1} \, \mathrm{COV}(\mathbf{X}, \mathbf{Y})\right). \tag{10.11}$$

Example 10.1   Estimation with Multiplicative Noise _____

A signal vector $S_1, \ldots, S_n$ is transmitted through a channel that attenuates the component S_i by a random factor N_i to yield an output $X_i = N_i S_i$. We assume we know that $E\mathbf{S} = \mathbf{m_S}$ and $\mathbb{C}_S = \mathrm{COV}(\mathbf{S}, \mathbf{S})$. If the total time taken to transmit the n signal components is small, then the random channel attenuation can be taken to be a single random variable N. Thus, we observe $\mathbf{X} = N\mathbf{S}$ at the channel output and wish to infer to the transmitted $\mathbf{S}$. Assume that the random amplitude N has $EN = m_N$ and $\text{VAR}(N) = \sigma_N^2$ and that it is independent of (unlinked with) the signal $\mathbf{S}$. As will be verified in Section 14.7, unlinkedness allows us to conclude that

$$(\forall j)\ E(NS_j) = ENES_j \text{ and } (\forall i, j)\ E(NS_iS_j) = ENE(S_iS_j).$$

We choose an affine estimator

$$\hat{\mathbf{S}} = \mathbb{A}\mathbf{X} + \mathbf{b}$$

and solve for $\mathbb{A}$ and $\mathbf{b}$ so as to minimize the MSE

$$\sum_{i=1}^{n} E(\hat{S}_i - S_i)^2.$$

From the preceding discussion, we know that

$$E\mathbf{X} = E(N\mathbf{S}) = ENE\mathbf{S} = m_N\mathbf{m_S}.$$

Therefore,

$$\mathbf{b} = E\mathbf{S} - \mathbb{A}E\mathbf{X} = E\mathbf{S} - m_N\mathbb{A}E\mathbf{S} = (\mathbb{I} - m_N\mathbb{A})E\mathbf{S},$$

$$\mathbb{A} = [\mathbb{C}_{\mathbf{X}}^{-1}\, \text{COV}(\mathbf{X}, \mathbf{S})]^T.$$

We first determine

$$\text{COV}(\mathbf{X}, \mathbf{S}) = [\text{COV}(X_i, S_j)],\ \ \text{COV}(X_i, S_j) = \text{COV}(NS_i, S_j)$$

$$= E(NS_iS_j) - E(NS_i)ES_j = ENE(S_iS_j) - ENES_iES_j,$$

where we have used the independence of N from $\mathbf{S}$ described earlier. Proceeding, we obtain

$$\text{COV}(X_i, S_j) = m_N\, \text{COV}(S_i, S_j) \iff \text{COV}(\mathbf{X}, \mathbf{S}) = m_N\mathbb{C}_{\mathbf{S}}.$$

Next, we determine

$$\mathbb{C}_{\mathbf{X}} = [\text{COV}(X_i, X_j)],\ \ \text{COV}(X_i, X_j) = E(X_iX_j) - (EX_i)(EX_j)$$

$$= E(N^2S_iS_j) - (m_NES_i)(m_NES_j) = EN^2E(S_iS_j) - m_N^2ES_iES_j$$

$$= EN^2\, \text{COV}(S_i, S_j) + \sigma_N^2ES_iES_j.$$

It follows that

$$\mathbb{C}_{\mathbf{X}} = EN^2\mathbb{C}_{\mathbf{S}} + \sigma_N^2\mathbf{m_S}\mathbf{m_S}^T.$$

Hence,

$$\mathbb{A} = [\mathbb{C}_{\mathbf{X}}^{-1}m_N\mathbb{C}_{\mathbf{S}}]^T = m_N\mathbb{C}_{\mathbf{S}}\mathbb{C}_{\mathbf{X}}^{-1}.$$

Finally, we obtain the minimum mean square error estimator from Eq. (10.10) as

$$\hat{\mathbf{S}} = m_N \mathbb{C}_\mathbf{S} \mathbb{C}_\mathbf{X}^{-1} (\mathbf{X} - m_N \mathbf{m_S}) + \mathbf{m_S}.$$

From Eq. (10.11), we have

$$E\left[(\hat{\mathbf{S}} - \mathbf{S})^T (\hat{\mathbf{S}} - \mathbf{S}) \right] = \text{Trace}(\mathbb{C}_\mathbf{S}) - \text{Trace}\left(\text{COV}(\mathbf{S}, \mathbf{X}) \mathbb{C}_\mathbf{X}^{-1} \text{COV}(\mathbf{X}, \mathbf{S}) \right)$$

$$= \sum_{i=1}^{n} \text{VAR}(S_i) - m_N^2 \text{Trace}\left(\mathbb{C}_\mathbf{S} \mathbb{C}_\mathbf{X}^{-1} \mathbb{C}_\mathbf{S} \right).$$

In the special case that the signals are uncorrelated with each other and are of mean zero, or

$$\mathbb{C}_\mathbf{S} = \sigma_\mathbf{S}^2 \mathbb{I}, \ E\mathbf{S} = \mathbf{0},$$

it follows that

$$\mathbb{C}_\mathbf{X} = \sigma_\mathbf{S}^2 EN^2 \mathbb{I},$$

and

$$\mathbb{A} = \frac{m_N}{EN^2} \mathbb{I}.$$

Thus, since $E\mathbf{X} = EN E\mathbf{S} = \mathbf{0}$,

$$\hat{\mathbf{S}} = \frac{m_N}{EN^2} \mathbf{X} = \frac{m_N}{m_N^2 + \sigma_N^2} \mathbf{X}.$$

Note that the "intuitive" thought that you might correct for the multiplicative random attenuation just by dividing $\mathbf{X} = n\mathbf{S}$ by $EN = m_N$ is correct only when N is nonrandom in that $\text{VAR}(N) = 0$.

10.6 APPLICATION TO KALMAN FILTERING

This section and its completion in Section 15.6 may be omitted without penalty in following the remainder of the material. Following the arguments presented next requires a high comfort level with matrix manipulations.

10.6.1 Setup

One of the greatest successes of applied probability has been the manifold applications (e.g., see Gelb [38], Brown and Hwang [16] , and Grewal and Andrews [41]) of what is known as Kalman filtering, first developed by R. E. Kalman [51]. Early applications were made to inertial navigation and stabilization systems used on submarines, ships, and aircraft, to process control for complex chemical engineering facilities, and to real-time control of other state-determined systems. Various generalized forms of the Kalman filter have continued to find application to all sorts of linear and nonlinear estimation and control problems.

Although both Wiener and Kalman filtering are applicable to continuous-time processes, such processes are out of the scope of this text. Hence, we continue to focus on discrete-time processes described by sequences $\{\mathbf{X}_k\}$ of n-dimensional vector-valued random variables, with $\mathbf{X}_k$ being a vector-valued response at time t_k, usually assumed to be linearly dependent upon the index k (i.e., $t_k = ak + b$). Kalman filtering provides a recursively formulated computational alternative to Wiener filtering that is appropriate in the common situation of tracking the state of a system described by a vector $\mathbf{X}$, with the state at time t_k given by $\mathbf{X}_k$. In a state-determined linear system, successive system states are related to each other through the state transition equation (see Section 10.2.3)

$$\mathbf{X}_{k+1} = \Phi_k \mathbf{X}_k + \mathbf{W}_k. \tag{10.12}$$

The matrix Φ_k describes the one-step transition from a state $\mathbf{X}_k$ at time t_k to one $\mathbf{X}_{k+1}$ at time t_{k+1} that would occur in the absence of external noise or forcing term $\mathbf{W}_k$. This iteration is initialized at some state $\mathbf{X}_0$. Note that $\mathbf{X}_k, k > 0$, is a linear function of $\mathbf{X}_0, \mathbf{W}_0, \dots, \mathbf{W}_{k-1}$. While Wiener filtering is applicable to all sequences of random variables, Kalman filtering is applicable only when the relations between successive terms evolve in a state-determined form. This restriction, however, confers the highly significant computational advantage of a recursive representation both for the minimum mean square estimator and for its performance as reflected in its covariance matrix. Such a representation is essential when we deal, as we frequently do, with a continuing stream of data—so-called *on-line processing*.

We simplify matters and assume that

$$E\mathbf{X}_0 = \mathbf{0}, \quad E\mathbf{X}_0\mathbf{X}_0^T = \mathbb{P}_0,$$

for some nonnegative definite Hermitian (symmetric if real-valued) matrix $\mathbb{P}_0$. Hence, $\mathbf{X}_0$ can be estimated simply by the zero vector $\mathbf{0}$ with resulting mean squared error covariance matrix $\mathbb{P}_0$. We assume further that, for a known nonnegative definite Hermitian matrix $\mathbb{Q}_k$,

$$E\mathbf{W}_k = \mathbf{0}, \; E\mathbf{W}_k\mathbf{W}_k^T = \mathbb{Q}_k.$$

Let $\mathbb{O}_{nm}$ denote the $n \times m$ matrix, all of whose entries are zeros. In what follows, we assume that the sequence of vector-valued "noises" $\{\mathbf{W}_k\}$ is uncorrelated in time in that, if $j \neq k$, then

$$E\mathbf{W}_j\mathbf{W}_k^T = \mathbb{O}_{nn}.$$

We are able to partially observe the state $\mathbf{X}_k, k > 0$, of this multidimensional linear system through an l-dimensional vector

$$\mathbf{Z}_k = \mathbb{H}_k\mathbf{X}_k + \mathbf{V}_k. \tag{10.13}$$

The matrix $\mathbb{H}_k$ is $l \times n$, typically with $l \leq n$, but is otherwise arbitrary. The l-dimensional vector $\mathbf{V}_k$ is a measurement noise vector about which we assume that

$$E\mathbf{V}_k = \mathbf{0}, \quad E\mathbf{V}_k\mathbf{V}_k^T = \mathbb{R}_k,$$

where $\mathbb{R}_k$ is known. We again assume uncorrelatedness of successive measurement noises in that, if $j \neq k$, then

$$E \mathbf{V}_j \mathbf{V}_k^T = \mathbb{O}_{ll}.$$

Finally, in conformity with much of the practice of Kalman filtering, we further assume that $\mathbf{X}_0, \mathbf{W}_1, \ldots, \mathbf{V}_1, \ldots,$ are all uncorrelated (recall that they all have zero expected values) with each other, or

$$(\forall j, k), \ E \mathbf{X}_0 \mathbf{W}_k^T = \mathbb{O}_{nn}, \ E \mathbf{X}_0 \mathbf{V}_k^T = \mathbb{O}_{nl}, \ E \mathbf{V}_j \mathbf{W}_k^T = \mathbb{O}_{ln}. \tag{10.14}$$

Our goal is to estimate the unobservable state $\mathbf{X}_k$ by an estimator $\hat{\mathbf{X}}_k$ that is a function of all of the observations $\mathbf{Z}_1, \ldots, \mathbf{Z}_k$. For convenience, we abbreviate these measurements by the $l \times k$ matrix

$$\mathbb{Z}_k^* = \left(\mathbf{Z}_1, \ldots, \mathbf{Z}_k \right).$$

Taking $\hat{\mathbf{X}}_k = \hat{\mathbf{X}}_k(\mathbb{Z}_k^*)$ formulates what is known as the filtering problem. Taking $\hat{\mathbf{X}}_k = \hat{\mathbf{X}}_k(\mathbb{Z}_j^*)$, for $j < k$ specifies the forecasting or prediction problem, while, if $j > k$, we have specified the smoothing or interpolation problem. The resulting random estimation error vector is given by

$$\mathbf{e}_k = \mathbf{X}_k - \hat{\mathbf{X}}_k.$$

Note that, in the filtering problem, $\hat{\mathbf{X}}_k(\mathbb{Z}_k^*), k > 0$, is a function of

$$\mathbf{X}_0, \mathbf{W}_0, \ldots, \mathbf{W}_{k-1}, \mathbf{V}_1, \ldots, \mathbf{V}_k.$$

In order to reduce the estimation error to a single numerical criterion (loss) that can be minimized, we introduce the weighted least mean square criterion. Let $\mathbb{M}$ be a positive definite matrix of our choosing (we shall in fact take it to be simply the identity matrix $\mathbb{I}$); construct the quadratic form $\mathbf{e}_k^T \mathbb{M} \mathbf{e}_k$, and then take its expectation to be the scalar quantity whose value we wish to minimize by choosing estimator $\hat{\mathbf{X}}_k(\mathbb{Z}_k^*)$.

10.6.2 Form of the Kalman Linear MMSE Estimator

In principle, we can solve for the linear function $\hat{\mathbf{X}}_k(\mathbb{Z}_k^*)$ of the $l \times k$ matrix $\mathbb{Z}_k^*$ that minimizes $E(\mathbf{X}_k - \hat{\mathbf{X}}_k)^T \mathbb{M}(\mathbf{X}_k - \hat{\mathbf{X}}_k)$. However, this calculation is lengthy and largely distracts us from the issues of expectation that we are examining. We will solve this problem in Chapter 15 when we introduce conditional expectation. In Section 15.4, we show that the minimizing solution without restriction to linearity is given by the conditional expectation (defined in Section 15.2)

$$\hat{\mathbf{X}}_k(\mathbb{Z}_j^*) = E(\mathbf{X}_k | \mathbb{Z}_j^*). \tag{10.15}$$

If we assume that $\mathbf{X}_0, \mathbf{W}_1, \ldots, \mathbf{V}_1, \ldots,$ are jointly normally distributed, then this conditional expectation becomes linear in the elements of the matrix $\mathbb{Z}_j^*$. (See Eq. (15.3).) Hence, this solution must also be the least mean square linear solution, without regard to assumptions of normality.

When $\mathbf{X}_k$ also satisfies a state equation, then this linear function can be written as a recursion in terms of a matrix $\mathbb{K}_k$, known as the *Kalman gain matrix*, as follows:

$$\hat{\mathbf{X}}_k(\mathbb{Z}_k^*) = \hat{\mathbf{X}}_k(\mathbb{Z}_{k-1}^*) + \mathbb{K}_k(\mathbf{Z}_k - \mathbb{H}_k\hat{\mathbf{X}}_k(\mathbb{Z}_{k-1}^*)). \tag{10.16}$$

Note the appearance of the one-step-ahead predictor $\hat{\mathbf{X}}_k(\mathbb{Z}_{k-1}^*)$, which, in what follows, will be denoted more compactly by $\hat{\mathbf{X}}_k^-$.

This conclusion bears closer examination. The best estimator $\hat{\mathbf{X}}_k(\mathbb{Z}_k^*)$ for the state vector $\mathbf{X}_k$ at time t_k is evaluated in terms of all of the available observations or measurements $\mathbb{Z}_k^* = (\mathbf{Z}_1, \ldots, \mathbf{Z}_k)$, up to and including time t_k. This evaluation is in terms of a linear combination of the best one-step-ahead predictor $\hat{\mathbf{X}}_k^-$ of the state vector $\mathbf{X}_k$, based upon only the observations $\mathbb{Z}_{k-1}^* = (\mathbf{Z}_1, \ldots, \mathbf{Z}_{k-1})$ prior to time t_k and the most current observation $\mathbf{Z}_k$. The term $\mathbf{Z}_k - \mathbb{H}_k\hat{\mathbf{X}}_k^-$ is the error we make in predicting the observation $\mathbf{Z}_k$ at time t_k from the information available at time t_{k-1} and represents the innovation in the observation process. By the orthogonality principle (see Section 10.5.2), this error term is orthogonal to linear combinations of this earlier information and hence is orthogonal to $\hat{\mathbf{X}}_k^-$.

We will be able to show easily in Section 15.6 that the one-step-ahead predictor $\hat{\mathbf{X}}_k^-$ is related to the estimator $\hat{\mathbf{X}}_{k-1}$ of the prior state $\mathbf{X}_{k-1}$ simply through the formula

$$\hat{\mathbf{X}}_k^- = \Phi_{k-1}\hat{\mathbf{X}}_{k-1}(\mathbb{Z}_{k-1}^*). \tag{10.17}$$

We can rewrite the expression for $\hat{\mathbf{X}}_k$ as

$$\hat{\mathbf{X}}_k(\mathbb{Z}_k^*) = \Phi_{k-1}\hat{\mathbf{X}}_{k-1}(\mathbb{Z}_{k-1}^*) + \mathbb{K}_k(\mathbf{Z}_k - \mathbb{H}_k\Phi_{k-1}\hat{\mathbf{X}}_{k-1}(\mathbb{Z}_{k-1}^*)). \tag{10.18}$$

The end result is a recursion relating the previous least mean square estimator $\hat{\mathbf{X}}_{k-1}$ of the previous state $\mathbf{X}_{k-1}$ to the current estimator $\hat{\mathbf{X}}_k$ of the current state $\mathbf{X}_k$, provided that we can determine the Kalman gain matrix $\mathbb{K}_k$.

10.6.3 Recursive Solution for the Kalman Gain Matrix

In the Kalman recursive approach to least mean square filtering, we need to recursively determine the sequence $\{\mathbb{K}_k\}$ of Kalman gain matrices, the sequence $\hat{\mathbf{X}}_k$ of estimators of $\mathbf{X}_k$, the sequence $\{\hat{\mathbf{X}}_k^-\}$ of one-step-ahead predictors of $\mathbf{X}_k$, and the performance of the resulting estimators in terms of the error covariance matrices

$$\mathbb{P}_k^- = E\mathbf{e}_k^-(\mathbf{e}_k^-)^T = E\left[(\mathbf{X}_k - \hat{\mathbf{X}}_k^-)(\mathbf{X}_k - \hat{\mathbf{X}}_k^-)^T\right]$$

and

$$\mathbb{P}_k = E\mathbf{e}_k\mathbf{e}_k^T = E\left[(\mathbf{X}_k - \hat{\mathbf{X}}_k)(\mathbf{X}_k - \hat{\mathbf{X}}_k)^T\right].$$

In order to evaluate $\mathbb{P}_k$ in terms of $\mathbb{P}_k^-$, write

$$\mathbf{e}_k = \mathbf{e}_k^- - \mathbb{K}_k(\mathbb{H}_k\mathbf{e}_k^- + \mathbf{V}_k) = (\mathbb{I} - \mathbb{K}_k\mathbb{H}_k)\mathbf{e}_k^- - \mathbb{K}_k\mathbf{V}_k.$$

Recall that $\mathbf{e}_k^-$ is a difference of $\mathbf{X}_k$ and $\mathbf{X}_k^-(\mathbb{Z}_{k-1}^*)$, both of which do not involve $\mathbf{V}_k$. Hence, $\mathbf{V}_k$ is orthogonal to $\mathbf{e}_k^-$. Invoking the results of Section 10.4, we obtain

$$
\begin{aligned}
\mathbb{P}_k &= (\mathbb{I} - \mathbb{K}_k \mathbb{H}_k)\mathbb{P}_k^-(\mathbb{I} - \mathbb{K}_k \mathbb{H}_k)^T + \mathbb{K}_k \mathbb{R}_k \mathbb{K}_k^T \\
&= \mathbb{P}_k^- - \mathbb{K}_k \mathbb{H}_k \mathbb{P}_k^- - (\mathbb{K}_k \mathbb{H}_k \mathbb{P}_k^-)^T + \mathbb{K}_k (\mathbb{H}_k \mathbb{P}_k^- \mathbb{H}_k^T + \mathbb{R}_k)\mathbb{K}_k^T .
\end{aligned} \tag{10.19}
$$

Note that this equation is valid for any choice of gain matrix $\mathbb{K}_k$. However, we wish to determine the best choice of such matrix. Our criterion is to choose $\mathbb{K}_k$ to minimize the sum of the mean squared errors made in estimating each of the components $\{\mathbf{X}_k\}$ at time t_k. Equivalently, we wish to choose $\mathbb{K}_k$ to minimize the sum of the diagonal terms in $\mathbb{P}_k$, which is known as the *trace* of $\mathbb{P}_k$. While this can be accomplished through what is likely to be unfamiliar methods of differentiation with respect to matrices, we choose to outline an approach that requires only familiar, if tedious, calculations. Note that the term $\mathbb{A} = \mathbb{H}_k \mathbb{P}_k^- \mathbb{H}_k^T + \mathbb{R}_k$ is a symmetric matrix that is nonnegative definite and positive definite if that is true for the covariance matrix $\mathbb{R}_k$. Assuming that this is the case, $\mathbb{A}$ has a matrix square root $\mathbb{A} = \mathbb{B}\mathbb{B}^T$. We can then complete the square in $\mathbb{K}_k$ for $\mathbb{P}_k$ through

$$
\mathbb{D} = \mathbb{P}_k^- \mathbb{H}_k^T (\mathbb{B}^T)^{-1}, \tag{10.20}
$$

$$
\mathbb{P}_k = (\mathbb{K}_k \mathbb{B} - \mathbb{D})(\mathbb{K}_k \mathbb{B} - \mathbb{D})^T + \mathbb{P}_k^- - \mathbb{D}\mathbb{D}^T . \tag{10.21}
$$

This equation can be verified by multiplying it out. The choice of gain matrix $\mathbb{K}_k$ that minimizes the diagonal elements of $\mathbb{P}_k$ is clearly

$$
\mathbb{K}_k \mathbb{B} = \mathbb{D},
$$

or

$$
\mathbb{K}_k = \mathbb{D}\mathbb{B}^{-1} = \mathbb{P}_K^- \mathbb{H}_k^T (\mathbb{B}^T)^{-1} \mathbb{B}^{-1} = \mathbb{P}_K^- \mathbb{H}_k^T (\mathbb{B}\mathbb{B}^T)^{-1} = \mathbb{P}_k^- \mathbb{H}_k^T (\mathbb{H}_k \mathbb{P}_k^- \mathbb{H}_k^T + \mathbb{R}_k)^{-1} . \tag{10.22}
$$

Substituting this solution for the optimal Kalman gain in Eq. (10.21) for the error covariance matrix yields

$$
\mathbb{P}_k = \mathbb{P}_k^- - \mathbb{D}\mathbb{D}^T . \tag{10.23}
$$

If we use the expression for $\mathbb{D}$ from Eq. (10.20) and Eq. (10.22) for the gain $\mathbb{K}_k$, then we can rewrite Eq. (10.23) as

$$
\mathbb{P}_k = (\mathbb{I} - \mathbb{K}_k \mathbb{H}_k)\mathbb{P}_k^- . \tag{10.24}
$$

It remains to evaluate $\mathbb{P}_k^-$ in terms of the earlier $\mathbb{P}_{k-1}$ so as to close the loop on a recursion between the various terms. Accordingly, write

$$
\begin{aligned}
\mathbf{e}_k^- &= \mathbf{X}_k - \hat{\mathbf{X}}_k^- = \mathbf{X}_k - \Phi_{k-1}\hat{\mathbf{X}}_{k-1} \\
&= \Phi_{k-1}\mathbf{X}_{k-1} + \mathbf{W}_{k-1} - \Phi_{k-1}\hat{\mathbf{X}}_{k-1} = \Phi_{k-1}\mathbf{e}_{k-1} + \mathbf{W}_{k-1} .
\end{aligned}
$$

Now note that $\mathbf{W}_{k-1}$ is orthogonal to $\mathbf{e}_{k-1}$ to conclude that

$$\mathbb{P}_k^- = \Phi_{k-1}\mathbb{P}_{k-1}\Phi_{k-1}^T + \mathbb{Q}_{k-1}. \tag{10.25}$$

Given the initial condition $\mathbb{P}_0$, we can determine $\mathbb{P}_1^-$ from Eq. (10.25). From Eq. (10.22), we can obtain $\mathbb{K}_1$ from $\mathbb{P}_1^-$. Equation (10.24) then enables us to determine $\mathbb{P}_1$ from both $\mathbb{K}_1$ and $\mathbb{P}_1^-$. This cycle can now be iterated to yield the succession of triples $(\mathbb{P}_k^-, \mathbb{K}_k, \mathbb{P}_k)$. We will derive a mathematically equivalent solution in Section 15.6 through the use of conditional expectation.

Example 10.2   Kalman Filtering

Consider a two-dimensional state vector ($n = 2$) $\mathbf{X}$ with time-invariant state transition matrix

$$\Phi = \frac{1}{4}\begin{pmatrix} 3 & 0 \\ 1 & 2 \end{pmatrix}.$$

Assume that the noise or forcing terms $\{\mathbf{W}_k\}$ are unlinked and normally distributed and that they satisfy the formula

$$E\mathbf{W}_k = \mathbf{0}, \quad E\mathbf{W}_k\mathbf{W}_k^T = \mathbb{Q}_k = \begin{pmatrix} 1 & .5 \\ .5 & 2 \end{pmatrix}.$$

The initial condition $\mathbf{X}_0$ is assumed to satisfy

$$E\mathbf{X}_0 = \mathbf{0}, \quad \mathbb{P}_0 = E\mathbf{X}_0\mathbf{X}_0^T = \mathbb{I} = \begin{pmatrix} 1 & 0 \\ 0 & 1 \end{pmatrix}.$$

We observe scalar Z_k ($l = 1$) given by

$$Z_k = \begin{pmatrix} 1 & 1 \end{pmatrix}\mathbf{X}_k + V_k,$$

where the measurement noises $\{V_k\}$ are unlinked and normally distributed; that is,

$$EV_k = 0, \quad \mathbb{R}_k = \text{VAR}(V_k) = 1.$$

A plot of the behavior of $\{\mathbf{X}_k\}$ is shown in Figure 10.1. On the left, we display the second coordinate $\mathbf{X}_k(2)$ on the y-axis and the first coordinate $\mathbf{X}_k(1)$ on the x-axis. On the right, we show both of the time series $\mathbf{X}_k(1)$ (solid) and $\mathbf{X}_k(2)$ (dotted) plotted against time k. Figure 10.2 displays the observation sequence Z_k plotted against time.

We demonstrate the Kalman recursive calculation by carrying out the first cycle starting with $\hat{\mathbf{X}}_0, \mathbb{P}_0$ and calculating, in order, $\mathbb{P}_1^-, \mathbb{K}_1, \hat{\mathbf{X}}_1$, and $\mathbb{P}_1$. We will then have what we need to carry out the second cycle. Our initial estimate is $\hat{\mathbf{X}}_0 = \mathbf{0}$ and $\mathbb{P}_0 = \mathbb{I}$.

From Eq. (10.25),

$$\mathbb{P}_1^- = \Phi_0\mathbb{P}_0\Phi_0^T + \mathbb{Q}_0 = \begin{pmatrix} 1.5625 & .6875 \\ .6875 & 2.3125 \end{pmatrix}.$$

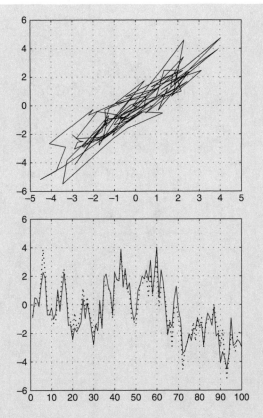

Figure 10.1 Gaussian vector markov process **X**(2) *vs.* **X**(1) and state components **X**(1) and **X**(2) (dotted).

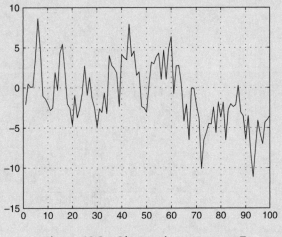

Figure 10.2 Observation sequence Z_k.

From Eq. (10.22),

$$\mathbb{K}_1 = \mathbb{P}_1^- \mathbb{H}_1^T (\mathbb{H}_1 \mathbb{P}_1^- \mathbb{H}_1^T + \mathbb{R}_1)^{-1}$$

$$= \begin{pmatrix} 1.5625 & .6875 \\ .6875 & 2.3125 \end{pmatrix} \begin{pmatrix} 1 \\ 1 \end{pmatrix} \left[(1 \quad 1) \begin{pmatrix} 1.5625 & .6875 \\ .6875 & 2.3125 \end{pmatrix} \begin{pmatrix} 1 \\ 1 \end{pmatrix} + 1 \right]^{-1} = \begin{pmatrix} .36 \\ .48 \end{pmatrix}.$$

From Eq. (10.18),

$$\hat{\mathbf{X}}_1 = \Phi_0 \hat{\mathbf{X}}_0 + \mathbb{K}_1 (\mathbf{Z}_1 - \mathbb{H}_1 \Phi_0 \hat{\mathbf{X}}_0)$$

$$= \frac{1}{4} \begin{pmatrix} 3 & 0 \\ 1 & 2 \end{pmatrix} \mathbf{0} + \begin{pmatrix} .36 \\ .48 \end{pmatrix} \left[\mathbf{Z}_1 - (1 \quad 1) \frac{1}{4} \begin{pmatrix} 3 & 0 \\ 1 & 2 \end{pmatrix} \mathbf{0} \right] = \begin{pmatrix} .36 \\ .48 \end{pmatrix} \mathbf{Z}_1.$$

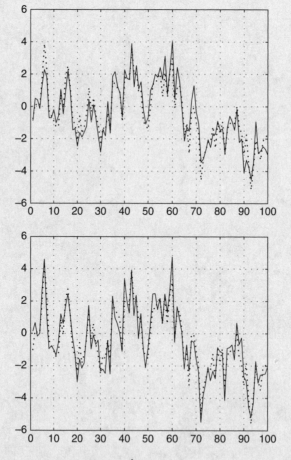

Figure 10.3 State $\mathbf{X}(1)$ and estimate $\hat{\mathbf{X}}(1)$ (dotted) and state $\mathbf{X}(2)$ and estimate $\hat{\mathbf{X}}(2)$ (dotted).

Finally, from Eq. (10.24),

$$\mathbb{P}_1 = (\mathbb{I} - \mathbb{K}_1\mathbb{H}_1)\mathbb{P}_1^- = \left[\begin{pmatrix} 1 & 0 \\ 0 & 1 \end{pmatrix} - \begin{pmatrix} .36 \\ .48 \end{pmatrix}(1 \quad 1)\right]\begin{pmatrix} 1.5625 & .6875 \\ .6875 & 2.3125 \end{pmatrix}$$

$$= \begin{pmatrix} .7525 & -.3925 \\ -.3925 & .8725 \end{pmatrix}.$$

The performance of the Kalman filter is shown in Figure 10.3 by plotting a state component (solid line) and its Kalman estimate (dotted line).

10.7 SUMMARY

We provided a brief background to the subject of deterministic linear systems, a subject that is well studied in every electrical engineering program. In support of goal G2, we took advantage of the introduction of moments in Chapter 9 to treat the response of linear systems to random inputs by studying the moments of such responses. Determining those moments does not require knowing a full specification of the input process. We then turned to our goal G3 and developed two forms of linear least mean square estimation. The first form is known as Wiener filtering; it was historically the first approach, and it applies to arbitrary sequences of random variables. The second form, known as Kalman filtering, applies to a wide class of random sequences, albeit not to all of them. However, Kalman filtering has the advantage of providing an iterative-in-time, computationally feasible solution to many problems of engineering importance, and it has a large literature reflecting its many applications. We will complete our discussion of Kalman filtering in Section 15.6.

EXERCISES

E10.1   a. A discrete-time (integer-valued) linear system is described by the impulse function

$$h(t, \tau) = \begin{cases} t^2 & \text{if } |t - \tau| \le 1.5 \\ 0 & \text{if otherwise} \end{cases}.$$

       If the system input $x_\tau = \beta^\tau$, what is the output y_t?

       b. Is the system of part (a) time varying?

       c. Is the system of part (a) causal?

       d. Is the system of part (a) stable?

E10.2   Examine the dynamics of the state-determined linear system given by

$$\mathbf{X}_0 = \begin{pmatrix} 0 \\ 0 \end{pmatrix}, \quad \mathbf{X}_{k+1} = \Phi\mathbf{X}_k + \mathbb{B}\mathbf{U}_k,$$

for

$$\mathbf{e}_1 = \frac{1}{\sqrt{2}}\begin{pmatrix} 1 \\ -1 \end{pmatrix}, \mathbf{e}_2 = \frac{1}{\sqrt{2}}\begin{pmatrix} 1 \\ 1 \end{pmatrix}, \Phi = \lambda\mathbf{e}_1\mathbf{e}_1^T + 0.5\mathbf{e}_2\mathbf{e}_2^T.$$

Using Matlab, select $\mathbb{B} = \text{randn}(2,2)$, but consider $\mathbb{B}$ to be fixed and nonrandom, and select $\mathbf{U}_k = \text{randn}(2,1)$.

    a. For each of the values $\lambda = -1.5 : 0.5 : 1.5$, use Matlab to plot both components, $\mathbf{X}_k(1)$ and $\mathbf{X}_k(2)$, for $k = 1 : 100$.

    b. If the observation $\mathbf{Z}_k$ of the state $\mathbf{X}_k$ is

$$\mathbf{Z}_k = \mathbb{H}\mathbf{X}_k + \mathbf{V}_k, \quad \mathbb{H} = \begin{pmatrix} 1 & -1 \\ 1 & 1 \end{pmatrix}, \quad \mathbf{V}_k = \text{randn}(2,1),$$

then use Matlab to plot $\mathbf{Z}_k(1)$ and $\mathbf{Z}_k(2)$ for $k = 1 : 100$.

E10.3 A dynamical system of the form given in the preceding problem has

$$E\mathbf{X}_0 = \mathbf{0}, \ \text{COV}(\mathbf{X}_0, \mathbf{X}_0) = \mathbb{C}_0 = \mathbb{I},$$

$$E\mathbf{U}_k = \mathbf{0}, \ \text{COV}(\mathbf{U}_k, \mathbf{U}_j) = \begin{cases} \mathbf{O} & \text{if } j \neq k \\ \sigma^2 \mathbb{I} & \text{if } j = k \end{cases}.$$

    a. Evaluate $E\mathbf{X}_k$.

    b. Evaluate $\mathbb{C}_k = \text{COV}(\mathbf{X}_k, \mathbf{X}_k)$ in terms of $\mathbb{C}_{k-1}$.

    c. Assuming that $E\mathbf{V}_k = \mathbf{0}$, evaluate $E\mathbf{Z}_k$.

    d. Assuming that $\mathbf{V}_k$ is uncorrelated with $\mathbf{X}_k$, and $\text{COV}(\mathbf{V}_k) = R$, evaluate $E\mathbf{Z}_k\mathbf{Z}_k^T$ and $\text{COV}(\mathbf{Z}_k, \mathbf{Z}_k)$.

E10.4   a. If the linear system

$$Y_n = \sum_{k=0}^{n} \beta^{n-k} X_k, \quad \{X_k\} \text{ pairwise uncorrelated}, \ X_k \sim \mathcal{N}(0, \sigma^2),$$

evaluate EY_n.

    b. For this linear system, evaluate $\text{COV}(Y_n, Y_m), n > m > 0$.

E10.5   a. If the linear system

$$Y_n = \sum_{k=-\infty}^{n} \beta^{n-k} X_k,$$

and we partially describe the input process by

$$EX_k = m, \ E(X_i X_j) = R_X(i, j) = e^{-|i-j|},$$

then evaluate EY_n.

    b. Provide an expression for the correlation function $R_Y(n, m)$ for the system output.

E10.6 For $i = 1, 2$, we observe $X_i = S + N_i$, where $S \sim \mathcal{N}(m, 1), N_i \sim \mathcal{N}(0, \sigma_i^2)$, and $\text{COV}(S, N_i) = \text{COV}(N_1, N_2) = 0$.

    a. Evaluate $\text{COV}(S, X_i)$.

    b. Evaluate the covariance matrix $\mathbb{C}_X = \text{COV}(\mathbf{X}, \mathbf{X})$.

    c. We wish to determine the minimum mean squared error linear estimator $\hat{S}(\mathbf{X})$ of S in the form $\hat{S}(\mathbf{X}) = \mathbf{a}^T(\mathbf{X} - E\mathbf{X}) + b$. Evaluate the optimum b.

    d. Evaluate the optimum $\mathbf{a}$ in $\hat{S}$.

E10.7   We observe $X_i = S + N_i$ and wish to estimate S from $\{X_i\}$. If $\{N_i\}$ are mean-zero, unit-variance, uncorrelated measurement noises that are also uncorrelated with the signal S, then determine the best estimator (minimum mean square) of the form $\hat{S}_n = \alpha \sum_{i=1}^{n} X_i + \beta$.

E10.8   We know that $\mathbf{X} \sim \mathcal{N}(\mathbf{0}, \mathbb{C}_\mathbf{X})$ with

$$\mathbb{C}_\mathbf{X} = \begin{pmatrix} 1 & 0 \\ 0 & 2 \end{pmatrix}.$$

a. If

$$Y = X_1^3 + X_2,$$

   evaluate EY and $\text{VAR}(Y)$.

b. Evaluate $\text{COV}(Y, \mathbf{X})$.

c. Determine the minimum mean square error estimator $\hat{Y}$ of Y when $\hat{Y} = \mathbf{a}^T \mathbf{X} + b$.

d. Evaluate the performance of $\hat{Y}$.

E10.9   We observe $\mathbf{X}$ and wish to estimate $\mathbf{Y}$ when we know that

$$E\mathbf{Y}^T = (-1\ 0), \ E\mathbf{X}^T = (0\ 0\ 1),$$

$$\mathbb{C}_\mathbf{X} = \begin{pmatrix} 5 & 6 & 0 \\ 6 & 14 & 3 \\ 0 & 3 & 8 \end{pmatrix} \text{ and } \mathbb{C}_\mathbf{Y} = \begin{pmatrix} 2 & 1 \\ 1 & 3 \end{pmatrix}.$$

a. Determine the minimum mean square error estimator $\hat{\mathbf{Y}}(\mathbf{X})$ of $\mathbf{Y}$ that is of the form $\mathbb{A}\mathbf{X} + \mathbf{b}$.

b. Determine the MSE $E[(\hat{\mathbf{Y}} - \mathbf{Y})^T(\hat{\mathbf{Y}} - \mathbf{Y})]$.

E10.10   A process $\{X_k\}$ satisfies the recursion

$$X_{k+1} = aX_k + W_k$$

with the initial conditions $EX_0 = 0$, $\text{VAR}(X_0) = \sigma_0^2$. The sequence of random variables $\{W_k\}$ satisfies

$$EW_k = 0, \ EW_k W_j = \sigma_w^2 \delta_{kj}.$$

We have available observations

$$Z_k = hX_k + V_k,$$

where

$$EV_k = 0, \ EV_k V_j = \sigma_v^2 \delta_{kj}, \ EV_k W_j = 0.$$

a. Determine the Kalman linear least mean square estimator $\hat{X}_k$ of X_k based upon $\mathbf{Z}_k = (Z_1, \ldots, Z_k)^T$.

b. Evaluate the error variance $P_k = E(X_k - \hat{X}_k)^2$.

Part III

Conditional Probability and Its Applications

11

Discrete Conditional Probability

11.1 PURPOSE, BACKGROUND, AND ORGANIZATION
11.2 CONDITIONAL PROBABILITY
11.3 PROPERTIES OF DISCRETE CONDITIONAL PROBABILITY
11.4 MULTIPLICATION/SEQUENCE THEOREM
11.5 TOTAL PROBABILITY THEOREM AND ITS APPLICATIONS
11.6 INVERTING CAUSE AND EFFECT: BAYES' THEOREM AND ITS APPLICATIONS
11.7 THREE PROBLEMS THAT CHALLENGE OUR INTUITION
11.8 SUMMARY
11.9 APPENDIX: DISCRETE CONDITIONAL PROBABILITY SUMMARY
EXERCISES

11.1 PURPOSE, BACKGROUND, AND ORGANIZATION

In this chapter, we organize our presentation to respond to our three goals and to account for the case in which the random variables are discrete (D). The chapter extends the introduction to conditional probability provided in Section 2.9 in the context of classical probability. The further extension to the important cases of continuous and mixed random variables is treated in Chapter 12. In an electrical engineering context, it is helpful to think of the basic system $Y = S(X)$ consisting of an input signal that is a random variable X, a (random) system S, and an output signal that is a random variable Y. Goal 1 has us develop mathematical models for random phenomena, and we have done so in Chapters 3, 4, 5, 6, and 7 for individual quantities like X, Y. We addressed Goal 2 in Chapter 8, but only for a nonrandom, deterministic system in which there is a unique response or output Y to a given excitation or input X. However, to describe a *random system* S, we must take into account the fact that the response or output Y depends in a probabilistic manner upon the excitation or input X. Hence, a description of a system must tell us what the probabilities of various outputs will be, given any input. Such a description requires the notion of conditional probability developed in this and the next chapters. The mathematical model constructed for S can then be combined with, say, the probability description for X to find the probability description for the response Y and thereby achieve Goal 2. This will be accomplished by the Total Probability Theorem of Section 11.5. Finally, we will explore aspects of Goal 3 when we discuss inference from the observed response Y (perhaps the received signal at the output of communications channel or the result of a measurement made on X using the noisy instrument S) to the excitation X (perhaps the transmitted signal or the true value of the quantity being measured) through use of Bayes' Theorem in Section 11.6. For ease of reference and to promote an understanding of the pattern of exposition, the results to be developed are summarized in Section 11.9.

We examine conditioning on an event B of positive probability. This immediately treats the case of discrete random variables. In Chapter 12, we examine the case of B having probability 0 as a limiting case of conditioning on suitable events of positive probability. This enables us to treat conditioning on continuous random variables. The expository pattern of this chapter will be the model for that of Chapter 12.

11.2 CONDITIONAL PROBABILITY

11.2.1 Motivation and Definition

Probabilities, on the frequentist, propensity, or belief interpretations (see Section 3.3), are based upon information or knowledge. Revising the information base should lead to revisions in the probabilities. While the information base for probability depends upon which of the many interpretations of probability is adopted, in all interpretations it is clear that knowledge of the occurrence of an event has an influence on the probabilities of the occurrences of the remaining events. To clarify this, we return to our motivating connection to relative frequencies of occurrence. Our setup is that of a collection of identical random experiments $\mathcal{E}_1, \ldots, \mathcal{E}_n$ that were performed in an "unlinked" fashion. If ω_i is the outcome of $\mathcal{E}_i$, then the relative frequency of occurrence of event A is

$$r_n(A) = \frac{1}{n} \sum_{i=1}^{n} I_A(\omega_i) = \frac{1}{n} N_n(A).$$

On our propensity view, the propensity $P(A)$ is displayed in such circumstances by the relationship $r_n(A) \approx P(A)$, for very large n. If we are now informed that event B has occurred, how are we to revise $P(A)$?

The revised probability is denoted "$P(A|B)$" and is read "the conditional probability of A given B."

Knowledge that B occurred directs us to consider the subsequence $\{\omega_i'\}$ of experimental outcomes in which $\omega_i' \in B$. This subsequence has length $N_n(B) \leq n$ that must be assumed to be nonzero (i.e., the event B should have been observed to occur at least once). The revised, or conditional, relative frequency is then

$$r_n(A|B) = \frac{N_n(A \cap B)}{N_n(B)} = \frac{N_n(A \cap B)/n}{N_n(B)/n};$$

the relative frequency of occurrences of A in the subsequence where B occurs is the ratio of the number of occurrences of both to the subsequence length. In the appropriate limit of large n, $r_n(A|B)$ approaches $P(A \cap B)/P(B)$.

We also came to this conclusion in Section 2.9, starting from the classical notion of probability in which

$$P(A) = \frac{||A||}{||\Omega||}$$

was revised to

$$P(A|B) = \frac{||A \cap B||}{||B||}.$$

The preceding reflections lead us to the next definition.

Definition 11.2.1   (Conditional Probability of Events)   If $P(B) > 0$, then the conditional probability of event A given event B is

$$P(A|B) = \frac{P(A \cap B)}{P(B)}.$$

If $P(B) = 0$, then the defining ratio becomes the indeterminate form $0/0$.

Example 11.1   Random Sampling

Consider a small silicon device of n atoms in which an unknown number N_0 of dopant atoms have been implanted. We wish to infer the number N_0 of dopant atoms in the device. In order to do so, we randomly select an even smaller volume of this device containing only s atoms and then explicitly count the number S_0 of dopant atoms in this smaller volume. Let A be the event that $S_0 = s_0$ and B be the event that $N_0 = n_0$. The number of arrangements of n_0 indistinguishable dopant atoms among the n indistinguishable silicon atoms is $\binom{n}{n_0}$, and this is the size of B. The number of arrangements in which s_0 dopant atoms are encountered in the s atoms that were selected when there were n_0 dopant atoms in the device is $\binom{s}{s_0}\binom{n-s}{n_0-s_0}$, and this is the size of $A \cap B$. Hence, under random sampling,

$$P(S_0 = s_0 | N_0 = n_0) = \frac{\binom{s}{s_0}\binom{n-s}{n_0-s_0}}{\binom{n}{n_0}}.$$

Notice that this probability is positive only for

$$s - (n - n_0) \leq s_0 \leq s.$$

11.2.2 Derivation of the Exponential Distribution

One application of the definition of conditional probability is a derivation of the exponential probability law $\mathcal{E}(\alpha)$ that was introduced in Section 6.6.1 and that is commonly used to describe the lifetimes of components that fail catastrophically or without aging (e.g., a glass windowpane can last essentially unchanged for centuries, but breaks when knocked sufficiently hard, with its fragility essentially independent of its age). We now use conditional probability to capture this notion precisely and thereby understand when the exponential law is applicable. We assert that the probability that the lifetime L will last at least an additional τ time units, given that it has already lasted t time units, does not depend upon t. That is, there is no aging effect that makes the item less likely to endure an additional τ time units, given that it has already survived some lengthy t time units. Formally,

$$(\forall t > 0, \tau > 0) \; P(L > t + \tau | L > t) = P(L > \tau).$$

To compare with the previous notation, adopt the correspondence between events of

$$A = \{L : L > t + \tau\} \subset B = \{L : L > t\}.$$

Rewritten, to eliminate conditioning, this claim becomes

$$(\forall t > 0, \tau > 0) \; P(L > t + \tau) = P(L > t)P(L > \tau),$$

where we have used the fact that the event $(L > t + \tau) \cap (L > t) = (L > t + \tau)$. Let $g(x) = \log P(L > x)$, as is legitimate for positive $P(L > x)$, to restate this conclusion as the identity

$$(\forall t > 0, \tau > 0) \; g(t + \tau) = g(t) + g(\tau).$$

This identity is known as the *Cauchy equation*. (See Aczel [1].) An evident solution for g is $g(x) = \beta x$. Other solutions exist, but they are highly pathological in that they are both unbounded in any finite interval and nonmeasurable. The only regular solution is the linear one, and it is the only possibility for $\log P$. However, if $g(x) = \beta x$, then $P(L > t) = e^{\beta t}$. The requirement that $P(L > t > 0) \leq 1$ implies that $\beta \leq 0$ or that

$$P(L > t) = e^{-\alpha t}, \text{ or } L \sim \mathcal{E}(\alpha).$$

We observe that the exponential distribution is the only family that can describe nonnegative real-valued lifetimes of systems that fail without aging.

11.2.3 Specialization to Discrete Random Variables

The application to discrete random variables X, Y comes from the identification of event A with, say, outcome $Y = y_j$ and event B with $X = x_i$. Hence,

$$P(Y = y_j | X = x_i) = \frac{P(X = x_i, Y = y_j)}{P(X = x_i)}.$$

One can rewrite these results in terms of probability mass functions. Defining joint and conditional pmfs

$$p_{X,Y}(x_i, y_j) = P(X = x_i, Y = y_j), \; p_{Y|X}(y_j | x_i) = P(Y = y_j | X = x_i),$$

we see that

$$p_X(x_i) = P(X = x_i) = \sum_j p_{X,Y}(x_i, y_j),$$

$$p_{Y|X}(y_j|x_i) = P(Y = y_j|X = x_i) = \frac{p_{X,Y}(x_i, y_j)}{p_X(x_i)}.$$

The conditional probability mass function $p_{Y|X}(y_j|x_i)$ provides a description for the random system $\mathcal{S}$ that transforms an input random variable X into an output random variable Y. A conditional cdf is

$$F_{Y|X}(y|x) = P(Y \leq y|X = x) = \sum_{y' \leq y} p_{Y|X}(y'|x).$$

Example 11.2   Digital Communications Channel _____

A *digital communications channel* $\mathcal{S}$ is a noisy channel described by conditional probabilities of system outputs given the system inputs. For example, $\mathcal{S}$ is a *binary symmetric channel (BSC)* when $X, Y \in \{0, 1\}$ and

$$p_{Y|X}(1|0) = P(Y = 1|X = 0) = P(Y = 0|X = 1) = p_{Y|X}(0|1),$$

$$p_{Y|X}(0|0) = P(Y = 0|X = 0) = P(Y = 1|X = 1) = p_{Y|X}(1|1).$$

We often abbreviate the error probabilities $P(Y = 1|X = 0), P(Y = 0|X = 1)$ by p and the probabilities of correct transmission $P(Y = 0|X = 0), P(Y = 1|X = 1)$ by $1 - p$.

Example 11.3   Pattern Class Features _____

Think of printed handwriting in which, each time we write, say, a lowercase "a," it comes out differently, or each time we pronounce a syllable, it comes out differently. We are thinking of a system $\mathcal{S}$ which transforms an input that is a member of a defined class of patterns into an output that is a specific realization of a member of the defined class. Let X represent a selected pattern class member (e.g., a specific alphanumeric character) and Y represent a feature measurement (e.g., a count of the number of dark pixels in the image or the energy in certain short-time speech spectrogram bands) that measures or describes the actual selected member of the pattern class. Say, X selects an optical signal source of intensity λ_X corresponding to some element of a pattern class. We then observe a photon count Y over some time interval T that satisfies

$$p_{Y|X}(y|x) = P(Y = y|X = x) = e^{-\lambda_x T} \frac{(\lambda_x T)^y}{y!}.$$

Example 11.4   Noisy Counter _____

X represents the actual count of, say, emissions of photons, electrons, or alpha particles over a given period, and Y is the measured or reported value. The measurement system $\mathcal{S}$, with input

X and output Y, is imperfect and occasionally drops or adds a single count. Hence, assuming that $\mathcal{S}$ is sufficiently well designed not to produce negative counts, we might model it by some p_c of correct count that does not depend upon the true value of X, for example,

$$p_{Y|X}(y|x) = P(Y = x|X = x) = p_c,$$

with errors occurring with probabilities

$$P(Y = 1|X = 0) = 1 - p_c,$$

and for $x > 0$,

$$P(Y = x + 1|X = x) = P(Y = x - 1|X = x) = \frac{1}{2}(1 - p_c).$$

11.3 PROPERTIES OF DISCRETE CONDITIONAL PROBABILITY

The results of this section repeat those of Section 2.9, where they were first established for the limited case of classical probability. The basic result will be that $P(\cdot|B)$; that is to say, $P(A|B)$, considered as a function of events A given a fixed event B, is a probability measure for the events A satisfying the Kolmogorov axioms. To verify this claim, first note that it is immediate from the definition of conditional probability that

$$P(A|B) \geq 0, \ P(\Omega|B) = 1, \ P(B|B) = 1, \ P(A|B) = P(A \cap B|B),$$

$$A \supset B \Rightarrow P(A|B) = 1.$$

Furthermore, if $A \perp C$ (disjoint), then

$$P(A \cup C|B) = \frac{P((A \cap B) \cup (C \cap B))}{P(B)} = \frac{P(A \cap B) + P(C \cap B)}{P(B)}$$

$$= P(A|B) + P(C|B).$$

Hence, $P(\cdot|B)$, thought of as a function of the "A" variable given a fixed event B, satisfies the Kolmogorov axioms K0–K3. Since K4 is also true, we have the following theorem.

Theorem 11.3.1   (Conditional Probability Is a Measure)   For any fixed event B with $P(B) > 0$, the function $P(\cdot|B)$ is a probability measure.

Thus, if we have a conditional pmf $p_{Y|X}$, then

$$P(Y \in A|X = x) = \sum_{y \in A} p_{Y|X}(y|x).$$

However, conditional probabilities are not probability measures when considered as functions $P(A|\cdot)$ of the "B" variable.

11.4 MULTIPLICATION/SEQUENCE THEOREM

We extend the use of conditional probability from evaluating

$$P(A \cap B) = P(B)P(A|B)$$

to evaluating the more complex $P(\cap_{i=1}^{n} A_i)$ (say, A_i is the ith word in a sentence or the order of cards in a deck after the ith shuffle of the deck) through the multiplication or sequence theorem.

   Theorem 11.4.1   (Multiplication or Sequence for Events)   Given a sequence of events $\{A_1, \ldots, A_n\}$,

$$P(\cap_{i=1}^{n} A_i) = P(A_1) \prod_{i=2}^{n} P(A_i \,|\, \cap_{j=1}^{i-1} A_j).$$

Thus,

$$P(A \cap B \cap C) = P(A)P(B|A)P(C|A \cap B).$$

   Proof. The proof is by induction on n. The claim is clearly correct for $n = 2$, by the definition of conditional probability. The induction hypothesis is to assume that it is true for all $k \leq n$ and then consider the next case,

$$P(\cap_{i=1}^{n+1} A_i) = P(A_{n+1} \cap B) \text{ for } B = \cap_{i=1}^{n} A_i.$$

By applying the $n = 2$ case, we see that

$$P(\cap_{i=1}^{n+1} A_i) = P(A_{n+1}|B)P(B).$$

Invoking the induction hypothesis, we have

$$P(B) = P(A_1) \prod_{i=2}^{n} P(A_i \,|\, \cap_{j<i} A_j).$$

Assembling results then yields the desired proof.   □

   We can use the Multiplication Theorem to analyze the probability of a sequence of events $\{A_i\}$ in terms of a product of perhaps simpler conditional probabilities.

   For example, suppose a device is placed into operation on day 1. The conditional probability that it will operate on day $k + 1$, given that it has been operating on each of the preceding days, is known to be p_{k+1}. What is the probability of the event B that it will operate for at least five consecutive days? Let A_i represent the event that the device is operating on day i. We have then been informed that $P(A_1) = 1$ and, for any $k \geq 1$,

$$P(A_{k+1}|\cap_{j \leq k} A_j) = p_{k+1}.$$

The event B of continuous operation for at least five days is $\cap_1^5 A_i$. Hence,

$$P(B) = P(\cap_{i=1}^5 A_i) = P(A_1) \prod_{i=2}^5 P(A_i \mid \cap_{j<i} A_j) = \prod_{i=2}^5 p_i.$$

What is the probability of the event C that it will operate for exactly five consecutive days? In this case, we have continuous operation for five days, as before, followed by failure on the sixth day. Hence, $C = A_6^c \cap (\cap_{i=1}^5 A_i)$, and

$$P(A_6^c \mid \cap_{i=1}^5 A_i) = 1 - P(A_6 \mid \cap_{i=1}^5 A_i) = 1 - p_6.$$

Thus, we find that

$$P(C) = (1 - p_6)P(B).$$

This technique is particularly useful if the sequence of events $\{A_i\}$ is a *causal chain* in which event A_i is caused (produced solely) by its temporal predecessor A_{i-1}. Such a sequence of events is called a *Markov process*, and its key property is given in the next definition.

Definition 11.4.1   (Markov Dependence)   The events $\{A_1, \ldots, A_n\}$ have Markov dependence if

$$(\forall i > 1) \; P(A_i \mid \cap_{j=1}^{i-1} A_j) = P(A_i \mid A_{i-1}).$$

When events in a particular enumeration have Markov dependence, we have the simplification of Theorem 11.4.1 given by

$$P(\cap_{i=1}^n A_i) = P(A_1) \prod_{i=2}^n P(A_i \mid A_{i-1}).$$

Example 11.5   Common Markov Processes

An example of a Markov process is provided by a cascade connection of possibly random electrical systems $\mathcal{S}_1, \ldots \mathcal{S}_n$. The response A_i of $\mathcal{S}_i$ depends only upon the excitation A_{i-1} provided by the response of system $\mathcal{S}_{i-1}$, and not upon the responses of earlier systems in the cascade. A gaming example is provided by the successive orderings $\{A_i\}$ of a deck of cards in repeat shuffles; the next ordering A_{i+1} depends only upon the current ordering A_i and the shuffling mechanism. A (too) simple model of natural language text production is to assume that the probability of the ith word w_i in a statement depends upon the earlier words $w_1, \ldots, w_{i-1}$ only through the most recent word w_{i-1}. This model can be made more accurate by considering word pairs, but then one is faced with assessing the very many conditional probabilities of one word pair given another word pair. As even a limited vocabulary will have 1,000 words, we now have 10^{12} probabilities to determine from samples of text in the language! Shannon [82], in his classic 1948 paper introducing information theory, provides examples of the generation of artificial text using this idea. An example of Markovian sentence production using a vocabulary of only seven words and a single punctuation mark (the period) can be found in Example 20.13.

In physical processes, if we can identify a system state, then we have a Markov process in the evolving sequence of system states. (For example, in a circuit composed of resistors, inductors, and capacitors, the system state is the inductor currents and capacitor voltages; in a mechanical particle system like a gas, unacted upon by external forces, the state would be the present position and momentum of each of the particles, or, more generally, the quantum mechanical wave function for a physical system.)

11.5 TOTAL PROBABILITY THEOREM AND ITS APPLICATIONS

Goal 2 has us treat the problem of the response of a possibly random system with random excitation. Unlike the case of deterministic systems, a description of an excitation or a system is given by providing the probabilities of events. Hence, the excitation, input, or source is characterized by specifying $P(B)$ for events B determined by the signal source. The *stochastic system* $\mathcal{S}$ (e.g., a communications channel transmitting inputs from a message source that are then received as outputs or an audio amplifier accepting as input a weak signal from a CD player and transforming it into a strong signal that can drive loudspeakers as its output) accepts the input event B and transforms it into an output event A. $\mathcal{S}$ is described or specified by providing $P(A|B)$ for all pairs of events B as before and events A determined by the system output. Achieving Goal 2 requires us to specify the response by calculating $P(A)$. Stated differently, we wish to assess the probability of an output, response, or "effect" A in terms of the probabilities of possible inputs, excitations, or "causes" $\{B_i\}$ and knowledge of the probabilistic *cause–effect mechanism*. For example, A could be the event of a set of possible outputs of a stochastic system (e.g., alphanumeric characters or ASCII symbols) and $\{B_i\}$ a list of the possible inputs (e.g., transmitted bytes generated by a keyboard) to the system. The probability of the effect A given the cause B_i is $P(A|B_i)$. Assuming that the list of causes $\{B_i\}$ is complete and without duplication in that one, and only one, cause can operate (e.g., all 256 possible bytes), the events $\{B_i\}$ form a partition Π (see Definition 1.6.3) of the sample space Ω.

Theorem 11.5.1   (Total Probability for Events)   If $\Pi = \{B_i\}$ is a partition of Ω, then

$$(\forall A \in \mathcal{A})\quad P(A) = \sum_i P(A|B_i)P(B_i),$$

where it is understood that $P(A|B_i)P(B_i) = 0$ when $P(B_i) = 0$.

Proof. If $\{B_i\}$ is a partition of Ω, then

$$A = A \cap \Omega = \cup_i(A \cap B_i),\ \text{ and }\ (\forall i \neq j)A \cap B_i \perp A \cap B_j.$$

Hence, from the (countable) additivity of P for disjoint sets,

$$P(A) = \sum_i P(A \cap B_i) = \sum_i P(A|B_i)P(B_i). \qquad \square$$

The Total Probability Theorem enables us to calculate the response (probabilities of outputs) of a stochastic system (characterized by a conditional probability function for outputs given inputs) driven by a stochastic signal (characterized by probabilities of inputs).

Example 11.6   Urn Model Output

An *urn model* consists of n urns $u_1, \ldots, u_n$, each urn having possibly different numbers of balls of different colors, and a method of choosing an urn and then a ball from that urn. For definiteness, let us assume that urn u_i has r_i red balls and g_i green balls, for a total of n_i balls. A random mechanism first selects an urn $U = u_i$ with probability p_i and then draws a ball B at random from the selected urn. The probability $P(B = r | U = u_i)$ of drawing a red ball from urn u_i is given, on the classical probability account, by r_i/n_i. We can use total probability to determine the overall probability $P(B = r)$ of drawing a red ball through

$$P(B = r) = \sum_{i=1}^{n} \frac{r_i}{n_i} p_i.$$

While literal urn models are of little interest to us, they are a simplified way of discussing two-stage compound random mechanisms (e.g., delays in communication networks due to the arrival of random numbers of customers, each requiring a random service time).

Example 11.7   Noisy Counter Output

The noisy counter was introduced in Example 11.4. Add the assumption that the true count $X \sim \mathcal{P}(\lambda)$. We calculate, for $y > 0$,

$$P(Y = y) = \sum_{x=0}^{\infty} P(Y = y | X = x) P(X = x)$$

$$= P(Y = y | X = y - 1) P(X = y - 1) + P(Y = y | X = y) P(X = y)$$

$$+ P(Y = y | X = y + 1) P(X = y + 1)$$

$$= \frac{1}{2}(1 - p_c) e^{-\lambda} \left(\frac{\lambda^{y-1}}{(y-1)!} + \frac{\lambda^{y+1}}{(y+1)!} \right) + p_c e^{-\lambda} \frac{\lambda^y}{y!}.$$

The case of $y = 0$ is a little simpler, in that $P(X = -1) = 0$ and there are only two terms in the sum and not three:

$$P(Y = 0) = P(Y = 0 | X = 0) P(X = 0) + P(Y = 0 | X = 1) P(X = 1)$$

$$= p_c e^{-\lambda} + (1 - p_c) \lambda e^{-\lambda}.$$

Digital communications are often binary at root. A binary-valued signal $X \in \{0, 1\}$ is transmitted through a communications channel $\mathcal{S}$ having output Y. The input X is viewed as a random variable with $\pi_i = P(X = i)$. The random/noisy channel $\mathcal{S}$ is characterized by specifying the conditional probabilities $P(Y = y | X = x)$. The two error probabilities are $P(Y = 1 | X = 0), P(Y = 0 | X = 1)$. Theorem 11.3.1 informs us that conditional probabilities are also probabilities in the variable that we are not conditioning upon. Hence, the corresponding probabilities of correct transmission follow by the rule for complementation as in $P(Y = 0|X = 0) = 1 - P(Y = 1|X = 0)$. In the free-space optical channel, it is more probable that a photon will be lost in transmission than that it will be reported present when it is absent. In this case $P(Y = 1|X = 0) < P(Y = 0|X = 1)$, assuming that "1" indicates the presence of a photon and "0" its absence. On the other hand, errors made over wired channels, with the two states $\{0, 1\}$ differing only in the polarity of the transmitted pulse, are likely to be equally probable. Such channels are modeled as *binary symmetric channels (BSC)*.

Example 11.8   Binary Symmetric Channel Output _____

Since conditional probabilities are also probabilities, it follows that, for a BSC ($x \in \{0, 1\}$),

$$P(Y = 1 - x|X = x) = p, \ P(Y = x|X = x) = 1 - p.$$

Letting $\pi_i = P(X = i)$, we find that the Total Probability Theorem enables us to describe the BSC response or output through

$$P(Y = 0) = P(Y = 0|X = 0)\pi_0 + P(Y = 0|X = 1)\pi_1 = (1 - p)\pi_0 + p(1 - \pi_0).$$

Of course, $P(Y = 1) = 1 - P(Y = 0)$.

Example 11.9   Binary Erasure Channel Output _____

In the *binary erasure channel* (BEC), $Y \in \{0, 1, e\}$, where we let e denote the event of erasure. Reception is either correct, $Y = X$, or an erasure $Y = e$. Hence,

$$P(Y = 1 - x|X = x) = 0, \ P(Y = x|X = x) = 1 - p, \ P(Y = e|X = x) = p.$$

A "0" is received either as a 0, with probability $1 - p$, or as an erasure e, with probability p. In this case, we conclude that

$$P(Y = e) = P(Y = e|X = 0)\pi_0 + P(Y = e|X = 1)\pi_1 = p,$$

$$P(Y = 1) = P(Y = 1|X = 0)\pi_0 + P(Y = 1|X = 1)\pi_1 = (1 - p)\pi_1,$$

$$P(Y = 0) = (1 - p)\pi_0.$$

11.6 INVERTING CAUSE AND EFFECT: BAYES' THEOREM AND ITS APPLICATIONS

11.6.1 Probabilistic Cause and Effect

Our Goal 3 addresses exposure to inference and decision-making uses of the probabilistic models that we learn to create in Goal 1 and to transform in Goal 2. In the language of "cause and effect," we observe an "effect" in the form of system response/output, and wish to infer back to the "causes" or excitations/inputs that could have given rise to the observed effect. Generally, there are many possible causes that are consistent with a given observed effect, because our systems are themselves random and not deterministic. Even if they are deterministic, they may be many-to-one functions. More specifically, in a communications system, the inputs are messages that are transmitted over a random channel that may distort the message and add noise (e.g., thermal noise, static) to it and then induce responses observed at a receiver. The function of the receiver is to infer or estimate the channel input or original message from the observed received signal at the channel output. In medical diagnostics (and other endeavors with reliability concerns), the effects are the observed symptoms and the causes are the diseases that could have given rise to the symptoms, taking into account the fact that not all patients respond in the same way to a given disease, nor need a given patient respond the same way to a given disease contracted at another time. As before, if we have countably many causes, then we can list them as a partition $\{B_i\}$ of the sample space, assuming that one, and only one, cause operates at a given time to produce an effect event A.

Theorem 11.6.1 (Bayes)   If (1) $\{B_i\}$ is a countable partition of Ω or of A, (2) $P(A) > 0$, and (3) we understand $P(A|B_k)P(B_k) = 0$ whenever $P(B_k) = 0$, then

$$P(B_i|A) = \frac{P(A|B_i)P(B_i)}{\sum_k P(A|B_k)P(B_k)}.$$

Proof. By the definition of conditional probability,

$$P(B_i|A) = \frac{P(A \cap B_i)}{P(A)} = \frac{P(A|B_i)P(B_i)}{P(A)}.$$

By the just-established Total Probability Theorem,

$$P(A) = \sum_k P(A|B_k)P(B_k).$$

Assembling these two results yields the theorem.   ☐

We can now assess the probability of a "cause" B_i given an "effect" A and infer the probabilities of inputs to a stochastic system from observing the outputs. This simple result has many other applications, including the recovery of signals in noise and the determination of the likely causes of the failure of a system. Many of its applications will require that it be generalized in Chapter 12 to conditioning on causes having zero probability.

Example 11.10   Urn Model Input from Output _____
Returning to our previous urn model of Example 11.6, we inquire into the probability $P(U = u_i|B = r)$ that the selected urn is u_i given that the observed draw was a red ball. From the previous results, we see that

$$P(U = u_i) = p_i, P(B = r | U = u_i) = \frac{r_i}{n_i},$$

$$P(B = r) = \sum_{j=1}^{n} \frac{r_j}{n_j} p_i.$$

Hence, applying Bayes' Theorem yields

$$P(U = u_i | B = r) = \frac{\frac{r_i}{n_i} p_i}{\sum_{j=1}^{n} \frac{r_j}{n_j} p_j}.$$

If the n urns were all equally likely to be selected, then this result simplifies to

$$P(U = u_i | B = r) = \frac{\frac{r_i}{n_i}}{\sum_{j=1}^{n} \frac{r_j}{n_j}}.$$

Example 11.11   Noisy Counter Input from Output

The noisy counter discussed in Example 11.7 leads us to ask what the probabilities are of the possible true count values given the measured count $Y = y$. Assuming that $y > 0$, Bayes' Theorem yields

$$P(X = x | Y = y) = \frac{P(Y = y | X = x) P(X = x)}{P(Y = y)} = 0 \text{ if } |y - x| > 1.$$

In particular,

$$P(X = y | Y = y) = \frac{p_c e^{-\lambda} \frac{\lambda^y}{y!}}{\frac{1}{2}(1 - p_c) e^{-\lambda} \left(\frac{\lambda^{y-1}}{(y-1)!} + \frac{\lambda^{y+1}}{(y+1)!} \right) + p_c e^{-\lambda} \frac{\lambda^y}{y!}}$$

$$= \frac{1}{\frac{1}{2} \frac{1 - p_c}{p_c} \left(\frac{y}{\lambda} + \frac{\lambda}{y+1} \right) + 1}.$$

Example 11.12   BSC Input from Output

A binary digital communications receiver observes a communications channel output (received signal) $Y = y$ and then infers the possible transmitted messages $X \in \{0, 1\}$, $P(X = i) = \pi_i$. In the BSC model of Example 11.8, $Y \in \{0, 1\}$ and

$$P(Y = x | X = x) = 1 - p, \ P(Y = 1 - x | X = x) = p.$$

Hence, applying Bayes' Theorem yields a description of the possible inputs given the observed output, say, $Y = 0$:

$$P(X = 0 | Y = 0) = \frac{\pi_0 (1 - p)}{\pi_0 (1 - p) + \pi_1 p}.$$

Of course, $P(X = 1 | Y = 0) = 1 - P(X = 0 | Y = 0)$.

Another application area is reliability and diagnoses of causes of failure. If A is the event of system failure in a particular manner (e.g., an ill individual with high temperature and rapid pulse rate) and $\{B_i\}$ is a list of possible causes of failure, of which one and only one can hold, then $P(B_i|A)$, as calculated through Bayes' Theorem, provides an assessment of the likelihoods of each of the possible causes. One might then proceed to repair or treat the system by examining it for failure modes in the order of the most probable ones being considered first. More generally, Bayes' Theorem is the basis of a school of statistical inference and decision making that has found particular favor in management applications.

11.7 THREE PROBLEMS THAT CHALLENGE OUR INTUITION

In the first two examples provided here, the results are often deemed paradoxical, and surveys that have been taken of actual responses show that people generally conclude wrongly.

11.7.1 Monty Hall's Game and Revision of Belief

A surprisingly controversial example is the so-called Monty Hall paradox (e.g., see Tierney [89]). Monty Hall was a popular television game show host whose game started with showing a contestant three closed doors, d_1, d_2, d_3, behind only one of which was a valuable gift. The contestant selected a door, say, d_1, but before the door was opened, Monty Hall, who knew which door hid the gift, say, d_2, opened a remaining door, d_3, that did not conceal the gift. (While Monty Hall did not always open a remaining door, the problem is of interest only in the cases in which he did so.) The contestant was then allowed to either stay with his original guess or change to the other closed door. The question is whether it is better to stay or to switch. The answer to this problem generated a lively controversy when it was correctly given by a columnist named Marilyn vos Savant, who is billed as the world's smartest woman on the basis of a very high IQ. Many qualified individuals, including mathematics teachers, criticized her correct answer. We now proceed to complete the model of this problem and to use Bayes' Theorem to analyze it. Let G be the random door concealing the gift; Y be the door that you, the contestant, chooses first; and M be the door Monty Hall opens. You have no prior knowledge as to the location of the gift, and this is modeled by

$$P(G = d_i | Y = d_j) = \frac{1}{3};$$

all doors are equally likely to conceal the gift, no matter what you guess. To model Monty Hall's constraint, assume that the $\{i_j\}$ are all distinct and that

$$P(M = d_{i_1} | Y = d_{i_2}, G = d_{i_3}) = 1.$$

However, if by chance you have chosen correctly, say, $Y = G = d_{i_2}$, then we assume that Monty Hall chooses at random between his two options:

$$P(M = d_{i_1} | Y = G = d_{i_2}) = \frac{1}{2} \text{ for } d_{i_1} \neq d_{i_2}.$$

The eventual answer does depend upon how we resolve this case, and that is often overlooked in discussions of the paradox.

To determine whether you should switch, we evaluate

$$P(G = d_1 | Y = d_2, M = d_3) = \frac{P(G = d_1, Y = d_2, M = d_3)}{P(Y = d_2, M = d_3)}$$

$$= \frac{P(M = d_3 | G = d_1, Y = d_2)P(G = d_1 | Y = d_2)P(Y = d_2)}{P(M = d_3 | Y = d_2)P(Y = d_2)}.$$

Cancelling the common factor of $P(Y = d_2)$, we find that the numerator is 1/3. To evaluate the denominator, use

$$P(M = d_3 | Y = d_2) = P(M = d_3 | G = d_1, Y = d_2)P(G = d_1 | Y = d_2)$$

$$+ P(M = d_3 | G = d_2, Y = d_2)P(G = d_2 | Y = d_2) + 0 = \left[1 + \frac{1}{2}\right]\frac{1}{3} = \frac{1}{2}.$$

Hence, $P(G = d_1 | Y = d_2, M = d_3) = 2/3$, and you, the contestant, should switch from your original choice d_2 to d_1!

11.7.2 False Positives on Diagnostic Tests

Let the event D be that a randomly selected individual in a population has a particular disease, and D^c is its complement. The probability that a randomly selected individual in this population has the disease is p_d. There is a test for the disease that always diagnoses correctly when an individual has the disease. However, when the individual does not have the disease, the test falsely reports this with probability p_t. Let TP denote the event that the test reports positively that the disease is present. Formalizing this discussion, we have

$$P(D) = p_d, \ P(TP|D) = 1, \ P(TP|D^c) = p_t.$$

An individual is interested in the probability $P(D|TP)$ of having the disease, given a positive test report. If, say, the disease is rare and $p_d = .001$, and the test falsely reports with small probability $P(TP|D^c) = .05$, then the intuitive response of many is that the individual is likely to have the disease. After all, the test is highly accurate. However, analysis demonstrates the weakness of this intuition. We evaluate the desired

$$P(D|TP) = \frac{P(TP|D)P(D)}{P(TP)},$$

where, by Total Probability,

$$P(TP) = P(TP|D)P(D) + P(TP|D^c)P(D^c).$$

Hence,

$$P(D|TP) = \frac{1}{1 + \frac{p_t(1-p_d)}{p_d}}.$$

One has to take into account the relative rareness of the disease, as well as the accuracy of the test. For the illustrative numbers provided, $P(D|TP) \approx .02$! For a rare disease, $P(D|TP) \approx p_d/p_t$.

11.7.3 Selecting a Partner

A final curious application of the Total Probability Theorem and classical probability is to the problem of "partner selection." We caricature this process by assuming that you have decided that you will interview or date n individuals in order to select a partner. These individuals are dated sequentially, and you assign a score X_i to the ith individual. We assume that there are no ties: $P(X_i = X_j) = 0$ for $i \neq j$. You wish to select the individual with the highest score in $X_1, \ldots, X_n$, but are forced to make your selection of an individual only while dating that individual. You cannot date them all and go back and select your most preferred one. Let M_l denote the index of the largest of $X_1, \ldots, X_l$, so that

$$X_{M_l} = \max_{1 \leq j \leq l} \{X_j\}.$$

We assume that the order of arrival of individuals is random and hence that

$$P(M_n = i) = \frac{1}{n}, \quad i = 1, \ldots, n.$$

Your goal is to adopt a strategy that will have a high probability of choosing X_{M_n} under the preceding constraints; we call this the success event S. The strategy will be to date τ individuals and then select the next individual i for which $X_i > X_{M_\tau}$, or $i = n$ if no such individual is encountered. That is, in the latter case, the best individual was unfortunately in the first τ that you dated. We can now use Total Probability to write

$$P(S) = \sum_{i=1}^{n} P(S|M_n = i) P(M_n = i).$$

Clearly, by our strategy, if $i \leq \tau$, then $P(S|M_n = i) = 0$. Hence,

$$P(S) = \frac{1}{n} \sum_{i=\tau+1}^{n} P(S|M_n = i).$$

For $i > \tau$,

$$P(S|M_n = i) = P(M_{i-1} \leq \tau) = \frac{\tau}{i-1},$$

from which we conclude that

$$P(S) = \frac{\tau}{n} \sum_{i=\tau+1}^{n} \frac{1}{i-1} = \frac{\tau}{n} \sum_{j=\tau}^{n-1} \frac{1}{j} \approx \frac{\tau}{n} \log\left(\frac{n}{\tau}\right).$$

The maximum in the large-n approximation is at $\tau/n = e^{-1}$, and it yields a maximum success probability of $P(S) = e^{-1}$. Your chance of choosing the best partner in a large search of n candidates is about 36%.

11.8 SUMMARY

A choice of probability measure is based upon knowledge or information, the form of knowledge required depending upon the interpretation chosen for the probability. Converting knowledge into probabilities is a thorny issue, a central role for statistics, and the subject of much controversy ranging from the philosophy of science to the practice of statistics. The least controversial use of knowledge to revise probability knowledge occurs when that knowledge is of a specific event, say, B occurring. We then revise our knowledge of $P(A)$ to $P(A|B)$. The classical and frequentist concepts of probability lead us to the same definition, namely,

$$P(A|B) = \frac{P(A \cap B)}{P(B)},$$

in the case that $P(B) > 0$. We explore the consequences of this definition for conditioning on any event of positive probability or any discrete random variable (one taking on only countably many distinct values). We are now able to derive the familiar exponential family of probability laws on the basis of a condition of "memorylessness" that can guide us in knowing when the exponential is the correct model for a random phenomenon.

Conditional probability enables us to describe a nondeterministic system S by specifying $P_S(Y \in A|X = x)$ for each possible x and set $A \in \mathcal{A}_Y$, thereby contributing to G1. From knowledge of $P(X \in B)$ and P_S, we can treat G2 by determining $P(Y \in A)$ through the Total Probability Theorem. Furthermore, Bayes' Theorem enables us to reverse cause and effect and to infer from system output back to the possible inputs by calculating $P(X \in B|Y \in A)$, thereby contributing to G3. We are able to infer, say, from the observed failure of a system to the probabilities of the possible causes, from a message received over a noisy channel to the probabilities of the possible transmitted messages, or from a measurement Y to the measured quantity X. The key definitions and theorems are summarized in the appendix following this section.

The chapter closed by applying these ideas of conditional probability to three examples, all of which are often poorly understood by the intuition of even individuals who are mathematically trained.

11.9 APPENDIX: DISCRETE CONDITIONAL PROBABILITY SUMMARY

- *System Characterization (Goal 1)*
  Definition of Conditional Probability in Discrete Case
  $P(A|B) = \frac{P(A,B)}{P(B)}$
  Probability of Effect A Given Cause B

- *System Response (Goal 2)*
  Total Probability Theorem
  $P(A) = \sum_i P(A|B_i)P(B_i)$
  Probability of Effect A

- *System Response (Goal 2)*
  Sequence Probability
  $$P(\cap_{i=1}^{n} A_i) = P(A_1) \prod_{2}^{n} P(A_k | A_1, \ldots, A_{k-1})$$
  Probability of Sequence/Cascade of Events $A_1, \ldots, A_n$

- *System Input from Output (Goal 3)*
  Bayes' Theorem
  $$P(B_i | A) = \frac{P(A|B_i)P(B_i)}{\sum_k P(A|B_k)P(B_k)}$$
  Probability of Cause B_i Given Effect A (Inference)

EXERCISES

E11.1   If $\Omega = \{0, 1, \ldots, 9\}$, $p(0) = .35, p(1) = .25$, $p(2) = \cdots = p(9) = .05$, $A = \{0, 4, 5\}$, and $B = \{2, 3, 5\}$, then evaluate $P(A|B)$.

E11.2   If $P(C|D) = .4$ and $P(D|C) = .5$, which of C, D is the more probable?

E11.3   If $P(E) = .3$ and $P(F) = .7$, what can you conclude about $P(E|F)$?

E11.4   If $\{B_1, B_2\}$ is a partition of Ω, with $P(B_1) = .2$, $P(A|B_1) = .5$, and $P(A|B_2) = .2$, then evaluate $P(B_1|A)$.

E11.5   Knowing nothing more, can you compare the values of $P(X \leq Y | Y = 2)$ and $P(X \leq 3 | Y = 2)$, and, if so, what is the comparison? (Explain your conclusion—do not give a yes/no answer.)

E11.6   Let $P(A) = 1/4$, $P(B|A) = 1/2$, and $P(A|B) = 1/4$.

   a. Is $A \perp B$?
   b. Does A contain B or B contain A?
   c. Evaluate $P(A^c | B^c)$.
   d. Is $P(A|B) + P(A|B^c) = 1$?
   e. Is $P(A|B) + P(A^c|B) = 1$?

E11.7   If $P(A) > 0$, is it true that

$$P(A \cap B | A) \geq P(A \cap B | A \cup B)?$$

E11.8   Event B is *corroborative* of event A, denoted $B \overset{+}{\to} A$, if $P(A|B) > P(A)$. Prove, or give a counterexample to, the following claims about being corroborative:

   a. symmetry: $B \overset{+}{\to} A \Rightarrow A \overset{+}{\to} B$.
   b. transitivity: $A \overset{+}{\to} B$, $B \overset{+}{\to} C \Rightarrow A \overset{+}{\to} C$.
   c. $B \overset{+}{\to} A$, $C \overset{+}{\to} A \Rightarrow B \cap C \overset{+}{\to} A$.

E11.9   Show that

   a. $P(A | B \cap C) = P(A \cap B | C)/P(B|C)$;
   b. $P(\cap_{k=1}^{n} A_k | B) = P(A_1|B) \prod_{m=2}^{n} P(A_m | B \cap (\cap_{j=1}^{m-1} A_j))$.

E11.10   Statistics suggest that if it has been snowing in Ithaca, New York, for exactly k days in a row, then the probability that it will not snow the next day is $\frac{1}{k+1}, k = 1, 2, \ldots$. Find the probability that a snowstorm that starts on January 25 will last through the end of the month.

E11.11 Message source M_1 produces a word that is a byte, and all bytes are equally probable. Message source M_2 produces a word of length eight characters, with the characters drawn from the ternary alphabet $\{0, 1, 2\}$, and all such words are equally probable.

    a. What is the probability that M_2 produces a word that looks like a byte (i.e., no appearance of "2")?

    b. If a word is equally likely to come from either source, what is the probability that it will be a byte?

    c. Given that we have observed a byte, how probable is it that it came from M_1?

E11.12 Let B refer to inputs to a binary channel and A refer to outputs. The source is described by $P(B = \{0\}) = .4$. The channel is described by

$$P(A = \{1\}|B = \{0\}) = .2, \quad P(A = \{1\}|B = \{1\}) = .7.$$

    a. Describe the outputs.

    b. If we observe $A = \{1\}$, what is the probability that $B = \{1\}$ occurred.

E11.13 Message source M_1 generates bytes having exactly 3 ones, with all arrangements being equally probable. Message source M_2 generates bytes having exactly 4 ones, with all arrangements being equally probable. If we are equally likely to receive a message from M_1 as from M_2, what is the probability of receiving a byte with a one in the first position?

E11.14 A message source is equally probable to be in any one of three states S_1, S_2, S_3. In state S_k, the source emits a byte having no more than k ones, with all such bytes being equally probable.

    a. How many different bytes can this source emit?

    b. What is the probability that the source generates the byte $B = 11000000$, given that the source is in state S_2?

    c. What is $P(B)$?

    d. Given that B was observed, what is the probability that the source was in state S_2?

E11.15 A signal $S \in \{s_1, s_2, s_3\}$ is the input to a noisy system having a nonnegative integer R as output. The three possible signals have respective probabilities of .2, .3, and .5. The system is characterized by

$$P(R = k|S = s_i) = (1 - \beta_i)\beta_i^k, \quad \beta_1 = \frac{1}{2}, \beta_2 = \frac{1}{3}, \beta_3 = \frac{1}{4}.$$

    a. What is the probability that the system output is 2?

    b. If the output $R = 2$, then which is the most likely signal to have been the input, and how likely is it?

E11.16 A particular circuit has 7 chips of type C_1, 3 of type C_2, and 5 of type C_3. The probability that a chip of type C_i will fail in the kth month of usage is $.25i(1 - .25i)^k, k = 0, 1, 2, \ldots$ What is the probability that a randomly selected chip will fail in the second month?

E11.17   A ternary-valued message with symbols drawn from the alphabet $\{a, b, c\}$ comes from one of the two message sources M_1, M_2. If the message source is M_1, then the probabilities of the symbols being drawn are .3, .3, .4, while if the source is M_2, the probabilities are .5, .25, .25, respectively. The probability that source M_1 is selected is .2.

   a. What is the probability of the message a?
   b. If we observe c, then what is the probability that it came from source M_1?

E11.18   A binary signal X with $P(X = 1) = .6$ is transmitted through a BSC with error probability of .1.

   a. Evaluate the probabilities of the possible outputs.
   b. Determine the most probable signal to have been sent if a one is received.

E11.19   An information source selects messages from the set $\mathcal{S}_M = \{0, 1, 2\}$ which are then transmitted over a memoryless noisy channel that has the set of outputs $\mathcal{S}_C = \{a, b\}$. The channel is characterized by specifying the conditional probability $P(C | M)$ for a channel output C, given a message M as input. We learn that

$$P(C = a | M = 0) = .2, \ P(C = a | M = 1) = .4,$$

$$P(C = a | M = 2) = .7.$$

The information source is characterized by specifying the probability $P(M)$ that message M is selected. We learn that

$$P(M = 0) = .3, \ P(M = 1) = .2.$$

   a. How probable is it that we will receive $C = b$?
   b. If we receive $C = b$, then how probable is it that $M = 1$ was sent?

E11.20   A binary symmetric channel (BSC) accepts an input X from the set $\{0, 1\}$ and produces an output Y in the same set. The probability of an error in transmission is p, no matter which error is made. An input is selected with probability $P(X = 0) = p_0$.

   a. Evaluate the probabilities for the possible outputs.
   b. Determine the probability that a one was sent, given that a one was received at the channel output.

E11.21   On a multiple-choice question having m alternative answers, the prior probability that the student knows the answer is p. If the student has to guess (event G), then all alternatives are equally probable. Find the probability that the student knew the answer to the question (event K) given that he or she answered it correctly (event C).

E11.22   A noisy counter observes nonnegative integer X and reports nonnegative integer Y such that

$$P(Y = x | X = x) = .75, \quad P(Y = x + 1 | X = x) = .25.$$

We know that $P(X = x) = .5^{x+1}$ for $x = 0, 1, \ldots$.

   a. Evaluate $P(Y = 2)$.
   b. Evaluate $P(X = x | Y = 2)$ for all nonnegative integers x.
   c. What is the most probable value of X, given that we observe $Y = 2$?

E11.23 A noisy counter observes $X \sim \mathcal{P}(\frac{1}{2})$ and reports nonnegative integer Y such that

$$P(Y = k|X = j) = \frac{e^{-1}}{(k-j)!} \ (= 0 \text{ if } j > k).$$

    a. Evaluate $P(Y = 1)$.
    b. Evaluate $P(X = k|Y = 1)$ for all nonnegative integer k.
    c. What is the MAP estimator of X (most probable value of X), given that we observe $Y = 1$?

E11.24 An attempt is made to classify a sample of radioactive material by means of a radiation count R over a fixed period of time. It is initially known that the sample is either a rare element A with $P(A) = .05$ or a more common element B with $P(B) = .95$. We know that the probability mass function for R is $\mathcal{P}(5)$ given that it is element A and $\mathcal{P}(3)$ given that it is element B.

    a. Evaluate $P(R = r)$.
    b. If $R = 5$ is observed, then what is the most probable source of the sample (i.e., compare $P(A|R = 5)$ with $P(B|R = 5)$), and how likely is it to be that element?

E11.25 There are three urns U_1, U_2, and U_3 with the following compositions of red, green, and blue balls:

$$r_1 = 5, g_1 = 7, b_1 = 0; \ \ r_2 = 0, g_2 = 10, b_2 = 15; \ \ r_3 = 5, g_3 = 5, b_3 = 5.$$

    a. If an urn is selected at random and then a ball is chosen at random from that urn, what is the probability P_b that a blue ball is chosen?
    b. Given the previous information and that a blue ball was chosen, what is the probability P_i, $i = 1, 2, 3$ that it was chosen from urn U_i, $i = 1, 2, 3$?
    c. If, now, one ball is chosen at random from each of the three urns, what is the probability that two of the balls are blue?
    d. Under the conditions of part (c), what is the probability that the third ball is red, given that two of the chosen balls are blue?

E11.26   a. For urn U_3, what is the probability of choosing a second ball that is red, given that the first ball chosen was blue and that there is no replacement?
    b. Repeat (a) for selection of balls with replacement.

E11.27 A binary-valued Markov source produces a random sequence $\{X_i\}$ with $X_i \in \{0, 1\}$, $P(X_1 = 0) = .4$, and

$$(\forall i > 1) \ P(X_i = x_i|X_{i-1} = x_{i-1}, \ldots, X_1 = x_1) = \begin{cases} .8 & \text{if } x_i = x_{i-1} \\ .2 & \text{if } x_i = 1 - x_{i-1} \end{cases}.$$

    a. Evaluate $P(X_2 = 1)$.
    b. Evaluate $P(X_1 = 1|X_2 = 1)$.
    c. If this source is the input to a binary symmetric channel with outputs $\{Y_i\}$ and error probability .1, and errors are made independently on successive inputs, then evaluate $P(Y_2 = 1)$.

E11.28   A binary erasure channel (BEC) has inputs $X \in \{0, 1\}$ and outputs $Y \in \{0, 1, E\}$. We know that $P(X = 0) = .35$ and that

$$P(Y = 0|X = 0) = .7, \; P(Y = E|X = 0) = .3,$$

$$P(Y = 0|X = 1) = 0, \; P(Y = E|X = 1) = .2.$$

   a. Evaluate the probability $P(Y = E)$ of an erasure.
   b. Evaluate $P(X = 0|Y = E)$.
   c. What is the most probable channel input if we observe an erasure as output?

E11.29   Given that at least one electron has been emitted in 1 μsec, the probability of emitting exactly one electron is twice its unconditional probability of being emitted in 1 μsec. Adopt an appropriate probability model.

   a. What is the probability of 0 electrons being emitted in 1 μsec?
   b. What is the probability of 0 electrons being emitted in 5 μsec?
   c. What is the average current?

12

Mixed Conditional Probability

12.1 PURPOSE, BACKGROUND, AND ORGANIZATION

As in Chapter 11, we organize our presentation of conditional probability to respond to our three goals. However, now we account for the three cases in which at least one of the random variables is continuous (C). Having treated the $Y|X = D|D$ case in Chapter 11, in this chapter we focus particularly on the case of conditioning on an event $X \in B$ having probability 0 and treat it as a limiting case of conditioning on suitable events of positive probability. We use this limiting case to define what is meant by conditional probability, although that is not the approach taken in more advanced treatments of conditional probability. Having defined conditional probability, we can also unproblematically define a conditional cdf. By analogy to our earlier work using density functions, we assume a representation of conditional probability by a conditional density and derive a relationship between the new object of a conditional density and the previously introduced object of a multivariate density. Conditional density functions enable us to treat conditioning on continuous random variables in a convenient manner.

For ease of reference and to promote an understanding of the pattern of exposition, the variety of results developed are organized and summarized in Sections 12.12 and 12.13. Unfortunately, the variety of results tends to obscure the simplicity of their pattern. In general, the results obtained in Chapter 11 for discrete random variables X and Y carry over in this chapter simply by making the following substitutions, as appropriate:

- replace a probability mass function p_Z by a probability density function f_Z when the variable Z is continuous;
- replace sums over pmfs by integrals over pdfs;
- when Y is continuous, replace $p_{Y|X}$ by $f_{Y|X}$.

12.2 CONDITIONAL PROBABILITY: $P(A|B)$ FOR $P(B) = 0$

When the sample space is a subset of $\mathbb{R}^n$, we often find ourselves attempting to condition on an event B that has zero probability. For example, we may model a pair of measurements, say, height H and weight W, and attempt to evaluate the probability that $H \le h$, given that we know $W = w$. The latter event will generally have probability zero. The general extension of the definition of conditional probability to include such circumstances relies upon mathematical notions (e.g., the Radon–Nikodym derivative and projections in Hilbert spaces) that are well beyond our prerequisites. For most of practice, an alternative is to proceed through a limiting process that approximates the conditioning event B by ones B_δ of positive probability. Our conclusion is presented in Definition 12.2.1, but is motivated by what follows.

We approximate an event $B \subset \mathbb{R}^n, P(B) = 0$, by a collection of events

$$\{B_\delta, \delta > 0\}, \ P(B_\delta) > 0, \ B_\delta \supset B, \ \cap_{\delta>0}B_\delta = B.$$

For instance, we might approximate each point $\mathbf{x}$ in B by a *neighborhood* containing $\mathbf{x}$ that might be a ball of radius δ or a hypercube centered at $\mathbf{x}$ and of side length 2δ to obtain, say,

$$B_\delta = \bigcup_{\mathbf{x}\in B}\{\mathbf{y} : (\forall i) \ |y_i - x_i| \le \delta\}, \quad B_\delta \supset B.$$

The set B_δ will have positive probability so long as the density f is positive in the neighborhood of some point of B. (If this is not the case, then $P(A|B)$ can be defined arbitrarily without real consequence.) In effect we increase B by taking the union of cubes of side 2δ centered at each of the points of B. Clearly, $\{B_\delta : \delta > 0\}$ is a collection of sets containing B and such that $\delta' > \delta \rightarrow B_{\delta'} \supset B_\delta$. If now $\cap_{\delta>0} B_\delta = B$, *which need not be the case,** then, by $K4$,

$$A_\delta = B_\delta - B \downarrow \emptyset \Rightarrow \lim_\delta P(A_\delta) = 0,$$

and by $K3$,

$$P(B_\delta) = P(A_\delta) + P(B) \downarrow P(B).$$

Thus, we can approximate $P(B) = 0$ by $P(B_\delta) > 0$, for small δ. Similarly, we can approximate $P(A \cap B)$ by $P(A \cap B_\delta)$. We then hope that a sensible definition of what we mean by conditional probability is the following:

---

*For example, if $\Omega = [0, 1]$ and B is the countably infinite set of rational numbers in the unit interval, then, for any $\delta > 0$, $B_\delta = [0, 1]$. Hence, $\cap_{\delta>0} B_\delta = [0, 1] \ne B$.

Definition 12.2.1 (Conditional Probability) The conditional probability of an event A given an event B, with $P(B)$ possibly 0, is

$$P(A|B) = \lim_{\delta \to 0} P(A|B_\delta),$$

provided that the limit exists and does not depend upon the choice of approximating sequence $\{B_\delta\}$ to B.

Example 12.1 Bivariate Conditional CDF _____

An important example of conditional probability is that of the conditional cdf. We examine this in the bivariate case where

$$F_{X_2|X_1}(x_2|x_1) = P(X_2 \le x_2 | X_1 = x_1).$$

If $P(X_1 = x_1) > 0$, then the conditional cdf is unproblematic. However, if $P(X_1 = x_1) = 0$, then, according to our approach,

$$F_{X_2|X_1}(x_2|x_1) = \lim_{\delta \to 0} P(X_2 \le x_2 | x_1 - \delta < X_1 \le x_1 + \delta).$$

$$P(X_2 \le x_2 | x_1 - \delta < X_1 \le x_1 + \delta) = \frac{P(x_1 - \delta < X_1 \le x_1 + \delta, X_2 \le x_2)}{P(x_1 - \delta < X_1 \le x_1 + \delta)}.$$

$$P(x_1 - \delta < X_1 \le x_1 + \delta, X_2 \le x_2) = \int_{-\infty}^{x_2} \int_{x_1-\delta}^{x_1+\delta} f_{X_1,X_2}(y_1, y_2)\, dy_1\, dy_2.$$

For small enough δ,

$$P(x_1 - \delta < X_1 \le x_1 + \delta, X_2 \le x_2) \approx 2\delta \int_{-\infty}^{x_2} f_{X_1,X_2}(x_1, y_2)\, dy_2.$$

Similarly,

$$P(x_1 - \delta < X_1 \le x_1 + \delta) \approx 2\delta f_{X_1}(x_1).$$

It follows that

$$F_{X_2|X_1}(x_2|x_1) = \lim_{\delta \to 0} P(X_2 \le x_2 | x_1 - \delta < X_1 \le x_1 + \delta) = \int_{-\infty}^{x_2} \frac{f_{X_1,X_2}(x_1, y_2)}{f_{X_1}(x_1)}\, dy_2.$$

This conclusion is meaningful, provided that $f_{X_1}(x_1)$ is positive throughout a sufficiently small neighborhood of x_1.

In the next section, we learn how to calculate $P(A|B)$ and the conditional cdf, using conditional densities.

12.3 BIVARIATE CONDITIONAL DENSITY

The bottom line of this section is that we can define a conditional density function $f_{X_2|X_1}(x_2|x_1)$ as the ratio of the unconditional densities, $f_{X_1,X_2}(x_1, x_2)/f_{X_1}(x_1)$ (when the denominator is nonzero).

The work to be done by this conditional density is that we can then evaluate a probability of the form $P(X_2 \in A | X_1 = x_1)$ as an integral, in the variable x_2, of the conditional density $f_{X_2|X_1}(x_2|x_1)$ over the set A. The discussion of this section is intended to motivate and justify that definition. A conditional density $f_{X_2|X_1}(x_2|x_1)$ is a function of x_1 and x_2 whose precise role is given by

$$P(X_2 \in A | X_1 = x_1) = \int_A f_{X_2|X_1}(x_2|x_1)\, dx_2. \tag{12.1}$$

The conditional density is the function that is integrated over the set A to calculate the conditional probability, and in this it parallels the role of the (unconditional) probability density function. In particular, for the conditional cdf

$$F_{X_2|X_1}(x_2|x_1) = P(X_2 \leq x_2 | X_1 = x_1),$$

we saw in the preceding example that our approach of Section 12.2, assuming that $f_{X_1}(x_1) > 0$, yielded

$$F_{X_2|X_1}(x_2|x_1) = \int_{-\infty}^{x_2} \frac{f_{X_1,X_2}(x_1, y_2)}{f_{X_1}(x_1)}\, dy_2.$$

However, from the defining condition for a bivariate conditional pdf given by Eq. (12.1),

$$F_{X_2|X_1}(x_2|x_1) = \int_{-\infty}^{x_2} f_{X_2|X_1}(y_2|x_1)\, dy_2.$$

Because the two expressions for the conditional cdf must agree for all x_2, we have

$$\frac{\partial F_{X_2|X_1}(x_2|x_1)}{\partial x_2} = \frac{f_{X_1,X_2}(x_1, x_2)}{f_{X_1}(x_1)} = f_{X_2|X_1}(x_2|x_1).$$

We conclude that, when $f_{X_1}(x_1) > 0$, it is necessary that

$$\frac{f_{X_1,X_2}(x_1, x_2)}{f_{X_1}(x_1)} = f_{X_2|X_1}(x_2|x_1).$$

Definition 12.3.1   (Bivariate Conditional PDF)   If the pdf $f_{X_1}(x_1)$ is piecewise continuous, then the bivariate conditional pdf for X_2 given X_1 is

$$f_{X_2|X_1}(x_2|x_1) = \frac{f_{X_1,X_2}(x_1, x_2)}{f_{X_1}(x_1)},$$

provided that x_1 is neither a point of discontinuity of f_{X_1} nor a point at which $f_{X_1}(x_1) = 0$. In the latter two cases, we can arbitrarily choose $f_{X_2|X_1}$ as any univariate pdf for X_2.

When $f_{X_1}(x_1) = 0$, arbitrarily specifying $f_{X_2|X_1}(x_2|x_1)$ has no impact on the unconditional probability of an event defined in terms of X_1 and X_2. The basic mathematical properties of a bivariate conditional density are summarized as follows: A function $g(x_1, x_2)$ is a conditional density for X_2 given X_1 if

$$\text{(i) } (\forall x_1, x_2) \; g(x_1, x_2) \geq 0;$$

$$\text{(ii) } (\forall x_1) \int_{-\infty}^{\infty} g(x_1, x_2) \, dx_2 = 1.$$

We do not state this as a formal lemma with proof, because such a lemma requires some additional technical restrictions that are beyond our interest.

Example 12.2   Bivariate Conditional PDF _____

Consider the bivariate cdf

$$F_{X,Y}(x, y) = \left[1 + e^{-x} + e^{-y}\right]^{-1}.$$

The univariate cdf

$$F_X(x) = \lim_{y \to \infty} F_{X,Y}(x, y) = \left[1 + e^{-x}\right]^{-1}.$$

The bivariate pdf

$$f_{X,Y}(x, y) = \frac{\partial^2 F_{X,Y}(x, y)}{\partial x \, \partial y} = 2 \frac{e^{-x-y}}{(1 + e^{-x} + e^{-y})^3}.$$

The univariate pdf

$$f_X(x) = \frac{\partial F_X(x)}{\partial x} = \frac{e^{-x}}{(1 + e^{-x})^2}.$$

Hence, from our definition,

$$f_{Y|X}(y|x) = \frac{f_{X,Y}(x, y)}{f_X(x)} = 2 \frac{e^{-y}(1 + e^{-x})^2}{(1 + e^{-x} + e^{-y})^3}.$$

Example 12.3   Conditional PDF and Independence _____

Consider a product bivariate pdf as described in Section 7.4.1 and defined in terms of two univariate pdfs f and g as

$$f_{X_1, X_2}(x_1, x_2) = f(x_1) g(x_2).$$

In this case,

$$f_{X_1}(x_1) = \int_{-\infty}^{\infty} f_{X_1, X_2}(x_1, x_2) \, dx_2 = f(x_1),$$

and, from our definition,

$$f_{X_2|X_1}(x_2|x_1) = \frac{f_{X_1,X_2}(x_1,x_2)}{f_{X_1}(x_1)} = g(x_2).$$

The conditional pdf for X_2 does not depend upon X_1, and it equals the unconditional pdf for X_2 in this case.

Example 12.4   Conditional PDF and Innovation Model

Consider an innovation-type pdf as described in Section 7.4.1 and given in terms of pdfs f and g and an arbitrary finite-valued function h by

$$f_{X_1,X_2}(x_1,x_2) = f(x_2 - h(x_1))g(x_1).$$

In this case,

$$f_{X_1}(x_1) = \int_{-\infty}^{\infty} f_{X_1,X_2}(x_1,x_2)\,dx_2 = g(x_1)\int_{-\infty}^{\infty} f(x_2 - h(x_1))\,dx_2 = g(x_1).$$

Hence, applying our definition, we have

$$f_{X_2|X_1}(x_2|x_1) = \frac{f_{X_1,X_2}(x_1,x_2)}{f_{X_1}(x_1)} = f(x_2 - h(x_1)).$$

The conditional pdf for X_2 does depend upon X_1.

Example 12.5   Bivariate Normal Conditional PDF

A *bivariate normal or Gaussian density* (see Section 7.4.4) is given in terms of two means m_1, m_2 written as components of a column vector $\mathbf{m}$ and a covariance matrix

$$\mathbf{m} = \begin{pmatrix} m_1 \\ m_2 \end{pmatrix}, \quad \mathbb{C} = \begin{pmatrix} \sigma_1^2 & \rho\sigma_1\sigma_2 \\ \rho\sigma_1\sigma_2 & \sigma_2^2 \end{pmatrix}, \quad \mathbf{x} = \begin{pmatrix} x_1 \\ x_2 \end{pmatrix},$$

with the parameter ρ known as the *correlation coefficient* (see Sections 9.11, 9.12), through

$$f_{X_1,X_2}(x_1,x_2) = \frac{1}{2\pi \det(\mathbb{C})^{\frac{1}{2}}} e^{-\frac{1}{2}(\mathbf{x}-\mathbf{m})^T \mathbb{C}^{-1}(\mathbf{x}-\mathbf{m})}.$$

Lemma 7.4.2 tells us that $X_1 \sim \mathcal{N}(m_1, C_{1,1}) = \mathcal{N}(m_1, \sigma_1^2)$, or

$$f_{X_1}(x_1) = \frac{1}{\sigma_1\sqrt{2\pi}} e^{-\frac{(x_1-m_1)^2}{2\sigma_1^2}}.$$

Letting

$$m_{2|1} = \frac{\rho\sigma_2}{\sigma_1}(x_1 - m_1) + m_2, \quad \sigma_{2|1}^2 = \sigma_2^2(1-\rho^2),$$

we compute

$$f_{X_2|X_1}(x_2|x_1) = \frac{f_{X_1,X_2}(x_1,x_2)}{f_{X_1}(x_1)} = \frac{1}{\sigma_{2|1}\sqrt{2\pi}} e^{-\frac{(x_2-m_{2|1})^2}{2\sigma_{2|1}^2}}.$$

In the bivariate normal case, the conditional density of X_2 given X_1 is again a normal with a mean $m_{2|1}$ revised from that of m_2 for X_2 alone and a variance $\sigma_{2|1}^2$ revised from that of σ_2^2.

Example 12.6   Another Bivariate Conditional PDF _____

As a further illustration of the possibilities for a bivariate conditional pdf, select any univariate pdfs f and g, and set

$$f_{X_2|X_1}(x_2|x_1) = \begin{cases} \dfrac{1}{x_1} f\left(\dfrac{x_2}{x_1}\right) & \text{if } x_1 > 0 \\ g(x_2) & \text{if } x_1 \le 0 \end{cases}.$$

Clearly, $f_{X_2|X_1}$ is nonnegative, and for any x_1, it integrates to unity in x_2.

12.4 MULTIVARIATE CONDITIONAL DENSITIES

Examples of multivariate conditional densities and of their applications will be given in the next section, following their definition and properties. For $n > m$, introduce the shorthand notation

$$x_m^n = (x_m, x_{m+1}, \ldots, x_n).$$

The extension of Definition 12.3.1 from the bivariate case to the multivariate case is given in the next definition.

Definition 12.4.1   (Multivariate Conditional PDF)   If the pdf $f_{X_1^m}(x_1^m) > 0$ in some sufficiently small neighborhood of the m-dimensional point x_1^m, then the multivariate conditional pdf for X_{m+1}^n given X_1^m is expressed in terms of the unconditional pdfs by

$$f_{X_{m+1}^n|X_1^m}(x_{m+1}^n|x_1^m) = \frac{f_{X_1^n}(x_1^n)}{f_{X_1^m}(x_1^m)}.$$

If $f_{X_1^m}(x_1^m) = 0$ in all sufficiently small neighborhoods of X_1^m, then arbitrarily choose $f_{X_{m+1}^n|X_1^m}$ as any $(n - m)$-variate pdf for X_{m+1}^n.

A function $g(x_1^n) = g(x_1, \ldots, x_m, \ldots, x_n)$ is a multivariate conditional pdf for X_{m+1}^n given X_1^m if

$$\text{(i) } (\forall x_1^n) \; g(x_1^n) \ge 0;$$

$$\text{(ii) } (\forall x_1^m) \int_{-\infty}^{\infty} \cdots \int_{-\infty}^{\infty} g(x_1^n) \, dx_{m+1} \ldots dx_n = 1.$$

Once again, we do not state this as a lemma with proof, for to do so would require additional technical conditions that lie beyond our interests.

The basic role of the conditional density is to enable us to evaluate conditional probabilities for an event $A \subset \mathbb{R}^{n-m}$, given information $X_1^m = x_1^m$.

Theorem 12.4.1   (Multivariate Conditional Density)   For any $(n-m)$-dimensional (Borel algebra) event A and any x_1^m for which $f_{X_1^m}(x_1^m) > 0$ in some sufficiently small neighborhood of X_1^m,

$$P(X_{m+1}^n \in A | X_1^m = x_1^m) = \int_A f_{X_{m+1}^n | X_1^m}(x_{m+1}^n | x_1^m)\, dx_{m+1} \ldots dx_n.$$

This result is essential for much of our subsequent work in this text. In particular, the conditional cdf

$$F_{X_{m+1}^n | X_1^m}(x_{m+1}^n | x_1^m) = P(X_{m+1} \leq x_{m+1}, \ldots, X_n \leq x_n | X_1^m = x_1^m)$$

$$= \int_{-\infty}^{x_{m+1}} \cdots \int_{-\infty}^{x_n} f_{X_{m+1}^n | X_1^m}(y_{m+1}^n | x_1^m)\, dy_{m+1} \ldots dy_n.$$

Conversely, we can recover the conditional density from the conditional cdf through

$$f_{X_{m+1}^n | X_1^m}(x_{m+1}^n | x_1^m) = \frac{\partial^{n-m} F_{X_{m+1}^n | X_1^m}(x_{m+1}^n | x_1^m)}{\partial x_{m+1} \cdots \partial x_n}.$$

An analogous result to the one given in Theorem 11.4.1 is the next theorem, which enables us to reverse the preceding and express a multivariate pdf in terms of the newly introduced conditional pdfs.

Theorem 12.4.2   (PDFs of Sequences)

$$f_{X_1^n}(x_1^n) = f_{X_1}(x_1) \prod_{j=2}^{n} f_{X_j | X_1^{j-1}}(x_j | x_1^{j-1}).$$

Proof. The proof is readily carried out by replacing $f_{X_j | X_1^{j-1}}(x_j | x_1^{j-1})$ by its definition $f_{X_1^j}(x_1^j)/f_{X_1^{j-1}}(x_1^{j-1})$ and evaluating the product by exploiting its many cancellations of terms. $\square$

12.5 APPLICATIONS OF CONDITIONAL DENSITIES

Discussions of applications can be found in several places in this chapter and later ones. The applications are located to support a particular topic in conditional probability. The short list given next focuses on different kinds of systems that require conditional probability for their description. The sections that follow note applications that rely upon the Total Probability Theorem (Theorem 12.7.1) and upon Bayes' Theorem (Theorem 12.9.1). In general, applications require all of these results on conditional probability for their formulation and subsequent development of optimal system design. Examples of applications requiring conditional probability for their formulation include the following:

- analog communications where S is the channel, X the transmitted signal and Y the received signal; in the common independent additive noise N model (see Section 13.5), S is defined by $f_{Y|X}(y|x) = f_N(y-x)$, where f_N is the channel noise density;
- measurement error model where S is the measuring instrument, X the true value of the quantity being measured, and Y the result of the measurement; S is described by $f_{Y|X}$ and is often also modeled by independent additive noise (see the same sections as for the previous example);
- control of a dynamical system in which $\mathbf{X}$ is the state variable (in this case it is likely to be multidimensional), $\mathbf{Y}$ is the observable, and S is the dynamical system, together with the observational system that relates present state to future state and thence to the observable, with our interest being in $f_{\mathbf{X}|\mathbf{Y}}$ as a description of what we know about $\mathbf{X}$, given that we have observed $\mathbf{Y}$ (e.g., see Kalman filtering in Section 15.6);
- converting a digital signal D into an analog signal A in which we are interested in $f_{A|D}$;
- hypothesis testing or target detection in which we observe X having a pdf $f_{X|H}$ that depends upon a hypothesis H about which one of two states of the world (H_0 is "target absent" and H_1 is "target present") is correct (see Section 13.4);
- pattern classification in which the distribution $f_{\mathbf{X}|\Theta}$ of a feature measurement $\mathbf{X}$ depends upon the value Θ of a pattern class (e.g., which of finitely many messages was actually sent, which of several types of vehicle is present in an image) (see Chapter 16);
- forecasting in which $\mathbf{X}$ is the observable (e.g., present and past stock prices), Y is the variable to be forecast (e.g., tomorrow's stock price) and S is the random system relating the two (e.g., the market) (see Chapter 16).

12.6 USEFUL FAMILIES OF CONDITIONAL DENSITIES

12.6.1 Independent or Product Model

Under the *product or independent model* (see Sections 7.4.1, 14.4) of a multivariate density modeling unlinked random variables $X_1, \ldots, X_n$ (e.g., the heights of unrelated students in a class, measurements made on physically uncoupled systems),

$$f_{X_1^n}(x_1^n) = \prod_{i=1}^{n} f_{X_i}(x_i).$$

In this case,

$$f_{X_{m+1}^n|X_1^m}(x_{m+1}^n|x_1^m) = \prod_{i=m+1}^{n} f_{X_i}(x_i).$$

Example 12.7   **Conditional PDF for Noise Given Signal** ——————————

In the signal-plus-noise model, which is also a standard model for measurement error, referred to earlier and treated more completely in Section 13.5, we observe a signal or make a measurement $X = S + N$, where S is the signal or true quantity and N is the random measurement perturbation. Typically, we assume a product model for

$$f_{S,N}(s, n) = f_S(s)f_N(n),$$

which in turn leads to the conditional model $f_{N|S} = f_N$ and $f_{X|S}(x|s) = f_N(x - s)$.

12.6.2 Markov Model

A simple and widely used first step in introducing dependence among random variables that are sequentially ordered (typically by time of occurrence) is provided by the notion of *Markov dependence*.

Definition 12.6.1   **(Markov Dependence)**   The random variables X_1^n are said to be Markov dependent if

$$(\forall i > 1)\, f_{X_i|X_1^{i-1}}(x_i|x_1^{i-1}) = f_{X_i|X_{i-1}}(x_i|x_{i-1}).$$

For $m \geq 1$, the sequence Theorem 12.4.2 followed by Markov dependence yields

$$f_{X_1^n}(x_1^n) = f_{X_1}(x_1)\prod_{k=2}^{n} f_{X_k|X_1^{k-1}}(x_k|x_1^{k-1}) = f_{X_1}(x_1)\prod_{k=2}^{n} f_{X_k|X_{k-1}}(x_k|x_{k-1}).$$

It is immediate from the definition of the conditional pdf that

$$f_{X_{m+1}^n|X_1^m}(x_{m+1}^n|x_1^m) = \prod_{i=m+1}^{n} f_{X_i|X_{i-1}}(x_i|x_{i-1}) = f_{X_{m+1}^n|X_m}(x_{m+1}^n|x_m).$$

The conditional density $f_{X_i|X_{i-1}}(x_i|x_{i-1})$ is known as the *one-step transition density*. In many applications, we find that $f_{X_i|X_{i-1}}$ does not depend on the (time) index i, and we have what is known as a *homogeneous Markov process*. (See Section 20.6.)

Example 12.8   **Natural Language Text Production** ——————————

An approximation to natural language text assumes that the ith word X_i is related to all preceding words X_1^{i-1} only through the most recent word X_{i-1}. A detailed, if contrived, example is provided in Example 20.13 in Chapter 20. That this Markov model cannot be correct for natural languages is evident when one considers the candidate next word continuations of "… probability is ??" and compares it with the continuations of "The father of modern probability is ??"

Example 12.9   Gambling Winnings _____

An analytically correct example of a Markov process is provided by the sequence $S_1, \ldots, S_n, \ldots$, where S_k is the cumulative amount won after gambling on the kth outcome of the toss of a coin or the spin of a roulette wheel. The actual amount won by betting on the outcome itself is $X_k = S_k - S_{k-1}$, or, rewritten,

$$S_n = \sum_1^n X_i = S_{n-1} + X_n.$$

This model is known as a *random walk* with step X_n. Formalizing what we have asserted is the Markov property

$$f_{S_n | S_1^{n-1}}(s_n | s_1^{n-1}) = f_{S_n | S_{n-1}}(s_n | s_{n-1}).$$

It can be shown (see Section 13.5 for the ideas) that

$$f_{S_n | S_{n-1}}(s_n | s_{n-1}) = f_{X_n}(s_n - s_{n-1}).$$

Example 12.10   Cascade of Systems _____

Consider a *cascade* (serial chain) of noisy systems $\mathcal{S}_1, \ldots, \mathcal{S}_n$ in which system $\mathcal{S}_k$ has input X_{k-1} and output X_k. Except for the output of $\mathcal{S}_{k-1}$ being the input of $\mathcal{S}_k$, these systems are otherwise physically unlinked. In this case, we know as much about X_k given X_1^{k-1} as we know from just X_{k-1} alone:

$$f_{X_k | X_1^{k-1}}(x_k | x_1^{k-1}) = f_{X_k | X_{k-1}}(x_k | x_{k-1}).$$

If, in addition, the system $\mathcal{S}_k$ is one that adds noise N_k to its input to determine its output, then

$$f_{X_k | X_{k-1}}(x_k | x_{k-1}) = f_{N_k}(x_k - x_{k-1}).$$

12.6.3 Multivariate Normal Model

We extend our results for the bivariate normal conditional density of Example 12.5 to the *multivariate normal* case by using the full joint density to evaluate $f_{X_k | X_1, \ldots, X_{k-1}}(x_k | x_1, \ldots, x_{k-1})$. Letting $\mathbb{Q}$ denote the inverse $\mathbb{C}^{-1}$ of the covariance matrix $\mathbb{C}$ of the full vector $\mathbf{X} = (X_1, \ldots, X_k)^T$ and $S = (\mathbf{x} - \mathbf{m})^T \mathbb{Q}(\mathbf{x} - \mathbf{m})$, we have

$$f_{\mathbf{X}}(\mathbf{x}) = (2\pi)^{-\frac{n}{2}} \sqrt{\det(\mathbb{Q})} \exp(-\tfrac{1}{2}(\mathbf{x} - \mathbf{m})^T \mathbb{Q}(\mathbf{x} - \mathbf{m}))$$

$$= (2\pi)^{-\frac{n}{2}} \sqrt{\det(\mathbb{Q})} \exp(-\tfrac{1}{2}S).$$

We focus on the exponent, S, as this is the only term that is a function of $\mathbf{x}$, and expand it as

$$S = (\mathbf{x} - \mathbf{m})^T \mathbb{Q}(\mathbf{x} - \mathbf{m}) = \sum_{i=1}^{k} \sum_{j=1}^{k} Q_{ij}(x_i - m_i)(x_j - m_j)$$

$$= Q_{kk}(x_k - m_k)^2 + \left[\sum_{j<k} (Q_{kj} + Q_{jk})(x_j - m_j) \right](x_k - m_k)$$

$$+ \sum_{i=1}^{k-1} \sum_{j=1}^{k-1} Q_{ij}(x_i - m_i)(x_j - m_j).$$

From the symmetry, $\mathbb{Q}^T = \mathbb{Q}$, of the inverse of a covariance matrix, we have $Q_{kj} = Q_{jk}$, and hence

$$S = Q_{kk}(x_k - m_k)^2 + 2\left[\sum_{j<k} Q_{kj}(x_j - m_j) \right](x_k - m_k)$$

$$+ \sum_{i=1}^{k-1} \sum_{j=1}^{k-1} Q_{ij}(x_i - m_i)(x_j - m_j).$$

Observing that S is a quadratic in $x_k - m_k$, we complete the square in this term to derive

$$S = S_1 + S_2, \quad S_1 = Q_{kk}\left[x_k - m_k + \frac{1}{Q_{kk}} \sum_{j<k} Q_{kj}(x_j - m_j) \right]^2,$$

$$S_2 = \sum_{i=1}^{k-1} \sum_{j=1}^{k-1} Q_{ij}(x_i - m_i)(x_j - m_j) - \frac{1}{Q_{kk}}\left[\sum_{j<k} Q_{kj}(x_j - m_j) \right]^2.$$

Note that S_2 is not a function of x_k. We can write, using the proportionality symbol $\propto$ to reflect that we are ignoring multiplicative constants (terms that are not functions of $\mathbf{x}$),

$$f_{\mathbf{X}}(\mathbf{x}) \propto \exp\left(-\frac{1}{2}S_1\right)\exp\left(-\frac{1}{2}S_2\right).$$

Inspecting the term S_1 reveals that in the variable x_k, the factor $\exp\left(-\frac{1}{2}S_1\right)$ is proportional to a density for X_k that is $\mathcal{N}\left(m_k - \frac{1}{Q_{kk}}\sum_{j<k} Q_{kj}(x_j - m_j), \frac{1}{Q_{kk}}\right)$. Thus, integration in $f_{\mathbf{X}}(\mathbf{x})$ over x_k yields the marginal density

$$f_{X_1,\dots,X_{k-1}}(x_1, \dots, x_{k-1}) \propto \exp\left(-\frac{1}{2}S_2\right),$$

which is a multivariate normal density as well. From

$$f_{X_k|X_1,\dots,X_{k-1}}(x_k|x_1, \dots, x_{k-1}) = \frac{f_{\mathbf{X}}(\mathbf{x})}{f_{X_1,\dots,X_{k-1}}(x_1, \dots, x_{k-1})},$$

we see that

$$f_{X_k|X_1,\ldots,X_{k-1}}(x_k|x_1,\ldots,x_{k-1}) \propto \exp\left(-\frac{1}{2}S_1\right).$$

Thus, we conclude that the conditional probability law of X_k given $(X_1,\ldots,X_{k-1})$ is

$$\mathcal{N}\left(m_k - \frac{1}{Q_{kk}}\sum_{j<k}Q_{kj}(x_j - m_j), \frac{1}{Q_{kk}}\right), \tag{12.2}$$

where $\mathbb{Q}$ is the inverse of the covariance matrix of $\mathbf{X}$.

Example 12.11   Trivariate Normal Conditional PDF _____

Let

$$\mathbf{m} = (-1\ 0\ 1)^T \text{ and}$$

$$\mathbb{C} = \begin{pmatrix} 7 & 4 & 5 \\ 4 & 4 & 3 \\ 5 & 3 & 7 \end{pmatrix}.$$

In this case,

$$\mathbb{Q} = \mathbb{C}^{-1} = \begin{pmatrix} 0.4634 & -0.3171 & -0.1951 \\ -0.3171 & 0.5854 & -0.0244 \\ -0.1951 & -0.0244 & 0.2927 \end{pmatrix} \text{ with } \frac{1}{Q_{3,3}} = 3.4167.$$

(Note that while $Q_{3,3}$ is the $(3,3)$ element of $\mathbb{C}^{-1}$, it is not the reciprocal of $C_{3,3}$.) Hence, X_3 given $X_1 = x_1$ and $X_2 = x_2$ has the conditional pdf

$$f_{X_3|X_1,X_2}(x_3|x_1,x_2) = \sqrt{\frac{3.4167}{2\pi}}e^{-\frac{3.4167}{2}(x_3-m_{3|12})^2}$$

with $m_{3|12} = 1 - 3.4167\left(-0.1951(x_1 + 1) - 0.0244x_2\right).$

The least mean square linear (and, as will be seen later, also nonlinear) predictor $\hat{X}_3(X_1, X_2)$ of X_3 given $X_1 = x_1$ and $X_2 = x_2$ is

$$\hat{X}_3(x_1, x_2) = 1 - 3.4167\left(-0.1951(x_1 + 1) - 0.0244x_2\right) = 1.6667 + .6667x_1 + .0833x_2.$$

The resulting mean square estimation error is $1/Q_{3,3} = 3.4167.$

12.7 EXTENSIONS OF THE TOTAL PROBABILITY THEOREM

The following extension of the Total Probability Theorem 11.5.1 covers the cases of conditioning either continuous or discrete random variables on either continuous or discrete random variables, but omits the case already treated in which both sets of variables are discrete. It is, of course, also possible to have arbitrary mixtures of discrete and continuous random variables. These can be treated by using impulse functions to incorporate pmfs in the pdfs. The theorem that follows is somewhat forbidding at first glance. However, the pattern established in Theorem 11.5.1 is just being repeated in three different settings.

Theorem 12.7.1   (Total Probability Extension)

a. If the variables in question are all continuous or real valued, then we have

$$f_{X^n_{m+1}}(x^n_{m+1}) = \int_{-\infty}^{\infty} \cdots \int_{-\infty}^{\infty} f_{X^n_1}(x^n_1)\, dx_1 \ldots dx_m$$

$$= \int_{-\infty}^{\infty} \cdots \int_{-\infty}^{\infty} f_{X^n_{m+1}|X^m_1}(x^n_{m+1}|x^m_1) f_{X^m_1}(x^m_1)\, dx_1 \ldots dx_m.$$

b. If the variables $X_1, \ldots, X_m$ are discrete valued, so that their density would contain impulse functions, while the remaining variables are continuous or real valued, then we can use the pmf $p_{X^m_1}$ to obtain

$$f_{X^n_{m+1}}(x^n_{m+1}) = \sum_{x_1} \cdots \sum_{x_m} f_{X^n_{m+1}|X^m_1}(x^n_{m+1}|x^m_1) p_{X^m_1}(x^m_1).$$

c. If the variables X^m_1 are continuous or real valued and the variables X^n_{m+1} are discrete valued, then we introduce a conditional pmf

$$p_{X^n_{m+1}|X^m_1}(x^n_{m+1}|x^m_1) = P(X^n_{m+1} = x^n_{m+1}|X^m_1 = x^m_1),$$

to obtain

$$p_{X^n_{m+1}}(x^n_{m+1}) = \int_{-\infty}^{\infty} \cdots \int_{\infty}^{\infty} p_{X^n_{m+1}|X^m_1}(x^n_{m+1}|x^m_1) f_{X^m_1}(x^m_1)\, dx_1 \ldots dx_m.$$

This theorem assists us in achieving Goal 2—calculating system response X^n_{m+1} to system excitation X^m_1.

12.8   APPLICATIONS OF THE EXTENDED TOTAL PROBABILITY THEOREM

12.8.1 Continuous Variables Conditioned on Continuous Variables

Example 12.12   Total Probability for Continuous Variables _____
Continuing with the first example in Section 12.3, we determined that

$$f_{Y|X}(y|x) = \frac{f_{X,Y}(x,y)}{f_X(x)} = 2\frac{e^{-y}(1+e^{-x})^2}{(1+e^{-x}+e^{-y})^3} \text{ and } f_X(x) = \frac{e^{-x}}{(1+e^{-x})^2}.$$

Applying Theorem 12.7.1(a) enables us to calculate

$$f_Y(y) = \int_{-\infty}^{\infty} 2 \frac{e^{-y}(1 + e^{-x})^2}{(1 + e^{-x} + e^{-y})^3} \frac{e^{-x}}{(1 + e^{-x})^2} \, dx.$$

Of course, in this case we know F_Y, and we know that the preceding integral evaluates as

$$f_Y(y) = \frac{e^{-y}}{(1 + e^{-y})^2}.$$

Further examples of continuous variables being conditioned on continuous variables were given in Section 12.5.

Example 12.13   Total Probability for Continuous Signal and Noise _____

If we consider the first two examples of communication of a signal S in the presence of independent additive noise N or of a measured value X of a true value S, then the received signal $X = S + N$ and

$$f_{X|S}(x|s) = f_N(x - s).$$

If we are given the density f_S describing the choice of signal S, then we can employ Theorem 12.7.1(a) to calculate a description of the received signal X:

$$f_X(x) = \int_{-\infty}^{\infty} f_N(x - s) f_S(s) \, ds.$$

More specifically, if $N \sim \mathcal{N}(0, \sigma_N^2)$ and $S \sim \mathcal{N}(m_S, \sigma_S^2)$ then, from the preceding integral, we can determine that $X \sim \mathcal{N}(m_S, \sigma_N^2 + \sigma_S^2)$.

Example 12.14   Total Probability for the Multivariate Normal _____

Another example is provided by the multivariate normal of the preceding section and its application to dynamical systems (see Section 10.6) in which, at time t, the n-dimensional state vector $\mathbf{X}_t \sim \mathcal{N}(\mathbf{0}, \mathbb{Q}_t)$ is multivariate normal. There is also the l-dimensional vector of observations $\mathbf{Z}_s$ on the behavior of the system at time s. The vectors $\mathbf{X}_t$ and $\mathbf{Z}_s$ are assumed to be related through

$$\mathbf{Z}_t = \mathbb{H}\mathbf{X}_t + \mathbf{V}_t,$$

where the $l \times n$ matrix $\mathbb{H}$ is nonrandom and known, the l-dimensional random vector $\mathbf{V}_t \sim \mathcal{N}(\mathbf{0}, \mathbb{R}_t)$, and $\mathbf{X}_t$ and $\mathbf{V}_t$ are jointly normal and uncorrelated [$E(\mathbf{X}_t \mathbf{V}_t^T) = \mathbb{O}_{nl}$]. What, then, is the description of $\mathbf{Z}_t$? By Theorem 12.7.1(a)

$$f_{\mathbf{Z}_t}(\mathbf{z}) = \int_{-\infty}^{\infty} \cdots \int_{-\infty}^{\infty} f_{\mathbf{Z}_t|\mathbf{X}_t}(\mathbf{z}|\mathbf{x}) f_{\mathbf{X}_t}(\mathbf{x}) \, dx_1 \ldots, dx_n.$$

The pdf $f_{\mathbf{X}_t}(\mathbf{x})$ is given as $\mathcal{N}(\mathbf{0}, \mathbb{Q}_t)$. We obtain the conditional pdf $f_{\mathbf{Z}_t|\mathbf{X}_t}(\mathbf{z}|\mathbf{x})$ as a version of signal plus noise. The uncorrelatedness of $\mathbf{X}_t$ and $\mathbf{V}_t$ when they are jointly normal implies that they are unlinked or independent, or

$$f_{\mathbf{X}_t, \mathbf{V}_t}(\mathbf{x}, \mathbf{v}) = f_{\mathbf{X}_t}(\mathbf{x}) f_{\mathbf{V}_t}(\mathbf{v}).$$

Hence,

$$f_{\mathbf{Z}_t|\mathbf{X}_t}(\mathbf{z}|\mathbf{x}) = f_{\mathbf{V}_t}(\mathbf{z} - \mathbb{H}\mathbf{x}).$$

We now have everything we need to apply Theorem 12.7.1(a). However, we do not pursue this further, as there are easier ways to determine that $\mathbf{Z}_t \sim \mathcal{N}(\mathbf{0}, \mathbb{M}_t)$. From Section 9.15,

$$\mathbb{M}_t = \mathbb{H}\mathbb{Q}_t\mathbb{H}^T + \mathbb{R}_t.$$

12.8.2 Continuous Variables Conditioned on Discrete Variables

The following are examples of discrete X_1^m and continuous X_{m+1}^n:

- pattern classification where X is the randomly selected pattern class (e.g., a phoneme or an alphanumeric character) and Y is the analog feature value (e.g., a short-time speech waveform, or an array of pixel amplitudes resulting from a scan of the character);
- reliability and diagnostic models in which X is the manufacturer of a component with lifetime Y or X is an illness and Y is a measured symptom (e.g., temperature, blood pressure), and the system S is the random relationship between manufacturing process and performance or between illness and its symptomatic display;
- communicating a digital signal X over an analog channel S yielding an analog received signal Y;
- mobile cellular communications where X is the cell membership of a caller, Y is the signal strength received at the base station antenna, and S is the propagation channel between the two.

Example 12.15   Total Probability for Continuous Conditioned on Discrete Variables __
We simplify the phoneme recognition problem to a Laplacian model in which

$$f_{Y|X}(y|x) = \frac{\alpha(x)}{2} e^{-\alpha(x)|y|};$$

unrealistically, we have reduced phonemes to amplitudes whose variance depends upon the choice X of phoneme. If we are given the pmf p_X for the likelihoods of the 44 phonemes of

English, then Theorem 12.7.1(b) allows us to describe the feature measurement Y by

$$f_Y(y) = \sum_x \frac{\alpha(x)}{2} e^{-\alpha(x)|y|} p_X(x).$$

Example 12.16   Total Probability for Continuous Conditioned on Discrete Variables (Continued)

A similar example assumes a digital signal X taking on K values $\{x_i\}$ and characterized by a pmf p_X that is observed over a channel characterized by independent additive Gaussian noise $N \sim \mathcal{N}(0, \sigma^2)$. We determine the description of the channel output or received signal Y given the input X through

$$f_{Y|X}(y|x) = \frac{1}{\sigma\sqrt{2\pi}} e^{-\frac{(y-x)^2}{2\sigma}}.$$

Recourse to Theorem 12.7.1(b) yields

$$f_Y(y) = \sum_{i=1}^{K} \frac{1}{\sigma\sqrt{2\pi}} e^{-\frac{(y-x_i)^2}{2\sigma^2}} p_X(x_i).$$

12.8.3 Discrete Variables Conditioned on Continuous Variables

The following are examples of continuous X_1^m and discrete X_{m+1}^n:

- quantizer or A/D converter in which the analog input X (e.g., speech, music) is quantized to one of finitely many levels Y by a possibly noisy quantizer $\mathcal{S}$;
- decision-making system $\mathcal{S}$, such as a receiver, that assigns one of finitely many pattern classes or possible original digital messages Y to an analog feature or received signal X.

Example 12.17   Total Probability for Discrete Conditioned on Continuous Variable

Let us model a binary-valued noisy A/D converter accepting analog X and outputting ± 1-valued Y as

$$Y = \text{sign}(X + N), \ N \sim \mathcal{N}(0, \sigma^2).$$

Then

$$P(Y = -1|X = x) = 1 - P(Y = 1|X = x) = P(X + N < 0|X = x)$$

$$= P(N < -x|X = x) = P(N < -x),$$

where the last conclusion follows if the noise N is independent of (unlinked to, $f_{N,X} = f_N f_X$) X. Hence,

$$P(Y = -1 | X = x) = \Phi\left(-\frac{x}{\sigma}\right).$$

In any event, we accept this conclusion for the purposes of the example. Given the description f_X of X, we can now compute the pmf describing Y, through Theorem 12.7.1(c), as

$$p_Y(-1) = 1 - p_Y(1) = \int_{-\infty}^{\infty} \Phi\left(-\frac{x}{\sigma}\right) f_X(x)\, dx.$$

Example 12.18   Laplace's Rule of Succession _____

A classical example is *Laplace's Rule of Succession*, used by him to infer the probability that the sun will rise tomorrow given that it had risen each day in the analysis of approximately 5,500 years since, according to the Bible, the world was created. We rephrase this question as the probability that a coin with unknown probability π for heads will produce an $(n + 1)$st head, given that it has already produced n heads. On this account, we have the conditional pmf

$$P(X_1 = x_1, \ldots, X_k = x_k | \pi = p) = p^{\sum_1^k I_H(x_i)}(1 - p)^{n - \sum_1^k I_H(x_i)}.$$

Laplace assumed that ignorance of the true value of π could be represented by the uniform distribution $\mathcal{U}(0, 1)$. (Modeling "ignorance" by a uniform distribution is a too frequent practice!) Applying Theorem 12.7.1(c) yields

$$P(X_1 = H, \ldots, X_k = H) = \int_0^1 p^k\, dp = \frac{1}{k + 1}.$$

Hence,

$$P(X_{n+1} = H | X_1 = H, \ldots, X_n = H) = \frac{P(X_1 = H, \ldots, X_{n+1} = H)}{P(X_1 = H, \ldots, X_n = H)} = \frac{n + 1}{n + 2}.$$

As there are approximately 2,000,000 days in 5,500 years, Laplace arrived at the comforting conclusion that the probability that the sun would rise on the next day was approximately $1 - .0000005$.

The use of the Beta prior (see Section 6.5.2) would give us a more flexible class of priors, but still one for which we could complete our calculations in closed form.

12.9 EXTENSIONS OF BAYES' THEOREM

The following extension of Bayes' Theorem 11.6.1 covers the three cases detailed in the preceding subsection. In order to preserve the ordering of the examples in that section, we needed to reverse the ordering of parts (b) and (c) of the theorem.

Theorem 12.9.1   (Bayes' Theorem Extension)

a. If the variables in question are all continuous or real valued, then we have

$$f_{X_1^m | X_{m+1}^n}(x_1^m | x_{m+1}^n) = \frac{f_{X_{m+1}^n | X_1^m}(x_{m+1}^n | x_1^m) f_{X_1^m}(x_1^m)}{f_{X_{m+1}^n}(x_{m+1}^n)},$$

where the denominator can be evaluated from the appropriate Total Probability Theorem as a multiple integral of the numerator over the variables x_1^m.

b. If the variables X_1^m are discrete valued, so that their density would contain impulse functions, while the remaining variables are real valued, then we can use a pmf $p_{X_1^m}$ to obtain

$$p_{X_1^m | X_{m+1}^n}(x_1^m | x_{m+1}^n) = \frac{f_{X_{m+1}^n | X_1^m}(x_{m+1}^n | x_1^m) p_{X_1^m}(x_1^m)}{f_{X_{m+1}^n}(x_{m+1}^n)},$$

where the denominator can be evaluated from the appropriate total probability theorem as a multiple sum of the numerator over the variables x_1^m.

c. If X_{m+1}^n is discrete and X_1^m continuous, we can use a conditional pmf to find

$$f_{X_1^m | X_{m+1}^n}(x_1^m | x_{m+1}^n) = \frac{p_{X_{m+1}^n | X_1^m}(x_{m+1}^n | x_1^m) f_{X_1^m}(x_1^m)}{p_{X_{m+1}^n}(x_{m+1}^n)}.$$

This theorem assists us in achieving Goal 3—inferring from observed X_{m+1}^n to the unobserved X_1^m when we know the description of X_{m+1}^n given X_1^m. The examples presented in the preceding section apply here as well. We illustrate each of the three types of Bayes' calculation by further developing the illustrations of the Total Probability Theorem given in the last section.

12.10   APPLICATIONS OF THE EXTENDED BAYES' THEOREM

12.10.1   Continuous Causes Conditioned on Continuous Effects

Example 12.19   Bayes' Theorem for Continuous Variables _____
In the example of a continuous or real-valued signal X received in the presence of independent additive noise N as $Y = X + N$, we now wish to determine $f_{X|Y}$, to specify what is known about X from observation of Y. From the work of the preceding section and the extended Bayes' Theorem, we find that

$$f_{X|Y}(x|y) = \frac{f_N(y - x) f_X(x)}{\int_{-\infty}^{\infty} f_N(y - z) f_X(z)\, dz}.$$

Example 12.20   Bayes' Theorem for Multivariate Normal Variables _____
In the dynamical system example of Section 12.8.1, we determined that the observations $\mathbf{Z}_t$, given the true state $\mathbf{X}_t$, had a conditional pdf $f_{\mathbf{Z}_t | \mathbf{X}_t}$ that was given by the $\mathcal{N}(\mathbf{0}, \mathbb{H}\mathbb{Q}_t\mathbb{H}^T + \mathbb{R}_t)$.

We also knew that $\mathbf{X}_t \sim \mathcal{N}(\mathbf{0}, \mathbb{Q}_t)$. Our interest in the measurement $\mathbf{Z}_t$ is in what it tells us about the state $\mathbf{X}_t$. In fact, the whole past history of such measurements tells us even more about $\mathbf{X}_t$, although we ignore this for now and leave it to our discussion of Kalman filtering. The information that $\mathbf{Z}_t$ conveys about $\mathbf{X}_t$ is given by Theorem 12.9.1(a):

$$f_{\mathbf{X}_t|\mathbf{Z}_t}(\mathbf{x}|\mathbf{z}) = \frac{f_{\mathbf{Z}_t|\mathbf{X}_t}(\mathbf{z}|\mathbf{x})f_{\mathbf{X}_t}(\mathbf{x})}{f_{\mathbf{Z}_t}(\mathbf{z})}.$$

All of the terms appearing in the ratio on the right-hand side have been evaluated in Section 12.8.1. The result of substitution is the confusing formula

$$f_{\mathbf{X}_t|\mathbf{Z}_t}(\mathbf{x}|\mathbf{z}) = \frac{1}{(2\pi)^{\frac{n}{2}}}\sqrt{\frac{\det(\mathbb{H}\mathbb{Q}_t\mathbb{H}^T + \mathbb{R}_t)}{\det(\mathbb{R}_t)\det(\mathbb{Q}_t)}}$$

$$\times \exp[-\tfrac{1}{2}((\mathbf{z} - \mathbb{H}\mathbf{x})^T\mathbb{R}_t^{-1}(\mathbf{z} - \mathbb{H}\mathbf{x}) + \mathbf{x}^T\mathbb{Q}_t^{-1}\mathbf{x} - \mathbf{z}^T(\mathbb{H}\mathbb{Q}_t\mathbb{H}^T + \mathbb{R}_t)^{-1}\mathbf{z})].$$

Simplifying this result is not essential to the point of this example as an illustration of the use of Bayes' Theorem. However, we provide a guide to such simplification leading to the conclusion that $\mathbf{X}_t$ conditional upon $\mathbf{Z}_t$ is again multivariate normal with a mean vector $\mathbf{m}_{\mathbf{X}_t|\mathbf{Z}_t}$ and covariance matrix $\mathbb{C}_{\mathbf{X}_t|\mathbf{Z}_t}$ that are given next. The first step is to notice that the exponent appearing in $f_{\mathbf{X}_t|\mathbf{Z}_t}(\mathbf{x}|\mathbf{z})$ has a term that is quadratic in $\mathbf{x}$ and given by

$$\mathbf{x}^T\mathbb{H}^T\mathbb{R}_t^{-1}\mathbb{H}\mathbf{x} + \mathbf{x}^T\mathbb{Q}_t^{-1}\mathbf{x} = \mathbf{x}^T\left(\mathbb{H}^T\mathbb{R}_t^{-1}\mathbb{H} + \mathbb{Q}_t^{-1}\right)\mathbf{x}.$$

From this, we can conclude that $\mathbf{X}_t$ given $\mathbf{Z}_t$ is again multivariate normal in $\mathbf{X}_t$, with a covariance matrix

$$\mathbb{C}_{\mathbf{X}_t|\mathbf{Z}_t} = \left(\mathbb{H}^T\mathbb{R}_t^{-1}\mathbb{H} + \mathbb{Q}_t^{-1}\right)^{-1}.$$

We can identify the mean of $\mathbf{X}_t$ given $\mathbf{Z}_t$ by looking at the terms in the exponent that are linear in $\mathbf{x}$ and "completing the square" in the quadratic. Doing so yields

$$\mathbf{X}_t|\mathbf{Z}_t \sim \mathcal{N}(\mathbb{H}^T\mathbb{R}_t^{-1}\mathbb{C}_{\mathbf{X}_t|\mathbf{Z}_t}^{-1}, \mathbb{C}_{\mathbf{X}_t|\mathbf{Z}_t}).$$

12.10.2 Discrete Causes Conditioned on Continuous Effects

We continue with the examples given in Section 12.8.2.

Example 12.21   Bayes' Theorem for Discrete Causes Conditioned on Continuous Effects

In the phoneme recognition problem, we determined that the measurement Y related to the phoneme X having a pmf $p_X(x)$ through

$$f_{Y|X}(y|x) = \frac{\alpha(x)}{2}e^{-\alpha(x)|y|} \quad \text{and} \quad f_Y(y) = \sum_x \frac{\alpha(x)}{2}e^{-\alpha(x)|y|}p_X(x).$$

Our interest in the measurement Y is in what it tells us about the true phoneme X. This information is given by the conditional pmf calculated from Theorem 12.9.1(b):

$$p_{X|Y}(x|y) = \frac{f_{Y|X}(y|x)p_X(x)}{f_Y(y)} = \frac{\frac{\alpha(x)}{2}e^{-\alpha(x)|y|}p_X(x)}{\sum_z \frac{\alpha(z)}{2}e^{-\alpha(z)|y|}p_X(z)}.$$

Example 12.22   Bayes' Theorem for Discrete Causes Conditioned on Continuous Effects (Continued) _____

In the example of a digital signal X characterized by a pmf p_X observed over an independent additive noise channel with $N \sim \mathcal{N}(0, \sigma^2)$, we have

$$f_{Y|X}(y|x) = \frac{1}{\sigma\sqrt{2\pi}}e^{-\frac{(y-x)^2}{\sigma^2}}.$$

The result from applying the Total Probability and Bayes' theorems is a conditional pmf

$$p_{X|Y}(x|y) = \frac{e^{-\frac{(y-x)^2}{\sigma^2}}p_X(x)}{\sum_{i=1}^{K}e^{-\frac{(y-x_i)^2}{2\sigma^2}}p_X(x_i)}.$$

12.10.3 Continuous Causes Conditioned on Discrete Effects

We continue with the examples of Section 12.8.3.

Example 12.23   Bayes' Theorem for Continuous Causes Conditioned on Discrete Effects _____

Finally, in the example of a binary-valued noisy A/D converter, the conditional pmf for the $\{-1, 1\}$-valued digital Y given the analog X was found to be

$$P(Y = -1|X = x) = \Phi(-\frac{x}{\sigma}).$$

The information about the analog X given the digital $Y = -1$ is

$$f_{X|Y}(x|-1) = \frac{\Phi(-\frac{x}{\sigma})f_X(x)}{\int_{-\infty}^{\infty}\Phi(-\frac{z}{\sigma})f_X(z)\,dz}.$$

Example 12.24   Laplace's Rule of Succession Revisited _____

Using Laplace's Rule of Succession and a uniform prior for the probability π that the sun would rise tomorrow, we determined the probability of a next success given a past run of n successes to be

$$P(X_{n+1} = H|X_1 = H, \ldots, X_n = H) = \frac{n+1}{n+2}.$$

However, we did not inquire into what this run of successes implied about the unknown probability π of an individual success. Recall our statement

$$P(X_1 = x_1, \ldots, X_n = x_n | \pi = p) = p^k (1 - p)^{n-k} \text{ for } k = \sum_{j=1}^n x_j.$$

Using the assumed $\pi \sim \mathcal{U}(0, 1)$, we can now determine

$$f_{\pi | K}(p|k) = \frac{p^k (1 - p)^{n-k}}{\int_0^1 x^k (1 - x)^{n-k} \, dx} U(p) U(1 - p).$$

From our knowledge of the Beta density (Section 6.5.2), we know that

$$\int_0^1 x^k (1 - x)^{n-k} \, dx = \frac{k!(n - k)!}{(n + 1)!}.$$

We conclude that

$$f_{\pi | K}(p|k) = \frac{(n + 1)!}{k!(n - k)!} p^k (1 - p)^{n-k}.$$

12.11 SUMMARY

The treatment of conditional probability given in Chapter 11 is extended to account for conditioning on certain events of probability zero. In particular, we wish to be able to evaluate the conditional probability $P(Y \in A | X = x)$ of a random variable $Y \in A$, given that a continuous random variable $X = x$, when this conditioning event has probability zero. The key is the introduction of a conditional density function $f_{Y|X}(y|x)$ and its multivariate extension $f_{\mathbf{Y}|\mathbf{X}}$. Important examples of multivariate conditional densities are used to define the product or independence case of unlinked random variables, Markov dependence in which the future is independent of the past conditional on the present, and the commonly encountered multivariate conditional normal. All of this contributes to achieving our Goal 1 of mathematical descriptions of random phenomena.

We restate the Total Probability, Sequence, and Bayes' theorems in this new context, although their forms remain the same as those of the versions given in Chapter 11. Theorem 12.7.1 enables us to achieve our Goal 2 of calculating the pdf or pmf of the output Y of a system S described by a conditional pdf or pmf of the output Y, given the input X described by a pdf or pmf. Theorem 12.9.1 similarly enables us to achieve our Goal 3 of using probability information to learn about causes (e.g., system inputs) from effects (e.g., system outputs).

These results are summarized and classified in two appendices following this section. The parallelism between the discrete and continuous cases is made evident. The two appendices are logically redundant, but the differences in exposition may help the reader.

12.12 APPENDIX 1: CONDITIONAL PROBABILITY SUMMARY I

The basic system configuration has an input random variable X transformed by a system S into an output random variable Y. G1 provides us with models for each component. G2 enables us to compute the description of Y, given that for X, S. G3 enables us to infer from knowledge of Y, S to X. In this summary, we organize results on the basis of the discrete or continuous classification of the random variables X and Y, first in the bivariate case where X and Y are scalar or real-valued random variables and second in the multivariate case of vector-valued $\mathbf{X}$ and $\mathbf{Y}$. As the fully discrete case was covered in Chapter 11 and summarized in Section 11.9, it is not repeated here.

12.12.1 Two Random Variables: Bivariate Case

X is continuous with pdf f_X and Y is continuous with pdf f_Y.

- System Characterization (Goal 1)
  Definition of Conditional Density
  $$f_{Y|X}(y|x) = \frac{f_{X,Y}(x,y)}{f_X(x)}$$
  Density of Effect Y Given Cause X

- System Response (Goal 2)
  Total Probability Theorem
  $$f_Y(y) = \int_{-\infty}^{\infty} f_{Y|X}(y|x) f_X(x)\, dx$$
  Density of Effect Y

- System Response (Goal 2)
  Sequence Density
  $$f_{X,Y}(x,y) = f_X(x) f_{Y|X}(y|x)$$
  Density of Sequence/Cascade of Events X, Y

- System Input from Output (Goal 3)
  Bayes' Theorem
  $$f_{X|Y}(x|y) = \frac{f_{Y|X}(y|x) f_X(x)}{\int_{-\infty}^{\infty} f_{Y|X}(y|x') f_X(x')\, dx'}$$
  Density of Cause Given Effect

  X is discrete with pmf $P(X = x_i) = p_i$ and Y is continuous with pdf f_Y.

- System Characterization (Goal 1)
  Definition of Conditional Density
  $$f_{Y|X}(y|x) = \frac{1}{p_i} \frac{\partial F_{Y|X}(Y|x_i)}{\partial y}$$
  Density of Effect Y Given Cause X

- System Response (Goal 2)
  Total Probability Theorem

$f_Y(y) = \sum_i f_{Y|X}(y|x_i)p_i$
Density of Effect Y

- System Response (Goal 2)
  Sequence Probability
  $P(Y \in A, X = x_i) = p_i \int_A f_{Y|X}(y|x_i)\,dy$
  Density of Sequence/Cascade of Events $X = x_i$, $Y \in A$

- System Input from Output (Goal 3)
  Bayes' Theorem
  $P(X = x_i|Y = y) = \frac{f_{Y|X}(y|x_i)p_i}{\sum_k f_{Y|X}(y|x_k)p_k}$
  Density of Cause Given Effect

  X is continuous with pdf f_X and Y is discrete with pmf $P(Y = y_i) = p_i$.

- System Characterization (Goal 1)
  Definition of Conditional Density
  $P(Y = y_i|X = x) = \frac{1}{f_X(x)} \frac{\partial P(Y=y_i, X \le x)}{\partial x}$
  Conditional PMF of Effect Y Given Cause X

- System Response (Goal 2)
  Total Probability Theorem
  $P(Y = y_i) = p_i = \int_{-\infty}^{\infty} P(Y = y_i|X = x)f_X(x)\,dx$
  PMF of Effect Y

- System Response (Goal 2)
  Sequence Probability
  $P(Y = y_i, X \in A) = \int_A P(Y = y_i|X = x)f_X(x)\,dx$
  Density of Sequence/Cascade of Events $X \in A$, $Y = y_i$

- System Input from Output (Goal 3)
  Bayes' Theorem
  $f_{X|Y}(x|Y = y_i) = \frac{P(Y=y_i|X=x)f_X(x)}{p_i}$
  Density of Cause Given Effect

12.12.2 Several Random Variables: Multivariate Case

Use the notation $X_j^k = \{X_j, X_{j+1}, \ldots, X_k\}$ to represent a sequence of random variables. Let the system $\mathcal{S}$ output or response Y correspond to X_{m+1}^n and the system input or excitation X correspond to X_1^m.

X_1^n is continuous with pdf $f_{X_1^n}$.

- System Characterization (Goal 1)
  Definition of Conditional Density

$$f_{X_{m+1}^n|X_1^m}(x_{m+1}^n|x_1^m) = \frac{f_{X_1^n}(x_1^n)}{f_{X_1^m}(x_1^m)}$$

Density of Effect $Y = X_{m+1}^n$ Given Cause $X = X_1^m$

- System Response (Goal 2)
  Total Probability Theorem
  $$f_{X_{m+1}^n}(x_{m+1}^n) = \int_{-\infty}^{\infty} \cdots \int_{-\infty}^{\infty} f_{X_{m+1}^n|X_1^m}(x_{m+1}^n|x_1^m) f_{X_1^m}(x_1^m)\, dx_1 \ldots dx_m$$
  Density of Effect $Y = X_{m+1}^n$

- System Response (Goal 2)
  Sequence Density
  $$f_{X_1^n}(x_1^n) = f_{X_1}(x_1) \prod_2^n f_{X_k|X_1^{k-1}}(x_k|x_1^{k-1})$$
  Density of Sequence/Cascade of Random Variables X_1^n

- System Input from Output (Goal 3)
  Bayes' Theorem
  $$f_{X_1^m|X_{m+1}^n}(x_1^m|x_{m+1}^n) = \frac{f_{X_{m+1}^n|X_1^m}(x_{m+1}^n|x_1^m) f_{X_1^m}(x_1^m)}{f_{X_{m+1}^n}(x_{m+1}^n)}$$
  Density of Cause X_1^m Given Effect X_{m+1}^n

X_1^m is discrete with pmf $p_{X_1^m}$ and X_{m+1}^n is continuous with pdf $f_{X_{m+1}^n}$.

- System Characterization (Goal 1)
  Definition of Conditional Density
  $$f_{X_{m+1}^n|X_1^m}(x_{m+1}^n|x_1^m) = \frac{f_{X_1^n}(x_1^n)}{f_{X_1^m}(x_1^m)},$$
  where the numerator and denominator pdfs contain impulses
  for the discrete random variables.
  In practice, the conditional pdf is specified directly through
  an understanding of the system $\mathcal{S}$.
  Density of Effect $Y = X_{m+1}^n$ Given Cause $X = X_1^m$

- System Response (Goal 2)
  Total Probability Theorem
  $$f_{X_{m+1}^n}(x_{m+1}^n) = \sum_{x_1^m} f_{X_{m+1}^n|X_1^m}(x_{m+1}^n|x_1^m) p_{X_1^m}(x_1^m)$$
  Density of Effect $Y = X_{m+1}^n$

- System Response (Goal 2)
  Sequence Density
  $$f_{X_1^n}(x_1^n) = f_{X_1}(x_1) \prod_2^n f_{X_k|X_1^{k-1}}(x_k|x_1^{k-1})$$
  Density of Sequence/Cascade of Random Variables X_1^n

The terms f_{X_1} and $f_{X_k|X_1^{k-1}}$ for $k \le m$ contain impulses because the random variables X_1^k are discrete.

- System Input from Output (Goal 3)
  Bayes' Theorem
  $$p_{X_1^m|X_{m+1}^n}(x_1^m|x_{m+1}^n) = \frac{f_{X_{m+1}^n|X_1^m}(x_{m+1}^n|x_1^m)p_{X_1^m}(x_1^m)}{f_{X_{m+1}^n}(x_{m+1}^n)}$$
  PMF of Cause X_1^m Given Effect X_{m+1}^n

  X_1^m is continuous with pdf $f_{X_1^m}$ and X_{m+1}^n is discrete with pmf $p_{X_{m+1}^n}$.

- System Characterization (Goal 1)
  Definition of Conditional Mass Function
  In practice, the conditional pmf $p_{X_{m+1}^n|X_1^m}(x_{m+1}^n|x_1^m)$ is specified directly through an understanding of the system $\mathcal{S}$.
  PMF of Effect $Y = X_{m+1}^n$ Given Cause $X = X_1^m$

- System Response (Goal 2)
  Total Probability Theorem
  $$p_{X_{m+1}^n}(x_{m+1}^n) = \int_{-\infty}^{\infty} \cdots \int_{-\infty}^{\infty} p_{X_{m+1}^n|X_1^m}(x_{m+1}^n|x_1^m)f_{X_1^m}(x_1^m)\,dx_1 \cdots dx_m$$
  PMF of Effect $Y = X_{m+1}^n$

- System Response (Goal 2)
  Sequence Density
  $$f_{X_1^n}(x_1^n) = f_{X_1}(x_1)\prod_2^n f_{X_k|X_1^{k-1}}(x_k|x_1^{k-1})$$
  Density of Sequence/Cascade of Random Variables X_1^n
  The terms f_{X_1} and $f_{X_k|X_1^{k-1}}$ for $k \le m$ contain impulses because the random variables X_1^k are discrete.

- System Input from Output (Goal 3)
  Bayes' Theorem
  $$f_{X_1^m|X_{m+1}^n}(x_1^m|x_{m+1}^n) = \frac{p_{X_{m+1}^n|X_1^m}(x_{m+1}^n|x_1^m)f_{X_1^m}(x_1^m)}{p_{X_{m+1}^n}(x_{m+1}^n)}$$
  PDF of Cause X_1^m Given Effect X_{m+1}^n

12.13 APPENDIX 2: CONDITIONAL PROBABILITY SUMMARY II

In this summary, we organize the presentation by the achievement of our three goals and not by the classification of X and Y as being either discrete (D) or continuous (C). We summarize only the bivariate case. We have 2×2 cases corresponding to the choices of D or C for each of X, Y. We proceed to treat each of our three goals under each of these four possibilities.

12.13.1 Goal 1:

Specifying the components of the probability model for X, Y, S.

$X, Y = D, D$: Use pmf $p_X(x) = P(X = x)$ for X, p_Y for Y, and $P(Y = y_i|X = x)$ for S, where

$$P(Y = y_i|X = x) \geq 0, \quad \sum_{y_i} P(Y = y_i|X = x) = 1.$$

$X, Y = C, C$: Use pdf f_X for X, f_Y for Y, and the conditional density $f_{Y|X}$ for S,

$$f_{Y|X}(y|x) \geq 0, \quad \int_{-\infty}^{\infty} f_{Y|X}(y|x)\,dy = 1.$$

$X, Y = D, C$: Use pmf p_X for X, pdf f_Y for Y, and $f_{Y|X}(y|x_i)$ for S.

$X, Y = C, D$: Use pdf f_X for input X, pmf p_Y for output Y, and $P(Y = y_i|X = x)$ for S, where

$$P(Y = y_i|X = x) \geq 0, \quad \sum_{y_i} P(Y = y_i|X = x) = 1.$$

12.13.2 Goal 2:

We use the Total Probability Theorem to compute the response pdf f_Y or pmf p_Y to excitation X passed through S:

$$X, Y = D, D: \quad P(Y = y_i) = \sum_{x_j} P(Y = y_i|X = x_j)p_X(x_j).$$

$$X, Y = C, C: \quad f_Y(y) = \int_{-\infty}^{\infty} f_{Y|X}(y|x)f_X(x)\,dx.$$

$$X, Y = D, C: \quad f_Y(y) = \sum_{x_j} f_{Y|X}(y|x_j)p_X(x_j).$$

$$X, Y = C, D: \quad P(Y = y_i) = \int_{-\infty}^{\infty} P(Y = y_i|X = x)f_X(x)\,dx.$$

12.13.3 Goal 3:

We use Bayes' Theorem to compute the posterior (so-called because we reassess what we know of X after observing Y) pdf $f_{X|Y}$ or pmf $p_{X|Y}$. The denominator in each right-hand side is evaluated by the expressions given under Goal 2:

$$X, Y = D, D: \quad P(X = x_j|Y = y_i) = \frac{P(Y = y_i|X = x_j)p_X(x_j)}{p_Y(y_i)}.$$

$$X, Y = C, C: \quad f_{X|Y}(x|y) = \frac{f_{Y|X}(y|x)f_X(x)}{f_Y(y)}.$$

$$X, Y = D, C: \quad P(X = x_j | Y = y) = \frac{f_{Y|X}(y|x_j) p_X(x_j)}{f_Y(y)}.$$

$$X, Y = C, D: \quad f_{X|Y}(x|y_i) = \frac{P(Y = y_i | X = x) f_X(x)}{p_Y(y_i)}.$$

EXERCISES

E12.1   If $f_{X,Y}(x, y) = 4xy$ on $[0, 1]^2$ (the unit square) and 0 otherwise, then evaluate $f_{Y|X}$ for $0 < x < 1$.

E12.2   For appropriately chosen constant c,

$$f_{X,Y}(x, y) = c e^{-x-y-xy} U(x) U(y).$$

Evaluate $P(X \leq 3 | Y = 2)$.

E12.3   A system $\mathcal{S}$ has input X and output Y.

    a. If the pdfs $f_X(x)$ and $f_Y(y)$ are known, then what can you say about $\mathcal{S}$?

    b. If, instead, you are given $f_{Y|X}(y|x)$ and $f_{X|Y}(x|y)$, what can you say about f_X and f_Y?

E12.4   A stochastic system with input $X \sim \mathcal{N}(0, 1)$ has response Y characterized by $f_{Y|X}(y|x) = \frac{1}{2} e^{-|y-x|}$. Provide expressions for $f_Y, f_{X|Y}$.

E12.5   A system $\mathcal{S}$ transforms X into Y such that

$$f_{X,Y}(x, y) = \frac{1}{2\pi} e^{-\frac{1}{2}(x^2 + y^2)}.$$

    a. Describe $\mathcal{S}$.

    b. What is known about Y?

    c. Given that we observe $Y = y$, what is known about X?

E12.6   a. If $f_{Y|X}(y|x) = |x + 1| e^{-|x+1|y} U(y)$ and $X \sim \mathcal{B}(2, \frac{1}{2})$, what can you say about Y?

    b. What is known about X given $Y = y$ for $y > 0$?

E12.7   A component is purchased from Company A or from Company B, each with equal probability. If it comes from A, then the probability that it will not fail prior to time t equals $e^{-t} U(t)$; if it comes from B, then the probability that it will not fail prior to time t is $e^{-2t} U(t)$.

    a. Express the probability that the component will fail prior to t_0.

    b. It was observed that the component failed at some time in (t_1, t_2). Find the probability that it was manufactured by A.

E12.8   An analog voltmeter can be thought of as providing a reading X that is the sum of the true voltage V and an unlinked (independent) noise N that is distributed as $\mathcal{N}(0, \sigma^2)$. In this case, $f_{X|V}(x|v)$ is $\mathcal{N}(v, \sigma^2)$. If the true voltage V is provided by a random source distributed as $\mathcal{N}(m, 1)$, then calculate $f_X, f_{V|X}$. (Hint: The answers will again be normal.)

E12.9   A signal $X \sim \mathcal{U}(-1, 1)$ is transmitted through a noisy channel and received as Y, with

$$f_{Y|X}(y|x) = \frac{1}{\sqrt{\pi}} e^{-(y-2x)^2}.$$

    a. Describe the response Y by providing an expression for f_Y.

    b. Given that we observe $Y = .5$, what value of X is the most likely (i.e., has the highest conditional density)?

E12.10   If

$$f_{X,Y}(x, y) = \begin{cases} 3 & \text{if } 0 \leq x \leq 1, \ 0 \leq y \leq x^2 \\ 0 & \text{if otherwise} \end{cases},$$

then determine $f_{Y|X}(y|x)$ for $0 < x < 1$, taking care to specify your answer for all values of the argument y.

E12.11   a. Show how to determine f_Y from $f_{Y|X}, f_{X|Y}$.

    b. If it is claimed that

$$f_{Y|X}(y|x) = \frac{1}{\sqrt{2\pi}} e^{-\frac{(y-x)^2}{2}}, \quad f_{X|Y}(x|y) = \frac{1}{\sqrt{\pi}} e^{-(x-\frac{y}{2})^2},$$

then can we determine f_Y?

E12.12   A system S has input X and output Y with joint density

$$f_{X,Y}(x, y) = U(x)U(y - x)U(1 + x - y)e^{-x}.$$

    a. Describe S through $f_{Y|X}$. If we observe $Y = 2$, what do we know about X?

E12.13   An amplitude-modulated (AM) signal has random phase angle $\Theta \in [-\pi, \pi]$ and random amplitude $A \geq 0$ having a joint density

$$f_{\Theta,A}(\theta, a) = \frac{1}{\pi} a e^{-a^2} U(a)U(\theta + \pi)U(\pi - \theta).$$

Evaluate the conditional density $f_{\Theta|A}$.

E12.14   We observe a voltage V that arises from noisy reception of a signal S drawn from $\{1, 2\}$. The voltage V has the conditional density

$$f_{V|S}(v|s) = (s + 1)v^s U(v)U(1 - v),$$

and the prior signal probabilities are given by $P(S = 1) = 1/3$. The MAP (maximum a posteriori) signal estimation rule requires us to determine $P(S = s|V = v)$ and then select that value s^* which maximizes this term.

    a. If we observe that $V = 1/2$, then what is the MAP estimate of S?

    b. Furthermore, how probable is it that this estimate is correct?

E12.15   A random system S has input X and output Y related through

$$f_{Y|X}(y|x) = \frac{1}{2} e^{-|y-x|}.$$

    a. If $X \sim \mathcal{U}(0, 1)$, then describe the system output in the regime $y > 1$.

    b. If we observe that $Y = 2$, then describe the system input.

E12.16   a. Verify that

$$f_{Y|X}(y|x) = \frac{1}{2}e^{-|y-x^2|}$$

is a pdf.

b. If $X \sim \mathcal{U}(0, 1)$ evaluate f_Y.

E12.17 A system $\mathcal{S}$ with input X and output Y is such that

$$f_{X,Y}(x, y) = \begin{cases} \dfrac{1}{\pi} & \text{if } x^2 + y^2 < 1 \\ 0 & \text{if otherwise} \end{cases}.$$

a. Describe the input X.

b. Describe the system $\mathcal{S}$ for $|x| < 1$.

c. If we observe $Y = 0.5$, then what do we know about X?

E12.18 $X \sim \mathcal{P}(\lambda)$ is input to a system $\mathcal{S}$ with output Y described by

$$f_{Y|X}(y|k) = \frac{1}{2}U(y - k)U(k + 2 - y) = \begin{cases} \dfrac{1}{2} & \text{if } k \leq y \leq k + 2 \\ 0 & \text{if } y \text{ other} \end{cases}.$$

a. Describe Y.

b. Describe what is known about X if we observe $Y = 3.5$.

E12.19 $X \sim \mathcal{U}(0, 1)$ is input to a system $\mathcal{S}$ with nonnegative integer output Y. $\mathcal{S}$ is described by

$$P(Y = y|X = x) = (1 - x)x^y.$$

a. Describe the output Y.

b. Describe what is known about the input X, given that we observe the output $Y = 2$.

E12.20 We have

$$F_{X,Y,Z}(x, y, z) = \left[1 + e^{-x} + e^{-y} + e^{-z}\right]^{-1}.$$

a. Evaluate $f_{X,Y,Z}$ through the triple partial differentiation of $F_{X,Y,Z}$.

b. Evaluate the cdf $F_{X,Y}(x, y)$.

c. Evaluate $f_{Z|X,Y}(z|x, y)$.

E12.21 A signal $X \sim \mathcal{E}(1)$ is transmitted optically by encoding it so that the number K of received photons has expected value X.

a. What is $P(K = k|X = x)$?

b. Show that $P(K = k)$ is $\mathcal{G}(\frac{1}{2})$.

c. Given that we have received $K = 2$ photons, what do we know about X?

E12.22 A system noisily reconstructs an analog signal S from a digital signal D. It is known that

$$P(D = d) = \frac{2}{3}3^{-d}, \ d = 0, 1, \ldots, \quad f_{S|D}(s|d) = \frac{1}{2}e^{-|s-d|}.$$

    a. Describe the analog signal S.

    b. If we observe $S = s$, what do we know about D?

    c. Given that we observed $S = s < 0$, what is the most probable value of D?

E12.23 A noisy 3-bit A/D converter accepts an input S with $f_S(s) = 3s^2 U(s) U(1 - s)$ and generates an output $D \sim \mathcal{B}(7, S)$.

    a. Evaluate the probability description of D and ED.

    b. What does observation of $D = d$ tell us about $S = s$?

    c. Given that we have observed $D = 2$, what is the most likely value of the original signal S?

E12.24 A system S has input X and output Y, and we know $f_{X,Y}$.

    a. Describe S.

    b. Describe what is known about the input when we observe the output.

E12.25   a. Given $f_{Y|X}, f_X$, what can you say about f_Y?

    b. Given $f_{Y|X}, f_Y$, what can you say about f_X?

E12.26 The state X of a finite-state system is measured by a noisy analog measurement Y, with the measurement related to the state by $Y|X \sim \mathcal{E}(X)$. (Y conditional upon $X = \alpha$ is $\mathcal{E}(\alpha)$.) You know that each of the four possible state values $1, 2, 3, 4$ is equally probable to occur.

    a. Describe Y.

    b. Describe what you know about X, given that we have observed Y.

    c. What is the most probable value of the state X, given that $Y = 1.5$?

E12.27 We observe $K \sim \mathcal{B}(n, \pi)$ with π unknown. If π is chosen to be $\beta(m, n)$, then evaluate what $K = k$ tells you about $\pi = p$.

E12.28 We observe a measurement Y of X, where we know that the conditional density of Y given $X = x$ is Laplacian with parameter $\alpha(x) = x$. We are told that $X \sim \mathcal{B}(n, \frac{1}{2})$. What does $Y = y$ tell you about $X = x$?

E12.29 You are given that the state vector $\mathbf{X} \sim \mathcal{N}(\mathbf{0}, \mathbb{Q})$, you observe $\mathbf{Z} = \mathbb{H}\mathbf{X} + \mathbf{V}$ with $\mathbf{V} \sim \mathcal{N}(\mathbf{0}, \mathbb{R})$, and the matrices are specified by

$$\mathbb{Q} = \begin{pmatrix} 1 & 0 \\ 0 & 2 \end{pmatrix}, \quad \mathbb{H} = \begin{pmatrix} 1 & 1 \\ 1 & -1 \end{pmatrix}, \quad \mathbb{R} = \begin{pmatrix} 1 & 0 \\ 0 & 1 \end{pmatrix}.$$

What does $\mathbf{Z} = \mathbf{z}$ tell you about $\mathbf{X} = \mathbf{x}$?

E12.30 Consider $\mathbf{X}$ with $E\mathbf{X} = \mathbf{m} = (1, 2, 3)^T$ and positive definite covariance matrix $\mathbb{C}$ that is randomly chosen as follows: Using Matlab,

$$\mathbb{A} = \mathrm{randn}(3,3), \quad \mathbb{C} = \mathbb{A}^T \mathbb{A} + .01\mathbb{I}.$$

Treating the selected $\mathbb{C}$ as nonrandom and assuming that $\mathbf{X} \sim \mathcal{N}(\mathbf{m}, \mathbb{C})$, explicitly calculate the conditional density $f_{X_3|X_1, X_2}$.

13

MAP, MLE, and Neyman–Pearson Rules

13.1 PURPOSE, BACKGROUND, AND ORGANIZATION

Our development of conditional probability in the previous two chapters enables us to advance significantly towards Goal 3 in our use of measurements or observations to make decisions, estimates, or inferences. Identifying an optimal function or rule for converting a measurement into a decision or estimate requires a performance or design criterion. Hitherto (e.g., Sections 9.12, 10.5, and 10.6) we have only used a criterion minimizing mean square error to identify an estimation rule, say, $\hat{Y}(X)$ of Y, based upon a measurement, X. In this chapter, we explore several other decision-making criteria and their resulting optimum decision rules. We treat first binary or dichotomous decision making, including the celebrated approach of Neyman and Pearson, and second, the fundamental model of additive measurement noise.

The basic problem of deciding between two hypotheses (e.g., which one of two possible signals was sent, target present or target absent) is treated first to identify a decision rule—the maximum a posteriori, or MAP, rule—that is optimum in that it yields a minimum error probability. Finding such a rule requires that we know the initial probabilities of the two hypotheses before we make any measurements. As such probabilities are often unknown (e.g., in radar or sonar target detection), we introduce next the approach of Neyman and Pearson, in which we choose a decision rule that minimizes the probability of a missed target detection, given that we limit the probability of a false alarm.

Finally, we address the problem of inferring from a noisy measurement X to a true real-valued S when the measurement noise N adds to S. We introduce maximum likelihood estimators as approximations to the MAP rule when prior probability information is missing. Significant additional progress towards achieving Goal 3 will be made in Chapter 15 for the mean square error criterion and in Chapter 16, where we develop the Bayesian approach to a large number of problems.

13.2 BINARY DECISION-MAKING SETUP

In consonance with our third goal, we consider the use of probabilistic information to make inferences and decisions. (See Ferguson [28] for the basics of statistical decision theory and Poor [73] and Scharf [81] for decision theory in the context of communications and signal processing.) We consider a setting in which there are two so-called *states of the world*, denoted H_0, H_1. The probabilities for these states of the world $\pi_i = P(H = H_i)$ are known as *prior probabilities*, because they represent what we know about the tendency for H_i to occur prior to our making a measurement on the world. For example, in a communications context, there could be two possible transmitted signals s_0, s_1, one of which is passed through a noisy channel and received as a signal X. In a radar or sonar target detection context, H_0 could represent the state of the world in which no target (e.g., aircraft in a section of sky under surveillance, or an underwater reef) is present and H_1 the state of the world in which a target is present. The radar or sonar emits a pulse, and the system then listens for an echo indicative of the presence of a reflector and hence a target. Thus, we make a measurement X whose relation to the state of the world is given by a conditional density

$$f_{X|H}(x|H = H_i) = f_i(x)$$

in more compact notation. These conditional densities are also known as *likelihood functions* in this problem setting. We can use Bayes' Theorem (Theorem 12.9.1(b)) to calculate the *posterior probabilities* (meaning that they are posterior to, or after, making the measurement):

$$P(H = H_i | X = x) = \frac{\pi_i f_i(x)}{\pi_0 f_0(x) + \pi_1 f_1(x)}. \tag{13.1}$$

Example 13.1   Dichotomous Posterior Probability _____

Assume that X given $H = H_i$ is distributed as $\mathcal{N}(m_i, \sigma_i^2)$. Assume further that $P(H = H_0) = P(H = H_1) = 0.5$. Then, after dividing the numerator and denominator by the numerator,

$$P(H = H_0 | X = x) = \left[1 + \frac{\sigma_0}{\sigma_1} \exp\left(\frac{(x - m_0)^2}{2\sigma_0^2} - \frac{(x - m_1)^2}{2\sigma_1^2} \right) \right]^{-1}.$$

Definition 13.2.1   (Dichotomous Decision Rule)   A decision-making rule ϕ is a $\{H_0, H_1\}$-valued function of a measurement X; $\phi(X) \in \{H_0, H_1\}$. Equivalently, if S_0, S_1, with $S_1 = S_0^c$, is a binary partition of the measurement space $X \in \Omega = \mathbb{R}^n$, then

$$\phi(x) = \begin{cases} H_1 & \text{if } x \in S_1 \\ H_0 & \text{if } x \in S_0 \end{cases}.$$

The *design problem* is to determine the "best" decision rule ϕ, or equivalently, the best set S_0.

13.3 MINIMUM ERROR PROBABILITY DESIGN: MAP RULE

While in Chapter 16 we will learn different explications of "best" based upon a variety of performance criteria, for the present we adopt the view that, by "best," we mean *minimum error probability*. Hence, we choose S_0 to minimize $P(\phi(X) \neq H)$. Writing this out with the use of the Total Probability Theorem yields the following expression for the error probability:

$$P(\phi(X) \neq H)$$
$$= P(H = H_0)P(\phi = H_1 | H = H_0) + P(H = H_1)P(\phi = H_0 | H = H_1)$$
$$= \pi_0 P(S_1 | H_0) + \pi_1 P(S_0 | H_1) = \pi_0(1 - P(S_0 | H_0)) + \pi_1 P(S_0 | H_1)$$
$$= \pi_0 + \int_{S_0} [\pi_1 f_1(x) - \pi_0 f_0(x)] \, dx.$$

Introduce the *likelihood ratio* $\Lambda(x) = f_1(x)/f_0(x)$ and *threshold* $\tau = \pi_0/\pi_1$ to write

$$P(\phi(X) \neq H) = \pi_0 + \pi_1 \int_{S_0} [\Lambda(x) - \tau] f_0(x) \, dx. \tag{13.2}$$

It is apparent that the choice of set S_0 that minimizes the error probability is the one that contains the measurements that give a nonpositive value to the square-bracketed term. Hence, the best decision rule is given by

$$S_0 = \{x : \Lambda(x) \leq \tau\}, \quad \phi(x) = \begin{cases} H_1 & \text{if } \Lambda(x) > \tau \\ H_0 & \text{if } \Lambda(x) \leq \tau \end{cases}.$$

Note that $P(\phi(X) \neq H)$ remains the same no matter to which case we assign $\Lambda(x) = \tau$. From Eq. (13.1), we see that

$$\frac{\Lambda}{\tau} = \frac{P(H = H_1 | X = x)}{P(H = H_0 | X = x)},$$

and an equivalent expression for the minimum error probability decision rule is

$$\phi(x) = \begin{cases} H_1 & \text{if } P(H = H_1 | X = x) > P(H = H_0 | X = x) \\ H_0 & \text{if otherwise} \end{cases}.$$

Summarizing, we have now established the following theorem.

Theorem 13.3.1   (Dichotomous MAP Rule)   The decision rule ϕ that maximizes the probability of a correct decision when there are only the two alternatives H_0 and H_1 is given by

$$\phi(x) = \begin{cases} H_1 & \text{if } \Lambda(x) > \tau = P(H = H_0)/P(H = H_1) \\ H_0 & \text{if } \Lambda(x) \le \tau \end{cases}$$

or by

$$\phi(x) = \begin{cases} H_1 & \text{if } P(H = H_1|X = x) > P(H = H_0|X = x) \\ H_0 & \text{if otherwise} \end{cases}.$$

As $P(H = H_i|X = x)$ is the *a posteriori* (after the observation) probability, this decision rule ϕ is also known as the *maximum a posteriori (MAP) rule*; it will be revisited in Section 16.5.3.

Example 13.2   MAP Rule in Dichotomous Case _____
Continuing with the preceding example, we can now calculate the MAP rule $\phi(x)$ by first calculating

$$\tau = \frac{\pi_0}{\pi_1} = \frac{0.5}{0.5} = 1$$

and

$$\Lambda(x) = \frac{f_1(x)}{f_0(x)} = \frac{\sigma_0}{\sigma_1} \exp\left(\frac{(x - m_0)^2}{2\sigma_0^2} - \frac{(x - m_1)^2}{2\sigma_1^2} \right).$$

The MAP rule $\phi(x) = H_1$ if and only if $\Lambda(x) > \tau$, and this occurs if and only if

$$\frac{(x - m_0)^2}{2\sigma_0^2} - \frac{(x - m_1)^2}{2\sigma_1^2} > \log\left(\frac{\sigma_1}{\sigma_0}\tau \right) = \log\left(\frac{\sigma_1}{\sigma_0} \right).$$

If $\sigma_0 = \sigma_1$, then the preceding condition simplifies to

$$(x - m_0)^2 > (x - m_1)^2.$$

Expanding the quadratics, cancelling the common term of x^2, and rearranging yields the condition

$$(m_0 - m_1)(m_0 + m_1) > 2(m_0 - m_1)x$$

for $\phi(x) = H_1$. The final simplification is

$$\phi(x) = H_1 \iff \begin{cases} x < \dfrac{m_0 + m_1}{2} & \text{if } m_0 > m_1 \\ x > \dfrac{m_0 + m_1}{2} & \text{if } m_0 < m_1 \end{cases}.$$

Having identified the best decision rule, we are also interested in its performance. From Eq. (13.2), the performance of the best decision rule is given by

$$P(\phi \neq H) = \pi_0 + \pi_1 \int_{\{x:\Lambda(x)\leq\tau\}} [\Lambda(x) - \tau]f_0(x)\,dx$$

$$= \pi_0 + \int_{\{x:\pi_1 f_1(x)\leq\pi_0 f_0(x)\}} [\pi_1 f_1 - \pi_0 f_0]\,dx.$$

If, when evaluated, this performance is inadequate to your needs, then the only recourse is to obtain a better measurement X (i.e., to change f_0, f_1, or both). We have found that the optimal way to process a given measurement and the prior probabilities π_i cannot be changed without changing the nature of the problem. For example, we may need to reduce channel noise (change f_0) or increase transmitter/radar/sonar signal power (change f_1).

Example 13.3 Performance of MAP Rule _____

To complete our example, we evaluate its performance assuming that $\pi_0 = \pi_1$, $\sigma_0 = \sigma_1 = \sigma$, and $m_0 > m_1$, and introducing the notation $\overline{m} = 0.5(m_0 + m_1)$. In this case, $\tau = 1$, and

$$2P(\phi \neq H) = 1 + \int_{\{x:\phi(x)=H_0\}} [f_1(x) - f_0(x)]\,dx = 1 + \int_{\overline{m}}^{\infty} [f_1(x) - f_0(x)]\,dx.$$

Using the $\mathcal{N}(0, 1)$ cdf Φ, we can write

$$\int_{\overline{m}}^{\infty} f_i(x)\,dx = 1 - \Phi\left(\frac{\overline{m} - m_i}{\sigma_i}\right)$$

to obtain

$$P(\phi \neq H) = 0.5 + 0.5\left[\Phi\left(\frac{\overline{m} - m_0}{\sigma}\right) - \Phi\left(\frac{\overline{m} - m_1}{\sigma}\right)\right].$$

Note that, by our assumption that $m_0 > m_1$, we have

$$\overline{m} - m_0 = 0.5(m_1 - m_0) < 0 < 0.5(m_0 - m_1) = \overline{m} - m_1.$$

Therefore, $\Phi(\frac{\overline{m}-m_0}{\sigma}) - \Phi(\frac{\overline{m}-m_1}{\sigma}) < 0$. We see that $P(\phi \neq H) < 0.5$, as it should be. We could have achieved an error probability of 0.5 simply by guessing.

13.4 HYPOTHESIS TESTING

13.4.1 Neyman–Pearson Rule

In the hypothesis-testing setting, we have two possible hypotheses H_0, H_1, one and only one of which describes the true state of affairs. In the approach pioneered by J. Neyman and E. Pearson in the 1930s, the hypothesis H_0 is called the _null hypothesis_ and is meant to describe the status quo (e.g., a patient is "well" or a target is absent in a radar/sonar setting). The hypothesis H_1 is called the *alternate hypothesis* and describes a unique alternative to the status quo (e.g., the patient has a specific illness or a particular target is present). Hence, their approach treated the two states of nature asymmetrically. We shall generally refer to H_0 *as the condition of target absent* and H_1 *as the condition of target present*. We have a measurement, observation, or datum X (e.g., diagnostics performed by a physician, measurements made by a radar receiver looking in a frequency band over a short period of time) whose density depends upon which hypothesis is correct. Under hypothesis H_i, we are given that X is described by the conditional density $f_{X|H}(x|i) = f_i(x)$. This time, however, we have neither information about the prior probabilities $\pi_i = P(H = H_i)$ nor data relevant to inferring these prior probabilities.

In the approach of Neyman and Pearson, we focus on the two types of error that a decision maker can make:

- Type I (*false alarm*)—decide H_1 when H_0 is true;
- Type II (*missed detection*)—decide H_0 when H_1 is true.

Neyman and Pearson suggested that a good decision rule would be one that achieves a minimum probability of Type II error (or, equivalently, a maximum probability of the complementary detection event), subject to an upper bound α on the probability of Type I error.

Definition 13.4.1 (Neyman–Pearson Rule) The Neyman–Pearson rule achieves the maximum detection probability P_D, subject to a given upper bound α on false-alarm probability P_{FA}.

In the statistical literature, the parameter α is called the *size* of the statistical test, and P_D, denoted by β, is the *power* of the test.

Neyman and Pearson determined the decision rule that is optimal under their formulation. As the full solution for the Neyman–Pearson rule is best presented in terms of the concept of a *randomized decision* or detection rule ($\phi(x)$ becomes the probability of deciding H_1 given that we have observed $X = x$), we will not present this rule in its full generality.

Lemma 13.4.1 (Neyman–Pearson) If the conditional cdfs $F_{\Lambda|H}(\lambda|H_i)$ are continuous in λ for each $i = 0, 1$ (equivalently, there are no jump discontinuities) and

$$(\forall i \forall c)\ P(\Lambda(X) = c \mid H_i) = 0,$$

then the decision rule

$$\phi(x) = \begin{cases} H_1 & \text{if } \Lambda(x) \geq \tau \\ H_0 & \text{if } \Lambda(x) < \tau \end{cases}$$

has the highest detection probability (power) $P_D = P(\phi(X) = H_1 \mid H_1)$ among all decision rules having false-alarm probability P_{FA} (size) no greater than $\alpha = P(\phi = H_1 \mid H_0)$. The case of $\Lambda = \tau$ can be assigned arbitrarily to either hypothesis.

The qualification about $P = 0$ is not needed in the general solution, but removing it introduces complexities that are irrelevant here. The rule ϕ is known as a *likelihood-ratio-threshold rule*, as was the MAP rule presented in the last section.

The form of ϕ given in the lemma is such that, by proper choice of τ, we can achieve any desired size α. Note that $P(\Lambda \geq \tau | H_0)$ is a continuous nonincreasing function of τ. Furthermore, when $\tau = 0$, it follows that $P(\Lambda \geq 0 | H_0) = 1$, and when $\tau = \infty$, it follows that $P(\Lambda \geq \tau | H_0) = 0$. Hence, for any $0 \leq \alpha \leq 1$, there is a choice of $\tau = \tau_\alpha$ such that $P(\Lambda \geq \tau_\alpha | H_0) = \alpha$. Indeed, a search for this value of threshold is materially assisted by knowledge of the monotonicity of $P(\Lambda \geq \tau | H_0)$.

Example 13.4   Neyman–Pearson Rule _____

Assume that, under hypothesis H_i, $X \sim \mathcal{E}(\beta_i)$ with $\beta_1 > \beta_0$. Then

$$\{x : \phi(x) = H_1\} = \{x : \Lambda(x) \geq \tau\} = \left\{ x : \frac{\beta_1}{\beta_0} e^{-(\beta_1 - \beta_0)x} \geq \tau \right\}$$

$$= \left\{ x : -(\beta_1 - \beta_0)x \geq \log\left(\frac{\beta_0}{\beta_1}\tau\right) \right\} = \left\{ x : x \leq -\frac{\log\left(\frac{\beta_0}{\beta_1}\tau\right)}{\beta_1 - \beta_0} \right\}.$$

It follows that

$$P(\phi(X) = H_1 \mid H = H_0) = \alpha = P\left(X \leq -\frac{\log\left(\frac{\beta_0}{\beta_1}\tau\right)}{\beta_1 - \beta_0} \mid H = H_0 \right)$$

$$= \int_0^{-\frac{\log\left(\frac{\beta_0}{\beta_1}\tau\right)}{\beta_1 - \beta_0}} \beta_0 e^{-\beta_0 x} \, dx = 1 - e^{\beta_0 \frac{\log\left(\frac{\beta_0}{\beta_1}\tau\right)}{\beta_1 - \beta_0}}.$$

To determine τ_α achieving the given α, we can solve

$$\log(1 - \alpha) = \beta_0 \frac{\log\left(\frac{\beta_0}{\beta_1}\tau_\alpha\right)}{\beta_1 - \beta_0} \iff \log\left(\frac{\beta_0}{\beta_1}\tau_\alpha\right) = \frac{\beta_1 - \beta_0}{\beta_0} \log(1 - \alpha)$$

$$\iff \tau_\alpha = \frac{\beta_1}{\beta_0} \exp\left[\frac{\beta_1 - \beta_0}{\beta_0} \log(1 - \alpha)\right].$$

Thus, for this two-hypothesis problem, we can calculate in closed form the threshold τ_α that will achieve a desired false-alarm probability α. In many cases, it may be necessary to use numerical methods to find the threshold.

13.4.2 Receiver Operating Characteristic (ROC)

Definition 13.4.2   (ROC) The receiver operating characteristic, abbreviated as *ROC*, is the function ρ giving the probability of detection P_D in terms of the false-alarm probability P_{FA},

$$P_D = \rho(P_{FA}).$$

Example 13.5   Receiver Operating Characteristic _____

Continuing with our last example, we see that determining P_D is the same calculation as determining $P_{FA} = \alpha$, with the appropriate substitution of f_1 for f_0. Hence, replacing the leading β_0 in the exponent by β_1, we find that

$$P_D = 1 - e^{\beta_1 \frac{\log\left(\frac{\beta_0}{\beta_1}\tau\right)}{\beta_1 - \beta_0}}.$$

We can now relate P_D to α by noting that, for the appropriately defined Σ,

$$P_D = 1 - e^{\beta_1 \Sigma} \text{ and } \alpha = 1 - e^{\beta_0 \Sigma}.$$

It follows that

$$P_D = \rho(\alpha) = 1 - (1 - \alpha)^{\frac{\beta_1}{\beta_0}}.$$

We have now determined the ROC. If we choose $\beta_1 = 2\beta_0$, then the ROC is shown in Figure 13.1.

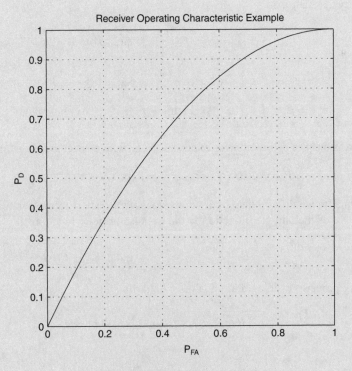

Figure 13.1   ROC example.

Our interest in ρ is that it allows the user to explore the consequences for P_D (power) of setting different false-alarm (size) levels. Basic properties of ρ are stated in the next lemma.

Lemma 13.4.2   (ROC of the Neyman–Pearson Rules)   Three properties describing the ROC of Neyman–Pearson rules are the following:

1. For a Neyman–Pearson rule ϕ,

$$P_D \geq P_{FA},$$

   and the ROC lies on or above the line $P_D = P_{FA}$.
2. The slope $\rho'(\alpha) = \tau_\alpha$ at $P_{FA} = \alpha$.
3. For a Neyman–Pearson (NP) rule, ρ is a concave function (i.e., $\rho'' \leq 0$).

Proof.   For property 1, consider the (randomized) decision rule $\theta(x) = \alpha$ that ignores X and that randomly selects H_1 with probability α. Then the false-alarm probability for this rule is α, and the detection probability is also α. Since an NP rule ϕ has the greatest power among all rules of its size, we see that the detection probability of it must be at least as large as the detection probability α achieved by $\theta(x)$.

For property 2, recall that

$$P_{FA} = \alpha = \int_{\{x:\Lambda(x)\geq\tau_\alpha\}}^{\infty} f_0(x)\,dx$$

and

$$P_D = \beta = \int_{\{x:\Lambda(x)\geq\tau_\alpha\}}^{\infty} f_1(x)\,dx = \int_{\{x:\Lambda(x)\geq\tau_\alpha\}}^{\infty} \Lambda(x)f_0(x)\,dx.$$

These two expressions can be rewritten as expectations E_0 with respect to f_0 of a unit-step function:

$$P_{FA} = E_0 U(\Lambda - \tau), \; P_D = E_0[U(\Lambda - \tau)\Lambda].$$

Proceeding formally, we differentiate both expressions with respect to the parameter τ, recalling that the derivative of the unit-step function is the impulse or delta function:

$$\frac{dP_{FA}}{d\tau} = -E_0\delta(\Lambda - \tau), \; \frac{dP_D}{d\tau} = -E_0[\delta(\Lambda - \tau)\Lambda]$$

From the sifting property of integration against a delta function (see Section 6.3), we know that

$$\frac{dP_D}{d\tau} = -E_0[\delta(\Lambda - \tau)\Lambda] = -\tau E_0\delta(\Lambda - \tau) = \tau\frac{dP_{FA}}{d\tau}.$$

Hence,

$$\frac{\frac{dP_D}{d\tau}}{\frac{dP_{FA}}{d\tau}} = \frac{dP_D}{dP_{FA}} = \tau.$$

If $P_{FA} = \alpha$ and $\tau = \tau_\alpha$ achieves this false-alarm probability, then we can rewrite the conclusion as

$$\frac{d\rho(\alpha)}{d\alpha} = \tau_\alpha.$$

We omit the proof of property 3, since it is best given when we allow randomized decision rules. An awkward proof can be given following the approach just taken to property 2.   □

It is generally the case that $\alpha = 0$ corresponds to $\beta = 0$ and always true that $\alpha = 1$ corresponds to $\beta = 1$. However, there are unusual situations in which a positive power is possible even when the size is 0.

Example 13.6   ROC Properties _____

In the last example, we determined that

$$\rho(\alpha) = 1 - (1 - \alpha)^2 = \alpha(2 - \alpha).$$

In the example before that, we determined that, for $\beta_1 = 2\beta_0$,

$$\tau_\alpha = 2(1 - \alpha).$$

The first claim of the last lemma is that $\rho(\alpha) \geq \alpha$. As α is a probability and lies between 0 and 1, it is immediate that this claim is correct. To check the second claim, note that

$$\rho'(\alpha) = 2(1 - \alpha),$$

and it follows that it equals τ_α. For the third claim, we see that

$$\rho''(\alpha) = -2 < 0,$$

and ρ is concave in α, as asserted.

13.5   ADDITIVE MEASUREMENT NOISE

13.5.1   Measurement Noise Model

The additive noise model arises in characterizing measurements and communication channels. Let S represent a "true" quantity—either the true value of a variable that is being measured or the signal/message that is being transmitted through a channel. The actual measurement or received signal X is generally a noise-corrupted version of S. In the additive noise model,

$$X = S + N,$$

where N is the noise. The standard description of such a situation is to provide the distribution for the true quantity, usually in the form of a pdf f_S and a description of the measurement process in terms of a conditional pdf $f_{X|S}$. This conditional density, in turn, is often derived from the description of the noise through f_N and an assertion that S, N are independent. Independence (to be discussed in Chapter 14) informs us that we should use the product model for the joint or bivariate density:

$$f_{S,N}(s, n) = f_S(s)f_N(n).$$

We can determine $f_{X|S}$ from $f_{S,N}$ by first determining $f_{S,X}$. We derive the latter through the techniques of Section 8.6.1 by observing the linear relation

$$\begin{pmatrix} S \\ X \end{pmatrix} = \begin{pmatrix} 1 & 0 \\ 1 & 1 \end{pmatrix} \begin{pmatrix} S \\ N \end{pmatrix}.$$

The determinant of the transformation matrix is 1, and thus

$$f_{S,X}(s, x) = f_{S,N}(s, x - s) = f_S(s)f_N(x - s).$$

It is immediate that

$$f_{X|S}(x|s) = \frac{f_{S,X}}{f_S} = f_N(x - s),$$

in the independent additive noise model described. We can also evaluate the measurement description through the use of the Total Probability Theorem to obtain

$$f_X(x) = \int_{-\infty}^{\infty} f_{S,X}(s, x) \, ds = \int_{-\infty}^{\infty} f_N(x - s)f_S(s) \, ds,$$

a convolution of the densities f_N and f_S.

Example 13.7   Additive Measurement Noise Model _____

Assume that $N \sim \mathcal{L}(\alpha)$ and $S \sim \mathcal{L}(\beta)$ are independent and that $X = S + N$. Then

$$f_X(x) = \int_{-\infty}^{\infty} f_N(x - s)f_S(s) \, ds = \frac{1}{4}\alpha\beta \int_{-\infty}^{\infty} e^{-\alpha|x-s|-\beta|s|} \, ds.$$

Note that f_X is a symmetric function of x ($f_X(x) = f_X(-x)$); this can be verified by replacing x by $-x$ and s by $-s$ in the given integral and seeing that its value is unchanged. For definiteness, we evaluate f_X for $x \geq 0$.

The presence of absolute values in the integrand means that we have to divide the range of integration into intervals within which the terms inside the absolute values have unchanging sign. The boundaries of these cases are identified by the values of s for which the terms in absolute value become zero. The absolute value terms are $|x - s|$, which becomes zero for $s = x$, and $|s|$, which becomes zero for $s = 0$. If, as assumed, $x \geq 0$, then the intervals of integration over which the terms inside the absolute values have constant sign are $s \leq 0$,

$0 < s \leq x$, and $s > x$. In this case,

$$\int_{-\infty}^{\infty} e^{-\alpha|x-s|-\beta|s|}\, ds$$

$$= \int_{-\infty}^{0} e^{-\alpha(x-s)-\beta(-s)}\, ds + \int_{0}^{x} e^{-\alpha(x-s)-\beta s}\, ds + \int_{x}^{\infty} e^{-\alpha(s-x)-\beta s}\, ds$$

$$= e^{-\alpha x} \int_{-\infty}^{0} e^{(\alpha+\beta)s}\, ds + e^{-\alpha x} \int_{0}^{x} e^{(\alpha-\beta)s}\, ds + e^{\alpha x} \int_{x}^{\infty} e^{-(\alpha+\beta)s}\, ds$$

$$= e^{-\alpha x} \left(\frac{1}{\alpha+\beta} + \frac{1}{\alpha-\beta}(e^{(\alpha-\beta)x} - 1) \right) + e^{\alpha x} \frac{1}{\alpha+\beta} e^{-(\alpha+\beta)x}$$

$$= \left(\frac{1}{\alpha+\beta} - \frac{1}{\alpha-\beta} \right) e^{-\alpha x} + \left(\frac{1}{\alpha-\beta} + \frac{1}{\alpha+\beta} \right) e^{-\beta x}.$$

In making this evaluation, we assumed that $\alpha \neq \beta$. Therefore, for $x \geq 0$,

$$4 f_X(x) = \frac{-2\alpha\beta^2}{\alpha^2 - \beta^2} e^{-\alpha x} + \frac{2\alpha^2\beta}{\alpha^2 - \beta^2} e^{-\beta x}$$

$$\text{or } f_X(x) = \frac{\alpha\beta}{2(\alpha^2 - \beta^2)} \left[-\beta e^{-\alpha x} + \alpha e^{-\beta x} \right].$$

While it is not immediate that $f_X(x)$ is always nonnegative for any positive α and β, it is the case.

From the symmetry of f_X noted at the outset, we conclude that, for all x,

$$f_X(x) = \frac{\alpha\beta}{2(\alpha^2 - \beta^2)} \left[-\beta e^{-\alpha|x|} + \alpha e^{-\beta|x|} \right].$$

13.5.2 Inferring to S from X: MLE Rule

We observe X, but would like to know S. The information about S contained in X is described by the conditional density $f_{S|X}$. Bayes' Theorem yields

$$f_{S|X}(s|x) = \frac{f_{X|S}(x|s) f_S(s)}{\int_{-\infty}^{\infty} f_{X|S}(x|s') f_S(s')\, ds'}.$$

In the case of the additive noise model,

$$f_{S|X}(s|x) = \frac{f_N(x-s) f_S(s)}{\int_{-\infty}^{\infty} f_N(x-s') f_S(s')\, ds'}.$$

Knowledge of $f_{S|X}$ enables us to design a receiver, say, to compensate for the noise introduced by the channel.

Example 13.8   Posterior of Signal Given Measurement _____
Continuing the additive noise example, we have

$$f_{S,X}(s,x) = f_S(s)f_N(x-s) = \frac{1}{4}\alpha\beta e^{-\beta|s|-\alpha|x-s|}$$

and

$$f_X(x) = \frac{\alpha\beta}{2(\alpha^2-\beta^2)}\left[-\beta e^{-\alpha|x|} + \alpha e^{-\beta|x|}\right],$$

enabling us to write

$$f_{S|X}(s|x) = \frac{f_{S,X}}{f_X} = \frac{\alpha^2-\beta^2}{2}\frac{e^{-\beta|s|-\alpha|x-s|}}{-\beta e^{-\alpha|x|} + \alpha e^{-\beta|x|}}.$$

A particular inference method (the *MAP* rule of Section 13.3) that is widely used is to infer S by that value $\hat{S}(X)$ which maximizes the conditional density of S given X:

$$\hat{S}(x) = \underset{s}{\operatorname{argmax}}[f_{S|X}(s|x)].$$

(The notation $\operatorname{argmax}_x g(x,y)$ stands for an argument in the variable x that maximizes the value of the function $g(\cdot,y)$.) When S is a discrete random variable, the *MAP* rule maximizes the probability of correct estimation. Note that

$$\underset{s}{\operatorname{argmax}}[f_{S|X}(s|x)] = \underset{s}{\operatorname{argmax}}[f_{X|S}(x|s)f_S(s)],$$

since the two terms differ only by a factor of $f_X(x)$ that does not affect the choice of maximizing s. Hence, more simply

$$\hat{S}(x) = \underset{s}{\operatorname{argmax}}[f_{X|S}(x|s)f_S(s)].$$

If $f_{X|S}(x|s)$ is a sharply peaked function of x for each s and $f_S(s)$ is slowly varying, then

$$\hat{S}(x) \approx \underset{s}{\operatorname{argmax}}[f_{X|S}(x|s)].$$

(In the additive noise case, $f_{X|S} = f_N(x-s)$ will be sharply peaked as needed if the noise pdf f_N is a unimodal function and the noise has a small variance.) This latter approximation has the advantage of not requiring us to know the prior f_S. The resulting inference rule is given in the next definition.

Definition 13.5.1   (Maximum Likelihood Estimator)   The *maximum likelihood (MLE) estimator* is given by

$$\tilde{S}(x) = \underset{s}{\operatorname{argmax}}[f_{X|S}(x|s)].$$

Example 13.9   MLE Rule for Signal Given Measurement
We determine the MLE $\tilde{S}(x)$ for our continuing example from

$$\tilde{S}(x) = \operatorname*{argmax}_s f_{S,X}(s,x) = \operatorname*{argmax}_s \left[\frac{1}{4}\alpha\beta e^{-\beta|s|-\alpha|x-s|}\right] = \operatorname*{argmax}_s [e^{-\beta|s|-\alpha|x-s|}]$$

$$= \operatorname*{argmax}_s [-\beta|s| - \alpha|x-s|].$$

The last line follows because the argument that maximizes $g(h(s))$ when g is an increasing function is the same argument that maximizes $h(s)$. To obtain the MLE, we assume that $\alpha \geq \beta$ and solve first, assuming that $x \geq 0$. In this case, we have the three cases of $s \leq 0$, $0 < s \leq x$, and $x < s$.

If $s \leq 0$, then

$$\operatorname*{argmax}_{s \leq 0}[-\beta|s| - \alpha|x-s|] = \operatorname*{argmax}_{s \leq 0}[\beta s - \alpha(x-s)] = \operatorname*{argmax}_{s \leq 0}[(\beta+\alpha)s] = 0,$$

and the tentative maximum value $\tilde{S}(x)$ is $-\alpha x$. If $0 < s \leq x$, then

$$\operatorname*{argmax}_{0<s\leq x}[-\beta|s| - \alpha|x-s|] = \operatorname*{argmax}_{0<s\leq x}[-\beta s - \alpha(x-s)] = \operatorname*{argmax}_{0<s\leq x}[(-\beta+\alpha)s] = x,$$

where we have used our assumption that $-\beta + \alpha > 0$). The resulting tentative maximum value $\tilde{S}(x)$ is $-\beta x$. The final case is $s > x$,

$$\operatorname*{argmax}_{s>x}[-\beta|s| - \alpha|x-s|] = \operatorname*{argmax}_{s>x}[\beta s - \alpha(s-x)] = \operatorname*{argmax}_{s>x}[(\beta-\alpha)s] = x,$$

and the resulting tentative maximum value $\tilde{S}(x)$ is $-\beta x$. As we have assumed that $\alpha > \beta$, the overall maximum is $-\beta x$ and is achieved by $\tilde{S}(x) = x$. This solution is also correct if $x < 0$.

13.5.3 The Discrete-Signal Case

If the signal or true variable S is discrete and described by a pmf

$$p_i = P(S = s_i),$$

then we can adapt the preceding results to obtain (for continuous noise)

$$f_X(x) = \sum_i f_{X|S}(x|s_i)p_i = \sum_i f_N(x-s_i)p_i,$$

$$P(S = s_i | X = x) = \frac{f_N(x-s_i)p_i}{\sum_j f_N(x-s_j)p_j}.$$

Inferences can then be made as before to choose the value of s_i, for a given x, that maximizes $P(S = s_i | X = x)$.

Example 13.10   MAP Rule for Discrete Signal _____

If

$$P(S = 0) = p_0, \; P(S = 1) = p_1 = 1 - p_0,$$

and $N \sim \mathcal{N}(0, \sigma^2)$, then

$$P(S = 1 | X = x) = \frac{p_1 e^{-\frac{(x-1)^2}{2\sigma}}}{p_1 e^{-\frac{(x-1)^2}{2\sigma}} + p_0 e^{-\frac{x^2}{2\sigma}}} = \left[1 + \frac{p_0}{p_1} e^{\frac{(x-1)^2}{2\sigma^2} - \frac{x^2}{2\sigma^2}} \right]^{-1}.$$

Hence, from $P(S = 1 | X = x) + P(S = 0 | X = x) = 1$, it follows that

$$P(S = 1 | X = x) > P(S = 0 | X = x) \iff \frac{p_0}{p_1} e^{\frac{(x-1)^2}{2\sigma^2} - \frac{x^2}{2\sigma^2}} < 1$$

$$\iff \frac{(x-1)^2}{2\sigma^2} - \frac{x^2}{2\sigma^2} < \log\left(\frac{p_1}{p_0}\right) \iff 1 - 2x < 2\sigma^2 \log\left(\frac{p_1}{p_0}\right)$$

$$\iff x > \frac{1}{2} - \sigma^2 \log\left(\frac{p_1}{p_0}\right).$$

Thus, the MLE

$$\tilde{S}(x) = \begin{cases} 1 & \text{if } x > \dfrac{1}{2} - \sigma^2 \log\left(\dfrac{p_1}{p_0}\right) \\ 0 & \text{if otherwise} \end{cases}.$$

13.6 SUMMARY

With the development of conditional probability in Chapters 11 and 12, we came into position to advance towards Goal 3 by solving several commonly encountered problems of inference under varying conditions of prior knowledge. First, we treated the case of making an optimal decision when faced with only two possibilities (e.g., binary digital communication receiver design or radar determination of the presence or absence of a target). Section 13.3 formulated and solved for the decision rule that yields the minimum error probability, a criterion that is used when the two types of errors that you can make have equal consequences for you. The solution introduces the important technique of a likelihood ratio-threshold rule that is encountered in other inference settings and the concept of a MAP decision rule to achieve minimum error probability. Digital communications problems fit this model.

Section 13.4 treated the same problems of binary decision making, but this time without assuming that we know the prior probabilities with which the two cases (say, target present or target absent) are presented. The celebrated formulation of this problem is due to Neyman and Pearson. Discussion of the Neyman–Pearson formulation is our first exposure to solving an inference or decision-making problem when we have incomplete probabilistic information—we do not know all that we need to. Radar and sonar target-detection problems fit this model.

Section 13.5 covered the important problem of signal plus noise. While the MAP rule can be used to solve this problem, we often have little knowledge of the prior probabilities required by the rule. We introduced the approximation by the maximum likelihood estimator (MLE) that is widely used in statistical and engineering practice. The MLE requires knowing only the conditional probability of the measurement X, given the true quantity S (i.e., knowing only the measurement process).

EXERCISES

E13.1 In a certain communications system, the received signal X is distributed $\mathcal{E}(1)$ if message m_1 is sent and is distributed $\mathcal{E}(2)$ if message m_2 is sent. Message m_1 is sent 40% of the time and m_2 is sent otherwise.

   a. Determine the receiver $\hat{m}(X)$ that has the minimum probability of error in deciding on the transmitted message.
   b. Evaluate the error probability for $\hat{m}$.

E13.2 In a given radar detection problem, if the target is absent, then we observe $X \sim \mathcal{N}(0, 1)$ whereas if a target is present, we observe $X \sim \mathcal{N}(1, 1)$. Professor Phynne has solved this problem for the decision rule $\hat{H}(X)$ having minimum error probability and found that

$$\hat{H}(x) = \begin{cases} \text{absent} & \text{if } x < 1 \\ \text{present} & \text{if } x \geq 1 \end{cases}.$$

   a. When is this the correct decision rule?
   b. Provide an expression for the error probability performance of this rule $\hat{H}$ when it is correct.

E13.3 a. If S is such that $P(Y = k | X = x)$ is $\mathcal{P}(\lambda)$ in k with $\lambda = 1 + x$ and $P(X = 0) = P(X = 1) = 0.5$, then describe Y.
   b. Solve for the MAP rule $\hat{X}(Y)$.

E13.4 A system S transforms X into Y such that

$$f_{X,Y}(x, y) = \frac{1}{2\pi} e^{-\frac{1}{2}(x^2 + y^2)}.$$

   a. Determine the MAP rule $\hat{X}$ for inferring X from Y.
   b. What is the error probability $P(\hat{X}(Y) \neq X)$?

E13.5 If $f_{Y|X}(y|x) = |x + 1| e^{-|x+1|y} U(y)$ and $X \sim \mathcal{B}(2, \frac{1}{2})$, determine the MAP rule $\hat{X}(Y)$ for $Y = 1$.

E13.6 Consider the so-called Z-channel system, used to model binary optical communications, in which the channel input X and output Y both take their values in $\{0, 1\}$, with the channel specified by

$$p_{Y|X}(y|x) = \begin{cases} 1 & \text{if } x = y = 0 \\ 1 - p & \text{if } x = y = 1 \\ p & \text{if } x = 1, y = 0 \end{cases}.$$

Note that a 0 is always received as a 0, but a 1 may be received either as a 0 or a 1. The input process is such that $P(X = 0) = \pi_0$, $P(X = 1) = \pi_1$.

   a. Describe the channel output.
   b. Evaluate the probability that the input is 0, given that the output is 0.
   c. Professor Phynne has proposed the decision rule $X^*(Y) = Y$ to recover the input from the observed output. Evaluate the error probability performance P_e^* of this rule.
   d. If $Y = 0$, then for what values of π_0 and p is the minimum error probability rule $\hat{X}(0) = 0$?

E13.7   A system $\mathcal{S}$ transforms X into Y such that

$$f_{X,Y}(x, y) = \begin{cases} 2e^{-x-y} & \text{if } 0 \le y \le x \\ 0 & \text{if otherwise} \end{cases}.$$

   a. Describe $\mathcal{S}$, being careful to specify correct ranges of variables x and y.
   b. Given that we observe $Y = y$, what is known about X?
   c. Determine the MAP rule $\hat{X}$ for inferring X from Y.
   d. What is the error probability $P(\hat{X}(Y) \ne X)$?

E13.8   In a certain communication system, the received signal X is described by a density function $e^{-x}U(x)$ if m_1 is sent and by density $2e^{-2x}U(x)$ if message m_2 is sent.

   a. Design a receiver so as to maximize the probability that we decide that m_1 was sent given that it was, subject to the restriction that the probability is no more than α of deciding that m_1 was sent when it was not.
   b. If message m_1 is sent 70% of the time and message m_2 is sent the remaining 30% of the time, then what is the probability of correctly deciding m_1 when we use the preceding design?

E13.9   If a system defect is present, then a monitor will yield a voltage X having a density $3x^2U(x)U(1 - x)$. If there is no defect, then the monitor voltage is described by the density $\mathcal{U}(0, 1)$.

   a. Determine the Neyman–Pearson ROC for the problem of detecting the presence or absence of a defect.
   b. If we desire a false-alarm probability of .2, then what is the corresponding detection probability?

E13.10   With a pseudonoise signal radar, if there is no reflected signal from a target, then we receive $Y \sim \mathcal{N}(0, 1)$. If a target is present and reflects the signal, then we receive $Y \sim \mathcal{N}(0, 2)$.

   a. Determine the Neyman–Pearson ROC for the problem of detecting the presence or absence of the target by specifying $P_D(\alpha)$ and $P_{FA}(\alpha)$.

   b. If you now learn that the prior probability of a target being present is .3, then evaluate the error probability when you use a Neyman–Pearson decision rule for a false-alarm probability of .1.

E13.11 A radar sends out a pseudorandom pulse and then waits for a given interval of time for a reflected attenuated pulse. If a certain target is absent, then the radar receiver output is described by $X \sim \mathcal{U}(0, 1)$. If the target is present, then $f_X(x) = 4x^3 U(x)U(1 - x)$.

   a. Determine the optimum receiver to use for a given false-alarm probability.

   b. Provide expressions for the false-alarm and detection probabilities.

   c. Calculate and sketch the receiver operating characteristic (ROC).

E13.12 We observe a voltage V that arises from noisy reception of a signal S drawn from $\{s_0, s_1\}$. V has the conditional density

$$f_{V|S}(v|s_i) = (i + 1)e^{-(i+1)v} U(v) = f_i(v).$$

   a. Identify the set $\{v : \text{decide } S = s_1\}$ for Neyman–Pearson hypothesis testing.

   b. Select the threshold τ so that the probability of deciding s_1 when s_0 was sent (false alarm or Type 1 error) is α.

   c. For this τ, what is the corresponding probability β of correctly deciding s_1 (detection probability or power of the test)?

   d. If we have the additional information that $P(S = s_0) = p_0$, then what is the probability of a wrong decision?

   e. Evaluate $P(S = s_0|V = v)$.

E13.13 A binary-valued signal $S \in \{-1, 1\}$ is measured in the presence of additive independent noise $N \sim \mathcal{N}(0, 1)$.

   a. Determine a decision rule based on the measurement X that incorrectly decides that $S = -1$ with a probability of .1 and maximizes the probability of correctly deciding that $S = -1$.

   b. Evaluate the probabilities of correctly deciding that $S = -1$ and of correctly deciding that $S = 1$.

E13.14 A binary-valued signal $S \in \{-1, 1\}$ is measured in the presence of additive independent noise N, $f_N(n) = \frac{1}{2}e^{-|n|}$.

   a. Determine a decision rule based on the measurement X that incorrectly decides that $S = 1$ with a probability of $\frac{1}{2}e^{-2}$ and maximizes the probability of correctly deciding that $S = 1$.

   b. Evaluate the probabilities of correctly deciding that $S = -1$ and of correctly deciding that $S = 1$ for a decision rule

$$d(x) = \begin{cases} 1 & \text{if } x > 0 \\ -1 & \text{if } x \le 0 \end{cases}.$$

It may help to note that

$$|x + a| - |x - a| = \begin{cases} 2a & \text{if } x > a \\ 2x & \text{if } |x| \le a \\ -2a & \text{if } x < -a \end{cases}.$$

E13.15   We observe $X \sim \mathcal{P}(\lambda)$ and wish to estimate λ from X.

   a. Can you determine the MAP estimator $\hat{\lambda}(X)$?
   b. Determine the MLE estimator $\tilde{\lambda}(X)$.
   c. If you are now informed that $\lambda \sim \mathcal{E}(1)$, provide an expression for the mean square error made by $\tilde{\lambda}(X)$.

E13.16   We observe $X = S + N$ arising from a signal $S \sim \mathcal{P}(\lambda_0)$ and additive, independent noise $N \sim \mathcal{P}(\lambda_1)$, with $\lambda_0 = \frac{3}{2}\lambda_1$.

   a. Specify $p_{X|S}(x|s)$.
   b. Provide an expression for $p_{S|X}(s|x)$.
   c. Determine the MAP rule $\hat{S}(2)$ for estimating S from X when $X = 2$. (Hint: Just try the different possible values of S.)
   d. Evaluate the error probability P_e incurred by using X itself to estimate S.

E13.17   We observe $X \sim \mathcal{B}(n, p)$ and wish to estimate p from X for given n.

   a. Can you determine the MAP estimator $\hat{p}(X)$?
   b. Determine the MLE estimator $\tilde{p}(X)$.
   c. If you are now informed that $p \sim \mathcal{U}(0, 1)$, provide an expression for the mean square error made by $\tilde{p}(X)$.

E13.18   We observe $X \sim \mathcal{G}(\beta)$ and wish to estimate β from X.

   a. Can you determine the MAP estimator $\hat{\beta}(X)$?
   b. Determine the MLE estimator $\tilde{\beta}(X)$.
   c. If you are now informed that $\beta \sim \mathcal{U}(0, 1)$, provide an expression for the mean square error made by $\tilde{\beta}(X)$.

E13.19   We observe $X \sim \mathcal{E}(\alpha)$ and wish to estimate α from X.

   a. Can you determine the MAP estimator $\hat{\alpha}(X)$?
   b. Determine the MLE estimator $\tilde{\alpha}(X)$.
   c. If you are now informed that $\alpha \sim \mathcal{E}(1)$, provide an expression for the mean square error made by $\tilde{\alpha}(X)$.

E13.20   We observe $X \sim \mathcal{N}(m, \sigma^2)$ and wish to estimate m from X for a given σ^2.

   a. Can you determine the MAP estimator $\hat{m}(X)$?
   b. Determine the MLE estimator $\tilde{m}(X)$.
   c. If you are now informed that $m \sim \mathcal{N}(0, 1)$, provide an expression for the mean square error made by $\tilde{m}(X)$.

E13.21   We observe $X \sim \mathcal{N}(0, v)$ and wish to estimate the variance v from X.

  a. Can you determine the MAP estimator $\hat{v}(X)$?
  b. Determine the MLE estimator $\tilde{v}(X)$.
  c. If you are now informed that $v \sim \mathcal{E}(1)$, provide an expression for the mean square error made by $\tilde{v}(X)$.

E13.22   We measure S and observe X. It is known that

$$f_{X|S}(x|s) = \frac{1}{2}e^{-|x-s|}, \ f_S(s) = e^{-s}U(s).$$

  a. Evaluate the MAP rule $\hat{S}(X)$.
  b. Evaluate the MLE rule $\tilde{S}(X)$.
  c. Evaluate the mean square error performances of $\hat{S}$ and $\tilde{S}$.

E13.23   We make a noisy measurement $Y = S + N$ of a signal S with pdf f_S in the presence of additive noise N having pdf f_N, and we know that $f_{S,N}(s, n) = f_S(s)f_N(n)$.

  a. Evaluate the conditional cdf $F_{Y|S}(y|s) = P(Y \le y|S = s)$ in terms of f_S and f_N.
  b. Evaluate $f_{S|Y}$ in terms of f_S and f_N.
  c. If $S \sim \mathcal{N}(m, 1)$ and $N \sim \mathcal{N}(0, \sigma^2)$, evaluate the MAP $\hat{S}(Y)$.
  d. Evaluate the MLE $\tilde{S}(Y)$.

E13.24   A voltage $V \sim \mathcal{E}(1)$ is measured by an instrument having response X that contains an additive independent normally distributed error of zero mean and variance σ^2.

  a. Evaluate the density $f_X(x)$.
  b. Evaluate the MLE $\tilde{V}(X)$.
  c. Evaluate the MAP $\hat{V}(X)$.
  d. We attempt to infer V from $\hat{V}(X) = aX + b$. What are the best choices of a, b to yield a minimum mean square estimate?

14

Independence

14.1 PURPOSE, BACKGROUND, AND ORGANIZATION

As noted by Kolmogorov in his epochal 1933 foundation for probability, as translated in [54], pp. 8, 9,

> The concept of mutual *independence* of two or more experiments holds, in a certain sense, a central position in the theory of probability.... Historically, the independence of experiments and random variables represents the very mathematical concept that has given the theory of probability its peculiar stamp.... We thus see, in the concept of independence, at least the germ of the peculiar type of problem in probability theory.... In consequence, one of the most important problems in the philosophy of the natural sciences is—in addition to the well-known one regarding the essence of the concept of probability itself—to make precise the premises which would make it possible to regard any given real events as independent.

The basic idea is that independence is an appropriate probability model when outcomes are *unlinked*. By "unlinked," we mean without a causal connection in a physical setting or without one outcome being informative about another outcome in an information-theoretic or belief-based setting.

We provided unmotivated definitions of independence for a pair of events in Section 2.10 in the classical probability setting and in Section 7.4.1 for independent random variables. We did so to afford earlier access to this important concept. However, as we believe that independence is best motivated and understood from conditional probability, a fuller discussion had to wait upon our development of conditional probability. By thus delaying our discussion of independence, we hope to discourage a common practice among students to assume that, in the absence of information to the contrary, events and random variables can be assumed to be independent. Identifying a relation of independence requires knowledge of the absence of possible linkages, and it is not just the "default" condition.

We develop the definition of independence by first giving it for pairs of events, then for n events, and, finally, for the infinitely many events usually involved in independent experiments or random variables. Mathematically, as we shall subsequently see, independence of events or random experiments or random variables is represented by an appropriate product factorization of the probability measure P, cdf $F_{X_1,\ldots,X_n}$, density $f_{X_1,\ldots,X_n}$, or pmf $p_{x_1,\ldots,x_n}$. We apply our concept of independence to derive the geometric and binomial probability models of Chapter 4, to calculate the cdfs of the maximum and minimum of several independent random variables, and to evaluate the expectation of a product of functions of independent random variables.

14.2 INDEPENDENT PAIRS OF EVENTS

14.2.1 Definition

The notation for independence

$$A \perp\!\!\!\perp B,$$

is read as "A is independent of B." By an assertion of independence, we mean that our knowledge of the tendency for A to occur given that we know that B has occurred has not been revised by knowing about B. From our concept of updating probabilities by using conditional probability, we can formalize this statement as

$$A \perp\!\!\!\perp B \Rightarrow P(A|B) = P(A).$$

This in turn would imply that

$$A \perp\!\!\!\perp B \Rightarrow P(A \cap B) = P(A)P(B).$$

This latter product form has the advantage of not requiring a caveat concerning $P(B) > 0$. Hence, we adopt the next definition.

Definition 14.2.1   (Independence of a Pair of Events)   The event A is independent of the event B, denoted

$$A \perp\!\!\!\perp B \iff P(A \cap B) = P(A)P(B).$$

Examples of pairs of independent events include the following:

- outcomes of successive die or coin tosses;
- the ordering of cards in a deck before and after thorough shuffling of the deck;
- heights or weights of two individuals who are not related;

- video frames or particular pixel values of such frames drawn from transmissions made at the same time by two broadcast TV channels carrying different programs;
- thermal noise voltage events $V_1 \in A$, $V_2 \in B$ measured at two different times;
- thermal noise voltage events $V_1 \in A$, $V_2 \in B$ measured at the same time in two different components;
- numbers N_1 and N_2 of photons emitted by a given light source when counted over nonoverlapping time intervals $[t_1, t_2]$, $[t_3, t_4]$, $t_2 < t_3$.
- numbers N_1 and N_2 of photons emitted by two different light sources, even when observed over the same time interval.

Examples of events that are not likely to be independent include the following:

- prices of a stock on successive days;
- heights or weights of two siblings;
- height and weight of a given individual;
- successive time samples of a speech signal;
- successive video frames or sequential pixel values in a broadcast TV program;
- successive letters or words in English or other natural language text.

We generally assume independence in practice when we cannot identify a physical linkage that could provide dependence. Information-based notions of independence using ideas of computational complexity have provided insights, but are intrinsically difficult (ineffectively computable) to apply in practice (e.g., see Li and Vitanyi [59]). However, in the end, it requires a study of data to determine independence, and even this is not done with complete confidence.

14.2.2 Basic Consequences of Independence

Of course,

$$P(B) > 0 \ \& \ A \perp\!\!\!\perp B \Rightarrow P(A|B) = P(A).$$

It is immediate from the symmetry of Definition 14.2.1 that

$$A \perp\!\!\!\perp B \Rightarrow B \perp\!\!\!\perp A.$$

Hence, we can also more symmetrically read $A \perp\!\!\!\perp B$ as "A and B are independent." Also,

$$A \perp\!\!\!\perp A \iff P(A) = P(A)^2 \iff P(A) = 0 \ \text{ or } \ 1, \tag{14.1}$$

$$P(A) = 0 \ \text{ or } \ 1 \Rightarrow (\forall B) \ A \perp\!\!\!\perp B. \tag{14.2}$$

Hence, when $P(A) = 0$ or 1, as is the case for ϕ and Ω, the event A is independent of all other events and we regard such independence involving an event of probability 0 or 1 as "trivial."

From $P(B) = P(A \cap B) + P(A^c \cap B)$, we see that $P(A^c \cap B) = P(B) - P(A \cap B)$. Hence, from $A \perp\!\!\!\perp B$, we conclude that

$$A \perp\!\!\!\perp B \Rightarrow A \perp\!\!\!\perp B^c \Rightarrow A^c \perp\!\!\!\perp B^c.$$

This is a first instance of a more general property of independence of Boolean functions of distinct sets of events. (See Theorem 14.3.1.)

The following properties show that two events are nontrivially independent only if they overlap properly:

$$A \perp B \ \& \ A \perp\!\!\!\perp B \Rightarrow P(A) = 0 \ \text{ or } \ P(B) = 0, \tag{14.3}$$

$$A \subset B \ \& \ A \perp\!\!\!\perp B \Rightarrow P(A) = 0 \ \text{ or } \ P(B) = 1. \tag{14.4}$$

Equations (14.3) and (14.4) help us to conclude that if $||\Omega|| < 4$, then there are no nontrivially independent events.

Example 14.1   Nontrivially Pairwise Independent Events _____

If $\Omega = \{1, 2, 3, 4\}$ and $P(\{\omega\}) = \frac{1}{4}$, then any two of $A = \{1, 2\}$, $B = \{2, 3\}$, and $C = \{1, 3\}$ are independent events. This is verified from

$$P(A \cap B) = P(\{2\}) = \frac{1}{4} = \frac{1}{2}\frac{1}{2} = P(A)P(B).$$

However, $D = \{1, 3, 4\}$ is not independent of any of $A, B,$ or C. Thus, $C \subset D$, and by Eq. (14.4), they could only be trivially independent and

$$P(A \cap D) = P(\{1\}) = \frac{1}{4} \neq P(A)P(D) = \frac{1}{2}\frac{3}{4} = \frac{3}{8}.$$

14.3   MUTUALLY INDEPENDENT EVENTS

When we turn to collections of more than two events, it is not enough for them to be independent when considered pairwise (two at a time). We also want several of them to be unlinked with, or uninformative about, several others of them.

14.3.1   Definition of Mutual Independence

We now turn to deal with more than two events. We let I, J denote index sets that are subsets of $\{1, \ldots, n\}$. Motivated by the desire for unlinkedness between a collection $\{A_1, \ldots, A_n\}$ of events to imply that any event A_j is independent of any collection $\{A_i, i \in I\}$ of events that does not contain A_j, or

$$(\forall I \subset \{1, 2, \ldots, n\})(\forall j \notin I) \ P(A_j | \cap_{i \in I} A_i) = P(A_j),$$

we accept the next definition.

Definition 14.3.1   (Mutually Independent Events)   The events $\{A_1, \ldots, A_n\}$ are mutually independent, denoted $\perp\!\!\!\perp_1^n A_i$ or $\perp\!\!\!\perp \{A_1, \ldots, A_n\}$, if

$$(\forall I \subset \{1, \ldots, n\}) \ P(\cap_{i \in I} A_i) = \prod_{i \in I} P(A_i).$$

Observe that, as this is trivially true for $||I|| < 2$, we need to verify $2^n - n - 1$ conditions, and we cannot reduce this number.

Example 14.2   Pairwise, but not Mutually, Independent Events _____

In the example at the close of Section 14.2 the events A, B, and C are pairwise independent. However, the probability of the triple

$$P(A \cap B \cap C) = P(\phi) = 0 \neq P(A)P(B)P(C) = \frac{1}{8},$$

and A, B, and C are not mutually independent.

14.3.2 Elementary Consequences of Mutual Independence

Clearly, the definition of mutual independence is symmetrical in that the ordering of events is irrelevant. If we let B_i denote either A_i or A_i^c, then

$$\coprod_{1}^{n} A_i \iff \coprod_{1}^{n} B_i.$$

In particular, we have Lemma 14.3.1.

Lemma 14.3.1   (Independence and Complements)

$$\coprod_{1}^{n} A_i \iff \coprod_{1}^{n} A_i^c.$$

Proof sketch. This lemma follows from a proof by induction on the number of events n and the observation that if $P(\cap_{i=1}^{n-1} A_i) = \prod_{i=1}^{n-1} P(A_i)$, then

$$P(\cap_{i=1}^{n} A_i) = \prod_{i=1}^{n} P(A_i) \iff P\left(A_n^c \cap \bigcap_{i=1}^{n-1} A_i\right) = P(A_n^c) \prod_{i=1}^{n-1} P(A_i). \qquad \square$$

A further result can be stated in terms of *Boolean functions of events*. By a Boolean function $f(A_1, \ldots, A_n) = F$ of the sets $A_1, \ldots, A_n$, we mean a set F that is constructed from these n sets through iterated use of the Boolean set operations of complementation, union, or intersection. Equivalently, we can replace sets by their $\{0, 1\}$-valued indicator functions and consider Boolean functions on n binary variables as functions that assign either 0 or 1 to each of the 2^n possible sequences of assignments to the n binary variables. Let $I = \{i_1, \ldots, i_k\}, J = \{j_1, \ldots, j_m\}$ be index sets that are subsets of $\{1, \ldots, n\}$, and let f, g denote two Boolean functions.

Theorem 14.3.1   (Independence and Boolean Functions)

$$I \perp J \ \& \ \coprod_{n=1}^{n} A_n \Rightarrow f(A_{i_1}, \ldots, A_{i_k}) = F \perp\!\!\!\perp G = g(A_{j_1}, \ldots, A_{j_m}).$$

Proof remark. We omit the proof, as this theorem is a special case of a more general result for mutually independent random variables.   □

In other words, given two nonoverlapping collections of events drawn from a larger collection of mutually independent events, the two new sets F, G, formed by choosing arbitrary Boolean functions defined on the two collections, will themselves be independent.

Example 14.3   Independence of Boolean Functions _____

If $\perp\!\!\!\perp_1^5 A_i$, then

$$(A_1 \cap A_2) \cup A_3^c = B \perp\!\!\!\perp C = A_4^c \cup A_5.$$

Theorem 14.3.1 also generalizes when we have more than two nonoverlapping collections of events and form more than two events through possibly different Boolean functions of each of them. Thus, in the preceding example, we also have the following three events being mutually independent:

$$D = A_1 \cap A_2, \quad E = A_3^c, \quad F = A_4^c \cup A_5.$$

14.3.3 Examples

Assume that $\perp\!\!\!\perp_1^n A_i$ and $P(A_k) = p_k$. We develop expressions for the probabilities of the following events:

a. Event A is the occurrence of all of these events; hence,

$$P(A) = P(\cap_1^n A_i) = \prod_1^n P(A_i) = \prod_1^n p_i$$

b. Event B is the nonoccurrence of all of them; thus,

$$P(B) = P(\cap_1^n A_i^c) = \prod_1^n (1 - p_i)$$

c. Event C is the occurrence of at least one of them; therefore,

$$P(C) = P(\cup_1^n A_i) = 1 - P(B^c) = 1 - P(\cap_1^n A_i^c) = 1 - \prod_1^n (1 - p_i)$$

d. Event D is the occurrence of exactly one event, no matter which one; consequently,

$$P(D) = P(\cup_{i=1}^n (A_i \cap (\cap_{j \neq i} A_j^c))) = \sum_{i=1}^n P(A_i \cap (\cap_{j \neq i} A_j^c))$$

$$= \sum_{i=1}^n p_i \prod_{j \neq i} (1 - p_j) = \left\{ \prod_1^n (1 - p_i) \right\} \sum_{k=1}^n \frac{p_k}{1 - p_k}$$

Example 14.4   Negative Binomial PDF

As a final example, consider repeated independent experiments in which the occurrence of event A on the ith experiment has probability p. Let N_r be the random variable that is the number of repetitions of the experiment (waiting time) to the rth occurrence of the event A. Clearly, if $n < r$, then $P(N_r = n) = 0$. Assume that $n \geq r$. The event $N_r = n$ occurs when A happens on the nth repetition of the experiment and A has occurred exactly $r - 1$ times in the previous $n - 1$ performances. The outcome of the nth repetition is independent of the previous outcomes.

There are $\binom{n-1}{r-1}$ choices for the $r - 1$ occurrences of A among the $n - 1$ experiment repetitions. From the independence of repetitions, each such pattern of $r - 1$ occurrences of event A and $n - 1 - (r - 1)$ nonoccurrences of A has probability $p^{r-1}(1 - p)^{n-r}$. Hence, for $n \geq r$,

$$P(N_r = n) = \binom{n-1}{r-1} p^r (1 - p)^{n-r}.$$

This probability mass function is known as the *negative binomial*.

14.4   INDEPENDENT RANDOM VARIABLES AND PRODUCT FACTORIZATION

Given random variables $\{X_1, \ldots, X_n\} = X_1^n$, we can define them to be independent if the individual random experiments $\mathcal{E}_1, \ldots, \mathcal{E}_n$ that generated them are independent. In order to provide more immediate access to the essential definition through product factorization of cdfs, pdfs, or pmfs, we proceed less formally. Select any n sets $A_i \subseteq \mathbb{R}$. Let $B_i = (X_i \in A_i)$ be the event that the random variable X_i takes a value in A_i. Consider the probability $P(\cap_1^n B_i)$ of the joint occurrence of the n events $X_i \in A_i$. If the random variables are to be independent, then so should be the events B_i. We can now use our preceding results on the mutual independence of events to conclude that

$$\underset{i=1}{\overset{n}{\amalg}} X_i \Rightarrow \underset{i=1}{\overset{n}{\amalg}} B_i,$$

$$P(X_1 \in A_1, \ldots, X_n \in A_n) = \prod_{i=1}^{n} P(X_i \in A_i).$$

As this should hold for all choices of events $\{A_i\}$, including the semi-infinite intervals $(-\infty, x]$, we are led to the next definition.

Definition 14.4.1   (Mutually Independent Random Variables)   The random variables $X_1, \ldots, X_n$ are mutually independent $\amalg_{i=1}^{n} X_i$ if and only if, for all $x_1, \ldots, x_n$,

$$F_{X_1, \ldots, X_n}(x_1, \ldots, x_n) = \prod_{i=1}^{n} F_{X_i}(x_i).$$

Note that this definition actually involves uncountably many conditions, as it must hold for all real-valued choices for each of the x_i. We have now provided a summary of the background to be found in the appendix, and that leads us to the next lemma.

Lemma 14.4.1   (Independence of Random Variables)   The following are equivalent conditions for the random variables $\mathbf{X} = (X_1, \ldots, X_n)$ to be *mutually independent*:

1. The joint cdf $F_{X_1,\ldots,X_n}$ has a product factorization in terms of the component cdfs given by

$$F_{X_1,\ldots,X_n}(x_1, \ldots, x_n) = \prod_1^n F_{X_i}(x_i).$$

2. The joint density

$$f_{\mathbf{X}}(x_1, \ldots, x_n) = \frac{\partial^n F_{\mathbf{X}}(x_1, \ldots, x_n)}{\partial x_1 \cdots \partial x_n} = \prod_{i=1}^n f_{X_i}(x_i).$$

3. In the case of discrete-valued random variables, the joint probability mass function factors as a product of the individual component experiment pmfs:

$$p_{X_1,\ldots,X_n}(x_1, \ldots, x_n) = P(X_1 = x_1, \ldots, X_n = x_n) = \prod_1^n p_{X_i}(x_i).$$

An alternative argument leading to product factorization of the pdf for independent random variables $X_1^n = (X_1, \ldots, X_n)$ starts from the conditional pdf $f_{X_{m+1}^n | X_1^m}(x_{m+1}^n | x_1^m)$. The conditional pdf tells us how to revise our knowledge of the tendency of X_{m+1}^n to occur, given knowledge that $X_1^m = x_1^m$ did occur. Hence, if the random variables are independent, we should not change our mind about this tendency, and we have

$$\overset{n}{\underset{1}{\perp\!\!\!\perp}} X_i \Rightarrow f_{X_{m+1}^n | X_1^m}(x_{m+1}^n | x_1^m) = f_{X_{m+1}^n}(x_{m+1}^n).$$

As this conclusion must hold for all m, n, we immediately derive the product factorization condition for the joint density.

An important special case occurs when the component experiments have the same probability description.

Definition 14.4.2   (i.i.d.)   The experiments $\mathcal{E}_1, \ldots, \mathcal{E}_n$ or the random variables $X_1, \ldots, X_n$ are independent and identically distributed—*i.i.d.*—if they are mutually independent and have a common cdf F_X:

$$(\forall i)\ F_{X_i}(x) = F_X(x).$$

In this case, the common cdf F_X yields a common density f_X and

$$f_{X_1,\ldots,X_n}(x_1, \ldots, x_n) = \prod_{i=1}^n f_X(x_i).$$

In the case of pmfs,

$$P(X_1 = x_1, \ldots, X_n = x_n) = p_{x_1, \ldots, x_n} = \prod_{i=1}^{n} p_{x_i}.$$

In the special case of X taking values in $\{0, 1\}$, we refer to the $i.i.d.$ model as *Bernoulli trials*. The $i.i.d.$ model is perhaps the single most commonly encountered joint experiment.

14.5 DERIVATIONS OF TWO PROBABILITY MODELS

14.5.1 Derivation of the Geometric Probability Model

Assume that we have mutually independent random variables $X_1, \ldots, X_n, \ldots$ and an event F defined for any individual X_i. For example, F might be the event of system failure due to a shock of magnitude X occurring in a unit time interval. Let T be the positive, integer-valued random variable that is the *waiting time* to the first occurrence of F. Hence, the event

$$(T = t) = \cap_1^{t-1}(X_i \in F^c) \cap (X_t \in F).$$

By the postulated independence,

$$P(T = t) = P(X_t \in F) \prod_1^{t-1} P(X_i \in F^c).$$

Making the assumption that there is no temporal evolution or aging over the trials, we find that all of the random variables have the same distribution. (They are now $i.i.d.$) Letting $\beta = P(X_1 \in F^c)$ yields

$$P(T = t) = (1 - \beta)\beta^{t-1}, \ t = 1, 2, \ldots.$$

Hence, we conclude that $T - 1 \sim \mathcal{G}(\beta)$. (See Section 4.4.3.)

14.5.2 Bernoulli Trials: Derivation of the Binomial Model

Let the random experiment $\mathcal{E}_i$ be described by

$$P(X_i = 1) = p_i = 1 - P(X_i = 0) \text{ or}$$

$$P(X_i = x) = p_i^x (1 - p_i)^{1-x} \text{ for } x \in \{0, 1\}.$$

X_i can be thought of as recording the outcome of the ith coin toss or marking the presence or absence of an error ($X_i = 1$ for error) on the ith symbol of text. Hence, under the assumption of mutually independent $\{X_i\}$,

$$P(X_1 = x_1, \ldots, X_n = x_n) = \prod_{i=1}^{n} p_i^{x_i} (1 - p_i)^{1-x_i}.$$

The special case of *i.i.d.* experiments is known as *Bernoulli trials*. In this case, $p_i = p$, and we have

$$P(X_1 = x_1, \ldots, X_n = x_n) = p^{\sum_{i=1}^{n} x_i}(1 - p)^{n - \sum_{i=1}^{n} x_i}.$$

Observe that $P(X_1 = x_1, \ldots, X_n = x_n)$ depends only upon $k = \sum_{i=1}^{n} x_i$; all sequences of outcomes that have the same sum k have the same probability

$$p^k(1 - p)^{n-k}, \quad k = 0, 1, \ldots, n.$$

Hence, if we now inquire into the description of

$$Y = \sum_{i=1}^{n} x_i,$$

the total number of errors made on the n symbols, we see that the event $Y = k$ is a disjoint union of $\binom{n}{k}$ possible outcome sequences, each having the sum k and the same probability. Thus, their sum is

$$P(Y = k) = \binom{n}{k} p^k(1 - p)^{n-k}, \quad k = 0, \ldots, n,$$

and we have derived the binomial probability law $\mathcal{B}(n, p)$ of Section 4.4.2.

Example 14.5   Even Number of Ones in Bernoulli Trials _____

In the case of Bernoulli trials, we evaluate the probability P_n that there is an even number (zero is even) of ones in the n trials. Clearly, $P_0 = 1$. If $n \geq 1$, then we have an even number of ones either if there was an odd number of ones in the first $n - 1$ trials and a one on the nth trial or if there was an even number of ones by trial $n - 1$ and a zero on the nth trial. Hence, using the independence of the nth trial from the preceding $n - 1$ trials, we obtain

$$P_n = p(1 - P_{n-1}) + (1 - p)P_{n-1} = p + (1 - 2p)P_{n-1}.$$

One can verify by substitution that this recursion, together with the initial condition for P_0, is uniquely satisfied by

$$P_n = \frac{1}{2}(1 + (1 - 2p)^n).$$

For $0 < p < 1$,

$$\lim_{n \to \infty} P_n = \frac{1}{2}.$$

Of course, we are not asserting that in the long run about half of the trials will be ones; rather, we assert only that the number of ones is equally likely to be even or odd.

14.6 EXAMPLE OF MAXIMA AND MINIMA OF RANDOM VARIABLES

We often encounter situations in which we are interested in either the largest or the smallest of a collection of $i.i.d.$ random variables. For example, in reliability, we might be interested in the shortest lifetime among a set of components or the maximum impact intensity that a system might be subject to in repeated uses. In networked communications, we are interested in maximum delays in packet transmission and in cellular communications in maximum numbers of simultaneous users. Assume that the random variables $X_1, \ldots, X_n$ are $i.i.d.$ with common cdf F_X and pdf f_X.

Proceeding to treat the maximum $Z = \max_{i \leq n} X_i$, we have

$$F_Z(z) = P(\max_{i \leq n} X_i \leq z) = P((\forall i \leq n)\, X_i \leq z) = P(\cap_i^n A_i),$$

where A_i is the event that $X_i \leq z$. The events $A_1, \ldots, A_n$ are $i.i.d.$ Hence,

$$P(\cap_i^n A_i) = \prod_i^n P(A_i) = F_X^n(z) \text{ and}$$

$$F_Z(z) = F_X^n(z).$$

We can enquire into the median μ_Z of Z,

$$F_Z(\mu_Z) = \frac{1}{2} \Rightarrow F_X(\mu_Z) = 2^{-\frac{1}{n}} = e^{-\frac{1}{n}\log 2} \approx 1 - \frac{\log 2}{n}.$$

We see that, for large n, $F_X(\mu_Z)$ is almost 1 and μ_Z is "large." Furthermore, by differentiating F_Z, we see that

$$f_Z(z) = nF_X^{n-1}(z)f_X(z).$$

Examining $F_Z(z)$, for large n, we have

$$F_Z(z) = F_X^n(z) = (1 - (1 - F_X(z)))^n \leq e^{-n(1-F_X(z))},$$

where we have used the inequality $1 - x \leq e^{-x}$. For small x, this inequality becomes a good approximation and offers

$$F_Z(z) \approx e^{-n(1-F_X(z))}$$

as a good approximation when $F_X(z)$ is close to 1 or z is large.

Example 14.6   CDF and Median of Maximum _____

If $X_i \sim \mathcal{U}(0, 1)$, then

$$F_Z(z) = z^n U(z)U(1 - z), \quad f_Z(z) = nz^{n-1}U(z)U(1 - z),$$

$$\mu_Z^n = \frac{1}{2} \text{ and } \mu_Z = 2^{-\frac{1}{n}}.$$

In this case, a large value of the median of Z is a value that is very close to 1.

If we inquire into the probability that the maximum exceeds a given value, we find that

$$P(Z > z) = 1 - F_Z(z) = 1 - F_X^n(z).$$

From the identity

$$1 - x^n = (1 - x) \sum_{k=0}^{n-1} x^k,$$

we see that

$$P(Z > z) = (1 - F_X(z)) \sum_{k=0}^{n-1} F_X^k(Z).$$

Since $F_X(z) \leq 1$, we have

$$P(Z > z) \leq n(1 - F_X(z)),$$

with the inequality only of interest for $F_X(z) > 1 - 1/n$ and becoming more accurate as z becomes larger, making $F_X(z)$ approach 1.

Example 14.7   Tail Probability for Maximum of Normals

If X_i are $i.i.d.$ $\mathcal{N}(m, \sigma^2)$, then the asymptotic approximation to the normal cdf given in Section 6.7.3 yields

$$P(\max_{i \leq n} X_i > z + m) \approx \frac{n\sigma e^{-\frac{z^2}{2\sigma^2}}}{z\sqrt{2\pi}},$$

or

$$P(\max_{i \leq n} X_i > z) \approx \frac{n\sigma e^{-\frac{(z-m)^2}{2\sigma^2}}}{(z - m)\sqrt{2\pi}}.$$

Turning now to examine the minimum $Y = \min_{i \leq n} X_i$, we have

$$F_Y(y) = P(\min X_i \leq y) = P((\exists i) \ X_i \leq y) = P(\cup_i A_i),$$

where A_i is the event that $X_i \leq y$. The events $A_1, \ldots, A_n$ are $i.i.d.$ Independence tells us directly only about intersections of independent events, not about their unions. However, de Morgan's laws allow us to convert a statement about unions into one about intersections:

$$P(\cup_i A_i) = 1 - P(\cap_i A_i^c) = 1 - \prod_i P(A_i^c)$$

$$= 1 - \prod_i (1 - P(X_i \leq y)) = 1 - (1 - F_X(y))^n.$$

Thus,

$$F_Y(y) = 1 - (1 - F_X(y))^n.$$

Furthermore, by differentiating F_Y, we see that

$$f_Y(y) = n(1 - F_X(y))^{n-1} f_X(y).$$

From $1 - x \le e^{-x}$,

$$F_Y(y) \ge 1 - e^{-nF_X(y)}.$$

For small y, meaning that $F_X(y)$ is close to zero,

$$F_Y(y) \approx 1 - e^{-nF_X(y)}.$$

We examine F_Y for large n by looking at the median value μ_Y of Y:

$$F_Y(\mu_Y) = \frac{1}{2},$$

For our present illustrative purposes, we assume that the median value is taken on uniquely. Hence,

$$1 - (1 - F_X(\mu_Y))^n = \frac{1}{2} \Rightarrow F_X(\mu_Y) = 1 - 2^{-\frac{1}{n}}.$$

The calculation done earlier for Z leads to

$$F_X(\mu_Y) \approx \frac{\log 2}{n},$$

and the median value for the minimum of n $i.i.d.$ random variables must be a "small" value as measured by the probability $F_X(\mu_Y)$ of an X_i not exceeding such a value.

Example 14.8   CDF and Median of Minimum _____

If $X \sim \mathcal{E}(\alpha)$, then we have the exact result

$$F_Y(y) = 1 - e^{-n\alpha y}.$$

In this case,

$$\mu_Y = \frac{\log 2}{\alpha n}.$$

Much is known about the asymptotic in n properties of the extreme value statistics Y and Z. Applying such results to a problem in radar was the subject of the author's first publication.

14.7 EXPECTATION AND INDEPENDENCE

Theorem 14.7.1   (Expectation of Products)   If $\perp\!\!\!\perp_1^n X_i$, then

$$E\left(\prod_1^n X_i\right) = \prod_1^n EX_i, \quad E\left(\prod_1^n h_i(X_i)\right) = \prod_1^n Eh_i(X_i).$$

Proof. Clearly, the first assertion is a special case of the second. To verify the second assertion, use independence and note that

$$E\left(\prod_1^n h_i(X_i)\right) = \int_{-\infty}^{\infty} dx_n \dots \int_{-\infty}^{\infty} dx_1 \left(\prod_1^n h_i(x_i)\right) \prod_1^n f_{X_i}(x_i)$$

$$= \prod_1^n \int_{-\infty}^{\infty} h_i(x_i) f_{X_i}(x_i)\, dx_i = \prod_1^n Eh_i(X_i). \qquad \Box$$

Hence, independence implies uncorrelatedness, but as we have seen in the discussion at the end of Section 9.12, the converse fails, and two random variables can be uncorrelated even though the one determines the other.

14.8 APPLICATION TO ESTIMATION OF THE MEAN AND VARIANCE

It is common to have available *i.i.d.* observations $\mathbf{X} = (X_1, \dots, X_n)$ from which to estimate a signal S or a parameter θ that governs the probability law of $\mathbf{X}$. We return to the MLE estimator introduced in Section 13.5.2 and apply it to the classical problem of the estimation of the unknown mean m and variance $v = \sigma^2$ of the normal law $X_i \sim \mathcal{N}(m, v)$. In this case,

$$f_{\mathbf{X}|m,v}(\mathbf{x}|m, v) = \prod_{i=1}^n f_{X_i|m,v}(x_i|m, v) = (2\pi v)^{-n/2} \prod_{i=1}^n e^{-\frac{1}{2v}(x_i - m)^2}.$$

The MLE $(\tilde{m}, \tilde{v})$ are chosen to maximize $f_{\mathbf{X}|m,v}(\mathbf{x}|m, v)$ for the given data $\mathbf{x}$. It is more convenient to maximize

$$\log f_{\mathbf{X}|m,v}(\mathbf{x}|m, v) = -\frac{n}{2}\log 2\pi - \frac{n}{2}\log v - \frac{1}{2v}\sum_{i=1}^n (x_i - m)^2.$$

We can ignore the initial constant term, as it does not involve the variable m or v. From the calculus, we take first partial derivatives with respect to m and v and set them to zero at the MLE values $\tilde{m}$ and $\tilde{v}$; thereby, we obtain the following equations for the two parameters:

$$\frac{1}{\tilde{v}}\sum_{i=1}^n (x_i - \tilde{m}) = 0;$$

$$-\frac{n}{2\tilde{v}} + \frac{1}{2\tilde{v}^2}\sum_{i=1}^n \left(x_i - \tilde{m}\right)^2 = 0.$$

We easily solve the first equation for the MLE estimator of the mean m,

$$\tilde{m} = \frac{1}{n} \sum_{i=1}^{n} x_i,$$

as the familiar sample average. Then, we just as easily solve the second equation for the MLE estimator of the variance v,

$$\tilde{v} = \frac{1}{n} \sum_{i=1}^{n} \left(x_i - \frac{1}{n} \sum_j x_j \right)^2.$$

These results for $\tilde{m}$ and $\tilde{v}$ can be compared with those in Section 1.7.2. Note that $\tilde{m}$ agrees with the sample mean. However, $\tilde{v}$ disagrees with the sample variance being smaller by a factor of $(n-1)/n$ that is negligible for large n.

If we had a prior distribution π over the parameters (m, v), then we could compute the MAP estimators $\hat{m}$ and $\hat{v}$. Typically, this would require numerical methods to carry out the required maximization of $\pi(m, v) f_{\mathbf{X}|m,v}(\mathbf{x}|m, v)$ with respect to m and v.

14.9 SUMMARY

It may be surprising to those who have had a prior introduction to probability that we have deferred a full discussion of the key concept of independence until now, with the exception of a brief mention in Section 2.10. We have done so because the soundest way to introduce independence is through conditional probability, although one can give the formal definition of independence without reference to conditional probability. However, following this common path provides the reader with so little insight into the meaning of independence that it encourages an excessive willingness to arbitrarily assume that events and random variables are independent. In his foundational treatise on the axioms of probability, Kolmogorov rightly claims that independence is a key to the distinction between probability theory and the mathematical theory of measure. (In his later contributions to algorithmic information theory (e.g., see Li and Vitanyi [59]) Kolmogorov provides another perspective on independence.) We attempt to understand independence through an informal notion of "unlinkedness." In practice, it is the informal recognition of this unlinkedness that leads one to postulate independence and thereby gain a substantially simplified and more tractable probability model for a random experiment.

We start with the simplest case of independent events and then extend the notion to the independence of random variables, which implies the independence of many events. (We defer to an appendix certain technical details about independent joint experiments that are needed to discuss independent random variables.) We point out the widely used special case of independent and identically distributed, *i.i.d.*, random variables. Our conceptual introduction is followed, as usual, by formal definitions, in this case derived from properties of conditional probability, and a study of the consequences of the definition. Among other results, we find ourselves able to clarify the conditions justifying the use of the binomial and geometric models by deriving them under appropriate assumptions of independence. The justification for the binomial leads to a justification for the Poisson based upon its connection as a rare events limit of the binomial. (See Section 4.5.)

Application is given to the properties of maxima and minima of $i.i.d.$ random variables and to expectation of products of independent random variables. This latter result will be used in Section 17.7 to provide an alternative definition of the independence of random variables that is computationally useful, but is an unintuitive starting point for independence.

EXERCISES

E14.1 If $\Omega = \{1, 2, 3, 4, 5, 6\}$, $A = \{1, 2, 4\}$, and $B = \{2, 3, 5\}$, then construct a pmf $p(k)$ such that $A \perp\!\!\!\perp B$ nontrivially.

E14.2 If $\Omega = \{0, 1, \ldots, 9\}$ and $p_k = .1$, what are the independence relations between the three events

$$A = \{0, 1, 2, 3, 4\}, \ \ B = \{2, 3, 4, 5, 6, 7\}, \ \ C = \{0, 1, 2, 6, 7, 8\}?$$

E14.3 The event F of system failure occurs if either A_1 or A_2 occurs, but A_3 does not occur. If $\perp\!\!\!\perp_i A_i$ and $P(A_1) = .4$, $P(A_2) = .35$, and $P(A_3) = .1$, then evaluate $P(F)$.

E14.4 Consider Bernoulli trials with arbitrary $P(X_i = 1) = p$. Group the odd and even trials $\{(X_{2k+1}, X_{2k})\}$. Let $N \geq 0$ be the index of the first k for which $X_{2N+1} \neq X_{2N}$.

    a. Determine $P(X_{2N} = 1)$.

    b. Can this process be used to create the equivalent of a truly fair coin?

E14.5 We are given that $\perp\!\!\!\perp_1^4 A_i$ with $P(A_i) = \frac{1}{1+i}$.

    a. If B is the event that at least one of $\{A_i\}$ occurs, then evaluate $P(B)$.

    b. If C is the event that either A_1 or A_2, but not A_4, occurs, then evaluate $P(C)$.

E14.6 Let X_i be 1 if there is an error in the ith bit and 0 if otherwise. Assume that $\mathbf{X} = \{X_i\}$ are $i.i.d.$ (independent and identically distributed) with $P(X_i = 1) = p$. Consider the events

$$A = \{\mathbf{X} : X_1 = X_3 = 1\}, \ \ B = \{\mathbf{X} : X_4 = 0\},$$

$$C = \{\mathbf{X} : X_2 = 0, \ X_5 = 1\}.$$

    a. Evaluate the probability P_a that all three events occur.

    b. Evaluate the probability P_b that none of them occur.

    c. Evaluate the probability P_c that at least one of them occurs.

    d. Evaluate the probability P_d that A and B, but not C, occur.

E14.7 Recall the network displayed in Figure 2.5. Let C_k, $k = 1 : 14$, be the events that the link between nodes k and $k + 1$ is intact. Assume that the events $\{C_k\}$ are $i.i.d.$ with probability p. Recall that all of the links are unidirectional from the lower numbered node to the higher.

    a. What is the probability $P_{1:6}$ of a connected path from node 1 to node 6?

    b. What is the probability $Q_{6:15}$ that there is no path from node 6 to node 15?

E14.8   Urn U_i has $r_i = i$ red balls and $g_i = 10 - i$ green balls. In an unlinked experiment $\mathcal{E}_0$, we select U_i with probability $\binom{5}{i-1}.5^5$ for $i = 1, \ldots, 6$. We then choose a ball at random from the selected urn. What is the probability that the ball is red?

E14.9   A given source produces an infinitely long sequence of messages $\{M_i\}$ that are chosen in an *i.i.d.* fashion with $P(M_i = m) = p_m$ for the choice of the specific message m on the ith trial. Consider two distinct possible messages m_0 and m_1.

    a. What is the probability that message m_0 will appear before message m_1 on the nth trial?

    b. Verify that the probability that eventually message m_0 will appear before message m_1 is $p_{m_0}/(p_{m_0} + p_{m_1})$.

E14.10 If

$$f_{X|Y}(x|y) = \begin{cases} 1 & \text{if } 0 \le x \le 1, \ y \ge 0 \\ 2x & \text{if } 0 \le x \le 1, \ y < 0 \end{cases},$$

then are X, Y independent?

E14.11 a. The input X to a system S with output Y is such that

$$f_X(x) = e^{-x}U(x) \text{ and } P(Y = y) = \left(\frac{1}{2}\right)^y \text{ for } y = 1, 2, \ldots.$$

    What can you say about S?

    b. For the setup of (a), if you also know that $X \perp\!\!\!\perp Y$, what can you say about S?

E14.12 A system is subjected to *i.i.d.* shocks $X_1, \ldots, X_n$ with $X_i \sim \mathcal{U}(0, S)$. The system fails if a shock exceeds γ. Find the probability of system failure.

E14.13 a. System A is composed of a "series" connection of components $C_1, \ldots, C_n$ and fails to operate if any component fails. The time to failure of C_i is T_i. If the $\{T_i\}$ are *i.i.d.* with density $\mathcal{E}(\alpha)$, then evaluate the density for the system lifetime T.

    b. System B is composed of a "parallel" connection of the same components as System A, and B fails to operate only if all of its components fail. Evaluate the density for the lifetime T of System B.

E14.14 A system is composed of three components C_1, C_2, and C_3, having respective lifetimes L_1, L_2, and L_3, with corresponding probability laws $L_i \sim \mathcal{E}(\alpha_i)$. If the individual components fail independently of each other and if the system fails if either of the components fail, then what is the density f_L for the overall system lifetime L?

E14.15 A system S having lifetime L is composed of three subsystems S_1, S_2, and S_3 having independent lifetimes L_1, L_2, and L_3 that are identically distributed with continuous cdf F. S fails if either S_1 fails or if both S_2 and S_3 fail.

    a. If A_i denotes the event that $L_i \le x$ and B denotes the event that $L \le x$, then express B in terms of $\{A_i\}$.

    b. Evaluate the cdf $F_L(x)$ for S.

    c. Evaluate the pdf f_L.

    d. Is $P(L_1 < L_2 < L_3) = P(L_3 < L_1 < L_2)$? (This is easy, with a little thought.)

    e. Evaluate the probability P_1 that S fails due to the failure of S_1 (i.e., $P_1 = P(L_1 < \max(L_2, L_3))$).

E14.16 A radar range detector works by considering the outputs $\{X_i\}$ from each of r "range bins," with a target being identified as at range r_j if X_j is large. When no target is present, the outputs are $i.i.d.$ $\mathcal{E}(\alpha)$. A detection occurs if any of these r outputs exceeds a threshold τ. What is the probability of false alarm?

E14.17 We observe $X_i, i = 1 : n, i.i.d.$ $\mathcal{U}(0, \theta)$. One way to infer θ is to form the estimator

$$Z_n = \frac{2}{n} \max_i X_i.$$

Evaluate the pdf $f_{Z_n}(z)$.

E14.18 If $X \sim \mathcal{E}(1)$, $Y \sim \mathcal{U}(0, 1)$ and they are independent, then evaluate $P(X > Y)$.

E14.19 A system composed of n components having individual lifetimes L_i that are $i.i.d.$ $\mathcal{E}(\alpha)$, fails if any component fails.

    a. What is the probability of the system lifetime T exceeding $\frac{1}{n\alpha}$?

    b. Evaluate the density function f_T of the system lifetime T.

E14.20 If X, Y are $i.i.d.$ $\mathcal{E}(1)$, then evaluate $P(Y > 2X)$.

E14.21 A random experiment $\mathcal{E}$ consists of choosing an integer from the set $\{1, 2, \ldots, N\}$, with each integer having equal probability of being chosen. Independent repetitions $\{\mathcal{E}_i\}$ of this experiment are conducted until the first time T that the number chosen equals one chosen in a previous experiment. Find $P(T = t)$.

E14.22 Calculate the probability that no two people in a group the size of your class section have the same birthday. (Neglect leap year and seasonal considerations.) Approximate $\log(1 - x)$ by $-x$ to obtain a simplified answer. See if any two people present share a common birthday.

E14.23 If $\perp\!\!\!\perp_1^3 X_i$, $X_i \sim \mathcal{N}(m_i, \sigma_i^2)$, then evaluate the joint density f_{X_1, X_2, X_3} and the conditional density $f_{X_1 | X_3}$.

E14.24 X_i, $i.i.d.$ $f_X(x) = \frac{1}{2}e^{-|x|}$ are inputs to a discrete-time integrator

$$Y_1 = X_1, \quad Y_k = X_k + Y_{k-1} \ (k > 1).$$

    a. Partially describe the system response by finding f_{Y_1, Y_2, Y_3}.

    b. Find $f_{X_3 | Y_2}$.

E14.25 X_1, X_2, X_3, $i.i.d.$ $\mathcal{E}(1)$, and

$$Y_1 = X_1 X_2 X_3, \quad Y_2 = X_1 X_2, \quad Y_3 = X_1.$$

Find $f_{Y_1, Y_2, Y_3}(y_1, y_2, y_3)$.

E14.26 In an $i.i.d.$ model for errors made in transmission of text symbols, T is the waiting time (number of symbols) to the first occurrence of an error. If the probability of an individual error is .1, what is the pmf $p_T(k)$?

E14.27 A photodetector receives a photon count Z arising from the sum of two independent light sources having photon counts $X_i \sim \mathcal{P}(\lambda_i)$, $i = 1, 2$. Evaluate $p_Z, p_{Z|X_1}, p_{X_1|Z}$.

E14.28 A binary symmetric channel has crossover (error) probability p and operates independently on its binary input symbols $\{X_i\}$ to produce its output symbols $\{Y_i\}$ (e.g., $(X_1, Y_1) \perp\!\!\!\perp (X_2, Y_2)$). The input message source can emit only one of the three binary sequences $m_1 = 00, m_2 = 01$, and $m_3 = 11$, with respective probabilities .5, .3, and .2.

    a. Find the probability that the output $Y_1 Y_2 = 10$.

    b. Determine the conditional probability of each of the three possible messages m_1, m_2, and m_3, given that we receive 10.

    c. Given any received sequence $y_1 y_2$, the minimum error probability decision as to the message that was sent is to choose the one with the largest posterior probability $P(M = m_i | Y_1 Y_2 = y_1 y_2)$. If $p = .2$ and 10 is received, then what should our decision be as to the transmitted message?

E14.29 An analog channel adds independent noise $N \sim \mathcal{N}(0, 2)$ to its input X to produce its output Y. The input is distributed with cdf $F_X(x) = .2U(x + 1) + .6U(x) + .2U(x - 1)$.

    a. Determine the channel characterization $f_{Y|X}$.

    b. Determine the channel output response f_Y.

    c. What is the most probable input value if the output is 1, and how probable is it?

E14.30 A system S having lifetime L is composed of two different components having individual lifetimes $L_1 \sim \mathcal{E}(1)$ and $L_2 \sim \mathcal{E}(3)$. These components of S fail independently of each other.

    a. If S fails if either component fails, then evaluate f_L.

    b. If S fails only if both components fail, then evaluate f_L.

E14.31 If $X \perp\!\!\!\perp Y$, $X \sim \mathcal{E}(1)$, $Y \sim \mathcal{U}(0, 1)$, and $Z = X + Y$, then determine $f_Z(z)$.

E14.32   a. If

$$X(t) = A \cos(\omega_0 t + \Theta), \ A \perp\!\!\!\perp \Theta, \ \Theta \sim \mathcal{U}(-\pi, \pi), \ A \sim \mathcal{E}(\alpha),$$

then evaluate $EX(t)$, $COV(X(t), X(s))$, and $VAR(X(t))$.

    b. Design the linear, least mean square predictor $\hat{X}(t)$ of $X(t)$ based upon observation of $X(s)$ for some $s < t$.

E14.33 A system S having lifetime L is composed of four subsystems $S_1, \ldots, S_4$ with lifetimes $L_1, \ldots, L_4$, respectively, that are $i.i.d.$ $\mathcal{E}(\alpha)$. S fails if both S_1 and S_2 fail or if both S_3 and S_4 fail.

    a. Letting $A = (L \leq \tau)$ and $A_i = (L_i \leq \tau)$, express A in terms of $A_1, \ldots, A_4$.

    b. Determine the cdf $F_L(\tau)$.

    c. Determine EL.

E14.34 We observe $X = S + N$,

$$S \perp\!\!\!\perp N, \ N \sim \mathcal{N}(0, \sigma^2), \ P(S = -1) = P(S = 1) = .25, \ P(S = 0) = .5.$$

    a. Evaluate $f_{X|S}$.

    b. Evaluate f_X.

    c. If $X = 1$, then what do we know about S?

E14.35 If all you know is that $P(X \leq 3 | Y = 2) < P(X \leq 2 | Y = 3)$, what can you conclude about the independence of X, Y? (Explain.)

E14.36 If $f_{X,Y}(x, y) = 1/\pi$ on the disk of unit radius and $f_{X,Y}(x, y) = 0$ otherwise, is $X \perp\!\!\!\perp Y$?

E14.37 If X, Y, Z are $i.i.d.$ and, individually, $\mathcal{N}(0, 1/2)$, what is $f_{X,Y,Z}$?

E14.38 A sequence of random variables $\{X_k\}$ is recursively generated through

$$X_0 = 0, \ X_1 = N_1, \ X_2 = aX_1 + N_2, \ X_k = aX_{k-1} + N_k,$$

where $\{N_k\}$ are $i.i.d.$ with common pdf f_N.

    a. Is $N_k \perp\!\!\!\perp X_{k-1}$?

    b. Is $N_k \perp\!\!\!\perp X_k$?

    c. Provide an expression for $f_{X_k|X_{k-1}}$ for $k \geq 2$.

E14.39   a. We know that $\perp\!\!\!\perp_1^3 A_i$, $P(A_1) = .4, P(A_2) = .4$, and $P(A_3) = .5$. Evaluate $P(B)$ for B the event that all three of these events do not occur.

    b. Repeat (a) for $P(C)$, where C is the event that either A_1 or A_3 occurs.

E14.40 The random variables $X_1, \ldots, X_n$ are $i.i.d.$ $\mathcal{U}(0, 1)$.

    a. If $Z = \max_{i \leq n} X_i$, provide an expression for EZ.

    b. If $Y = \min_{i \leq n} X_i$, provide an expression for EY.

E14.41 The random variables $X_1, \ldots, X_n$ are $i.i.d.$ $\mathcal{E}(\alpha)$.

    a. If $Z = \max_{i \leq n} X_i$, provide an expression for EZ.

    b. If $Y = \min_{i \leq n} X_i$, provide an expression for EY.

E14.42 We observe $\mathbf{X} = (X_1, \ldots, X_n)$ $i.i.d.$ $\mathcal{P}(\lambda)$ and wish to estimate λ from $\mathbf{X}$.

    a. Determine the MLE estimator $\tilde{\lambda}(\mathbf{X})$.

    b. If you are now informed that $\lambda \sim \mathcal{E}(1)$, provide an expression for the mean square error made by $\tilde{\lambda}(\mathbf{X})$.

E14.43 We observe $\mathbf{X} = (X_1, \ldots, X_n)$ $i.i.d.$ $\mathcal{B}(n_0, p)$ and wish to estimate p from $\mathbf{X}$ for given n_0.

    a. Determine the MLE estimator $\tilde{p}(\mathbf{X})$.

    b. If you are now informed that $p \sim \mathcal{U}(0, 1)$, provide an expression for the mean square error made by $\tilde{p}(\mathbf{X})$.

E14.44 We observe $\mathbf{X} = (X_1, \ldots, X_n)$ $i.i.d.$ $\mathcal{G}(\beta)$ and wish to estimate β from $\mathbf{X}$.

    a. Determine the MLE estimator $\tilde{\beta}(\mathbf{X})$.

    b. If you are now informed that $\beta \sim \mathcal{U}(0, 1)$, provide an expression for the mean square error made by $\tilde{\beta}(\mathbf{X})$.

E14.45 We observe $\mathbf{X} = (X_1, \ldots, X_n)$ $i.i.d.$ $\mathcal{E}(\alpha)$ and wish to estimate α from $\mathbf{X}$.

    a. Determine the MLE estimator $\tilde{\alpha}(\mathbf{X})$.

    b. If you are now informed that $\alpha \sim \mathcal{E}(1)$, provide an expression for the mean square error made by $\tilde{\alpha}(\mathbf{X})$.

E14.46 We observe $\mathbf{X} = (X_1, \ldots, X_n)$ $i.i.d.$ $\mathcal{N}(m, \sigma^2)$ and wish to estimate m from $\mathbf{X}$ for given σ^2.

    a. Determine the MLE estimator $\tilde{m}(\mathbf{X})$.

    b. If you are now informed that $m \sim \mathcal{N}(0, 1)$, provide an expression for the mean square error made by $\tilde{m}(\mathbf{X})$.

E14.47   We observe $\mathbf{X} = (X_1, \ldots, X_n)$ *i.i.d.* $\mathcal{N}(0, v)$ and wish to estimate the variance v from $\mathbf{X}$.

    a. Determine the MLE estimator $\tilde{v}(\mathbf{X})$.

    b. If you are now informed that $v \sim \mathcal{E}(1)$, provide an expression for the mean square error made by $\tilde{v}(\mathbf{X})$.

E14.48   We observe $Y = X + N$ as the output of a measurement system $\mathcal{S}$ having input $X \sim \mathcal{P}(1)$ and additive independent $N \sim \mathcal{E}(\alpha)$.

    a. Evaluate $P(X \leq Y)$. (This is easy, given a little thought.)

    b. Specify what we know about X, given that we have observed Y.

    c. Provide an expression for the minimum error probability MAP estimator $\hat{X}(Y)$ of X given Y that is valid for all Y.

    d. Evaluate $\hat{X}(Y)$ for $Y = 1.5$.

15

Conditional Expectation

15.1 PURPOSE, BACKGROUND, AND ORGANIZATION

We define a concept of conditional expectation $E(Y|\mathbf{X})$ of Y given $\mathbf{X}$ through the use of conditional pmfs and pdfs, in a manner parallel to our definition of the expectation EY in Chapter 9. $E(Y|\mathbf{X} = \mathbf{x})$ is to be the mean of Y, given the information that another random variable $\mathbf{X} = \mathbf{x}$. We introduce the concept of conditional expectation, explore its basic properties, and then apply it to two problems of minimum mean square estimation (MMSE), as well as revisiting Kalman filtering. (See Section 10.6.) We are now able to find the most general minimum mean square estimator of Y given $\mathbf{X}$ without needing the restriction to linearity imposed in Chapter 10.

15.2 CONDITIONAL EXPECTATION BASICS

Conditional expectation is given as an extension of expectation through the following definition.

Definition 15.2.1 (Conditional Expectation) The conditional expectation

$$E(Y|\mathbf{X} = \mathbf{x}) = \int_{-\infty}^{0} y f_{Y|\mathbf{X}}(y|\mathbf{x})\, dy + \int_{0}^{\infty} y f_{Y|\mathbf{X}}(y|\mathbf{x})\, dy,$$

provided that at least one of the two integrals is finite.

The conditional expectation $E(\mathbf{Y}|\mathbf{X} = \mathbf{x})$ of a vector $\mathbf{Y}$ given $\mathbf{X} = \mathbf{x}$ is just the vector of conditional expectations of the components Y_i of $\mathbf{Y}$, or

$$E(\mathbf{Y}|\mathbf{X} = \mathbf{x}) = [E(Y_i|\mathbf{X} = \mathbf{x})].$$

Example 15.1   Calculation of Conditional Expectation, Innovations Case _____
Consider the innovations joint pdf

$$f_{X,Y}(x, y) = e^{-2|y-x^2|-x}U(x).$$

We recover

$$f_X(x) = \int_{-\infty}^{\infty} e^{-2|y-x^2|-x}U(x)\,dy = e^{-x}U(x).$$

The conditional pdf

$$f_{Y|X}(y|x) = \frac{f_{X,Y}(x, y)}{f_X(x)} = e^{-2|y-x^2|}.$$

Applying Definition 15.2.1, we obtain

$$E(Y|X = x) = \int_{-\infty}^{\infty} yf_{Y|X}(y|x)\,dy = \int_{-\infty}^{\infty} ye^{-2|y-x^2|}\,dy$$

$$= \int_{-\infty}^{\infty} (z + x^2)e^{-2|z|}\,dz = 0 + x^2,$$

a nonlinear function of x.

A more abstract and more general approach to defining conditional expectation (see Section 15.8) is available from such sources as Billingsley [10] and Loeve [60]. However, our definition suffices for most of engineering practice.

Let

$$h(\mathbf{x}) = E(Y|\mathbf{X} = \mathbf{x}), \text{ and define } E(Y|\mathbf{X}) = h(\mathbf{X}).$$

In other words,

> $E(Y|\mathbf{X})$ is a function only of the variable $\mathbf{X}$ and a functional of (average over) the variable Y.

As was the case for conditional probability $P(A|B)$ acting like a probability measure in the A variable, conditional expectation $E(Y|\mathbf{X})$ acts like an expectation in the Y variable. The next three theorems are stated informally, in that they omit certain needed technical restrictions (e.g., equality holds only for _almost_ all values of x). Heuristically, their proofs closely parallel those given for unconditional expectation in Chapter 9.

Theorem 15.2.1   (Linearity of Conditional Expectation)

$$E(aY_1 + bY_2 + c|\mathbf{X}) = aE(Y_1|\mathbf{X}) + bE(Y_2|\mathbf{X}) + c.$$

Theorem 15.2.2   (Conditional Expectation of Functions)

$$E(g(Y, \mathbf{X})|\mathbf{X} = \mathbf{x}) = \int_{-\infty}^{\infty} g(y, \mathbf{x}) f_{Y|\mathbf{X}}(y|\mathbf{x}) \, dy = E(g(Y, \mathbf{x})|\mathbf{X} = \mathbf{x}).$$

In particular,

$$E(g(\mathbf{X})|\mathbf{X} = \mathbf{x}) = g(\mathbf{x}).$$

Theorem 15.2.3   (Conditional Expectation and Independence)   If $Y \perp\!\!\!\perp \mathbf{X}$, then

$$E(Y|\mathbf{X}) = EY.$$

Thus, $E(Y|\mathbf{X})$ is a *linear functional* in the Y variable and is a version of an integral. This is not true for the $\mathbf{X}$ variable.

A counterpart to the Total Probability Theorem is given by the next theorem.

Theorem 15.2.4   (Expectation from Conditional Expectation)

$$EY = E\{E(Y|\mathbf{X})\},$$

where the outer expectation in the right-hand side is just over the remaining variable $\mathbf{X}$.

Proof.

$$EY = \int_{-\infty}^{\infty} y f_Y(y) \, dy = \int_{-\infty}^{\infty} y \left\{ \int_{-\infty}^{\infty} f_{Y|\mathbf{X}}(y|\mathbf{x}) f_{\mathbf{X}}(\mathbf{x}) d\mathbf{x} \right\} dy$$

$$= \int_{-\infty}^{\infty} E(Y|\mathbf{X} = \mathbf{x}) f_{\mathbf{X}}(\mathbf{x}) \, d\mathbf{x},$$

where we changed the order of integration from y being the last variable of integration to its being the first variable. $\square$

An interesting illustration of the use of this result is given in Section 17.10 on random sums.

Example 15.2   Unconditional from Conditional Expectation _____
Continuing with the previous example, we have

$$EY = E\{E(Y|\mathbf{X})\} = E(X^2) = \int_0^{\infty} x^2 e^{-x} \, dx = 2! = 2.$$

An extension of this property called the *property of subconditioning*, also known as the *property of iterated expectation*, is given by Theorem 15.2.5.

Theorem 15.2.5   (Subconditioning)

$$E(Y|\mathbf{X} = \mathbf{x}) = E\{E(Y|\mathbf{X}, \mathbf{Z})|\mathbf{X} = \mathbf{x}\}$$

$$= \int_{-\infty}^{\infty} f_{\mathbf{Z}|\mathbf{X}}(\mathbf{z}|\mathbf{x}) E(Y|\mathbf{Z} = \mathbf{z}, \mathbf{X} = \mathbf{x}) \, d\mathbf{z}.$$

Proof. The proof is similar to that of Theorem 15.2.4. For simplicity, we use scalar, rather than vector, notation:

$$h(x) = E(Y|X = x) = \int_{-\infty}^{\infty} y f_{Y|X}(y|x)\, dy = \int_{-\infty}^{\infty} y \frac{f_{X,Y}(x, y)}{f_X(x)}\, dy$$

$$= \int_{-\infty}^{\infty} y \left[\int_{-\infty}^{\infty} \frac{f_{X,Y,Z}(x, y, z)}{f_X(x)}\, dz \right] dy$$

$$= \int_{-\infty}^{\infty} y \left[\int_{-\infty}^{\infty} f_{Y|X,Z}(y|x, z) f_{Z|X}(z|x)\, dz \right] dy$$

$$= \int_{-\infty}^{\infty} f_{Z|X}(z|x) \left[\int_{-\infty}^{\infty} y f_{Y|X,Z}(y|x, z)\, dy \right] dz$$

$$= \int_{-\infty}^{\infty} f_{Z|X}(z|x) E(Y|X = x, Z = z)\, dz = E\{E(Y|X, Z)|X = x\}. \qquad \square$$

Example 15.3   Subconditioning _____

Assume that

$$Y = N + XZ, \quad \perp\!\!\!\perp (N, X, Z), \quad N \sim \mathcal{N}(0, \sigma^2), \quad Z \sim \mathcal{N}(m, 1).$$

Hence, by Theorem 15.2.2,

$$E(Y|X, Z) = E(N|X, Z) + E(XZ|X, Z) = E(N|X, Z) + XZ.$$

By Theorem 15.2.3,

$$E(N|X, Z) = EN = 0.$$

We have

$$E(Y|X) = E(XZ|X) = XE(Z|X) = XEZ = mX.$$

15.3 EXAMPLE OF THE MULTIVARIATE NORMAL

If X, Y are jointly normally distributed with means m_X and m_Y, variances σ_X^2 and σ_Y^2, and normalized correlation coefficient ρ, then we can use our prior derivation in Section 12.6.3 of $f_{Y|X}$ as normal to show that the conditional expectation

$$E(Y|X = x) = \frac{\rho \sigma_Y}{\sigma_X}(x - m_X) + m_Y$$

and the conditional variance

$$\mathrm{VAR}(Y|X = x) = \sigma_Y^2(1 - \rho^2).$$

Note that the conditional variance, which is a function of x, turns out to be a constant.

If X, Y are jointly normal, are not linearly dependent ($|\rho| < 1$), and have equal variances $\sigma_X^2 = \sigma_Y^2 = \sigma^2$, then

$$|E(Y|X) - m_Y| = |\rho||X - m_X| < |X - m_X|.$$

The conditional expected value of Y given X lies closer to the mean of Y than does X to the mean of X, a phenomenon known as *Galton's regression to the mean*. If we make the reasonable assumption of a jointly normal distribution of the intelligence X of a parent and the intelligence Y of a child, with a positive correlation between the two and equal means and variances, then we can conclude that the average intelligence of a child of an intelligent parent, while above average, is below that of the parent. As it is clear that the parents of university students are of above-average intelligence, we are led with seeming mathematical inexorability to a sad conclusion about the students themselves. (Of course, those students lacking in respect for their elders might wonder if this argument couldn't be reversed.)

The linearity of conditional expectation $E(Y|\mathbf{X})$ continues to hold for vector-valued random variable $\mathbf{X}$ in the jointly Gaussian case. Change notation so that $Y = X_k$, $\mathbf{X} = (X_1, \ldots, X_{k-1})^T$, and $\mathbb{Q}$ denotes the inverse $\mathbb{C}^{-1}$ of the covariance matrix for all of $X_1, \ldots, X_{k-1}, X_k$. It is then immediate from the calculations of Section 12.6.3 that

$$E(X_k|X_1, \ldots, X_{k-1}) = m_k - \frac{1}{Q_{kk}} \sum_{j < k} Q_{kj}(X_j - m_j).$$

Furthermore, the conditional variance

$$\text{VAR}(X_k|X_1, \ldots, X_{k-1}) = \frac{1}{Q_{kk}}$$

and is a constant function of the conditioning variables. Hence, the conditional expectation for the multivariate normal is again linear in the variables being conditioned upon.

15.4 MMSE ESTIMATION

Given data $\mathbf{X}$, we may wish to infer to a random variable Y or random vector $\mathbf{Y}$ that cannot be observed directly. Examples include inferences

- from the time-sampled characteristics $\mathbf{X}$ of a received signal to the transmitted signal $\mathbf{Y}$;
- from a series of noisy measurements $\mathbf{X}$ to the value Y of the true quantity being measured;
- from a series of noisy measurements $\mathbf{X}$ of the past positions of a vehicle to its present position Y in an inertial navigation system;
- from measurements of certain state variables $\mathbf{X}$ in a network we wish to infer to other network characteristics Y;
- from observations $\mathbf{X}$ on temperature and barometric pressure at a variety of geographical sites to a future temperature Y at a given site;
- from observations on the past behavior $\mathbf{X}$ of the Dow Jones closing stock prices to tomorrow's closing Dow Jones price Y;

- from power measurements $\mathbf{X}$ made by an array of sensors to the locations $\mathbf{Y}$ of callers in a mobile cellular environment.

15.4.1 Scalar Y

The inference is made by a *nonlinear estimator* $\hat{Y}(\mathbf{X})$ that is a function of the data and is intended to approximate to the unknown Y. A measure of degree of closeness or approximation is provided by mean square error (MSE) $E(\hat{Y} - Y)^2$. We wish to identify the estimator that yields a minimum mean square error (MMSE) among all possible functions of the data and thereby generalize the work on linear estimators of Sections 9.12, 10.5, and 10.6.

    Theorem 15.4.1 (Scalar MMSE Estimator)   The (potentially nonlinear) function $\hat{Y}(\mathbf{X})$ that uniquely achieves the minimum of $E(Y - \hat{Y}(\mathbf{X}))^2$ is given by

$$Y^*(\mathbf{X}) = E(Y|\mathbf{X}).$$

The resulting performance of $\hat{Y} = Y^*$ is given by the conditional variance $E(Y - E(Y|\mathbf{X}))^2$.

    Proof. For notational convenience, we use

$$Y^*(\mathbf{X}) = E(Y|\mathbf{X}).$$

For $\hat{Y} = \hat{Y}(\mathbf{X})$, an arbitrary function of the data, we add and subtract $Y^*(\mathbf{X})$ to obtain

$$
\begin{aligned}
E(Y - \hat{Y})^2 &= E(Y - Y^*(\mathbf{X}) + Y^*(\mathbf{X}) - \hat{Y})^2 \\
&= E(Y - Y^*(\mathbf{X}))^2 + E(Y^*(\mathbf{X}) - \hat{Y})^2 + 2E\left[(Y^*(\mathbf{X}) - \hat{Y}(\mathbf{X}))(Y - Y^*(\mathbf{X}))\right].
\end{aligned}
$$

We use subconditioning (Theorem 15.2.5) to evaluate the last, cross-product, term:

$$E\left[(Y^*(\mathbf{X}) - \hat{Y}(\mathbf{X}))(Y - Y^*(\mathbf{X}))\right] = EZ = E\{E(Z|\mathbf{X})\},$$

$$E(Z|\mathbf{X} = \mathbf{x}) = (Y^*(\mathbf{x}) - \hat{Y}(\mathbf{x}))E(Y - Y^*(\mathbf{X})|\mathbf{X} = \mathbf{x}).$$

The second factor is

$$E(Y - Y^*(\mathbf{X}))|\mathbf{X} = \mathbf{x}) = E(Y|\mathbf{X} = \mathbf{x}) - Y^*(\mathbf{x}) = 0.$$

Hence,

$$E(Y - \hat{Y})^2 = E\left(Y - Y^*(\mathbf{X})\right)^2 + E\left(Y^*(\mathbf{X}) - \hat{Y}\right)^2,$$

where the first term does not depend upon $\hat{Y}$ and the second nonnegative term can be minimized (set to zero) by choosing

$$\hat{Y}(\mathbf{X}) = Y^*(\mathbf{X}) = E(Y|\mathbf{X}).$$

The resulting MMSE is

$$E\left(Y - \hat{Y}(\mathbf{X})\right)^2 = E\left[(Y - E(Y|\mathbf{X}))^2\right].$$

This last term is the expected *conditional variance* VAR$(Y|\mathbf{X})$.   □

Example 15.4   Calculation of MMSE Estimator and Performance _____
In our first example using an innovations bivariate pdf, we found that

$$E(Y|X = x) = x^2.$$

Hence, the MMSE estimator of Y is $Y^*(X) = X^2$. The performance of this estimator is

$$E(Y - X^2)^2 = \int_{-\infty}^{\infty} dx \int_{-\infty}^{\infty} dy\,(y - x^2)e^{-2|y-x^2|-x}U(x).$$

Evaluating the integral on y yields

$$\int_{-\infty}^{\infty} (y - x^2)^2 e^{-2|y-x^2|-x}U(x)\,dy = e^{-x}U(x)\int_{-\infty}^{\infty} (y - x^2)^2\,dy e^{-(2|y-x^2|)}\,dy$$

$$= e^{-x}U(x)\int_{-\infty}^{\infty} z^2 e^{-2|z|}\,dz = 2e^{-x}U(x)\int_{0}^{\infty} z^2 e^{-2z}\,dz$$

$$= \frac{1}{2}e^{-x}U(x).$$

We find that the estimation error performance of $Y^* = X^2$ is given by

$$E(Y - X^2)^2 = \int_{0}^{\infty} \frac{1}{2}e^{-x}\,dx = \frac{1}{2}.$$

15.4.2 Vector Y

These results generalize to the case of vector-valued $\mathbf{Y}$ estimated by $\hat{\mathbf{Y}}(\mathbf{X})$ so as to minimize the slightly generalized weighted least mean square criterion $E[(\mathbf{Y} - \hat{\mathbf{Y}})^T \mathbb{M}(\mathbf{Y} - \hat{\mathbf{Y}})]$. The weighting matrix $\mathbb{M}$ is arbitrary so long as it is symmetric and positive definite, an example being the identity matrix. For notational convenience, let

$$\mathbf{Y}^*(\mathbf{X}) = E(\mathbf{Y}|\mathbf{X}) \iff Y_i^* = E(Y_i|\mathbf{X}).$$

We will show that the choice $\hat{\mathbf{Y}} = \mathbf{Y}^*$ is optimal in that it yields the minimum possible value of the weighted error.

Theorem 15.4.2   (Vector MMSE Estimator)   The MSE

$$E[(\mathbf{Y} - \hat{\mathbf{Y}}(\mathbf{X}))^T \mathbb{M}(\mathbf{Y} - \hat{\mathbf{Y}}(\mathbf{X}))]$$

for symmetric and positive definite $\mathbb{M}$ is uniquely minimized by the choice of estimator

$$\hat{\mathbf{Y}}(\mathbf{X}) = \mathbf{Y}^*(\mathbf{X}) = E(\mathbf{Y}|\mathbf{X}).$$

The resulting mean square error is the scalar value

$$E[(\mathbf{Y} - \mathbf{Y}^*)^T \mathbb{M} (\mathbf{Y} - \mathbf{Y}^*)].$$

Proof. As in the previous proof, we add and subtract $\mathbf{Y}^*$, group terms, carry out the indicated products, and show that two of the resulting four terms are zero:

$$E[(\mathbf{Y} - \hat{\mathbf{Y}})^T \mathbb{M} (\mathbf{Y} - \hat{\mathbf{Y}})] = E[(\mathbf{Y} - \mathbf{Y}^* + \mathbf{Y}^* - \hat{\mathbf{Y}})^T \mathbb{M} (\mathbf{Y} - \mathbf{Y}^* + \mathbf{Y}^* - \hat{\mathbf{Y}})]$$

$$= E[(\mathbf{Y} - \mathbf{Y}^*)^T \mathbb{M} (\mathbf{Y} - \mathbf{Y}^*)] + E[(\mathbf{Y}^* - \hat{\mathbf{Y}})^T \mathbb{M} (\mathbf{Y}^* - \hat{\mathbf{Y}})]$$

$$+ E[(\mathbf{Y} - \mathbf{Y}^*)^T \mathbb{M} (\mathbf{Y}^* - \hat{\mathbf{Y}})] + E[(\mathbf{Y}^* - \hat{\mathbf{Y}})^T \mathbb{M} (\mathbf{Y} - \mathbf{Y}^*)].$$

We use the property of subconditioning,

$$E\mathbf{Z} = EE[\mathbf{Z}|\mathbf{X}],$$

to show that the two terms in the third line of this equation are zero. As the second of the two terms is the transpose of the first ($\mathbb{M}$ is symmetric), and each term is a scalar, it is immediate that their transposes are equal. It suffices to show that

$$E[(\mathbf{Y}^* - \hat{\mathbf{Y}})^T \mathbb{M} (\mathbf{Y} - \mathbf{Y}^*)] = EE[(\mathbf{Y}^* - \hat{\mathbf{Y}})^T \mathbb{M} (\mathbf{Y} - \mathbf{Y}^*)|\mathbf{X}]$$

is zero by subconditioning on $\mathbf{X}$. Since $\mathbf{Y}^*$ and $\hat{\mathbf{Y}}$ are both functions of $\mathbf{X}$,

$$E[(\mathbf{Y}^* - \hat{\mathbf{Y}})^T \mathbb{M} (\mathbf{Y} - \mathbf{Y}^*)|\mathbf{X}] = (\mathbf{Y}^* - \hat{\mathbf{Y}})^T \mathbb{M} E[(\mathbf{Y} - \mathbf{Y}^*)|\mathbf{X}].$$

However,

$$E[(\mathbf{Y} - \mathbf{Y}^*)|\mathbf{X}] = E(\mathbf{Y}|\mathbf{X}) - \mathbf{Y}^* = \mathbf{Y}^* - \mathbf{Y}^* = \mathbf{0}.$$

Hence, we have established that

$$E[(\mathbf{Y} - \hat{\mathbf{Y}})^T \mathbb{M} (\mathbf{Y} - \hat{\mathbf{Y}})] = E[(\mathbf{Y} - \mathbf{Y}^*)^T \mathbb{M} (\mathbf{Y} - \mathbf{Y}^*)] + E[(\mathbf{Y}^* - \hat{\mathbf{Y}})^T \mathbb{M} (\mathbf{Y}^* - \hat{\mathbf{Y}})].$$

The first of the two terms on the right-hand side of the equation does not depend upon our choice of $\hat{\mathbf{Y}}$. Hence, to minimize the mean squared error, we must choose $\hat{\mathbf{Y}}$ to minimize the remaining term

$$E[(\mathbf{Y}^* - \hat{\mathbf{Y}})^T \mathbb{M} (\mathbf{Y}^* - \hat{\mathbf{Y}})] = EE[(\mathbf{Y}^* - \hat{\mathbf{Y}})^T \mathbb{M} (\mathbf{Y}^* - \hat{\mathbf{Y}})|\mathbf{X}].$$

Noting that $\mathbf{Y}^* - \hat{\mathbf{Y}}$ is a function of $\mathbf{X}$, we see that

$$E[(\mathbf{Y}^* - \hat{\mathbf{Y}})^T \mathbb{M} (\mathbf{Y}^* - \hat{\mathbf{Y}})|\mathbf{X}] = (\mathbf{Y}^* - \hat{\mathbf{Y}})^T \mathbb{M} (\mathbf{Y}^* - \hat{\mathbf{Y}}).$$

Since $\mathbb{M}$ is assumed to be positive definite, this term is positive unless the vector $\mathbf{Y}^* - \hat{\mathbf{Y}}$ is zero. Hence, the uniquely best choice for $\hat{\mathbf{Y}}$ is $\mathbf{Y}^*$. $\square$

Clearly, the vector-valued MMSE estimator is the vector of the individual scalar MMSE estimators. The only generalization is the inclusion of the positive definite weighting matrix $\mathbb{M}$, and this is often just taken to be the identity $\mathbb{I}$.

Example 15.5   Emitter Location _____

Four power-measuring sensors $\mathbf{X} = (X_1, \ldots, X_4)^T$ are located at the respective vertices $\{(0, 0),$ $(1, 0), (0, 1), (1, 1)\}$ of a square of unit length. An emitter is located at position $\mathbf{Y} = (Y_1, Y_2)^T$ that is distributed uniformly over the unit square. The emitter power $\Pi \sim \mathcal{E}(\alpha)$ and is independent of the emitter location. The sensors report received power that is proportional to the random emitter power Π and attenuated inversely as the distance $d_i(\mathbf{Y})$ between the sensor X_i and the emitter at $\mathbf{Y}$. For example,

$$d_2^2(\mathbf{y}) = (y_1 - 1)^2 + y_2^2.$$

The received power propagation is assumed to satisfy an inverse-square law, with some $d_0^2 > 0$ added to deal with the "near field," and with additive noise

$$X_i = \frac{\Pi}{d_0^2 + d_i^2(\mathbf{Y})} + N_i.$$

(In cellular systems, received power can attenuate more rapidly with a higher exponent of distance.) We assume that $\perp\!\!\!\perp \{N_1, \ldots, N_4, \mathbf{Y}, \Pi\}$ and that the $i.i.d.$ additive noises are distributed as $\mathcal{U}(0, \epsilon)$. The conditional pdf

$$f_{\mathbf{X}|\mathbf{Y},\Pi}(\mathbf{x}|\mathbf{y}, \pi) = \prod_{i=1}^{4} f_N\left(x_i - \frac{\pi}{d_0^2 + d_i^2(\mathbf{y})}\right)$$

$$= \prod_{i=1}^{4} U\left(x_i - \frac{\pi}{d_0^2 + d_i^2(\mathbf{y})}\right) U\left(\epsilon + \frac{\pi}{d_0^2 + d_i^2(\mathbf{y})} - x_i\right).$$

From the Total Probability Theorem and the independence between Π and $\mathbf{Y}$,

$$f_{\mathbf{X}|\mathbf{Y}}(\mathbf{x}|\mathbf{y}) = \int_{-\infty}^{\infty} f_{\mathbf{X}|\mathbf{Y},\Pi}(\mathbf{x}|\mathbf{y}, \pi) f_{\Pi|\mathbf{Y}}(\pi|\mathbf{y}) \, d\pi = \int_{-\infty}^{\infty} f_{\mathbf{X}|\mathbf{Y},\Pi}(\mathbf{x}|\mathbf{y}, \pi) f_{\Pi}(\pi) \, d\pi$$

$$= \int_{0}^{\infty} \alpha e^{-\alpha\pi} \prod_{i=1}^{4} U\left(x_i - \frac{\pi}{d_0^2 + d_i^2(\mathbf{y})}\right) U\left(\epsilon + \frac{\pi}{d_0^2 + d_i^2(\mathbf{y})} - x_i\right) d\pi$$

$$= \int_{0}^{\infty} \alpha e^{-\alpha\pi} I_A(\pi) \, dp.$$

In the last integral, the term $I_A(\pi)$ summarizes the product of the unit-step function with

$$A = \{\pi : \min_{i}\left[x_i(d_0^2 + d_i^2(\mathbf{y}))\right] \geq \pi \geq \max_{i}\left[(x_i - \epsilon)(d_0^2 + d_i^2(\mathbf{y}))\right]\}.$$

The last integral is now readily evaluated to yield

$$f_{\mathbf{X}|\mathbf{Y}}(\mathbf{x}|\mathbf{y}) = e^{-\alpha \max_i \left[(x_i - \epsilon)(d_0^2 + d_i^2(\mathbf{y})) \right]} - e^{-\alpha \min_i \left[x_i (d_0^2 + d_i^2(\mathbf{y})) \right]}.$$

Calculating the MMSE estimator $\hat{\mathbf{Y}}(\mathbf{X})$ requires the use of Bayes' Theorem,

$$f_{\mathbf{Y}|\mathbf{X}}(\mathbf{y}|\mathbf{x}) = \frac{f_{\mathbf{X}|\mathbf{Y}}(\mathbf{x}|\mathbf{y}) f_{\mathbf{Y}}(\mathbf{y})}{f_{\mathbf{X}}(\mathbf{x})}.$$

The term $f_{\mathbf{X}|\mathbf{Y}}(\mathbf{x}|\mathbf{y})$ is available from the last integral. The term

$$f_{\mathbf{Y}}(\mathbf{y}) = \begin{cases} 1 & \text{if } 0 \leq y_1 \leq 1 \text{ and } 0 \leq y_2 \leq 1 \\ 0 & \text{if otherwise} \end{cases}$$

corresponds to an emitter location that is chosen to be uniformly distributed over the unit square. The Total Probability Theorem yields

$$f_{\mathbf{X}}(\mathbf{x}) = \int_0^1 \int_0^1 \left(e^{-\alpha \max_i \left[(x_i - \epsilon)(d_0^2 + d_i^2(\mathbf{y})) \right]} - e^{-\alpha \min_i \left[x_i (d_0^2 + d_i^2(\mathbf{y})) \right]} \right) dy_1 \, dy_2.$$

Finally, from Theorem 15.4.2, the MMSE estimator of location coordinate Y_j is

$$\hat{Y}_j(\mathbf{x}) = \frac{1}{f_{\mathbf{X}}(\mathbf{x})} \int_0^1 \int_0^1 y_j \left(e^{-\alpha \max_i \left[(x_i - \epsilon)(d_0^2 + d_i^2(\mathbf{y})) \right]} - e^{-\alpha \min_i \left[x_i (d_0^2 + d_i^2(\mathbf{y})) \right]} \right) dy_1 \, dy_2.$$

We leave matters at this point.

15.5 TWO APPLICATIONS OF MMSE ESTIMATION

15.5.1 Signal-Plus-Additive Independent Noise

Consider the now-familiar additive noise model $X = S + N$, where S, N are independent with individual density functions f_S, f_N, respectively. In our analysis of independent signal-plus-noise in Section 13.5, we found that

$$f_{S|X}(s|x) = \frac{f_N(x - s) f_S(s)}{\displaystyle\int_{-\infty}^{\infty} f_N(x - s) f_S(s) \, ds}.$$

Hence, the desired MMSE $\hat{S}(X)$ of signal S observed as X is given by

$$\hat{S}(x) = E(S|X = x) = \frac{\displaystyle\int_{-\infty}^{\infty} s f_N(x - s) f_S(s) \, ds}{\displaystyle\int_{-\infty}^{\infty} f_N(x - s) f_S(s) \, ds}.$$

We study this result from two different vantage points.

We start by assuming that S, N are individually normally distributed with

$$ES = m_S, \quad EN = 0, \quad \text{VAR}(S) = \sigma_S^2, \quad \text{VAR}(N) = \sigma_N^2.$$

Note that X, S are jointly normally distributed with

$$EX = m_S + 0, \quad \text{VAR}(X) = \sigma_S^2 + \sigma_N^2,$$

and with the correlation coefficient ρ calculated from

$$COV(X, S) = E(X - EX)(S - m_S) = \text{VAR}(S) = \sigma_S^2,$$

$$\rho = \frac{\sigma_S^2}{\sigma_S \sqrt{\sigma_S^2 + \sigma_N^2}}$$

$$= \frac{1}{\sqrt{1 + \frac{\sigma_N^2}{\sigma_S^2}}}.$$

From Theorem 15.4.1, the mean square estimator and its error are given by

$$\hat{S}(X) = E(S|X) = m_S + \frac{1}{1 + \frac{\sigma_N^2}{\sigma_S^2}}(X - m_S), \quad E(S - S^*)^2 = \frac{\sigma_N^2}{1 + \frac{\sigma_N^2}{\sigma_S^2}}.$$

If we introduce the commonly useful notion of *signal-to-noise ratio*

$$SNR = \frac{\sigma_S^2}{\sigma_N^2},$$

then

$$\hat{S}(X) = m_S + \frac{SNR}{1 + SNR}(X - m_S) = \frac{SNR}{1 + SNR}X + \frac{m_S}{1 + SNR}.$$

It is evident that if $SNR \gg 1$, then we have a very good observation and we see that $\hat{S}(X) \approx X$. However, if $SNR \ll 1$, then we have a very bad measurement and $\hat{S}(X) \approx m_S$, our prior information about S.

A more general analysis for independent signal and noise, proceeding from the expression for $\hat{S}(X) = E(S|X)$, starts by substituting $u = x - s$ for s and then $du = -ds$ to write

$$\hat{S}(x) = \frac{\int_{-\infty}^{\infty} (x - u)f_N(u)f_S(x - u)\, du}{\int_{-\infty}^{\infty} f_N(u)f_S(x - u)\, du}$$

$$= x - \frac{Num = \int_{-\infty}^{\infty} uf_N(u)f_S(x - u)\, du}{Den = \int_{-\infty}^{\infty} f_N(u)f_S(x - u)\, du}.$$

Assume that

$$EN = EN^3 = 0, EN^2 = \sigma_N^2,$$

properties of the noise N that would be true if $N \sim \mathcal{N}(0, \sigma_n^2)$, but also if N had finite third moment and symmetrical density ($f_N(-u) = f_N(u)$). Assume further that the noise standard deviation σ_N is small relative to the spread σ_S of the signal S, so that the noise density is sharply peaked relative to the signal density. Introduce a Taylor's series, in u about x, for f_S that ignores terms beyond the quadratic:

$$f_S(x - u) = f_S(x) - f_S'(x)u + \frac{1}{2}f_S''(x)u^2 + \cdots$$

Hence,

$$Num \approx \int_{-\infty}^{\infty} f_N(u)u[f_S(x) - f_S'(x)u + \frac{1}{2}f_S''(x)u^2]\,du$$

$$= f_S(x)EN - f_S'(x)EN^2 + \frac{1}{2}f_S''(x)EN^3 = -\sigma_N^2 f_S'(x),$$

Assuming that $f_X(x) > 0$,

$$Den \approx \int_{-\infty}^{\infty} f_N(u)[f_S(x) - f_S'(x)u + \frac{1}{2}f_S''(x)u^2]\,du$$

$$= f_S(x) - f_S'(x)EN + \frac{1}{2}f_S''(x)EN^2 = f_S(x) + \frac{1}{2}\sigma_N^2 f_S''(x).$$

If the previous approximation is nonzero, then we conclude with the approximate MMSE estimator

$$\hat{S}(x) \approx x + \frac{\sigma_N^2 f_S'(x)}{f_S(x) + \frac{1}{2}\sigma_N^2 f_S''(x)} \approx x + \sigma_N^2 \frac{\partial \log(f_S(x))}{\partial x}.$$

Example 15.6   Approximate MMSE Estimation of Signal in Additive Noise _____
Assume that $S \perp\!\!\!\perp N$, $N \sim \mathcal{L}(\alpha)$, and $S \sim \Gamma(2, 1)$. We have

$$f_S(s) = se^{-s}U(s).$$

In this case, the argument for $\hat{S}(x)$ applies, with

$$f_S'(s) = (1 - s)e^{-s}U(s), \quad f_S''(s) = (s - 2)e^{-s}U(s).$$

Hence, if $x \geq 0$,

$$\hat{S}(x) \approx x + \frac{\frac{2}{\alpha^2}(1 - x)}{x + \frac{x-2}{\alpha^2}} = x + \frac{2(1 - x)}{\alpha^2 x + x - 2}.$$

This approximation to $\hat{S}(x)$ is more accurate for larger α when $\hat{S}(x)$ is closer to x.

15.5.2 Quantization

As a second application, we consider the problem of reconstructing an analog signal S from its quantized version given by a digital codeword c. For example, music is time sampled at a rate of 44.1 KHz to generate 44,100 analog samples each second for each stereo channel. Each analog sample is then quantized into one of $2^{16} = 65,536$ levels and each level indexed by a 16-bit codeword. A *staircase quantizer* is a function defined by an increasing sequence $\{\tau_i\}$ of thresholds, with $\tau_0 = -\infty$, $\tau_i < \tau_{i+1}$, $\tau_{2^{16}} = \infty$. The codeword $c = c_i$ if $\tau_i \leq S < \tau_{i+1}$. For convenience, let

$$q_i = \frac{\tau_i + \tau_{i+1}}{2}, \quad l_i = \tau_{i+1} - \tau_i,$$

denote the midpoint and length of the ith quantization interval. In a CD player, the reconstruction of the analog signal is accomplished by a DAC. We address the optimal reconstruction of S when it has been quantized and assigned codeword c_i corresponding to the information that $|S - q_i| \leq l_i/2$. For clarity, we drop the subscript i. Hence, the optimal reconstruction

$$\hat{S}(c) = E\left(S \bigg| |S - q| \leq \frac{l}{2}\right).$$

In this case, we are conditioning on an event of positive probability and can write

$$\hat{S}(c) = \frac{\displaystyle\int_{q-l/2}^{q+l/2} s f_S(s)\, ds}{P(|S - q| \leq l/2)}.$$

To proceed further, we assume that S has been well quantized in that the first and last infinite quantization intervals have a negligible probability of containing S and the remaining finite-length intervals are short compared with the rate of variation of f_S. A first-order Taylor's series for f_S about $s = q$ is

$$f_S(s) \approx f_S(q) + (s - q)f_S'(q).$$

Inserting this approximation, we then find that

$$P(|S - q| \leq l/2) = \int_{q-l/2}^{q+l/2} f_S(s)\, ds \approx l f_S(q) + \frac{l^2}{4} f_S'(q),$$

$$\int_{q-l/2}^{q+l/2} s f_S(s)\, ds \approx q l f_S(q) + \frac{l^3}{12} f_S'(q).$$

The ratio of terms yields

$$\hat{S}(x) \approx \frac{q + \dfrac{l^2}{12} \dfrac{d \log f_S(s)}{ds}\bigg|_{s=q}}{1 + \dfrac{l}{4} \dfrac{d \log f_S(s)}{ds}\bigg|_{s=q}}.$$

Example 15.7   Approximate MMSE D/A Conversion _____

Recalling our Laplacian model for speech and applying it as well to music, we have

$$f_S(s) = \frac{\alpha}{2} e^{-\alpha|s|}.$$

This yields the conclusion that

$$\hat{S}(c) \approx \frac{q - \alpha \frac{l^2}{12} \text{sign}(x)}{1 - \alpha \frac{l}{4} \text{sign}(x)}.$$

The MMSE digital-to-analog (D/A) converter $\hat{S}(c)$ in this case adjusts the midpoint q of the interval containing S, depending upon the algebraic sign of x and the other parameters.

15.6 APPLICATION TO KALMAN FILTERING REVISITED

This section can be omitted without any consequence for the material to follow. It is more demanding than the general level of this text. We return to our examination of Kalman filtering against the background provided in Section 10.6. We are now in a position to derive the results that were assumed in our earlier discussion. The key results established in this chapter are that the conditional expectation is the optimal weighted, least mean square estimator (established in Theorem 15.4.2) and that, when all random variables are jointly normal, the conditional expectation is linear in the variables being conditioned on. (See Section 15.3.)

The one-step-ahead MMSE estimator $\hat{\mathbf{X}}_k^-$ of the state $\mathbf{X}_k$, based upon all of the past measurements $\mathbb{Z}_{k-1}^*$, recorded in the matrix as $k-1$ columns, is $E(\mathbf{X}_k|\mathbb{Z}_{k-1}^*)$. Recalling the dynamical system equation from Section 10.6, given in terms of the state transition matrix Φ_{k-1}, we see that

$$E(\mathbf{X}_k|\mathbb{Z}_{k-1}^*) = E(\Phi_{k-1}\mathbf{X}_{k-1}|\mathbb{Z}_{k-1}^*) + E(\mathbf{W}_{k-1}|\mathbb{Z}_{k-1}^*).$$

Uncorrelatedness of normally distributed random variables implies their independence. Hence, $\mathbf{W}_{k-1} \perp\!\!\!\perp \mathbb{Z}_{k-1}^*$ and $E(\mathbf{W}_{k-1}|\mathbb{Z}_{k-1}^*) = E\mathbf{W}_{k-1} = \mathbf{0}$. We conclude that

$$\hat{\mathbf{X}}_k^- = E(\mathbf{X}_k|\mathbb{Z}_{k-1}^*) = \Phi_{k-1}E(\mathbf{X}_{k-1}|\mathbb{Z}_{k-1}^*) = \Phi_{k-1}\hat{\mathbf{X}}_{k-1},$$

as claimed in Section 10.6 but now without having to assume linearity.

While the preceding remarks defend those made earlier in the discussion of Kalman filtering, they do not establish the precise linear forms assumed for the estimators. We undertake this task by using the assumed normality and the linear form of the conditional expectation. Recalling that $\mathbb{Z}_k^* = (\mathbb{Z}_{k-1}^*, \mathbf{Z}_k)$, elementary use of the definition of conditional density enables us to write the following expression:

$$f_{\mathbf{X}_k|\mathbb{Z}_k^*} = \frac{f_{\mathbf{Z}_k|\mathbf{X}_k, \mathbb{Z}_{k-1}^*} f_{\mathbf{X}_k|\mathbb{Z}_{k-1}^*}}{f_{\mathbf{Z}_k|\mathbb{Z}_{k-1}^*}}.$$

The optimum estimator of $\mathbf{X}_k$ given $\mathbb{Z}_k^*$ is $\hat{\mathbf{X}}_k$, and it has error covariance matrix $\mathbb{P}_k$. Recall that linear combinations of normal random vectors are normally distributed and that conditional densities of normal random vectors are again normal. (This reasoning will be implicitly repeated shortly.) Hence, we have the probability law

$$\mathcal{L}(\mathbf{X}_k | \mathbb{Z}_k^*) = \mathcal{N}(\hat{\mathbf{X}}_k, \mathbb{P}_k).$$

Recalling the elements of the Kalman filtering problem from Section 10.6, we have $\mathbf{V}_k \perp\!\!\!\perp \mathbb{Z}_{k-1}^*$, $\mathbf{Z}_k = \mathbb{H}_k \mathbf{X}_k + \mathbf{V}_k$, and $COV(\mathbf{V}_k, \mathbf{V}_k) = \mathbb{R}_k$, yielding

$$\mathcal{L}(\mathbf{Z}_k | \mathbf{X}_k, \mathbb{Z}_{k-1}^*) = \mathcal{N}(\mathbb{H}_k \mathbf{X}_k, \mathbb{R}_k).$$

The optimum estimator of $\mathbf{X}_k$ given $\mathbb{Z}_{k-1}^*$ is the one-step-ahead estimator $\hat{\mathbf{X}}_k^-$. The estimation error covariance matrix is $\mathbb{P}_k^-$ for $\hat{\mathbf{X}}_k^-$. Hence,

$$\mathcal{L}(\mathbf{X}_k | \mathbb{Z}_{k-1}^*) = \mathcal{N}(\hat{\mathbf{X}}_k^-, \mathbb{P}_k^-).$$

The optimum estimate $\hat{\mathbf{Z}}_k$ of $\mathbf{Z}_k$ given $\mathbb{Z}_{k-1}^*$ is

$$\begin{aligned}
\hat{\mathbf{Z}}_k &= E(\mathbf{Z}_k | \mathbb{Z}_{k-1}^*) = E(\mathbb{H}_k \mathbf{X}_k + \mathbf{V}_k | \mathbb{Z}_{k-1}^*) \\
&= \mathbb{H}_k E(\mathbf{X}_k | \mathbb{Z}_{k-1}^*) + E(\mathbf{V}_k | \mathbb{Z}_{k-1}^*) = \mathbb{H}_k \hat{\mathbf{X}}_k^- + E(\mathbf{V}_k) = \mathbb{H}_k \hat{\mathbf{X}}_k^-,
\end{aligned}$$

where we have used $\mathbf{V}_k \perp\!\!\!\perp \mathbb{Z}_{k-1}^*$ and $E\mathbf{V}_k = 0$. The corresponding error covariance matrix

$$E(\mathbf{Z}_k - \hat{\mathbf{Z}}_k)(\mathbf{Z}_k - \hat{\mathbf{Z}}_k)^T = \mathbb{H}_k \mathbb{P}_k^- \mathbb{H}_k^T + \mathbb{R}_k.$$

Hence,

$$\mathcal{L}(\mathbf{Z}_k | \mathbb{Z}_{k-1}^*) = \mathcal{N}(\mathbb{H}_k \hat{\mathbf{X}}_k^-, \mathbb{H}_k \mathbb{P}_k^- \mathbb{H}_k^T + \mathbb{R}_k).$$

We now have specified each of the conditional densities required to compute $f_{\mathbf{X}_k | \mathbb{Z}_k^*}$. We introduce the following temporary notation:

$$f_{\mathbf{X}_k | \mathbb{Z}_k^*} = c_0 \exp\left(-\frac{1}{2} S_0\right),$$

$$f_{\mathbf{Z}_k | \mathbf{X}_k, \mathbb{Z}_{k-1}^*} = c_1 \exp\left(-\frac{1}{2} S_1\right),$$

$$f_{\mathbf{X}_k | \mathbb{Z}_{k-1}^*} = c_2 \exp\left(-\frac{1}{2} S_2\right),$$

$$f_{\mathbf{Z}_k | \mathbb{Z}_{k-1}^*} = c_3 \exp\left(-\frac{1}{2} S_3\right).$$

Furthermore, we introduce the following temporary notation for the inverses of particular matrices:

$$\mathbb{Q}_0 = \mathbb{P}_k^{-1},$$

$$\mathbb{Q}_1 = \mathbb{R}_k^{-1},$$

$$\mathbb{Q}_2 = (\mathbb{P}_k^-)^{-1},$$

$$\mathbb{Q}_3 = (\mathbb{H}_k \mathbb{P}_k^- \mathbb{H}_k^T + \mathbb{R}_k)^{-1}.$$

Evaluating each of these conditional densities by a normal density, we obtain

$$S_0 = (\mathbf{x}_k - \hat{\mathbf{x}}_k)^T \mathbb{Q}_0 (\mathbf{x}_k - \hat{\mathbf{x}}_k),$$

$$S_1 = (\mathbf{z}_k - \mathbb{H}_k \mathbf{x}_k)^T \mathbb{Q}_1 (\mathbf{z}_k - \mathbb{H}_k \mathbf{x}_k),$$

$$S_2 = (\mathbf{x}_k - \hat{\mathbf{x}}_k^-)^T \mathbb{Q}_2 (\mathbf{x}_k - \hat{\mathbf{x}}_k^-),$$

$$S_3 = (\mathbf{z}_k - \mathbb{H}_k \hat{\mathbf{x}}_k^-)^T \mathbb{Q}_3 (\mathbf{z}_k - \mathbb{H}_k \hat{\mathbf{x}}_k^-).$$

In this new notation, the expression for the conditional density $f_{\mathbf{X}_k | \mathbb{Z}_k^*}$ leads us to

$$S_0 = S_1 + S_2 - S_3.$$

We proceed by completing the square in $\mathbf{x}_k$ in the expression $S_1 + S_2 - S_3$. The quadratic terms are

$$\mathbf{x}_k^T (\mathbb{H}_k^T \mathbb{Q}_1 \mathbb{H}_k + \mathbb{Q}_2) \mathbf{x}_k = \mathbf{x}_k^T \mathbb{A} \mathbf{x}_k.$$

The linear terms (using the symmetry of the covariance matrices) are

$$[-2 \mathbf{z}_k^T \mathbb{Q}_1 \mathbb{H}_k - 2 (\hat{\mathbf{x}}_k^-)^T \mathbb{Q}_2] \mathbf{x}_k = -2 \mathbf{b}^T \mathbf{x}_k.$$

Completing the square yields

$$\mathbf{x}_k^T \mathbb{A} \mathbf{x}_k - 2 \mathbf{b}^T \mathbf{x}_k + c = (\mathbf{x}_k - \mathbb{A}^{-1} \mathbf{b})^T \mathbb{A} (\mathbf{x}_k - \mathbb{A}^{-1} \mathbf{b}) + c - \mathbf{b}^T \mathbb{A}^{-1} \mathbf{b},$$

with all of the terms depending upon $\mathbf{x}_k$ being displayed explicitly.

Since $S_0 = S_1 + S_2 - S_3$, it follows immediately that

$$\mathbb{A} = \mathbb{P}_k^{-1} \quad \text{or} \quad \mathbb{P}_k = \mathbb{A}^{-1},$$

$$\hat{\mathbf{x}}_k = \mathbb{A}^{-1} \mathbf{b} = \mathbb{P}_k [\mathbb{H}_k^T \mathbb{Q}_1 \mathbf{z}_k + \mathbb{Q}_2 \hat{\mathbf{x}}_k^-] = \mathbb{P}_k [(\mathbb{P}_k^-)^{-1} \hat{\mathbf{x}}_k^- + \mathbb{H}_k^T \mathbb{R}_k^{-1} \mathbf{z}_k].$$

If we recall that $\hat{\mathbf{x}}_k^- = \Phi_{k-1} \hat{\mathbf{x}}_{k-1}$ and therefore that

$$\mathbb{P}_k^- = \Phi_{k-1} \mathbb{P}_{k-1} \Phi_{k-1}^T + \mathbb{Q}_k,$$

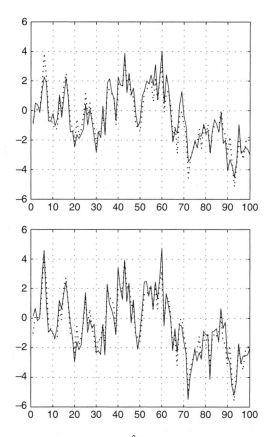

Figure 15.1   State **X** and Estimate $\hat{\mathbf{X}}$ (dotted), $X(1)$ (top), $X(2)$ (bottom).

then we can eliminate $\hat{\mathbf{x}}_k^-$ and $\mathbb{P}_k^-$ from the preceding and have a recursion for $\hat{\mathbf{X}}_k$ (restoring the random variable notation) and $\mathbb{P}_k$ in terms of $\hat{\mathbf{X}}_{k-1}$ and $\mathbb{P}_{k-1}$:

$$\mathbb{P}_k = \mathbb{A}^{-1} = \left[\mathbb{H}_k^T \mathbb{R}_k^{-1} \mathbb{H}_k + (\Phi_{k-1} \mathbb{P}_{k-1} \Phi_{k-1}^T + \mathbb{Q}_k)^{-1} \right]^{-1},$$

$$\hat{\mathbf{X}}_k = \mathbb{P}_k \left[(\Phi_{k-1} \mathbb{P}_{k-1} \Phi_{k-1}^T + \mathbb{Q}_k)^{-1} \Phi_{k-1} \hat{\mathbf{X}}_{k-1} + \mathbb{H}_k^T \mathbb{R}_k^{-1} \mathbf{Z}_k \right].$$

Solutions to the Kalman filtering problem come in a wide variety of forms, and we have now derived one of them. Figure 15.1 illustrates the solution we have derived by running it on the same problem, as described in Section 10.6, with the same result.

15.7 SUMMARY

The concept of conditional expectation stands in the same relation to conditional probability that expectation stands to probability. Having introduced expectation in Chapter 9 and conditional probability in Chapters 11 and 12, we were then in a position to introduce conditional expectation. We deferred this introduction to gain greater familiarity both with expectation and conditional probability and their roles. Once again, we motivate conditional expectation, provide a formal definition using a conditional density in place of the unconditional density in the definition of (unconditional) expectation, and explore the consequences of this definition. We can now discuss averages when we restrict the population to members satisfying a conditioning proposition or event. We illustrate conditional expectation by evaluating it for the multivariate normal in Section 15.3.

Section 15.4 applies conditional expectation to solve what is perhaps the most widely encountered estimation problem, that of minimum mean square estimation. The minimum mean square error estimation problem, when the estimator was restricted to be linear in the observations, was discussed in Section 10.5 (on Wiener filtering) and Section 10.6 (on Kalman filtering). We are now able to remove the restriction that our estimator be linear in the observations. We use our new insights in Section 15.5 to examine the common problem of estimating a signal embedded in additive independent noise and to analyze quantization (analog-to-digital conversion). We return to Kalman filtering in Section 15.6 and complete our derivation of the Kalman filtering equations first presented in Section 10.6.

15.8 APPENDIX: CONDITIONING ON EVENT ALGEBRAS

This section attempts to convey a more general approach to defining conditional expectation. However, it can be omitted without harm to the remainder of the text. As you recall from Chapter 3, probability starts from a triple of sample space (set) $\Omega = \{\omega\}$, σ-algebra $\mathcal{A}$ of selected subsets of Ω, and a probability measure P assigning real numbers to the sets in $\mathcal{A}$ in a manner satisfying the Kolmogorov axioms. We say that the collection of events $\mathcal{B}$ is a sub-σ-algebra of the collection of events $\mathcal{A}$ if $\mathcal{B} \subset \mathcal{A}$ and $\mathcal{B}$ is itself a σ-algebra.

The general approach to defining conditional expectation of a random variable Y is as an expectation conditioned on a given sub-σ-algebra $\mathcal{B}$ of $\mathcal{A}$, rather than as conditioned on a random variable X. Given X, we can instead condition on the sub-σ-algebra $\mathcal{B}$ of $\mathcal{A}$ that is the smallest σ-algebra containing events of the form $\{\omega : X(\omega) \leq x\}$. Denote this conditional expectation either as $E(Y|\mathcal{B})$ or as $Y_{\mathcal{B}}$. In keeping with our original approach, $Y_{\mathcal{B}}$ is to be a $\mathcal{B}$-measurable random variable (likely, unlike Y). That is, for any constant a, the event $\{\omega : Y_{\mathcal{B}}(\omega) \leq a\} \in \mathcal{B}$. In order to distinguish the random variable (it is not unique) that is a conditional expectation given $\mathcal{B}$ from all other such random variables, we now incorporate the key property established in Theorem 15.2.4 directly into the following definition of conditional expectation.

Definition 15.8.1   (Conditional Expectation)   The conditional expectation $E(Y|\mathcal{B})$ is any random variable $Y_{\mathcal{B}}$ that is $\mathcal{B}$-measurable and that satisfies

$$(\forall B \in \mathcal{B})\ E(YI_B) = E(Y_{\mathcal{B}}I_B).$$

Since the defining conditions of the conditional expectation involve only measurability and behavior under unconditional expectation, then any other random variable, say, Z, that is also

$\mathcal{B}$-measurable and that agrees almost surely with $Y_{\mathcal{B}}$ ($P(Z = Y_{\mathcal{B}}) = 1$) will also satisfy the definition of being a conditional expectation. Hence, there are many versions of conditional expectation, but each pair agrees almost surely. Observe how the awkward case of conditioning on an event, say, $B_0 \in \mathcal{B}$, that is null ($P(B_0) = 0$) is handled with apparent ease in this definition of conditional expectation—there is no division by zero. It is clear from the definition that any choice of $Y_{\mathcal{B}}(\omega)$ for $\omega \in B_0$ is acceptable. Conditional expectation $E(Y|\mathcal{B})$ can be defined arbitrarily on a null set in $\mathcal{B}$. However, we still need constructive means to identify $Y_{\mathcal{B}}$.

EXERCISES

E15.1   a. If $f_{X,Y}(x, y) = e^{-x-y}U(x)U(y)$, then evaluate $E(Y|X)$.
              b. If $f_{X,Y}(x, y) = \frac{x+y}{2}e^{-x-y}U(x)U(y)$, then evaluate $E(Y|X)$.

E15.2   If $(\forall i)EX_i^2 < \infty$ and $(\forall i \neq j)E(X_i|X_j) = 0$, then evaluate EX_1 and $E(X_1 X_2)$.

E15.3   We observe voltages V_1 and V_2 having a joint density that is uniform over the unit disk. Evaluate $E(V_2|V_1)$.

E15.4   a. If we know that $E(V_2|V_1) = V_1^3$ and that $V_1 \sim \mathcal{U}(-1, 1)$, then evaluate the correlation between V_1 and V_2.
              b. Are V_1 and V_2 independent?

E15.5   a. If $X \sim \mathcal{U}(0, 1)$ and $E(Z|X) = X^2$, then evaluate EZ.
              b. For the specification of part (a), evaluate the correlation $E(XZ)$.

E15.6   a. If, in a system with output Y and input $X \sim \mathcal{E}(\alpha)$, we know that $E(Y|X) = X^2$, evaluate EY.
              b. Evaluate the input–output correlation $E(YX)$.

E15.7   It is known that $X \sim \mathcal{N}(0, 1)$ and $E(Y|X) = X^4$. Giving reasons for your answers, are X and Y independent or uncorrelated?

E15.8   It is known that, for all x, $E(Y|X = x) > x$.

              a. If EX is finite, what can you conclude about EY and EX?
              b. If EX is infinite, what can you conclude about EY?

E15.9   A receiver whose antenna is pointed at the sky yields output voltage measurements V_1 and V_2, made at different times, that are jointly normally distributed with mean zero, equal variances of 1/2, and a covariance of 1/4.

              a. Evaluate the conditional density $f_{V_2|V_1}$.
              b. Evaluate the minimum mean square estimator $\hat{V}_2(V_1)$ of V_2.

E15.10   If

$$f_{X,Y}(x, y) = (x + y)U(x)U(y)U(1 - x)U(1 - y),$$

              then evaluate the minimum mean square estimator $\hat{Y}(X)$ of Y.

E15.11   We observe the product $X = SN$, where the signal $S \sim \mathcal{U}(1, 2)$ and the independent noise N has the Pareto model

$$f_N(n) = \frac{3}{n^4}U(n - 1).$$

a. Determine $f_{X|S}(x|s)$. (Hint: Consider the cdf $F_{X|S}$.)

b. Determine $f_{S|X}(s|x)$. Be explicit about the range of s and the cases $x < 2$ and $x \geq 2$.

c. What is the minimum mean square estimator $\hat{S}(X)$ of the signal given the observation, for observations $x \geq 2$?

d. What is the performance of this estimator?

e. Determine the asymptotic behavior (large sample size n) of $E(S_n^* - S)^2$.

E15.12   a. If $Y = e^X + N$ and $X \perp\!\!\!\perp N$, with $X \sim \mathcal{N}(m, \sigma^2)$, $N \sim \mathcal{N}(0, 1)$, provide an expression for the MMSE estimator $\hat{X}(Y)$ of X given Y.

b. What is $E\hat{X}(Y)$? (This can be done with very little calculation.)

E15.13   A thermal noise voltage V appears across an independent random resistor $R \sim \mathcal{N}(R_0, 1)$, $R_0 \gg 1$, and it induces a noise current I.

a. If $EV^2 = 2$, what is f_V?

b. Evaluate the joint pdf $f_{V,R}(v, r)$.

c. Evaluate the average power $E(V^2/R)$ dissipated in R.

d. Evaluate $f_{V,I}(v, i)$ for the current I through R.

e. Are V and I independent?

f. What is the MMSE $\hat{R}(V)$ that estimates R from V?

E15.14   a. If

$$f_S(x) = 2sU(s)U(1 - s), N \perp\!\!\!\perp S, N \sim \mathcal{L}(1), X = S + N, Y = S^2,$$

then find the minimum mean square estimator $\hat{Y}(X)$ of Y.

b. Evaluate the performance of this estimator.

E15.15   A system transforms its input $X \sim \mathcal{E}(\alpha)$ into an output Y through

$$f_{Y|X}(y|x) = \frac{1}{2}e^{-|y-\cos(x)|}.$$

We wish to predict the output Y from the input X.

a. Evaluate the minimum mean square error predictor $\hat{Y}(X)$.

b. Evaluate the mean square error performance of this estimator.

E15.16   a. If

$$f_{Y|X_1,X_2}(y|x_1, x_2) = \frac{1}{\sqrt{\pi}}e^{-(y-x_1 x_2)^2},$$

with $\{X_i\}$ *i.i.d.* $\mathcal{N}(0, \sigma^2)$, then determine the least mean square estimator $\hat{Y}(X_1, X_2)$ of Y.

b. Evaluate the performance of $\hat{Y}$. (Hint: Consider subconditioning.)

E15.17   a. A noisy measurement $Y = X + N$ of $X \sim \mathcal{P}(\lambda_1)$ arises from additive independent noise $N \sim \mathcal{P}(\lambda_0)$. Evaluate $P(Y = y)$.

b. Describe what we know about X after observing Y.

c. Evaluate the least mean square estimator $\hat{X}(Y)$ of X given Y.

d. Provide an expression for the performance of $\hat{X}(Y)$.

E15.18 a. If $X \perp\!\!\!\perp Y$, $X \sim \mathcal{N}(1, 2)$, $Y \sim \mathcal{N}(-1, 1)$ and $Z = XY$, then evaluate EZ and VAR(Z).

b. Evaluate $\text{COV}(X, Z)$.

c. Evaluate the linear least mean square estimator $\hat{X}_L(Z)$ of X.

d. Evaluate the least mean square estimator $\hat{Z}(X)$ of Z given X. (Caution: Note the changes from part (c).)

E15.19 If $Y = \prod_{i=1}^{N} X_i$, where $N \perp\!\!\!\perp \{X_i\}$, $\{X_i\}$ are $i.i.d.$, $X_1 \sim \mathcal{U}(0, 1)$, and $N \sim \mathcal{P}(\lambda)$, then evaluate $E(Y|N)$ and EY.

E15.20 a. Using the matrix operation of `trace`, express the error criterion $E(\mathbf{X} - \hat{\mathbf{X}})^T \mathbb{M}$ $(\mathbf{X} - \hat{\mathbf{X}})$ in terms of the positive definite weighting matrix $\mathbb{M}$ and the unweighted error covariance matrix $\mathbb{P} = E(\mathbf{X} - \hat{\mathbf{X}})^T(\mathbf{X} - \hat{\mathbf{X}})$.

b. Simplify the expression when $\mathbb{M}$ is a diagonal matrix with diagonal elements $\{d_i\}$.

E15.21 Consider the emitter location problem of Example 15.5, only now we have just a single sensor X_1 at $(0, 0)$. Provide an expression for the MMSE estimator $\hat{\mathbf{Y}}$ of the emitter location $\mathbf{Y}$.

E15.22 A dynamical system with state vector $\mathbf{X}_k$ at time t_k is specified by

$$\Phi = \texttt{randn(3,3)}, \; \mathbb{B} = \texttt{randn(3,3)},$$

with initial condition $\mathbf{X}_0$ satisfying

$$E\mathbf{X}_0 = \mathbf{0}, \; COV(\mathbf{X}_0, \mathbf{X}_0) = \mathbb{P}_0 = \mathbb{I}$$

and input process $\{\mathbf{U}_k\}$ satisfying

$$E\mathbf{U}_k = \mathbf{0}, \; COV(\mathbf{U}_j, \mathbf{U}_k) = \begin{cases} \mathbb{O} & \text{if } j \neq k \\ \mathbb{I} & \text{if } j = k \end{cases}.$$

The observation process $\mathbf{Z}_k$, $k \geq 1$, is specified by

$$\mathbb{H} = \texttt{randn(2,3)}, \; E\mathbf{V}_k = \mathbf{0}, \; COV(\mathbf{V}_k, \mathbf{V}_k) = \begin{cases} \sigma^2 \mathbb{I} & \text{if } j = k \\ \mathbb{O} & \text{if } j = k \end{cases}.$$

Once selected, $\Phi, \mathbb{B}, \mathbb{H}$ should be treated as nonrandom. The initial condition $\mathbf{X}_0$, measurement noises, and input noises are also uncorrelated with each other.

a. What is the least mean square estimator $\hat{\mathbf{X}}_0$ of $\mathbf{X}_0$?

b. Using Matlab or any other computing environment, for $k = 1 : 100$, iteratively evaluate

$$\mathbb{P}_k = COV(\mathbf{X}_k - \hat{\mathbf{X}}_k, \mathbf{X}_k - \hat{\mathbf{X}}_k).$$

c. Using your computing environment, for $k = 1 : 100$, iteratively evaluate $\hat{\mathbf{X}}_k$ $(\mathbf{Z}_k, \ldots, \mathbf{Z}_1)$.

d. For each of the three components of the state vector, plot the true state and its Kalman estimate, both as a function of the time index k.

16

Bayesian Inference

16.1 PURPOSE, BACKGROUND, AND ORGANIZATION

Our newly gained knowledge of conditional expectation enables us to make significant progress towards achieving our Goal 3 of using probabilities to make good inferences, estimates, and decisions. We can now develop the Bayesian approach to decision making. The Bayesian approach is the "gold standard" in decision making, and it can be followed confidently when all of the information it requires (listed subsequently) is available. Bayesian statistics is rooted in preferences between estimators or decision rules that admit an expected utility representation (e.g., see Ferguson [28], Fishburn [32], and Savage [80]). If your preferences satisfy the rationality and structural axioms of utility theory, then there exists a loss function L such that it is correct to choose a decision rule d^* which minimizes the expected loss. We have already accomplished much in this direction. Our previous work on MMSE estimation in Sections 9.12, 10.5, 10.6, 15.4, 15.5, and 15.6 and our work on minimum error probability estimation in Section 13.3 are all examples of the Bayesian approach for specific loss functions.

If you are not provided with objective probabilities, then there are extensions that enable you to take the same Bayesian approach, using subjective or epistemic probabilities described in Section 3.3. However, controversies in statistics arise when it appears that we lack some of the information needed for the Bayesian approach. Those statisticians known variously as Bayesians, subjectivists, or personalists believe that the necessary information is always available from individuals and, furthermore, that these individuals are capable of making arbitrarily precise judgements through introspection. Those statisticians known as objectivists/frequentists believe that the information should be objective and typically based upon plentiful frequentist data. In the absence of such frequentist data, the objective frequentists would follow such other approaches to decision making as that of Neyman–Pearson treated in Section 13.4 or to estimation as that of maximum likelihood touched on in Section 13.5. A smaller school believes that objective approaches can still be taken in the absence of extensive frequentist data by resorting to epistemic principles that are questioned by others (e.g., see Jaynes [50]). In the end, these controversies amount to debates about the meaning of probability. Alternative meanings of probability were outlined in Section 3.3.

A new approach based upon *interval-valued probability* or *sets of measures*, noted in Section 3.2, in principle, allows for the rational incorporation of imprecise beliefs, as on the subjectivist account. While this approach promises to eliminate the more objectionable assumptions of previous work on subjective Bayesian decision making, it is still in an early stage of development (Walley [96, 97]).

After setting up the Bayesian decision-making framework, we define the Bayes principle of selecting rules that minimize expected loss. We then examine three special cases in which we can solve explicitly for the form of the optimal decision rules.

16.2 SETUP AND NOTATION FOR BAYESIAN DECISION MAKING

We proceed to specify the components of a Bayesian decision problem. We are considering decision making under uncertainty, where this uncertainty is about the unknown true state of nature or of the small world of our problem that determines the consequences of our actions. Our choice of action is to be based upon information or data that provide partial information about that state. The components of a decision problem (describing our prior knowledge about the setting of the problem and specifying our preferences for the consequences of our actions) are as follows:

1. $S = \{\theta\}$, set of *states of nature*
   Possible parameter values, set of transmitted messages (e.g., JPEG images), set of pattern categories (e.g., alphanumeric ASCII characters), etc. The "true," but unknown, state of nature is presumed to be represented by some element of a sample space S.

2. Θ denotes the random variable on S corresponding to the actual or true state of nature. The message actually transmitted or pattern selected.

3. $A = \{a\}$, *set of actions* or decisions available to the decision maker
   When we wish to infer to Θ, then examples of A include the sets of "true" messages, categories, or parameter values. However, A need not be the same as S. In medical treatment, S could be a set of states of health (illnesses) for an individual, while A could be the set of treatments; our primary objective is curing the patient, not making an accurate diagnosis. A could be the set, say, of control actions (e.g., positions of the gas pedal, the brake pedal, and the steering wheel while driving in a winter storm with impaired road visibility) to be taken, given the information in the received message (e.g., about our approximate location on the road, other traffic, our speed, and the traction condition of the road surface).

4. $L : S \times A \to \mathbb{R}$, *loss function*

   $L(\theta, a)$ is a numerical measure of the consequences to the decision maker of adopting action $a \in A$ when the true state of nature is $\theta \in S$. Smaller losses are preferred to larger ones. The appendix to this chapter discusses the meaning and source of a loss function.

5. $X = \{x\}$, *observation space* or measurement set

   For example, these might be the sets of possible received signals, laboratory instrument readings, measurements of a pattern feature, other data relevant to learning Θ.

6. X denotes the random variable on X corresponding to the observation.

7. $d : X \to A$, *decision function* or rule

   If we use rule d, then, when we observe x, we decide or take action $d(x) = a \in A$. We will assume that d is an ordinary deterministic single-valued function, although the general theory of decision making requires the use of randomized decision rules (where $d(a|x)$ is now the probability of choosing action a, given that we have observed x) in such problems as Neyman–Pearson hypothesis testing.

8. $D = \{d\}$, set of possible, allowable, constructible, or practical decision rules

   D might be all (measurable) functions from X to A or a restricted class, such as the linear functions, or all neural networks of a given architecture on X if, say, X is $\mathbb{R}^n$. D depends upon available technology and resources.

9. π, *prior probability* measure on S describing Θ

   π represents our probabilistic knowledge as to the states of nature prior to making an observation or measurement. This prior knowledge may be based upon subjective belief or upon frequentist data. An alternative notation is $p_\Theta(\theta)$, a probability mass function on S, or $f_\Theta(\theta)$, a density function.

10. $p_{X|\Theta}(x|\theta)$ or $f_{X|\Theta}(x|\theta)$, *likelihood function*

    These are conditional probability descriptions of the measurement system, relating observations and states of nature. For example, if message θ is sent, then the probability of receiving discrete-valued signal x is $p_{X|\Theta}(x|\theta)$.

11. $\pi(\theta|x)$ or $\pi_{\Theta|X}(\theta|x)$, *posterior conditional probability* of state of nature θ, given observation x

    This describes what is known about the state of nature after (posterior to) the observation. An alternative notation is $p_{\Theta|X}$, $f_{\Theta|X}$.

12. $EL(\Theta, d)$, *expected loss* incurred by use of the rule $d \in D$

    Chapters 11 and 12 enable us to evaluate EL in a variety of circumstances. In the case of discrete X and Θ,

$$EL(\Theta, d(X)) = \sum_{\theta \in S} \sum_{x \in X} \pi(\theta) p_{X|\Theta}(x|\theta) L(\theta, d(x)).$$

In the case of continuous Θ and discrete X,

$$EL(\Theta, d(X)) = \sum_{x \in X} \int_S \pi(\theta) p_{X|\Theta}(x|\theta) L(\theta, d(x)) \, d\theta.$$

In the case of discrete Θ and continuous X,

$$EL(\Theta, d(X)) = \sum_{\theta \in S} \pi(\theta) \int_X f_{X|\Theta}(x|\theta) L(\theta, d(x)) \, dx.$$

In the case of continuous $\mathcal{X}$ and Θ,

$$EL(\Theta, d(X)) = \int_{\mathcal{S}} \int_{\mathcal{X}} \pi(\theta) f_{X|\Theta}(x|\theta) L(\theta, d(x)) \, dx \, d\theta.$$

$L(\Theta, d(X))$ is the random loss associated with Θ, X. EL is the numerical value of the performance of rule d.

16.3 BAYES APPROACH TO DECISION MAKING

Definition 16.3.1   (Bayes Principle)   Decision rule d is *preferred to* (better than) rule d' if $EL(\Theta, d) < EL(\Theta, d')$.

The Bayes principle and the construction of the required loss function are grounded in utility theories of rational preference (e.g., see Savage [80], Ferguson [28], Fishburn [32], and the appendix to this chapter). The better decision rule is the one incurring the lower expected loss. This principle does not take into account the cost of computing or implementing the decision rule.

Definition 16.3.2   (Bayes Risk)   The Bayes risk

$$r^* = \inf_{d \in \mathcal{D}} EL(\Theta, d).$$

The inf, or infimum, is the greatest lower bound and is not necessarily achievable. However, in our applications, it will generally be achievable and therefore identical to the minimum. The Bayes risk r^* is the greatest lower bound to the best achievable performance.

Definition 16.3.3   (Bayes Rule) d^* is a Bayes rule if

$$EL(\Theta, d^*(X)) = r^*.$$

Lemma 16.3.1   (Equivalent Loss Functions)   If L and L' are loss functions linearly related through $L' = aL + b$, for $a > 0$, then a Bayes rule d^* for L is also a Bayes rule for L'.

Linear scale changes in the loss function affect neither the choice of best decision rule nor the comparison between any two decision rules.

16.4 CALCULATING BAYES RULES

Employing the subconditioning result that, for any random variables Y and X,

$$EY = E\{E\{Y|X\}\},$$

and the fact that the minimum with respect to a parameter of a sum or integral is greater than or equal to the corresponding sum or integral of the minima with respect to a parameter of the summands or integrands,

$$\inf_\lambda \sum_x g(x, \lambda) \geq \sum_x \inf_\lambda g(x, \lambda),$$

$$\inf_\lambda Eg(X, \lambda) \geq E\left[\inf_\lambda g(X, \lambda)\right],$$

enables us to write (we use minima in place of infima for clarity)

$$r^* = \min_{d \in \mathcal{D}} EL(\Theta, d(X)) = \min_{d \in \mathcal{D}} E\{E\{L(\Theta, d(X))|X\}\}$$

$$\geq E\{ \min_{d(x) \in \mathcal{A}} E\{L(\Theta, d(X))|X = x\}\}.$$

Whereas the first minimization is over choices of a function $d \in \mathcal{D}$, the last (interior) minimization is much easier to evaluate, as it is only over the value in $\mathcal{A}$ of the decision rule d evaluated at x. The inequality is an equality,

$$r^* = E\{\min_{a \in \mathcal{A}} E\{L(\Theta, a)|X = x\}\},$$

if we have not restricted the class $\mathcal{D}$; any collection (graph) of values $\{(x, d(x)) : x \in \mathcal{X}\}$ assembles into some function in $\mathcal{D}$. (This raises technical issues of the measurability of the resulting function $d^* \in \mathcal{D}$ that are beyond our scope.) Hence, the expectation of the minima averaged over X is achieved by some rule d^* (not necessarily unique) in $\mathcal{D}$. We summarize the preceding in the following somewhat informally stated theorem.

Theorem 16.4.1   (Extensive Form Bayes Calculation)   A Bayes rule d^* is given by

$$(\forall x \in \mathcal{X}) \; d^*(x) = a^* \iff E\{L(\Theta, a^*)|X = x\} = \min_{a \in \mathcal{A}} E\{L(\Theta, a)|X = x\}.$$

Our process, then, for calculating a Bayes rule d^* for a given loss function L is to determine the best action $a^* \in \mathcal{A}$ for each possible observation $x \in \mathcal{X}$. This is known as an *extensive form analysis*.

A so-called *normal form analysis* would have us enumerate each decision rule $d \in \mathcal{D}$, assign to that rule its expected loss $EL(\Theta, d)$, and then choose the rule having the smallest expected loss. This approach is generally intractable, but, on occasion, proves simplest. A normal form approach may be preferred if $\mathcal{D}$ is restricted, say, to a finite list of decision rules.

16.5 PARTICULAR BAYES SOLUTIONS

In what follows, we calculate Bayes rules for several important cases of decision making and estimation.

16.5.1 Ideal Observer Loss

Our first loss function assumes that $\mathcal{S} = \mathcal{A}$.

Definition 16.5.1   (Ideal Observer Loss Function)   The *ideal observer (IO)* case is defined by

$$S = \mathcal{A}, \quad L(\theta, a) = 1 - \delta_{\theta,a} = \begin{cases} 1 & \text{if } \theta \neq a \\ 0 & \text{otherwise} \end{cases}.$$

Hence, all correct decisions incur a loss of 0 and all incorrect decisions incur a loss of 1. This loss function is common in digital communication system analyses (e.g., see the minimum error probability design in Section 13.3) and in some pattern classification or hypothesis-testing problems. It is not used when S is uncountably infinite (e.g., an interval of numbers). Moreover, it is not often the case that all correct decisions are equally desirable or that all incorrect decisions are equally undesirable. A correct decision that diagnoses a dangerous illness is preferred to one that misses the diagnosis, but is not preferred to a correct decision of being in good health.

Theorem 16.5.1   (IO Bayes rule)   The Bayes rule d^* for the IO case is the so-called *maximum a posteriori (MAP) rule* wherein

$$d^*(x) = \text{argmax}_a P(\Theta = a | X = x);$$

$d^*(x) = a^*$ if a^* maximizes $P(\Theta = a | X = x)$.

Proof. For the IO loss function

$$EL(\Theta, d^*) = P(d^*(X) \neq \Theta) = 1 - P(d^*(X) = \Theta),$$

$$E(L(\Theta, d^*)|X = x) = 1 - P(d^*(X) = \Theta | X = x).$$

Hence, by Theorem 16.4.1,

$$d^*(x) = a^* \iff a^* = \underset{a \in \mathcal{A}}{\text{argmax}} \, P(\Theta = a^* | X = x). \qquad \square$$

Thus, the action $d^*(x)$ is a state (it need not be unique) that has the highest conditional probability of being correct given the measurement; it maximizes the *a posteriori* probability. The MAP rule was first derived in Chapter 13.

Example 16.1   IO Loss and MAP Rule _____

Consider a light source that has 10 levels of brightness and is used to signal the digits 0 through 9. We assume that the source, as observed by a photon-counting detector X, is described by a Poisson with parameter λ_i representing the average brightness of digit i. We assume further that each of the 10 digits is chosen equally probably, so that $\pi_i = .1$. Finally, we assume that the IO loss is appropriate such that confusing, say, a 0 with a 1 is no worse than confusing it with a 9. The posterior

$$P(\Theta = i | X = k) = \frac{P(X = k | \Theta = i)\pi_i}{P(X = k)} = \frac{e^{-\lambda_i} \frac{\lambda_i^k}{k!} .1}{\sum_{j=0}^{9} e^{-\lambda_j} \frac{\lambda_j^k}{k!} .1}$$

Note that, as we are only interested in the value of i that maximizes this posterior probability, it will be the same value that maximizes the simpler expression based solely on the numerator:

$$d^*(k) = i^* = \operatorname*{argmax}_{i \in \{0,...,9\}} [e^{-\lambda_i} \lambda_i^k].$$

For a given k, we can evaluate each of the 10 terms of the form given in the square brackets, identify the largest one, and assign its index i^* to $d^*(k)$.

16.5.2 Quadratic Loss

If S and $\mathcal{A}$ are both subsets of the reals $\mathbb{R}$, then the most commonly used loss function is the *quadratic loss* $L(\theta, a) = (\theta - a)^2$, as discussed in Section 15.4. We solved a version of this problem, for $\mathcal{A} = \mathbb{R}$, in Sections 15.4 and 15.5 under the heading of MMSE estimation.

Theorem 16.5.2   (Quadratic Loss Bayes Rule)   The Bayes rule in the quadratic loss case is given by

$$(\forall x \in \mathcal{X}) \; d^*(x) = a^*(x) \in \mathcal{A} \text{ where } |a^*(x) - E(\Theta|X = x)| = \min_{a \in \mathcal{A}} |a - E(\Theta|X = x)|;$$

that is, choose $a^* \in \mathcal{A}$ as close to $E(\Theta|X = x)$ as you can.

If $d^*(X) = E(\Theta|X)$ is allowed by $\mathcal{A}$, then

$$r^* = E\{VAR(\Theta|X)\}.$$

Proof.   We redo the end of the calculation in Section 15.4 to allow for $\mathcal{A} \subset \mathbb{R}$, and we solve for the Bayes rule d^* as follows:

$$E\{L(\Theta, a)|X = x\} = E\{(\Theta - E(\Theta|X))^2|X = x\} + E\{(E(\Theta|X) - a)^2|X = x\}.$$

Here, the cross-product term

$$E\left[(\Theta - E(\Theta|X))(E(\Theta|X) - a)\Big|X = x\right] = E(\Theta - E(\Theta|X = x))(E(\Theta|X = x) - a) = 0,$$

for the reasons given in Section 15.4. Note that the first term, which we write as

$$VAR(\Theta|X = x) = E\{(\Theta - E(\Theta|X))^2|X = x\},$$

is independent of the choice of action a. Hence,

$$E\{L(\Theta, a)|X = x\} = VAR(\Theta|X = x) + E\{(E(\Theta|X) - a)^2|X = x\}$$
$$= VAR(\Theta|X = x) + (E(\Theta|X = x) - a)^2.$$

It is now evident that $\min_{a \in \mathcal{A}} E\{L(\Theta, a)|X = x\}$ is achieved by the a^* in $\mathcal{A}$ that is closest to $E(\Theta|X = x)$ and equal to it if possible. $\square$

Example 16.2   Quadratic Loss

Assume that $S = \mathbb{R}$, $\mathcal{A} = Z$ (the set of all integers), the prior π is $\mathcal{N}(0, 1)$, and $f_{X|\Theta}$ is $\mathcal{N}(\Theta, \sigma^2)$. In this case, X and Θ are jointly normally distributed, and we can think of $X = \Theta + N$ for some "noise" N that is independent of Θ and is distributed $\mathcal{N}(0, \sigma^2)$. From the results obtained previously for the bivariate normal,

$$E(\Theta|X = x) = \frac{\rho \sigma_\Theta}{\sigma_X}(x - EX) + E\Theta = \frac{x}{1 + \sigma^2}.$$

Hence, $d^*(x)$ is the integer (element of **Z**) that is closest to $\frac{x}{1+\sigma^2}$. If, to the contrary of our assumption, $\mathcal{A} = \mathbb{R}$, then

$$d^*(x) = \frac{x}{1 + \sigma^2}$$

and

$$r^* = (1 - \rho^2)\sigma_\Theta^2 = \frac{\sigma^2}{1 + \sigma^2}.$$

16.5.3 Binary or Dichotomous Decision Problem

In this problem, first treated in Chapter 13, there are only two states of nature, $S = \{\theta_0, \theta_1\}$—for example, two possible messages as in binary digital communications, two states of health ("well" and "ill"), target absent and target present, etc. The states of nature are often referred to as "hypotheses," with θ_0 representing the so-called null hypothesis and θ_1 representing the so-called alternate hypothesis. The null hypothesis is thought of as representing the status quo and the alternate as a deviation from the status quo; for example, the null hypothesis is that you are "well" or that the "target is absent," and the alternate is that you are "ill" or that the "target is present." We encountered this terminology previously in our discussion of Neyman–Pearson hypothesis testing in Section 13.4.

- Given that there are only two states of nature, it is reasonable (but not necessarily the case) to suppose that we have only two actions, $\mathcal{A} = \{a_0, a_1\}$.
- We assume, without loss of generality, that a_i is the correct action to take if you know that the state of nature is θ_i.
- Our loss function $L(\theta, a)$ is specified in the form $L(\theta_i, a_j) = L_{ij}$, and from our immediately preceding assumption, $L_{ii} < L_{ij}$ for $j \neq i$.
- The prior probability mass function is specified through $\pi(\theta_i) = \pi_i$, where $\pi_0 + \pi_1 = 1$.
- Finally, we are provided with the conditional densities or probability mass functions. For convenience, we let $p_{X|\Theta}(x|\theta_i) = p_i(x)$ and $f_{X|\Theta}(x|\theta_i) = f_i(x)$.

Using Bayes' formula yields the posterior

$$\pi(\theta_i|x) = \frac{\pi_i p_i(x)}{\pi_0 p_0(x) + \pi_1 p_1(x)}$$

in the case of a discrete measurement, and it equals

$$\frac{\pi_i f_i(x)}{\pi_0 f_0(x) + \pi_1 f_1(x)}$$

in the case of a continuous measurement.

For a given decision rule d, we define the set of observations corresponding to d choosing $a_j \in \mathcal{A}$ as

$$D_j = \{x : d(x) = a_j\},$$

and we observe that $D_1 = D_0^c$.

We can now write

$$E\{L(\Theta, a_j) | X = x\} = L_{0j} \pi(\theta_0 | x) + L_{1j} \pi(\theta_1 | x).$$

Since we wish to minimize $E\{L(\Theta, a_j) | X = x\}$ by a choice of j, the optimal or Bayes decision rule d^* is given in terms of D_j^* by

$$D_0^* = \{x : E\{L(\Theta, a_0) | X = x\} \le E\{L(\Theta, a_1) | X = x\}\},$$

$$D_1^* = D_0^{*c} = \{x : E\{L(\Theta, a_0) | X = x\} > E\{L(\Theta, a_1) | X = x\}\}.$$

(Note that how we assign x when there is equality in the above is immaterial; it does not affect the resulting expected loss.) Hence,

$$D_0^* = \{x : L_{00} \pi(\theta_0 | x) + L_{10} \pi(\theta_1 | x) \le L_{01} \pi(\theta_0 | x) + L_{11} \pi(\theta_1 | x)\}.$$

We summarize the preceding in the next theorem.

Theorem 16.5.3   (Dichotomous Decision Bayes Rule)   The Bayes decision rule in the binary case is given by

$$D_0^* = \{x : d^*(x) = a_0\} = \{x : \Lambda(x) \le \tau\}, \quad D_1^* = \{x : \Lambda(x) > \tau\},$$

where the *likelihood ratio*

$$\Lambda(x) = \frac{p_1(x)}{p_0(x)} \quad \text{or} \quad = \frac{f_1(x)}{f_0(x)},$$

and the decision *threshold*

$$\tau = \frac{\pi_0(L_{01} - L_{00})}{\pi_1(L_{10} - L_{11})}.$$

This solution to the binary decision problem is known as the *likelihood-ratio-threshold rule*.

A decision rule of this form, but restricted to the IO loss function, was previously encountered in our discussion of the MAP rule for minimum error probability in Section 13.3 and the Neyman–Pearson decision rule for hypothesis testing in Section 13.4. In the current Bayes setup, we can express the threshold as a function of the given prior probabilities and losses, and we are not restricted to the special case of minimum error probability. In the Neyman–Pearson setup, the threshold was determined by the tolerable false-alarm probability. This approach is applicable when we have the IO loss, but do not know the prior π.

16.5.4 Examples of Binary Bayesian Decision Making

Example 16.3   Multivariate Gaussian Observations _____

A frequently encountered two-category pattern classification application of the preceding has a d-dimensional, normally distributed feature vector $\mathbf{X} \sim \mathcal{N}(\mathbf{m}, \mathbb{C})$, with $\mathbb{C}$, $\mathbf{m}$ that depend upon the pattern category. For example, we may have two possible Gaussian d-dimensional signals $\mathbf{S}_0$ and $\mathbf{S}_1$, with $\mathbf{S}_i \sim \mathcal{N}(\mathbf{m}_i, \mathbb{D}_i)$, that are observed in independent noise $\mathbf{N} \sim \mathcal{N}(\mathbf{0}, \mathbb{G})$. The resulting observation $\mathbf{X} = \mathbf{S} + \mathbf{N}$ depends upon which of the two signals is transmitted. If it is $\mathbf{S}_i$, then, conditional upon the choice of signal class (note that this is not conditional upon the value of the signal itself), $\mathbf{X} \sim \mathcal{N}(\mathbf{m}_i, \mathbb{D}_i + \mathbb{G})$. In what follows, we let $\mathbb{C}_i = \mathbb{D}_i + \mathbb{G}$. In this case, twice the logarithm of the likelihood ratio of multivariate normal densities is given by

$$2 \log \Lambda(\mathbf{x}) = \log\left(\frac{\det \mathbb{C}_0}{\det \mathbb{C}_1}\right) + (\mathbf{x} - \mathbf{m}_0)^T \mathbb{C}_0^{-1}(\mathbf{x} - \mathbf{m}_0) - (\mathbf{x} - \mathbf{m}_1)^T \mathbb{C}_1^{-1}(\mathbf{x} - \mathbf{m}_1)$$

$$= \mathbf{x}^T(\mathbb{C}_0^{-1} - \mathbb{C}_1^{-1})\mathbf{x} + \mathbf{x}^T(\mathbb{C}_1^{-1}\mathbf{m}_1 - \mathbb{C}_0^{-1}\mathbf{m}_0) + (\mathbf{m}_1^T \mathbb{C}_1^{-1} - \mathbf{m}_0^T \mathbb{C}_0^{-1})\mathbf{x}$$

$$+ \log\left(\frac{\det \mathbb{C}_0}{\det \mathbb{C}_1}\right) + \mathbf{m}_0^T \mathbb{C}_0^{-1}\mathbf{m}_0 - \mathbf{m}_1^T \mathbb{C}_1^{-1}\mathbf{m}_1.$$

Note that the log-likelihood ratio depends upon the observation $\mathbf{x}$ only through the three terms given on the second line of the preceding equation. The three terms in the third line depend upon the known covariance matrices and mean vectors. From the transpose of a scalar being the same scalar, and from the symmetry of covariance matrices and their inverses, it follows that the scalar

$$\mathbf{x}^T \mathbb{C}^{-1}\mathbf{m} = \mathbf{m}^T \mathbb{C}^{-1}\mathbf{x}.$$

Hence, we can simplify the foregoing expression for the $\mathbf{x}$-dependent terms as

$$\log \Lambda(\mathbf{x}) = \tfrac{1}{2}\mathbf{x}^T(\mathbb{C}_0^{-1} - \mathbb{C}_1^{-1})\mathbf{x} + (\mathbb{C}_1^{-1}\mathbf{m}_1 - \mathbb{C}_0^{-1}\mathbf{m}_0)^T\mathbf{x} + \dots.$$

For ease of reference, we introduce the quadratic function

$$Q(\mathbf{x}) = \tfrac{1}{2}\mathbf{x}^T(\mathbb{C}_0^{-1} - \mathbb{C}_1^{-1})\mathbf{x} + (\mathbb{C}_1^{-1}\mathbf{m}_1 - \mathbb{C}_0^{-1}\mathbf{m}_0)^T\mathbf{x}$$

and a bias term b given by

$$2b(\mathbb{C}_0, \mathbb{C}_1, \mathbf{m}_0, \mathbf{m}_1) = \log\left(\frac{\det \mathbb{C}_0}{\det \mathbb{C}_1}\right) + \mathbf{m}_0^T \mathbb{C}_0^{-1}\mathbf{m}_0 - \mathbf{m}_1^T \mathbb{C}_1^{-1}\mathbf{m}_1.$$

Hence, we can now write

$$\log \Lambda(\mathbf{x}) = Q(\mathbf{x}) + b(\mathbb{C}_0, \mathbb{C}_1, \mathbf{m}_0, \mathbf{m}_1).$$

Returning to the optimal decision rule, we see that the set D_1^* of observations for which we should decide that the transmitted signal was $\mathbf{S}_1$ is given by

$$D_1^* = \{\mathbf{x} : \log \Lambda(\mathbf{x}) > \log \tau\} = \{\mathbf{x} : Q(\mathbf{x}) > c = \log \tau - b\}.$$

The equation $Q = c$ defines a quadratic surface in $\mathbb{R}^d$. If $\mathbf{x}$ lies above this surface, then our optimal decision is that $\mathbf{S}_1$ was transmitted.

If the covariance matrices are both equal to $\mathbb{C}$, then we have the further simplifications

$$Q(\mathbf{x}) = \mathbb{C}^{-1}(\mathbf{m}_1 - \mathbf{m}_0)^T \mathbf{x},$$

$$b(\mathbb{C}, \mathbb{C}, \mathbf{m}_0, \mathbf{m}_1) = \tfrac{1}{2}[\mathbf{m}_0^T \mathbb{C}^{-1} \mathbf{m}_0 - \mathbf{m}_1^T \mathbb{C}^{-1} \mathbf{m}_1].$$

It is still the case that $\log \Lambda(\mathbf{x}) = Q(\mathbf{x}) + b$. Letting $\mathbf{w} = \mathbf{m}_1 - \mathbf{m}_0$, we see that the log-likelihood ratio depends upon $\mathbf{x}$ only through the linear function (dot or scalar product) $\mathbf{w}^T \mathbf{x}$. The optimal decision rule is now given by

$$D_1^* = \{\mathbf{x} : \mathbf{w}^T \mathbf{x} > c = \log \tau - b(\mathbb{C}, \mathbb{C}, \mathbf{m}_0, \mathbf{m}_1)\}.$$

This is the well-known case of *Fisher's linear discriminant function*. It also corresponds to the operation of a *perceptron*, a single-linear-threshold-element neural network. Geometrically, the parameters $(\mathbf{w}, c)$ define a hyperplane in $\mathbb{R}^d$. D_1^* is the half-space of points lying above this hyperplane. If $\mathbf{x} \in D_1^*$, then we decide that $\mathbf{S}_1$ was transmitted.

Example 16.4   Binomial Observations

We now examine an example of decision making with discrete data where we have two possible coins $\Theta = \{\theta_0, \theta_1\}$, with coin θ_i having probability p_i of heads. While coins per se are not of great interest to us, they schematically represent a number of more interesting situations. Coin θ_i has been selected with prior probability π_i, and we assume that $p_1 > p_0$. We are allowed to toss the selected coin n times with $\{0, 1\}$-valued outcomes $\mathbf{x} = (x_1, \ldots, x_n)$. We then have to choose an action $\mathcal{A} = \{a_0, a_1\}$, where action a_i means that we decide that it is coin θ_i, and this is represented by a choice of loss function $L(\theta_i, a_j) = L_{ij}$ for which $L_{ii} < L_{i,1-i}$. We assume that we can ascertain our loss function L_{ij}. In this case, the likelihood ratio is given by

$$k = \sum_1^n x_i, \quad \Lambda(\mathbf{x}) = \frac{p_1^k (1-p_1)^{n-k}}{p_0^k (1-p_0)^{n-k}} = \left(\frac{p_1(1-p_0)}{p_0(1-p_1)}\right)^k \left(\frac{1-p_1}{1-p_0}\right)^n.$$

For convenience, let

$$\gamma = \frac{p_1(1-p_0)}{p_0(1-p_1)} \quad \text{and} \quad \tau = \frac{\pi_0}{\pi_1} \frac{L_{01} - L_{00}}{L_{10} - L_{11}},$$

and note that $\gamma > 1$ follows from the assumption that $p_1 > p_0$. We can now write the Bayes decision rule

$$D_1^* = \left\{\mathbf{x} : \log \Lambda(\mathbf{x}) = k \log \gamma + n \log \frac{1-p_1}{1-p_0} > \log \tau\right\}.$$

Hence,

$$D_1^* = \left\{ \mathbf{x} : \sum_1^n x_i > \frac{\log \tau - n \log \frac{1-p_1}{1-p_0}}{\log \gamma} \right\}$$

is the set of observations for which we choose action a_1 that the coin has probability p_1 for heads. A reasonable outcome is an outcome in which we decide it is the coin with a higher probability of heads if we observe a large number of heads. Of course, our result is more precise than this statement.

16.6 SUMMARY

This chapter systematically presented the Bayesian approach to achieving Goal 3 under the circumstance that all required probability information is available. Conditional expectation is the key to making significant progress. We first define a decision problem by listing its components. We then define the Bayes approach of minimizing expected loss for an arbitrarily selected loss function and show how to solve for the Bayes decision rule in a large family of estimation and decision problems. After deriving general results on extensive form calculation of Bayes decision rules, we examine three important loss functions of ideal-observer loss corresponding to a criterion of minimum error probability (see Section 13.3), quadratic loss, and the general loss function when there are only two states of the world (e.g., target present and target absent). The topic of quadratic loss and its applications was previously considered in Chapters 9, 10, and 15, and the binary case of two states of the world was treated in Chapter 13. The systematic Bayesian approach critically assumes knowledge of the prior probability measure π. While there is an outspoken Bayesian school of statisticians which urges that all of inference be cast in this mold of a known prior, we find this claim too confining. We need to respect the limits of our knowledge, and prior probabilities represent a significant degree of knowledge. The material on Neyman–Pearson decision making in Section 13.4 and maximum likelihood estimation in Section 13.5 indicate what can be achieved in the absence of a prior probability measure.

16.7 APPENDIX: AN INFORMAL PRIMER ON LOSS FUNCTIONS

16.7.1 Setting the Stage

Systems, including decision-making systems, are designed at the expense of time and material resources to achieve a purpose defined by more or less clearly understood performance criteria. But who provides these criteria and in what form? We follow convention and refer to the individual or group that is to benefit from the system as the "User." While much of decision theory and utility theory contemplates a User that is an individual, the User is much more often a group of individuals. However, the members of this group might not be known to the system designer at the time the system is being considered. Nor need this group be a well-defined one. How can the designer solicit the preferences for quality of service (e.g., balances between accuracy, reliability, cost, and ease of use of the system) of such user groups as the Internet, World Wide Web, cellular and public-switched telephony users, as well as consumers of electric power, in advance of the deployment of the system? In these cases, the system designers may consult only their own preferences and thereby become both designer and User. In this primer, it is easiest to think of the User as a single known individual—perhaps yourself.

What, then, are the User's preferences among possible decision-making systems sharing such elements of our initial list as set $\Theta = \{\theta\}$ of states of the world, actions $\mathcal{A} = \{a\}$, and observations/data $\mathcal{X} = \{x\}$? There usually is also a technological constraint that limits the set $\mathcal{D} = \{d\}$ of decision rules to ones that can be implemented in a given technology at a given cost and within a given time. How can the User's preferences among potential systems or decision rules be expressed mathematically so that optimization techniques can be employed to find an acceptable rule? We do not ask the User, who may be a "user" and not a technical expert, to jump directly to the choice of system. Rather, we ask the User to contemplate the consequences of using a decision rule in the sound belief that the consequences lie closer to the needs and understanding of the User of the rule.

16.7.2 Preference Orderings

The mathematical tool for representing preferences is that of binary order relations. Given, say, the set $\mathcal{D}$ of possible decision rules or systems, we consider a binary relation $\succeq_{\mathcal{D}}$ with the meaning that if $d_0, d_1 \in \mathcal{D}$, then $d_1 \succeq_{\mathcal{D}} d_0$ is read, "Rule d_1 is at least as desirable to the User as is rule d_0." We subscript the relation by $\mathcal{D}$ to emphasize that it is a relation between elements of $\mathcal{D}$. Subsequently, we will consider such preference relations between elements of other sets of objects. A related notion is $\succ_{\mathcal{D}}$, where $d_1 \succ_{\mathcal{D}} d_0$ is read as "Rule d_1 is strictly preferred to rule d_0." The link between $\succeq_{\mathcal{D}}$ and $\succ_{\mathcal{D}}$ is defined by

$$d_1 \succ_{\mathcal{D}} d_0 \iff d_1 \succeq_{\mathcal{D}} d_0 \text{ and false } d_0 \succeq_{\mathcal{D}} d_1.$$

Finally, we have the binary relation $\approx_{\mathcal{D}}$, where $d_1 \approx_{\mathcal{D}} d_0$ is read, "The User is indifferent between d_0 and d_1." $\approx_{\mathcal{D}}$ is defined by

$$d_1 \approx_{\mathcal{D}} d_0 \iff d_1 \succeq_{\mathcal{D}} d_0 \text{ and } d_0 \succeq_{\mathcal{D}} d_1,$$

and it has the symmetry property that

$$d_1 \approx_{\mathcal{D}} d_0 \iff d_0 \approx_{\mathcal{D}} d_1.$$

Further constraints on the nature of these binary relations, provided by axioms discussed shortly, are needed to develop loss functions.

16.7.3 Consequences and Gambles

A difficulty with the given scenario is that it presumes the User has enough technical skill to be able to express preferences for the rules themselves. It is more realistic to focus on the consequences of the use of these rules and preferences among those consequences.

When nature selects as state $\Theta = \theta$ and the decision rule d selects an action $a = d(x)$, then consequences ensue from the pair (θ, a). Let $\mathcal{C}$ denote a set $\{c\}$ of "elementary" consequences (e.g., awards of certain amounts of money or loss of a certain amount of time); it may help to think of $\mathcal{C}$ as a range of monetary payoffs, say, from $-\$10^4$ (loss) to $+\$10^7$ (gain). We assume that there is a function

$$C : \Theta \times \mathcal{A} \to \mathcal{C}, \ C(\theta, a) = c,$$

that assigns a particular consequence c to each pair (θ, a). Thus, if the User was informed that she would be facing (θ, a), then she would know that this amounted to facing a consequence c. We assume that the User has a preference ordering $\succeq_C$ between the pairs of consequences in $\mathcal{C}$.

However, the choice of a system or decision rule d actually confronts the User, not with a single elementary consequence, but with a probability distribution over the consequences. Nature selects θ and the observation x with certain probabilities. The decision rule then selects the action $a = d(x)$, and $C(\theta, d(x))$ becomes the consequence. Let Δ denote the random variable that is the consequence produced by the system. We assume, for simplicity, that $\Theta, \mathcal{X}, \mathcal{A}$, and $\mathcal{C}$ are countable sets. We evaluate the probability mass function $p_d(c)$ for consequences, when we use decision rule d, by

$$p_d(c) = P(\Delta = c) = P(\{(\theta, x) : d(x) = a \text{ and } C(\theta, a) = c\})$$

$$= \sum_{\theta \in \Theta} \pi(\theta) \sum_{\{a : C(\theta, a) = c\}} \sum_{\{x : d(x) = a\}} p_{X|\Theta}(x|\theta). \tag{16.1}$$

Hence, the choice of d exposes the User to a probability distribution with pmf p_d over the set $\mathcal{C}$ of consequences. We call this distribution a "gamble" offered to the User.

Let $\mathcal{C}^* = \{c^*\}$ denote a set of distributions over consequences in $\mathcal{C}$. We require that the degenerate gamble $\mathbf{1}_c$ that assigns probability 1 to the particular consequence c lie in $\mathcal{C}^*$ for all choices of c. Furthermore, we require that finite convex combinations of gambles in $\mathcal{C}^*$ also lie in $\mathcal{C}^*$. Finite convex combinations of gambles or probability measures were discussed in Section 3.9. Hence, the gambles c^* in $\mathcal{C}^*$ are probability distributions of the form

$$c^*(c) = \sum_i \lambda_i \mathbf{1}_{c_i}(c),$$

where the weights $\lambda_i > 0$ and they sum to 1; gamble c^* selects elementary consequence c_i with probability λ_i. Henceforth, we assume that

$$(\forall d \in \mathcal{D}) \ p_d \in \mathcal{C}^*.$$

The User is now asked to express her preferences between pairs of elements in $\mathcal{C}^*$ through $\succeq_{C^*}$, and we can again define strict preference $\succ_{C^*}$ and indifference $\approx_{C^*}$ as we did earlier. Thus, $c_1^* \succ_{C^*} c_0^*$ is read, "The gamble over consequences c_1^* is strictly preferred by the User to the gamble c_0^*."

16.7.4 Numerical Representation of $\succeq_{C^*}$

Our task is to render tractable and assessable the rather abstract User's preference relation $\succeq_{C^*}$ between gambles over consequences of the kind produced by the choice of a system or a decision rule. In order to do this, and to eventually arrive at an understanding of loss functions, we need to impose several axioms to constrain what can be a User preference relation. Some of these are *axioms of rationality*, in that they are close to our intuitive understanding of preference. Thus, we assume that $\succeq_{C^*}$ is *transitive*. That is, for any gambles $c_0^*, c_1^*,$ and c_2^*, if $c_2^* \succeq_{C^*} c_1^*$ and $c_1^* \succeq_{C^*} c_0^*$, then it must also be the case that $c_2^* \succeq_{C^*} c_0^*$. Restated, if I prefer apples to oranges and oranges to pears, then I prefer apples to pears. We also assume *comparability* in that, for any c_0^* and c_1^*, the User will assert either $c_1^* \succeq_{C^*} c_0^*$ or $c_0^* \succeq_{C^*} c_1^*$ or both. The User can always compare any two gambles. In addition, to reach the end we are approaching, we also need to impose what have been called *structural axioms*. These are technical axioms without which we cannot achieve the desired representation. Presentations of such sets of axioms and discussions of their implications can be found in Ferguson [28], Fishburn [32], Krantz, et al. [56], and Savage [80].

When $\succeq_{C^*}$ satisfies these (unstated) axioms, there exists a numerical representation in the sense that each gamble $c^* \in C^*$ can be assigned a real number, say, $r(c^*)$ such that

$$r : C^* \to \mathbb{R}, \quad c_1^* \succeq_{C^*} c_0^* \iff r(c_1^*) \geq r(c_0^*).$$

Preference between gambles is reduced to numerical inequality between representations. What greatly simplifies the structure of $\succeq_{C^*}$ and of its representation by r is that we need only have an appropriate representation of the simpler set C of the consequences themselves. The axioms imply that there exists a *value or utility function*

$$v : C \to \mathbb{R}$$

such that, for

$$c^* = \sum_i \lambda_i \mathbf{1}_{c_i},$$

$$r(c^*) = \sum_i \lambda_i v(c_i) = E_{c^*} v(c).$$

Note that v need be defined only over the simpler set C and not over the set C^* of gambles with payoffs in C. The numerical representation of the gamble c^* is simply the expectation of the value function v taken with respect to the probability distribution c^* in C! Restated, there exists a value function assigning real numbers to each of the elementary consequences in C such that

$$c_1^* \succeq_{C^*} c_0^* \iff E_{c_1^*} v \geq E_{c_0^*} v.$$

As in Lemma 16.3.1, it is evident that if we change the value function v to

$$v' = \alpha v + \beta,$$

with positive α, then we will induce exactly the same preference ordering between pairs of gambles. Hence, the value function v is not unique to within positive affine transformations. Moreover, it is highly unlikely that a nonlinear transformation of v to v' will maintain the same ordering when we compute expected values. Under the appropriate axioms for preference, every nonlinear transformation will yield a new value function that represents a different preference ordering.

16.7.5 Loss Function

Assembling the previous ideas, under axioms unspecified in this primer, rational User preferences between gambles have an expected value numerical representation. If we return to the choice between decision rules, we have

$$d_1 \succeq_\mathcal{D} d_0 \iff E_{p_{d_1}} v \geq E_{p_{d_0}} v,$$

where p_d was defined in Eq. (16.1). Thus, we can numerically represent the preference ranking of a decision rule d through

$$r(d) = \sum_{\theta \in \Theta} \pi(\theta) \sum_{\{a : C(\theta, a) = c\}} \sum_{\{x : d(x) = a\}} p_{X|\Theta}(x|\theta) v(C(\theta, a)),$$

with

$$d_1 \succeq_\mathcal{D} d_0 \iff r(d_1) \geq r(d_0).$$

The *loss function* $L(\theta, a)$ can now be identified as

$$L(\theta, a) = -v(C(\theta, a)),$$

with the negative sign showing that L represents loss and not gain. Restated,

$$d_1 \succeq_\mathcal{D} d_0 \iff E_{p_{d_1}} L(\theta, d_1(X)) \leq E_{p_{d_0}} L(\theta, d_0(X)).$$

We have succeeded in developing the Bayes principle of Definition 16.3.1 that a rational User's preferences for decision rules is such that the preferable rule is the one with the smaller expected loss. Moreover, we have done so via a path that renders the loss function L meaningful in terms of a User's preferences between the consequences of gambles to which she is exposed by a choice of system or decision rule d.

16.7.6 Ascertaining the Value or Loss Function

The preceding discussion argued for the existence of a loss or value function that can represent a User's preferences. How can we elicit or learn the User's actual value function? One such process that is easiest to explain, although it may tax the User's psychological resources and patience to conduct it, is the following: Given that we have determined the set $\mathcal{C}$ of consequences of the choices $\{(\theta, a)\}$, select a consequence $\overline{c}$ that the User finds as desirable as any other consequence in $\mathcal{C}$ and another consequence $\underline{c}$ that is no more desirable than any consequence in $\mathcal{C}$. Arbitrarily, and without loss of generality given the nonuniqueness of the value function with respect to positive affine transformations, assign the value function

$$v(\overline{c}) = 1 > v(\underline{c}) = 0,$$

consistent with $\mathbf{1}_{\overline{c}} \succ_{\mathcal{C}^*} \mathbf{1}_{\underline{c}}$. Given any $c \in \mathcal{C}$, ask the User to identify a $0 \leq \lambda \leq 1$ such that

$$\mathbf{1}_c \approx_{\mathcal{C}^*} \lambda \mathbf{1}_{\overline{c}} + (1 - \lambda) \mathbf{1}_{\underline{c}}.$$

We are asking the user to identify a probability λ such that she is indifferent between receiving the consequence c as a sure thing or receiving a gamble that yields the desirable $\overline{c}$ with probability λ and the undesirable $\underline{c}$ with the remaining probability of $1 - \lambda$. Hence,

$$v(c) = \lambda v(\overline{c}) + (1 - \lambda)v(\underline{c}) = \lambda,$$

and the User has revealed $v(c)$ through her choice of λ. This process must then be repeated for the other consequences in $\mathcal{C}$, a likely arduous task.

Example 16.5   An Experiment in Valuation
It will help to fix ideas if the reader tries the preceding elicitation process on himself or herself in the role of User. Assume, as usual, that more money is preferred to less, take $\mathcal{C}$ to be the range of dollar amounts from 0 to $+10^7$ dollars, and choose $\overline{c} = 10^7$ and $\underline{c} = 0$. Setting $v(10^7) = 1$ and $v(0) = 0$, evaluate, through your personal introspection, the values for the following eight monetary consequences in US dollars:

$$1, \ 10, \ 100, \ 10^3, \ 10^4, \ 10^5, \ 10^6, \ 5 \times 10^6.$$

Most individuals will find that plotting $v(c)$ vs. c will yield an "S"-shaped curve. An example is shown in Figure 16.1. The x-axis is logarithmic and starts at 10 rather than 1. The dotted curve is actually a straight line when not plotted semilogarithmically and corresponds to a value function that is linear with dollars. For this curve, the individual is indifferent between a 50–50 gamble between $\$10^7$ and $\$0$ or getting a little more than $\$10^5$ as a sure thing. Value is nonlinear in dollars.

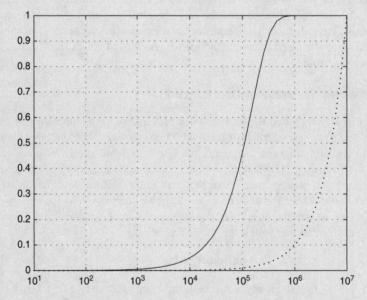

Figure 16.1 Sigmoidal curve typical of value/utility functions.

Few people who are not very wealthy would prefer a bet that has .5 probability of yielding 10^7 and .5 probability of yielding 0 to one that gives them 2.5×10^6 as a sure thing, even though the expected dollar winnings in the first bet is twice that in the second bet. We tend to prefer the sure thing to risky gambles when large amounts of money are involved, but we are willing to take very unfavorable bets (e.g., state lotteries) when the sure thing is a small amount of money.

EXERCISES

E16.1 $S = \mathcal{X} = \mathbb{R}$, $\mathcal{A} = \{0, 1, 2, \ldots\}$ and $L, f_{X|\Theta}, \pi$ are such that

$$E(L(\theta, a)|X = x) = |x - a| + x^2.$$

a. Find a Bayes rule d^*;
b. Evaluate the Bayes risk r^* when $X \sim \mathcal{U}(-.5, .5)$.

E16.2 In a given decision problem with $S = \mathcal{A} = \mathcal{X} = \mathbb{R}$ and $X \sim \mathcal{N}(0, \sigma^2)$, calculation based upon $L(\theta, a), f_{X|\Theta}, \pi$, reveals that

$$E\{L(\Theta, d)|X = x\} = d^2(x) + x^2 + (d(x) - x)^2.$$

a. Find a Bayes rule d^*.
b. Evaluate the Bayes risk r^*.

E16.3 In a certain decision problem with $S = \mathcal{A} = \mathcal{X} = \{0, 1, 2, \ldots\}$, calculation reveals that

$$E[L(\Theta, a)|X = x] = (a - x^2)^2 + a + x.$$

We also know that

$$\pi_\Theta(\theta) = 2^{-1-\theta}, \quad p_X(x) = \frac{e^{-1}}{x!}.$$

a. Find a Bayes rule d^*.
b. Evaluate the Bayes risk r^*.

E16.4 If $S = \mathcal{A} = \mathbb{R}$, $L(\theta, a) = (\theta - a)^2 + a + \theta$, $f_{X|\Theta}(x|\theta) = \frac{1}{2}e^{-|x-\theta|}$, and $\Theta \sim \mathcal{U}(-1, 1)$, then evaluate the Bayes rule d^* for $x > 1$.

E16.5 If $S = \mathcal{X} = \mathbb{R}$, $\mathcal{A} = Z$, $L(\theta, a) = (\theta - a)^2$, $X|\Theta \sim \mathcal{N}(\Theta, 1)$, and $\Theta \sim \mathcal{N}(0, 1)$, then evaluate the Bayes rule.

E16.6 Θ, X are jointly normally distributed with $E\Theta = 1$, $EX = 0$, $VAR(\Theta) = 4$, $VAR(X) = 2$, and $COV(X, \Theta) = 1$. The loss function $L(\theta, a) = (\theta - a)^2$ and $\mathcal{A} = \mathbb{R}$.

a. Find a Bayes rule d^*.
b. Evaluate its performance r^*.

E16.7 If a light source is present, then a photon counter yields a count $N \sim \mathcal{P}(2)$. However, if the source is absent, which it is with probability $\frac{2}{3}$, then $N \sim \mathcal{P}(1)$.

What is the optimum decision rule to use to detect the presence or absence of this light source if the loss to the user of this system is 5 whenever an error is made and 3 whenever no error is made?

E16.8   We receive a binary signal S in additive noise N that is independent of S. We know that $P(S = 0) = 1/3 = 1 - P(S = 1)$ and $N \sim \mathcal{N}(0, 1)$. We wish to estimate the signal S on the basis of the received sum X in such a manner that we minimize the expected squared difference between S and its estimate $d(X)$.

Calculate the best decision rule d^*.

E16.9   We receive a binary signal S in additive noise N that is independent of S. We know that $P(S = 0) = 1/3 = 1 - P(S = 1)$ and $N \sim \mathcal{G}(\frac{1}{2})$. We wish to estimate the signal S on the basis of the received sum X in such a manner that we minimize the expected squared difference between S and its estimate $d(X) \in \{0, 1\}$.

Calculate the best decision rule d^* for $X > 0$.

E16.10   In an analog communications channel, the conditional density of the received output X given the input signal S is given by

$$f_{X|S}(x|s) = \begin{cases} 1 & \text{if } |x - s| \leq \dfrac{1}{2} \\ 0 & \text{if otherwise} \end{cases}.$$

The probability density $f_S(s) = \frac{1}{2}e^{-|s|}$.

a. Find the estimator $\hat{S}(X)$ of S that minimizes $E(S - \hat{S})^2$.

b. Find the estimation error.

E16.11   In a digital communications channel, the conditional pmf of the $\{0, 1\}$-valued received output X given the $\{0, 1\}$-valued input signal S is given by

$$p_{X|S}(x|s) = \begin{cases} 1 & \text{if } x = s = 1 \\ .6 & \text{if } x = 1, s = 0 \end{cases}.$$

a. If all inputs are equally probable, then determine a receiver $\hat{S}(X)$ to identify the most probable input for a given output X.

b. Evaluate the probability of erroneous decoding of the channel output.

E16.12   An optical radar ("lidar") sends out a pulse of light and then waits for a given interval $[t_0, t_0 + T]$ of time for a reflected beam. If a target is absent, then a photodetector operating over the observation interval will count $N \sim \mathcal{P}(T)$ photons. If a target is present, then $N \sim \mathcal{P}(2T)$. A target can be expected to be present with probability .2.

a. Calculate the best detection rule to use to determine the status of the target when our loss is as follows:

5 if the target is present, but we do not detect it (missed detection);

0 if the target is present and we do detect it;

2 if the target is absent and we claim otherwise (false alarm);

0 if the target is absent and we realize this.

b. Provide an expression for the performance of this detection rule.

E16.13   If $S = A = X = \mathbb{R}$ and $L(\theta, a) = |\theta - a|$ (*absolute value loss*), show that the Bayes rule $d^*(x)$ is given by the median of the conditional cdf $F_{\Theta|X}(\theta|x)$.

E16.14   An observation X given Θ is described by

$$f_{X|\Theta}(x|\theta) = \theta e^{-\theta x} U(x),$$

$\Theta \sim \mathcal{E}(1)$, and $\mathcal{A} = \mathbb{R}$, and we use absolute value loss $L(\theta, a) = |\theta - a|$.

    a. Assuming the solution provided in the preceding exercise, evaluate the Bayes estimator of Θ for absolute value loss.

    b. Evaluate the MMSE estimator $\hat{\Theta}$ for the same specification.

    c. Do the two estimators agree?

E16.15   Undertake the introspection process discussed in Section 16.7.6 to evaluate your present value for U.S. dollars in the range from losing \$100 to gaining \$1,000. Sketch your value function v.

Part IV

Characteristic Functions and Probability Bounds

17

Characteristic Functions

17.1 PURPOSE, BACKGROUND, AND ORGANIZATION

In mathematics and in its applications, it is always valuable to have alternative ways of representing the same mathematical object. An analogy might be a set of vectors that can be represented in infinitely many coordinate systems. In our case, our fundamental concept is that of a probability measure P assigning numerical probability to each event A in an appropriate algebra of events $\mathcal{A}$. In Chapter 5, we represented P for a random variable X and events of the form $X \in A$ through a cumulative distribution function (cdf) F_X. In Chapters 4, 5, and 7, we represented the cdfs through probability mass functions (pmf) and probability density functions (pdf). Each of these representations enabled us to reconstruct P for events of the form $X \in A$. Each representation was simpler to describe than was the measure P. Our final representation is given by the characteristic function ϕ_X for a random variable X (or a random vector $\mathbf{X}$). The characteristic function will be seen to be a Fourier transform of the pdf f_X. As such, it will enable us to calculate moments by differentiating ϕ_X rather than integrating f_X, which is generally more difficult. It will also give us a particularly easy way to deal with the responses of general linear systems. The characteristic function provides access to the family of celebrated Central Limit Theorems [60] that help

459

to explain the prevalence in nature of the normal or Gaussian distribution. Finally, combining conditional expectation and characteristic functions enables us to obtain detailed results for the complex problem of a sum of independent random variables in which the number of summands is itself a random variable.

17.2 CHARACTERISTIC FUNCTION DEFINITION AND EXAMPLES

We have seen that, in the case of real- or vector-valued random variables, the probability measure P can be equivalently specified by either a cdf F_X, density f_X, or pmf p_X. We now introduce another means of specifying P: through the *characteristic function*.

Definition 17.2.1 (Characteristic Function) The characteristic function ϕ_X of a random variable X is

$$\phi_X(u) = Ee^{iuX} = E\cos(uX) + iE\sin(uX) = \int_{-\infty}^{\infty} e^{iux} f_X(x)\,dx.$$

This definition holds for the complex-valued argument u for which the expectation exists.

Example 17.1 Characteristic Function Calculations _____
If X has a continuous distribution described by a pdf, say, $\mathcal{U}(a,b)$, then

$$\phi_X(u) = \int_a^b \frac{e^{iux}}{b-a}\,dx = \frac{e^{iub} - e^{iua}}{iu(b-a)}.$$

If X has a discrete distribution described by a pmf, say, $\mathcal{G}(\beta)$, then the expectation is evaluated as the sum

$$\phi_X(u) = \sum_{k=0}^{\infty}(1-\beta)\beta^k e^{iuk} = (1-\beta)\sum_{k=0}^{\infty}(\beta e^{iu})^k = \frac{1-\beta}{1-\beta e^{iu}},$$

where we have used the summation formula for a geometric series.

There is a one-to-one relationship between characteristic functions and pdfs or pmfs.

Theorem 17.2.1 (Inversion of a Characteristic Function) Given a characteristic function ϕ_X, we can recover the pdf f_X through

$$f_X(x) = \frac{1}{2\pi}\int_{-\infty}^{\infty} e^{-iux}\phi_X(u)\,du.$$

Proof. It is clear from its definition that the characteristic function ϕ_X is also a variant (replace u by $-u$ or take the complex conjugate) of the usual *Fourier transform* of the density f_X. Hence, we appeal to the well-known Fourier inversion. If one wishes to avoid unit-impulse functions, that can be accomplished by the use of Fourier–Stieltjes transforms. In this approach, we establish the desired one-to-one relationship between a cdf F_X and ϕ_X. □

Hence, we can describe the probability measure P by a characteristic function as well as by a density or distribution function. Indeed, from ϕ_X, we can recover f_X through Fourier inversion, and from f_X, we can recover both $F_X(x)$ and $P(X \in A)$ by integrating over appropriate sets. The advantage of having several equivalent representations of the probability measure P is that, for each representation, there are problems for which it provides the simplest or most tractable approach. The following table gives the characteristic functions for common probability laws or models:

Model	$\mathcal{R}(n)$	$\mathcal{B}(n,p)$	$\mathcal{G}(\beta)$	$\mathcal{P}(\lambda)$
$\phi_X(\mathbf{u})$	$\dfrac{e^{iu}(1-e^{iun})}{n(1-e^{iu})}$	$(pe^{iu}+1-p)^n$	$\dfrac{1-\beta}{1-\beta e^{iu}}$	$e^{\lambda(e^{iu}-1)}$

Model	$\mathcal{U}(a,b)$	$\mathcal{E}(\alpha)$	$\mathcal{L}(\alpha)$	$\mathcal{N}(m,\sigma^2)$
$\phi_X(\mathbf{u})$	$e^{iu\frac{b+a}{2}}\dfrac{sin(u\frac{b-a}{2})}{u\frac{b-a}{2}}$	$\dfrac{\alpha}{\alpha-iu}$	$\dfrac{\alpha^2}{\alpha^2+u^2}$	$e^{ium-\frac{1}{2}\sigma^2u^2}$

Model	$\Gamma(\alpha,\lambda)$
$\phi_X(\mathbf{u})$	$\left(1-\dfrac{iu}{\lambda}\right)^{-\alpha}$

17.3 PROPERTIES OF THE CHARACTERISTIC FUNCTION

Elementary properties of a characteristic function include the following:

1.
$$\phi_X(0) = E^{i0X} = E1 = 1.$$

2. We now use the fact that the magnitude of an integral or a sum is less than or equal to the integral or sum of the magnitudes. From

$$|\phi(u)| = \left| \int_{-\infty}^{\infty} e^{iux}f_X(x)\,dx \right| \le \int_{-\infty}^{\infty} \left| e^{iux}f_X(x) \right| dx = \int_{-\infty}^{\infty} f_X(x)\,dx = 1,$$

we conclude that, for real-valued u,

$$|\phi_X(u)| \le 1 = \phi_X(0).$$

3. By the usual changes of variable in an integral or basic properties of expectation, we find that

$$\phi_{aX+b}(u) = Ee^{iu(aX+b)} = e^{iub}Ee^{iuaX} = e^{iub}\phi_X(au).$$

4. Recalling that the complex conjugate of $e^{i\alpha}$ is $e^{-i\alpha}$, we readily arrive at

$$\phi_X(-u) = Ee^{-iuX} = \left(Ee^{iuX}\right)^* = \phi_X^*(u).$$

5. Note that

$$|\phi_X(u + \epsilon) - \phi_X(u)| \le \int_{-\infty}^{\infty} |e^{i\epsilon x} - 1| f_X(x)\, dx.$$

We would like to take the limit of the left-hand side as ϵ goes to 0 to prove the continuity of the function $\phi_X(u)$. That conclusion will be correct if we can show that the right-hand side also has a limit of 0. Proving this last claim requires us to interchange the limit on ϵ with the integral so as to take the limit of the integrand. Such an interchange can be justified by the Dominated Convergence Theorem (see Section 0.8), since the integrand has an upper bound of $2f_X(x)$ that does not depend upon ϵ and that integrates to a finite value of 2. Hence, we conclude that

$$\phi_X(u) \text{ is a continuous function of } u.$$

6. A less obvious property, but one that essentially defines the class of characteristic functions, is that the function $\phi_X(u)$ is a *nonnegative definite* (related to the definition given in Section 9.14 for matrices).

Definition 17.3.1   **(Nonnegative Definite Function)**   The function $g(x)$ is nonnegative definite if

$$(\forall n)(\forall x_1, \ldots, x_n)(\forall a_1, \ldots, a_n) \sum_{i=1}^{n} \sum_{j=1}^{n} a_i a_j^* g(x_i - x_j) \ge 0,$$

where a^* is the complex conjugate of a.

These properties define the class of characteristic functions.

The next theorem is somewhat technical. However, what it tells us is that, under reasonable conditions, if a sequence of cdfs converges to a cdf F, then the corresponding characteristic functions also converge to a characteristic function ϕ that is the characteristic function of F and vice versa.

Relations between convergence of sequences of cdfs and convergence of the corresponding sequence of characteristic functions that will be needed to establish a Central Limit Theorem in Section 17.9 are given by Levy's Continuity Theorem.

Theorem 17.3.1   **(Levy's Continuity Theorem)**   Let $\{F_n\}$ be a sequence of cdfs with corresponding characteristic functions $\{\phi_n\}$. If there is a cdf F with characteristic function ϕ for which, at all x that are points of continuity of F (i.e., not jump points of F),

$$\lim_{n \to \infty} F_n(x) = F(x),$$

then, for all u,

$$\lim_{n \to \infty} \phi_n(u) = \phi(u).$$

Conversely, if there is some function ϕ that is continuous at 0 and is such that, for all u,

$$\lim_{n \to \infty} \phi_n(u) = \phi(u),$$

then ϕ is a characteristic function for some cdf F, and at all points x of continuity of F,

$$\lim_{n \to \infty} F_n(x) = F(x).$$

Proof citation. Proofs are available from many texts on advanced probability (e.g., Billingsley [10]). □

17.4 RELATIONSHIP BETWEEN CHARACTERISTIC FUNCTIONS AND MOMENTS

It is easy to calculate finite moments from a characteristic function.

Theorem 17.4.1   (Moments from Characteristic Functions)   If $|EX^k| < \infty$, then

$$EX^k = \frac{1}{i^k} \frac{d^k \phi_X(u)}{du^k} \bigg|_{u=0}.$$

Proof. The proof, which has some technical issues needed to verify the existence of the kth derivative of the characteristic function, is accomplished by justifying interchanging derivatives and expectation (which is like an integral) and then calculating. Thus,

$$\frac{d^k \phi_X(u)}{du^k} = \frac{d^k E e^{iuX}}{du^k} = E\left(\frac{d^k e^{iuX}}{du^k}\right) = E\left((iX)^k e^{iuX}\right).$$

Evaluating at $u = 0$ yields

$$\frac{d^k \phi_X(u)}{du^k}\bigg|_{u=0} = i^k EX^k. \qquad \qquad □$$

Example 17.2   Mean from Characteristic Function _____
From the table of characteristic function,

$$X \sim \Gamma(\alpha, \lambda) \Rightarrow \phi_X(u) = \left(1 - \frac{iu}{\lambda}\right)^{-\alpha}.$$

Hence,

$$\frac{d\phi_X(u)}{du} = \frac{i\alpha}{\lambda}\left(1 - \frac{iu}{\lambda}\right)^{-\alpha-1}.$$

$$EX = \frac{1}{i}\left.\frac{d\phi_X(u)}{du}\right|_{u=0} = \frac{1}{i}\frac{i\alpha}{\lambda} = \frac{\alpha}{\lambda}.$$

We have some words of caution. When the moment does not exist finite, applying the above can yield a finite and erroneous result. Furthermore, failure to divide by i^k yields either imaginary moments or even moments (e.g., the second moment) that are negative when they must be positive. Finally, failure to evaluate at $u = 0$ results in a function of u rather than the expected constant value for a moment.

It is sometimes more convenient to evaluate the moments from the logarithm of the characteristic function. Note that

$$\left.\frac{\partial \log(\phi_X(u))}{\partial u}\right|_{u=0} = \frac{1}{\phi_X(0)}\left.\frac{\partial \phi_X(u)}{\partial u}\right|_{u=0} = \left.\frac{\partial \phi_X(u)}{\partial u}\right|_{u=0} = iEX,$$

$$\frac{\partial^2 \log(\phi_X(u))}{\partial u^2} = -\frac{1}{(\phi_X(u))^2}\left(\frac{d\phi_X(u)}{du}\right)^2 + \frac{1}{\phi_X(u)}\frac{\partial^2 \phi_X(u)}{\partial u^2}.$$

Evaluating at $u = 0$ and using Theorem 17.4.1 yields

$$\left.\frac{\partial^2 \log(\phi_X(u))}{\partial u^2}\right|_{u=0} = (EX)^2 - EX^2 = -\text{VAR}(X).$$

17.5 EXTENSION TO RANDOM VECTORS

Definition 17.5.1 (Multivariate Characteristic Function) The multivariate characteristic function is defined by

$$\phi_{\mathbf{X}}(\mathbf{u}) = Ee^{i\mathbf{u}^T\mathbf{X}} = \int_{-\infty}^{\infty} dx_n \ldots \int_{-\infty}^{\infty} dx_1 e^{i\sum_{k=1}^{n} u_k x_k} f_{\mathbf{X}}(\mathbf{x}).$$

The Fourier inversion in this case is given by

$$f_{\mathbf{X}}(x_1, \ldots, x_n) = \frac{1}{(2\pi)^n}\int_{-\infty}^{\infty} du_n \ldots \int_{-\infty}^{\infty} du_1 e^{-i\sum_{k=1}^{n} u_k x_k} \phi_{\mathbf{X}}(u_1, \ldots, u_n).$$

A parallel to the property that we can find the cdf for a lower dimensional random vector that contains some of the components of a higher dimensional random vector from the cdf for the higher dimensional vector by setting some of the arguments to infinity (Section 7.2) (or densities for lower dimensional random vectors by integrating out on the unwanted components in the higher order density as in Section 7.3) is provided by setting selected arguments of the characteristic function to zero. We illustrate this property with the formula

$$\phi_{X_1, X_2}(u_1, u_2) = E^{i(u_1 X_1 + u_2 X_2)} = E^{i(u_1 X_1 + u_2 X_2 + 0 X_3)} = \phi_{X_1, X_2, X_3}(u_1, u_2, 0).$$

Thus, lower dimensional characteristic functions are easily constructed from higher dimensional ones without the need for integration, as was required for pdfs.

Higher order correlations are also easily calculated by repeated differentiation. Thus,

$$EX_k X_l = -\left. \frac{\partial^2 \phi_{X_k, X_l}(u_k, u_l)}{\partial u_k \partial u_l} \right|_{u_k = u_l = 0} = -\left. \frac{\partial^2 \phi_{\mathbf{X}}(\mathbf{u})}{\partial u_k \partial u_l} \right|_{\mathbf{u} = \underline{0}}.$$

Example 17.3   Correlation from Characteristic Function _____

Consider the bivariate characteristic function arising from a linear transformation on two *i.i.d.* $\mathcal{E}(1)$ random variables:

$$\phi_{X_1, X_2}(u_1, u_2) = (1 - i(u_1 + u_2))^{-1} (1 - i(u_1 - u_2))^{-1}$$

$$= g(u_1 + u_2) g(u_1 - u_2) \text{ for } g(v) = (1 - iv)^{-1},$$

$$\frac{\partial^2 \phi_{X_1, X_2}(u_1, u_2)}{\partial u_1 \partial u_2} = g^{(2)}(u_1 + u_2) g(u_1 - u_2) + g^{(1)}(u_1 + u_2) g^{(1)}(u_1 - u_2)$$

$$-g^{(1)}(u_1 + u_2) g^{(1)}(u_1 - u_2) - g(u_1 + u_2) g^{(2)}(u_1 - u_2),$$

$$E(X_1 X_2) = -\left. \frac{\partial^2 \phi_{X_1, X_2}(u_1, u_2)}{\partial u_1 \partial u_2} \right|_{u_1 = u_2 = 0}$$

$$= g^{(2)}(0) g(0) + [g^{(1)}(0)]^2 - [g^{(1)}(0)]^2 - g(0) g^{(2)}(0) = 0.$$

We did not need to evaluate the derivatives for the given function g to determine cancellation.

More generally, letting $p = \sum_1^n p_k$ for integer-valued $p_k \geq 0$, we have

$$E\left(\prod_1^n X_k^{p_k}\right) = \frac{1}{i^p} \left. \frac{\partial^p \phi_{\mathbf{X}}(\mathbf{u})}{\partial u_1^{p_1} \cdots \partial u_n^{p_n}} \right|_{\mathbf{u} = \underline{0}}.$$

17.6 LINEAR TRANSFORMATIONS AND THE MULTIVARIATE NORMAL REVISITED

The behavior of the multivariate characteristic function under linear transformations is given by

$$\mathbf{Y} = \mathbb{A}\mathbf{X} + \mathbf{b} \Rightarrow \phi_{\mathbf{Y}}(\mathbf{u}) = E e^{i\mathbf{u}^T \mathbf{Y}} = E e^{i\mathbf{u}^T (\mathbb{A}\mathbf{X} + \mathbf{b})}$$

$$= E\left(e^{i\mathbf{u}^T \mathbf{b}} e^{i(\mathbb{A}^T \mathbf{u})^T \mathbf{X}}\right) = e^{i\mathbf{b}^T \mathbf{u}} \phi_{\mathbf{X}}(\mathbb{A}^T \mathbf{u}),$$

where we have used the facts that $(\mathbb{A}\mathbb{B})^T = \mathbb{B}^T\mathbb{A}^T$ and that $\exp(i\mathbf{u}^T\mathbf{b})$ is nonrandom and can be brought outside the expectation operation.

Example 17.4   Characteristic Function for Independent Variables _____

If X_1 and X_2 are $i.i.d.$ $\mathcal{E}(1)$, then, using Theorem 14.7.1 to evaluate the expected value of a product of independent random variables as the product of the expected values, we obtain

$$\phi_{X_1,X_2}(u_1, u_2) = E e^{iu_1 X_1 + iu_2 X_2} = E e^{iu_1 X_1} E e^{iu_2 X_2} = \phi_{X_1}(u_1)\phi_{X_2}(u_2).$$

As they are both $\mathcal{E}(1)$,

$$\phi_{X_1,X_2}(u_1, u_2) = (1 - iu_1)^{-1}(1 - iu_2)^{-1}.$$

Let

$$\mathbb{A} = \begin{pmatrix} 1 & 1 \\ 1 & -1 \end{pmatrix}, \text{ and } \mathbf{Y} = \mathbb{A}\mathbf{X}.$$

We then find that $\phi_{Y_1,Y_2}(u_1, u_2)$ is as given in the preceding example.

If $E\mathbf{X} = \mathbf{m}_X$ and the covariance matrix $\mathbb{C}_X = E(\mathbf{X} - \mathbf{m})(\mathbf{X} - \mathbf{m})^T$, then determining means and correlations from the characteristic function $\phi_\mathbf{Y}$, as described at the close of the preceding section, yields

$$E\mathbf{Y} = \mathbb{A}\mathbf{m}_X + \mathbf{b}, \ E(\mathbf{Y} - E\mathbf{Y})(\mathbf{Y} - E\mathbf{Y})^T = \mathbb{A}\mathbb{C}_X\mathbb{A}^T,$$

as expected from Section 9.15.

We can now define $\mathbf{X}$ as being *multivariate normal* in a simpler fashion than that of the multivariate normal density of Section 7.4.4.

Definition 17.6.1   (Multivariate Normal)   The random vector $\mathbf{X}$ is multivariate normal if and only if every linear transformation $Y = \mathbf{a}^T\mathbf{X} + b$ to a scalar random variable Y results in $Y \sim \mathcal{N}(m_Y, \sigma_Y^2)$.

In this case, applying the immediately preceding results, we determine that

$$m_Y = \mathbf{a}^T\mathbf{m}_X + b, \ \text{VAR}(Y) = \sigma_Y^2 = \mathbf{a}^T\mathbb{C}_X\mathbf{a} = \sum_i \sum_j a_i a_j C_X(i, j).$$

Knowing $\mathbf{m}_X$ and $\mathbb{C}_X$ suffices to determine the mean and variance of Y, and the assumed normality of Y requires us only to specify that mean and variance. Proceeding further with the consequences of our definition of $\mathbf{X}$ being multivariate normal, we see that Y being scalar normal means

$$\phi_Y(u) = E e^{iuY} = e^{ium_Y} e^{-\frac{1}{2}u^2\sigma_Y^2}.$$

However, an alternative evaluation is

$$Ee^{iuY} = Ee^{iu(\mathbf{a}^T\mathbf{X}+b)} = e^{iub}\phi_{\mathbf{X}}(u\mathbf{a}).$$

Equating the two evaluations produces

$$e^{iub}\phi_{\mathbf{X}}(u\mathbf{a}) = e^{ium_Y}e^{-\frac{1}{2}u^2\sigma_Y^2} = e^{iu(\mathbf{a}^T\mathbf{m}_X+b)}e^{-\frac{1}{2}u^2\mathbf{a}^T\mathbb{C}_X\mathbf{a}}.$$

Setting $u = 1$, we discover that

$$\phi_{\mathbf{X}}(\mathbf{a}) = e^{i\mathbf{a}^T\mathbf{m}_X}e^{-\frac{1}{2}\mathbf{a}^T\mathbb{C}_X\mathbf{a}}$$

must hold for any choice of vector $\mathbf{a}$. Bowing to notational convention, we can replace the arbitrary vector $\mathbf{a}$ by $\mathbf{u}$ to obtain the characteristic function of the multivariate normal

$$\phi_{\mathbf{X}}(\mathbf{u}) = e^{i\mathbf{u}^T\mathbf{m}_X}e^{-\frac{1}{2}\mathbf{u}^T\mathbb{C}_X\mathbf{u}}.$$

We restate this conclusion as Lemma 17.6.1.

Lemma 17.6.1   (Multivariate Normal)   **X**, as specified in Definition 17.6.1 for a nonnegative definite matrix $\mathbb{C}_X$ and vector m_X, if and only if

$$\phi_{\mathbf{X}}(\mathbf{u}) = e^{i\mathbf{m}_X^T\mathbf{u}-\frac{1}{2}\mathbf{u}^T\mathbb{C}_X\mathbf{u}}.$$

This representation by characteristic function has the advantage over the density of not requiring the inverse of the covariance matrix $\mathbb{C}_X$ and thus being applicable in all cases. In addition, the definition does not require a confusing multiplicative factor that is the reciprocal of $(2\pi)^{n/2}\sqrt{\det\mathbb{C}_X}$. We exercise the results of the preceding section to verify that, for the given characteristic function of **X**, the mean is the vector $\mathbf{m}_X$ and the covariance matrix is $\mathbb{C}_X$, as expected for the multivariate normal:

$$EX_j = i^{-1}\left.\frac{\partial\log(\phi_{\mathbf{X}}(\mathbf{u}))}{\partial u_j}\right|_{\mathbf{u}=0} = m_j,$$

$$EX_jX_k = i^{-2}\left.\frac{\partial^2\phi_{\mathbf{X}}(\mathbf{u})}{\partial u_j\,\partial u_k}\right|_{\mathbf{u}=0} = m_jm_k + C_{jk}.$$

Hence, $\text{COV}(X_j, X_k) = C_{j,k}$, as expected.

Our defining condition that a random vector **X** is multivariate normal if and only if every linear transformation $Y = \mathbf{a}^T\mathbf{X} + b$ of **X** to a scalar yields a normally distributed scalar random variable has the following important practical implication:

Jointly normally distributed inputs to an arbitrary linear system yield normally distributed outputs.

For example, normally distributed noise (say, thermal noise) input and passed through a linear circuit or system is output as normally distributed noise.

17.7 CHARACTERISTIC FUNCTIONS AND INDEPENDENCE

Theorem 17.7.1   (Independence and Characteristic Functions) $\perp\!\!\!\perp_1^n X_i$ if and only if

$$\phi_{\mathbf{X}}(\mathbf{u}) = \prod_1^n \phi_{X_i}(u_i).$$

> **Remark.**   This is just a corollary to Theorem 14.7.1 about independence and expectation.

> *Product factorization of the multivariate characteristic function is an alternative definition of independence for random variables.*

Example 17.5   Laplacian Derivation _____
As an illustration, we provide a derivation of the Laplacian probability law $\mathcal{L}(\alpha)$. Assume that X, Y are $i.i.d.$ $\mathcal{E}(\alpha)$ with $Z = X - Y$. From the table, we find that

$$\phi_X(u) = \phi_Y(u) = \frac{\alpha}{\alpha - iu}.$$

Hence,

$$\phi_Z(u) = Ee^{iu(X-Y)} = Ee^{iuX} Ee^{-iuY} = \frac{\alpha^2}{\alpha^2 + u^2},$$

which is precisely the characteristic function of $\mathcal{L}(\alpha)$. This result suggests that if we identify a phenomenon as having a Laplacian probability law, then we might consider whether it arises as a difference of independent exponentially distributed variables.

17.8 CHARACTERISTIC FUNCTIONS OF SUMS

We demonstrate the simplicity of the calculation of a characteristic function for a sum of random variables. The sole requirement is that you can calculate the joint characteristic function of the summands, but doing so may be difficult when the summands are dependent. In this approach, you end up with a characteristic function as an answer and not a probability statement. Chapter 18 tells us how to easily convert characteristic functions into upper bounds on probabilities.

Lemma 17.8.1   (Characteristic Function of a Sum)   Define a weighted sum

$$S_n = \sum_1^n a_k X_k = \mathbf{a}^T \mathbf{X}.$$

1. For general random variables $\mathbf{X} = (X_1, \ldots, X_n)$,

$$\phi_{S_n}(u) = \phi_{\mathbf{X}}(\mathbf{a}u) = \phi_{X_1,\ldots,X_n}(a_1 u, \ldots, a_n u).$$

2. If the random variables are also independent, then

$$\phi_{S_n}(u) = \prod_1^n \phi_{X_k}(a_k u).$$

3. If the random variables are $i.i.d.$ and the sum has constant weights $a_k = a$, then

$$\phi_{S_n}(u) = \phi_{X_1}^n(au).$$

Proof. In the general case, we start from $\phi_{X_1,\dots,X_n}(u_1,\dots,u_n) = \phi_{\mathbf{X}}(\mathbf{u})$. From the definition of the multivariate characteristic function,

$$\phi_{S_n}(u) = Ee^{iuS_n} = Ee^{iu\sum_{k=1}^n a_k X_k} = \phi_{X_1,\dots,X_n}(a_1 u,\dots,a_n u).$$

If, in addition, the random variables are independent,

$$\phi_{S_n}(u) = E\prod_{k=1}^n e^{iu a_k X_k} = \prod_{k=1}^n \phi_{a_k X_k}(u) = \prod_1^n \phi_{X_k}(a_k u),$$

where we have used the independence of the summands to write an expectation of a product as a product of expectations of functions of independent random variables. If, in addition, the random variables are $i.i.d.$, then

$$\phi_{S_n}(u) = \prod_1^n \phi_{X_1}(a_k u).$$

If, further, the weights $a_k = a$, then

$$\phi_{S_n}(u) = \prod_1^n \phi_{X_1}(au) = \left(\phi_{X_1}(au)\right)^n. \qquad \square$$

The simplicity of this expression for ϕ_{S_n}, especially compared with that for F_{S_n}, reveals the advantages of the characteristic function representation for sums.

Example 17.6   Characteristic Function of a Sum _____
If $S_n = \mathbf{a}^T\mathbf{X}$ and $\mathbf{X} \sim \mathcal{N}(\mathbf{m}, \mathbb{C})$, then

$$\phi_{\mathbf{X}}(\mathbf{u}) = e^{i\mathbf{m}^T\mathbf{u} - \frac{1}{2}\mathbf{u}^T\mathbb{C}\mathbf{u}}.$$

From Lemma 17.8.1(1),

$$\phi_{S_n}(u) = \phi_{\mathbf{X}}(\mathbf{a}u) = e^{iu\mathbf{m}^T\mathbf{a} - \frac{1}{2}u^2\mathbf{a}^T\mathbb{C}\mathbf{a}}.$$

Note that this is a function of the single variable u.

If, in the previous equation, the random variables are also independent, so that $\mathbb{C}$ is diagonal with entry $C_{k,k} = \sigma_k^2$, then, from Lemma 17.8.1(2),

$$\phi_{S_n}(u) = \prod_1^n e^{iua_k m_k - \frac{1}{2}u^2 a_k^2 \sigma_k^2} = e^{iu \sum_1^n a_k m_k - \frac{1}{2}u^2 \sum_1^n a_k^2 \sigma_k^2}.$$

We recognize that $S_n \sim \mathcal{N}(\sum_1^n a_k m_k, \sum_1^n a_k^2 \sigma_k^2)$.

If $X_k \sim \mathcal{E}(\alpha)$ and they are $i.i.d.$ and $S_n = \sum_1^n X_k$, then, from Lemma 17.8.1(3),

$$\phi_{S_n}(u) = \left(\phi_{X_1}(u)\right)^n = \left(\frac{\alpha}{\alpha - iu}\right)^n.$$

Examining the table in Section 17.2 reveals that $S_n \sim \Gamma(n, \alpha)$.

17.9 CENTRAL LIMIT THEOREM

There are a variety of central limit theorems, and we discuss only the basic case of sums of $i.i.d.$ random variables. It is remarkable how such sums, satisfying just a condition of bounded variance summands, will tend towards a normal distribution when the number of summands grows and when the sum is properly scaled.

We now restrict the preceding result pertaining to the characteristic function of a sum to $i.i.d.$ mean zero summands (or achieve mean zero in what follows by replacing X_k by $X_k - EX_k$) with a constant weight a_n that depends only upon the number of summands:

$$S_n = a_n \sum_1^n X_k, \qquad \phi_{S_n}(u) = \phi_{X_1}^n(a_n u).$$

If we let $a_n \to 0$ as $n \to \infty$, then, for large numbers of $i.i.d.$ summands having a finite second moment, we can use a Taylor's series with remainder (o abbreviates "zero order"):

$$\phi_{X_1}(a_n u) = \phi_{X_1}(0) + \phi_{X_1}'(0)a_n u + \frac{1}{2}\phi_{X_1}''(0)a_n^2 u^2 + o(a_n^2 u^2).$$

Then we can use the relations between moments and derivatives of a characteristic function at zero (e.g., $EX_1 = 0$ implies $\phi_X'(0) = 0$ and removes the term in u) to obtain

$$\phi_{X_1}(a_n u) = 1 + 0 - \frac{1}{2}EX_1^2 a_n^2 u^2 + o(a_n^2 u^2).$$

The existence of a finite second moment ensures the existence of a continuous second derivative of the characteristic function. Furthermore, because the mean is zero, $EX_1^2 = \text{VAR}(X_1) = \sigma^2$. Hence,

$$\phi_{S_n}(u) = \left[1 - \frac{1}{2}\sigma^2 a_n^2 u^2 + o(a_n^2 u^2)\right]^n.$$

If we recall that

$$\lim_{n \to \infty} \left[1 + \frac{x}{n} \right]^n = e^x,$$

then we see that

$$\lim_{n \to \infty} \phi_{S_n}(u) = e^{-\frac{1}{2}\sigma^2 u^2 \lim_{n \to \infty} n a_n^2}.$$

The choice $a_n^2 = \frac{1}{n}$ yields

$$\lim_{n \to \infty} \phi_{S_n}(u) = e^{-\frac{1}{2}\sigma^2 u^2}.$$

In other words, the characteristic functions of $\{s_n\}$ converge to the characteristic function for a $\mathcal{N}(0, \sigma^2)$ random variable. We sum up the preceding in the so-called "equal components case" of the next theorem.

Theorem 17.9.1   (Central Limit Theorem)   If $\{X_k\}$ are $i.i.d.$ with $EX_k = m$, $VAR(X_k) = \sigma^2$, and

$$S_n = \frac{1}{\sqrt{n}} \sum_{k=1}^{n} (X_k - m),$$

then

$$\lim_{n \to \infty} \phi_{S_n}(u) = e^{-\frac{1}{2}\sigma^2 u^2},$$

and by the Levy Continuity Theorem,

$$\lim_{n \to \infty} F_{S_n}(s) = \int_{-\infty}^{s} \frac{1}{\sigma \sqrt{2\pi}} e^{-\frac{1}{2}\frac{x^2}{\sigma^2}} \, dx = \Phi\left(\frac{s}{\sigma} \right).$$

Proof remark. The formal proof of this requires the Levy Continuity Theorem (Theorem 17.3.1) to justify the transition from the convergence of characteristic functions to a function continuous at the origin ($u = 0$) to the convergence of the corresponding cdfs to the cdf corresponding to the limiting characteristic function. $\square$

Hence, properly normalized sums of $i.i.d.$ random variables of finite mean square have distributions which closely approximate that of the normal distribution. Figure 17.1 illustrates convergence to the normal distribution by computing and plotting the exact densities for normalized sums of $i.i.d.$ random variables that are uniformly distributed over $[-1, 1]$ and are themselves clearly nonnormal. It is evident that, while the densities for a few summands are far from normal, the density for 16 summands is similar to that of the normal density plotted by a dotted line.

The Central Limit Theorem suggests an explanation for the prevalence of normally distributed random variables in nature. Many macroscopic (directly observable) phenomena are a resultant (summation) of a large number of identically distributed random microscopic phenomena (e.g.,

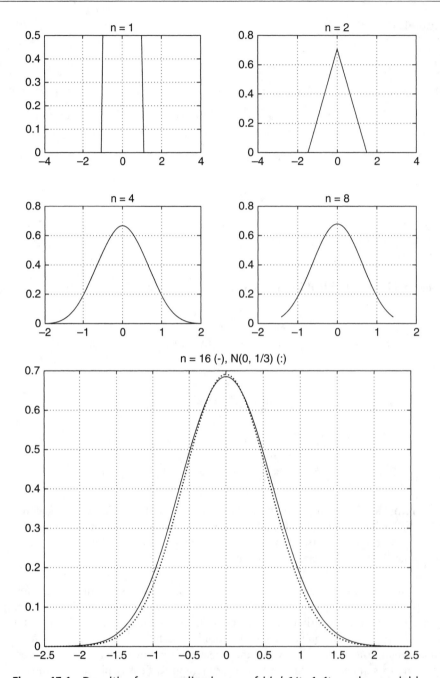

Figure 17.1   Densities for normalized sums of *i.i.d.* $\mathcal{U}(-1, 1)$ random variables.

signals emitted by individual electrons) acting independently. For example, the current flowing in a photodetector exposed to a nonnegligible light source will have a normal distribution for the fluctuations from the mean (DC) current. The photodetector current is the sum of current pulses produced by individual electrons that are released by the independent arrival of photons at the photodetector. As the photons arrive at random and act independently of each other, and because the observed current is the sum of their effects, the (properly normalized) current flowing at any particular time has a normal distribution, a phenomenon known as *shot noise*. For strong light sources, the mean of the current is so much larger than the standard deviation that we do not notice this shot noise effect.

17.10 RANDOM NUMBERS OF SUMMANDS

We refer to sums of a random number of random variables, or

$$S = \sum_{i=0}^{N} X_i,$$

where now N is a nonnegative, integer-valued random variable assumed to be independent of the summands $\{X_i\}$. For example, N could be the number of customers, packets, or jobs arriving in a queue in a given time interval and X_i might be the variable service time required to process the ith job. S then would be the total service time. Another application has N as the number of photons arriving at a photodetector in time T and X_i as the number of electrons released by the arrival of the ith photon; X_i might be 0 if the photon frequency is too low to release an electron, or X might be some random positive integer. S/T would be the average current generated by the photodetector exposed to this light source. In our applications, we will assume that $N = 0$ means that $S = 0$ or, equivalently, $X_0 = 0$ with a characteristic function $\phi_{X_0}(u) = 1$. From subconditioning,

$$ES = E\{E(S|N)\},$$

and

$$E(S|N = n) = \sum_{0}^{n} E(X_i|N = n).$$

If $N \perp\!\!\!\perp X_i$, then

$$E(S|N = n) = \sum_{0}^{n} EX_i.$$

If the random variables $\{X_i, i > 0\}$ have a common expectation/mean m, then

$$E(S|N = n) = nm, \quad ES = mEN.$$

For more detailed information about S, we calculate its characteristic function ϕ_S by assuming that $\perp\!\!\!\perp\{N, X_1, \ldots, X_n, \ldots\}$ and again using subconditioning:

$$\phi_S(u) = Ee^{iuS} = E\{E(e^{iuS}|N)\}, \quad E(e^{iuS}|N = n) = \prod_0^n \phi_{X_i}(u),$$

$$\phi_S(u) = P(N = 0) + \sum_{n=1}^{\infty} P(N = n) \prod_{i=1}^n \phi_{X_i}(u).$$

If the summands are also identically distributed (except for X_0) with common characteristic function ϕ_X, then

$$\phi_S(u) = \sum_{n=0}^{\infty} P(N = n)\phi_X^n(u),$$

where we have used $\phi_X^0(u) = 1 = \phi_{X_0}(u)$. Note that the characteristic function

$$\phi_N(u) = \sum_{n=0}^{\infty} P(N = n)e^{iun} = \sum_{n=0}^{\infty} P(N = n)[e^{iu}]^n.$$

Comparing the expressions for ϕ_S and ϕ_N, we see that choosing u in $\phi_N(u)$ so that $e^{iu} = \phi_X$ enables us to write

$$\phi_S(u) = \phi_N(-i \log \phi_X(u)).$$

We have now proven the following theorem.

Theorem 17.10.1   (Random Numbers of Summands)   If N is a nonnegative, integer-valued random variable, $S = \sum_{i=0}^N X_i$, $X_0 = 0$, the $\{X_i, i \geq 1\}$ are $i.i.d.$ with common characteristic function ϕ_X, and they are independent of the random number of summands N described by characteristic function ϕ_N, then

$$\phi_S(u) = \phi_N(-i \log \phi_X(u)).$$

From this equation, we can conclude that

$$ES = EXEN,$$

although this conclusion was derived at the outset without the assumption that the $\{X_i\}$ are independent of each other.

Example 17.7   Random Numbers of Summands _____
In many cases, $N \sim \mathcal{P}(\lambda)$; therefore, from

$$\phi_N(u) = e^{-\lambda + \lambda e^{iu}},$$

we obtain

$$\phi_S(u) = e^{-\lambda + \lambda \phi_X(u)}.$$

As a check on our answer,

$$\phi_S(0) = e^{-\lambda + \lambda \phi_X(0)} = e^{-\lambda + \lambda} = 1.$$

17.11 GENERATING FUNCTIONS

When the random variable X is integer valued, as is the case in many counting applications, then we sometimes use a function related to the characteristic function and given in the next definition.

Definition 17.11.1   (Generating Function)   The generating function (GF)

$$G_X(s) = Es^X = \sum_{k=0}^{\infty} p_k s^k,$$

where p_k is the pmf for the nonnegative integer-valued X.

Example 17.8   Calculation of a Generating Function _____

If $X \sim \mathcal{B}(n, p)$, then its generating function

$$G_X(s) = \sum_{k=0}^{n} \binom{n}{k} p^k (1-p)^{n-k} s^k = (1-p)^n \sum_{k=0}^{n} \binom{n}{k} \left(\frac{ps}{1-p} \right)^k$$

$$= (1-p)^n \left(1 + \frac{ps}{1-p} \right)^n = (1 - p + ps)^n.$$

$G_X(s)$ is finite for any $s \in [-1, 1]$. If $G_X(s) < \infty$ for some $s > 1$, then all moments of X exist finite. Evidently, the GF

$$G_X(e^{iu}) = \phi_X(u)$$

and corresponds to a z transform of the pmf in place of a Fourier transform of the pdf. Generating functions have been well studied (e.g., see Feller [27]) and are widely used as analytical tools in combinatorial analysis.

There is a one-to-one relationship between pmfs for nonnegative integer-valued random variables and generating functions.

Theorem 17.11.1   (Inversion of Generating Function)   We can recover the pmf $\{P(X = k) = p_k\}$ from $G_X(s)$ through a Taylor's series (power series) expansion of G_X about $s = 0$ with

$$p_k = \frac{G_X^{(k)}(0)}{k!}.$$

Proof. Equating the Taylor's series

$$G_X(s) = \sum_{k=0}^{\infty} \frac{G_X^{(k)}(0)}{k!} s^k$$

to the definition

$$G_X(s) = \sum_{k=0}^{\infty} p_k s^k$$

immediately yields (by the uniqueness of power series representations)

$$p_k = \frac{G_X^{(k)}(0)}{k!}.$$

Of course, this result requires that, at $s = 0$, the GF have derivatives of all orders. Indeed, the GF has derivatives of all orders so long as $|s| < 1$. $\square$

Example 17.9   PMF from Generating Function _____
If

$$G_X(s) = \frac{1}{8}(s + 1)^3,$$

then, expanding the formula in a power series

$$G_X(s) = \frac{1}{8}(1 + 3s + 3s^2 + s^3),$$

we see that G_X is the generating function for a pmf

$$P(X = 0) = p_0 = \frac{1}{8}, \ p_1 = \frac{3}{8}, \ p_2 = \frac{3}{8}, \ p_3 = \frac{1}{8}.$$

Basic properties of G_X, analogous to the ones provided in Section 17.3, are as follows:

1. From $\sum_k p_k = 1$,

$$G_X(1) = 1.$$

2. For $s \in [0, 1]$, the nonnegativity of the coefficients $\{p_k\}$ implies that $G_X(s)$ is increasing in s; therefore,

$$G_X(s) \le G_X(1).$$

3. $G_X(s)$ is a continuous function of $s \in [0, 1]$.

4. For integer-valued X,

$$EX = G_X^{(1)}(1),$$

possibly infinite. Higher order moments are also computable from higher order derivatives, albeit with the need for a little additional algebra.
5. It is again true that if X and Y are independent, nonnegative integer-valued random variables and $Z = X + Y$, then

$$G_Z(s) = G_X(s)G_Y(s).$$

A novel property of GFs is the ease with which we can determine the GF for the probabilities of exceeding a given integer. Let $T_X(s)$ be the GF of the tail probabilities

$$t_k = P(X > k), \ T_X(s) = \sum_{k=0}^{\infty} t_k s^k.$$

We have the following lemma.

Lemma 17.11.1   (Tail Probabilities)

$$T_X(s) = \frac{1 - G_X(s)}{1 - s}.$$

Proof. For $k = 0$, $t_0 = P(X > 0) = 1 - p_0$. Following Feller [27], p. 249, observe that, for $k \geq 1$, the coefficient of s^k in $(1 - s)T_X(s) = t_k - t_{k-1} = P(X > k) - P(X > k - 1) = -P(X = k) = -p_k$. Hence, $(1 - s)T_X(s) = 1 - G_X(s)$. $\qquad\qquad\square$

Example 17.10   Tail Probabilities from Generating Function _____
The GF for $X \sim \mathcal{P}(\lambda)$ is

$$G_X(s) = Es^X = \sum_{k=0}^{\infty} e^{-\lambda} \frac{(s\lambda)^k}{k!} = e^{-\lambda + s\lambda}.$$

Hence, the tail probabilities for the Poisson are given by

$$T_X(s) = \frac{1 - e^{-\lambda(1-s)}}{1 - s}.$$

Recalling that $EX = G_X^{(1)}(1)$ and using L'Hôpital's rule reveals the alternative expression

$$EX = T_X(1).$$

Multivariate generating functions are easily defined, with the bivariate case being

$$G_{X_1,X_2}(s_1, s_2) = E(s_1^{X_1} s_2^{X_2}).$$

17.12 SUMMARY

The introduction of expectation in Chapter 9 and our treatment of functions of random variables in Chapter 8 together enable us to introduce the notion of a characteristic function $\phi_X(u)$ as yet another description of a cdf F_X. Characteristic functions provide a convenient way to calculate moments. As shown in Section 17.4, moments can now be calculated from derivatives of the characteristic function evaluated at the origin and do not require the often difficult integration associated with the pdf description f_X of a random variable X.

Characteristic functions also provide a useful tool for dealing with commonly encountered sums of random variables (e.g., the output of integrating systems, the macroscopic effect of the collective action of many microscopic random variables). Furthermore, characteristic functions provide an analytical tool that is well adapted to the study of sums of a very large number of independent random variables. The practical importance of this can be seen in that the typical behavior of the collective processes lying at the root of many random phenomena can be understood as the macroscopically observable response being a summation of very many independently acting unobservable phenomena of small magnitude. More formally, the response random variable is a sum of independent random variables. Understanding and describing the response random variable requires us to calculate all or part of the probabilistic description of such sums. Section 8.6.1 provided one such approach through the analysis developed there to determine the density of a linear transformation of a random vector. Section 10.4 provided the partial description of such a sum through its moments. A convenient full probabilistic description awaited our introduction of the characteristic function of a sum. The use of characteristic functions also enables us to derive powerful asymptotic results for cases of very many summands—results that depend only weakly on the precise distributions of the individual summands. These results include a (very) weak form of the law of large numbers for averages and the basic case of the Central Limit Theorem.

The Central Limit Theorem offers an argument for the prevalence of the normal or Gaussian distribution in macroscopic natural phenomena. Indeed, as discussed in Section 17.6, characteristic functions provide the easiest means for defining the important family of multivariate normal models, a means that does not require the covariance matrix to have an inverse.

Conditional expectation combined with characteristic functions enables us to treat the novel problem of a sum of a random number of random variables; hitherto, our sums of random variables had a fixed and known number of summands. Such problems of a random number of summands arise, for example, when a random number of events occur (e.g., the number of photons arriving in a given time interval from a light source or the number of customers or digital communication packets arriving in a queue in a given time interval) and each of them produces a possibly random response (e.g., the number of electrons emitted by a photodetector in response to an individual photon or the service time required by a given customer).

The chapter closed with a special case of a characteristic function, known as a generating function, for random variables that take on only nonnegative integer values. Such "counting" random variables arise in combinatorial problems (e.g., see Chapter 2), and generating functions provide a useful technique for their solution.

EXERCISES

E17.1   If $P(X = 1) = 1 - P(X = 0) = p$, evaluate $\phi_X(u)$.

E17.2   a. Compute the characteristic function $\phi_X(u)$ for $X \sim \mathcal{U}(-a, a)$.

        b. Compute the characteristic function for the Laplacian, $X \sim \mathcal{L}(\alpha)$.

E17.3   a. Compute the characteristic function for $X \sim \mathcal{B}(n, \frac{1}{2})$.

        b. Compute the characteristic function for $X \sim \mathcal{P}(\lambda)$.

E17.4   If X is an integer-valued random variable with $P(X = k) = p_k$ and characteristic function ϕ_X, then show that ϕ_X is periodic by evaluating $\phi_X(u) - \phi_X(u + 2\pi)$. (Carry out any simplifications.)

E17.5   If $\phi_X(u) = 1/(1 + u^2)$, evaluate VAR(X).

E17.6   If

$$\phi_X(u) = \left(\frac{1}{1 + u^2}\right)^2,$$

then evaluate $VAR(X)$.

E17.7   Determine EX and $VAR(X)$ from the characteristic function

$$\phi_X(u) = (1 - iu)^{-2}.$$

E17.8   a. Evaluate EX, EX^2, and VAR(X) for $X \sim \mathcal{B}(n, p)$ from its characteristic function.

        b. Repeat part (a) for $X \sim \mathcal{P}(\lambda)$.

E17.9   Evaluate EX, EX^2, VAR(X), and EX^3 from the characteristic function for $X \sim \mathcal{N}(m, \sigma^2)$.

E17.10   a. If a random variable X has the characteristic function

$$\phi_X(u) = \frac{(1 - \alpha)^2}{1 + \alpha^2 - 2\alpha \cos(u)}, \quad \text{for } 0 < \alpha < 1,$$

then evaluate EX and VAR(X).

        b. If $\{X_1, X_2\}$ are $i.i.d.$ as X of part (a), with $\alpha = .5$, then evaluate the joint characteristic function $\phi_\mathbf{X}(\mathbf{u}) = \phi_{X_1, X_2}(u_1, u_2)$.

E17.11   a. Professor Phynne claims that

$$\phi_X(u) = \frac{1}{1 + u^4}$$

is a characteristic function. Evaluate VAR(X) from ϕ_X.

        b. Is Professor Phynne correct that ϕ is a characteristic function? (Give precise reasons for your answer.)

E17.12   a. Show that $\phi_X(u) = e^{-u^4}$ cannot be a characteristic function by evaluating EX^2.

        b. Which of the properties of characteristic functions stated in Section 17.3 do not hold?

E17.13   A Brownian motion particle with mass m moving with a velocity $\mathbf{V}$ has kinetic energy

$$K = \frac{1}{2} \sum_{i=1}^{3} m V_i^2,$$

where the components $\{V_i\}$ of $\mathbf{V}$ are $i.i.d.$ $\mathcal{N}(0, \sigma^2)$. Evaluate the characteristic function ϕ_K in terms of $\phi_{V_1^2}$. (Note that the density for $2K/m\sigma$ is chi squared with three degrees of freedom, or χ_3^2.)

E17.14   If

$$\phi_{\mathbf{X}}(\mathbf{u}) = e^{i2u_1 - u_1^2 - u_2^2 + u_1 u_2},$$

then evaluate $EX_1 X_2$, and $\phi_{X_1}(u)$.

E17.15   A photodetector is exposed to the light from three unrelated sources N_1, N_2, and N_3 of respective intensities I_1, I_2, and I_3.

   a. Making appropriate assumptions, provide the characteristic function ϕ_N for the total number $N = N_1 + N_2 + N_3$ of incident photons in time T.
   b. What is the probability that zero photons will be observed?
   c. What is the most probable number of photons to be observed?

E17.16   a. The random variables $\{X_1, \ldots, X_n\}$ are $i.i.d.$ Assuming knowledge of ϕ_X, determine the characteristic function $\phi_{S_n}(u)$ for the weighted sum

$$S_n = \frac{1}{\sqrt{n}} \sum_{k=1}^{n} X_k.$$

   b. Determine the characteristic function for $S_n - ES_n$.

E17.17   A system $\mathcal{S}$ having input X adds independent noise $N \sim \mathcal{L}(1)$ to X to produce its output Y.

   a. Characterize the system by calculating $f_{Y|X}$.
   b. If the characteristic function of the output is

$$\phi_Y(u) = \frac{e^{-u^2}}{1 + u^2},$$

   then what is the input density f_X?

E17.18   A discrete-time integrator accepts inputs $\{X_i\}$ and yields outputs

$$Y_n = \sum_{j=n-L}^{n} a_j X_j.$$

   a. If $\{X_i\}$ are independent with $\phi_{X_i}(u) = a_i^2/(a_i^2 + u^2)$ for $a_i > 0$ (the Laplacian), then evaluate ϕ_{Y_n}.
   b. Evaluate EY_n and $\text{VAR}(Y_n)$ by using ϕ_{Y_n}.

E17.19   In a certain system $\mathcal{S}$, the input X and output Y have a joint characteristic function

$$\phi_{X,Y}(u, v) = e^{-\lambda_0 + \lambda_0 e^{iu}} e^{-\lambda_1 + \lambda_1 e^{i(u+v)}}.$$

   a. Describe the output Y by providing its characteristic function.
   b. Evaluate EY.
   c. Evaluate the correlation $E(XY)$.
   d. Determine the characteristic function $\phi_Z(u)$ for $Z = Y - X$.

E17.20 The two voltage sources in a circuit have amplitudes V_1 and V_2, with means of 1 and a covariance matrix

$$\mathbb{C}_{\mathbf{V}} = \begin{pmatrix} 2 & 1 \\ 1 & 3 \end{pmatrix}.$$

There are two circuit currents I_1 and I_2, related to $\mathbf{V}$ through $\mathbf{I} = \mathbb{G}\mathbf{V}$, where

$$\mathbb{G} = \begin{pmatrix} 5 & -2 \\ -2 & 3 \end{pmatrix}.$$

    a. Calculate $E\mathbf{I}$ and the covariance matrix $\mathbb{C}_{\mathbf{I}}$.

    b. If V_1 and V_2 are jointly normally distributed, then evaluate the characteristic function $\phi_{\mathbf{I}}$.

E17.21   a. If Y_1, Y_2 is the response of a linear system given by

$$\mathbf{Y} = \begin{pmatrix} Y_1 \\ Y_2 \end{pmatrix} = \begin{pmatrix} 1 & 2 \\ 3 & 1 \end{pmatrix} \begin{pmatrix} Z_1 \\ Z_2 \end{pmatrix}, \; Z_1 \perp\!\!\!\perp Z_2, \; Z_1 \sim \mathcal{N}(1,2), \; Z_2 \sim \mathcal{N}(0,6),$$

    then evaluate $E\mathbf{Y}$.

    b. Evaluate the joint characteristic function $\phi_{\mathbf{Y}}(\mathbf{u}) = \phi_{Y_1,Y_2}(u_1, u_2)$ for $\mathbf{Y}$ in the preceding equation.

E17.22 In a given circuit $\mathbf{V} = \mathbb{Z}\mathbf{I}$, we know that

$$\phi_{\mathbf{I}}(\mathbf{u}) = e^{-u_1^2 - u_2^2 + u_1 u_2}, \quad \mathbb{Z} = \begin{pmatrix} 4 & -2 \\ -2 & 3 \end{pmatrix}.$$

    a. Evaluate $COV(I_1, I_2)$.

    b. Evaluate $\phi_{\mathbf{V}}(\mathbf{u})$.

E17.23 In a given circuit, the mesh currents $\mathbf{I}$ are related to the source voltages $\mathbf{V}$ through a conductance matrix

$$\mathbb{G} = \begin{pmatrix} 2 & -1 \\ -1 & 3 \\ 0 & -1 \end{pmatrix}, \quad \mathbf{I} = \mathbb{G}\mathbf{V}.$$

    a. If the source voltages are $i.i.d.$ with common characteristic function $\phi_V(u) = e^{-|u|}$, then evaluate $\phi_{\mathbf{V}}(\mathbf{u})$.

    b. Evaluate $\phi_{\mathbf{I}}$.

E17.24 In a given circuit, the mesh currents $\mathbf{I}$ are related to the source voltage $V \sim \mathcal{N}(m, \sigma^2)$ through

$$\mathbf{I} = \begin{bmatrix} 2 \\ 1 \end{bmatrix} V.$$

Determine the characteristic function $\phi_{\mathbf{I}}$.

E17.25 In a given circuit, the node voltages $\mathbf{V}$ are related to the source currents $\mathbf{I}$ through the impedance matrix

$$\mathbf{V} = \mathbb{Z}\mathbf{I}, \; \mathbb{Z} = \begin{pmatrix} 3 & -1 \\ -1 & 2 \end{pmatrix}.$$

We are informed that

$$\phi_{\mathbf{I}}(\mathbf{u}) = \frac{e^{-u_1^2}}{1 + u_2^2}.$$

a. Evaluate EI_2.
b. Evaluate $\phi_{\mathbf{V}}(\mathbf{u})$.
c. Are V_1 and V_2 independent?

E17.26   In a linear, discrete-time system the d-dimensional vector of state variables $\mathbf{X}_k$ at time k is related to the vector of state variables $\mathbf{X}_{k+1}$ at the next time $k + 1$ through

$$\mathbf{X}_0 = \mathbf{Z}_0, \quad \mathbf{X}_{k+1} = \mathbb{A}\mathbf{X}_k + \mathbf{Z}_{k+1} \text{ for } k \geq 0,$$

where $\mathbb{A}$ is a known nonrandom square matrix and the random vectors $\{\mathbf{Z}_k\}$ are $i.i.d.$ with common characteristic function

$$\phi_{\mathbf{Z}}(\mathbf{u}) = e^{-\mathbf{u}^T \mathbf{u}}.$$

Evaluate the characteristic function $\phi_{\mathbf{X}_1}$ of $\mathbf{X}_1$.

E17.27   We have

$$\phi_{Y_1, Y_2}(u_1, u_2) = \frac{e^{-u_1^2 - u_2^2}}{1 + (u_1 + u_2)^2}.$$

a. Evaluate ϕ_{Y_1} and ϕ_{Y_2}.
b. Is $Y_1 \perp\!\!\!\perp Y_2$?
c. Evaluate EY_1.
d. Evaluate $VAR(Y_1)$.
e. If

$$\mathbf{Z} = \begin{pmatrix} Z_1 \\ Z_2 \end{pmatrix} = \begin{pmatrix} Y_1 - Y_2 \\ Y_1 + 2Y_2 \end{pmatrix},$$

evaluate $\phi_{Z_1, Z_2}(v_1, v_2)$.

E17.28   a. If the characteristic function $\phi_{\mathbf{X}}(\mathbf{u})$ for a d-dimensional random vector $\mathbf{X}$ is given by

$$\phi_{\mathbf{X}}(\mathbf{u}) = \exp\left(i \sum_{k=1}^{d} k u_k - \sum_{k=1}^{d} k u_k^2\right),$$

then evaluate the characteristic function for the second component X_2 of $\mathbf{X}$.
b. For the setup of (a), evaluate EX_2 and $VAR(X_2)$.

E17.29   If X, Y are $i.i.d.$ Cauchy, verify that $2X = X + X$ has the same characteristic function as $Z = X + Y$. (This is a curious instance in which the sum of two $i.i.d.$ random variables is identically distributed to a sum in which the two variables are completely dependent.)

E17.30   A motion-detecting, narrow-beam infrared (IR) detector looking at low-level background optical radiation produces an output Y that is the difference $X_1 - X_2$ of two $i.i.d.$ photon sources. What is the characteristic function $\phi_Y(u)$?

E17.31 Verify that the Laplacian $\mathcal{L}(\alpha)$ is the correct probability law for the difference $D = I_1 - I_2$ in intensity levels between adjacent pixels by postulating that the intensity levels are $i.i.d.$ $\mathcal{E}(\alpha)$.

E17.32 In an Aloha protocol, two callers whose packets have collided are assigned $i.i.d.$ retransmission times R_1 and R_2 that are distributed as $\mathcal{U}(0, \tau)$. Retransmission is successful if $|R_1 - R_2| > \theta$.

    a. Calculate the characteristic function ϕ_{R_1}.

    b. Calculate the characteristic function ϕ_Z for $Z = R_1 - R_2$.

E17.33 If $\{X_i\}$ are $i.i.d.$ Cauchy with characteristic function $e^{-|u|}$ and $S_n = (1/n)\sum_1^n X_i$, then evaluate the characteristic function ϕ_{S_n}. (Note from this that, in the absence of a finite mean, the average of even a very large number of $i.i.d.$ random variables need not stabilize or converge to a constant!)

E17.34 $\{X_i\}$ are $i.i.d.$ with $P(X = 1) = p = 1 - P(X = 0)$.

    a. Consider the average

$$A_n = \frac{1}{n}\sum_{i=1}^{n} X_i,$$

       and evaluate its characteristic function $\phi_{A_n}(u)$.

    b. Determine the limiting value of $\phi_{A_n}(u)$ as n grows, and conclude that the distribution of the average converges to a particular constant.

    c. Consider the normalized sum

$$S_n = \frac{1}{\sqrt{n}}\sum_{i=1}^{n}(X_i - p),$$

       and evaluate its characteristic function $\phi_{S_n}(u)$.

    d. Determine the limiting value of $\phi_{S_n}(u)$ as n grows, and identify the corresponding distribution.

E17.35 In a compound Poisson model for Unix netstat, we have a number C of client processes, each of which generates numbers N_i of packets that are $i.i.d.$ $\mathcal{P}(\lambda)$ and are independent of C. The total number of packets $N = \sum_1^C N_i$ if $C > 0$, and $N = 0$ if $C = 0$. The characteristic function $\phi_{N_i}(u) = e^{-\lambda(1-e^{iu})}$.

    a. If $C = c > 0$, determine ϕ_N and repeat for $C = 0$.

    b. If $C \sim \mathcal{P}(\gamma)$, then evaluate $P(N = n)$.

E17.36 The number N of γ-rays incident on a particular material in time T is random and described by a $\mathcal{G}(\beta)$ distribution. The ith incident γ-ray produces a random number X_i of α-particles, where X_i is described by a characteristic function $\phi_X(u)$. Assuming independent effects from different γ-rays and making reasonable assumptions, evaluate the characteristic function ϕ_A for the random number A of α-particles emitted in time T.

E17.37 The random number X_i of electrons emitted from a particular material due to the ith incident photon is described by

$$P(X_i = 0) = 1/3, \ P(X_i = 1) = 2/3,$$

and the responses to different photons are independent. The number N of photons incident in time T is random and described by the Poisson $\mathcal{P}(\lambda T)$ distribution.

    a. Evaluate the characteristic function of the number S of emitted electrons in time T.

    b. Evaluate the expected value of the time-average current

$$I = \frac{1}{T} \sum_{i=1}^{N} X_i$$

    flowing in time T.

E17.38 In a given period T, a random number N of symbols are transmitted by a terminal. The probability of exactly n symbols being transmitted is $(\frac{1}{2})^{n+1}$. Errors are made independently from symbol to symbol with probability .01, and the errors are made independently of the number of symbols transmitted in T.

    a. Evaluate the characteristic function $\phi_S(u)$ for the total number S of symbols in error that have been transmitted in T. (It may help to let X_i be 1 if there is an error in the ith symbol and be 0 if otherwise.)

    b. Evaluate ES.

E17.39 A random number N of packets of varying bit-lengths $\{L_i\}$ are received in time T in a certain digital communication network, where

$$N \sim \mathcal{P}(2T), \quad \{L_i\} \ i.i.d. \ \phi_L(u) = \frac{1}{n} \frac{e^{iun} - 1}{e^{iu} - 1}.$$

Evaluate the characteristic function for the total number B of bits received in time T.

E17.40 For the design of a cache size, we need to know the amount of storage that might be required to cover a period of T seconds. Assume that the number N_T of file sizes requested is Poisson with an average of λ files per second. Assume further that files sizes S_i in bytes are $i.i.d.$ with $S \sim \mathcal{G}(\beta)$ (the Zeta/Zipf would be more realistic, but the characteristic function is awkward) and independent of the number of requests. Let B_T denote the total number of bytes received in time T.

    a. Determine the expected number EB_T of bytes arriving to be stored in time T.

    b. Determine the characteristic function for B_T.

E17.41 We have

$$Y = \sum_{1}^{N} X_i, \ \perp\!\!\!\perp \{N, X_1, \ldots\}, \ N \sim \mathcal{P}(\lambda_0), X_i \sim \mathcal{P}(\lambda_1).$$

    a. Evaluate EY.

    b. Evaluate ϕ_Y.

    c. Is Y Binomial, Poisson, or Geometric?

E17.42 In a certain situation, the random opportunities for performing tasks are such that the number N of such tasks that can be performed in an allotted time T is described by $N \sim \mathcal{R}_{100}$ (a uniform distribution over $0, 1, \ldots, 99$). The ith task has a payoff X_i with $P(X_i = -1) = p = 1 - P(X_i = 1)$. N and $X_1, X_2, \ldots$ are mutually independent.

a. Evaluate $\phi_Y(u)$ for the cumulative payoff

$$Y = \sum_{i=1}^{N} X_i .$$

b. Evaluate VAR(Y) from ϕ_Y.

E17.43 If $P(X = 1) = 1 - P(X = 0) = p$, evaluate the generating function $G_X(s)$.

E17.44 Using the table of characteristic functions, determine the GF $G_X(s)$ for the following and check your results by evaluating $G_X(1)$:

a. $X \sim \mathcal{R}(n)$;
b. $X \sim \mathcal{B}(n, p)$;
c. $X \sim \mathcal{P}(\lambda)$.

18

Probability Bounds and Sums

18.1 PURPOSE, BACKGROUND, AND ORGANIZATION

The techniques introduced in this chapter provide useful approximate answers to questions about evaluating probabilities that are either too difficult to compute by exact methods or ones in which the information needed for an exact answer is lacking. The moments and characteristic functions that we have introduced are computed from knowledge of P, F_X or f_X, but do not have a direct interpretation in terms of probabilities. It is frequently the case—for example, in dealing with sums of many random variables—that we know only selected moments or can compute the characteristic function, but are unable to determine the complete probability specification. The bounds we introduce will enable us to convert information on moments and characteristic functions into upper bounds on probabilities of certain events. As we shall see, for instance, in our discussion of the laws of large numbers in Chapter 19, these upper bounds enable us to draw useful conclusions about the behavior of random variables and their sums without requiring full knowledge of the probability specification. In much of practice, we have only partial information concerning P and need to be able to convert that partial information into bounds on, or estimates of, the probabilities of events of interest. We illustrate the utility of these bounds by evaluating them in the important case of sums of independent random variables. The bounds are used in a wide variety of applications, including information theory and recent work on statistical machine learning (e.g., Fine [31]).

18.2 GENERALIZED CHEBYCHEV BOUNDS

Theorem 18.2.1   (Generalized Chebychev Bound)   Given a set A, select a function $g(x)$ such that

$$(\forall x)\; g(x) \geq I_A(x).$$

It then follows that

$$P(X \in A) \leq \min(1, Eg(X)).$$

Proof. We recall the following two facts:

$$P(A) = EI_A(X) \text{ and } P(Y \geq X) = 1 \Rightarrow EY \geq EX.$$

Hence, if we select a function g with the hypothesized property, then we can conclude with the upper bound

$$Eg(X) \geq EI_A(X) = P(X \in A).$$

However, because the upper bound can exceed 1,

$$\min(1, Eg(X)) \geq P(X \in A). \qquad \square$$

It is easy to verify, for a given event A and function g, whether the first condition of the theorem holds. The issues are whether we can evaluate $Eg(X)$ and how close the upper bound is to the probability.

We focus on the following two types of events:

- One-Sided Case: $A = \{x : x \geq \tau\}$;
- Two-Sided Case: $A = \{x : |x| \geq \tau\}$.

Example 18.1   Original Chebychev Bounds _____

Our first example of a suitable function is

$$g(x) = \frac{x^2}{\tau^2}.$$

In this case, we find from the argument just outlined, that

$$\frac{EX^2}{\tau^2} \geq P(|X| \geq \tau) \geq P(X \geq \tau).$$

Equivalently, if we replace X by $X - EX$, then

$$\frac{\text{VAR}(X)}{\tau^2} \geq P(|X - EX| \geq \tau) \geq P(X - EX \geq \tau).$$

These bounds are the original heavily used *Chebychev bounds* and convert knowledge of a second moment or variance into an upper bound on a probability that a random variable X is "large" ($\geq \tau$).

The original Chebychev bounds decrease relatively slowly with increasing τ; their so-called "tail behavior" is that of slow algebraic (inverse quadratic) convergence to zero. Given what we have assumed to this point—just the existence of a finite second moment—it is not possible to improve on this result. The Chebychev bound is *tight*; that is to say, for each value of τ, there is a distribution P that achieves equality in the bound. Consider

$$P(X = \tau) = P(X = -\tau) = \frac{1}{2}.$$

Then

$$EX = 0, \ \ \text{VAR}(X) = \tau^2, \ \ P(|X| \geq \tau) = 1 = \frac{\text{VAR}(X)}{\tau^2}.$$

If we have additional knowledge that, say, $P(X \geq \tau)$ decreases at least exponentially fast in τ, then we can use Theorem 18.2.1 to find an upper bound with more rapidly decreasing tail behavior.

Example 18.2   Exponential Chebychev Bounds _____

Choose

$$g(x) = e^{\lambda(x-\tau)} \ \text{ for some } \lambda \geq 0.$$

This choice implies that

$$e^{-\lambda\tau} E e^{\lambda X} \geq P(X \geq \tau).$$

(Note that if $P(X \geq \tau)$ decreased less rapidly than $e^{-\lambda\tau}$, then $Ee^{\lambda X}$ would be infinite and the upper bound useless and better replaced by 1.) Of course, in terms of the characteristic function ϕ_X,

$$Ee^{\lambda X} = \phi_X(-i\lambda)$$

yields

$$e^{-\lambda\tau} \phi_X(-i\lambda) \geq P(X \geq \tau).$$

A version of the exponential bound in the two-sided case can be given in terms of the hyperbolic cosine function

$$\cosh(x) = \frac{e^x + e^{-x}}{2} :$$

$$g(x) = \frac{\cosh(\lambda x) - 1}{\cosh(\lambda \tau) - 1}.$$

Note that

$$E \cosh(\lambda X) = \frac{1}{2} (\phi_X(-i\lambda) + \phi_X(i\lambda)).$$

Hence,

$$\frac{E \cosh(\lambda X) - 1}{\cosh(\lambda \tau) - 1} \geq P(|X| \geq \tau).$$

Knowledge of the characteristic function can yield exponentially rapidly decaying upper bounds on probabilities of particular events.

Example 18.3   Bounding the Normal

As we know, probabilities for normally distributed random variables are not expressible in terms of finite combinations of elementary functions. Consider the case of $X \sim \mathcal{N}(0, \sigma^2)$:

$$\phi_X(u) = e^{-\frac{1}{2}\sigma^2 u^2}, \quad e^{-\lambda\tau}e^{\frac{1}{2}\sigma^2\lambda^2} \geq P(X \geq \tau).$$

While this bound is valid for any $\lambda \geq 0$, λ is of our choosing and it behooves us to make the best choice: Choose $\lambda \geq 0$ to minimize the upper bound. Equivalently, taking logarithms, we see that

$$-\lambda\tau + \frac{1}{2}\sigma^2\lambda^2$$

is minimized by the choice

$$\lambda^* = \frac{\tau}{\sigma^2}$$

for $\tau > 0$, yielding the upper bound

$$e^{-\frac{1}{2}\frac{\tau^2}{\sigma^2}} \geq P(X \geq \tau), \text{ for } X \sim \mathcal{N}(0, \sigma^2),$$

which decays faster than exponentially in τ.

Example 18.4   Bounding the Poisson

Consider the Poisson

$$X \sim \mathcal{P}(\gamma), \quad EX = \gamma, \quad \text{VAR}(X) = \gamma, \quad \phi_X(u) = e^{-\gamma + \gamma e^{iu}}.$$

Hence,

$$\frac{\gamma}{\tau^2} \geq P(|X - \gamma| \geq \tau) \geq P(X - \gamma \geq \tau),$$

$$e^{-\lambda\tau}e^{-\gamma + \gamma e^{\lambda}} \geq P(X \geq \tau).$$

For this latter one-sided exponential bound to be nontrivial (less than 1), we require that

$$\lambda\tau + \gamma > \gamma e^{\lambda}.$$

Note that, for $\lambda \geq 0$,

$$\frac{e^{\lambda} - 1}{\lambda} \geq 1.$$

Now, rearrange terms to conclude that

$$\tau > \gamma\frac{e^{\lambda} - 1}{\lambda} \Rightarrow \tau > \gamma$$

is necessary if this upper bound is to be nontrivial. In effect, the upper bound is useless, unless we are in the region where $\tau > EX$. If we now identify the best choice of λ to yield the lowest upper bound, we see from differentiation that, for $\tau > \gamma$,

$$e^{\lambda} = \frac{\tau}{\gamma}, \quad e^{-\gamma} \left(\frac{e\gamma}{\tau} \right)^{\tau} \geq P(X \geq \tau).$$

Since

$$\tau^{-\tau} = e^{-\tau \log \tau},$$

this upper bound to the tail probability of the Poisson decays more rapidly than exponentially in τ.

18.3 CHERNOFF BOUNDS

Sums of random variables, including sums in which the random variables are multiplied by weights, as in time averages, are a common occurrence. Because these sums generally involve large (possibly infinite) numbers of summands, a direct calculation of their distribution is usually out of the question. Nonetheless, what appears to be a difficult problem is capable of a resolution in terms of much more easily calculated upper bounds to the probabilities in question. These upper bounds are often sufficient for applications. *Chernoff bounds* are upper bounds to sums of *i.i.d.* random variables that start out in the form $e^{-\lambda\tau}$ and then choose λ optimally and dependent upon τ. The final function of τ may decrease more rapidly than exponentially.

From our previous study of

$$S_n = \sum_1^n X_i,$$

with $EX_1 = m$, $\text{VAR}(X_1) = \sigma^2$, and $\{X_i\}$ *i.i.d.*, we know that

$$ES_n = nm, \quad \text{VAR}(S_n) = n\sigma^2, \quad \phi_{S_n}(u) = \phi_{X_1}^n(u).$$

Hence,

$$\frac{n\sigma^2}{\tau^2} \geq P(|S_n - ES_n| \geq \tau), \quad e^{-\lambda\tau}\phi_{X_1}^n(-i\lambda) \geq P(S_n \geq \tau).$$

We can tighten the exponential bound if we can select $\lambda > 0$ to minimize the upper bound. From the calculus, we set to zero the first partial derivative with respect to λ in order to find the optimal λ^*. The governing equation is

$$\left. \frac{\partial \log \left(\phi_{X_1}(-i\lambda) \right)}{\partial \lambda} \right|_{\lambda=\lambda^*} = \frac{\tau}{n}.$$

Example 18.5   Bounding the Binomial $\mathcal{B}(n,p)$ _____

As an illustration, we consider the case of

$$X_1, ..., X_n, \quad i.i.d., \quad P(X_i = 1) = p = 1 - P(X_i = 0).$$

In this case,

$$S_n = \sum_1^n X_i \sim \mathcal{B}(n,p),$$

although we will analyze the bounds from the viewpoint of sums. From our earlier work, letting $q = 1 - p$, we have

$$ES_n = np, \ \ \mathrm{VAR}(S_n) = npq, \ \ \phi_{S_n}(u) = [pe^{iu} + q]^n.$$

Hence,

$$\frac{npq}{\tau^2} \geq P(|S_n - np| \geq \tau), \ \ \ e^{-\lambda\tau}[pe^\lambda + q]^n \geq P(S_n \geq \tau).$$

We refine the exponential upper bound by looking for the best choice of λ. Taking a derivative with respect to λ and setting it to zero yields

$$e^{\lambda^*} = \frac{q\tau}{p(n-\tau)}.$$

Substituting into the upper bound, letting $\rho = \frac{\tau}{n}$, and rearranging yields

$$\left\{\frac{p^\rho q^{1-\rho}}{\rho^\rho (1-\rho)^{1-\rho}}\right\}^n \geq P(S_n \geq \tau).$$

Figure 18.1 shows the discrepancy between the exponential upper bound just derived and the true tail probabilities for $\mathcal{B}(50, .3)$.

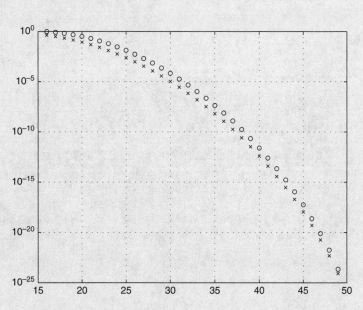

Figure 18.1 Semilog plot of Chernoff bound ("o") and tail probabilities ("x") for $\mathcal{B}(50, .3)$.

Central concepts in information theory (see Cover and Thomas [20]) are the Boltzmann/ Shannon entropy function (see Section 2.5) for a discrete pmf $\mathbf{p} = \{p_i\}$,

$$H(\mathbf{p}) = -\sum_i p_i \log p_i,$$

and the *Kullback–Leibler (KL) discrimination* or divergence D (see Section 9.9) between two pmfs $\mathbf{p}^0, \mathbf{p}^1$,

$$D(\mathbf{p}^0 \| \mathbf{p}^1) = \sum_i p_i^0 \log \frac{p_i^0}{p_i^1}.$$

Note that KL discrimination is asymmetric in that, in general, $D(\mathbf{p}^0 \| \mathbf{p}^1) \neq D(\mathbf{p}^1 \| \mathbf{p}^0)$.

In the binary case, these quantities simplify to

$$H(\mathbf{p}) = -p_1 \log p_1 - (1 - p_1) \log(1 - p_1),$$

$$D(\mathbf{p}^0 \| \mathbf{p}^1) = p_1^0 \log \frac{p_1^0}{p_1^1} + (1 - p_1^0) \log \frac{1 - p_1^0}{1 - p_1}.$$

In the binary case, we can simplify our notation to $D(p_1^0 \| p_1^1)$. In the special case of $p_1^1 = \frac{1}{2}$ (discriminating against the coin being fair),

$$D(p_1^0 \| 1/2) = \log 2 - H(\mathbf{p}^0).$$

We can use these concepts to write our upper bound to $P(S_n \geq \tau)$ as

$$e^{-nD(\frac{\tau}{n}, p)} \geq P(S_n \geq \tau) = \sum_{i=\tau}^{n} \binom{n}{i} p^i q^{n-i}.$$

The upper bound is nontrivial for $\frac{\tau}{n} > p$.

Example 18.6   U.S. Presidential Elections

The results just derived can be applied to the 1960 and 2000 U.S. presidential elections. In the closely contested 1960 election between Kennedy and Nixon, Kennedy won with 34,227,096 votes and Nixon lost with 34,108,546 votes, a difference of 118,550 votes out of 68,335,642 votes cast. Kennedy won by only 59,275 votes more than half of the votes cast—less than .1%. Can this outcome be construed as the voters tossing a fair coin to determine the president? Of course, any specific outcome of the toss of 68 million fair coins is going to have very low probability. Instead, we ask about the probability of the number of heads exceeding the number of tails by the observed margin and use our upper bounds to estimate this quantity:

$$n = 68,335,642, \ p = \frac{1}{2}, \ ES_n = \frac{n}{2}, \ \text{VAR}(S_n) = \frac{n}{4}, \ \tau = 59,275,$$

$$P(S_n - ES_n \geq \tau) = \sum_{k=\frac{n}{2}+\tau}^{n} \binom{n}{k} 2^{-n} \leq \frac{n}{4\tau^2} = .0049.$$

Hence, it is unlikely that this outcome was just a random result. Indeed, that it is far more unlikely than it appears even from the preceding bound can be seen by evaluating an exponential bound

$$P(S_n - ES_n \geq \tau) \leq e^{-nK(\frac{\tau + \frac{n}{2}}{n}, p = \frac{1}{2})} = e^{-n(\log 2 - H(\frac{\tau}{n} + \frac{1}{2}))}.$$

In our case, $\frac{\tau}{n} = .000867$, and we use the Taylor's series expansion

$$H(x) = \log 2 - 2(x - \frac{1}{2})^2 + \ldots$$

to obtain

$$P(S_n - ES_n \geq \tau) \leq e^{-103} \approx 2 \times 10^{-45}.$$

One can repeat these analyses to study the 2000 presidential election between Bush and Gore. Gore won the popular vote (but lost in the electoral vote by a close contest in Florida) by 50,996,064 to Bush's 50,456,167; there were about 4,000,000 votes for other candidates. Hence, Gore won the popular vote by 269,948 votes more than half of the votes cast for both of them, yielding a $\frac{\tau}{n}$ of .00266. This ratio is three times greater than that observed in the 1960 election and implies that there was an even stronger preference for Gore over Bush than for Kennedy over Nixon.

18.4 HOEFFDING BOUNDS

An upper bound due to Hoeffding applies to sums of independent random variables $\{X_i\}$ when these variables are, in addition, individually bounded, or

$$P(a_i \leq X_i - EX_i \leq b_i) = 1,$$

but not necessarily identically distributed.

Theorem 18.4.1 (Hoeffding Inequality) If $\{X_i\}$ are independent and $P(a_i \leq X_i - EX_i \leq b_i) = 1$, then

$$P\left(\left|\sum_{i=1}^{n}(X_i - EX_i)\right| \geq \epsilon\right) \leq 2e^{-2\epsilon^2/\sum_{i=1}^{n}[b_i - a_i]^2}.$$

In particular, if all of the mean-centered random variables have the common range $[a, b]$, then

$$P\left(\left|\frac{1}{n}\sum_{i=1}^{n}(X_i - EX_i)\right| > \epsilon\right) \leq 2e^{-\frac{2n\epsilon^2}{(b-a)^2}}.$$

Proof remark. The Hoeffding bound can be derived by starting from the one-sided exponential Chernoff bound

$$P\left(\sum_{i=1}^{n}(X_i - EX_i) \geq \epsilon\right) \leq e^{-\lambda\epsilon}\prod_{i=1}^{n}\phi_{X_i - EX_i}(-i\lambda).$$

The key is to establish the connection that if $P(a \leq X - EX \leq b) = 1$, then

$$\phi_{X-EX}(-i\lambda) \leq e^{\frac{\lambda^2}{8}(b-a)^2}.$$

Hence,

$$P\left(\sum_{i=1}^{n}(X_i - EX_i) \geq \epsilon\right) \leq e^{-\lambda\epsilon + \frac{\lambda^2}{8}\sum_{i=1}^{n}(b_i-a_i)^2}.$$

Following the Chernoff bound approach, we minimize this upper bound by the choice of $\lambda > 0$. The minimizing λ for the exponent is

$$\lambda^* = \frac{4\epsilon}{\sum_{i=1}^{n}(b_i - a_i)^2}.$$

Substituting λ^* yields the one-sided Hoeffding bound

$$P\left(\sum_{i=1}^{n}(X_i - EX_i) \geq \epsilon\right) \leq e^{-2\epsilon^2/\sum_{i=1}^{n}(b_i-a_i)^2}.$$

The double-sided bound is just twice as large, since the same argument can be made for $-\sum_{i=1}^{n}(X_i - EX_i)$ as for the term without the negative sign. (See Pollard [72], p. 192, for a complete proof.)   □

Thus, if $X_i \in [0, 1]$ then $b_i - a_i = 1$, and the Hoeffding bound yields

$$P\left(\left|\frac{1}{n}\sum_{i=1}^{n}(X_i - EX_i)\right| > \epsilon\right) \leq 2e^{-2n\epsilon^2}.$$

Example 18.7   U.S. Presidential Elections Revisited

Applying the Hoeffding bound to the Kennedy–Nixon election of 1960 with $n = 68,335,642$ and $n\epsilon = 59,275$, yields an upper bound to the probability of $2e^{-106}$ that is only slightly better than the one found by using Chernoff bounds.

18.5 SUMMARY

Prior to this chapter, we considered methods that were, in principle, exact for determining probabilistic quantities. In many cases, these methods cannot be carried out in practice, either due to analytical limitations or, quite commonly, due to a lack of sufficient knowledge of the true probability model. Analytical limitations often arise in dealing with functions of a large number of (even independent) random variables. The simplest such function, that of a weighted summation, frequently yields an intractable problem if we attempt to determine the cdf of the resulting summation. In practice, our knowledge of the true probability model is often incomplete; we may only know a few moments that have been estimated from data. This chapter explains methods for finding upper bounds to probabilities that have proven their worth countless times in facing the problems we have indicated. We are often able to derive useful results and insights even when we cannot solve for the true probability distribution, either in practice or in principle. Typically, we are led to upper bounds for probabilities of the forms $P(X \geq \tau) \leq f(\tau)$ (single-sided bounds) and $P(|X| \geq \tau) \leq f(\tau)$ (double-sided bounds). Generalized Chebychev bounds for both the single- and double-sided cases are derived in Sections 18.2 and specialized to exponential functions and refined in the Chernoff bounds of Section 18.3. Hoeffding bounds are applicable to sums of bounded independent random variables and are described in Section 18.4. They have found useful application to analyses of machine learning algorithms.

The generalized Chebychev bounds are of the form

$$P(|X| \geq \tau) \leq \frac{Eg(X)}{g(\tau)},$$

for a suitably chosen function g that is also one about which we know enough to evaluate $Eg(X)$. Examples of such a g include $g(x) = |x|^k$, where $Eg(X)$ is the kth moment, and in the single-sided case, $g(x) = e^{\lambda x}$ for positive λ, where $Eg(X)$ is easily determined from the characteristic function $\phi_X(-i\lambda)$. Generalized Chebychev bounds apply to any random variable for which we can evaluate the needed expectation. In particular, in Section 18.3, we discuss their application to sums of independent random variables. Hoeffding bounds exploit an additional assumption: that all of the summands are strictly bounded to lie in a known finite interval, which enables them to provide slightly sharper upper bounds for sums of independent random variables. We illustrated the application of these bounds by analyzing the outcome of the 1960 presidential election, in which there was a seemingly very slight margin in favor of Kennedy over Nixon. We found that this slight margin is actually indicative of a statistically highly significant voter preference for Kennedy, albeit the preference itself is only slight.

EXERCISES

E18.1   a. Use the basic Chebychev inequality to show that the probability that a random variable X differs from its mean by at least three standard deviations is no more than 1/9.

   b. For what distribution is the probability equal to the upper bound?

E18.2   We are attempting to estimate the cdf $F_X(1)$ on the basis of $i.i.d.$ $\{X_i\}$ with $X_i \sim F_X$. In order to do so, we employ the relative frequency $r_A(n) = \frac{1}{n} \sum_{i=1}^{n} I_A(X_i)$, for A the event $X \leq 1$.

   Evaluate a Chebychev upper bound to $P(|r_A(n) - P(A)| \geq \epsilon n^{-\frac{1}{2}})$ in terms of $F_X(1)$.

E18.3   How reliable an estimator of the probability of heads can we form from the relative frequency of occurrence in the first 1,000 tosses of a coin with probability $P(H) = 1/2$? Estimate the probability that the relative frequency will differ from 1/2 by at least .05.

E18.4   In Exercise E17.39, if the number of bits B received in time T exceeds a buffer length q, then bits will be lost. Provide a nontrivial upper bound to $P(B \geq q)$ of the form $\frac{c}{q^2}$.

E18.5   In a communications setting, we receive 100 symbols, each with a probability of error of .1. If errors are made independently from symbol to symbol, then use both a quadratic and an exponential Chebychev inequality to bound the probability that there will be at least 20 errors in the received symbols.

E18.6   If $X \sim \mathcal{N}(m, \sigma^2)$, then provide a good exponential upper bound to $P(X - m \geq \tau)$, using ϕ_X.

E18.7   For $n > 1$,

$$\phi_{R_n}(u) = \left(\frac{1}{1 + \frac{u^2}{n^2}} \right)^n.$$

Provide an upper bound to $P(R_n > \tau)$ of the form $ce^{-\tau}$.

E18.8   If $X \sim \Gamma(\alpha, \lambda)$, then use the characteristic function to provide an exponential upper bound to $P(X \geq \tau)$.

E18.9   A radar receiver designed to determine the presence or absence of a target in its surveillance region calculates the received energy over a fixed time interval. When no target is present, the inputs $\{X_i\}$ to the decision-making part of the receiver may be taken to be $i.i.d.$ $\mathcal{N}(0, \sigma^2)$. The decision maker squares and integrates (sums) these inputs to yield outputs $Z_n = \frac{1}{\sigma^2} \sum_{i=1}^{n} X_i^2$. At time T, Z_T is compared with a threshold value τ, and we decide a target is absent if $Z_T < \tau$ and present if $Z_T \geq \tau$ (Neyman–Pearson decision making).

   a. Partially analyze the performance of this radar receiver by calculating the mean, variance, and characteristic function of Z_T. Does the form of the characteristic function enable you to identify the pdf of Z_T?

   b. Use a Chernoff-type bound to estimate the probability $P(Z_T \geq \tau | \text{absent})$ of a false alarm (deciding that the target is present when it is absent).

E18.10   *Markov's inequality* is of the form

$$P(X \geq \tau > 0) \leq \frac{c}{\tau}.$$

Use the generalized Chebychev approach to determine c.

E18.11   A signal detector based upon an energy measurement observes that

$$Y = \sum_{i=1}^{n} X_i^2 \geq 0.$$

If the $\{X_i\}$ are $i.i.d.$ $\mathcal{N}(0, 1)$, then evaluate the constant α in the bound

$$P(Y \geq \epsilon) \leq \frac{\alpha}{\epsilon}.$$

E18.12   A random signal $X \sim \mathcal{N}(m, \sigma^2)$ is propagated through n layers of attenuating materials with final output

$$Y = X \prod_{1}^{n} A_i = X(A_1 \cdots A_n).$$

The random attenuations are $i.i.d.$, $A_1 \sim \mathcal{U}(0, 1)$, and are independent of the signal X.

a. Evaluate VAR(Y).

b. Evaluate an upper bound to $P(Y \geq \tau)$ and indicate the values of τ for which it is of interest.

E18.13 Given that $\{X_i\}$ are $i.i.d.$ and $Y = \sum_{i=1}^n X_i$, find constants c_1 and c_2 that satisfy

$$P(|Y| \geq \tau) \leq \frac{c_1}{\tau^2}$$

$$P(Y \geq \tau) \leq c_2 e^{-\tau}$$

nontrivially when $\phi_{X_1}(u) = e^{-u^2}$.

E18.14 $S = X_1 + X_2 + X_3$, where $\{X_i\}$ are $i.i.d.$ $\mathcal{L}(1)$.

a. Provide an upper bound of the form $c_1 \tau^{-2}$ to $P(S \geq \tau)$.

b. Repeat part (a) for an upper bound of the form $c_2 e^{-\lambda \tau}$.

c. For $\tau = 6$ and $\lambda = 1/2$, which bound should you use?

d. For $\tau = 5$, can λ be chosen to make the bound of (b) better than the bound of (a)?

E18.15 The waiting time W to the arrival of the third photon is $W = T_1 + T_2 + T_3$, where T_1, T_2, T_3 are $i.i.d.$ $\mathcal{E}(\alpha)$.

a. Evaluate $\phi_W(u)$.

b. Provide an upper bound of the form $c e^{-\lambda \tau}$ to $P(W \geq \tau)$ by evaluating c.

c. For what values of λ and τ is this bound nontrivial?

d. Provide an upper bound of the form c/τ^2 to $P(|W - EW| \geq \tau)$ by evaluating c.

E18.16 a. Provide a Chernoff upper bound to $P(X \geq \tau)$ when $X \sim \mathcal{P}(\mu)$.

b. For what values of τ is this optimized bound less than 1?

E18.17 If $X_1, \ldots, X_n$ are $i.i.d.$ $\beta(3, 4)$ random variables, use the Hoeffding inequality to provide an upper bound to

$$P\left(\left|\sum_{i=1}^n (X_i - EX_i)\right| \geq \epsilon\right).$$

E18.18 (Bernstein's inequality) When $EX = 0$, $\text{VAR}(X) = \sigma^2$, and

$$(\forall n \geq 2) \ E|X|^n \leq 0.5\sigma^2 n! c^{n-2},$$

it can be shown that, for $\lambda > 0$,

$$Ee^{\lambda X} \leq \exp\left(\frac{\sigma^2 \lambda^2}{2(1 - \lambda c)}\right).$$

Under these circumstances, show that you can select $\lambda > 0$ so that

$$P(X \geq \tau) \leq e^{-\frac{\tau^2}{2(\sigma^2 + c\tau)}}.$$

Part V

Infinitely Many Random Variables

19

Limits and Laws of Large Numbers

19.1 PURPOSE, BACKGROUND, AND ORGANIZATION

A highly influential viewpoint is that

Probability is about what happens in the long run.

More specifically, probability is about the long-run fraction of time that an event occurs. The mathematics of the long run is that of limits. There are many different kinds of limits, particularly when we talk not just about limits of sequences of real numbers, but about limits of functions; random variables are real-valued functions on the sample space. We would like to know if the fraction of times that an event is seen to occur will, in the actual repeated experiment we are performing, converge to the probability. Moreover, we would like to ask that this "convergence" be seen to happen quickly—that it will not take many tosses of the coin to get a pretty good estimate of what the long-run fraction of heads will be. After all, in other experimental situations, such as repeated measurement of the length of an object or of an electrical resistance or voltage, we quickly come up with a number that is believed to be precise to within so many significant figures. Unfortunately, in this, we are asking for more than what probability can provide:

Probability can provide only probabilistic guarantees

—guarantees only as to what the probability is of convergence to within a given precision after a given number of repeated, unlinked trials. Such guarantees go by the name of "laws of large numbers," the "large numbers" being the number of repetitions of the random experiment. Probability will not provide guarantees that, after 1,000 tosses of a coin, we can be certain that the probability of heads is $.5 \pm .1$.

How, then, can we relate probabilities, discussed at great length in the earlier chapters, to observed outcomes of random experiments? The interpretation of probability we favor is that $P(A)$ is the *propensity* for event A to occur in a performance of a random experiment $\mathcal{E}$; note that the probability is relative to a particular experiment. When there is opportunity for repeated unlinked performances $\mathcal{E}_1, ..., \mathcal{E}_n$ of $\mathcal{E}$, this propensity (tendency, disposition) can display itself through the relative frequency

$$r_n(A) = \frac{1}{n} \sum_1^n I_A(X_i) \approx P(A)$$

with which A is observed to occur among the outcomes $X_1, ..., X_n$ of the random experiments. Our goal is to expand upon this frequentist link to probability, originally offered in Sections 3.3 and 3.4.

In the theory of probability, the frequentist link is captured in *laws of large numbers*. These "laws" describe a mathematically idealized asymptotic long-run behavior of averages of outcomes (these may be general random variables $\{X_i\}$ or the $\{0, 1\}$-valued indicators $\{I_A(X_i)\}$ for some event A that appear in relative frequency) of repeated unlinked random experiments which are mathematically modeled as having outcomes that are independent and identically distributed random variables. Note that

Laws of large numbers are theorems in the mathematical theory of probability and not physical laws governing what must happen in actual repeated random experiments.

As there are a number of different mathematical methods for construing the approximation of the random sequence of relative frequencies to nonrandom probability, there are a corresponding number of different laws of large numbers. This is a fairly abstract subject, and to maintain interest and insight, we open our discussion by examining the late 17th-century founding views of Jacob Bernoulli on this subject. We then proceed to develop a first law of large numbers based upon the simple notion of convergence in mean square. Having established that, we review notions of convergence for ordinary functions, then introduce parallel notions of convergence for random variables (in probability, with probability one, in rth mean, and in distribution), and provide the corresponding laws of large numbers.

Another reason for the material in this chapter is that it is of importance in the study of random or stochastic processes. (See Chapter 20.) Such processes are the probabilistic mathematical models for signals that persist either in time or in space (e.g., a speech or video signal observed over the time interval $[0, T]$ or the temperature or electromagnetic fields established over a region of space). Such random-process models involve considering infinitely (either countable or uncountable) many random variables representing the values of signal amplitudes at various times or field vector values at points in space. The presence of infinitely many random variables leads to the need for analytical methods capable of providing probability distributions for functions (e.g., averages calculated over time intervals or over cells in space) of these infinitely many random variables. The analysis of functions of infinitely many variables requires a consideration of limits. It is the very notion of limits of random sequences that we address in this chapter.

A word of caution: While the study of the asymptotic (in the number of observations n) behavior of statistics (functions of the random observations that are meant to estimate or infer aspects of the probability law governing the observations) has long been a central subject and part of the defense of the choice of particular statistics (e.g., the sample mean), purely asymptotic results have little to no bearing on our real-world concerns, which are always with not very large finite values of n. We share the following opinion expressed by Kolmogorov [55]:

> (1) The frequency concept based on the notion of *limiting frequency* as the number of trials increases to infinity does not contribute anything to substantiate the applicability of the results of probability theory to real practical problems, where we have always to deal with a finite number of trials.

19.2 JACOB BERNOULLI ON THE LAWS OF LARGE NUMBERS

The first law of large numbers is due to Jacob (also called "James") Bernoulli, professor of mathematics at the University of Basel, working at the close of the 17th century. (See Figure 19.1.) Bernoulli struggled to formulate what he could expect to be the case and then to prove it with the limited methods of his times.

Historians (e.g., Hacking [43]) estimate that Bernoulli produced the first such law in about 1692, although his work containing it, *Ars Conjectandi, Pars Quarta*, was published posthumously (deceased 1705), in 1713. At the time of writing of this text, Bernoulli's masterpiece in the development of probability has been translated to English only in a 1966 technical report sponsored by the Office of Naval Research and produced by Bing Sung for the Dept. of Statistics at Harvard [9]. Bernoulli was working against a background of classical probability (Chapter 2) originating from models of symmetrical games of chance. In this model, probability is to be assessed as the ratio of the number of favorable "cases" for an event's occurrence to the total number of "cases."

Figure 19.1   Jacob Bernoulli and LLN on Swiss stamp.

In effect, there is a uniform distribution over the set of "cases," but we are left with the problem of determining this set. In the instance of a symmetrical die, we may determine the cases simply as the six different faces that may appear uppermost. What, though, of an asymmetrical die? What are the "cases" that determine the weather or our own mortality? Bernoulli, properly despairing of enumerating the cases in such phenomena, observed that what could not be counted *a priori* might yet be inferred by observing outcomes of the experiments determined by such phenomena—that is, from relative frequencies of occurrence. We now refer to the translation of Bernoulli's own remarks on these issues.

> And there we concluded that for correctly forming conjectures about anything at all, nothing is required other than that the numbers of these cases be accurately determined and that it be found out how much more easily some cases can happen than others. But here, finally, we seem to have met our problem, since this may be done only in a very few cases and almost nowhere other than in games of chance the inventors of which, in order to provide equal chances for the players, took pains to set up so that the numbers of cases would be known and fixed for which gain or loss must follow, and so that all these cases could happen with equal ease. For in several other occurrences which depend upon either the work of nature or the judgment of men, this is by no means the situation. (pp. 34, 35)

> But what mortal will ever determine, for example, the number of diseases [i.e., the number of cases] which are able to seize upon the uncountable parts of the human body... (p. 36)

> But indeed, another way is open to us here by which we may obtain what is sought; and what you cannot deduce *a priori,* you can at least deduce *a posteriori* [i.e., you will be able to make a deduction from many observed outcomes of similar events]. For it must be presumed that every single thing is able to happen and not to happen in as many cases as it was previously observed to have happened... (p. 37)

> And so, if anyone has observed the weather for the past several years and has noted how many times it was calm or rainy: or if anyone has judiciously watched two players and has seen how many times this one or that one has emerged victorious in this way he has detected what the ratio probably is between the number of cases in which the same events, with similar circumstances prevailing, are able to happen and not to happen later on.

> And this empirical way of determining the numbers of cases by trials is neither new nor unusual, for the celebrated author [A. Arnauld] of the *Ars Cogitandi*, a man of great insight and intelligence, prescribes a similar method in Chapter 12 ... it is not enough to use one or two trials, but rather a great number of trials is required. (pp. 37, 38)

> It certainly remains to be inquired whether after the number of observations has been increased, the probability is increased of attaining the true ratio between the numbers of cases in which some event can happen and in which it cannot happen, so that this probability finally exceeds any given degree of certainty; or whether the problem has, so to speak, its own asymptote—that is, whether some degree of certainty is given which one can never exceed, so that however many observations are made, we can never be more than 1/2 or 2/3 or 3/4 certain that we have detected the true ratio of cases. (p. 39)

Therefore, this is the problem which I now set forth and make known after I have already pondered over it for twenty years. (p. 42)

Whence, finally, this one thing seems to follow: that if observations of all events were to be continued throughout all eternity, (and hence the ultimate probability would tend toward perfect certainty), everything in the world would be perceived to happen in fixed ratios and according to a constant law of alternation, so that even in the most accidental and fortuitous occurrences we would be bound to recognize, as it were, a certain necessity and, so to speak, a certain fate. (pp. 65, 66)

It is clear from the closing remarks that in the very long run there is a lawlike aspect to even the "most accidental and fortuitous occurrences." This lawlike aspect is to be captured in probabilities based on relative frequencies. Bernoulli's proof of his law of large numbers occupies pages 44 to 64 of the translation.

19.3 LLN FOR MEAN SQUARE CONVERGENCE

19.3.1 MS Convergence Basics

We have a sequence of random variables and a possible limit that is another random variable. We need to determine whether the terms in the sequence "approach" the limit random variable. We start with the classical case of the terms in the sequence being independent of each other and identically distributed. The easiest notion of convergence to calculate with is that of convergence in mean square: It requires just taking a limit of the sequence of real numbers that is the expectation of the square of the difference between the nth term in the sequence and its limit.

We note from our work in Chapters 17 and 18 on sums of $i.i.d.$ random variables that, for the relative frequency $r_n(A)$,

$$Er_n(A) = P(A), \quad \text{VAR}(r_n(A)) = \frac{P(A)(1 - P(A))}{n} \leq \frac{1}{4n},$$

with the last inequality a consequence of the maximum of $x(1 - x)$ being .25 for $0 \leq x \leq 1$. More generally, if $\{X_i\}$ are $i.i.d.$ with finite mean square ($EX_i^2 < \infty$), then they have a common finite mean, say, m, and common finite variance, say, σ^2. Hence,

$$E\left(\frac{1}{n}\sum_1^n X_i\right) = m, \quad \text{VAR}\left(\frac{1}{n}\sum_1^n X_i\right) = \frac{\sigma^2}{n}.$$

Thus, "on average," the relative frequency is the probability $Er_n(A) = P(A)$, and the fluctuation about the average, as measured by the variance $\text{VAR}(r_n(A))$, decreases with increasing sample size n. Since the relative frequency is a random variable, we cannot expect it to equal the probability, except by chance. However, if we let n approach infinity, then we might expect the sequence of random variables $r_1(A), ..., r_n(A), ...$ to "converge" to $P(A)$. If this is the case, then we will have shown a consistency between the mathematical notion of probability and an interpretation based upon real-world observations. It is evident from the preceding calculations that

$$\lim_{n \to \infty} E(r_n(A) - P(A))^2 = 0,$$

$$\lim_{n \to \infty} E\left(\frac{1}{n}\sum_1^n X_i - m\right)^2 = 0.$$

Definition 19.3.1   (Mean Square Convergence) A sequence of random variables $\{X_n\}$ converges in mean square to a random variable X if

$$\lim_{n \to \infty} E(X_n - X)^2 = 0.$$

Such convergence is denoted by $X_n \xrightarrow{ms} X$.

Thus, there is a sense of a vanishingly small mean-squared discrepancy in which the sequence of relative frequencies converges to the probability, denoted

$$r_n(A) \xrightarrow{ms} P(A),$$

or the averages to the mean. As only some finite initial segment of this sequence is ever observable, we have established just a partial bridge between what is observable and its probability.

19.3.2 MS Convergence and Wide-Sense Stationarity

Now that we know about mean square convergence, we can weaken the condition that all the terms in the sequence are independent and identically distributed. Calculating the mean square requires only knowledge of means and covariances and not of the full distribution. By making the assumption that the sequence is wide-sense stationary, we know enough from the mean function and covariance function to determine whether the time average of such terms converges to their mean or expected value. We can establish laws of large numbers even when the $\{X_i\}$ are not independent by making other assumptions. If we focus on sufficient conditions for mean square convergence, then the assumptions, restated next (see also Section 20.10), of wide-sense stationarity (WSS) and asymptotic uncorrelatedness suffice.

Definition 19.3.2   (Wide-Sense Stationary) A sequence of random variables $\{X_i\}$ is wide-sense stationary if both of the following conditions hold:

$$\textbf{(constant mean)} \quad (\exists m)(\forall i)\ EX_i = m,$$

$$\textbf{(time difference)} \quad (\exists c : Z \to \mathbb{R})\ \text{COV}(X_i, X_j) = c(i - j).$$

Hence, under WSS, the covariance between two random variables depends only upon their time difference—that is, only upon how far apart they are in the sequence. It is immediate from $\text{COV}(X, Y) = \text{COV}(Y, X)$ that

$$c(k) = c(-k)$$

and, from the Schwarz inequality of Section 9.11, that

$$c(0) \geq |c(k)|.$$

A sequence of random variables is *asymptotically uncorrelated* if $\lim_{k \to \infty} c(k) = 0$. We can now state the means square law of large numbers.

Theorem 19.3.1   (Mean Square LLN)   If the sequence $\{X_i\}$ is wide-sense stationary and asymptotically uncorrelated, then

$$\frac{1}{n} \sum_1^n X_i \xrightarrow{ms} m.$$

Proof. We need to establish

$$\lim_{n \to \infty} E \left[\frac{1}{n} \sum_1^n (X_i - m) \right]^2 = 0.$$

Proceeding with the evaluation of the mean square yields

$$E \left[\frac{1}{n} \sum_1^n (X_i - m) \right]^2 = \frac{1}{n^2} \sum_{i=1}^n \sum_{j=1}^n \mathrm{COV}(X_i, X_j) = \frac{1}{n^2} \sum_{i=1}^n \sum_{j=1}^n c(i - j)$$

$$= \frac{1}{n^2} \sum_1^n c(0) + \frac{2}{n^2} \sum_{1 \leq j < i \leq n} c(i - j).$$

We reorder the double summation as a summation over the indices $\{i, k\}$, with $i, k = i - j$, which have a 1:1 relationship to the original indices i, j:

$$\frac{2}{n^2} \sum_{1 \leq j < i \leq n} c(i - j) = \frac{2}{n^2} \sum_{k=1}^{n-1} c(k) \sum_{i=k+1}^n 1 = \frac{2}{n^2} \sum_{k=1}^{n-1} c(k)(n - k).$$

Hence,

$$E \left[\frac{1}{n} \sum_1^n (X_i - m) \right]^2 = \frac{c(0)}{n} + \frac{2}{n} \sum_{k=1}^{n-1} c(k) \left(1 - \frac{k}{n} \right).$$

Consider the upper bound

$$E \left[\frac{1}{n} \sum_1^n (X_i - m) \right]^2 \leq \frac{c(0)}{n} + \frac{2}{n} \sum_{k=1}^{n-1} |c(k)|.$$

The first term in the upper bound has a limit of 0. The second term has an upper bound of

$$\frac{2}{n} \sum_{k=1}^{n-1} |c(k)| \leq \frac{2}{n} \sum_{k=1}^{\sqrt{n}} c(0) + \frac{2}{n} \sum_{k=\sqrt{n}+1}^{n-1} |c(k)| \leq \frac{2}{\sqrt{n}} c(0) + 2 \max_{\sqrt{n} < k < n} |c(k)|.$$

We now invoke the assumption of asymptotic uncorrelatedness, $\lim_k c(k) = 0$, to conclude that

$$\lim_{n \to \infty} E \left[\frac{1}{n} \sum_1^n (X_i - m) \right]^2 = 0,$$

as is needed to establish convergence in mean square.   □

Example 19.1   WSS and Convergence in MS _____

Consider the $w.s.s.$ correlated sequence $X_1, \ldots$, where $EX_i = m$ and

$$\text{COV}(X_i X_j) = \frac{1}{1 + (i - j)^2}.$$

If we proceed in accordance with the definition of mean square convergence, then we have to evaluate

$$\lim_{n \to \infty} E \left(\frac{1}{n} \sum_1^n X_i - m \right)^2 = \lim_{n \to \infty} \frac{1}{n^2} \sum_{i=1}^n \sum_{j=1}^n \text{COV}(X_i, X_j)$$

$$= \lim_{n \to \infty} \frac{1}{n^2} \sum_{i=1}^n \sum_{j=1}^n \frac{1}{1 + (i - j)^2}.$$

However, thanks to Theorem 19.3.1, we do not have to engage in lengthy calculations to know that

$$\lim_{k \to \infty} \text{COV}(X_i, X_{i+k}) = \lim_{k \to \infty} \frac{1}{1 + k^2} = 0$$

implies that $X_i \xrightarrow{ms} m$.

19.4 CONVERGENCE OF SEQUENCES OF FUNCTIONS

Limits of sequences of random variables are limits of sequences of functions; a random variable is a real-valued function with arguments drawn from the sample space. Thus, the sequence $\{X_n(\omega)\}$ depends upon the index n along which we will be taking limits and upon the argument ω of the function that can give us different values at the same n. Limits of sequences of functions are statements about limits of the generally infinitely many values of the functions for each of their possible arguments ω. Hence, we have to be able to determine the limiting behavior of infinitely many real-valued sequences. There are several ways to look at all of these cases at once.

We prepare to generalize the laws of large numbers based on mean square convergence to other forms of convergence. Limits are an essential concept in analysis and one that also has several extensions to sequences of random variables. It is often the case in probability and statistics that we deal with "large" numbers of outcomes for which it is impractical to calculate precise joint densities. In such circumstances, we may be satisfied with knowledge of large-scale general characteristics of the data or phenomena (e.g., the Central Limit Theorem of Section 17.9).

We first review the concepts of convergence of sequences of functions and then provide probabilistic analogs for convergence of sequences of random variables, as needed, to understand and develop the laws of large numbers. This discussion will also provide an introduction to dealing with infinite collections of random variables, the subject of random/stochastic processes that provide probability models for signals persisting in time or space.

The basic notion of convergence is the familiar *convergence of sequences of real numbers*, defined by

$$\lim_{n \to \infty} a_n = a \iff (\forall \epsilon > 0)(\exists N_\epsilon)(\forall n > N_\epsilon) \; |a_n - a| < \epsilon.$$

If we inquire into *convergence of sequences of functions*, then the meaning of

$$\lim_{n \to \infty} f_n = f$$

is less clear and can be given several interpretations, three of which we provide. We say that we have a *pointwise limit* if the following biconditional holds:

$$\lim_{n \to \infty} f_n = f \text{ (pointwise)} \iff (\forall x \in \mathbb{R}) \; \lim_{n \to \infty} f_n(x) = f(x).$$

For example,

$$f_n(x) = xe^{\frac{x}{n}} \to x \quad \text{pointwise.}$$

Pointwise convergence of functions amounts to ordinary convergence at each argument of the function and is the strongest form of convergence.

A weaker form that is very useful allows for divergence on an exceptional set of points D that is very small (i.e., has length 0). This form is known as *convergence almost everywhere (a.e.)* and is defined by

$$\lim_{n \to \infty} f_n = f \text{ (a.e.)} \iff (\forall x \in D^c) \; \lim_{n \to \infty} f_n(x) = f(x).$$

For example,

$$f_n(x) = \frac{1}{n \sin(x)} \to 0 \text{ (a.e.),}$$

with D the countably infinite set $\{k\pi\}$ of integer multiples of π.

Finally, we consider a form of convergence in which the differences between functions are measured by integrals of positive powers of the magnitudes of those differences. This form is known as *convergence in rth mean*, for $r > 0$, and is defined by

$$f_n \xrightarrow{r} f \iff \lim_{n \to \infty} \int_{-\infty}^{\infty} |f_n(x) - f(x)|^r \, dx = 0.$$

Example 19.2 Convergence of Functions in rth Mean _____

Consider the sequence of square pulse functions

$$f_n(x) = \alpha_n \left[U(x) - U\left(x - \frac{1}{n}\right) \right] = \begin{cases} \alpha_n & \text{if } 0 \le x \le \frac{1}{n} \\ 0 & \text{if otherwise} \end{cases}.$$

If we expect the limiting function to be 0, then, for convergence in rth mean, we need

$$f_n \xrightarrow{r} 0 \iff \lim_{n \to \infty} \int_{-\infty}^{\infty} |f_n(x)|^2 \, dx = \lim_{n \to \infty} \frac{1}{n} |\alpha_n|^r = 0.$$

Hence, we have convergence in the rth mean if and only if

$$\lim_{n \to \infty} \frac{|\alpha_n|}{n^{\frac{1}{r}}} = 0.$$

If

$$\alpha_n = n^{\frac{1}{r}} \log n,$$

then we do not have convergence in the rth mean; dividing by the term $\log n$, instead of multiplying by it, would ensure convergence in the rth mean. Depending upon the positive value of r, we may or may not have convergence.

19.5 CONVERGENCE OF SEQUENCES OF RANDOM VARIABLES

19.5.1 Definitions of Types of Convergence

In this context, random variables are best thought of as functions from an underlying sample space Ω to the real numbers $\mathbb{R}$. With this interpretation, issues of convergence are parallel to convergence questions for functions. Pointwise convergence is too strict to be useful for random variables. However, the notion of *a.e.* convergence has a clear counterpart as "almost sure" (*a.s.*) convergence when we take a set with probability zero for the "small" exceptional set D.

Definition 19.5.1 (Convergence *a.s./wp*1) The sequence of random variables $\{X_n\}$ converges almost surely (*a.s.*) or with probability 1 (*wp1*) to a random variable X:

$$X_n \to X \text{ a.s.} \iff P(\{\omega : \lim_{n \to \infty} X_n(\omega) = X(\omega)\}) = 1.$$

In this definition, $\lim_{n \to \infty} X_n(\omega)$ is the ordinary limit of a sequence $\{X_n(\omega)\}$ of real numbers; for each $\omega \in \Omega$, we have a likely different sequence of real numbers, some of them having limits and some not.

Example 19.3 Convergence Almost Surely of Random Variables _____

If, for a given random variable Z,

$$P(\{\omega : 0 \le |Z(\omega)| < 1\}) = 1, \quad X_n(\omega) = Z^n(\omega),$$

then X_n converges to 0 wp1; the exceptional set $D \subset \Omega$ is the set $D = \{\omega : |Z(\omega)| \geq 1\}$ and has $P(D) = 0$. Hence, in this notion of convergence, $\{X_n\}$ converges to a random variable $X = 0$ everywhere, except on a possible set of probability 0.

Convergence in the rth mean has a counterpart in the next definition.

Definition 19.5.2   (Convergence in the rth Mean) X_n converges in the rth mean to X for some r is defined as follows:

$$X_n \xrightarrow{r} X \iff \lim_{n \to \infty} E|X_n - X|^r = 0.$$

We often set $r = 2$ and refer to this as *convergence in mean square (m.s.)*, as given in Definition 19.3.1. We denote such convergence by $X_n \xrightarrow{ms} X$.

Example 19.4   Convergence in Mean Square and Almost Surely _____
For example, if, for a given random variable Z,

$$X_n = \frac{n}{n+1} Z,$$

then $X_n \xrightarrow{ms} Z$ if $EZ^2 < \infty$, but not if $EZ^2 = \infty$. In either case, X_n converges to Z *a.s.* or *wp*1 so long as Z is almost surely finite (i.e., $P(|Z| < \infty) = 1$).

A concept of convergence for random variables that is weaker than the two just defined is that of convergence in probability.

Definition 19.5.3   (Convergence in Probability) X_n converges in probability to X is defined as follows:

$$X_n \xrightarrow{P} X \iff (\forall \epsilon > 0) \lim_{n \to \infty} P(|X_n - X| > \epsilon) = 0.$$

Example 19.5   Convergence in Probability _____
The previous example of $X_n = (n/(n+1))Z$ is also one for which we have convergence in probability to Z. For another example, consider X and $X_1, \ldots$, where they are jointly normally distributed with common mean 0 and covariance matrix partially defined by

$$\text{COV}(X, X) = \text{COV}(X_n, X_n) = 1, \text{COV}(X, X_n) = 1 - \frac{1}{n}.$$

Letting $Y_n = X_n - X$, we know from our previous work on normally distributed random variables that

$$Y_n = X_n - X \sim \mathcal{N}\left(0, \frac{2}{n}\right).$$

Hence,

$$P(|X_n - X| \geq \epsilon) = P(|Y_n| \geq \epsilon) = 2 \int_{\epsilon \frac{n}{2}}^{\infty} \frac{1}{\sqrt{2\pi}} e^{-x^2} \, dx.$$

It follows that

$$(\forall \epsilon > 0) \lim_{n \to \infty} P(|Y_n| \geq \epsilon) = 0,$$

and $Y_n = X_n - X \overset{P}{\longrightarrow} 0$ or $X_n \overset{P}{\longrightarrow} X$.

Our final concept of convergence for random variables is actually not a notion of convergence for the random variables, but a much weaker notion of the convergence of their individual cumulative distribution functions to a cumulative distribution function.

Definition 19.5.4   (Convergence in Distribution) X_n converges in distribution to X is defined as follows:

$$X_n \overset{\mathcal{D}}{\longrightarrow} X \iff (\forall u) \lim_{n \to \infty} \phi_{X_n}(u) = \phi_X(u).$$

This definition is more commonly given in terms of convergence of the cdfs $\{F_{X_n}\}$ at the points of continuity of the limiting cdf F, with the equivalence following from the Levy Continuity Theorem 17.3.1. One should be aware that convergence in distribution actually implies nothing about the more conventional senses of convergence of random variables or functions. A sequence converging in distribution to some random variable $X \sim F$ also converges in distribution to any other random variable $Z \sim F$ sharing the same cdf.

Example 19.6   Convergence in Distribution _____
If $\{X_n\}$ *i.i.d.* F, then, for all n, $F_n = F$, and hence we have convergence in distribution to any random variable $X \sim F$. Of course, since the sequence is *i.i.d.*, values taken on by successive terms are independent and do not exhibit any usual behavior of convergence.

19.5.2 Further Examples of Convergence

As an initial illustration of the differences between these several definitions consider the following construction: We first put a uniform distribution on the unit interval sample space and then define a sequence of random variables over this interval, resulting in

$$\Omega = [0, 1], \; P([a, b]) = b - a \text{ for } 0 \leq a \leq b \leq 1, \; \omega \sim \mathcal{U}(0, 1).$$

Now define

$$X_n(\omega) = \begin{cases} n & \text{if } 0 \leq \omega \leq \frac{1}{n} \\ 0 & \text{if otherwise} \end{cases}.$$

Note that we have the two variables n and ω to deal with. In this case,

$$\{\omega : \lim_{n \to \infty} X_n(\omega) = 0\} = \{0\}^c.$$

The exceptional set is just the single point 0, and clearly, since $P(\omega = 0) = 0$, it follows that

$$X_n \to X \ a.s.$$

With regard to convergence in rth mean, we note that

$$E|X_n|^r = EX_n^r = n^r \left(\frac{1}{n}\right).$$

If we then evaluate $\lim_{n \to \infty} E|X_n|^r$, we see that the limit is infinite if $r > 1$, 1 if $r = 1$, and 0 if $r < 1$. Hence, X_n converges to 0 almost surely, but not in mean square if $r \geq 1$. It is easy to see that we have convergence in probability to 0 from

$$P(|X_n - X| > \epsilon) = P(X_n > \epsilon) = P\left(0 \leq \omega \leq \frac{1}{n}\right) = \frac{1}{n}.$$

Finally,

$$\phi_{X_n}(u) = e^{iun} \int_0^{\frac{1}{n}} dx + \int_{\frac{1}{n}}^1 dx = 1 + \frac{1}{n} e^{iun} \to 1 = \phi_0(u),$$

and $X_n \xrightarrow{\mathcal{D}} 0$.

A more complex example of convergence in distribution is provided by our analysis of the long-run behavior of

$$S_n = \frac{1}{\sqrt{n}} \sum_1^n (X_i - EX_i)$$

for $i.i.d.$ X_i. In this case, we established convergence in distribution to the $\mathcal{N}(0, \sigma^2)$ when $\text{VAR}(X_i) = \sigma^2 < \infty$ and referred to this result as a Central Limit Theorem (Section 17.9). We will see shortly that, in this case, $\{S_n\}$ does not converge in any of the other senses we introduced.

Our last example, to show that we can have convergence in rth mean and in probability, but not with probability 1, is again an explicit construction of a sequence $\{Y_n(\omega)\}$ of random variables with argument ω uniformly distributed along the unit interval and Y_n being simply $\{0, 1\}$ valued. We are going to arrange things so that $Y_n(\omega)$ takes the value 1 only over a short interval of argument ω, with the interval decreasing to zero length as the index n goes to infinity. This will assure convergence in rth mean and in probability. However, to frustrate convergence with probability 1, we need to successively choose these intervals of ω so that they keep changing and wandering over all of the unit interval sample space as n increases. Again, take $\omega \sim \mathcal{U}(0, 1)$. Introduce a sequence of intervals $\{I_n = [a_n, b_n]\}$ that will provide successively more refined partitions of the unit interval, defined as follows: Any integer index $n \geq 1$ has a unique representation (based on $\sum_{j=0}^{k-1} j = k(k-1)/2$) as

$$n = \frac{1}{2} k_n (k_n - 1) + j_n, \ k_n \geq 1, \ 1 \leq j_n \leq k_n.$$

For example, $n = 1 \iff k_1 = 1, j_1 = 1$, and $n = 8 \iff k_8 = 4, j_8 = 2$, etc. The number of integers n sharing the same value of k_n is exactly k_n. Define I_n through

$$a_n = \frac{j_n - 1}{k_n}, \ b_n = \frac{j_n}{k_n}.$$

Note that, in all cases, $0 \le a_n < b_n \le 1$ and $b_n - a_n = 1/k_n$. Given any $\omega \in [0, 1] = \Omega$, there will be infinitely many n for which $\omega \in I_n$. Define

$$Y_n(\omega) = \begin{cases} 1 & \text{if } \omega \in I_n \\ 0 & \text{if otherwise} \end{cases}.$$

If we turn to the sequence of random variables Y_n, then we might again expect convergence to 0. Indeed,

$$P(\{\omega : Y_n(\omega) > 0\}) = P(\omega \in I_n) = \frac{1}{k_n}, \quad EY_n^r = \frac{1}{k_n},$$

and thus,

$$Y_n \xrightarrow{P} 0, \quad Y_n \xrightarrow{r} 0.$$

Hence, we have convergence both in probability and in rth mean. However, in this case we do not have convergence almost surely. If you fix attention on any $\omega^0 \in [0, 1]$, then infinitely often there will be indices n such that $Y_n(\omega^0) = 1$ and infinitely often there will be indices such that $Y_n(\omega^0) = 0$. Hence,

$$\{\omega : Y_n(\omega) \text{ converges }\} = \phi,$$

and the sequence diverges pointwise and thus almost surely.

19.5.3 Convergence Relationships

The next theorem establishes the interrelationships between the four modes of convergence of random sequences.

Theorem 19.5.1   (Implications of Convergences)

$$X_n \to X \; a.s. \; \Rightarrow X_n \xrightarrow{P} X \tag{C1};$$

$$X_n \xrightarrow{r} X \Rightarrow X_n \xrightarrow{P} X \tag{C2};$$

$$X_n \xrightarrow{P} X \Rightarrow X_n \xrightarrow{\mathcal{D}} X \tag{C3};$$

$$(\exists Z \ge 0, EZ^r < \infty)(\forall n)(|X_n| \le Z), X_n \xrightarrow{P} X \Rightarrow X_n \xrightarrow{r} X \tag{C4};$$

$$X_n \xrightarrow{P} X \Rightarrow (\exists \{n_k\}) \; X_{n_k} \to X \; a.s. \tag{C5}.$$

Implications that are absent were omitted because they do not hold without additional hypotheses.

The proofs of $C1$ and $C2$ are provided in an appendix. A weaker form of $C3$, that convergence in mean square implies convergence in distribution, is established in the appendix. Our example of X_n shows that we can have convergence *a.s.* without having convergence in rth mean, and our example Y_n shows the converse. $C5$ asserts that a sequence $\{X_n\}$ that converges in probability to a random variable X has a subsequence that converges *a.s.* to X. In our example of $\{Y_n\}$, such a subsequence is given by, say, all of the n having a representation with $j_n = 1$.

19.6 MUTUAL CONVERGENCE

Thus far, our discussion of convergence assumed that we knew or could hypothesize the limiting random variable. What do we do when we do not know this limiting random variable? Furthermore, if the sequence does not converge and a limiting random variable does not exist, how can we find this out? In the event that we wish to discuss the convergence of a sequence of random variables, but do not know the limiting random variable, we resort to the notion of *mutual*, or *Cauchy, convergence.*

Definition 19.6.1   (Mutual Convergence)   The sequence $\{X_n\}$ mutually converges in the sense $\rightarrow$, where this arrow represents any of the three preceding notions of convergence in probability, rth mean, or $wp\,1$, if

$$|X_n - X_m| \rightarrow 0,$$

no matter how we allow n and m to diverge.

Remark.   A showing of mutual convergence requires that convergence to 0 holds for any conceivable way in which n and m approach infinity. If we took $n = m$, then it would always be the case that $X_n - X_m = 0$!

Theorem 19.6.1   (Mutual Convergence) $\{X_n\}$ mutually converges in either of the three senses given if and only if it converges to some random variable in the respective sense.

Proof.   See Grimmett and Stirzaker [42], Chapter 7, and Loeve [60], I, p. 114.   □

If we wish to show that a particular sequence converges, say, in mean square and are unsure of the limit, if any, then we need only show that

$$\lim_{\substack{n \to \infty \\ m \to \infty}} E|X_n - X_m|^2 = 0.$$

Should this condition fail to hold for some choice of n, m going to infinity, then the sequence is not convergent. Mutual convergence is particularly important in showing that sequences are divergent. In that case, there is no limiting random variable and we cannot exhaustively go through all random variables, showing that convergence fails for each!

19.7 LAWS OF LARGE NUMBERS

We can now apply what we have learned about different modes of convergence to the original problem of the laws of large numbers. We will see that we can justify our concept of the expected value of a random variable as a limiting time average of an appropriate sequence of random variables. Hence, what is "expected" from EX is that this will be the limiting time average of $i.i.d.$ trials. The various forms of convergence of random variables elaborated upon in the preceding section generate a corresponding variety of laws of large numbers. These laws always refer to convergence of averages of random variables, with the random variables occasionally restricted to be $\{0, 1\}$-valued indicator functions for an event. If we eliminate convergence in distribution as too weak to serve as an explanation for the link between relative frequency and probability, then we can establish a law of large numbers based upon the next weakest convergence notion.

Theorem 19.7.1   (Weak Law of Large Numbers)   If $\{I_A(X_i)\}$ are $i.i.d.$, then

$$r_n(A) \xrightarrow{P} P(A).$$

Proof.

$$r_n(A) = \frac{1}{n} \sum_1^n I_A(X_i) \;\Rightarrow\; Er_n(A) = EI_A(X_1) = P(A),$$

$$\mathrm{VAR}(r_n(A)) = \frac{P(A)(1 - P(A))}{n} \le \frac{1}{4n}.$$

Invoking a Chebychev inequality (Section 16.2) yields

$$\frac{1}{4n\epsilon^2} \ge P(|r_n(A) - P(A)| > \epsilon).$$

The theorem follows by letting $n \to \infty$. $\qquad\qquad\qquad\qquad\qquad\qquad\qquad\qquad$ $\square$

Of course, we have also again proven that there is mean square convergence,

$$r_n(A) \xrightarrow{ms} P(A).$$

The stronger form of $a.s./wp1$ convergence is the basis of the next LLN.

Theorem 19.7.2   (Strong Law of Large Numbers (SLLN))   If $\{I_A(X_i)\}$ are $i.i.d.$, then

$$r_n(A) \longrightarrow P(A) \; a.s.$$

This theorem is originally due to E. Borel early in the 20th century, and it asserts the stronger property of convergence $a.s.$ rather than just convergence in probability. We will not prove the theorem; it was generalized by Kolmogorov to Theorem 19.7.3.

Theorem 19.7.3   (Kolmogorov's SLLN))   If $\{X_i\}$ $i.i.d.$ and (a possibly infinite) EX_1 exists, then

$$\frac{1}{n} \sum_1^n X_i \to EX_1 \; (a.s.).$$

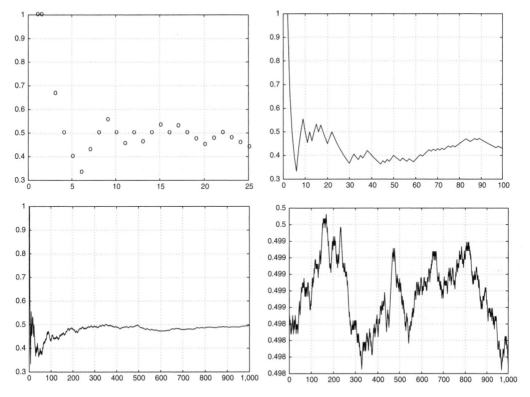

Figure 19.2   Plots of relative frequency for fair coin vs. number of tosses.

Proofs can be found in Billingsley [10], p. 250, Grimmett and Stirzaker [42], p. 326, and Loeve [60], I, p. 251.

Hence, the relative frequency connection to probability is that of probabilistic convergence, an idealization that cannot be observed. The first three plots in Figure 19.2 show the results of a simulation of relative frequency for 25, 100, and 1,000 tosses of a fair coin. The fourth plot in the figure (note the change in vertical scale) shows the relative frequency for trials 9,001 to 10,000.

19.8 AN EXAMPLE OF DIVERGENCE OF TIME AVERAGES

A random variable $X \sim \mathcal{C}(a)$ is Cauchy distributed if, for $a > 0$,

$$f_X(x) = \frac{a}{a^2 + x^2}.$$

Assume that $\{X_i\}$ are $i.i.d.$ $\mathcal{C}(a)$. Let

$$S_n = \frac{1}{n} \sum_1^n X_i.$$

The results on characteristic function of Section 17.8 enable us to conclude that

$$\phi_X(u) = e^{-a|u|}, \quad \phi_{S_n}(u) = e^{-a|u|}.$$

Hence, the average S_n is distributed exactly as an individual summand. As we are attempting to prove that the averages do not converge, we have recourse to the notion of mutual convergence and consider, for $n \geq m$,

$$S_n - S_m = \left(1 - \frac{m}{n}\right)\left(\left[Z_{n,m} = \frac{1}{n-m}\sum_{i=m+1}^n X_i\right] - \left[Y_{n,m} = \frac{1}{m}\sum_{i=1}^m X_i\right]\right).$$

Observe that, because $Y_{n,m}$ and $Z_{n,m}$ are averages of nonoverlapping sets of independent random variables, they are independent of each other. Furthermore, by the results of ϕ_{S_n}, it is the case that they are also identically distributed with common characteristic function $e^{-a|u|}$. Letting $W_{n,m} = Z_{n,m} - Y_{n,m}$, we see that

$$S_n - S_m = \left(1 - \frac{m}{n}\right)W_{n,m}.$$

However,

$$\phi_{W_{n,m}}(u) = \phi_{Z_{n,m}}(u)\phi_{Y_{n,m}}(-u) = e^{-2a|u|}.$$

Hence, $W_{n,m}$ has a fixed, nondegenerate distribution that is independent of n and m. As we are free to choose how n and m go to infinity, we can take $n = 2m$ to derive

$$\phi_{S_{2m}-S_m}(u) = e^{-a|u|}.$$

Clearly, as m goes to infinity, $S_{2m} - S_m$ not only does not converge to 0, but for all m, it has a $\mathcal{C}(a)$ distribution. The first three plots in Figure 19.3 show the results of a simulation of averages of 25, 100, and 1,000 Cauchy random variables. The fourth plot in the figure shows these averages for between 5,001 and 10,000 Cauchy random variables.

How, then, has the Kolmogorov SLLN failed to yield convergence of $\{S_n\}$? The Cauchy random variable X has no expected value; that is,

$$EX = -\int_{-\infty}^0 \frac{a|x|}{a^2 + x^2}\,dx + \int_0^\infty \frac{ax}{a^2 + x^2}\,dx,$$

is undefined, since neither integral is finite. We see that the existence of EX is necessary for an SLLN, and this expectation need not be finite. It is with such an example in mind that, in Definition 9.3.1 of expectation, we required at least one of these two integrals to be finite

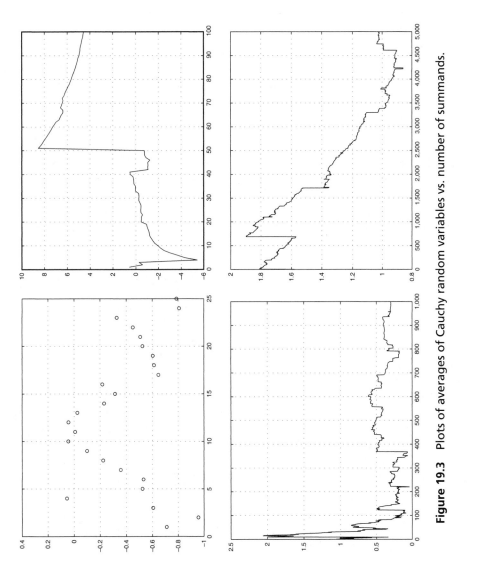

Figure 19.3 Plots of averages of Cauchy random variables vs. number of summands.

for EX to be defined. The tempting alternative of a principal value definition ($\lim_{a \to \infty} \int_{-a}^{a} x f_X$ $(x) \, dx$) was not used because, in this case, it would yield an expectation of 0 that does not correspond to the long-run time average.

19.9 SUMMARY

In Section 3.4, we discussed a possible relation between long-run relative frequencies of occurrences of an event and the probability of that event. This relation was meant to contribute towards an interpretation of probability and towards a choice of axioms for probability. At that point in our discussion, we were unable to move from the qualitative ideas to quantitative ones. The celebrated laws of large numbers, explicated in this chapter, undergird the reasonableness of a frequentist interpretation of probability and defend the link to propensity discussed in Section 3.4.

The laws of large numbers study limits of sequences of random variables—what are often referred to as "asymptotics." Jacob Bernoulli, working at the close of the 17th century, carried out the first examination of asymptotics in making the first formulation of laws of large numbers. In Section 19.2, we quoted from the celebrated writings of Bernoulli, as he wrestled to create an understanding of probability that went beyond the classical one (Chapter 2) in which it is assumed that we can recognize equally probable cases. Bernoulli wanted to know what can justify the use of long-run data to estimate probabilities when we do not have access to the classical probability argument. In our modern notation, Bernoulli wanted to prove that

$$(\forall \epsilon > 0) \quad \lim_{n \to \infty} P\left(\left| \frac{1}{n} \sum_{i=1}^{n} I_A(X_i) - P(A) \right| \geq \epsilon \right) = 0.$$

This becomes the notion of convergence in probability defining a convergence of long-run averages of random variables, which are themselves random variables, to their expectation. An examination of the laws of large numbers requires us to consider countably infinitely many independent random variables $\{X_i\}$.

While this consideration also arose in our discussion of the Central Limit Theorem in Section 17.9, we were able to sidestep it by focusing instead on traditional issues of the convergence of deterministic functions such as characteristic functions. Section 19.3 adopts a similar tactic when it focuses on the asymptotic behavior of the mean square difference between the average of the first n random variables and their common mean m. We use the generalized Chebychev bounds of Section 18.2 and 18.3, in the form

$$P\left(\left| \frac{1}{n} \sum_{i=1}^{n} I_A(X_i) - P(A) \right| \geq \epsilon \right) \leq \frac{E\left(\frac{1}{n} \sum_{i=1}^{n} I_A(X) - P(A) \right)^2}{\epsilon^2},$$

to provide conditions under which convergence of the mean square difference to zero implies that the probability of a significant discrepancy also goes to zero.

In Section 19.4, we prepare for a more direct confrontation with the convergence of random variables by first reviewing traditional notions of the convergence of sequences of numbers and several notions of the convergence of functions. Section 19.5 then translates the convergence-of-functions concepts into the random variable convergence notions of almost surely (also known as "with probability 1"), in probability, in rth mean (especially the case $r = 2$), and in distribution. We ignore the case of pointwise convergence of functions as too strong a notion for random variables. The notion of convergence in distribution has no counterpart in convergence of functions. Implications are pointed out between the convergence notions; some are stronger than others and some cannot be compared with others.

Section 19.6 recalls from analysis the notion of mutual convergence and adapts it readily to the convergence of sequences of random variables. The importance of mutual convergence is that it enables us to discuss convergence without first having to know the limiting random variable, if any.

The variety of convergence notions gives rise to a corresponding variety of laws of large numbers that are discussed in Section 19.7. In the end, laws of large numbers cannot "verify" that long-run relative frequencies provide "the," or even "a" correct interpretation of probability. However, they do establish a mutual consistency that supports the use of long-run relative frequencies (when these make sense) to estimate probabilities.

19.10 APPENDIX: PROOF OF THEOREM 19.5.1 C1, C2

The proof that convergence with probability 1 implies convergence in probability returns us to the techniques of Chapter 3.

Proof of C1. In order to establish that convergence $a.s.$ implies convergence in probability, we introduce the family of events

$$A_{n,\epsilon} = \{\omega : |X_n(\omega) - X(\omega)| \le \epsilon\}.$$

Then, by the meaning of pointwise convergence,

$$C = \{\omega : X_n(\omega) \to X(\omega)\} = \cap_{\epsilon > 0} \cup_{N=1}^{\infty} \cap_{n > N} A_{n,\epsilon}.$$

If $X_n \to X$ $a.s./wp1$, then $P(C) = 1$. Equivalently, by de Morgan's laws,

$$D = C^c = \cup_{\epsilon > 0} D_\epsilon \text{ where } D_\epsilon = \cap_{N=1}^{\infty} \cup_{n > N} A_{n,\epsilon}^c,$$

and

$$P(\cup_{\epsilon > 0} D_\epsilon) = P(D) = 0.$$

Hence, $a.s.$ convergence implies that

$$(\forall \epsilon > 0) \ P(D_\epsilon) = 0.$$

Furthermore, by the probability axiom K4 of monotone continuity,

$$P(\cap_{N=1}^{\infty} \cup_{n > N} B_n) = \lim_{N \to \infty} P(\cup_{n > N} B_n),$$

and we see that

$$0 = P(D_\epsilon) = \lim_{N\to\infty} P(\cup_{n>N} A^c_{n,\epsilon}) \ge$$

$$\lim_{N\to\infty} P(A^c_{N,\epsilon}) = \lim_{N\to\infty} P(|X_N(\omega) - X(\omega)| > \epsilon) = 0.$$

We have, therefore, concluded that *a.s.* convergence implies convergence in probability.   □

Proof of C2. Use the Chebychev inequality in the form

$$\frac{E|X_n - X|^r}{\epsilon^r} \ge P(|X_n - X| \ge \epsilon),$$

to conclude that

$$\lim_{n\to\infty} E|X_n - X|^r = 0 \Rightarrow \lim_{n\to\infty} P(|X_n - X| \ge \epsilon) = 0.$$ □

Instead of proving that convergence in probability implies convergence in distribution, we prove the weaker statement that convergence in mean square implies convergence in distribution. This claim is weaker because convergence in mean square does imply convergence in probability.

Proof. We can show convergence in distribution for the sequence $\{X_n\}$ to X if we can show that

$$\lim_{n\to\infty} |\phi_{X_n}(u) - \phi_X(u)| = 0$$

for all real u. Observe that, from the definition of the characteristic function, it follows that

$$|\phi_{X_n}(u) - \phi_X(u)| = \left|E(e^{iuX_n} - e^{iuX})\right| \le E\left|e^{iuX_n} - e^{iuX}\right|,$$

where we have once again used the fact that $|EZ| \le E|Z|$, as is true for definite integrals and sums. Note that, by the elementary properties of the complex exponential,

$$\left|e^{iuX_n} - e^{iuX}\right| = \left|e^{iu(X_n - X)} - 1\right|.$$

We now make use of the analytical fact that, for real-valued θ,

$$\left|e^{i\theta} - 1\right| \le |\theta|.$$

Thus, for real-valued random variables and real u, we have

$$|\phi_{X_n}(u) - \phi_X(u)| \le E|u(X_n - X)| = |u|E|X_n - X|.$$

From the Schwarz inequality (see Section 9.11), we have

$$(E|Z|)^2 \le EZ^2.$$

Hence,

$$|\phi_{X_n}(u) - \phi_X(u)|^2 \le u^2 E(X_n - X)^2.$$

It is now immediate that convergence in mean square of X_n to X shows that the upper bound to $|\phi_{X_n}(u) - \phi_X(u)|$ goes to zero, and convergence in distribution follows from convergence in mean square. $\qquad\square$

EXERCISES

E19.1 Given a rational $p = m/q$, with integers $0 < m < q$, provide an example of a binary-valued sequence $\{x_n\}$, $x_n \in \{0, 1\}$, such that the sequence of relative frequencies

$$r_n(1) = \frac{1}{n} \sum_{j=1}^{n} x_n$$

converges to p.

E19.2 Repeat Problem 1 for $0 < p < 1$, but with irrationals (not a ratio of integers).

Hint: There exist $0 < m_n < q_n$, q_n increasing, such that

$$p = \lim_{n \to \infty} \frac{m_n}{q_n}.$$

E19.3 Provide an example of a binary-valued sequence $\{x_n\}$ such that the sequence $\{r_n(1)\}$ of relative frequencies diverges.

E19.4   a. Given that $X_n = a_n Z$, where Z is *a.s.* finite, provide nontrivial conditions on the sequence of numbers a_n to ensure that X_n converges almost surely.

  b. What nontrivial conditions on Z and a_n will guarantee convergence in mean square?

E19.5 Given *i.i.d.* random variables $\{X_j\}$ and a fixed set A, show that the formula

$$r_n(A) = \frac{1}{n} \sum_{1}^{n} I_A(X_j)$$

for the relative frequencies of occurrence converges in all four senses (distribution, in probability, in mean square, almost surely).

E19.6 Given *i.i.d.* observations $X_n \sim F$, show that the empirical distribution function (see Section 5.5)

$$\hat{F}_n(x) = \frac{1}{n} \sum_{1}^{n} U(x - X_k)$$

converges in all four senses to the true value $F(x)$ of the cdf F at x.

(Note: It is also true that, simultaneously in all x, $F_n(x)$ converges to $F(X)$ in the sense that $\sup_x |F_n(x) - F(x)|$ converges to 0 $wp1$. This is known as the *Glivenko–Cantelli Theorem* [10], p. 232, [60], I, p. 20)

E19.7   The measurements $\{X_n\}$ are independent with common mean m, but possibly different variances (measurement accuracies) $\{\sigma_n^2\}$. Provide a condition on the sequence of

$$V_n = \sum_1^n \sigma_k^2$$

to ensure that the average of the measurements converges in probability to m.

E19.8   If $\{X_k\}$ are $i.i.d.$ $EX_k = 0$, $\mathrm{VAR}(X_k) = \sigma^2 < \infty$, and

$$S_n = \frac{1}{\sqrt{n}} \sum_{k=1}^n X_k,$$

then use mutual convergence to show that $\{S_n\}$ does not converge in mean square. (Note that, by the Central Limit Theorem, this sequence does converge in distribution to $\mathcal{N}(0, \sigma^2)$.)

E19.9   Consider

$$\Omega = \mathbb{R}, \ f_Z(z) = \frac{1}{2} e^{-|z|}, \ X_n(\omega) = \begin{cases} \sqrt{n} & \text{if } n \le Z(\omega) \le n + \sqrt{n} \\ Z(\omega) & \text{if otherwise} \end{cases}.$$

Using each of the four senses of convergence ($a.s./wp\,1$, mean square, in probability, in distribution), verify whether $\{X_n\}$ converges to some random variable X and identify this limiting random variable (it will be the same in each case). Do this directly by applying the appropriate definition in each case and not by recourse to the theorem on implications between modes of convergence. You will need to look ahead to the Borel–Cantelli Lemma of Section 20.3 to verify convergence almost surely.

E19.10   If $\{X_i\}$ are mean zero and have $\mathrm{COV}(X_i, X_j) = c(i - j)$, where $c(0) = 1$, $c(1) = c(-1) = \frac{1}{2}$, and $c(k) = 0$ for $|k| > 1$, then what can you conclude about the statistical stability of the long-run time averages, $\{\frac{1}{n} \sum_1^n X_i\}$, of these variables by examining convergence in mean square and in probability?

20

Random Processes

20.1 PURPOSE, BACKGROUND, AND ORGANIZATION

With the exception of Chapter 19, we have hitherto concerned ourselves just with finitely many (usually one or two) random variables that could be considered to be components of a finite-dimensional random vector. Countably or uncountably infinite collections of random variables are the subject of *random processes* or random fields. In this chapter we provide definitions, common uses, and elementary properties for a number of useful families of random processes that are encountered in practice. The subject of random processes is a rich one, well developed in its theory and having a myriad of applications of engineering and scientific significance. Our purpose here is to provide a glimpse into the subject that shows you a number of important examples and prepares you for additional studies in this subject. Much more comprehensive treatments of random processes are available from a number of sources, including the classic and still relevant works of Doob [25], Parzen [70], Rosenblatt [78], and Yaglom [104], and the later texts by Gallager [37], Gray and Davisson [40], Grimmett and Stirzaker [42], Karlin and Taylor [52], and Resnick [75]. There exist many monographs devoted to each of the families of random processes we introduce here and to other random processes, as well.

Illustrative examples of random processes and random fields are listed next; throughout, the set T is the set indexing the individual random variables and is often just time.

1. A binary message source in which the random amplitude X_t takes values in $\{0, 1\}$ at times in the set T that may be taken to be the integers.
2. A speech signal (with continuous amplitude X_t) recorded over a time set $T = [a, b]$.
3. A speech signal recorded, time-sampled (T the integers), and digitized (X_t ranging over the finite set of quantization values).
4. The arrival pattern of photons at a detector, where X_t is the integer number of photons that have arrived at time t for $T = [0, \infty)$. Arrival times are the times where there are jumps in the values of X_t.
5. The surface locations $X_{y,z}$ of ions that have been implemented in a layer of a small device. $X_{y,z}$ is binary valued: "1" denotes the presence of an ion at surface location (y, z) and "0" denotes otherwise. The index set $T = [a, b]^2$ is the set of coordinates for the surface of this layer.
6. The ambient electromagnetic field observed at a point by an antenna and consisting of a three-dimensional vector electric field $\mathbf{E}_t$ and a three-dimensional vector magnetic field $\mathbf{H}_t$ that are evolving in times from a set $T = \mathbb{R}$.
7. A three-dimensional electric field $\mathbf{E}_{\mathbf{x},t}$ observed at a point in space $\mathbf{x}$ and a time t, with the index set $T = \mathbb{R}^3 \times \mathbb{R}$ consisting of three real-valued space coordinates and one real-valued time coordinate.
8. Turbulent fluid flow in a region of space where we record the temperature, pressure, and velocity at each point in a cubical region of space and at each time. The index set $T = \mathbb{R} \times \mathbb{R} \times \mathbb{R}^3 \times [a, b]^3 \times \mathbb{R}$.
9. The responses $\mathbf{X}_n$, where $\mathbf{X}_n \in \mathbb{R}^s$ is an s-dimensional measurement, from a field of s sensors, each reporting at integer times in T.

It is characteristic of these examples that the processes or signals in question cannot be expressed as a fixed function of the time/space index $t \in T$ and a given finite number of random variables. Any attempt at such a representation must involve infinitely many random variables. It is in this sense that the subject of random processes departs from what has occupied our attention in all but the last chapter. We organize our development of random processes first by generalities applicable to all instances. We then proceed with separate sections devoted to individual families of random processes sharing some defining characteristic. For each such family, we discuss how to find the joint distributions of any finite number of its random variables, some of its basic properties that provide insight into its applicability, and examples of its applications.

20.2 DEFINITION OF A RANDOM PROCESS

Random processes are interchangeably called *stochastic processes*. Random processes provide us with the mathematical descriptions of random signals existing in time, in space, or in both. Random processes are collections of infinitely many random variables, each one corresponding perhaps to a signal amplitude at any of the uncountably infinitely many times in the continuous time interval $T = [0, \tau]$. We start with the following definition.

Definition 20.2.1   (Random Process)   A random process is a collection $\{X_t : t \in T\}$ of (typically, infinitely many) random variables defined on a probability space $(\Omega, \mathcal{A}, P)$. This collection is indexed by a parameter t drawn from an index set T such that we can consistently specify the multivariate cumulative distribution $F_{X_{t_1}, \ldots, X_{t_n}}$ for any finite subset of random variables in the collection.

Definition 20.2.2   (Random Sequence)   A random process $\{X_t : t \in T\}$ for which the set T is countably infinite is also known as a *random sequence*.

By citing a "consistent" specification, we draw attention to the fact that the same random variable (e.g, X_{t_1}) can appear in more than one collection (e.g., $\{X_{t_1}, X_{t_{10}}\}$ or $\{X_{t_0}, X_{t_1}, X_{t_8}\}$) and the multivariate distributions for the different collections must agree in how they describe the random variables common to these different collections.

Examples of different index sets were provided in the previous section. Typically, the index set T is a set of integers or real numbers representing the times, either discrete or continuous, at which we are interested in the amplitudes of a signal. Thus, X_{t_0} is the signal amplitude considered at time t_0. The index set T might also represent position in space for a random process (now known as a *random field*) defined in space, such as the vector values of electric and magnetic fields distributed in space. Chellappa and Jain [19] collect a number of discussions on the use of *Markov random fields* to model and classify images and to model visual textures.

A random process is a function $X_t(\omega)$, also written as $X(t, \omega)$, of the two variables $t \in T$ and $\omega \in \Omega$. For a given t_0, $X_{t_0}(\omega) = X(t_0, \omega)$ is a random variable as ω ranges over the probability sample space Ω. For a given ω_0, $X_t(\omega_0)$ is a function of t called a *sample function*.

Example 20.1   Explicit Random Processes

A somewhat trivial example of a random process $\{X_t : t \in \mathbb{R}\}$ can be given constructively by taking any n random variables $Y_1(\omega), \ldots, Y_n(\omega)$ that may have any joint distribution and any n nonrandom functions $g_1(t), \ldots, g_n(t)$ defined on $\mathbb{R}$ and setting

$$X_t(\omega) = \sum_{i=1}^{n} Y_i(\omega) g_i(t).$$

If n is large compared with the number of time points at which X_t is examined, then this will appear to be a nondegenerate random process. However, once X_t is examined at sufficiently many time points, the underlying structure will become apparent. An example of such a random process arises in the context of representing a narrowband communications signal in its quadrature components:

$$X_t = Y_1 \cos(\omega_0 t) + Y_2 \sin(\omega_0 t).$$

An alternative representation of the same signal can be given in terms of an envelope A and phase Θ as

$$X_t = A \cos(\omega_0 t + \Theta),$$

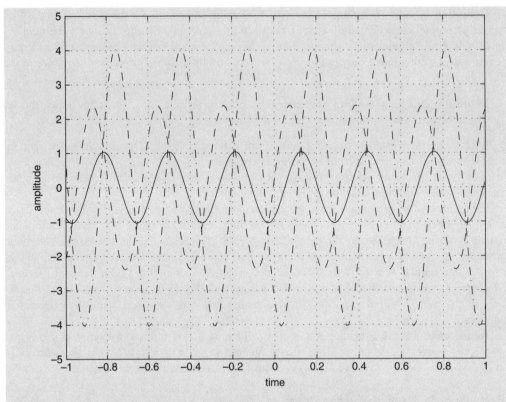

Figure 20.1   Two-parameter signal, three sample functions.

but now the parameter Θ enters nonlinearly in X_t. We illustrate the behavior of this model in Figure 20.1 by plotting the sample functions (time histories of a realization of the random process) for three randomly selected $i.i.d.$ $\mathcal{N}(0, 4)$ choices for Y_1 and Y_2.

A more accurate model would have the $Y_i(t)$ be random processes that are slowly (compared with the frequency ω_0) varying in time. This is discussed more fully in Section 20.11.3.

In each of these examples in which X_t is a function of t and of a finite number of random variables $\{Y_1, \ldots, Y_n\}$ or A and Θ, the methods of Chapter 8, particularly Section 8.5, enable us to compute the required joint distribution of $X_{t_1}, \ldots, X_{t_k}$ from the given joint distribution of, say, the $\{Y_i\}$.

A nontrivial version of the same idea is the *Karhunen–Loeve representation*. (See Loeve [60], II, pp. 144.) In this representation, a random process $\{X_t\}$ is expanded in a generally infinite series in which the expansion coefficient random variables are uncorrelated and the expansion functions are the eigenfunctions of an integral equation with the covariance function as kernel.

In previous chapters, we studied at length a trivial example of random processes—trivial in that T is taken to be a finite set such as $\{1, \ldots, n\}$. In this case, we merely have a random vector $\mathbf{X}$ with components $\{X_i\}$. The use of the term "random process" is reserved to the case of the index set T being either a countably infinite subset of the real numbers or an uncountable interval of real numbers. When the index set is drawn from the multidimensional subsets of $\mathbb{R}^k$ for $k > 1$, we speak of random fields. Henceforth, we assume that T is an appropriate subset of $\mathbb{R}$. Furthermore, if we do not say otherwise, then, in this chapter, $T = \mathbb{R}$.

20.3 BOREL–CANTELLI LEMMA

The celebrated Borel–Cantelli Lemma enables us to draw conclusions about the probabilities of certain events, defined in terms of infinitely many other events involving individual random variables, from the univariate probabilities of these individual random variables. As the lemma focuses on properties of an infinite collection of events, its concerns were not ours until we reached Chapter 19 and this chapter. Let A_k denote an event involving the outcome of the random variable X_{t_k}; this can be generalized to have A_k be determined by the outcomes of several random variables. Given a countably infinite sequence of events $\{A_k\}$, the event

$$B = \cap_{n=1}^{\infty} \cup_{k=n}^{\infty} A_k$$

(read, "For all n, there is a $k \geq n$ such that A_k occurs") is the event that infinitely many of these events occur, and it is denoted either "A_k i.o." or "lim sup A_k." The complement of B (use the de Morgan laws) is the event

$$F = \cup_{n=1}^{\infty} \cap_{k=n}^{\infty} A_k^c.$$

(read, "There exists an n such that, for all $k \geq n$ the event A_k does not occur") that only finitely many of these events occur, and it is denoted either "A_k f.o." or "lim inf A_k."

Lemma 20.3.1   (Borel–Cantelli)

(f.o.) If $\sum_{k=1}^{\infty} P(A_k) < \infty$, then $P(A_k \text{ i.o.}) = 0$ or $P(A_k \text{ f.o.}) = 1$.

(i.o.) If $\sum_{k=1}^{\infty} P(A_k) = \infty$ and $\{A_k\}$ are mutually independent, then $P(A_k \text{ i.o.}) = 1$ or $P(A_k \text{ f.o.}) = 0$.

Proof. We prove the (f.o.) part first. Note that, for any $m > 0$,

$$\cap_{n=1}^{\infty} \cup_{k \geq n} A_k \subset \cup_{k \geq m} A_k.$$

Hence,

$$P(\cap_{n=1}^{\infty} \cup_{k \geq n} A_k) \leq P(\cup_{k \geq m} A_k).$$

By Boole's Inequality, extended to countable unions through K4,

$$P(\cup_{k \geq m} A_k) \leq \sum_{k=m}^{\infty} P(A_k).$$

However, by the hypothesis that $\sum_{k=1}^{\infty} P(A_k)$ is convergent, it follows that, for any $\epsilon > 0$, we can take m_ϵ large enough that

$$\sum_{k=m_\epsilon}^{\infty} P(A_k) < \epsilon.$$

Hence,

$$(\forall \epsilon > 0) \; P(A_k \text{ i.o.}) < \epsilon,$$

and the first half of the lemma follows.

Turning now to the proof of the (i.o.) part under the additional assumption that the events are mutually independent, consider

$$B_n = \cap_{k \geq n} A_k^c.$$

By the hypothesis of mutual independence of $\{A_k\}$, and therefore of $\{A_k^c\}$,

$$P(B_n) = \prod_{k \geq n} P(A_k^c) = \prod_{k \geq n} (1 - P(A_k)).$$

Recall that

$$1 - x \leq e^{-x},$$

to identify the upper bound

$$P(B_n) \leq \prod_{k \geq n} e^{-P(A_k)} = e^{-\sum_{k=n}^{\infty} P(A_k)}.$$

The assumed divergence of $\sum_1^{\infty} P(A_k)$ implies the divergence of $\sum_n^{\infty} P(A_k)$. Hence, we find that $P(B_n) \leq 0$ or that, for any n, $P(B_n) = 0$. Combining this conclusion with the Kolmogorov axiom K4$'$, we have

$$P(\cup_{n=1}^{\infty} B_n) = 0,$$

or, taking complements,

$$P(\cap_{n=1}^{\infty} \cup_{k=n}^{\infty} A_k) = 1,$$

and the second half of the lemma is proven.   □

Example 20.2   Borel–Cantelli

If we know that, for a given collection of events $\{A_k\}$, their individual probabilities satisfy $P(A_k) \leq 1/k^2$, then we can conclude that infinitely many of those events occur with probability 0 or, conversely, that finitely many of those events occur with probability 1. Restated, in this case, there exists a random time N such that, with probability 1, none of the A_k occur for $k > N$. Furthermore, we can reach this conclusion without needing to know about the interactions between events such as those expressed by pairwise probabilities $P(A_i \cap A_j)$. However, if we know only that $P(A_k) \geq 1/k$, then we cannot reach any conclusion based upon the first portion of the Borel–Cantelli Lemma.

From the second portion of the Borel–Cantelli Lemma, if we know also that $\{A_k\}$ are mutually independent, then, from $P(A_k) \geq 1/k$, we can conclude that $\{A_k\}$ occurs infinitely often with probability 1.

Example 20.3   Boundary Exceedance

Consider a random process $\{X_k\}$ with an integer time index set T. We are interested in how often the sample function $X_k(\omega)$ from this process exceeds a prescribed boundary function $\{b_k\}$. For example, we might be interested in whether X_k exceeds a certain maximum value τ and would then take $b_k = \tau$, a constant boundary function. Borel–Cantelli will not tell us the probability of a boundary exceedance occurring with a given time interval. However, it can tell us about whether, considered for all time, the boundary will be exceeded finitely often or not. For ease of calculation, assume that X_k is a speech process with univariate $\mathcal{L}(1)$ marginal pdf. In this case, the probability of exceeding a positive b_k at time k is just

$$P(X_k > b_k) = \int_{b_k}^{\infty} \frac{1}{2} e^{-x} \, dx = \frac{1}{2} e^{-b_k}.$$

From the first portion of the Borel–Cantelli Lemma, for a positive boundary function b_k,

$$\sum_{k=1}^{\infty} P(X_k > b_k) = \frac{1}{2} \sum_{k=1}^{\infty} e^{-b_k}.$$

If this sum is convergent, then no matter what the joint distributions for collections of random variables, the border will be exceeded only finitely often with probability 1. The sum will be convergent if b_k grows rapidly enough, and $b_k = \log k + 2 \log \log k$ will suffice. If, instead, the sum is divergent, then we would need additional information about the joint distributions to know what happens to border exceedances.

20.4 INDEPENDENT RANDOM VARIABLES

20.4.1 Basics of an Independent Random-Variables Process

The simplest example of a random process $\{X_t, t \in T\}$ is one in which for any finite subset $t_1, \ldots, t_n$ of T, we have the mutual independence $\perp\!\!\!\perp_{i=1}^{n} X_{t_i}$ of the random variables. Hence,

$$F_{X_{t_1}, \ldots, X_{t_n}}(x_1, \ldots, x_n) = \prod_{i=1}^{n} F_{X_{t_i}}(x_i).$$

Equivalently, we can express this result in terms of the probability density functions

$$f_{X_{t_1}, \ldots, X_{t_n}}(x_1, \ldots, x_n) = \prod_{i=1}^{n} f_{X_{t_i}}(x_i),$$

and in terms of the characteristic functions

$$\phi_{X_{t_1},\ldots,X_{t_n}}(u_1,\ldots,u_n) = \prod_{i=1}^{n}\phi_{X_{t_i}}(u_i).$$

Completing the specification of the random process simply requires that, for all $t \in T$, we specify the cdfs F_{X_t}. If all of the cdfs are equal to a common cdf F_X, then we have a process of $i.i.d.$ random variables.

A random process of $i.i.d.$ random variables is stationary in the sense to be discussed in Section 20.5.1; for all n and $t_1,\ldots,t_n$ drawn from T, and for any τ such that $t_1 + \tau,\ldots,t_n + \tau$ are also in T, the two cdfs

$$F_{X_{t_1},\ldots,X_{t_n}}(x_1,\ldots,x_n) \text{ and } F_{X_{t_1+\tau},\ldots,X_{t_n+\tau}}(x_1,\ldots,x_n)$$

are identical functions of $x_1,\ldots,x_n$. Shifting each time t_i by the same amount τ does not change the joint cdf; the origin of the time scale is immaterial. We use stationary random-process models for physical phenomena in which the underlying physical mechanism does not depend upon when we started to measure time, although it typically does depend upon time differences; from the process values, we may be able to determine events concerned with duration, but not ones concerned with the time of initiation.

Example 20.4   Boundary Exceedance (Continued) _____

In Example 20.3, we were unable to conclude anything about infinitely often exceedances of the boundary b_k by the random process X_k with $\mathcal{L}(1)$ univariate distributions. If X_k is a speech process that has been time sampled at such a rate that the samples are mutually independent, then we have a process of independent random variables and can access the second portion of the Borel–Cantelli Lemma. If

$$\sum_{k=1}^{\infty}P(X_k > b_k) = \sum_{k=1}^{\infty}e^{-b_k} = \infty,$$

then we can conclude that, with probability 1, X_k will exceed b_k for infinitely many times k. This will be true if the boundary b_k is a slowly growing function of time k, such as any constant τ or the slowly growing $\log k$.

20.4.2 Bernoulli Process

If the index set T is discrete—say, the integers—then this construction makes sense. Unlimited coin tossing provides an example of a simple process of $i.i.d.$ random variables that is known as a Bernoulli process.

Definition 20.4.1   (Bernoulli Process)   A countable collection $\{X_n, n \geq 1\}$ of random variables is a Bernoulli process with parameter $0 \leq p \leq 1$ if the random variables are $i.i.d.$ with common distribution given by

$$P(X_n = 1) = p = 1 - P(X_n = 0).$$

From our earlier work, we know that, say, the random variable

$$Y_n = \sum_{i=1}^{n} X_{t_i},$$

with the $\{t_i\}$ being distinct, has a distribution given by the binomial $\mathcal{B}(n, p)$:

$$P(Y = k) = \binom{n}{k} p^k (1 - p)^{n-k}.$$

If the process random variables $\{X_n\}$ are not restricted to be binary valued, but do have a finite expected value $EX_n = m$, then we know from Section 19.7 that the time averages converge with probability 1 to their common mean

$$P\left(\lim_{n \to \infty} \frac{1}{n} \sum_{1}^{n} X_i = m \right) = 1.$$

If the process random variables have finite mean square, and therefore finite $EX = m$ and $VAR(X) = \sigma^2$, then the convergence of the time averages is also in the mean square:

$$\lim_{n \to \infty} E \left(\frac{1}{n} \sum_{1}^{n} X_i - m \right)^2 = 0.$$

A sample function from an $i.i.d.$ process X_t, where each value is distributed as $\mathcal{P}(1)$, is shown in Figure 20.2.

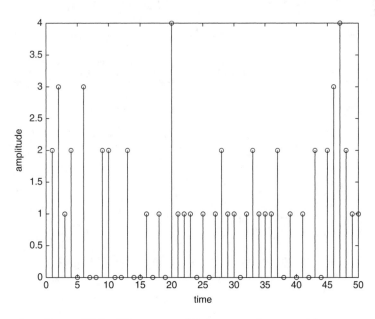

Figure 20.2   IID process of Poisson random variables.

Example 20.5 Pulse Code Modulation (PCM)
We revise the definition of a Bernoulli process slightly to allow ± 1-valued X_n and to allow the time index set $T = Z = \{0, \pm 1, \pm 2, \ldots\}$:

$$P(X_n = 1) = p = 1 - P(X_n = -1).$$

Define $\Theta \sim \mathcal{U}(0, \tau)$, independent of $\{X_n\}$. Introduce a pulse function $p(t)$ that is bounded in magnitude and square integrable:

$$\sup_t |p(t)| < \infty \text{ and } \int_{-\infty}^{\infty} |p(t)| \, dt < \infty.$$

The pulse p is usually thought of as being 0 outside of the interval $[0, \tau)$, but it need not be. Define the message signal random process, resulting from pulse code modulation by X_n, as

$$Y_t = \sum_{n=-\infty}^{\infty} X_n p(t - n\tau + \Theta).$$

We evaluate EY_t, and show that it does not depend upon t. We proceed by first subconditioning on $\{X_n\}$:

$$EY_t = E\left(E\left(Y_t | \{X_n\}\right)\right),$$

$$E\left(Y_t | \{X_n\}\right) = E\left(\sum_{n=-\infty}^{\infty} X_n p(t - n\tau + \Theta) | \{X_n\}\right) = \sum_{n=-\infty}^{\infty} X_n E(p(t - n\tau + \Theta) | \{X_n\}).$$

Since Θ is independent of $\{X_n\}$, it follows that

$$\sum_{n=-\infty}^{\infty} X_n E(p(t - n\tau + \Theta) | \{X_n\}) = \sum_{n=-\infty}^{\infty} X_n \frac{1}{\tau} \int_0^{\tau} p(t - n\tau + \theta) \, d\theta.$$

Note that, by a change of variables,

$$\int_0^{\tau} p(t - n\tau + \theta) \, d\theta = \int_{t-n\tau}^{t-(n-1)\tau} p(u) \, du.$$

Now take expectations over $\{X_n\}$, with $EX_n = 2p - 1$, to find

$$EY_t = \frac{2p-1}{\tau} \sum_{n=-\infty}^{\infty} \int_{t-n\tau}^{t-(n-1)\tau} p(u) \, du = \frac{2p-1}{\tau} \int_{-\infty}^{\infty} p(u) \, du.$$

The integral of the pulse function is finite, since p was assumed to be absolutely integrable. Observe that EY_t does not depend upon the time t.

We continue and evaluate the autocorrelation by making use of the independence of $\{X_n\}$ and Θ to write

$$E(Y_t Y_s) = R_Y(t,s) = \sum_{n=-\infty}^{\infty} \sum_{m=-\infty}^{\infty} E(X_n X_m) E\left(p(t - n\tau + \Theta) p(s - m\tau + \Theta)\right).$$

Of course,

$$E(X_n X_m) = \begin{cases} 1 & \text{if } n = m \\ (2p-1)^2 & \text{if } n \neq m \end{cases},$$

$$E\left(p(t - n\tau + \Theta) p(s - m\tau + \Theta)\right) = \frac{1}{\tau} \int_0^{\tau} p(t - n\tau + \theta) p(s - m\tau + \theta)\, d\theta$$

$$= \frac{1}{\tau} \int_{s-m\tau}^{s-(m-1)\tau} p(t - s - (n-m)\tau + u) p(u)\, du.$$

Assembling results yields

$$R_Y(t,s) = \frac{1}{\tau} \sum_{m=-\infty}^{\infty} \int_{s-m\tau}^{s-(m-1)\tau} p(t - s + u) p(u)\, du$$

$$+ \frac{(2p-1)^2}{\tau} \sum_{n \neq m} \int_{s-m\tau}^{s-(m-1)\tau} p(t - s - (n-m)\tau + u) p(u)\, du.$$

It is now apparent that $R_Y(t,s)$ depends upon t and s only through their difference $t - s$. Given that EY_t does not depend on t, we can conclude that the PCM random process Y_t is wide-sense stationary. (See Section 20.10.1.)

We simplify the previous result for R_Y by making the usual assumption that $p = 1/2$. In this case, $EY_t = 0$ and

$$R_Y(t,s) = \frac{1}{\tau} \sum_{m=-\infty}^{\infty} \int_{s-m\tau}^{s-(m-1)\tau} p(t - s + u) p(u)\, du = \frac{1}{\tau} \int_{-\infty}^{\infty} p(t - s + u) p(u)\, du.$$

The finiteness of R_Y follows from the uniform boundedness of $p(t - s + u)$ and the absolute integrability of $p(u)$. In the specific case of PCM with nonoverlapping rectangular pulses of height a and duration from 0 to τ $(p(t) = a(U(t) - U(t - \tau)))$,

$$R_Y(t,s) = a^2 \begin{cases} 1 - \left|\frac{t-s}{\tau}\right| & \text{if } |t - s| < \tau \\ 0 & \text{if otherwise} \end{cases}.$$

20.4.3 White Noise

If $T = [a, b]$ is an interval of the reals or even if it is all of the reals, then using a process of independent random variables means that we are modeling some phenomenon by claiming that its amplitudes X_t and X_s are independent at times t and s that may be arbitrarily close to each other. This lack of dependence between amplitudes at nearly identical times suggests a pathological process, violating physical continuity for arbitrarily nearby times. Curiously, there is one such process of $i.i.d.$ random variables for $T = \mathbb{R}$, and it is for (idealized) thermal noise mentioned in Section 6.7.2! This idealized model for thermal noise is known as *white noise*. White noise is a process of $i.i.d.$ random variables having a Gaussian cdf with mean zero and infinite variance. The correlation function $R_X(t, s) = E(X_t X_s)$ must then be 0 for $t \neq s$ and infinite for $t = s$, or

$$R_X(t, s) = E(X_t X_s) = \sigma^2 \delta(t - s),$$

where δ is the unit impulse function. (See Section 6.3.) This is a rather pathological state of affairs. At each instant, the probability is 0 that the amplitude is bounded by any finite number! White noise, like Medusa of Greek mythology, is not meant to be looked at directly. We do not measure its individual amplitudes. Rather, white noise makes sense when viewed through filters, much as Medusa could be looked at indirectly. If one considers the response Y_t of a linear system characterized by an impulse response function $h(t, \tau)$ (see Section 10.2.2) and excited by white noise, then

$$Y_t = \int_{-\infty}^{\infty} h(t, \tau) X_\tau \, d\tau.$$

In this case, we can use the methods of Section 10.4 to determine that $EY_t = 0$, as a consequence of $EX_\tau = 0$, and that

$$R_Y(t, s) = E(Y_t Y_s) = \sigma^2 \int_{-\infty}^{\infty} h(t, \tau) h(s, \tau) \, d\tau.$$

The correlation function R_Y will now be well behaved, provided that the impulse response h is square integrable. Furthermore, as we know from Section 17.6, linear combinations of Gaussian random variables result in Gaussian random variables. While this result will be extended in Section 20.11.2 from the linear combinations of finitely many random variables to those of infinitely many random variables, the extension goes through and the $\{Y_t\}$ are jointly Gaussian random variables with zero mean and finite variances.

White-noise processes are also discussed in terms of the concept of a power spectral density, to be introduced in Section 20.10.

20.5 STATIONARY AND ERGODIC PROCESSES

20.5.1 Stationarity

In many applications, we have reason to believe that the physical mechanism generating the observed random signal is a mechanism which has defining characteristics that are constant over time. Thus, thermal noise generated by a resistor is defined by the ambient temperature T and the value R of the resistance, and in many circumstances these two parameters are constant over reasonable periods of immediate interest. Radioactive decay is determined by the fixed half-life of the element involved and by the number of atoms of the element that are present. The number of atoms changes only very slowly over periods that are short compared with the half-life. Hence, for such periods, a fixed Poisson distribution governs the observed decay counts. The number of calls placed to a cellular wireless base station will depend upon the time of day, as well as upon the occurrence of unusual events (e.g., anomalous weather, holidays). However, over a time scale of minutes, we can assume that the environmental characteristics governing call initiation are essentially unchanged. Random signals modeled over a period $T = [\tau_0, \tau_1]$ whose length $t_1 - t_0$ is short enough that the underlying generating mechanisms have unchanged characteristics are said to be *stationary processes* and should be described by joint distributions that depend only upon time differences and not upon the absolute, or true, time. Notwithstanding our remarks, stationary random-process models commonly assume stationarity for all time, using either $T = \mathbb{R}$ for continuous-time models or $T = Z = \{0, \pm 1, \pm 2, \ldots\}$ for discrete-time models or the nonnegative parts of such choices.

Definition 20.5.1   (Stationary Random Process)   A random process $\{X_t, t \in T\}$ is *stationary* if, for any n, any choice of time points $t_1, \ldots, t_n$ lying in the index set T and any τ chosen so that, for all t_i, $\tau + t_i \in T$, the joint cdf satisfies

$$(\forall x_1, \ldots, x_n) \ F_{X_{t_1}, \ldots, X_{t_n}}(x_1, \ldots, x_n) = F_{X_{\tau+t_1}, \ldots, X_{\tau+t_n}}(x_1, \ldots, x_n).$$

In particular, we see that if X_t is stationary and $n = 1$, then, for any $t \in T = [\tau_0, \tau_1]$, take $\tau = \tau_0 - t$ to conclude that

$$F_{X_t}(x) = F_{X_{\tau_0}}(x),$$

and the univariate cdf is constant over time t. If τ_0 is infinite, then any other finite choice in T will do. We often assume that the time $0 \in T$. In that case, we usually choose $\tau = -t$ to conclude that

$$F_{X_t}(x) = F_{X_0}(x).$$

It follows that, for X_t stationary, the mean function

$$EX_t = m_X(t) = m_X(\tau_0),$$

and it is constant over the period T. Furthermore, if X_t is stationary and $n = 2$, then, for any times $s < t$, both in T, take $\tau = \tau_0 - s$ to conclude that

$$F_{X_s, X_t}(x_1, x_2) = F_{X_{\tau_0}, X_{\tau_0 + t - s}}(x_1, x_2).$$

Hence, the cdf for X_s, X_t depends upon the two times only through their time difference. It follows that when X_t is stationary, the autocorrelation function

$$E(X_t X_s) = R_X(t, s) = R_X(\tau_0, \tau_0 + |t - s|) = R(t - s)$$

depends upon the two times t and s only through their difference $t - s$.

Example 20.6   Stationary Process of Independent Variables _____
A stationary process X_t of independent random variables must, by the previous discussion, have all of its univariate distributions identical. Hence, the only stationary process of independent variables is the *i.i.d.* process.

A useful, but abstract, way of discussing stationarity, which will prove useful in the next subsection on ergodicity, is to define a *time-shift operator* T_τ that works as follows: Assume that the index set T is either $\mathbb{R}$ or Z. Given a single random variable X_t,

$$T_\tau X_t = X_{t+\tau},$$

where, if $T = Z$, then τ is restricted to be an integer in Z. Hence, the application of T_τ maps any given random variable into another one that is time shifted from the first by τ. We then extend the definition of T_τ to collections of random variables through

$$T_\tau \{X_{t_1}, \ldots, X_{t_k}\} = \{X_{t_1+\tau}, \ldots, X_{t_k+\tau}\};$$

each random variable in the collection is transformed into one with a time index shifted by τ. Given any event A whose outcome is determined by the collection of random variables $\{X_{t_1}, \ldots, X_{t_k}\}$, we define $T_\tau A$ as the corresponding new event now determined by the time-shifted $\{X_{t_1+\tau}, \ldots, X_{t_k+\tau}\}$. Finally, if the event A is determined by infinitely many random variables (e.g., A involves limits of sequences of random variables), then $T_\tau A$ is the corresponding event when the random variables involved in the definition of A are time shifted (e.g., time shifted in the sequences whose limits are then taken). For example,

$$A = \{\omega : \lim_{n \to \infty} \frac{1}{n} \sum_{k=1}^{n} X_{2k}(\omega) < 0\}$$

is the event that the limit of the time averages along the even times is less than 0. However,

$$T_{-1}A = \{\omega : \lim_{n \to \infty} \frac{1}{n} \sum_{k=1}^{n} X_{2k-1}(\omega) < 0\}$$

is the event that the limit of the time averages along the odd times is less than 0. These might be very different sets, depending upon the behavior of $\{X_k\}$ at even and odd times.

We can redefine a stationary random process $\{X_t : t \in T\}$ to be a random process for which, for all τ and all events A,

$$P(T_\tau A) = P(A).$$

Given any event A defined in terms of the random variables of the process, time shifting A by any amount τ produces a new event $T_\tau A$ with the same probability. Restated, a stationary process measure P is invariant under shift.

20.5.2 Ergodicity

When one thinks of a random process $X_t(\omega)$ (now used as a shorthand for the process and not just as a single random variable) as a function of t and ω, it becomes natural to ask, To what extent are the probability properties describing subsets of Ω displayed in the time behavior through the sample functions of the process? By examining a single sample function, are we likely to be able to learn the expected values or the joint cdfs of the random variables? The general answer is that this will not be the case. A single sample function need not provide full information about the joint cdfs.

Example 20.7 Nonergodic Processes

Consider

$$X_t = A \cos(\omega_c t + \Theta),$$

where $A \perp\!\!\!\perp \Theta$ with $\Theta \sim \mathcal{U}(-\pi, \pi)$. As can be seen in Figure 20.1, a single sample function tells us nothing about the distributions of amplitude and phase. You would have to examine a large number of such sample functions to learn that Θ was uniformly distributed over a period and that A was perhaps Rayleigh.

As a second example, consider a Bernoulli process in which p is first chosen according to a Beta prior. If $p = p_0$ is chosen for the given sample function, then the Strong Law of Large Numbers (see Section 19.7) would assure us that, in the limit of any infinitely long sample function, we could estimate $p = p_0$ correctly with probability 1. However, this would tell us nothing about the prior used to select p. To complete the process description, we would need to see many sample functions and to infer the prior from the observed choices of p.

The condition of ergodicity we will introduce provides a constraint on a random process such that, with probability 1, an individual sample function that is infinitely long can be used to correctly infer the distributions defining the full random process. To define ergodicity, we first need to define an *invariant event*. An event/set A is invariant if

$$(\forall \tau) \; T_\tau A = A;$$

arbitrary time shifts of A return the same set A. A is invariant if and only if every point in A has all of its shifts also in A.

Example 20.8 An Invariant Event or Set

$$A_m = \{\omega : m - 1 < \lim_{n \to \infty} \frac{1}{n} \sum_{k=1}^{n} X_k(\omega) \leq m\}$$

$$= T_\tau A_m = \{\omega : m - 1 < \lim_{n \to \infty} \frac{1}{n} \sum_{k=1+\tau}^{n+\tau} X_k(\omega) \leq m\}.$$

The truth of this claim follows from the limit not depending upon any initial segment $X_1, \ldots, X_\tau$.

Lemma 20.5.1 (Invariant σ-algebra) The collection of invariant events forms an invariant σ-algebra $\mathcal{I}$.

Definition 20.5.2 (**Ergodic Process**) A stationary random process $\{X_t : t \in T\}$ is an *ergodic process* if every invariant event A has $P(A)$ either 0 or 1.

Example 20.9 Ergodic Process and Time Averages _____

For an ergodic process X_t, each of the events A_m in the preceding example must have probability either 0 or 1. If they all have probability 0, then, by the countable additivity axiom,

$$P\left(\cup_{m=-\infty}^{\infty} A_m\right) = \sum_{m=-\infty}^{\infty} P(A_m) = 0.$$

However,

$$\cup_{m=-\infty}^{\infty} A_m = \{\omega : -\infty < \lim_{n\to\infty} \frac{1}{n} \sum_{k=1}^{n} X_k(\omega) < \infty\} = \{\omega : \lim_{n\to\infty} \frac{1}{n} \sum_{k=1}^{n} X_k(\omega) \text{ exists}\}.$$

Thus, if each of the invariant events A_m has probability 0, then, for the ergodic process X_t, with probability 1 the limit of time averages does not exist. However, if, for some $m = m_0$, $P(A_{m_0})$ is positive, then $P(A_{m_0}) = 1$. In this case, with probability 1 the limit exists and takes values in the interval $(m_0 - 1, m_0]$.

We assert the following theorem without proof:

Theorem 20.5.1 (**Ergodicity of *i.i.d.* Processes**) A random process (sequence) of *i.i.d.* random variables is an ergodic process.

The Bernoulli process is a stationary, ergodic process, and the set of sample functions whose time averages converge to p is a set of probability 1. In this sense, almost every sample function provides full information about the joint distributions of the process; given p, we can compute any desired joint probabilities. An ergodicity condition for Gaussian random processes is provided in Section 20.11.1.

20.5.3 Stationary Ergodic Processes

Having discussed stationarity and ergodicity, we are now ready for these properties to do their work. We state the next theorem without proof.

Theorem 20.5.2 (**Convergence for Stationary Ergodic Processes**) If $\{X_t : t \in \mathbb{R}\}$ or $\{X_t : t \in Z\}$ is a stationary and ergodic process and h is an arbitrary real-valued function such that $Eh(X_0) = m_h$ is finite, then, with probability 1,

$$\lim_{n\to\infty} \frac{1}{n} \sum_{k=1}^{n} h(X_k) = Eh(X_0);$$

$$\lim_{a\to\infty} \frac{1}{2a} \int_{-a}^{a} h(X_t)\,dt = Eh(X_0).$$

Long-run time averages of functions of a stationary ergodic process converge with probability 1 to a constant (a consequence of ergodicity) that is their expected value $Eh(X_0)$. This being the case, with probability 1 we are guaranteed that averages calculated from a single sample function will converge to the statistical average over the ensemble of such sample functions. The importance of stationary ergodic processes is that we can estimate elements of the probability description of the process from appropriate long-run time averages of sample functions. (Further information on stationarity and ergodicity is available from many sources, including Grimmett and Stirzaker [42] and Loeve [60].)

20.6 MARKOV CHAINS IN DISCRETE TIME

20.6.1 Defining a Markov Chain in Discrete Time

A very useful class of random processes is those in which the random variables take values only in a given finite or countably infinite set $\mathcal{X} = \{\xi_i\}$ known as their state set. Furthermore, chains in discrete time have a countably infinite index set T that we will take to be the integers. In this digital age, much of the information we encounter is prepared in this state of a process taking on only finitely many values at regular, discrete times.

Definition 20.6.1   (Markov Chain)   A random process $\{X_t : t \in T\}$ is a *Markov chain* in discrete time if

1. The index set T is countably infinite and can be taken to be the integers $0, 1, \ldots$.
2. There is a countable set $\mathcal{X}$ such that, for all $t \in T$,

$$P(X_t \in \mathcal{X}) = 1.$$

3. For all $t > 0$ and for all $x_0, \ldots, x_t$ in $\mathcal{X}$,

$$P(X_{t+1} = x_{t+1} | X_t = x_t, \ldots, X_0 = x_0) = P(X_{t+1} = x_{t+1} | X_t = x_t). \qquad (20.1)$$

Equation (20.1) is the key Markovian property that

The future X_{t+1} at time $t + 1$ is independent of the past $X_{t-1}, \ldots, X_0$, given knowledge of the present X_t.

Treatment of such Markov chains is available from many sources, including Ethier and Kurtz [26], Gallager [37], and Kemeny, Snell, and Knapp [53]. The function $P(X_{t+1} = x_{t+1} | X_t = x_t)$ is known as the *one-step transition probability*.

Definition 20.6.2   (Homogeneous Markov Chain)   A Markov chain is homogeneous if its one-step transition probability matrix between states does not depend upon the time at which the transition occurs.

In this case, for all $t > 0$ and x and y,

$$P(X_t = y | X_{t-1} = x) = P(X_1 = y | X_0 = x).$$

Example 20.10   Homogeneous Markov Chain

The simplest example of a homogeneous Markov chain is a Markov chain having only a single state and a transition probability of return to that state of unity. A much more interesting example that can be used to describe a binary message source has two states, one for each of the possible transmitted bits, which we take to be $\{0, 1\}$. The transition is that between successive bits; we assume that the last bit sent can influence the next bit, but that earlier bits are irrelevant. In this case, there are two probabilities: $p_{0,0}$, that a "0" will be followed by another "0," and $p_{1,1}$, that a "1" will be followed by another "1":

$$P(X_t = 1|X_{t-1} = 1) = p_{1,1}, \quad P(X_t = 0|X_{t-1} = 1) = 1 - p_{1,1},$$

$$P(X_t = 0|X_{t-1} = 0) = p_{0,0}, \quad P(X_t = 0|X_{t-1} = 1) = 1 - p_{0,0}.$$

We can also apply this model to a *burst error* model. In this case, say, you are keyboarding and $X_t = 1$ if you enter the tth symbol correctly and $X_t = 0$ if you make an error in entering this symbol. It is more likely that you will make an error on the $t + 1$st symbol if you have made an error on the tth symbol than if you had entered the tth symbol correctly. For example, we might have $p_{1,1} = .98$ and $p_{0,0} = .1$. Given that you just made an error, the probability is .1 that you will follow it with another error. Given that you have just entered a symbol correctly, the probability is .98 of entering the next symbol correctly. A sample function from such a Markov process will typically show long runs of correct symbols followed by short bursts of errors.

An intuitively useful way to describe a homogeneous Markov chain is through a state transition diagram. Each possible state is a node on a weighted, directed graph, typically shown as a circle containing the state label. The directed links between the nodes are assigned probability values (weights) with $P(X_1 = y|X_0 = x) = p_{x,y}$ denoting the one-step transition probability. The link is omitted when its weight would be zero (i.e., if there is zero probability of a transition in the direction of the link being made in one step).

Example 20.11   Markov State Transition Diagram

Figure 20.3 shows such a state transition diagram for a homogeneous Markov chain having five nodes.

A computationally useful way to describe a homogeneous Markov chain when the state set $\mathcal{X}$ is of finite size s is by an $s \times s$ matrix

$$\mathbb{P} = [p_{i,j}], \quad p_{i,j} = P(X_1 = \xi_j|X_0 = \xi_i)$$

of the one-step transition probabilities. No matter what state ξ the chain is currently in, it is required to make a transition to some state (possibly itself) in $\mathcal{X}$. The immediate implication is that

$$\sum_{\xi_j \in \mathcal{X}} P(X_1 = \xi_j|X_0 = \xi) = 1;$$

it follows that

The row sums of this one-step transition matrix $\mathbb{P}$ must be unity.

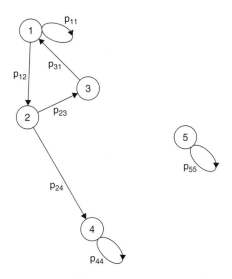

Figure 20.3   Markov state transition diagram.

Of course, all of the elements of $\mathbb{P}$, being probabilities, must lie in $[0, 1]$. In terms of the state transition diagram, all of the weights on the links must be numbers in $[0, 1]$ and, for each node, the sum of the weights must be unity on links leaving that node.

Example 20.12   Markov State Transition Diagram (Continued) _____

For the state transition graph of Figure 20.3, we must have

$$p_{1,1} + p_{1,2} = 1, \quad 0 < p_{1,1} < 1,$$

$$p_{2,3} + p_{2,4} = 1, \quad 0 < p_{2,3} < 1,$$

$$p_{3,1} = 1,$$

$$p_{4,4} = 1,$$

$$p_{5,5} = 1.$$

20.6.2 Dynamics: Evolution in Time

Much is known about how homogeneous Markov chains evolve in time. Let $\pi^{(n)}$, with $\pi^{(n)}(\xi) = P(X_n = \xi)$, be a row vector specifying the probability mass function over the states $\mathcal{X}$ at time n. $\pi^{(0)}$ is the initial pmf—the one that starts the process at time 0. We can calculate the pmf at time 1 from the one-step transition probabilities and the Total Probability Theorem:

$$\pi^{(1)}(\xi') = \sum_{\xi \in \mathcal{X}} \pi^{(0)}(\xi) P(X_1 = \xi' | X_0 = \xi).$$

We rewrite this result in matrix–vector form as

$$\pi^{(1)} = \pi^{(0)} \mathbb{P}.$$

This result generalizes to arbitrary time t:

$$\pi^{(t)} = \pi^{(t-1)} \mathbb{P}. \tag{20.2}$$

Given these two results, and applying them iteratively, we see that

$$\pi^{(t)} = \pi^{(0)} \mathbb{P}^t. \tag{20.3}$$

The tth power $\mathbb{P}^t$ of the one-step transition matrix $\mathbb{P}$ is the t-step transition matrix for a homogeneous Markov chain.

A version of this result for a Markov chain that is not necessarily homogeneous is to let $\mathbb{P}^{(s,t)}$ denote the transition (conditional probability) matrix taking the process from where it is at time s to where it will be at time $t > s$:

$$\mathbb{P}^{(s,t)} = [P^{(s,t)}_{i,j}], \quad P^{(s,t)}_{i,j} = P(X_t = \xi_j | X_s = \xi_i).$$

Theorem 20.6.1   (Chapman–Kolmogorov)   If $t > s > r$, then

$$p^{(r,t)}_{i,j} = \sum_{\xi_k \in \mathcal{X}} p^{(r,s)}_{i,k} p^{(s,t)}_{k,j}$$

or, equivalently, for all $t > s > r$,

$$\mathbb{P}^{(r,t)} = \mathbb{P}^{(r,s)} \mathbb{P}^{(s,t)}.$$

Proof.   By the Total Probability Theorem,

$$p^{(r,t)}_{i,j} = \sum_{\xi_k \in \mathcal{X}} p^{(r,s)}_{i,k} P(X_t = \xi_j | X_s = \xi_k, X_r = \xi_i).$$

However, by the Markov property of Eq. (20.1),

$$P(X_t = \xi_j | X_s = \xi_k, X_r = \xi_i) = P(X_t = \xi_j | X_s = \xi_k) = p^{(s,t)}_{k,j}.$$

This proves the first claim. The second claim just rewrites the first claim in matrix form.   □

If the Markov chain is homogeneous, then the transition matrix $\mathbb{P}^{(s,t)}$ depends upon the times $t > s$ only through their difference $t - s$, and we can change notation to reflect this through $\mathbb{P}^{(t-s)}$. We have shown in Eq. (20.3), or it can be shown by iteration in the Chapman–Kolmogorov Theorem, that

$$\mathbb{P}^{(t-s)} = \mathbb{P}^{t-s};$$

the transition matrix from time s to time t is the $(t - s)$th power of the one-step transition matrix.

Example 20.13   English Text Production _____
Consider the following very rudimentary model for English text production that might be of interest in communications, information processing, and artificial intelligence. Our vocabulary consists only of the following seven words and the period punctuation mark:

    the, gray, cat, is, in, a, bag, .

These seven words and the punctuation mark "." are the elements of our state space $\mathcal{X}$, with ξ_i being the ith such word in the order given; thus, ξ_4 is the word "is". We will model text production with these seven words as a homogeneous Markov chain. By using a Markov chain model, we can impose some constraints upon how words immediately follow each other. It is true that in natural language the constraints between words can reach over much longer distances than just the adjacent word. Hence, our model is a toy one to illustrate the operation of Markov chains, rather than a realistic one of text production.

After a little thought, we adopted the following 8×8 one-step transition matrix:

$$
\mathbb{P} = \begin{pmatrix}
0 & .3 & .4 & 0 & 0 & 0 & .3 & 0 \\
0 & 0 & .5 & 0 & 0 & 0 & .4 & .1 \\
0 & 0 & 0 & .6 & 0 & 0 & .1 & .3 \\
.3 & .2 & 0 & 0 & .25 & .25 & 0 & 0 \\
.5 & 0 & 0 & 0 & 0 & .5 & 0 & 0 \\
0 & .3 & .4 & 0 & 0 & 0 & .3 & 0 \\
0 & 0 & 0 & .3 & 0 & 0 & 0 & .7 \\
0 & 0 & 0 & 0 & 0 & 0 & 0 & 1
\end{pmatrix}.
$$

Note that the row sums are all unity, as is necessary for a transition matrix. There are zeros on the main diagonal, with the exception of the $(8,8)$ element. The zeros prevent a word from following itself. The exception is for the terminal punctuation period.

For the purposes of this example, we choose a prior probability row vector

$$
\pi^{(0)} = \begin{pmatrix} .4 & 0 & 0 & .3 & 0 & .3 & 0 & 0 \end{pmatrix}.
$$

This prior allows a sentence to begin only with the, is, or a.

A Matlab simulation of the operation of this Markov chain was run with a limit of nine words. As can be seen from the one-step transition matrix, once a period is selected, only periods will be selected from then on. Hence, sample sentences may end with multiple periods. The following 15 sample "sentences" were the first produced by the simulation:

```
'a'    'gray'  'bag'   '.'    '.'    '.'    '.'    '.'    '.'
'is'   'in'    'a'     'bag'  '.'    '.'    '.'    '.'    '.'
'is'   'a'     'gray'  'cat'  '.'    '.'    '.'    '.'    '.'
'is'   'the'   'gray'  'cat'  '.'    '.'    '.'    '.'    '.'
'a'    'cat'   'is'    'in'   'a'    'gray' 'bag'  '.'    '.'
'is'   'in'    'the'   'cat'  'is'   'the'  'bag'  '.'    '.'
'the'  'bag'   '.'     '.'    '.'    '.'    '.'    '.'    '.'
'is'   'a'     'cat'   'is'   'the'  'cat'  'is'   'in'   'the'
'a'    'cat'   'is'    'in'   'a'    'cat'  'bag'  '.'    '.'
'a'    'cat'   '.'     '.'    '.'    '.'    '.'    '.'    '.'
'the'  'gray'  'bag'   'is'   'the'  'bag'  'is'   'gray' 'cat'
'a'    'cat'   'is'    'in'   'the'  'cat'  'is'   'in'   'a'
'is'   'in'    'the'   'cat'  '.'    '.'    '.'    '.'    '.'
```

```
'the'  'bag'  'is'  'gray'  'cat'    '.'     '.'     '.'      '.'
'a'    'bag'   '.'    '.'     '.'     '.'     '.'     '.'      '.'
```

Two of these sentences are grammatically correct and meaningful.

Example 20.14   Markov State Probabilities _____

We see that most of our "sentences" in the previous example end with a period. How probable are they to end that way? To answer this question, we need to calculate the state distribution vector $\pi^{(8)}$, where we take into account the fact that we start at $t = 0$ and generate a string of nine words. From Eq. (20.3), we have

$$\pi^{(8)} = \pi^{(0)} \mathbb{P}^8.$$

Allowing Matlab to do the algebra, we find that

$$\mathbb{P}^8 = \begin{pmatrix} .0202 & .0259 & .0369 & .0257 & .0083 & .0185 & .0372 & .8272 \\ .0140 & .0231 & .0385 & .0287 & .0044 & .0131 & .0389 & .8392 \\ .0157 & .0203 & .0308 & .0503 & .0103 & .0136 & .0331 & .8259 \\ .0325 & .0299 & .0315 & .0433 & .0201 & .0285 & .0334 & .7808 \\ .0354 & .0305 & .0293 & .0341 & .0211 & .0312 & .0306 & .7877 \\ .0213 & .0277 & .0398 & .0278 & .0087 & .0196 & .0401 & .8148 \\ .0074 & .0094 & .0141 & .0241 & .0050 & .0064 & .0152 & .9184 \\ 0 & 0 & 0 & 0 & 0 & 0 & 0 & 1 \end{pmatrix},$$

and the vector describing the probability mass function for the selection of states at time 8 is

$$\pi^{(8)} = (.0242 \quad .0276 \quad .0362 \quad .0316 \quad .0120 \quad .0218 \quad .0370 \quad .8096).$$

Hence, the probability is .81 of a terminal period in our sentences of nine words. In a sample of 15 sentences, the expected number of sentences that do not end with a period is 3, and this is actually what we observed in our sample in the previous example.

20.6.3 Communicating Classes

While we can calculate the state distribution $\pi^{(n)}$ for large n, depending upon how large the state set $\mathcal{X}$ is, eventually we will exceed our computational patience. A detailed picture is available concerning the long-run behavior of homogeneous Markov chains. However, before we can present elements of this picture, we need to better understand the kinds of motion between states in a Markov chain.

The states in a finite-state Markov chain in discrete time can be divided into groups according to whether they are accessible to each other. In a countable-state Markov chain, we have a further distinction as to whether a state recurs infinitely often or not.

Definition 20.6.3   (State Classification)   A variety of states and classes of states are identified:

1. State ξ_j is *accessible* from a state ξ_i if there is an n such that the n-step transition probability $p_{i,j}^{(n)} > 0$.

2. States ξ_i and ξ_j *communicate* if each is accessible from the other.
3. State ξ_i is *absorbing* if $p_{i,i}^{(1)} = 1$.
4. State ξ_i is *persistent* or *recurrent* if the probability (denoted by $f_{i,i}$) of eventually returning to ξ_i is 1.
5. State ξ_i is *transient* if the probability $f_{i,i}$ of eventually returning to ξ_i is less than 1.
6. The set $C(j)$ of states that communicate with state ξ_j is said to be the *communicating class* of state ξ_j.
7. If $C(j)$ is empty, then state ξ_j is said to be a *nonreturn* state and is not accessible from itself.
8. A nonempty set C of states is said to be *closed* if no state outside of C is accessible from any state in C; otherwise, it is *open*.

It is easily verified that if $C(i)$ and $C(j)$ are two communicating classes, then either they are the same set or they are disjoint (have no states in common).

Example 20.15   Classes of Markov States _____

Consider the Markov chain for our text production model with one-step transition probability matrix $\mathbb{P}$ given in Example 20.13. If one examines its eighth power, given later in Example 20.14, one sees that there are no zero terms in the first seven rows. Hence, each of the first seven states are accessible from each of the others in eight steps (actually in fewer, but we have already made the calculation whose results we are using), and they all form one communicating class. This class is open, since the eighth state can be reached from the class and there is no return from the eighth state. The eighth state is absorbing and therefore does not communicate with any of the other states in the chain. The eighth state is persistent, and the other seven states are transient.

Example 20.16   Classes of Markov States (Continued) _____

Consider the Markov chain with state transition diagram given in Figure 20.3. States 1, 2, and 3 are accessible from states 1, 2, or 3 and from no others. State 4 is accessible from state 2 and from itself. State 5 is accessible only from itself.

States 1, 2, and 3 form a communicating class that is open. (A transition to state 4 can happen.) State 4 is in a communicating class with itself, as is state 5.

States 4 and 5 are absorbing states; once entered, they cannot be left. They each form a closed communicating class. These two states are persistent, whereas the other states are all transient.

An example of a nonreturn state is state 1 for the one-step transition 3×3 matrix

$$\mathbb{P} = \begin{pmatrix} 0 & .5 & .5 \\ 0 & .4 & .6 \\ 0 & .6 & .4 \end{pmatrix}.$$

In this case, state 1, if entered, transitions with equal probability to either state 2 or state 3. Once states 2 or 3 have been entered, future motion is restricted to them. Thus, there is a probability of .4 of staying in the state and a probability of .6 of moving to the other one.

We need one further classification tool: periodicity. The greatest common divisor (gcd) of a set of integers is the largest integer that divides each of them; the gcd will be 1 if the integers are relatively prime.

Definition 20.6.4   (Periodicity)   A state ξ_i in a Markov chain has *period* $d(i)$ if $d(i)$ is the greatest common divisor of the set of times

$$\{n : P_{i,i}^{(n)} > 0\}$$

in which there is a positive probability of returning to ξ_i.

Consider the trivial two-state chain with

$$\mathbb{P} = \begin{pmatrix} 0 & 1 \\ 1 & 0 \end{pmatrix}.$$

The dynamics of this closed communicating class is simple. The chain oscillates between the two states. It will return to a given state in exactly an even number of transitions and only in those. These two states have a periodicity of 2. What we have just seen, the two states having the same periodicity, is a fact of closed communicating classes: All states in the same closed communicating class have the same period. If the period is 1, then we say that the Markov chain is *aperiodic*. The examples of Markov chains given up to now were all aperiodic.

20.6.4 First-Passage Probabilities

Let $f_{i,j}^{(n)}$ denote the probability of a *first passage* in n steps from state ξ_i to state ξ_j. $f_{i,j}^{(n)}$ is the conditional probability of a path that is conditioned to start at ξ_i and then end in n steps at ξ_j without having reached ξ_j in any fewer steps. If $n = 1$, then a first passage occurs in one step, and

$$f_{i,j}^{(1)} = p_{i,j}.$$

If $n \geq 2$, then a first passage from ξ_i to ξ_j occurs by taking a first step to a state, say, ξ_k that is other than ξ_j, followed by a first passage in $n-1$ steps from ξ_k to the goal of ξ_j. From the basic Markov property, we easily write the recursion for $n > 1$:

$$f_{i,j}^{(n)} = \sum_{k \neq j} p_{i,k} f_{k,j}^{(n-1)} = \sum_{k=1}^{s} p_{i,k} f_{k,j}^{(n-1)} - p_{i,j} f_{j,j}^{(n-1)}.$$

Let diag($\mathbb{A}$) denote a diagonal matrix formed out of the diagonal of a square matrix $\mathbb{A}$. This operation is linear. Let

$$\mathbb{F}^{(n)} = [f_{i,j}^{(n)}].$$

We can now rewrite the recursion for the first passage probabilities as

$$\mathbb{F}^{(1)} = \mathbb{P}, \ \ \mathbb{F}^{(n)} = \mathbb{P}\left(\mathbb{F}^{(n-1)} - \operatorname{diag}(\mathbb{F}^{(n-1)})\right).$$

Alternatively, given that we start at state ξ_i, we can reach state ξ_j in n steps, not necessarily for the first time, by reaching ξ_j for the first time in $m \leq n$ steps and then returning to ξ_j in the remaining $n - m$ steps. Adopting the convention that $\mathbb{P}^{(0)} = \mathbb{I}$, the identity matrix, we can write the recursion

$$p_{i,j}^{(n)} = \sum_{m=1}^{n} f_{i,j}^{(m)} p_{j,j}^{(n-m)}.$$

Or we can solve for

$$f_{i,j}^{(n)} = p_{i,j}^{(n)} - \sum_{m=1}^{n-1} f_{i,j}^{(m)} p_{j,j}^{(n-m)}.$$

The probability $f_{i,j}$ that we can ever make a transition from state ξ_i to state ξ_j is

$$f_{i,j} = \sum_{n=1}^{\infty} f_{i,j}^{(n)}.$$

If $f_{i,j} = 0$, then state ξ_j is not accessible from state ξ_i, and the two states are not in the same communicating class. If $f_{i,j} > 0$, then ξ_j is accessible from ξ_i. If it is also true that $f_{j,i} > 0$, then the two states are in the same communicating class. If $f_{i,j} = f_{j,i} = 1$, then the states are in a closed communicating class.

The expected waiting time $T_{i,j}$ to a first transition from state ξ_i to ξ_j is infinite if $f_{i,j} < 1$; there is a positive probability of never making this transition. If $f_{i,j} = 1$, then

$$ET_{i,j} = \sum_{n=1}^{\infty} n f_{i,j}^{(n)}.$$

Introduce the generating function (see Section 17.10)

$$F_{i,j}(s) = \sum_{n=1}^{\infty} f_{i,j}^{(n)} s^n.$$

Clearly,

$$F_{i,j}(1) = f_{i,j}.$$

If $F_{i,j}(1) = 1$, then, in terms of the generating function,

$$ET_{i,j} = \left. \frac{dF_{i,j}(s)}{ds} \right|_{s=1} = \lim_{s \uparrow 1} \frac{1 - F(s)}{1 - s}.$$

An alternative means of evaluating the mean waiting time $ET_{j,j}$ to return to a given state ξ_j will be provided in Section 20.6.5.

Example 20.17   First Transitions _____

We return to the English text production presented in Example 20.13 and ask for the probability that the word bag first appears as the third word in a sentence beginning with the word the.

In the enumeration of words used in that example, the is the first word and bag is the seventh word. Hence, we are interested in $f_{1,7}^{(2)}$. We can evaluate

$$\mathbb{F}^{(2)} = \mathbb{P}\left(\mathbb{F}^{(1)} - \mathrm{diag}(\mathbb{F}^{(1)})\right) = \mathbb{P}\left(\mathbb{P} - \mathrm{diag}(\mathbb{P})\right)$$

$$= \begin{pmatrix}
0 & 0 & .15 & .33 & 0 & 0 & .16 & .36 \\
0 & 0 & 0 & .42 & 0 & 0 & .05 & .43 \\
.18 & .12 & 0 & .03 & .15 & .15 & 0 & .07 \\
.125 & .165 & .32 & 0 & 0 & .125 & .245 & .02 \\
0 & .3 & .4 & 0 & 0 & 0 & .3 & 0 \\
0 & 0 & .15 & .33 & 0 & 0 & .16 & .36 \\
.09 & .06 & 0 & 0 & .075 & .075 & 0 & 0 \\
0 & 0 & 0 & 0 & 0 & 0 & 0 & 0
\end{pmatrix}.$$

Hence, $f_{1,7}^{(2)} = .16$. In Example 20.13, we found that 3 of the 15 sentences began with the, and one of those had bag as the third word for the first time. The sample is too small for reliable inference to $f_{1,7}^{(2)}$.

Observe that the matrix $\mathbb{F}^{(2)}$ is not a one-step transition matrix, in that its row sums do not add to 1. Thus, you cannot go from the state of a "period" back to a "period" for the first time in two steps. Once you enter the state of a "period," you return to it immediately.

20.6.5 Long-Run Dynamics

Returning to the question of the long-run behavior of a finite-state Markov chain, we see that, as the number n of transitions grows, the probability approaches 1 that the chain is no longer in any open communicating class. The greater is the number of transitions, the greater is the probability that there will be a transition out of the open class. However, once this transition is made, the chain can never return to the members of the open class, by its very definition. Therefore, asymptotically in n, we can find ourselves only in the closed communicating classes. As any two communicating classes either are the same or are disjoint, eventually the chain will enter and remain in one of its closed communicating classes. Its further dynamics will be determined just by the dynamics of the closed communicating class it entered. The essence of the long-run behavior of a finite-state Markov chain is the behavior of a closed communicating class. A Markov chain that is composed of a single closed communicating class is said to be *irreducible*.

Theorem 20.6.2   (Asymptotic Distribution)   A finite-state homogeneous Markov chain that is irreducible and aperiodic has a unique limiting distribution π, written as a row vector, over its state space $\mathcal{X}$ that is given by

$$\pi = \pi\mathbb{P},\tag{20.4}$$

where $\mathbb{P}$ is the one-step transition matrix.

Proofs of this theorem and its straightforward extension to the periodic case are available in many texts, including Doob [25], Karlin and Taylor [52], Parzen [70], and Resnick [75]. A corollary to the theorem enables us to determine the mean waiting times to return to the individual states.

Corollary 20.6.1   (Mean Waiting Time to Return)   A finite-state homogeneous Markov chain that is irreducible and aperiodic has mean waiting times to return to states given by

$$ET_{j,j} = \frac{1}{\pi_j},$$

where π is the asymptotic or stationary distribution for the chain.

Example 20.18   Stationary Prior

Consider the general two-state case

$$\mathbb{P} = \begin{pmatrix} p_{1,1} & p_{1,2} \\ p_{2,1} & p_{2,2} \end{pmatrix},$$

where the row sums are unity. The unique row vector solution as a probability vector to

$$\pi = \pi \mathbb{P}$$

is given by

$$\pi = (\pi_1, \pi_2), \quad \pi_2 = 1 - \pi_1, \quad \pi_1 = \frac{p_{2,1}}{1 + p_{2,1} - p_{1,1}}. \tag{20.5}$$

20.6.6 Stationary Markov Chain

Thus far, we have focused on finite-state, discrete-time parameter Markov processes that are homogeneous in that the one-step transition matrix $\mathbb{P}$, or $\mathbb{P}^{(1)}$, does not depend upon the time t at which the transition is made. The notion of stationarity, discussed in Section 20.5.1, is that any collection of finitely many random process random variables $X_{t_1}, \ldots, X_{t_n}$, all of $t_i \in T$, has exactly the same joint distribution as any collection $X_{t_1+\tau}, \ldots, X_{t_n+\tau}$ where the times have been shifted by any common τ such that the shifted times are also in T. The physical meaning of this condition is that the mechanism generating this process is not changing in time; all that matters are time differences, but not the absolute time origin itself. The homogeneous Markov chains we have discussed are not, in general, stationary. This can be seen from Eq. (20.3), where we calculated the state distribution at time t in terms of the state distribution at time 0:

$$\pi^{(t)} = \pi^{(0)} \mathbb{P}.$$

In general, $\pi^{(t)}$ varies with t.

A necessary condition to ensure that a homogeneous Markov chain is stationary is suggested by Theorem 20.6.2. The distribution of states at time t will be independent of t if and only if the initial distribution is chosen such that

$$\pi^{(0)} = \pi^{(0)} \mathbb{P}. \tag{20.6}$$

It is not difficult to verify that this condition is also necessary.

Theorem 20.6.3 (Markov Chain Stationarity) A finite-state Markov chain is stationary if and only if it is homogeneous and its initial state distribution satisfies Eq. (20.6).

Theorem 20.6.4 (Markov Chain Ergodicity) A finite-state Markov chain is ergodic if it is irreducible and aperiodic.

20.7 INDEPENDENT INCREMENT RANDOM PROCESSES

Given a random process $\{X_t : t \in T\}$, we define its increments through

$$D_{s,t} = X_t - X_s.$$

A random process of *independent increments* is defined in Definition 20.7.1.

Definition 20.7.1 (Independent Increments) The random process $\{X_t\}$ with associated increments $\{D_{s,t}\}$ is a process of independent increments if, for all n and for all times $t_0 < t_1 < \ldots < t_n$, the random variables $\{D_{t_0,t_1}, \ldots, D_{t_{n-1},t_n}\}$ are mutually independent.

Example 20.19 Independent Increments Process _____

As an example, take T to be the set of nonnegative integers, let $\{Y_i\}$ be a set of mutually independent random variables, and define the process (where, for clarity, we use n as an integer-valued time index)

$$X_n = \sum_{k=0}^{n} Y_k.$$

It is immediate that the increments

$$D_{m,n} = \sum_{k=m+1}^{n} Y_k,$$

as sums of mutually independent random variables, are mutually independent for time intervals that do not overlap (in that they have no summands in common).

Note that a process of independent random variables can be a process of independent increments only in the degenerate case of constant random variables.

We further specialize to a random process of *stationary independent increments* through the next definition.

Definition 20.7.2 The random process $\{X_t\}$ with associated increments $\{D_{s,t}\}$ is a process of stationary independent increments if it is one of independent increments and, in addition, for each $s < t$, the distribution of $D_{s,t}$ depends upon the times s and t only through their difference $t - s$.

Example 20.20 Stationary Independent Increments _____

Take the preceding example, but now require that $\{Y_k\}$ be *i.i.d.*

In the next two subsections, we define random processes in continuous time, $t = [0, \infty)$, that have stationary independent increments. In each case, for any set of time points $t_0 < t_1 < \ldots < t_n$, we show how to determine the joint distributions for $X_{t_0}, \ldots, X_{t_n}$.

20.7.1 Poisson Process

We discussed Poisson random variables at some length in Sections 4.4 and 4.5, because such random variables occur frequently in engineering practice. A Poisson random variable X represents a count taken over a defined period T in which particles (e.g., photons, gamma rays, electrons) or customers arrive or events occur with an average rate of λ occurrences per unit time. In this case,

$$P(X = k) = e^{-\lambda T} \frac{(\lambda T)^k}{k!}.$$

However, this formula does not tell us about the pattern of arrivals or occurrences that would be found by examining the specific random times of these occurrences, $\tau_1, \ldots, \tau_k$. The random variable X_t for a Poisson process is often denoted instead by N_t, to emphasize the integer nature of the counting process. N_t is the number of events (e.g., particle arrivals) that occurred in the time interval $[0, t]$. Clearly, $N_t - N_s$, for $t > s$, is the number of events that occurred in the time interval $(s, t]$.

Definition 20.7.3   (Poisson Process)   The collection of random variables $\{N_t : t \geq 0\}$ is a Poisson process if the following are true:

1. $P(N_0 = 0) = 1$.
2. There exists an arrival rate $\lambda > 0$ such that, for all $0 \leq s < t$,

$$P(N_t - N_s = k) = e^{-\lambda(t-s)} \frac{(\lambda(t - s))^k}{k!}.$$

3. For all n and for all time points $0 \leq t_1 < t_2 < \ldots < t_n$, the random variables $\{D_{t_i, t_{i+1}} = N_{t_{i+1}} - N_{t_i}\}$ are mutually independent.

Thus, the Poisson process is defined as a random process with stationary independent increments. The numbers of particles arriving in disjoint intervals are mutually independent random variables.

Having defined the Poisson process, we can address the probabilistic properties of the random times of arrival of particles or customers. Let τ_1 be the waiting time to the arrival of the first particle. Evidently, for $x > 0$,

$$P(\tau_1 \leq x) = P(N_x \geq 1);$$

if a particle arrived prior to time x, then the particle count N_x at time x must be at least 1. However,

$$P(N_x \geq 1) = 1 - P(N_x = 0) = 1 - e^{-\lambda x}.$$

Hence, we conclude that the waiting time to the arrival of the first particle is exponentially distributed:

$$P(\tau_1 \leq x) = F_{\tau_1}(x) = 1 - e^{-\lambda x}.$$

Proceeding similarly, we see that the arrival time τ_n for the nth particle satisfies the formula

$$P(\tau_n \leq x) = P(N_x \geq n) = 1 - P(N_x < n) = 1 - e^{-\lambda x} \sum_{k=0}^{n-1} \frac{(\lambda x)^k}{k!}.$$

A complete description of a random process requires that, for any collection of its random variables, we can specify their joint distribution. We show how to do this for the Poisson process by specifying its joint probability mass function. Consider the bivariate case of $0 \leq t_0 < t_1$. Note that

$$P(N_{t_0} = n_0, N_{t_1} = n_1) = P(N_{t_1} = n_1 | N_{t_0} = n_0)P(N_{t_0} = n_0).$$

Earlier, we showed how to determine $P(N_{t_0} = n_0)$ as $\mathcal{P}(\lambda t_0)$. The conditional probability can be rewritten as

$$P(N_{t_1} = n_1 | N_{t_0} = n_0) = P(N_{t_1} - N_{t_0} = n_1 - n_0 | N_{t_0} = n_0).$$

Note that the increment $D_{t_0, t_1} = N_{t_1} - N_{t_0}$ is independent of the increment $D_{0, t_0} = N_{t_0}$. Hence,

$$P(N_{t_1} = n_1 | N_{t_0} = n_0) = P(N_{t_1} - N_{t_0} = n_1 - n_0) = P(N_{t_1 - t_0} = n_1 - n_0).$$

We conclude that N_{t_1} given N_{t_0} is distributed $\mathcal{P}(\lambda(t_1 - t_0))$, and we can now calculate the bivariate pmf.

The extension to the multivariate case follows by induction on the number k of time points $0 \leq t_0 < t_1 < \ldots < t_{k-1}$. The key observation is that the conditional distribution of $N_{t_{k-1}}$, given all of the earlier values $N_{t_0}, \ldots, N_{t_{k-2}}$, is just the distribution of the most recent increment $N_{t_{k-1}} - N_{t_{k-2}}$, namely,

$$P(N_{t_{k-1}} = n_{k-1} | N_{t_0} = n_0, \ldots, N_{t_{k-2}} = n_{k-2}) = P(N_{t_{k-1}} - N_{t_{k-2}} = n_{k-1} - n_{k-2})$$

and is $\mathcal{P}(\lambda(t_{k-1} - t_{k-2}))$.

Implicit in the preceding discussion is the fact that the random waiting times $\{W_i\}$ between unit-height jumps in the amplitude of N_t are $i.i.d.$ and distributed as $\mathcal{E}(\lambda)$, as was shown for $W_1 = \tau_1$. This observation enables us to simulate a Poisson process by generating these waiting times and setting

$$N_t = k \iff \sum_{i=1}^{k} W_i \leq t < \sum_{i=1}^{k+1} W_i.$$

A plot of a sample function from a Poisson process with $\lambda = 1$ is shown in Figure 20.4.

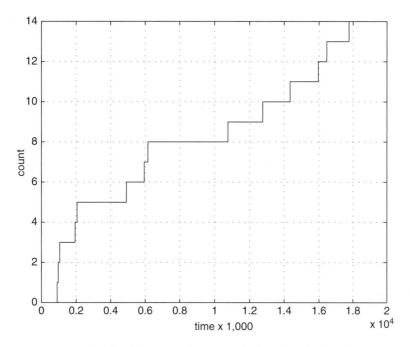

Figure 20.4   Poisson process sample function for $\lambda = 1$.

20.7.2 Brownian Motion

Brownian motion was named for Robert Brown, the early 19th-century experimentalist who first observed the highly irregular movement of microscopic particles. The particles he observed were so small that their motion was affected significantly by collisions with invisible gas molecules. Other phenomena modeled by Brownian motion include integrated white-noise processes (see the Exercises section at the end of the chapter) and movement of certain financial quantities. Brownian motion is both a process of stationary independent increments and a *Gaussian random process (GRP)*. It is typically defined for a time index set $T = [0, \infty)$.

Definition 20.7.4   (Brownian Motion)   Let $\sigma^2 > 0$ be given. A collection of random variables $\{X_t : t \in T = [0, \infty)\}$ is a Brownian motion random process if the following are true:

1. Every increment $D_{t,t+\tau} = X(t + \tau) - X(t)$ is distributed $\mathcal{N}(0, \sigma^2 \tau)$.
2. Increments calculated over nonoverlapping time intervals are independent and distributed as indicated in item 1.
3. $X(0) = 0$ and $X(t)$ is continuous at 0.

Hence, $EX_t = 0$, $\mathrm{VAR}(X_t) = \sigma^2 t$, and, for $t > s$,

$$\mathrm{COV}(X_s, X_t) = E(X_s X_t) = E\left(X_s(X_t - X_s + X_s)\right) = \mathrm{VAR}(X_s) + 0,$$

where we have used the independence of the increment $X_t - X_s$ from X_s or $X_s - X_0$ and X_s being mean zero. For any sequence of times for which $t_1 < \ldots < t_n$,

$$P(X_{t_1} = x_1, \ldots, X_{t_n} = x_n)$$

$$= P\left(D_{0,t_1} = x_1, D_{t_1,t_2} = x_2 - x_1, \ldots, D_{t_{n-1},t_n} = x_n - \sum_{k<n} x_k\right).$$

As the increments are independent and individually normally distributed, it follows that the $X_{t_1}, \ldots, X_{t_n}$ are jointly normally distributed. Hence, for Brownian motion,

$$\text{COV}(X_s, X_t) = \sigma^2 \min(s, t) \text{ and } EX_t = 0.$$

We now have a complete specification of the Gaussian random process that is Brownian motion. From our previous study of the multivariate normal or from the results in Section 17.6, we can explicitly write the joint pdf for any collection of random variables produced by Brownian motion. A sample function from a Brownian motion random process is shown in Figure 20.5.

Clearly, the marginal density

$$f_{X_t}(x) = \frac{1}{\sigma\sqrt{2\pi t}} e^{-\frac{x^2}{2\sigma^2 t}}$$

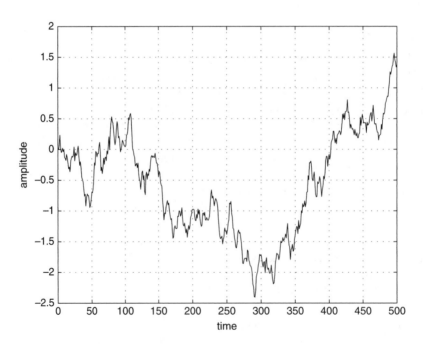

Figure 20.5 Brownian motion sample function.

follows from $X_t = D_{0,t}$ and $X(0) = 0$. If $t > s$, the conditional density

$$f_{X_t|X_s}(y|x) = \frac{1}{\sigma\sqrt{2\pi(t-s)}}e^{-\frac{(y-x)^2}{2\sigma^2(t-s)}}$$

follows from

$$X_t = X_s + D_{s,t} = D_{0,s} + D_{s,t}.$$

In our discussion of the Borel–Cantelli Lemma and its applications to random processes, we considered the question of how often a random process sample function exceeded a given boundary function. We now ask a related question of Brownian motion. Let T_b denote the first time that $X_t \geq b$. We take $b > 0$; otherwise $T_b = 0$, due to our initializing the Brownian motion process at 0 at time 0. We derive the distribution of the random *hitting time* T_b from that of X_t by starting with the Total Probability Theorem:

$$P(X_t \geq b) = P(X_t \geq b|T_b \leq t)P(T_b \leq t) + P(X_t \geq b|T_b > t)P(T_b > t).$$

From the definition of T_b as the *first* time that the process reaches b,

$$P(X_t \geq b|T_b > t) = 0.$$

By the independent increments property, the distribution of the increment $D_{T_b,t}$ when $T_b \leq t$ depends only upon $t - T_b$. We are omitting some details, since now T_b is a random variable and not a fixed time. However, the event $T_b = s$ is determined by the Brownian motion process sample path in the time interval $[0, s]$ and does not depend upon future values of the process. Thus, if $T - b \leq t$, then $X_{T_b} = b$, by the continuity of the sample paths of X_t and because the future increment $D_{T_b,t}$, being mean 0 and Gaussian, has equal probabilities of $1/2$ of being positive or negative. Therefore,

$$P(X_t \geq b|T_b \leq t) = \frac{1}{2}.$$

We now conclude that

$$P(T_b \leq t) = 2P(X_t \geq b) = \frac{\sqrt{2}}{\sigma\sqrt{t\pi}}\int_t^\infty e^{-\frac{x^2}{2\sigma^2 t}}\,dx = 2\left(1 - \Phi\left(\frac{\sqrt{t}}{\sigma}\right)\right),$$

where Φ is the $\mathcal{N}(0, 1)$ cdf.

20.8 MARTINGALES

20.8.1 Basics of Martingales

Martingales are a class of random processes that were first studied in connection with defining fair games of chance and defining a random sequence as a sequence that was resistant to exploitation by a gambler who observes its past outcomes and predicts its future outcomes. The basic idea is a random process X_t, with index set T, such that the expected value of a future amplitude X_t is independent of whatever can be known through observation of past amplitudes $\{X_s : s \leq r < t\}$. (See Doob [25], Grimmett and Stirzaker [42], Karlin and Taylor [52], and Neveu [65].) This idea motivates the next definition.

Definition 20.8.1   (Martingale)   The random process $\{X_t : t \in T\}$ is a martingale if, for every $t > s$,

$$E|X_t| < \infty$$

and

$$E(X_t | \{X_r : r \leq s\}) = X_s \text{ with probability 1.}$$

In the definition, we condition on the complete past, all of the X_r for times r that are less than or equal to a given s that is in turn less than t. The next theorem shows us that the martingale property holds even if we omit some of the conditioning variables.

Theorem 20.8.1   (Martingale Generalization)   If X_t is a martingale and S is any subset of times prior to and including s, with $s < t$, then

$$E(X_t | X_s, \{X_r : r \in S\}) = X_s \text{ with probability 1.}$$

Proof. We use the property of subconditioning for conditional expectation as given in Section 15.2. We subcondition on the full set of variables on or before time s, $\{X_r : r \leq s\}$, and then take the conditional expectation (average) over the variables we introduced, given the ones we had at the outset:

$$E(X_t | X_s, \{X_r : r \in S\}) = E\big([E(X_t | \{X_r : r \leq s\})] | X_s, \{X_r : r \in S\}\big).$$

From the definition of a martingale, the interior term in square brackets

$$E(X_t | \{X_r : r \leq s\}) = X_s \text{ with probability 1.}$$

Hence,

$$E(X_t | X_s, \{X_r : r \in S\}) = E(X_s | X_s, \{X_r : r \in S\}) = X_s \text{ with probability 1.} \qquad \square$$

A further generalization of the concept of a martingale is to define X_n as a martingale when we condition on another family of random variables $\{Y_n : n \in N\}$, provided that

$$E(X_n | Y_1, \ldots, Y_{n-1}) = X_{n-1}.$$

We use this important generalization shortly in our construction of Doob's martingale.

Much is known about the behavior of martingales. A satisfyingly simple result on convergence is given in the next theorem.

Theorem 20.8.2   (Martingale Convergence)   Let $\{X_n, n \in N\}$ be a martingale satisfying

$$\sup_{n \geq 0} E|X_n| < \infty.$$

Then there exists a random variable X_∞ to which the martingale converges with probability 1; that is,

$$P(\lim_{n \to \infty} X_n = X_\infty) = 1.$$

Furthermore, if the sequence of mean squares $\{EX_n^2\}$ is uniformly bounded by finite b, then there is also convergence in mean square:

$$\lim_{n \to \infty} E(X_n - X_\infty)^2 = 0.$$

Proof reference.   See Doob [25] or Karlin and Taylor [52]. □

Note that the theorem does not assert convergence to a constant.

20.8.2 Examples and Applications of Martingales

We provide four examples of classes of martingale processes. Our first example is to specialize homogeneous Markov chains so that they are also martingales. Given that we have a Markov chain with only finitely many states, we automatically satisfy the requirement of a martingale that $E|X_n|$ be finite for every n. (If the Markov chain has countably many states, then this condition becomes nontrivial.) The remaining condition is that

$$E(X_{n+1}|X_0, \ldots, X_n) = X_n.$$

Evaluating the required conditional expectation is simplified by the Markov property

$$E(X_{n+1}|X_0, \ldots, X_n) = E(X_{n+1}|X_n) = \sum_{\xi \in \mathcal{X}} \xi P(X_{n+1} = \xi | X_n) = X_n.$$

A specific way to satisfy this condition is to take the state set $\mathcal{X} = \{\underline{k}, 1, \ldots, \overline{k} - 1, \overline{k}\}$ of $\overline{k} - \underline{k} + 1$ elements and satisfy the extreme state conditions that

$$p_{\underline{k},j} = \delta_{\underline{k},j}, \quad p_{\overline{k},j} = \delta_{\overline{k},j},$$

and intermediate state conditions for $\underline{k} < i < \overline{k}$ that

$$p_{i,j} = \begin{cases} p_{i,2i-j} & \text{if } \underline{k} \leq 2i - j \leq \overline{k} \\ 0 & \text{otherwise} \end{cases}.$$

If X_n is one of the two extreme values $\underline{k}$ or $\overline{k}$, then X_{n+1} can only equal X_n; there is no way to "average" other states in $\mathcal{X}$ and obtain the extreme values. If X_n is an intermediate value, then, by assigning exactly the same transition probability to a state j that is larger than $X_n = i$ by $j - i$ as is assigned to a state $2i - j$ that is smaller than state i by the same $j - i$, the average of this pair of states j and $2i - j$ will be precisely i. In conclusion, there are homogeneous Markov chains that are martingales, but they must satisfy special conditions on their one-step transition probabilities.

For convenience, enumerate the states with positive and negative indices such that $\xi_{-i} = -\xi_i$. Given any $X_n = \xi_i$, we require that the one-step transition probabilities satisfy

$$p_{i,i} > 0 \text{ to ensure that } X_n = \xi_i \text{ has a positive probability of being chosen, or}$$

$$p_{i,j} = p_{i,-j}.$$

It is easy to see that, for any i,

$$E(X_{n+1}|X_n = \xi_i) = \sum_j \xi_j p_{i,j} = \xi_i.$$

As a second example, we show that a class of random processes that we have just examined can also be martingales.

Theorem 20.8.3   **(Independent Increments and Martingales)**   If $\{X_t : t \in T\}$ is a process of independent increments and these increments have mean zero, then it is also a martingale.

Proof. Note that, for $t > s \geq r$ and X_t a process of independent increments, $D_{s,t} = X_t - X_s$ is independent of X_r. We evaluate

$$E(X_t|\{X_r : r \leq s\}) = E(D_{s,t} + X_s|\{X_r : r \leq s\})$$

$$= E(D_{s,t}|\{X_r : r \leq s\}) + E(X_s|\{X_r : r \leq s\}).$$

However, by independence and $ED_{s,t} = 0$ by hypothesis,

$$E(D_{s,t}|\{X_r : r \leq s\}) = ED_{s,t} = 0,$$

and

$$E(X_s|\{X_r : r \leq s\}) = X_s \text{ with probability 1.}$$

We conclude that

$$E(X_t|\{X_r : r \leq s\}) = X_s \text{ with probability 1,}$$

and X_t is a martingale, as asserted.   □

Therefore, Brownian motion is a martingale. Thus, Figure 20.5 is a plot of a sample function of a martingale. The Poisson process has independent increments with nonzero mean,

$$ED_{s,t} = \lambda(t - s),$$

and is not a martingale. Nonetheless, it is easily corrected in that $Y_t = X_t - \lambda t$ is a martingale when X_t is a Poisson process. The increments $Y_t - Y_s$ are again independent of $\{Y_r : r \leq s\}$ and now have mean zero, as needed to apply Theorem 20.8.3.

A third class of martingales that is easily constructed is given in terms of an arbitrary random process $\{Y_n : n \in N\}$ and a random variable X that need only have $E|X| < \infty$. *Doob's martingale* $\{X_n, n \in N\}$ is then defined by

$$X_n = E(X|Y_0, \dots, Y_n).$$

To verify that we have indeed defined a martingale, we need to establish the two conditions given in Definition 20.8.1. First,

$$E|X_n| = E\left[|E(X|Y_0, \dots, Y_n)|\right] \leq E\left[E(|X||Y_0, \dots, Y_n)\right] = E|X| < \infty,$$

where we used the special case of Jensen's inequality

$$E|X| \geq |EX|.$$

Second,

$$E(X_{n+1}|Y_0, \dots, Y_n) = E\left(\left[E(X|Y_0, \dots, Y_n, Y_{n+1}\right]|Y_0, \dots, Y_n\right)$$
$$= E(X|Y_0, \dots, Y_n) = X_n,$$

where we have used the conditional expectation property of subconditioning. We thus have verified that Doob's process is a martingale.

One use of Doob's process is in studying the behavior of least mean square estimators. As we know from Section 15.4, the least mean square estimator $\hat{X}_n$ of X, given observations $Y_0, \dots, Y_n$, is

$$\hat{X}_n(Y_0, \dots, Y_n) = E(X|Y_0, \dots, Y_n).$$

We now know that the sequence $\{\hat{X}_n\}$ for an increasing number of observations forms a martingale, and this makes available all that is known about martingales to help us understand the behavior of this sequence of estimators.

As a fourth example, we consider a sequence of observations $X_1, \dots, X_n, \dots$ and the sequence of likelihood ratios $\Lambda_n = \Lambda_n(X_1, \dots, X_n)$ formed when we have two hypotheses, described by the multivariate density functions f_0 and f_1, about the source of the observations,

$$\Lambda_n = \frac{f_1(X_1, \dots, X_n)}{f_0(X_1, \dots, X_n)}.$$

Likelihood ratios appear in optimal decision-making rules for deciding which of the two hypotheses applies best to the observations; they were discussed in Chapter 13 and in Section 16.5. We claim that if, in fact, the observations come from f_0, then the sequence of likelihood ratios forms a martingale. In order to remind us that f_0 governs the observations, we will replace the usual notation of E for expectations by E_0. To verify this claim, we use subconditioning to write

$$E_0(\Lambda_n|\Lambda_1, \dots, \Lambda_{n-1}) = E_0([E_0(\Lambda_n(X_1, \dots, X_{n-1})]|\Lambda_1, \dots, \Lambda_{n-1}).$$

Letting $f_0(x_n | X_1, \ldots, X_{n-1})$ denote the conditional density of X_n given $X_1, \ldots, X_{n-1}$, we find that the square-bracketed term

$$
\begin{aligned}
E_0 \Lambda_n (X_1, \ldots, X_{n-1}) &= \int_{-\infty}^{\infty} \Lambda_n (X_1, \ldots, X_{n-1}, x_n) f_0(x_n | X_1, \ldots, X_{n-1})\, dx_n \\
&= \int_{-\infty}^{\infty} \frac{f_1(X_1, \ldots, X_{n-1}, x_n)}{f_0(X_1, \ldots, X_{n-1}, x_n)} \frac{f_0(X_1, \ldots, X_{n-1}, x_n)}{f_0(X_1, \ldots, X_{n-1})}\, dx_n \\
&= \frac{1}{f_0(X_1, \ldots, X_{n-1})} \int_{-\infty}^{\infty} f_1(X_1, \ldots, X_{n-1}, x_n)\, dx_n = \frac{f_1(X_1, \ldots, X_{n-1})}{f_0(X_1, \ldots, X_{n-1})} \\
&= \Lambda_{n-1}(X_1, \ldots, X_{n-1}).
\end{aligned}
$$

Returning, we see that

$$
E_0(\Lambda_n | \Lambda_1, \ldots, \Lambda_{n-1}) = E_0(\Lambda_{n-1} | \Lambda_1, \ldots, \Lambda_{n-1}) = \Lambda_{n-1} \text{ with probability one.}
$$

We have established that the sequence of likelihood ratios generated by adding new observations is a martingale when evaluated according to the hypothesis f_0. This conclusion is of value in studying the asymptotic behavior of likelihood ratio tests as the number of observations grows. If, for example, the likelihood ratios are known to be uniformly upper bounded, then the Martingale Convergence Theorem 20.8.2 applies immediately to guarantee the convergence of $\{\Lambda_n\}$, both in mean square and with probability 1.

20.8.3 Optional Stopping for Martingales

Not only do martingales have a number of important applications, but they have a number of properties that enable us to understand and calculate with these random-process models. One of these properties concerns the concept of an *optional stopping time*. For example, if a sequence of gambles is fair, as it should be, then, no matter how we examine the past, we cannot do better in the future. An optional stopping time S is an abstraction of a rule to "know when to quit."

Definition 20.8.2   (Optional Stopping Time)   Given a random process $\{X_t, t \in N\}$, the random variable S is an optional stopping time for the process if, for every $n \in N = \{0, 1, \ldots\}$, the event $S = n$ is determined by the random variables $X_0, \ldots, X_n$.

In effect, we decide to stop at time n, depending only upon what we could have observed up to that time.

Theorem 20.8.4   (Optional Stopping for Martingales)   Assume that $\{X_t, t \in N\}$ is a martingale and S is an optional stopping time for this martingale. If $P(S < \infty) = 1$ (so that we stop somewhere with probability 1), and

$$
E\left(\sup_{n \geq 0} |X_{\min(S, n)}| \right) < \infty,
$$

($X_{\min(S,n)}$ means we evaluate the process either at S or at n, whichever occurs first), then

$$EX_S = EX_0.$$

Proof reference. See Karlin and Taylor [52] Theorem 3.1 for a proof.   □

The import of the Optional Stopping Theorem for martingales is that, putting aside some technical conditions, no matter how you choose a stopping rule or random variable S, the expected value of the stopped random variable X_S is equal to that of any other random variable. You cannot profit from, or be penalized by, optional stopping, at least with respect to expected values. An important caveat is that S be finite with probability 1. If this condition is violated, then the conclusions do not follow. For example, we could take as our stopping time in a fair gambling system that we stop when we have won a large amount of money M. This would violate the conclusion of the theorem that $EX_S = EX_0$, since EX_S would appear to be M and larger than $EX_0 = 0$. However, there is a positive probability of losing all we have before we reach our goal of M, and it is not true that S is finite with probability 1.

Example 20.21   Optional Stopping _____

We can use the Optional Stopping Theorem to determine expected waiting times to events that will eventually occur with probability 1. Thus, choose some boundary function b_n, and let $S = n$ if $X_n = b_n$. If b_n is chosen so that $P((\exists N)X_N = b_N) = 1$, then we can apply optional stopping. Consider the example of

$$X_t = N_t - \lambda t,$$

which is a martingale if N_t is a Poisson process with parameter λ. Select a threshold $\tau > 0$ and rate $\rho > 0$, and let S be the first time that $X_t \geq \tau - \rho t$. From Theorem 20.8.4,

$$EX_S = EX_0 = E(N_0 - \lambda 0) = 0.$$

Hence,

$$0 \geq \tau - \rho ES \text{ or } ES \leq \frac{\tau}{\rho}.$$

We easily identify an upper bound to the expected waiting time for X_t to reach or exceed the boundary $\tau - \rho t$.

20.9 CALCULUS

20.9.1 Continuity

When are the sample functions, realizations of the random process considered as functions of time, continuous? Continuity requires limits. Consider the basic question of continuity for a random process $\{X_t, t \in T\}$: Is

$$\lim_{s \to t} X_s = X_t?$$

The condition $s \to t$ makes the most sense when the index set T is an interval or all of the real numbers, and we assume this to be true. If the random process does not satisfy continuity, then there are pathological possibilities that are well beyond our mathematical scope. However, if we have continuity with probability 1 for all $t \in T$, with the possible exception of a subset of T of length 0, or

$$(\forall t \in T) \; P\left(\lim_{s \to t} X_s = X_t\right) = 1, \tag{20.7}$$

then the random process will have sufficient regularity (technically, its sample functions will be measurable; see Doob [25], Theorem 2.5) for the usual issues of analysis. The condition just given, however, does not mean that the sample functions of the process are themselves continuous functions. The Poisson process is an example in which the sample functions will have jump discontinuities over an interval with positive probability. Of course, these jump discontinuities occur at random times, and the probability is still 1 that at any preselected time point, the probability of a discontinuity is 0.

A weaker continuity condition is continuity in mean square at each t:

$$\lim_{s \to t} E(X_t - X_s)^2 = 0.$$

A random process satisfying this condition at each t in T is said to be *mean square continuous* on T. Expansion of the quadratic yields

$$E(X_t - X_s)^2 = R_X(t, t) + R_X(s, s) - 2R_X(t, s).$$

Hence, mean square continuity on T is ensured when the autocorrelation function R_X is continuous along the diagonal $\{(t, s) : t = s \in T\}$.

20.9.2 Differentiation

In our study of (finite-dimensional) random variables and vectors, we carefully analyzed the effects of linear transformations on such collections of random variables (e.g., see Section 8.6.1 and topics in Chapter 17). Now that we are looking at processes defined for any real-valued time, we have access to linear transformations that involve infinitely many random variables, typically through limiting processes. Basic examples are provided by differentiation and integration.

We inquire into what happens when we try to define

$$\left.\frac{dX_t}{dt}\right|_{t=t_0} = \lim_{\epsilon \to 0} \frac{X_{t+\epsilon} - X_t}{\epsilon}. \tag{20.8}$$

Theorem 20.9.1   (Mean Square Derivative)   The random process X_t is differentiable at t_0, in the sense of mean square convergence, if EX_t is differentiable at t_0 and

$$\frac{\partial^2 R_X(t, s)}{\partial t \, \partial s}$$

exists and is continuous in $t = s = t_0$.

Proof (sketch). It is easiest to examine the existence of such a derivative through mutual convergence in mean square. The question is, Is

$$\lim_{\epsilon \to 0, \delta \to 0} E \left(\frac{X_{t+\epsilon} - X_t}{\epsilon} - \frac{X_{t+\delta} - X_t}{\delta} \right)^2 = 0?$$

Expanding and examining the terms would reveal that the limit is zero if and only if both of the following limits exist:

$$\lim_{\epsilon \to 0} \frac{EX_{t+\epsilon} - EX_t}{\epsilon}, \quad \lim_{\epsilon \to 0, \delta \to 0} \text{COV} \left(\frac{X_{t+\epsilon} - X_t}{\epsilon}, \frac{X_{t+\delta} - X_t}{\delta} \right).$$

The statement of the theorem contains sufficient conditions for these limits to exist.   □

The correlation function for the derivative process X_t' is given by

$$E \left(X_t' X_s' \right) = R_{X'}(t, s) = \frac{\partial^2 R_X(t, s)}{\partial t \, \partial s}.$$

This result is understandable if we interchange differentiation and expectation:

$$E \left(X_t' X_s' \right) = E \left(\frac{\partial X_t}{\partial t} \frac{\partial X_s}{\partial s} \right) = \frac{\partial}{\partial t} \frac{\partial}{\partial s} E \left(X_t X_s \right) = \frac{\partial}{\partial t} \frac{\partial}{\partial s} R_X(t, s) = \frac{\partial^2 R_X(t, s)}{\partial t \, \partial s}.$$

Example 20.22   Mean Square Derivative

Consider a random process with mean 0 and covariance or autocorrelation function

$$R_X(\tau) = \frac{\sigma^2}{1 + \tau^2}.$$

Clearly, the mean function is everywhere differentiable. Furthermore, setting $\tau = t - s$, we have

$$\frac{\partial^2 R_X(t - s)}{\partial t \, \partial s} = - \frac{d^2 R_X(\tau)}{d\tau^2} \bigg|_{\tau = t - s} = \frac{2(3\tau^2 - 1)}{(1 + \tau^2)^3} \bigg|_{\tau = t - s},$$

and the derivative of X_t exists at all t, since the differentiability of the mean and covariance functions is established.

If we now consider a random process with mean 0 and covariance given by

$$R_X(\tau) = e^{-\alpha |\tau|},$$

then R_X is not even once continuously differentiable, due to the cusp at $\tau = 0$;

$$\frac{d R_X(\tau)}{d\tau} = -\alpha e^{-\alpha |\tau|} \text{sign}(\tau).$$

In this case, the random process nowhere satisfies the sufficient condition for differentiability. It does not have a mean square derivative.

20.9.3 Riemann Integration

The Riemann integral familiar to engineering and physics students defines an integral

$$\int_a^b f(x)\, dx$$

as a limit of approximating sums of finitely many terms of the form $f(x_i)(x_{i+1}^{(n)} - x_i^{(n)})$, where $\{x_i^{(n)}\}$ are n points, arranged in increasing order $a = x_1 < x_2 < \cdots < x_n = b$, in the interval $[a, b]$ and that partition it so that

$$\lim_{n \to \infty} \max_i (x_{i+1}^{(n)} - x_i^{(n)}) = 0.$$

Hence, if we attempt to define a Riemann integral for a random process X_t, namely,

$$\int_a^b X_t\, dt,$$

we immediately encounter an issue of the existence of a limit of a sequence of sums of increasingly many weighted random variables. As we learned in Chapter 19, there are many probabilistic senses of convergence. A common approach is to prove mutual convergence in the mean square sense. A sufficient (but not necessary) condition is then available from the next theorem.

Theorem 20.9.2   (Riemann Integration)   If $\{X_t : t \in [a, b]\}$ is of finite mean square with a mean function $m_X(t) = EX_t$ and a covariance function $C(t, s) = \mathrm{COV}(X_t, X_s)$, then the Riemann integral

$$\int_a^b X_t\, dt$$

exists as a mean square limit of Riemann approximating sums if m_X and $C(t, s)$ are continuous functions.

The advanced theory of probability and random processes is based largely on Lebesgue integration and not on Riemann. Doob [25] Theorem 2.7, stated informally, informs us that if the random process satisfies the continuity condition of Eq. (20.7), then the sample functions of the process will be Lebesgue integrable over a time set T_0, with probability 1, provided that

$$\int_{T_0} E(|X_t|)\, dt < \infty.$$

This is an easily verified condition for almost sure integrability of the sample functions of a random process.

20.10 WIDE-SENSE STATIONARY PROCESSES AND POWER SPECTRAL DENSITY

20.10.1 Basics of Wide-Sense Stationary Processes

A weaker notion than that of stationarity as discussed in Section 20.5.1, but one that has proven to be of great value in practice, is that of the wide-sense stationary process.

Definition 20.10.1   (Wide-Sense Stationary)   A random process $X(t)$ of finite mean square is *wide-sense stationary* (abbreviated *w.s.s.*) if it has a constant mean function

$$(\exists m)(\forall t)\ EX(t) = m_X(t) = m$$

and an autocorrelation function $E(X(t)X(s)) = R_X(t,s)$ that is a function R of only the time difference $t - s$, or

$$(\forall t, s)\ R_X(t,s) = R(t - s).$$

Note that we can replace the statement about R_X by one about the covariance function $C_X(t,s) = \text{COV}(X(t)X(s))$.

Example 20.23   WSS Autocorrelation _____
An example of a *w.s.s.* autocorrelation function is given by

$$R_X(t,s) = \begin{cases} \sigma^2(1 - \frac{|t-s|}{\alpha}) & \text{if } |t - s| \le \alpha \\ 0 & \text{otherwise} \end{cases}.$$

A similar second example is

$$R_X(t,s) = \sigma^2 e^{-\alpha|t-s|}, \text{ for } \alpha > 0.$$

As the multivariate normal density (see Sections 7.4.4, 9.15, 12.6.3, or 17.6) is defined solely in terms of the mean function and the autocovariance function, a little consideration reveals that a wide-sense stationary normal or Gaussian random signal is also strictly stationary. The advantages of the wide-sense stationary class of random processes are (1) that we need know only the generally available mean function and covariance function and not all of the joint distributions and (2) that this class of processes is preserved under transformation through time-invariant linear systems. If X_t is a wide-sense stationary process and is the input to a time-invariant linear system—say, one characterized by an impulse response $h(t,\tau) = h(t - \tau)$—then the output process Y_t is also wide-sense stationary. To verify this, assume that $EX_t = m$ and $R_X(t,s) = R(t - s)$. Then

$$Y_t = \int_{-\infty}^{\infty} h(t-s)X_s\, ds \Rightarrow EY_t = \int_{-\infty}^{\infty} h(t-s)EX_s\, ds = m \int_{-\infty}^{\infty} h(u)\, du$$

and

$$
R_Y(t, s) = E(Y_t Y_s) = E \left(\int_{-\infty}^{\infty} \int_{-\infty}^{\infty} h(t - \tau_1) X_{\tau_1} h(s - \tau_2) X_{\tau_2} \, d\tau_1 d\tau_2 \right)
$$

$$
= \int_{-\infty}^{\infty} \int_{-\infty}^{\infty} h(t - \tau_1) h(s - \tau_2) R(\tau_1 - \tau_2) \, d\tau_1 d\tau_2
$$

$$
= \int_{-\infty}^{\infty} \int_{-\infty}^{\infty} h(u) h(v) R(v - u + t - s) \, du dv.
$$

Clearly, $R_Y(t, s)$ depends only upon $t - s$, and Y_t is also a $w.s.s.$ process when X_t is.

20.10.2 Basics of Power Spectral Density

An important characteristic of $w.s.s.$ random processes is how they distribute their average/expected power over the frequency domain. The instantaneous power of a random process $\{X_t, t \in \mathbb{R}\}$, when X_t is thought of as a voltage or current across or through a resistance, is proportional to X_t^2 and equal to it if the resistance is 1 ohm. It is common in engineering and statistics to think of X_t^2 as power at time t, without regard to whether it is a voltage or current. As we are thinking of X_t as an amplitude in a random process, this power would also be a random variable. We are interested in the expected value EX_t^2 of this instantaneous power and note that it is given in terms of the $w.s.s.$ autocorrelation function by $R_X(t, t) = R_X(t - t) = R_X(0)$. (We abuse notation by having R_X stand for both a function of two variables and a function of one variable, and in the $w.s.s.$ case R_X, it is constant in time.) We first propose a definition of the distribution of power in frequency, rather than in time, and then defend it after a digression into time-invariant linear systems. (See also Sections 10.2 and 10.3.) The expected power allocation in frequency is given by the function (psd) $S_X(\omega)$ and has physical units of watts per hertz.

Definition 20.10.2   (Power Spectral Density)   The power spectral density (psd) function S_X for a wide-sense stationary random process is given in terms of the autocorrelation function R_X for this process through the Fourier transform

$$
S_X(\omega) = \int_{-\infty}^{\infty} R_X(\tau) e^{-i\omega\tau} \, d\tau. \tag{20.9}
$$

This Fourier transform will exist if R_X is absolutely integrable. That is,

$$
\int_{-\infty}^{\infty} |R_X(\tau)| \, d\tau < \infty.
$$

Note that, for real random processes, the autocorrelation function $R_X(t, s) = R_X(s, t)$, and in the $w.s.s.$ case $R_X(t - s) = R_X(s - t)$. Hence, we can rewrite Eq. (20.9) as

$$
S_X(\omega) = 2 \int_0^{\infty} R_X(\tau) \cos(\omega\tau) \, d\tau.
$$

It is evident from this result that S_X is a real-valued function for real-valued X_t. Furthermore, it also follows that S_X is an even function ($S_X(\omega) = S_X(-\omega)$). The remaining property of the psd is that it is nonnegative, or $S_X(\omega) \geq 0$. The nonnegativity of S_X is a mathematical consequence of a correlation matrix or function being nonnegative definite, as discussed in Section 9.14, although we do not demonstrate this here. We sum up these observations and strengthen them by including their converse in the next theorem.

Theorem 20.10.1   (Power Spectral Density Characterization)   If X_t is a real-valued *w.s.s.* random process for all real-valued t that has finite mean square and an absolutely integrable auto-correlation function $R_X(\tau)$, then the psd $S_X(\omega)$ exists. Furthermore, a function S_X of ω is a psd if and only if it has the following properties:

1. S_X is real valued.
2. S_X is an even function.
3. S_X is nonnegative.

If the psd S_X is also absolutely integrable, then we have an inverse Fourier transform that recovers R_X from S_X through

$$R_X(\tau) = \frac{1}{2\pi} \int_{-\infty}^{\infty} e^{i\omega\tau} S_X(\omega)\, d\omega. \tag{20.10}$$

Our definition of the psd requires discussion to give it meaning as expected power (power is time-averaged energy) per unit of frequency (thus, the meaning of the usage "density" as density in frequency and not as a probability density) of the random process.

20.10.3 PSD and Linear Systems

We begin our explanation of S_X with a summary of results from linear system theory applied to wide-sense stationary random processes. Let X_t be an input *w.s.s.* random process and Y_t the corresponding output random process, both defined for the time index set $\mathbb{R}$. Let

$$m_X = EX_t = m, \quad R_X(\tau) = E(X_t X_{t+\tau}), \quad m_Y(t) = EY_t, \quad R_Y(t,s) = E(Y_t Y_s).$$

The time-invariant linear (LTI) system is characterized by an impulse response function h through

$$Y_t = \int_{-\infty}^{\infty} h(t-\tau) X_\tau\, d\tau.$$

In order to avoid unnecessary complications, we will assume throughout that the impulse response is square integrable, or

$$\int_{-\infty}^{\infty} h^2(t)\, dt < \infty.$$

Assuming that the expectation operator E can be interchanged with integration over time, we see that

$$EY_t = E\left(\int_{-\infty}^{\infty} h(t-\tau) X_\tau\, d\tau\right) = \int_{-\infty}^{\infty} h(t-\tau) m_X\, d\tau = m \int_{-\infty}^{\infty} h(\tau)\, d\tau.$$

Hence, EY_t is constant in time. Continuing, we have

$$R_Y(t, s) = E(Y_t Y_s) = E\left(\int_{-\infty}^{\infty} h(t - \tau) X_\tau \, d\tau \int_{-\infty}^{\infty} h(s - \sigma) X_\sigma \, d\sigma\right)$$

$$= \int_{-\infty}^{\infty} \int_{-\infty}^{\infty} h(t - \tau) h(s - \sigma) R_X(\tau - \sigma) \, d\tau d\sigma.$$

We change variables from τ and σ to u and v through $u = t - \tau$ and $v = s - \sigma$, to write

$$R_Y(t, s) = \int_{-\infty}^{\infty} \int_{-\infty}^{\infty} h(u) h(v) R_X(t - s + v - u) \, du dv.$$

We see that Y_t is also w.s.s., since EY_t is constant and $R_Y(t, s)$ depends only upon $t - s$. With the usual abuse of notation, we write

$$R_Y(\tau) = \int_{-\infty}^{\infty} \int_{-\infty}^{\infty} h(u) h(v) R_X(\tau + v - u) \, du dv. \tag{20.11}$$

We have now proven the following theorem.

Theorem 20.10.2   (WSS and LTI)   If X_t is a *w.s.s.* random process, with absolutely integrable autocorrelation function R_X, that is the input to an LTI system with square-integrable impulse response function h, then the output Y_t is also a *w.s.s.* process with mean function and autocorrelation function as we have calculated.

We pursue Eq. (20.11) further to derive an expression for the psd S_Y of the output process Y_t that is now known to be *w.s.s.* We know from Eq. (20.9) that we can define S_Y from R_Y by a Fourier transform. Hence, applying this result to that of Eq. (20.11), we find that

$$S_Y(\omega) = \int_{-\infty}^{\infty} e^{-i\omega\tau} R_Y(\tau) \, d\tau$$

$$= \int_{-\infty}^{\infty} \int_{-\infty}^{\infty} \int_{-\infty}^{\infty} e^{-i\omega\tau} h(u) h(v) R_X(\tau + v - u) \, du dv d\tau$$

$$= \int_{-\infty}^{\infty} \int_{-\infty}^{\infty} \int_{-\infty}^{\infty} e^{-i\omega u} h(u) e^{i\omega v} h(v) e^{-i\omega(\tau + v - u)} R_X(\tau + v - u) \, du dv d\tau.$$

Making a change of variables in which $\sigma = \tau + v - u$ replaces τ yields

$$S_Y(\omega) = \left[\int_{-\infty}^{\infty} e^{-i\omega u} h(u) \, du\right] \left[\int_{-\infty}^{\infty} e^{i\omega v} h(v) \, dv\right] \left[\int_{-\infty}^{\infty} e^{-i\omega\sigma} R_X(\sigma) \, d\sigma\right].$$

The Fourier transform

$$H(\omega) = \int_{-\infty}^{\infty} e^{-i\omega t} h(t) \, dt$$

of the impulse response h is known as the *transfer function*. Note that the complex-conjugate transfer function

$$H^*(\omega) = \int_{-\infty}^{\infty} e^{i\omega v} h(v)\, dv.$$

Hence, we conclude with the important result that, for *w.s.s.* random processes passed through LTI systems,

$$S_Y(\omega) = |H(\omega)|^2 S_X(\omega). \tag{20.12}$$

20.10.4 Interpreting the PSD

The expected power EY_t^2 at time t in the output Y_t of a LTI system with transfer function H is

$$R_Y(t - t) = R_Y(0) = \frac{1}{2\pi} \int_{-\infty}^{\infty} S_Y(\omega)\, d\omega$$

and is independent of t. In terms of the transfer function H, we can use Eq. (20.12) to write

$$EY_t^2 = R_Y(0) = \frac{1}{2\pi} \int_{-\infty}^{\infty} |H(\omega)|^2 S_X(\omega)\, d\omega. \tag{20.13}$$

Equation (20.13) suggests that if we wish to determine the expected power in the input random process X_t in a frequency band $[\omega_0, \omega_1]$, then we can do so by evaluating the mean square output of a system, with input X_t, that is a filter passing just those frequencies. An ideal band-pass filter—one that passes uniformly only those frequency components that lie in the given band and uniformly rejects those outside that band—is described by the transfer function

$$H_{\omega_0,\omega_1}(\omega) = U(|\omega| - \omega_0)U(\omega_1 - |\omega|).$$

We conclude that the expected power in the input random process X_t that lies in the frequency band between ω_0 and ω_1 is given simply by

$$2\pi R_Y(0) = \int_{\omega_0}^{\omega_1} S_X(\omega)\, d\omega + \int_{-\omega_1}^{-\omega_0} S_X(\omega)\, d\omega.$$

If we recall from Theorem 20.10.1 that S_X is an even function, we can rewrite this result just in terms of positive frequencies:

$$\pi R_Y(0) = \int_{\omega_0}^{\omega_1} S_X(\omega)\, d\omega.$$

If we focus on a very narrow band $[\omega_0 - \delta, \omega_0 + \delta]$ of positive frequencies about a particular frequency ω_0, then we see that the expected power in that band is approximately

$$\pi R_Y(0) \approx 2\delta S_X(\omega_0). \tag{20.14}$$

This equation then justifies our calling S_X a power spectral *density* function. It is analogous to our interpretation of a probability density function $f_X(x)$ (see Section 6.2) through

$$P(x_0 - \delta \le X < x_0 + \delta) \approx 2\delta f_X(x_0).$$

For white noise, it is immediate from the properties of the delta function in the autocorrelation function that $S_X(\omega) = \sigma^2/2\pi$ and is constant over all radian frequencies ω. Radiation, in the visible range, that has a constant distribution of power with frequency appears white. Hence, the alternative name of "white noise" for thermal noise.

The power spectral density function is an important statistical characteristic of a wide-sense stationary random process, particularly when such processes are to be filtered or passed through linear systems. It is also of value in classifying random processes. White or thermal noise was discussed in Sections 6.7.2 and 20.4.3. White noise is a random process with a power spectral density $S_X(\omega)$ that is constant (positive) over all frequencies. Thermal noise is white noise that is a Gaussian random process. If the thermal noise is generated by a device with resistance R ohms and it is operating at a temperature of $T°$ Kelvin, then the psd has the value $2kTR$, where k is Boltzmann's constant. In Section 6.7.2, we also noted flicker, or semiconductor, or $1/f$ noise prevalent in semiconductor devices. Such noise has a power spectral density function

$$S_X(\omega) \propto \frac{1}{|\omega|}$$

and has most of its power at low frequencies. (There is something problematic with this expression for a psd, since it fails to be integrable both at zero and at infinity. S_X in this case can be understood as valid over a wide finite range of frequencies that avoids zero.)

Another approach to interpreting the psd S_X is through the Wiener–Khinchin Theorem.

Theorem 20.10.3   (Wiener–Khinchin)   If the autocorrelation function $E(X_t X_s) = R_X$ $(t - s)$ is such that

$$\int_{-\infty}^{\infty} |R_X(\tau)|\, d\tau < \infty,$$

then

$$S_x(\omega) = \lim_{a \to \infty} \frac{1}{2a} E \left| \int_{-a}^{a} X_t e^{-i\omega t}\, dt \right|^2.$$

While a stationary random process X_t does not have a Fourier transform (it is not integrable over all $t \in \mathbb{R}$), we can estimate the energy at a given frequency by

$$\left| \int_{-a}^{a} X_t e^{-i\omega t}\, dt \right|^2.$$

This integral will be finite when the limits of integration are finite. We then divide by the total time $2a$ over which we integrated the energy and thereby have a power. The outer expectation then yields the expected power at the frequency ω. Finally, we let $a \to \infty$, with the division by

$2a$ encouraging a finite limiting value. The absolute integrability of R_X suffices for this to work as desired. We do not carry out the details. Under ergodicity of X_t this Wiener–Khinchin Theorem enables us to estimate the psd from a finite-length sample function.

20.11 GAUSSIAN RANDOM PROCESS

20.11.1 Basics

A Gaussian random process is a process in which, for all finite subsets of indices in T, the joint distribution of the amplitudes at those times is multivariate normal.

Definition 20.11.1   (Gaussian Random Process) $\{X_t, t \in T\}$ is a Gaussian random process (GRP) if there exists a function $m(t)$ defined on T and a covariance function $C_X(t, s) = \text{COV}(X_t, X_s)$ defined on pairs of indices t and s in T such that, for any n and index points $t_1, \ldots, t_n$ in T, the random variables $X_{t_1}, \ldots, X_{t_n}$ are multivariate normal with mean $EX_t = m(t)$ and covariance matrix $\mathbb{C} = [C_{i,j}]$ with $C_{i,j} = C_X(t_i, t_j)$.

Recall from Section 17.6 that the multivariate normal characteristic function for $\mathbf{X} = (X_{t_1}, \ldots, X_{t_n})^T$ is given by

$$\mathbf{m} = (m(t_1), \ldots, m(t_n))^T, \quad \mathbb{C} = [C_{i,j}], \quad C_{i,j} = \text{COV}(X_{t_i}, X_{t_j}),$$

$$\phi_{\mathbf{X}}(\mathbf{u}) = \exp\left(i\mathbf{u}^T\mathbf{m} - \frac{1}{2}\mathbf{u}^T\mathbb{C}\mathbf{u}\right).$$

If the GRP is *w.s.s.*, then its mean function $m(t)$ will be constant in t and its covariance function $C_X(t, s)$ will be a function only of $t - s$. In the *w.s.s.* case, we can alternatively specify the GRP by its mean and psd S_X. Furthermore, since the GRP is defined in terms of its mean and covariance functions, if it is *w.s.s.*, then all of its multivariate distributions depend only upon time differences and not upon the time origin.

Theorem 20.11.1   (WSS GRP)   A wide-sense stationary Gaussian random process is a strictly stationary random process.

Although we cannot prove this here, the next theorem provides a useful sufficient condition for a GRP to be ergodic.

Theorem 20.11.2   (Ergodicity of a GRP)   A wide-sense stationary GRP is ergodic if its covariance function is absolutely integrable, or

$$\int_{-\infty}^{\infty} |\text{COV}(X_t, X_{t+\tau})| \, d\tau < \infty.$$

The class of GRPs is very important in electrical engineering because of its prevalence in a variety of electrical noise sources. (See Section 6.7.2.) It is important in physics, particularly in the case of thermal noise, because it is rooted in fundamental statistical mechanical, physical processes.

20.11.2 Linear Transformations of a GRP

A key property of being a GRP is that this class is closed under linear transformations. We saw in Section 17.6 that a linearly transformed multivariate normal random vector yields another multivariate normal random vector, and this property can even be used to define the multivariate normal characteristic function. However, these were only linear transformations between finite-dimensional Gaussian random vectors. If we are to understand the results of such operations as differentiation and integration, then, as noted in Section 20.9, we need to account for the distribution of quantities defined by limiting processes. It is clear from our discussion of differentiation and Riemann integration that the sequences of which we are taking limits involve only linear combinations of finitely many random variables. Hence, from the results just cited, we know that each term in these sequences will have a Gaussian distribution if the random process is a GRP. We will have our conclusion if we can determine that a convergent sequence of normally distributed random variables is itself normally distributed.

The key result comes from Theorem 19.5.1 where it is asserted that convergence in mean square implies convergence in probability and convergence in probability implies convergence in distribution; actually, the proof given in the appendix establishes directly the implication that convergence in mean square implies convergence in distribution. Our remarks in Section 20.9 established conditions guaranteeing mean square convergence for differentiation and integration. Hence, combining results, we obtain the following statement: When each term in the approximating sequence of either the derivative or the integral has a Gaussian distribution and the sequence itself is mean square convergent, the limit exists and also has a Gaussian distribution.

Theorem 20.11.3 (**GRP and Calculus**) If the conditions for mean square convergence to derivatives and Riemann integrals hold for a GRP, then these derivatives and integrals are also normally distributed.

In brief, the derivative of a GRP is a GRP, and under appropriate regularity conditions,

$$Y_t = \int_{-\infty}^{\infty} h(t, \tau) X_\tau \, d\tau$$

is a GRP if X_t is.

20.11.3 Narrowband GRP

A random process of particular importance in communications is a *narrowband* GRP. By a narrowband process, we refer to a *w.s.s.* process having a psd S_X that is zero for $|\omega| = 2\pi f$ outside of a frequency (measured in Hz and not in radian frequency) interval $[f_c - B/2, f_c + B/2]$. For a process to be narrowband, we must have the center frequency f_c be much larger than the bandwidth B. In AM radio, the typical center frequency f_c is 1 MHz and the bandwidth B is 20 KHz (of which the signal is designed to occupy only 16 KHz). In FM radio, the typical center frequency f_c is 100 MHz and the bandwidth B is 200 KHz (of which the signal is designed to occupy only 150 KHz). In mobile cellular communications, f_c is on the order of 1 GHz and the signal bandwidth only on the order of several kilohertz, although signal bandwidths of 1 MHz are becoming available. In all of these cases, the center frequency is orders of magnitude greater than the signal bandwidth. When one tunes a receiver to a channel (bandpasses to the bandwidth in which one is looking for a signal), one receives the desired signal together with additive noise introduced in the transmission channel. Such additive noise in a free-space propagation channel always includes thermal noise X_t that is a mean zero, w.s.s. GRP with constant psd. Once we have tuned, the signal passed to other stages in the receiver contains a bandpass-filtered additive noise term Y_t that is a narrowband GRP. The filtered thermal noise Y_t, having been passed through a band-pass filter with transfer function H, then has a psd S_Y that is itself bandpass:

$$S_Y(\omega) = |H(\omega)|^2 S_X(\omega).$$

In what follows, we do not assume that Y_t is just narrowband thermal noise, but allow it to be any w.s.s. narrowband GRP that could have arisen from band-passing any other GRP with mean zero. A nice representation for such a mean-zero, narrowband GRP Y_t is in terms of *quadrature components*. There exist U_t and V_t, mean-zero, lowpass GRPs, and a random phase θ that is independent of U_t and V_t, such that

$$Y_t = U_t \cos(\omega_c t + \theta) + V_t \sin(\omega_c t + \theta). \qquad (20.15)$$

Furthermore, we claim that, in this representation of Y_t, we can take the quadrature components to have psds S_U and S_V that are identical to some psd S, to be determined, and lowpass in that they are zero outside of the base-band $[-B/2, B/2]$, provided that $f_c > B/2$.

To substantiate these claims, we express the quadrature components in terms of Y_t through a process of linear filtering. First, multiply (modulate) Y_t by $2 \cos(\omega_c t + \theta)$ and recall the following trigonometric identities:

$$2 \cos(A) \cos(B) = \cos(A - B) + \cos(A + B),$$

$$2 \sin(A) \sin(B) = \cos(A - B) - \cos(A + B),$$

$$2 \sin(A) \cos(B) = \sin(A + B) + \sin(A - B).$$

These identities are easily derived from the Euler formula

$$e^{i\theta} = \cos(\theta) + i \sin(\theta).$$

Write

$$2 Y_t \cos(\omega_c t + \theta) = U_t[1 + \cos(2\omega_c t + 2\theta)] + V_t \sin(2\omega_c t + 2\theta).$$

The terms $U_t \cos(2\omega_c t + 2\theta)$ and $V_t \sin(2\omega_c t + 2\theta)$ have both been modulated to become narrowband signals at a center frequency of $2f_c$. If we now lowpass $2Y_t \cos(\omega_c t + \theta)$, keeping only the part of the signal whose Fourier transform lies in a band below a frequency of $f_c - B/2$, then we are left with only U_t. Thus, U_t can be recovered from Y_t by a process of linear filtering through a lowpass filter with impulse response h_B:

$$U_t = 2 \int_{-\infty}^{\infty} h_B(t - \tau) Y_\tau \cos(\omega_c \tau + \theta) \, d\tau.$$

Similarly,

$$V_t = 2 \int_{-\infty}^{\infty} h_B(t-\tau)Y_\tau \sin(\omega_c\tau+\theta)\,d\tau.$$

Recall now that a linearly filtered GRP is again a GRP. Hence, U_t and V_t are GRPs. In addition, since Y_t has mean zero, the same holds for its linearly filtered versions U_t and V_t.

We establish that U_t and V_t are w.s.s. if Y_t is w.s.s.:

$$E(U_tU_s) = 4 \int_{-\infty}^{\infty}\int_{-\infty}^{\infty} h_B(t-\tau)h_B(s-\sigma)E\left[Y_\tau Y_\sigma \cos(\omega_c\tau+\theta)\cos(\omega_c\sigma+\theta)\right]d\tau d\sigma$$

$$= 2 \int_{-\infty}^{\infty}\int_{-\infty}^{\infty} h_B(t-\tau)h_B(s-\sigma)R_Y(\tau-\sigma)$$

$$\times E[\cos(\omega_c(\tau-\sigma))+\cos(\omega_c(\tau+\sigma)+2\theta)]\,d\tau d\sigma.$$

We were able to factor the expectation of the product of a term involving Y and one involving cosine terms due to the postulated independence of Y and θ. Since the random phase $\theta \sim \mathcal{U}(-\pi,\pi)$, it follows that

$$E\cos(\omega_c(\tau+\sigma)+2\theta) = 0,$$

and we are left with

$$E(U_tU_s) = 2 \int_{-\infty}^{\infty}\int_{-\infty}^{\infty} h_B(t-\tau)h_B(s-\sigma)R_Y(\tau-\sigma)\cos(\omega_c(\tau-\sigma))\,d\tau d\sigma.$$

Make the one-to-one change of variables

$$u = \tau - \sigma \text{ and } v = t - \tau,$$

to achieve

$$R_U(t,s) = E(U_tU_s) = 2 \int_{-\infty}^{\infty}\int_{-\infty}^{\infty} h_B(v)h_B(s-t+v+u)R_Y(u)\cos(\omega_c u)\,du\,dv.$$

Inspecting the right-hand side of the equation reveals that $R_U(t,s)$ depends upon t and s only through $t-s$. Hence, U_t is indeed w.s.s. A parallel argument shows that V_t is also w.s.s. and $R_V(t,s) = E(V_tV_s)$ is given by an identical expression. Thus, we also conclude that the psds S_U and S_V are identical.

We identify this common psd (call it S_Q) by taking the Fourier transform of R_U to recover S_U:

$$S_Q(\omega) = \int_{-\infty}^{\infty} e^{-i\omega\tau}R_U(\tau)\,d\tau$$

$$= \int_{-\infty}^{\infty}\int_{-\infty}^{\infty}\int_{-\infty}^{\infty} h_B(v)h_B(-\tau+v+u)R_Y(u)e^{-i\omega\tau}(e^{i\omega_c u}+e^{-i\omega u})\,du\,dv\,d\tau.$$

Make the change of variables $u = u$, $v = v$, and $\sigma = -\tau+v+u$ to obtain

$$S_Q(\omega) = \int_{-\infty}^{\infty}\int_{-\infty}^{\infty}\int_{-\infty}^{\infty} h_B(v)h_B(\sigma)R_Y(u)e^{-i\omega(v+u-\sigma)}(e^{i\omega_c u}+e^{-i\omega u})\,du\,dv\,d\sigma.$$

We see that this expression now factors into three univariate integrals:

$$S_Q(\omega) = \left[\int_{-\infty}^{\infty} h_B(v) e^{-i\omega v} \, dv \right]$$

$$\times \left[\int_{-\infty}^{\infty} h_B(\sigma) e^{i\omega\sigma} \, d\sigma \right]$$

$$\times \left[\int_{-\infty}^{\infty} R_Y(u)(e^{-i(\omega-\omega_c)u} + e^{-i(\omega+\omega_c)u}) \, du \right].$$

The preceding expression enables us to conclude that

$$S_Q(\omega) = |H(\omega)|^2 \left[S_Y(\omega - \omega_c) + S_Y(\omega + \omega_c) \right].$$

In words, the psd S_Q of a quadrature component of a narrowband w.s.s. GRP is the lowpass part of the sum of the psd of Y_t, shifted to the left by ω_c, plus the psd of Y_t, shifted to the right by the same center frequency ω_c. Since S_Y is an even function, we can rewrite this as

$$S_Q(\omega) = |H(\omega)|^2 \left[S_Y(-\omega + \omega_c) + S_Y(\omega + \omega_c) \right]. \tag{20.16}$$

In terms of the representation of S_Y given in Eq. (20.16), we find that

$$S_Q = Q(-\omega) + Q(\omega).$$

The function Q is low-pass and 0 if $|\omega| > \pi B$. Thus, although Q need not be symmetric about the origin, S_Q is symmetric, as is required for a psd.

The quadrature component GRPs U_t and V_t need not be independent. As they are jointly GRPs, we need only calculate their cross-correlation $E(U_t V_s)$ to investigate their relationship:

$$E(U_t V_s) = 4 \int_{-\infty}^{\infty} \int_{-\infty}^{\infty} h_B(t-\tau) h_B(s-\sigma) E\left[Y_\tau Y_\sigma \cos(\omega_c \tau + \theta) \sin(\omega_c \sigma + \theta) \right] d\tau d\sigma$$

$$= 2 \int_{-\infty}^{\infty} \int_{-\infty}^{\infty} h_B(t-\tau) h_B(s-\sigma) R_Y(\tau-\sigma)$$

$$\times E[\sin(\omega_c(\tau - \sigma)) + \sin(\omega_c(\tau + \sigma) + 2\theta)] \, d\tau d\sigma.$$

As we did previously, we conclude from the presence of the random phase angle that

$$E[\sin(\omega_c(\tau - \sigma)) + \sin(\omega_c(\tau + \sigma) + 2\theta)] = \sin(\omega_c(\tau - \sigma)).$$

Making the same change of variables as before yields

$$E(U_t V_s) = 2 \int_{-\infty}^{\infty} \int_{-\infty}^{\infty} h_B(v) h_B(s - t + v + u) R_Y(u) \sin(\omega_c u) \, du dv.$$

We see that $E(U_t V_s)$ depends upon t and s only through their difference. We calculate the cross-psd S_{UV} by taking a Fourier transform of the cross-correlation. As this calculation is similar (a sine replaces a cosine in the integrand) to the one we have just done, we leap to the conclusion that

$$S_{UV}(\omega) = \frac{i}{\pi} |H|^2 \left[S_Y(\omega - \omega_c) - S_Y(\omega + \omega_c) \right].$$

The cross-psd will be zero if the left shift and right shift of S_Y by the same center frequency ω_c are equal. This cancellation requires that the function Q appearing in the representation of the narrowband S_Y given in Eq. (20.16) be symmetric about 0. If Q is symmetric, then the quadrature components U_t and V_t are independent GRPs. If Q is not symmetric, then the quadrature components in the representation of the narrowband w.s.s. GRP Y_t are dependent.

20.12 SUMMARY

Although the focus of this text has been on finitely many random variables, in Chapter 19 and in this chapter we have provided a transition to the study of infinitely many random variables. Random processes can be thought of as signals in time or space with the (possibly vector-valued) amplitude X_t at location t being a random variable. As there are typically infinitely many times or locations, we are dealing with an infinite collection of random variables. Indeed, this is the concern of many systems in electrical engineering. The study of random processes is usually the study of important families of such processes that share some characteristic.

In our discussion we made a distinction between processes proceeding in discrete time and those taking place in continuous time. We discussed both discrete- and continuous-time processes of independent random variables, with white noise being the often-used, but pathological, continuous-time version. Stationary and ergodic processes were defined. Processes that are both stationary and ergodic have the desirable property that, with probability 1, long-run sample function time averages converge to statistical averages or expected values, and we can learn these statistical quantities from the sample function data. We then discussed the very useful model of discrete-time Markov chains. Examples of continuous-time Markov processes were provided by the independent increment Poisson and Brownian motion processes. Martingales—models for fair games and for many other phenomena of interest in mathematical finance and in statistical analysis—were discussed in discrete time, although Brownian motion is an example of a martingale in continuous time.

In moving to continuous-time processes, we first reviewed the elements of stochastic calculus. This subject is a complex one and best left to more advanced courses in random processes. However, a focus on mean square concepts renders the issues accessible to us. With this calculus background, we proceeded to the class of wide-sense stationary processes, their frequency-domain representation through the power spectral density function, and their passage through linear time-invariant systems characterized either by an impulse response function $h(\tau)$ or by a transfer function $H(\omega)$.

We closed our discussion with the Gaussian random process, of which Brownian motion was an early example. The GRP is important in electrical engineering, for it provides the mathematical models for many physical noise sources.

The discussions of the families of random processes in this chapter serve only as an introduction to their subjects. You will need to continue your studies of probability begun here with additional study of random processes. You do so having been provided with a solid foundation.

EXERCISES

E20.1 Consider the random process

$$X_t = A\cos(\omega_c t + \Theta),$$

where the amplitude A has a Rayleigh density

$$f_A(x) = 2xe^{-x^2}U(x),$$

the phase $\Theta \sim \mathcal{U}(-\pi, \pi)$, and A and Θ are independent.

    a. Evaluate EX_t.
    b. Evaluate the autocorrelation $R_X(t, s)$ and autocovariance $C_X(t, s)$.
    c. Show that X_t is normally distributed.

E20.2 Consider the random process $\{X_t, t \in \mathbb{N}\}$ of *i.i.d.* random variables given by $P(X_t = 1) = p = 1 - P(X_t = 0)$ for $0 < p < 1$.

    a. What is the probability that we will never see a t for which $X_t = 1$?
    b. What is the probability that we will see no more than k ones?
    c. What is the probability that, for any time t_0, there exists $t > t_0$ with $X_t = 1$?
    d. What is the probability that we will see infinitely many ones?
    e. Using the results of Chapter 19, for $\epsilon > 0$, evaluate

$$\lim_{n \to \infty} P\left(\left|\frac{1}{n}\sum_1^n X_t - p\right| < \epsilon\right).$$

E20.3 Consider the random process $\{X_t, t \in \mathbb{N}\}$ of independent random variables with

$$E(X_t) = m(t), \quad \text{VAR}(X_t) = \sigma_t^2 \le b < \infty.$$

    a. Show that

$$S_n = \frac{1}{n}\sum_{k=1}^n \left(X_{t_k} - m(t_k)\right)$$

    converges to 0 in mean square.
    b. Provide an example of $m(t)$ and choice of times $\{t_k\}$ such that

$$Y_n = \frac{1}{n}\sum_{k=1}^n X_{t_k}$$

    converges to 0 in mean square.
    c. Provide an example of $m(t)$ and choice of distinct times $\{t_k\}$ such that $\{Y_n\}$ does not converge in mean square.

E20.4 Consider the independent random variables process $\{T_k\}$, where T_k is the waiting time between the arrival of the $(k-1)$st and the kth photon. However, we now allow the average light intensity α_k to vary (perhaps it is a very weak signal that is being transmitted) so that $T_k \sim \mathcal{E}(\alpha_k)$.

   a. Determine the joint characteristic function $\phi_{T_1,\ldots,T_k}(u_1,\ldots,u_k)$.

   b. Determine the cdf of the largest waiting time,

$$T_k^* = \max_{i \leq k} T_i,$$

up to the arrival of the kth photon.

   c. Determine the cdf of the smallest waiting time,

$$T_k' = \min_{i \leq k} T_i,$$

up to the arrival of the kth photon.

   d. Show that you can choose the signal $\{\alpha_k\}$ such that, for a given $t > 0$,

$$\sum_{k=1}^{\infty} P(T_k > t) < \infty.$$

For this choice of signal, use the Borel–Cantelli Lemma to show that the event of a waiting time between arrivals exceeding t in duration will occur only finitely often with probability 1.

   e. Show that you can choose the signal $\{\alpha_k\}$ such that, for a given $t > 0$,

$$\sum_{k=1}^{\infty} P(T_k > t) = \infty.$$

   f. For this choice of signal, use the converse of the Borel–Cantelli Lemma to show that the event of a waiting time between arrivals exceeding t in duration will occur infinitely often with probability 1.

E20.5 Which of the following random processes or random sequences are stationary or ergodic?

   a. $X_t = \Theta \sim \mathcal{U}(0,1)$.

   b. X_t for $T = [0,\infty)$ is a GRP with

$$EX_t = 0, \quad R_X(t,s) = \min(t,s).$$

   c. X_t is a GRP on $T = \mathbb{R}$ (see Section 20.11.1)

$$EX_t = t, \quad \mathrm{COV}(t,s) = C_X(t,s) = e^{-|t-s|}.$$

   d. X_k is a GRP with $T = Z$ and

$$EX_k = 0, \quad E(X_j X_k) = \begin{cases} 1 & \text{if } j = k \\ 0 & \text{if otherwise} \end{cases}.$$

E20.6 X_t for $T = [0,\infty)$ is a stationary and ergodic process.

   a. Explain how to estimate EX_5^2 from data provided by a sample function.

   b. Explain how to estimate $\mathrm{VAR}(X_5)$ from data provided by a sample function.

E20.7 A Markov binary message source has the one-step transition probabilities $p_{0,0} = .9$ and $p_{1,1} = .05$.

   a. Provide the one-step transition matrix $\mathbb{P}$ and sketch the state transition diagram.

b. Determine the limiting distribution of states, $\lim_{t \to \infty} \pi^{(t)}$.

c. What is the expected waiting time between successive ones?

d. Given that the chain starts in state 0, provide an expression for the probability that the first 1 will not appear in the next k symbols.

E20.8 a. For the Markov chain of the preceding problem, if the initial state distribution is

$$\pi^{(1)} = \begin{pmatrix} .7 & .3 \end{pmatrix},$$

determine the state distributions $\pi^{(n)}$ for $n = 2$ and $n = 5$.

b. If you have access to computational resources, plot both components of $\pi^{(n)}$ for n ranging from 5 to 20. Does $\pi^{(n)}$ appear to be converging to a limit?

E20.9 Consider the Markov state transition graph of Figure 20.3 with transition probabilities specified by

$$p_{1,1} = .1, \ p_{1,2} = .3, \ p_{2,3} = .5.$$

a. Specify the remaining positive transition probabilities.

b. If the Markov chain starts in state ξ_1 at $t = 1$, evaluate the state distribution $\pi^{(2)}$.

c. Evaluate $\lim_{n \to \infty} \pi^{(n)}$.

d. By considering the state transition diagram, evaluate the probability $f_{1,2}^{(2)}$ of returning to state ξ_1 for the first time in two steps.

e. What is the mean waiting time to return to each of the states?

E20.10 Draw state transition diagrams for a Markov chain having five states such that the following statements hold individually:

a. Every state is in the same communicating class.

b. There are three nonempty communicating classes.

c. There are two nonempty communicating classes, and one of them is open.

d. State ξ_2 is transient.

e. State ξ_3 is absorbing.

f. State ξ_1 is nonreturn.

g. State ξ_1 is accessible from state ξ_5, but not conversely.

E20.11 Argue from the Chapman–Kolmogorov equation that if a homogeneous Markov chain has a transition probability matrix $\mathbb{P}$ such that, for some r, $\mathbb{P}^r$ has all positive entries, then, for any $t > r$, the matrix $\mathbb{P}^t$ will also have all positive entries.

E20.12 Consider a reliability problem of cascading faults in which the system has s states. State ξ_1 is the normal operating state and is the state we are in at time 1. State ξ_s is the absorbing state of catastrophic failure from which recovery cannot be achieved. The remaining states of successively impaired operation are such that they communicate only with their nearest neighbor; that is, for $2 \le k \le s - 1$, state ξ_k communicates only with states ξ_{k-1} and ξ_{k+1}. Simplify our model by assuming that, for $k < s$, $p_{k,k+1} = p_1$, and for $1 < k$, $p_{k,k-1} = p_0$.

a. Sketch the state transition diagram.

b. Provide an expression for the probability that we will find ourselves on the verge of disaster (in state ξ_{s-1}) at time n given that we started in ξ_1 at time 1.

    c. Provide an expression for the probability $f_{1,s}^{(n)}$ of failure first occurring at time n.

    d. What is the probability $f_{1,s}$ of eventual failure?

    e. Provide an expression for the mean waiting time to catastrophic failure.

E20.13 For the preceding Markov chain for cascading faults, what is the mean time to return to the normal operating state?

E20.14 If a Poisson process $\{N_t, t \geq 0\}$ is described by a mean number of jumps per unit time of $\lambda = 5$, provide an explicit expression for the trivariate joint probability

$$P(N_{t_1} = n_1, N_{t_2} = n_2, N_{t_3} = n_3) \text{ with } t_3 > t_2 > t_1.$$

E20.15 Consider a Poisson process $\{X_t, t \geq 0\}$ with time measured in seconds.

    How should you choose the parameter λ to maximize the probability that $P(X_5 = 2)$?

E20.16   a. Determine EN_t for N_t a Poisson process with parameter λ.

    b. Calculate the autocorrelation function $R_N(t, s) = E(N_t N_s)$ for this Poisson process.

    c. Show that the autocovariance function $C_N(t, s) = E(N_t - EN_t)(N_s - EN_s)$ is $\lambda \min(t, s)$.

E20.17 We define a random process $\{X_t, t \geq 0\}$ known as the *random telegraph wave* as taking on only the binary values $\{-1, 1\}$, starting with $P(X_0 = 1) = 1$, and changing sign at those times when a given Poisson process N_t increases in value by 1. Alternatively, $X_t = 1$ if N_t is even and $X_t = -1$ if N_t is odd.

    a. Show that

$$P(X_t = 1) = \cosh(\lambda t)e^{-\lambda t} \text{ and } P(X_t = 1) = \sinh(\lambda t)e^{-\lambda t}.$$

    b. Evaluate EX_t and $\text{VAR}(X_t)$.

    c. Observe that, for $t > s$, the event that $X_t = 1$ given that $X_s = -1$ is that $N_t - N_s$ is odd. Use this condition to determine $P(X_t = 1 | X_s = -1)$.

    d. Show that the correlation function

$$R_X(t, s) = e^{-2\lambda(t-s)} \text{ for } t > s.$$

E20.18 If $\{X_n, n \in N\}$ is a martingale, then evaluate $E\left(X_j(X_m - X_k)\right)$ for any $m > k \geq j$.

E20.19 If

$$S_n = \sum_0^n X_n$$

is a martingale, then are any X_j and X_k orthogonal for $j \neq k$?

E20.20   a. Provide an example of a Markov chain having five states such that its one-step transition matrix $\mathbb{P}$ is not trivially a diagonal matrix and the chain is also a martingale.

    b. Classify the states of this Markov chain according to the categories of Definition 20.6.3.

    c. Evaluate the asymptotic state distribution π for this chain.

E20.21 If $N_t, t \geq 0$, is a Poisson process, define the random process

$$Y_t = (N_t - \lambda t)^2 - \lambda t.$$

Observe that there is a one-to-one relationship between N_t and Y_t.

     a. Show that Y_t is a martingale by applying the definition.

     b. What is the mean waiting time ES for $Y_S = a > 0$ for the first time?

E20.22 Consider a very long buffer initially containing K packets. At each unit time, we either receive a new packet with probability p or forward a packet with probability $1 - p$. Assume that these events at each time are $i.i.d.$ Let $Y_m = 1$ denote the event that we receive a new packet and $Y_m = -1$ the event that we forward a packet. Then

$$P(Y_m = 1) = p = 1 - P(Y_m = -1);$$

The cumulative number of packets in the buffer after the nth time are

$$B_n = K + \sum_{m=1}^{n} Y_m.$$

     a. Rewrite

$$B_n = M_n - K - n(1 - 2p).$$

     Is $\{M_n\}$ a martingale?

     b. If S is the waiting time to first emptying the buffer ($B_n = 0$ for the first time), evaluate ES. You will need to be careful about p.

E20.23 For all n, we have estimates of X distributed as $\mathcal{N}(m, \sigma^2)$ based upon observations $Y_0, \ldots, Y_n$. In addition, we know that the random variables X and $Y_0, \ldots, Y_n$ are jointly normally distributed for any n. Consider the sequence of least mean square estimates of X given by

$$Z_n = E(X|Y_0, \ldots, Y_n),$$

with mean square estimation error e_n.

     a. Evaluate EZ_n.

     b. Using the martingale property, evaluate the autocorrelation $E(Z_m Z_n)$ for $n \geq m$.

     c. Write the univariate distribution for Z_n.

     d. Write the bivariate distribution for Z_m and Z_n.

E20.24   a. For the process of least mean square estimates $\{Z_n\}$ defined in the preceding problem, does it converge in mean square or with probability 1?

     b. If it does converge, evaluate $\lim_{n \to \infty} Z_n$.

E20.25   a. If a mean-zero Gaussian random process $\{X_n, n \in N\}$ in discrete time is a martingale, then what can you say about the least mean square estimator $\hat{X}_n(X_0, \ldots, X_{n-1})$ of X_n?

     b. Determine its autocovariance function $\text{COV}(X_m, X_n)$.

E20.26 Let X_t be a Gaussian white-noise process having mean 0 and autocorrelation function $R_X(t, s) = \sigma^2 \delta(t - s)$. Let

$$Y_t = \int_0^t X_s \, ds.$$

     a. Ignoring the subtleties regarding the integration of such a pathological process, we conclude that Y_t is also a Gaussian random process. Evaluate EY_t and $R_Y(t, s)$.

     b. Is Y_t Brownian motion?

E20.27
   a. Determine the bivariate pdf for Brownian motion variables X_t and X_s, assuming that $t > s$.

   b. Determine the trivariate pdf for Brownian motion variables X_t, X_s, and X_r, assuming that $t > s > r$.

   c. Evaluate $E(X_t X_s X_r)$.

E20.28 Let X_t be a wide-sense stationary random process with common mean m, autocorrelation function $R_X(\tau)$, and power spectral density S_X. For each version of Y_t defined in the following list, evaluate R_Y and the psd S_Y:

   a. $Y_t = aX_t + b$;

   b. $Y_t = X_{t-t_0}$, for a given t_0;

   c. $Y_t = \frac{dX_t}{dt}$;

   d. $Y_t = X_t \cos(\omega_c t + \Theta)$, with Θ uniformly distributed over $[-\pi, \pi]$ and independent of the X_t process.

E20.29 If

$$S_X(\omega) = e^{-a|\omega|},$$

determine EX and $\mathrm{VAR}(X)$.

E20.30 If X_t is w.s.s. with

$$R_X(\tau) = e^{-a|\tau|}$$

and

$$Y = \int_0^1 X_t \, dt,$$

then determine EY_t^2.

E20.31 If X_t and Y_t are orthogonal w.s.s. random processes (i.e., for all s and t, $E(X_s Y_t) = 0$) and $Z_t = X_t + Y_t$, then express the autocorrelation R_Z and psd S_Z in terms of the autocorrelations and psds for X_t and Y_t.

E20.32 Let X_n be a message sequence that is an $i.i.d.$ ± 1-valued Bernoulli process with

$$P(X_n = 1) = p = 1 - P(X_n = -1).$$

Define $\Theta \sim \mathcal{U}(-\frac{\pi}{2}, \frac{\pi}{2})$ and independent of $\{X_n\}$. The modulated signal is

$$Y_t = \cos\left(\omega_c t + \Theta + \frac{\pi}{2}X_n\right) \text{ for } n\tau \leq t \leq (n+1)\tau.$$

Hence, the random process Y_t is a phase-shift-keyed version of the message sequence X_n.

   a. Evaluate EY_t and the autocorrelation $R_Y(t, s)$.

   b. If Y_t is wide-sense stationary, then evaluate the psd S_Y.

E20.33 Returning to Exercise 20.5, assume that the pulse shape is rectangular, or

$$p(t) = U(t) - U(t - \tau).$$

Show that the process Y_t is now wide-sense stationary.

E20.34 Define a thermal noise process generated by a resistance of 1,000 ohms at a room temperature of $20°$ C. Specify the autocorrelation function and the power spectral density.

E20.35 If the thermal noise source of the preceding problem is the input X_t to a linear, time-invariant filter with transfer function

$$H(\omega) = \frac{1}{1 + i\omega},$$

describe the filter output process Y_t.

E20.36 Consider the random process

$$X_t = \cos\left((\omega_c + \Omega_t)t + \Theta\right),$$

where the instantaneous frequency $\omega_c + \Omega_t$ is random with the message random process Ω_t being a Gaussian random process with

$$E\Omega_t = 0 \text{ and } R_\Omega(t, s) = e^{-\alpha|t-s|}.$$

The phase $\Theta \sim \mathcal{U}(-\pi, \pi)$, and Ω_t and Θ are independent. X_t is a frequency-modulated signal carrying the message Ω_t.

  a. Evaluate EX_t. (*Hint*: Using subconditioning, first take the conditional expectation of Θ given Ω.)
  b. Evaluate the autocorrelation $R_X(t, s)$ and autocovariance $C_X(t, s)$. (*Hint*: Use the trigonometric identities to rewrite a product of cosines as a sum of cosines. Relate an expectation of a cosine to characteristic functions.)

E20.37 X_t is a Gaussian random process with mean 0 and autocorrelation function $R_X(t, s)$. If $Y_t = [X_t]^2$, then evaluate R_Y. (*Hint*: See Section 17.4 and apply it to the bivariate normal.)

E20.38 $X_t, t \geq 0$, is a Gaussian random process with mean function $\sin(t)$ and autocovariance function

$$C_X(t, s) = \min(t, s).$$

Completely describe the random process

$$Y_t = \int_0^t X_s \, ds.$$

E20.39 If a mean-zero Gaussian random process X_t with autocorrelation function

$$R_X(t, s) = \frac{1}{1 + (t - s)^2},$$

then describe the derivative process $Y_t = X_t'$.

E20.40 If a mean-zero Gaussian random process X_t has autocorrelation function

$$R_X(t, s) = t^2 s^2 e^{-(t-s)^2},$$

then describe the derivative process $Y_t = X_t'$.

E20.41 This exercise concerns the Karhunen–Loeve representation for a random process. Choose $T = [0, \tau]$ and define

$$X_t = \sum_{k=1}^{\infty} \sqrt{\lambda_k} \xi_k \phi_k(t),$$

where $\lambda_k > 0$, $\{\xi_k\}$ are *i.i.d.* $\mathcal{N}(0, 1)$ random variables and $\{\phi_k\}$ are orthonormal functions on T in that

$$\int_0^\tau \phi_k(t)\phi_j(t)\,dt = \delta j, k = \begin{cases} 1 & \text{if } j = k \\ 0 & \text{otherwise} \end{cases}.$$

a. Show that $\{X_t, t \in T\}$ is a Gaussian random process.
b. Evaluate EX_t and the autocorrelation function $R_X(t, s) = E(X_t X_s)$.
c. Show that $\{\phi_k(t)\}$ are the eigenfunctions and $\{\lambda_k\}$ the eigenvalues for the integral equation

$$\int_0^\tau R_X(t, s)f(s)\,ds = \gamma f(t).$$

d. Starting with a Gaussian random process on $T = [0, \tau]$ with mean 0 and autocorrelation function $R_X(t, s)$, how can we determine its Karhunen–Loeve representation?

Bibliography

[1]   Aczel, J. [1966], *Lectures on Functional Equations and Their Applications,* translated by Scripta Technica, Inc., Academic Press, New York.

[2]   Adler, R. J., R. E. Feldman, and M. S. Taqqu, eds. [1998], *A Practical Guide to Heavy Tails: Statistical Techniques and Applications,* Birkhauser, Boston.

[3]   Arnauld, A., and P. Nicole [1662], *La Logique, ou l'art de Penser,* translated by J. Dickoff, and P. James [1964], *The Art of Thinking,* Bobbs–Merrill, Indianapolis.

[4]   Ash, R. [1972], *Real Analysis and Probability,* Academic Press, New York.

[5]   Axtell, R. [2001], Zipf distribution of U.S. firm sizes, *Science,* **293**, 7 September, 1818–1820.

[6]   Barabasi, A. [2002], *Linked: The New Science of Networks,* Perseus Publishing, Cambridge, MA.

[7]   Berger, J. [1985], *Statistical Decision Theory and Bayesian Analysis, 2d ed.,* Springer-Verlag, New York.

[8]   Bernoulli, Daniel [1738], Specimen theoriae novae de mensura sortis, *Commentarii Academiae Scientarum Imperialis Petropolitanae,* **5**, 175–192. Translated by L. Sommer [1954], Exposition of a new theory on the measurement of risk, *Econometrica,* **22**, 23–36.

[9]   Bernoulli, James/Jacob [1713], *Ars Conjectandi, Pars Quarta,* translated by Bing Sung [1966], *Translations from James Bernoulli,* Dept. of Statistics, Harvard University, Cambridge.

[10]   Billingsley, P. [1979], *Probability and Measure,* 1st ed., J. Wiley & Sons: New York.

[11]   Bollobas, B. [1985], *Random Graphs,* Academic Press: London.

[12]   Blackwell, D. [1969], *Basic Statistics,* McGraw–Hill, New York.

[13]   Boole, G. [1854], *An Investigation of the Laws of Thought,* reprinted in 1951 by Dover, New York.

[14]   Born, M. [1926], Zur quantenmechanik der stoßvorgange, *Zeitschrift fur Physik,* **XXXVII**, 863–867.

[15]   Broder, A., R. Kumar, F. Maghoul, P. Raghavan, S. Rajagopalan, R. Stata, A. Tomkins, and J. Wiener [2000], Graph structure in the web, Ninth International World Wide Web Conference, Amsterdam.

[16]   Brown, R., and P. Hwang [1992], *Introduction to Random Signals and Applied Kalman Filtering,* 2d ed., Wiley, New York.

[17]   Cardano, G. [1564/1663], *Liber de Ludo Alea,* trans. S. H. Gould [1961], *The Book on Games of Chance,* Holt, Rinehart & Winston, New York.

[18]   Carnap, R. [1962], *The Logical Foundations of Probability,* Univ. of Chicago Press.

[19]   Chellappa, R., and A. Jain, eds. [1993], *Markov Random Fields: Theory and Application,* Academic Press Inc., San Diego.

[20]   Cover, T., and J. Thomas [1991], *Elements of Information Theory,* Wiley, New York.

[21]   Dagpunar, J. [1988], *Principles of Random Variate Generation,* Oxford Science Publications, Clarendon Press: Oxford.

[22]   David, F. N. [1962], *Games, Gods and Gambling,* Griffin, London.

[23]   de Finetti, B. [1974], *Theory of Probability*, translation by A. Machi and Adrian Smith, Wiley Classics, New York (restatement and further development of much earlier ideas).

[24]   de Finetti, B. [1990], *Theory of Probability: A Critical Introductory Treatment*, translated by A. Machi and A. Smith, Wiley Classics Library, Interscience Pubs., New York.

[25]   Doob, J. [1953], *Stochastic Processes*, Wiley, New York.

[26]   Ethier, S., and T. Kurtz [1986], *Markov Processes*, Wiley-Interscience, New York.

[27]   Feller, W. [1950,1957], *An Introduction to Probability Theory and Its Applications, I*, Wiley, New York.

[28]   Ferguson, T. [1967], *Mathematical Statistics: A Decision Theoretic Approach*, Academic Press, New York.

[29]   Fine, T. L. [1973], *Theories of Probability: An Examination of Foundations*, Academic Press, New York.

[30]   Fine, T. L. [1999], Foundations of probability: an update, in S. Kotz, C. Read, and D. Banks, eds. *Encyclopedia of Statistical Sciences: Update Volume 3*, Wiley, New York, pp. 246–254.

[31]   Fine, T. L. [1999], *Feedforward Neural Network Methodology*, Springer-Verlag, New York.

[32]   Fishburn, P. C. [1970], *Utility Theory for Decision Making*, Wiley, New York.

[33]   Franklin, J. [2001], *The Science of Conjecture: Evidence and Probability before Pascal*, The Johns Hopkins Press, Baltimore and London.

[34]   Freedman, D., et al. [1991], *Statistics, 2d ed.*, W.W. Norton, New York.

[35]   Fry, T. C. [1928], *Probability and Its Engineering Uses*, Bell Telephone Laboratories Series, Van Nostrand, New York.

[36]   Galambos, J., and I. Simonelli [1996], *Bonferroni-Type Inequalities with Applications*, Springer, New York.

[37]   Gallager, R. [1996], *Discrete Stochastic Processes*, Kluwer Academic Publishers, Boston.

[38]   Gelb, A., ed. [1974], *Applied Optimal Estimation*, M.I.T. Press, Cambridge.

[39]   Gnedenko, B. V. [1967], *The Theory of Probability*, 4th ed., Chelsea Publ. Co., Bronx, New York.

[40]   Gray, R., and L. Davisson [1986], *Random Processes: A Mathematical Approach for Engineers*, Prentice Hall, Englewood Cliffs, NJ.

[41]   Grewal, M., and A. Andrews [1993], *Kalman Filtering: Theory and Practice*, Prentice Hall, Englewood Cliffs, NJ.

[42]   Grimmett, G., and D. Stirzaker [2001], *Probability and Random Processes*, Oxford University Press Inc., New York.

[43]   Hacking, I. [1975], *The Emergence of Probability*, Cambridge Univ. Press.

[44]   Hald, A. [1990], *A History of Probability and Statistics: and Their Applications before 1750*, Wiley, New York.

[45]   Halmos, P. [1974], *Naive Set Theory*, Springer-Verlag, New York.

[46]   Horn, R. A., and C. R. Johnson [1985], *Matrix Analysis*, Cambridge Univ. Press, New York.

[47]   Humphreys, P. [1985], Why propensities cannot be probabilities, *The Philosophical Review*, **94**, 557–570.

[48]   Huyghens, C. [1657], *De Ratiociniis in Ludo Aleae*, in F. van Schooten, *Exercitationum Mathematicarum Libri Quinque, V*.

[49]   Jauch, J. M. [1968], *Foundations of Quantum Mechanics*, Addison-Wesley, Reading, MA.

[50]   Jaynes, E. T. [2003], *Probability Theory: The Logic of Science*, Cambridge University Press, Cambridge.

[51]   Kalman, R. [1960], A new approach to linear filtering and prediction problems, *Trans. of the ASME J. Basic Engineering, Series D*, **82**, 35–45.

[52]   Karlin, S., and H. Taylor [1975], *A First Course in Stochastic Processes*, 2d ed., Academic Press, New York.

[53] Kemeny, J., J. L. Snell, and A. W. Knapp [1976], *Denumerable Markov Chains,* Springer-Verlag, New York.

[54] Kolmogorov, A. N. [1933], *The Foundations of Probability*, translated by N. Morrison and reprinted 1956 in the second English edition, Chelsea Pub. Co. New York.

[55] Kolmogorov, A. N. [1963], On tables of random numbers, *Sankhya: The Indian Journal of Statistics, Series A,* p. 369.

[56] Krantz, D., R. Luce, P. Suppes, and A. Tversky [1971], *Foundations of Measurement,* **I**, Academic Press, New York.

[57] Laplace, Pierre-Simon, Marquis de [1825], *Philosophical Essay on Probabilities,* translated in 1995 from fifth French edition by A. I. Dale, Springer-Verlag, New York.

[58] Leon-Garcia, A. [1994], *Probability and Random Processes for Electrical Engineering,* Second ed., Addison-Wesley, Reading, MA.

[59] Li, M., and P. Vitanyi [1997], *An Introduction to Kolmogorov Complexity and Its Applications,* Springer, New York.

[60] Loeve, M. [1977], *Probability Theory I,* Springer, New York.

[61] Maistrov, L. E. [1974], *Probability Theory: A Historical Sketch*, Academic Press, New York.

[62] Mandelbrot, B. [1983], *The Fractal Geometry of Nature,* W.H. Freeman.

[63] von Mises, R., and H. Geiringer [1964], *The Mathematical Theory of Probability and Statistics,* Academic Press, New York (has a discussion of the much earlier work by von Mises on the Kollektiv).

[64] von Mises, R. [1957], *Probability, Statistics, and Truth,* H. Geiringer, translator, third edition of 1951 version in German, Allen and Unwin, London.

[65] Neveu, J. [1975], *Discrete-Parameter Martingales*, trans. T. Speed, North-Holland Pub.: Amsterdam.

[66] Oppenheim, A. V., A. Willsky, and S. H. Nawab [1997], *Signals and Systems,* 2d ed., Prentice Hall, Upper Saddle River, NJ.

[67] Pais, A. [1982], *'Subtle is the Lord...': The Science and Life of Albert Einstein,* Chapter 4, Oxford University Press, New York.

[68] Pais, A. [1982], Max Born's statistical interpretation of quantum mechanics, *Science,* **218**, 1193–1198.

[69] Palmer, E. M. [1985], *Graphical Evolution: An Introduction to the Theory of Random Graphs,* John Wiley & Sons, Inc., New York.

[70] Parzen, E. [1962], *Stochastic Processes*, Holden-Day, Inc., San Francisco.

[71] Pisacane, V. L., and R. C. Moore, eds. [1994], *Fundamentals of Space Systems,* Oxford Univ. Press, New York, Sec. 9.4, 583–588.

[72] Pollard, D. [1984], *Convergence of Stochastic Processes*, Springer.

[73] Poor, H. [1994], *An Introduction to Signal Detection and Estimation, 2d ed.*, Springer-Verlag, New York.

[74] RAND Corporation [1955], *A Million Random Digits with 100,000 Normal Deviates,* Free Press, Glencoe, IL.

[75] Resnick, S. [1992], *Adventures in Stochastic Processes,* Birkhauser, Boston.

[76] Riordan, J. [1958], *An Introduction to Combinatorial Analysis,* Wiley, New York.

[77] Robbins, H. [1955], A remark on Stirling's formula, *Amer. Math. Monthly,* **62**, 26–29.

[78] Rosenblatt, M. [1974], *Random Processes*, 2d ed., Springer-Verlag, New York.

[79] Sambursky, S. [1956], On the possible and the probable in ancient Greece, *Osiris,* **12**, 35–48.

[80] Savage, L. J. [1954], *The Foundations of Statistics,* Wiley, New York.

[81] Scharf, L. [1991], *Statistical Signal Processing: Detection, Estimation, and Time Series Analysis,* Addison-Wesley, Reading, MA.

[82] Shannon, C., and W. Weaver [1949], *A Mathematical Theory of Communication*, Univ. of Illinois Press, reprint of 1948 BSTJ articles by Shannon accompanied by an essay by Weaver.

[83] Stanley, R. [1986], *Enumerative Combinatorics*, vol. I, Chapman & Hall.

[84] Stigler, S. [1986], *The History of Statistics,* The Belknap Press of Harvard University Press, Cambridge, MA.

[85] Strichartz, R. [1995], *The Way of Analysis,* Jones and Bartlett, Boston.

[86] Strogatz, S. [2001], Exploring complex networks, *Nature,* **410**, 8 March 2001, 268–276.

[87] Suppes, P. [1972], *Axiomatic Set Theory,* Dover, New York.

[88] Suppes, P. [1987], Propensity representations of probability, *Erkenntnis,* **26**, 335–358.

[89] Tierney, John [1991], Behind Monty Hall's doors: puzzle, debate and answer?, *The New York Times,* July 21, 1991, Section 1, Part 1, Page 1, Col. 5.

[90] Todhunter, I. [1865], *A History of the Mathematical Theory of Probability from the Time of Pascal to That of Laplace*, reprinted in 1949 by Chelsea Pub., New York.

[91] Tonomura, A., J. Endo, T. Matsuda, T. Kawasaki, and H. Ezawa [1989], Demonstration of single-electron buildup of an interference pattern, *Amer. Jour. of Physics,* **57**, pp. 117–120.

[92] Tufte, E. [1983], *The Visual Display of Quantitative Information,* Graphics Press, Cheshire, CT, 143.

[93] Tutte, W. [1984], *Graph Theory,* Addison–Wesley Pub.: Menlo Park, CA.

[94] Venn, J. [1888], *The Logic of Chance,* 3d ed., Macmillan, New York.

[95] Walley, P., and T. L. Fine [1979], Varieties of modal (classificatory) and comparative probability, *Synthese,* **41**, 321–374.

[96] Walley, P. [1991], *Statistical Reasoning with Imprecise Probabilities*, Chapman and Hall Pub.

[97] Walley, P. [1996], Inferences from multinomial data: learning about a bag of marbles, *J. Roy. Statist. Soc. Ser. B*, **58**, 3–57.

[98] Welsh, D. [1988], *Codes and Cryptography*, Oxford University Press.

[99] Wiener, N. [1947], *Extrapolation, Interpolation and Smoothing of Stationary Time Series,* Wiley, New York.

[100] Wilks, S. S. [1962], *Mathematical Statistics,* Wiley & Sons, Inc.: New York.

[101] Williams, D. [1991], *Probability with Martingales*, Cambridge Univ. Press.

[102] Wong, E. [1983], *Introduction to Random Processes,* Springer-Verlag, New York.

[103] Wong, E., and B. Hajek [1984], *Stochastic Processes in Engineering Systems,* Springer-Verlag, New York.

[104] Yaglom, A. [1962], *An Introduction to the Theory of Stationary Random Processes,* trans. R. Silverman, Prentice Hall, Englewood Cliffs, NJ.

Solution Notes for Selected Exercises

These selected detailed solution notes to 153 of the 650 exercises were prepared by teaching assistants Pablo I. Fierens, Chin-Jen Ku, Ivan Lysiuk, and Leandro Rego, while carrying out their duties in Cornell's ECE310 between 2002 and 2004, and by the author. The author prepared solutions to exercises in Chapters 13, 16, 18, 19, and 20, and to a few other exercises in other chapters. Lysiuk and Rego prepared the most recent versions of most of the other solutions included here.

CHAPTER 0

3.

$$a = \sum_{n=0}^{\infty} 3^{-n} = \sum_{n=0}^{\infty} \left(\frac{1}{3}\right)^n = \frac{1}{1 - \frac{1}{3}} = \frac{3}{2},$$

$$b = \sum_{n=1}^{\infty} 2^{-n} = (1 - 1) + \sum_{n=1}^{\infty} \left(\frac{1}{2}\right)^n = \sum_{n=0}^{\infty} \left(\frac{1}{2}\right)^n - 1 = 2 - 1 = 1,$$

$$c = \sum_{n=0}^{\infty} \frac{2^n}{n!} = e^2.$$

5.

$$\mathbf{x} \cdot \mathbf{y} = (-1) \times 1 + 0 \times 1 + 1 \times (-1) = -2.$$

Not orthogonal because $\mathbf{x} \cdot \mathbf{y} \neq 0$.

8.

$$q = \begin{pmatrix} 0 & 2 & 1 \end{pmatrix} \begin{pmatrix} 3 & 1 & 2 \\ 0 & 1 & 2 \\ 1 & 1 & 4 \end{pmatrix} \begin{pmatrix} 0 \\ 2 \\ 1 \end{pmatrix} = \begin{pmatrix} 0 & 2 & 1 \end{pmatrix} \begin{pmatrix} 4 \\ 4 \\ 6 \end{pmatrix} = 14.$$

14.

$$f'(x) = \left. \frac{d\Phi(zx)}{dz} \right|_{z=\sqrt{x}} (\sqrt{x})' = \frac{1}{\sqrt{2\pi}} e^{-\frac{x}{2}} \frac{1}{2\sqrt{x}} = \frac{e^{-\frac{x}{2}}}{2\sqrt{2\pi x}}.$$

15.

$$\frac{d}{dx} \int_x^{x^2} [\log(x+y)]^2 \, dy = 2x \log(x+x^2) - \log(2x) + \int_x^{x^2} \frac{2\log(x+y)}{x+y} \, dy.$$

16.

$$\int_A f(x,y) \, dx \, dy = \int_{y=0}^1 \int_{x=0}^y f(x,y) \, dx \, dy = \int_{y=0}^1 e^y \left[\int_{x=0}^y e^x \, dx \right] dy = \int_{y=0}^1 e^y [e^y - 1] \, dy$$

$$= \int_{y=0}^1 (e^{2y} - e^y) \, dy = \frac{1}{2}(e^2 - 1) - (e - 1) = \frac{e^2 - 2e + 1}{2} = \frac{(e-1)^2}{2}.$$

18. Since

$$I_k = \int_0^\infty x^k e^{-x} \, dx = (-1) \int_0^\infty x^k \, de^{-x} = -e^{-x} x^k \Big|_0^\infty + \int_0^\infty e^{-x} \, dx^k$$

$$= k \int_0^\infty x^{k-1} e^{-x} \, dx = k I_{k-1}.$$

for $\forall k \geq 1$ and $I_0 = 1$, then by induction $I_k = k(k-1) \cdots 1 = k!$

19. a. The result of the integration cannot be a function of the dummy variable.

b.

$$\sum_{n=0}^{100}\sum_{m=0}^{100}(-2)^{n+m} = \sum_{n=0}^{100}(-2)^n \sum_{m=0}^{100}(-2)^m = \left(\sum_{m=0}^{100}(-2)^m\right)^2 > 0.$$

c. The size of the resulting matrix must be 4×3.

CHAPTER 1

1. a. $\Omega = \{x \in R : 50 < x < 500\}$, where x is measured in lbs. (answers are not unique)
   b. $\Omega = N$, where N are all nonnegative integers.
   c. $\Omega = L \times L$, where L is the set of all letters in the English alphabet.
   d. $\Omega = N$, where N are all nonnegative integers or the number A of atoms in one micro-gram of radium (determinable from Avogadro's number and the atomic number of radium).

2. $\Omega = N_{10} = \{0, 1, 2, \ldots, 9\}$,
   $A = \{0, 3, 6, 9\}$ and $B = \{0, 4, 8\}$, therefore
   $A^c = \{1, 2, 4, 5, 7, 8\}$; $A \cup B = \{0, 3, 4, 6, 8, 9\}$ and $A \cap B = \{0\}$.

3. C is equal to the part inside the disc with radius 1 and center at $(0, 0)$ which is in the first and third quadrant of the cartesian plane.
   D is equal to the union of the first and third quadrant of the cartesian plane and the part inside the disc with radius 1 and center at $(0, 0)$ which is in the second and fourth quadrant of the cartesian plane.
   (Figures are omitted.)

9. $I_{(A^c \cup B^c)^c} = 1 - I_{A^c \cup B^c} = 1 - (\max\{I_{A^c}, I_{B^c}\}) = 1 - (\max\{1 - I_A, 1 - I_B\}) = 1 - (1 - \min\{I_A, I_B\}) = \min\{I_A, I_B\} = I_{A \cap B}$.

15. a. If $I_A I_B$ is identically zero, there should be no $\omega \in \Omega$ such that $\omega \in A$ and $\omega \in B$; this implies A and B must be disjoint sets, $A \perp B$.
    b. If $I_A - I_B \geq 0$, this implies that if $I_B(\omega) = 1$ for some $\omega \in \Omega$, then $I_A(\omega) = 1$; therefore, if $\omega \in B \Longrightarrow \omega \in A \Longrightarrow B \subseteq A$.
    c. If $I_A(\omega) - I_B(\omega) = 1$ for all $\omega \in \Omega$, this implies $I_A(\omega) = 1$ and $I_B(\omega) = 0$ for all $\omega \in \Omega$, and therefore $A = \Omega$ and $B = \emptyset$.
    d. If $A \cap B = A \cup B$, then $\min\{I_A, I_B\} = \max\{I_A, I_B\}$, which implies $I_A = I_B$, and therefore $A = B$.
    e. Using indicator functions, $I_{A \cap B^c} = I_A(1 - I_B) = I_{B \cap A^c} = I_B(1 - I_A)$. Therefore $I_A = I_B$ or $A = B$.

16. a. $A \cup B \cup C$
    d. $A^c \cap B^c \cap C^c$
    f. B
    i. $(A \cap B \cap C)^c$
    j. $A \cap B \cap C$

19. As $\{a\}$, $\{b\}$ and $\{c,d,e,f\}$ form a partition of the sample space Ω, according to Example 1.5, the smallest algebra $\mathcal{A}$ of subsets of Ω containing A and B is:
    $\mathcal{A} = \{\emptyset, \Omega, \{a\}, \{b\}, \{a,b\}, \{a,c,d,e,f\}, \{b,c,d,e,f\}, \{c,d,e,f\}\}$.
    Therefore, $C \in \mathcal{A}$ but $D \notin \mathcal{A}$.

20. $\mathcal{A} = \{\emptyset, \Omega, A, A^c\}$, where $A = \{a,b\}$.

22. We will answer this exercise with a smaller sample, so that you can understand what you were supposed to do. Consider the following text from *The Ithaca Journal*:

    a. "A poem that began as a second-grade assignment to encourage writing through rapping, became the theme to commemorate the third anniversary of Rashad Richardson's death." The corresponding binary sequence is to be created according to the first letter in the words (a, p, t, b, a, a, s, a, ..., d) and therefore is the following (1, 0, 0, 0, 1, 1, 0, 1, ..., 0).

    b. The occurrence of vowels and consonants in the text is random. There is no simple formula that predicts the location of the next vowel given the location of the preceding ones.

    c. Mean = .28; median = 0; standard = .458.

    d. The sequence corresponding to the occurrence of vowels in the letters is given by: (1, 0, 1, 1, 0, 0, 0, 1, 0, 0, 1, 0, 1, 0, 1, 0,—(first 5 words before this point)—1, 0, 1, 0, 1, 0, 0, 0, 0, 1, 0, 1, 1, 0, 0, 1, 0, 0, 0, 1, 0, 0, 0, 1, 1, 0, 0, 1, 1, 0, 1, 0, 1,—(first 10 words before this point)—...). So that plot should include the points: (5, 7/16) and (10, 21/49).

CHAPTER 2

1. a. Using sampling with replacement formula from Section 2.3.1: $\mu_{26,2} = 26^2 = 676$.

   b. Let A be the set of all Scrabble words of length 2 and Ω be all possible pairs of letters. Then, $\|A\| = 96$ and $\|\Omega\| = 26^2$.

   $$\implies P(A) = \frac{96}{26^2} = .142.$$

   c. Let C be the set of all Scrabble words of length 3 that begin with c and Ω as before. Then, $\|C\| = 37$ and $\|\Omega\| = 26^2$.

   $$\implies P(C) = \frac{37}{26^2} \approx 0.055.$$

   d. Let A be the set of all words of length 4, and B be the set of all possible words of length 4 that contain exactly three vowels. Then: $\|A \cap B\| = 62$, $\|B\| = 4 \times 21 \times 5^3 = 10,500$, where

   4—positions of a consonant,
   21—possible consonants,
   5^3—possible triples of vowels.

   $$\implies P(A|B) = \frac{62}{10,500} \approx 5.905 \times 10^{-3}.$$

3. Let A be the event that M produces a byte. Then

$$P(A) = \frac{2^8}{3^8} = 0.039.$$

7. For the given length n, there are 4^n different messages, so we want

$$4^n > 1,000 \implies n > 4.982;$$

thus, the minimum number required is 5.

12. Ω is the set of all possible 64-bit sequences and A is the set of all possible 64-bit sequences that start with 00 and have exactly 4 zeros. Then, $\|A\| = \binom{62}{2}$ and $\|\Omega\| = 2^{64}$.

$$\implies P(A) = \frac{\binom{62}{2}}{2^{64}}.$$

15. a. If A is the event of receiving 10101010, then $P(A) = \frac{1}{2^8}$.

   b. Let B be the event $b_1 = 1$; then $\|B\| = 2^7$. Therefore, $P(B) = \frac{2^7}{2^8} = \frac{1}{2}$.

   c. Let C be the event of receiving 11111111; then $P(C) = \frac{1}{2^8}$.

   d. Let D be the event of at least one bit has a value 1; then D^c is the event of receiving 00000000. Therefore

$$P(D) = 1 - P(D^c) = 1 - \frac{1}{2^8}.$$

   e. Let E be the event that $b_1 + b_2 = 1$; then $\|E\| = 2 \times 2^6 = 2^7$, where

   2—possible outcomes for $b_1 b_2$, either 10 or 01;
   2^6—possible outcomes for the remaining 6 bits.

$$\implies P(E) = \frac{2^7}{2^8} = \frac{1}{2}.$$

21. a. $\|\Omega\| = \binom{1,000}{300}$. Let A be the event of generating all ones in the initial 300 positions; then

$$P(A) = \frac{1}{\binom{1,000}{300}}.$$

   b. Let B the event of generating all ones in the first 600 positions; then $\|B\| = \binom{600}{300}$. So,

$$P(B) = \frac{\binom{600}{300}}{\binom{1,000}{300}} = \frac{700! \, 600!}{1000! \, 300!}.$$

   c. Since there are 2^m possible messages of length m, $2^m \geq \binom{1,000}{300}$. Therefore, the smallest possible m is $\lceil \log_2 \binom{1,000}{300} \rceil \approx 1000 \, H(.3)$.

d. A source has bit-per-symbol rate of H if the set of messages of length n produced by the source has size 2^{nH}. (See Section 2.5.) Therefore,

$$2^{nH} = \binom{1,000}{300} \implies H = \frac{1}{n} \log_2 \binom{1,000}{300} = \frac{1}{1,000} \log_2 \binom{1,000}{300}.$$

22. a. There are $\binom{10}{2}$ possible links, and each of them either belongs or does not belong to the graph. So, there are $2^{\binom{10}{2}}$ undirected labeled graphs.

b. There are $\binom{\binom{10}{2}}{5} = \binom{45}{5}$ such graphs possible.

c. The number of graphs with n nodes and m links is given by

$$\binom{\binom{n}{2}}{m} = \binom{\frac{n(n-1)}{2}}{m}.$$

So, $m^* = \lceil \frac{n(n-1)}{4} \rceil$ or $m^* = \lfloor \frac{n(n-1)}{4} \rfloor$.

26. a. Let A be the event of interest. Then, $\|A\| = \binom{6}{2}$ and $\|\Omega\| = 2^{\binom{6}{2}} = 2^{15}$.

$$\implies P(A) = \frac{\binom{15}{10}}{2^{15}}.$$

41. a. Let Ω be the set of all possible arrangements of k indistinguishable photons in N intervals. By Bose–Einstein statistics, $\|\Omega\| = \binom{N-1+k}{k}$. Let A be the set of possible arrangements, at which no two photons occupy the same interval; in other words, each interval contains either one photon or zero photons. Then $\|A\| = \binom{N}{k}$ and

$$P_a = P(A) = \frac{\|A\|}{\|\Omega\|} = \frac{\binom{N}{k}}{\binom{N-1+k}{k}}.$$

b. Let B be the event that at least two photons share some unit interval. Then $B = A^c$ and

$$P_b = P(B) = 1 - P(A) = 1 - \frac{\binom{N}{k}}{\binom{N-1+k}{k}}.$$

c. Let C be the event that two of k photons turned out to be in the first interval or, in other words, $k - 2$ remaining photons were arranged in $N - 1$ intervals. Hence, $\|C\| = \binom{N-4+k}{k-2}$ and

$$P_c = P(C) = \frac{\binom{N-4+k}{k-2}}{\binom{N-1+k}{k}}.$$

42. a. According to Bose–Einstein statistics (page 78), there are $m = \binom{10-1+3}{3} = \binom{12}{3} = 220$ distinguishable arrangements.

b. The event A corresponds to one of m possible arrangements; thus, $\|A\| = 1$ and $P(A) = \frac{1}{\binom{12}{3}} = 0.0045(45)$.

c. The event B is that none of 3 photons are in the cell q_1; hence, all 3 photons are in the remaining 9 cells $\{q_2, \ldots, q_{10}\}$. Therefore, $\|B\| = \binom{9-1+3}{3} = \binom{11}{3}$.

$$\Longrightarrow P(B) = \frac{\binom{11}{3}}{\binom{12}{3}} = 0.75.$$

43. a. According to the Fermi–Dirac Statistics (page 79), there are $m = \binom{10}{3}$ physically distinguishable arrangements.

b. Since the event A corresponds to one of m possible arrangements, $P(A) = \frac{1}{\binom{10}{3}}$.

c. Event B is that all 3 electrons are distributed by Fermi–Dirac in the remaining 9 cells. Hence,

$$P(B) = \frac{\binom{9}{3}}{\binom{10}{3}} = .7.$$

47. a. Let Ω be the set of all possible images of the size $n = 10 \times 10$, composed of 20 red pixels, 30 green pixels, and 50 blue pixels. Then the total number of different images one can form is $\|\Omega\| = \binom{100}{20,30,50} = \frac{100!}{20!30!50!}$.

b. Let B the set of all possible pictures where the first pixel is red. Therefore, $\|B\|$ is the number of different pictures one can form by fixing one red pixel at the first position and placing 19 red pixels, 30 green pixels, and 50 blue pixels into the remaining 99 positions. Thus, $\|B\| = \binom{99}{19,30,50}$ and

$$P(B) = \frac{\binom{99}{19,30,50}}{\binom{100}{20,30,50}} = \frac{1}{5}.$$

c. Let C be the set of possible pictures with 50 blue pixels placed in the last 9 rows. Hence, $\|C\| = \binom{90}{50}\binom{50}{20}$, where

$\binom{90}{50}$—possible arrangements of 50 blue pixels in the last 90 positions,

$\binom{50}{20}$—possible arrangements of 20 red pixels in the remaining 50 positions.

$$\Longrightarrow P(C) = \frac{\binom{90}{50}\binom{50}{20}}{\binom{100}{20,30,50}}.$$

d. Similarly to (b), D is the set of all pictures with 19 red pixels, 29 green pixels, and 50 blue pixels arranged in 98 positions. $\|D\| = \binom{98}{19,29,50}$.

$$\Longrightarrow P(D) = \frac{\binom{98}{19,29,50}}{\binom{100}{20,30,50}} = \frac{2}{33}.$$

49. a. From Section 2.9.3:

$$P(A) = \sum_{i=1}^{3} P(A|B_i)P(B_i) = 0.23.$$

b. By Bayes' Theorem:

$$P(B_2|A) = \frac{P(A|B_2)P(B_2)}{\sum_{i=1}^{3} P(A|B_i)P(B_i)} = \frac{P(A|B_2)P(B_2)}{P(A)} = 0.5217.$$

51. Set A contains all possible bytes with $b_1 b_2 = 11$. Therefore, $\|A\| = 2^6$ and

$$P(A) = \frac{2^6}{2^8} = 0.25.$$

Since B is the event that a byte contains an odd number of ones, $\|B\| = \binom{8}{1} + \binom{8}{3} + \binom{8}{5} + \binom{8}{7} = 128$. Hence,

$$P(B) = \frac{128}{2^8} = 0.5.$$

$A \cap B$ is the event that first two bits are 11 and the sum of the remaining 6 bits is odd, therefore $\|A \cap B\| = \binom{6}{1} + \binom{6}{3} + \binom{6}{5} = 32$, and by definition of the conditional probability,

$$P(B|A) = \frac{\|A \cap B\|}{\|A\|} = \frac{32}{2^6} = 0.5,$$

$$P(A|B) = \frac{\|A \cap B\|}{\|B\|} = \frac{32}{128} = 0.25.$$

52. a. $\|A\| = 4$, since only 4 outcomes lead to the sum of the spots equal to 9: $6 + 3$, $5 + 4$, $4 + 5$, $3 + 6$. $A \cap B$ is the event that the first die shows 4, and the sum of the spots on both dice is equal to 9; thus, the second die shows 5. Hence, $\|A \cap B\| = 1$. By definition of the conditional probability,

$$P(B|A) = \frac{\|A \cap B\|}{\|A\|} = \frac{1}{4}.$$

b. In the case of the event B, the second die may show one of its six sides, thus $\|B\| = 6$. Let Ω be all possible pairs of spots on two dice; then $\|\Omega\| = 36$. Since $\|A\|\|B\| = 4 \times 6 = 24$ and $\|A \cap B\|\|\Omega\| = 1 \times 36 = 36$, we conclude that A and B are *not independent*.

53. a. Let A be the set of all bytes with exactly 3 ones; then $\|A\| = \binom{8}{3}$. Ω is the set of all possible bytes; thus, $\|\Omega\| = 2^8$.

$$\implies P(A) = \frac{\binom{8}{3}}{2^8} = \frac{7}{32}.$$

b. Let A be as before, and B be the set of all bytes with $b_1 = 1$; then $\|B\| = 2^7$. $A \cap B$ is the set of all bytes such that $b_1 = 1$ and $\sum_{i=2}^{8} b_i = 2$. Then, $\|A \cap B\| = \binom{7}{2}$. By definition of the conditional probability,

$$P(B|A) = \frac{\|A \cap B\|}{\|A\|} = \frac{\binom{7}{2}}{\binom{8}{3}} = 0.375.$$

c. $P(B) = \frac{\|B\|}{\|\Omega\|} = \frac{1}{2}$. Since $P(B|A) \neq P(B)$, the events A and B are *not independent*.

CHAPTER 3

5. a. From Section 3.4,

$$\left(1 - \frac{1}{n}\right) r_{n-1}(A) + \frac{1}{n} I_A(\omega_n) = \frac{n-1}{n} \times \frac{1}{n-1} \sum_{i=1}^{n-1} I_A(\omega_i) + \frac{1}{n} I_A(\omega_n)$$

$$= \frac{1}{n} \sum_{i=1}^{n} I_A(\omega_i) = r_n(A).$$

   b. From part (a), $r_n(A) - r_{n-1}(A) = \frac{1}{n}(I_A(\omega_n) - r_{n-1}(A))$.

$$\Rightarrow |r_n(A) - r_{n-1}(A)| = \frac{1}{n}|I_A(\omega_n) - r_{n-1}(A)| \le \frac{1}{n}.$$

   c. No. Consider the following construction of a binary sequence as a counterexample:

   Let $x_1 = 1$ and $r_i = \frac{1}{i} \sum_{j=1}^{j=i} x_j$.
   For $i \ge 2$

   if $r_i \ge 2/3$, **then** $x_{i+1} = 0$
   else if $r_i \le 1/3$, **then** $x_{i+1} = 1$
   else if $1/3 < r_i < 2/3$ and $x_i = 1$, **then** $x_{i+1} = 1$
   else if $1/3 < r_i < 2/3$ and $x_i = 0$, **then** $x_{i+1} = 0$.

   The sequence of relative frequencies will oscillate between 1/3 and 2/3.

8. The polynomial has the following roots: $p_1 = 2, p_2 = 2\sqrt{-1}, p_3 = -2\sqrt{-1}, p_4 = -0.5$, and **$p_5 = 0.5$**. The fifth root, $p_5 = 0.5$, is the only that belongs to the interval $[0, 1]$; therefore, $p = 0.5$ is the answer.

9.

$$P(\{a\}) = P(\{a, b\} \cap \{a, c\}) = P(\{a, b\}) + P(\{a, c\}) - P(\{a, b, c\}) = .5 + .7 - 1 = .2.$$

   Similarly, we find $P(\{b\}) = 0.3$ and $P(\{c\}) = 0.5$.

13. a. $P(A^c) = 1 - P(A) = .3$.
    b. $P(A \cup B) = P(A) + P(B) - P(A \cap B) \Rightarrow P(A \cup B) + P(A \cap B) = P(A) + P(B) = .3$.
    c. By the union bound, $P(A \cup B) \le P(A) + P(B) = .6$. In addition, $P(A \cup B) \ge \max(P(A), P(B)) = .4$.

14. a. The answer is $P(B) = .5$ and occurs for the case $A \subset B$ with $P(A) = .1$.
    b.

$$P(A \cap B) \ge P(A) + P(B) - 1 \Rightarrow P(A) + P(B) \le 1.2.$$

   So, to maximize $P(A)P(B)$ we must take

$$P(A) = P(B) = .6 \Rightarrow P(A)P(B) = .36.$$

20. a. Yes. Consider: $P(\{0\}) = P(\{5\}) = \frac{1}{20}$; $P(\{1\}) = P(\{2\}) = P(\{3\}) = \frac{3}{10}$; $P(\{4, 6, 7, 8, 9\}) = 0$. Then, $P(A) = \frac{7}{20}$, $P(B) = \frac{13}{20}$, and $P(C) = \frac{9}{10}$.

   b. Yes. Consider: $P(\{1, 2, 3\}) = 0$ and $P(\{i\}) = \frac{1}{7}$ for $i \in \{0, 4, 5, 6, 7, 8, 9\}$. Then, $P(A) = \frac{4}{7}$, $P(B) = \frac{3}{7}$, and $P(C) = 0$.

29. By the Inclusion–Exclusion Principle,

$$P(A \cup B \cup C) = P(A) + P(B) + P(C) - P(A \cap B) - P(B \cap C) - P(C \cap A)$$

$$+ P(A \cap B \cap C)$$

$$= .3 + .4 + .5 - .2 - .1 - .15 + .1 = .85.$$

CHAPTER 4

1. $p_3 = 1 - P(\{\omega_1, \omega_2\}) = 0.35$; $p_1 = 1 - P(\{\omega_2, \omega_3\}) = 0.15$; $p_2 = 1 - p_1 - p_3 = 0.5$; $P(\{\omega_1, \omega_3\}) = p_1 + p_3 = 0.5$.

4. Let A be the event that fewer than two errors occurred: $P(A) = \sum_{k=0}^{1} \binom{5}{k}(0.1)^k(0.9)^{5-k} = (0.9)^5 + 5(0.1)(0.9)^4 \approx 0.92$.

10. a.

$$p_c = \sum_{k=0}^{2} \binom{100}{k}(0.01)^k(0.99)^{100-k} \approx 0.92.$$

   b. The most probable number of errors is $\lfloor (101)(0.01) \rfloor = 1$.

16. Lifetime L should be modeled with geometric distribution; we know $\beta = \frac{m}{m+1} = \frac{10}{11}$, thus: $P(L = n) = (1 - \frac{10}{11})(\frac{10}{11})^n = \frac{10^n}{11^{n+1}}$, $n \in \mathbb{N}$.

19. We know that T has geometric distribution:

$$P(T = k) = (1 - \beta)\beta^{k-1}, \; k = 1, 2, \dots.$$

For $k = 1$, we know $P(T = 1) = P(\{5, 6\}) = \frac{1}{3}$, so: $(1 - \beta) = \frac{1}{3} \Rightarrow \beta = \frac{2}{3} \Rightarrow$

$$P(T = k) = \frac{1}{3}\left(\frac{2}{3}\right)^{k-1} = \frac{2^{k-1}}{3^k}, \; k = 1, 2, \dots.$$

20. We must use $Poisson(100)$. Then: $P(s) = \sum_{k=80}^{120} e^{-100}\frac{100^k}{k!} \approx 0.96$.

21. a. Using the Poisson model, let A be the desired event. Then

$$P(A) = e^{-\gamma T}\frac{(\gamma T)^0}{0!} = e^{-\gamma T}.$$

   b. Using the result of Section 4.4.4:

$$k^* = \lfloor \gamma T \rfloor \Rightarrow P_{k^*} = e^{-\gamma T}\frac{(\gamma T)^{\lfloor \gamma T \rfloor}}{(\lfloor \gamma T \rfloor)!} \approx \frac{1}{\sqrt{2\pi \gamma T}}.$$

24. Using the Poisson model,

$$P_0 = e^{-\lambda}\frac{\lambda^0}{0!} = 0.1 \Rightarrow \lambda = \ln(10).$$

Let A be the event that at least 2 photons are emitted in time T. Hence,

$$P(A) = 1 - P(A^c) = 1 - \sum_{k=0}^{1}(0.1)\frac{(\log(10))^k}{k!} = 1 - 0.1 - 0.1\log(10) \approx 0.67.$$

37. a.

$$\frac{p_{1K}}{p_{1M}} = \frac{(10^3)^{-\alpha}}{(10^6)^{-\alpha}} = 10^{3\alpha} = 10^4 \Rightarrow 3\alpha = 4 \Rightarrow \alpha = \frac{4}{3}.$$

  b.

$$\frac{p_{1M}}{p_{1G}} = \frac{(10^6)^{-\frac{4}{3}}}{(10^9)^{-\frac{4}{3}}} = 10^4.$$

CHAPTER 5

1. a. Yes, it is a cdf for $\Omega = R$ (see Example 5.4).
   b. No, because (via) fails:

$$\lim_{x \to -\infty} U(x) + [1 - U(x)]\frac{1 + e^x}{2} = 1/2$$

   c. No, because (ii) and (vib) fail.
   d. Yes.
   e. No, it is not even a function, because for $x = 0$, it assumes two values: 0 and 1.

2. Since K_T must be nonnegative integer,

$$F_K(x) = \sum_{i=0}^{\infty} P(K_T = i)U(x - i).$$

   a. $P(K_T = 1) = F_K(1) - F_K(1^-) = F_K(1) - F_K(1/2) = 0.1$.
   b. Assuming K_T is distributed according to Poisson, we can find λT using the result of the part (a): $P(K_T = 1) = 0.1 = e^{-\lambda T}\lambda T$. Solving numerically, we obtain two solutions, $\lambda_1 T \approx 0.1118$ and $\lambda_2 T \approx 3.5772$. Therefore,

$$F_K(1) = P(K_T = 0) + P(K_T = 1) = e^{-\lambda T} + 0.1,$$

   and substituting the two possible values of λ, we get

$$F_K^1(1) \approx 0.9942$$

$$F_K^2(1) \approx 0.1280.$$

With the information given in the exercise, it is not possible to decide which of these two values is the "true" one.

8. Left plot:

$$P(X < -3) \approx 0.05,$$

$$P(X \le -1) \approx 0.27,$$

$$P(X = 0) = 0,$$

$$P(X > 2) = 1 - P(X \le 2) \approx 0.12.$$

Right plot:

$$P(X < -3) = 0,$$

$$P(X \le -1) = 0.2,$$

$$P(X = 0) = 0,$$

$$P(X > 2) = 1 - P(X \le 2) = 0.$$

10.

$$F_X(x) = \sum_{i=0}^{3} b(i, 3, 1/3) U(x - i) = \sum_{i=0}^{3} \binom{3}{i} \left(\frac{1}{3}\right)^i \left(\frac{i}{3}\right)^{3-i} U(x - i).$$

12. a.

$$P(Y \le -1) = F_Y(-1) = 0.1,$$

$$P(Y = 1) = F_Y(1) - F_Y(1^-) = 0.2,$$

$$P(Y = 2) = F_Y(2) - F_Y(2^-) = 0.$$

b. There is a single jump point of $F(y)$ at $y = 1$ and of height $.5 - .3 = .2$. Hence,

$$\lambda_d F_d(y) = .2U(y - 1), \quad \lambda_d = .2, \quad \text{and } F_d(y) = U(y - 1).$$

Thus, $\lambda_{ac} = 1 - \lambda_d = .8$ and

$$.8F_{ac}(y) = F(y) - .2U(y - 1),$$

$$F_{ac}(y) = \begin{cases} 0 & \text{if } y < -2 \\ \dfrac{1}{8}(y + 2) & \text{if } -2 \le y < 1 \\ \dfrac{3}{8} & \text{if } 1 \le y < 2 \\ \dfrac{5}{16}y - \dfrac{1}{4} & \text{if } 2 \le y < 4 \\ 1 & \text{otherwise} \end{cases}.$$

21. a.

$$P(X \leq 1) = F_X(1) = 2/3,$$

$$P(X \geq 2) = 1 - P(X < 2) = 1 - F_X(2^-) = 1 - 2/3 = 1/3.$$

b.

$$P(1/2 < X \leq 3/4) = F_X(3/4) - F_X(1/2) = 0.$$

c.

$$P(X = 0) = F_X(0) - F_X(0^-) = 0,$$

$$P(X = 1) = F_X(1) - F_X(1^-) = 2/3 - 1/3 = 1/3.$$

d. A sketch of

$$F_X(x) = \begin{cases} \dfrac{1}{3}e^x & \text{if } x < 0 \\[2mm] \dfrac{1}{3} & \text{if } 0 \leq x < 1 \\[2mm] \dfrac{2}{3} & \text{if } 1 \leq x < 2 \\[2mm] 1 - \dfrac{1}{4}e^{2-x} & \text{if } 2 \leq x \end{cases}$$

reveals that F_X has two jump points of height $\frac{2}{3} - \frac{1}{3} = \frac{1}{3}$ at $x = 1$ and of height $\frac{3}{4} - \frac{2}{3} = \frac{1}{12}$ at $x = 2$. Therefore,

$$\lambda_d = \frac{1}{3} + \frac{1}{12} = \frac{5}{12},$$

$$\lambda_d F_d(x) = \frac{1}{3}U(x-1) + \frac{1}{12}U(x-2),$$

$$\lambda_{ac} = 1 - \lambda_d = \frac{7}{12},$$

$$\lambda_{ac}F_{ac}(x) = F_X(x) - \lambda_d F_d(x),$$

and

$$F_{ac}(x) = \begin{cases} \dfrac{4}{7}e^x & \text{if } x < 0 \\[2mm] \dfrac{4}{7} & \text{if } 0 \leq x < 2 \\[2mm] 1 - \dfrac{3}{7}e^{2-x} & \text{if } 2 \leq x \end{cases}.$$

CHAPTER 6

1. The cdf

$$
F_X(x) = \begin{cases} e^{x-2} & \text{if } x < 1 \\ \dfrac{1}{2} & \text{if } 1 \le x < 2 \\ 1 & \text{otherwise} \end{cases}
$$

has a jump discontinuity of height $\frac{1}{2} - e^{-1}$ at $x = 1$ and another one of height $1 - \frac{1}{2} = \frac{1}{2}$ at $x = 2$. Hence, it can be represented as a weighted sum of a discrete cdf and an absolutely continuous cdf.

$$
\lambda_d = \frac{1}{2} - e^{-1} + \frac{1}{2} = 1 - e^{-1} \text{ and } \lambda_{ac} = 1 - \lambda_d = e^{-1},
$$

$$
\lambda_d F_d(x) = \left(\frac{1}{2} - e^{-1}\right) U(x-1) + \frac{1}{2} U(x-2),
$$

$$
\lambda_{ac} F_{ac}(x) = F_x(x) - \lambda_d F_d(x) = e^{x-2}[1 - U(x-1)] + e^{-1} U(x-1),
$$

and

$$
F_X(x) = \lambda_{ac} F_{ac}(x) + \lambda_d F_d(x), \text{ where } \lambda_{ac} = 1 - \lambda_d.
$$

Thus the desired pdf f_X is given by

$$
f_X(x) = \lambda_d f_d(x) + \lambda_{ac} f_{ac}(x),
$$

where

$$
\lambda_{ac} f_{ac}(x) = \frac{\lambda_{ac}\, dF_{ac}(x)}{dx} = e^{x-2}[1 - U(x-1)],
$$

$$
\lambda_d f_d(x) = \frac{\lambda_d\, dF_d(x)}{dx} = \left(\frac{1}{2} - e^{-1}\right)\delta(x-1) + \frac{1}{2}\delta(x-2),
$$

and

$$
f(x) = e^{x-2}[1 - U(x-1)] + \left(\frac{1}{2} - e^{-1}\right)\delta(x-1) + \frac{1}{2}\delta(x-2).
$$

3. Phase is uniformly distributed in the interval $(-180, 180)$; therefore,

$$
f_P(x) = \frac{1}{360} U(x + 180)(1 - U(x - 180)).
$$

Let A be the event that the received phase is within $(-30, 30)$. Then,

$$
P(A) = \int_A f_P(x)\, dx = \int_{-30}^{30} \frac{1}{360} U(x + 180)(1 - U(x - 180))\, dx
$$

$$
= \int_{-30}^{30} \frac{1}{360}\, dx = \frac{60}{360} = 1/6.
$$

8. a. Assuming the lifetime is exponentially distributed, $F(x) = (1 - e^{-\alpha x})U(x)$:

$$F(5{,}000) = \frac{1}{2} \Rightarrow \frac{1}{2} = (1 - e^{-5{,}000\alpha}) \Rightarrow \alpha = \frac{\log 2}{5{,}000}.$$

Therefore, the mean chip lifetime is $\frac{1}{\alpha} = \frac{5{,}000}{\log 2}$.

b.

$$P(\{X < 1{,}000\} \cup \{X > 10{,}000\}) = P(X < 1{,}000) + P(X > 10{,}000)$$
$$= P(X \le 1{,}000) + (1 - P(X \le 10{,}000))$$
$$= F(1{,}000) + 1 - F(10{,}000)$$
$$= 1 - e^{-\frac{\log 2}{5}} + 1 - (1 - e^{-2\log 2})$$
$$= 1 - \left(\frac{1}{2}\right)^{1/5} + \left(\frac{1}{2}\right)^{2}.$$

11. Assuming that the waiting time is exponentially distributed, $F(t) = (1 - e^{-\alpha t})U(t)$:

$$P(T > 1) = 1 - P(T \le 1) = 1 - F(1) = e^{-\alpha} = 0.1 \Rightarrow \alpha = \log 10.$$

So,

$$F(t) = \left[1 - \left(\frac{1}{10}\right)^{t}\right]U(t).$$

16. a. Assuming the waiting time is exponentially distributed, if the mean value is 1 ns, then $\lambda = 1$ (for time measured in *nanoseconds*). Therefore, $T_2 \sim \Gamma(2, 1)$. So, the pdf is

$$f_{T_2}(x) = \frac{1}{\Gamma(2)}xe^{-x}U(x), \; \Gamma(2) = 1.$$

b. The cdf is: $F_{T_2}(x) = F_{2,1}(x) = (1 - (1+x)e^{-x})U(x)$.

18. Assuming a Pareto distribution with $\tau = 1$, $F(x) = (1 - x^{-\alpha})U(x - 1)$:

$$P(R > 10) = 1 - P(R \le 10) = 1 - F(10) = 10^{-\alpha} = 0.5$$
$$\Rightarrow \alpha = \log_{10} 2.$$

Hence,

$$P(R > 1{,}000) = 1 - F(1{,}000) = 1{,}000^{-\log_{10} 2} = \frac{1}{8}.$$

19. Assuming a Pareto distribution with $\tau = 1$, $F(x) = (1 - x^{-\alpha})U(x - 1)$:

$$P(X > 20) = 1 - P(X \le 20) = 20^{-\alpha} = 0.5$$
$$\Rightarrow \alpha = \log_{20} 2$$

and

$$P(X > 60) = 1 - F(60) = 60^{-\log_{20} 2}.$$

23. Assuming a Laplacian distribution, $F(x) = \frac{1}{2}e^{\alpha x}(1 - U(x)) + (1 - \frac{1}{2}e^{-\alpha x})U(x)$:

$$P(A > 1) = 1 - P(A \le 1) = 1 - F(1) = \frac{1}{2}e^{-\alpha} = 0.3$$

$$\Rightarrow \alpha = -\log(0.6)$$

and

$$P(A > 2) = 1 - F(2) = \frac{1}{2}e^{2\log(0.6)} = \frac{1}{2}(0.6)^2 = 0.18.$$

29. We know that $V \sim N(0, \sigma^2)$, so:

$$P(-10^{-3} < V < 10^{-3}) = \Phi\left(\frac{10^{-3}}{\sigma}\right) - \Phi\left(\frac{-10^{-3}}{\sigma}\right) \approx \Phi'(0)\frac{2 \cdot 10^{-3}}{\sigma}.$$

Since this probability is given to be 6×10^{-3} and $\Phi'(0)$ is equal to the standard normal pdf evaluated at zero,

$$\frac{1}{\sqrt{2\pi}}\frac{2 \cdot 10^{-3}}{\sigma} \approx 6 \times 10^{-3} \Rightarrow \sigma \approx \frac{1}{3\sqrt{2\pi}} \approx 0.1330.$$

So,

$$P(0.999 < V < 1.001) = \Phi\left(\frac{1.001}{\sigma}\right) - \Phi\left(\frac{0.999}{\sigma}\right) \approx \Phi'\left(\frac{1}{\sigma}\right)\frac{2 \cdot 10^{-3}}{\sigma}$$

$$\approx \frac{1}{\sqrt{2\pi}} \times e^{-\frac{1}{2\sigma^2}} \times \frac{2 \cdot 10^{-3}}{\sigma} \approx 6 \times 10^{-3} \times e^{-9\pi}.$$

32. a. We know that $V \sim N(0, \sigma^2)$, where $\sigma^2 = 4kTRB$. We have

$$P(|V| < 2 \times 10^{-6}) \approx \Phi'(0)\frac{4 \cdot 10^{-6}}{\sigma}.$$

Since this probability is given to be 10^{-3} and $\Phi'(0)$ is equal to the standard normal pdf evaluated at zero:

$$\frac{1}{\sqrt{2\pi}}\frac{4 \cdot 10^{-6}}{\sigma} \approx 10^{-3} \Rightarrow \sigma \approx \frac{4 \cdot 10^{-3}}{\sqrt{2\pi}} \approx 1.596 \times 10^{-3}.$$

Therefore,

$$T \approx \frac{\left(\frac{4 \cdot 10^{-3}}{\sqrt{2\pi}}\right)^2}{4 \cdot 1.38 \cdot 10^{-23} \cdot 10^6 \cdot 10^9} \approx 46.132 \,^{\circ}\text{K}.$$

b.
$$P(|V| < 10^{-3}) = \Phi\left(\frac{10^{-3}}{\sigma}\right) - \Phi\left(\frac{-10^{-3}}{\sigma}\right) \approx \Phi'(0)\frac{2\cdot 10^{-3}}{\sigma} \approx \frac{1}{\sqrt{2\pi}}\frac{2\cdot 10^{-3}}{\sigma} = 0.5.$$

37.
$$P(X > 700) = 1 - P(X \le 700) = 1 - \Phi\left(\frac{700-500}{100}\right) = 1 - \Phi(2) = 0.023.$$

CHAPTER 7

1. a.
$$\lim_{x,y\to\infty} F(x,y) = 0.5 + 0.5 = 1.$$

b. From Section 5.3
$$F_Y(y) = \lim_{x\to\infty} F(x,y) = 0.5F_2(y) + 0.5F_4(y)$$

is a univariate cdf.

c.
$$f(x,y) = \frac{\partial^2 F(x,y)}{\partial x\,\partial y} = 0.5f_1(x)f_2(y) + 0.5f_3(x)f_4(y).$$

d. Yes, since $f(x,y) \ge 0$ and $\int_{\mathbb{R}^2} f(x,y) = 1$.

e. Yes.

5. a.
$$f_Y(y) = \int_{-\infty}^{\infty} f_{X,Y}(x,y)\,dx$$

$$= \begin{cases} \int_0^1 3x^2\,dx & \text{if } 0 \le y \le 1 \\ 0 & \text{otherwise} \end{cases}$$

$$= \begin{cases} 1 & \text{if } 0 \le y \le 1 \\ 0 & \text{otherwise} \end{cases}$$

$$= U(y)U(1-y).$$

b.
$$F_Y(y) = Pr(Y \le y) = \int_{-\infty}^{y} f_Y(z)\,dz = yU(y)[1 - U(y-1)] + U(y-1).$$

c.
$$P(X \ge Y) = \iint_{y\le x} f_{X,Y}(x,y)\,dx\,dy = \int_{x=0}^{1}\int_{y=0}^{x} 3x^2\,dy\,dx$$

$$= \int_{x=0}^{1} 3x^2[y\,|_0^x]\,dx = \int_0^1 3x^3\,dx = \frac{3x^4}{4}\bigg|_0^1 = \frac{3}{4}.$$

d.

$$f_X(x) = 3x^2 U(x)U(1-x) \Rightarrow f_{X,Y}(x,y) = f_X(x)f_Y(y) \Rightarrow \text{ product construction.}$$

12.

$$P\left(Y \geq X + \frac{1}{2}\right) = \iint_{x+\frac{1}{2} \leq y} f_{X,Y}(x,y)\, dx\, dy = \int_{x=0}^{\frac{1}{2}} \int_{y=x+\frac{1}{2}}^{1} (x+y)\, dy\, dx$$

$$= \int_0^{\frac{1}{2}} \left(xy + \frac{y^2}{2}\right)\Big|_{x+\frac{1}{2}}^{1} dx = \int_0^{\frac{1}{2}} \left[\left(x + \frac{1}{2}\right)\right.$$

$$\left. - \left(x^2 + \frac{x}{2} + \frac{1}{2}\left(x^2 + x + \frac{1}{4}\right)\right)\right] dx$$

$$= \int_0^{\frac{1}{2}} \left(\frac{3}{8} - \frac{3x^2}{2}\right) dx = \left(\frac{3x}{8} - \frac{x^3}{2}\right)\Big|_0^{\frac{1}{2}} = \frac{3}{16} - \frac{1}{16} = \frac{1}{8}.$$

20. a. Omitted.

b. From Section 7.4.4,

$$\mathbb{C}_{1,3} = \begin{pmatrix} 6 & 2 \\ 2 & 4 \end{pmatrix}$$

and

$$\mathbf{m}_{1,3} = \begin{pmatrix} 0 \\ 2 \end{pmatrix}.$$

Then, we have:

$$\mathbb{C}_{1,3}^{-1} = \begin{pmatrix} 0.2 & -0.1 \\ -0.1 & 0.3 \end{pmatrix},$$

and the corresponding pdf:

$$f_{X_1,X_3}(x_1, x_3) = \frac{1}{4\pi\sqrt{5}} \exp\left(-\frac{1}{2}[0.2x_1^2 - 0.2x_1(x_3 - 2) + 0.3(x_3 - 2)^2]\right).$$

c. $\mathbb{C}_2 = \sigma_2^2 = 10$ and $m_2 = -1 \Rightarrow f_{X_2}(x_2) = \frac{1}{\sqrt{2\pi}\sqrt{10}} e^{\frac{-(x_2+1)^2}{20}}$.

d. We can express these probabilities in terms of the $\mathcal{N}(0, 1)$ cdf Φ:

$$P(0 < X_2 < \infty) = \Phi(\infty) - \Phi\left(\frac{-m_2}{\sigma_2}\right) = 1 - \Phi\left(\frac{1}{\sqrt{10}}\right),$$

and

$$P(0 < X_3 < \infty) = 1 - \Phi\left(\frac{-m_3}{\sigma_3}\right) = 1 - \Phi\left(\frac{-2}{2}\right) = 1 - \Phi(-1).$$

Thus,

$$P(X_3 > 0) - P(X_2 > 0) = \Phi\left(\frac{1}{\sqrt{10}}\right) - \Phi(-1) > 0 \Rightarrow P(X_3 > 0) > P(X_2 > 0).$$

21. a. We have to check whether $f_{X,Y,Z}(x, y, z)$ is nonnegative for all triples (x, y, z) and its integral over $\mathbb{R}^3$ is equal to 1. The nonnegativity follows from the fact that g_1, g_2, and g_3 are univariate pdfs. Hence, they are nonnegative and their product is nonnegative also. Now we check the second property:

$$\iiint_{\mathbb{R}^3} f_{X,Y,Z}(x, y, z)\, dx\, dy\, dz = \iiint_{\mathbb{R}^3} g_1(z)g_2(y - z^2)g_3(x - yz)\, dx\, dy\, dz$$

$$= \iint_{\mathbb{R}^2} g_1(z)g_2(y - z^2)\left(\int_{-\infty}^{\infty} g_3(x - yz)\, dx\right) dy\, dz.$$

Using substitution: $a = x - yz$, and $da = dx$, we get

$$= \iint_{\mathbb{R}^2} g_1(z)g_2(y - z^2)\left(\int_{-\infty}^{\infty} g_3(a)\, da\right) dy\, dz = \iint_{\mathbb{R}^2} g_1(z)g_2(y - z^2)\, dy\, dz.$$

Substituting $c = y - z^2$ and $dc = dy$, we obtain:

$$= \int_{-\infty}^{\infty} g_1(z)\left(\int_{-\infty}^{\infty} g_2(c)\, dc\right) dz = \int_{-\infty}^{\infty} g_1(z)\, dz = 1.$$

Thus, it is a trivariate pdf, and it is innovations model by construction.

b. By Section 7.3:

$$f_{Y,Z}(y, z) = \int_{-\infty}^{\infty} f_{X,Y,Z}(x, y, z)\, dx$$

$$= g_1(z)g_2(y - z^2) \int_{-\infty}^{\infty} g_3(x - yz)\, dx$$

$$= g_1(z)g_2(y - z^2).$$

c. From Section 7.3:

$$f_X(x) = \iint_{\mathbb{R}^2} g_1(z)g_2(y - z^2)g_3(x - yz)\, dy\, dz.$$

22. a. As in Section 7.5:

$$F_L(x) = P(L < x) = P(\max(L_1, L_2, L_3) \le x) = P((L_1 \le x) \cap (L_2 \le x) \cap (L_3 \le x))$$

$$= P(L_1 \le x, L_2 \le x, L_3 \le x) = F_{L_1, L_2, L_3}(x, x, x) = U(x)\Pi_{i=1}^{3}(1 - e^{-\alpha_i x}).$$

Differentiating $F_L(x)$ with respect to x, we get the corresponding pdf:

$$f_L(x) = (F_L(x))' = f_{L_1}(x)F_{L_2}(x)F_{L_3}(x) + f_{L_2}(x)F_{L_1}(x)F_{L_3}(x) + f_{L_3}(x)F_{L_1}(x)F_{L_2}(x),$$

where $F_{L_i}(x) = U(x)(1 - e^{-\alpha_i x})$ and $f_{L_i}(x) = \alpha_i e^{-\alpha_i x}$.

b.

$$F_L(x) = P(L \leq x) = P(\min(L_1, L_2, L_3) \leq x) = P((L_1 \leq x) \cup (L_2 \leq x) \cup (L_3 \leq x)).$$

Using the Inclusion–Exclusion Principle,

$$\begin{aligned} F_L(x) = {} & P(L_1 \leq x) + P(L_2 \leq x) + P(L_3 \leq x) - P((L_1 \leq x) \cap (L_2 \leq x)) \\ & - P((L_1 \leq x) \cap (L_3 \leq x)) - P((L_2 \leq x) \cap (L_3 \leq x)) \\ & + P((L_1 \leq x) \cap (L_2 \leq x) \cap (L_3 \leq x)) \\ = {} & F_{L_1}(x) + F_{L_2}(x) + F_{L_3}(x) - F_{L_1}(x)F_{L_2}(x) - F_{L_1}(x)F_{L_3}(x) - F_{L_2}(x)F_{L_3}(x) \\ & + F_{L_1}(x)F_{L_2}(x)F_{L_3}(x), \end{aligned}$$

and pdf:

$$\begin{aligned} f_L(x) = (F_L(x))' = {} & \sum_{i=1}^{3} f_{L_i}(x) + f_{L_1}(x)[F_{L_2}(x)F_{L_3}(x) - F_{L_2}(x) - F_{L_3}(x)] \\ & + f_{L_2}(x)[F_{L_1}(x)F_{L_3}(x) - F_{L_1}(x) - F_{L_3}(x)] \\ & + f_{L_3}(x)[F_{L_1}(x)F_{L_2}(x) - F_{L_1}(x) - F_{L_2}(x)]. \end{aligned}$$

c. Omitted.

CHAPTER 8

4. a.

$$Y = X^2 \Rightarrow X = \sqrt{Y} \Rightarrow \frac{dx}{dy} = \frac{1}{2\sqrt{y}},$$

$$f_Y(y) = f_X(\sqrt{y}) \cdot \frac{1}{2\sqrt{y}} = \alpha \tau^\alpha y^{\frac{-\alpha-1}{2}} U(\sqrt{y} - \tau) \cdot \frac{1}{2\sqrt{y}} = \frac{\alpha}{2}(\tau^2)^{\frac{\alpha}{2}} y^{-\frac{\alpha}{2}-1} U(y - \tau^2).$$

So,

$$Y \sim Par\left(\frac{\alpha}{2}, \tau^2\right). \Rightarrow \beta = \frac{\alpha}{2}.$$

b.

$$Y = X^n \Rightarrow X = Y^{1/n} \Rightarrow \frac{dx}{dy} = \frac{1}{ny^{\frac{n-1}{n}}},$$

$$f_Y(y) = f_X(y^{1/n}) \cdot \frac{1}{ny^{\frac{n-1}{n}}} = \alpha \tau^\alpha y^{\frac{-\alpha-1}{n}} U(y^{1/n} - \tau) \cdot \frac{1}{ny^{\frac{n-1}{n}}} = \frac{\alpha}{n}(\tau^n)^{\frac{\alpha}{n}} y^{-\frac{\alpha}{n}-1} U(y - \tau^n).$$

Therefore, $f_Y(y) \sim Par(\alpha/n, \tau^n)$.

8. a.

$$p(q_2) = P(X > \tau) = \int_\tau^\infty \frac{1}{2} e^{-x}\, dx = -\frac{1}{2} e^{-x}\Big|_\tau^\infty = \frac{1}{2} e^{-\tau},$$

$$p(q_1) = P(0 < X \le \tau) = \int_0^\tau \frac{1}{2} e^{-x}\, dx = -\frac{1}{2} e^{-x}\Big|_0^\tau = \frac{1}{2}(1 - e^{-\tau}).$$

By symmetry of the pdf, we get:

$$p(-q_1) = p(q_1) = \frac{1}{2}(1 - e^{-\tau}),$$

$$p(-q_2) = p(q_2) = \frac{1}{2} e^{-\tau}.$$

b. We need $\frac{1}{2} e^{-\tau} = \frac{1}{4} \Rightarrow \tau = \ln 2$.

9. a. $P(Y = 0) = P(-\frac{1}{3} \le x \le 0) = \frac{1}{3}$.

b. $F_Y(y) = P(Y \le y) = \begin{cases} 0 & \text{if } y < 0 \\ \dfrac{1}{3} & \text{if } y = 0 \end{cases}$.

c. $F_Y(y) = P(Y \le y) = P(X \le \sqrt{y}) = \begin{cases} \sqrt{y} + \dfrac{1}{3} & \text{if } 0 < y \le \dfrac{4}{9} \\ 1 & \text{if } y > \dfrac{4}{9} \end{cases}$.

d. $f_Y(y) = \frac{1}{3}\delta(y) + \frac{1}{2\sqrt{y}} U(y) U(\frac{4}{9} - y)$.

16. We know that $V \sim N(0, \sigma^2)$, where $\sigma^2 = 4kTW$ and $P = V^2$, so

$$F_\rho(x) = Pr(10 \log_{10} V^2 \le x) = Pr\left(|V| \le \left(10^{\frac{x}{10}}\right)^{\frac{1}{2}}\right) = \Phi\left(\frac{10^{\frac{x}{20}}}{\sigma}\right) - \Phi\left(\frac{-10^{\frac{x}{20}}}{\sigma}\right).$$

Therefore,

$$f_\rho(x) = \frac{dF_p(x)}{dx} = \frac{1}{20}(\log 10) 10^{\frac{x}{20}} \cdot \frac{1}{\sigma} \cdot \frac{1}{\sqrt{2\pi}} \exp\left(\frac{-10^{\frac{x}{10}}}{2\sigma^2}\right) \cdot 2$$

$$= \frac{1}{\sigma\sqrt{2\pi}} (\ln 10) 10^{(\frac{x}{20} - 1)} \exp\left(\frac{-10^{\frac{x}{10}}}{2\sigma^2}\right).$$

35. a.

$$f_{X,Y} = \frac{\partial^2 F_{X,Y}(x, y)}{\partial x\, \partial y},$$

$$\frac{\partial F_{X,Y}(x, y)}{\partial y} = \frac{e^{-y}}{(1 + e^{-2x} + e^{-y})^2},$$

$$f_{X,Y}(x,y) = \frac{\partial^2 F_{X,Y}(x,y)}{\partial x \partial y} = \frac{\partial}{\partial x}\left(\frac{\partial F_{X,Y}(x,y)}{\partial y}\right)$$

$$= \frac{4e^{-y}e^{-2x}(1+e^{-2x}+e^{-y})}{(1+e^{-2x}+e^{-y})^4} = \frac{4e^{-y}e^{-2x}}{(1+e^{-2x}+e^{-y})^3}.$$

b. $V = X^2 - Y^2 = (X+Y)(X-Y) = W(X-Y) \Rightarrow X - Y = V/W$ and as $X + Y = W$, adding both equations, we have:

$$2X = \frac{V}{W} + W \Rightarrow X = \frac{1}{2}\left(\frac{V}{W} + W\right) \Rightarrow Y = \frac{1}{2}\left(W - \frac{V}{W}\right).$$

c.

$$\mathbb{J} = \begin{pmatrix} \frac{1}{2}\left(1 - \frac{v}{w^2}\right) & \frac{1}{2w} \\ \frac{1}{2}\left(1 + \frac{v}{w^2}\right) & -\frac{1}{2w} \end{pmatrix} \Rightarrow \det(\mathbb{J})$$

$$= \left| -\frac{1}{4w}\left(1 - \frac{v}{w^2}\right) - \frac{1}{4w}\left(1 + \frac{v}{w^2}\right) \right| = \left| \frac{1}{2w} \right|,$$

$$f_{W,V}(w,v) = \left| \frac{1}{2w} \right| f_{X,Y}\left(\frac{1}{2}\left(\frac{v}{w} + w\right), \frac{1}{2}\left(w - \frac{v}{w}\right)\right)$$

$$= \left| \frac{1}{2w} \right| \frac{4e^{-\left(\frac{3w}{2} + \frac{v}{2w}\right)}}{\left(1 + e^{-\left(\frac{v}{w} + w\right)} + e^{-\frac{1}{2}\left(w - \frac{v}{w}\right)}\right)^3}.$$

d. $f_V(v) = \int_{-\infty}^{\infty} f_{W,V}(w,v)\,dw$, where $f_{W,V}(w,v)$ is as given above.

42. We have $X_1 = Y_1 Y_2$ and $X_2 = Y_1 - Y_1 Y_2$.

$$\mathbb{J} = \begin{pmatrix} y_2 & y_1 \\ 1 - y_2 & -y_1 \end{pmatrix} \quad |\det(\mathbb{J})| = |-y_2 y_1 - y_1(1 - y_2)| = |-y_1| = y_1,$$

since y_1 is nonnegative. Therefore,

$$f_{Y_1,Y_2}(y_1,y_2) = y_1 f_{X_1,X_2}(y_1 y_2, y_1 - y_1 y_2)$$

$$= y_1 \cdot \alpha e^{-\alpha y_1 y_2} U(y_1 y_2) \cdot \alpha e^{-\alpha(y_1 - y_1 y_2)} U(y_1 - y_1 y_2)$$

$$= \alpha^2 y_1 e^{-\alpha y_1} U(y_1 y_2) U(y_1 - y_1 y_2)$$

$$= \alpha^2 y_1 e^{-\alpha y_1} U(y_1) U(1 - y_2) U(y_2).$$

43. a. Let $Z = X_2$; then $X_1 = \frac{Y - Z^2}{3}$.

$$|\det(\mathbb{J})| = \left| \det\begin{pmatrix} \frac{1}{3} & -\frac{2z}{3} \\ 0 & 1 \end{pmatrix} \right| = \frac{1}{3}.$$

$$f_{Y,Z}(y,z) = \frac{1}{3}f_{X_1,X_2}\left(\frac{y - z^2}{3}, z\right) = \frac{1}{3\pi}e^{-\left(\frac{y-z^2}{3}\right)^2 - z^2}.$$

b. $f_Y(y) = \int_{-\infty}^{\infty} f_{Y,Z}(y,z)\,dz$, where $f_{Y,Z}(y,z)$ is as given previously.

47. Let $Z = A \sin\theta$. From Section 8.5:

$$|\det(\mathbb{J})| = \left| \det \begin{pmatrix} \cos\theta & -A\sin\theta \\ \sin\theta & A\cos\theta \end{pmatrix} \right|^{-1} = |A^{-1}| = \frac{1}{\sqrt{s^2 + z^2}}$$

$$f_{S,Z}(s,z) = \frac{1}{\sqrt{s^2 + z^2}} f_{A,\theta}\left(\sqrt{s^2 + z^2}, \arctan\frac{z}{s}\right)$$

$$= \frac{1}{\sqrt{s^2 + z^2}} \cdot \sqrt{s^2 + z^2}\, e^{-\frac{s^2 + z^2}{2}} \cdot \frac{1}{2\pi}$$

$$= \frac{1}{2\pi} e^{-\frac{s^2 + z^2}{2}}, \quad \forall s, z \in \mathbb{R}.$$

CHAPTER 9

1.

$$E\,X = \int_0^\infty x f_X(x)\,dx = \int_0^1 x \cdot 2x\,dx = \left.\frac{2x^3}{3}\right|_0^1 = \frac{2}{3},$$

$$E(X^2) = \int_0^1 x^2 \cdot 2x\,dx = \left.\frac{2x^4}{4}\right|_0^1 = \frac{1}{2}.$$

Therefore,

$$\mathrm{VAR}(X) = E(X^2) - (E\,X)^2 = \frac{1}{2} - \frac{4}{9} = \frac{1}{18}.$$

2.

$$E\,X = \int_0^\infty x f_X(x)\,dx = \int_1^\infty \frac{2}{x^2}\,dx = \left.-\frac{2}{x}\right|_1^\infty = 2,$$

$$E(X^2) = \int_{-\infty}^\infty x^2 f_X(x)\,dx = \int_1^\infty \frac{2}{x}\,dx = 2\ln x\Big|_1^\infty = \infty.$$

Therefore,

$$\mathrm{VAR}(X) = E(X^2) - (E\,X)^2 = \infty.$$

3.

$$E\,X = \sum_{i=0}^{i=2} i\,p_i = 0 \cdot 0.2 + 1 \cdot 0.5 + 2 \cdot 0.3 = 1.1,$$

$$\mathrm{VAR}(X) = E(X - E\,X)^2 = \sum_{i=0}^{2} p_i \cdot (i - E\,X)^2$$

$$= .2 \cdot (-1.1)^2 + .5 \cdot (-0.1)^2 + .3 \cdot (0.9)^2 = .49.$$

13. Using properties E4 and E5,

$$E\left(\frac{1}{2}X + Y\right) = \frac{1}{2}EX + EY = \frac{1}{2}\int_0^1 x\,dx + 2 = 2.25.$$

15.

$$p_X(x) = \sum_{y=0}^{1} p_{X,Y}(x, y) = p_{X,Y}(x, y = 0) + p_{X,Y}(x, y = 1).$$

Therefore,

$$p_X(0) = 0.4 \text{ and } p_X(1) = 0.6,$$

$$E(XY) = \sum_{x=0}^{1}\sum_{y=0}^{1} xyp_{X,Y}(x, y) = 1 \cdot 0.4 = 0.4.$$

27. a.

$$p(q_0) = P(X \leq \tau) = \int_{-\infty}^{\tau} f_X(x)\,dx = \int_0^{\tau} e^{-x}\,dx = 1 - e^{-\tau},$$

$$p(q_1) = P(X > \tau) = \int_{\tau}^{\infty} f_X(x)\,dx = \int_{\tau}^{\infty} e^{-x}\,dx = e^{-\tau}.$$

Hence,

$$F_Y(y) = (1 - e^{-\tau})U(y - q_0) + e^{-\tau}U(y - q_1),$$

and

$$EY = q_0(1 - e^{-\tau}) + q_1 e^{-\tau},$$
$$E(Y^2) = q_0^2(1 - e^{-\tau}) + q_1^2 e^{-\tau}.$$

Therefore,

$$\text{VAR}(Y) = E(Y^2) - (EY)^2 = e^{-\tau}(1 - e^{-\tau})(q_1 - q_0)^2.$$

b. First, we rewrite $Y = q(X)$ as

$$q(X) = q_0 U(\tau - X) + q_1(1 - U(\tau - X)).$$

Denoting $g(X) = Xq(X)$ and using property E8,

$$E(XY) = E(Xq(X)) = E(g(X)) = \int_{-\infty}^{\infty} g(x)f_X(x)\,dx$$

$$= \int_0^{\infty} x \cdot [q_0 U(\tau - x) + q_1(1 - U(\tau - x))] \cdot e^{-x}\,dx$$

$$= q_0 \int_0^{\infty} xU(\tau - x)e^{-x}\,dx + q_1 \int_0^{\infty} x(1 - U(\tau - x))e^{-x}\,dx$$

$$= q_0 \int_0^\tau x e^{-x}\, dx + q_1 \int_\tau^\infty x e^{-x}\, dx$$

$$= -q_0 \int_0^\tau x\, de^{-x} - q_1 \int_\tau^\infty x\, de^{-x}$$

$$= -q_0 \left[x e^{-x} \Big|_0^\tau - \int_0^\tau e^{-x}\, dx \right] - q_1 \left[x e^{-x} \Big|_\tau^\infty - \int_\tau^\infty e^{-x}\, dx \right]$$

$$= -q_0 (\tau e^{-\tau} - 1 + e^{-\tau}) - q_1 (-\tau e^{-\tau} - e^{-\tau})$$

$$= q_0 + \tau e^{-\tau}(q_1 - q_0) + e^{-\tau}(q_1 - q_0).$$

Therefore,

$$E(Y - X)^2 = E(Y^2 + X^2 - 2XY) = E(Y^2) + E(X^2) - 2E(XY)$$

$$= E(Y^2) + \text{VAR}(X) + (EX)^2 - 2E(XY)$$

$$= q_0^2(1 - e^{-\tau}) + q_1^2 e^{-\tau} + 2 - 2(q_0 + \tau e^{-\tau}(q_1 - q_0) + e^{-\tau}(q_1 - q_0)).$$

c. We need to find τ such that $\dfrac{dE(Y - X)^2}{d\tau} = 0$. Thus, τ can be found from the following equation:

$$(q_0^2 - q_1^2)e^{-\tau} - 2e^{-\tau}(q_1 - q_0) + 2\tau e^{-\tau}(q_1 - q_0) + 2e^{-\tau}(q_1 - q_0) = 0.$$

After dividing it by $e^{-\tau}$ and by $(q_1 - q_0)$ we get

$$-(q_0 + q_1) - 2 + 2\tau + 2 = 0 \Rightarrow \tau = \frac{q_0 + q_1}{2}.$$

29. a. We can write $f_{X,Y}(x, y)$ in more convenient form:

$$f_{X,Y}(x, y) = 3x^2 U(x)U(1 - x)U(y)U(1 - y).$$

Then, we can find the marginal pdfs:

$$f_X(x) = \int_{-\infty}^\infty f_{X,Y}(x, y)\, dy = 3x^2 U(x)U(1 - x) \int_0^1 dy = 3x^2 U(x)U(1 - x),$$

$$f_Y(x) = \int_{-\infty}^\infty f_{X,Y}(x, y)\, dx = U(y)U(1 - y) \int_0^1 3x^2\, dx = U(y)U(1 - y).$$

Having found the marginal pdfs, we can calculate the moments:

$$E X^k = \int_{-\infty}^\infty x^k f_X(x)\, dx = \int_0^1 x^k 3x^2\, dx = \frac{3x^{k+3}}{k + 3} \Big|_0^1 = \frac{3}{k + 3},$$

$$E Y^k = \int_{-\infty}^\infty y^k f_Y(y)\, dy = \int_0^1 y^k\, dy = \frac{y^{k+1}}{k + 1} \Big|_0^1 = \frac{1}{k + 1}.$$

b. Using the results of the part (a),

$$\text{VAR}(X) = E(X^2) - (E\,X)^2 = \frac{3}{5} - \left(\frac{3}{4}\right)^2 = \frac{3}{80},$$

$$\text{VAR}(Y) = E(Y^2) - (E\,Y)^2 = \frac{1}{3} - \left(\frac{1}{2}\right)^2 = \frac{1}{12}.$$

c.

$$E(XY) = \int_{-\infty}^{\infty} \int_{-\infty}^{\infty} xy f_{X,Y}(x,y)\,dx\,dy = \int_{0}^{1} \int_{0}^{1} xy\,3x^2\,dx\,dy$$

$$= \int_{0}^{1} 3x^3\,dx \cdot \int_{0}^{1} y\,dy = \frac{3}{4} \cdot \frac{1}{2} = \frac{3}{8}.$$

d. Using the results of (a) and (c),

$$\text{COV}(X, Y) = E(XY) - E\,X \cdot E\,Y = \frac{3}{8} - \frac{3}{4} \cdot \frac{1}{2} = 0.$$

e.

$$E\left(a(X - EX) - (Y - E\,Y)\right)^2$$

$$= E\left(a^2(X - EX)^2 + (Y - EY)^2 - 2a(X - EX)(Y - EY)\right)$$

$$= a^2 E(X - EX)^2 + E(Y - EY)^2 - 2aE\left((X - EX)(Y - EY)\right)$$

$$= a^2 \underbrace{\text{VAR}(X)}_{=\frac{3}{80}} + \underbrace{\text{VAR}(Y)}_{=\frac{1}{12}} - 2a \underbrace{\text{COV}(X, Y)}_{=0}$$

$$= \frac{3a^2}{80} + \frac{1}{12}.$$

f. We are going to construct a linear estimator in the form: $\hat{Y}(X) = aX + b$. Having defined $c = b - E\,Y + aE\,X$, we know (from page 273) that the estimator yields the minimum mean square error (MMSE) if $c = 0$ and

$$a = \frac{\text{COV}(X, Y)}{\text{VAR}(X)}.$$

Using the result of (d),

$$\text{COV}(X, Y) = 0 \Rightarrow a = \frac{\text{COV}(X, Y)}{\text{VAR}(X)} = 0,$$

and

$$c = 0 \Rightarrow b - EY + aEX = 0 \Rightarrow b = EY = \frac{1}{2}.$$

Thus, the MMSE affine estimator is $\hat{Y}(X) = \frac{1}{2}$, and the reason why the random variable X is not used in the estimator is because X and Y are uncorrelated.

34. a. We are looking for the estimator in the form $\hat{S}(X) = aX + b$ that has the minimum mean square error. Denote $c = b - ES + aEX$. From Section 9.12.1, we know that MMSE must have $c = 0$ and $a = \frac{COV(X,S)}{VAR(X)}$. At first, we find a:

$$COV(X,S) = E[(X - EX)(S - ES)] = E[(S + N - ES - EN)(S - ES)]$$
$$= E(S - ES)^2 + E[(N - EN)(S - ES)]$$
$$= VAR(S) + COV(N,S) = \frac{1}{12} + 0 = \frac{1}{12},$$

and

$$VAR(X) = E(S + N - ES - EN)^2 = VAR(S) + VAR(N) + 2\,COV(S,N)$$
$$= \frac{1}{12} + \sigma^2 + 0 = \frac{1}{12} + \sigma^2.$$

Thus, $a = \frac{COV(X,S)}{VAR(X)} = \frac{\frac{1}{12}}{\frac{1}{12} + \sigma^2} = \frac{1}{1 + 12\sigma^2}.$

$$c = 0 \Rightarrow b = ES - aEX = \frac{1}{2} - \frac{1}{1 + 12\sigma^2} \cdot \frac{1}{2} = \frac{6\sigma^2}{1 + 12\sigma^2}.$$

So, the linear MMSE for S is $\hat{S}(X) = \dfrac{X}{1 + 12\sigma^2} + \dfrac{6\sigma^2}{1 + 12\sigma^2}.$

b. Using the results in Section 9.12.1,

$$E(S - \hat{S})^2 = VAR(S) - \frac{COV(X,S)^2}{VAR(X)} = \frac{1}{12} - \frac{1}{12} \cdot \frac{\frac{1}{12}}{\frac{1}{12} + \sigma^2} = \frac{\sigma^2}{1 + 12\sigma^2}.$$

37. a. From Section 9.15.1, we know (as $\mathbf{b} = \mathbf{0}$):

$$R_Y = AR_X A^T = \begin{pmatrix} 7 & 1 \\ 1 & 3 \end{pmatrix}.$$

b. No, because we don't have any information about the mean of either the input or the output.

42. a. $COV(S, X_i) = E[(S - m)(S + N_i - m)] = VAR(S) + COV(S, N_i) = 1.$

b. $COV(X_i, X_j) = E[(S + N_i - m)(S + N_j - m)] = VAR(S) + COV(S, N_j) + COV(N_i, S) + COV(N_i, N_j)$. So,

$$COV(X_i, X_j) = \begin{cases} 1 & \text{if } i \neq j \\ 1 + \sigma_i^2 & \text{if } i = j \end{cases}.$$

43. a.

$$EX = \mathbf{m} = \begin{pmatrix} 0 \\ 1 \end{pmatrix}, \quad \text{VAR}(\mathbf{X_1}) = 2, \quad \text{VAR}(\mathbf{X_2}) = 3, \quad \text{and} \quad \text{COV}(\mathbf{X_1}, \mathbf{X_2}) = -1.$$

b. Using the result of the previous item, we know:

$$a^* = \frac{\text{COV}(X_1, X_2)}{\text{VAR}(X_1)} = \frac{-1}{2},$$

$$b^* = EX_2 - a^* EX_1 = 1.$$

c. Again, from Section 9.12.1, we know:

$$E(X_2 - \hat{X}_2)^2 = \text{VAR}(X_2) - \frac{\text{COV}(X_1, X_2)^2}{\text{VAR}(X_1)} = 3 - \frac{1}{2} = 2.5.$$

d. From Section 9.15.1, we know:

$$EY = \mathbb{A}EX = \begin{pmatrix} 2 \\ -1 \\ 2 \end{pmatrix},$$

$$\text{COV}(\mathbf{Y}, \mathbf{Y}) = \mathbb{A}\,\text{COV}(\mathbf{X}, \mathbf{X})\mathbb{A}^T = \begin{pmatrix} 10 & -5 & 10 \\ -5 & 27 & -26 \\ 10 & -26 & 28 \end{pmatrix}.$$

CHAPTER 10

4. a. $EY_n = E(\sum_{k=0}^n \beta^{n-k} X_k) = \sum_{k=0}^n \beta^{n-k} EX_k = 0.$
   b. For $n > m > 0$,

$$\text{COV}(Y_n, Y_m) = E\left[\left(\sum_{k=0}^n \beta^{n-k} X_k\right)\left(\sum_{k=0}^m \beta^{m-k} X_k\right)\right]$$

$$= \sum_{j=0}^n \sum_{k=0}^m \beta^{m+n-j-k} EX_j X_k.$$

And since $\{X_k\}$ are pairwise uncorrelated, we have:

$$\text{COV}(Y_n, Y_m) = \sum_{j \neq k}^{n} \sum^{m} \beta^{m+n-j-k} EX_j EX_k + \sum_{k=0}^m \beta^{m+n-2k} EX_k^2$$

$$= 0 + \sigma^2 \beta^{m+n} \sum_{k=0}^m \beta^{-2k} = \sigma^2 \beta^{m+n} \left(\frac{1 - \beta^{-2(m+1)}}{1 - \beta^{-2}}\right).$$

6. a. $\mathrm{COV}(S, X_i) = E[(S - m)(S + N_i - m)] = \mathrm{VAR}(S) + \mathrm{COV}(S, N_i) = 1$.

   b. $\mathrm{COV}(X_i, X_j) = E[(S + N_i - m)(S + N_j - m)] = \mathrm{VAR}(S) + \mathrm{COV}(S, N_j) +$
      $\mathrm{COV}(N_i, S) + \mathrm{COV}(N_i, N_j)$. So,

   $$\mathrm{COV}(X_i, X_j) = \begin{cases} 1 & \text{if } i \neq j \\ 1 + \sigma_i^2 & \text{if } i = j \end{cases}$$

   c. Define $\mathbf{Z} = \mathbf{X} - E\mathbf{X}$. Using the result of Section 10.5.4, we know that:

   $$\mathbf{b}^* = E\mathbf{S} - \mathbb{A}E\mathbf{Z} = \mathbf{m}.$$

   d. Since $\mathrm{COV}(\mathbf{S}, \mathbf{Z}) = \mathrm{COV}(\mathbf{S}, \mathbf{X})$ and since $C_{\mathbf{Z}} = C_{\mathbf{X}}$, using the result of Section 10.5.4, we get

   $$\mathbf{a}^* = \mathrm{COV}(\mathbf{S}, \mathbf{Z})\mathbb{C}_{\mathbf{Z}}^{-1} = \begin{pmatrix} 1 & 1 \end{pmatrix} \frac{-1}{\sigma_1^2 + \sigma_2^2 + \sigma_1^2\sigma_2^2} \begin{pmatrix} 1 & -(1 + \sigma_1^2) \\ -(1 + \sigma_2^2) & 1 \end{pmatrix}.$$

   So,

   $$\mathbf{a}^* = \frac{1}{\sigma_1^2 + \sigma_2^2 + \sigma_1^2\sigma_2^2} \begin{pmatrix} \sigma_2^2 & \sigma_1^2 \end{pmatrix}.$$

8. a.

   $$EY = EX_1^3 + EX_2 = 0,$$

   $$\mathrm{VAR}(Y) = EY^2 = EX_1^6 + 2EX_1^3X_2 + EX_2^2.$$

   Note that $EX_2^2 = \mathrm{VAR}(X_2)$ because $EX_2 = 0$. Since X_1 and X_2 are normal random variables and are uncorrelated, they are also independent. Also since X_1 is a standard normal, using the results in Section 9.10.2, we get $EX_1^6 = 15$. So,

   $$\mathrm{VAR}(Y) = 15 + 0 + 2 = 17.$$

   b.

   $$\mathrm{COV}(Y, X_1) = E[(X_1^3 + X_2)X_1] = EX_1^4 + EX_1X_2.$$

   Again, using independence of X_1 and X_2, and the results in Section 9.10.2, on standard normal that $EX_1^4 = 3$, we get

   $$\mathrm{COV}(Y, X_1) = 3,$$

   $$\mathrm{COV}(Y, X_2) = E[(X_1^3 + X_2)X_2] = EX_1^3X_2 + EX_2^2.$$

   Again, using independence of X_1 and X_2, we get

   $$\mathrm{COV}(Y, X_2) = 2.$$

c. Using the result of Section 10.5.4, we know that:

$$b^* = E\mathbf{Y} - A E\mathbf{X} = 0$$

and

$$\mathbf{a}^* = \mathrm{COV}(Y, X) C_X^{-1}$$

$$= \begin{bmatrix} 3 & 2 \end{bmatrix} \begin{bmatrix} 1 & 0 \\ 0 & 1/2 \end{bmatrix} = \begin{bmatrix} 3 & 1 \end{bmatrix}.$$

Therefore, $\hat{Y} = \begin{bmatrix} 3 & 1 \end{bmatrix} \mathbf{X}$.

d. Using the result of Section 10.5.4, we get

$$E(Y - \hat{Y})^2 = \mathrm{VAR}(Y) - \mathrm{COV}(Y, \mathbf{X}) C_X^{-1} \mathrm{COV}(\mathbf{X}, Y)$$

$$= 17 - \begin{bmatrix} 3 & 2 \end{bmatrix} \begin{bmatrix} 1 & 0 \\ 0 & 1/2 \end{bmatrix} \begin{bmatrix} 3 \\ 2 \end{bmatrix} = 6$$

CHAPTER 11

1. Since $A \cap B = \{5\}$, then by definition of conditional probability,

$$P(A|B) = \frac{P(A \cap B)}{P(B)} = \frac{p(5)}{p(2) + p(3) + p(5)} = \frac{0.05}{0.15} = \frac{1}{3}.$$

4. Notice that $P(B_2) = 1 - P(B_1) = 0.8$. By Bayes' Theorem,

$$P(B_1|A) = \frac{P(A|B_1)P(B_1)}{P(A|B_1)P(B_1) + P(A|B_2)P(B_2)} = \frac{0.5 \cdot 0.2}{(0.5 \cdot 0.2) + (0.2 \cdot 0.8)} = \frac{5}{13}$$

12. a. By Total Probability Theorem:

$$P(A = \{0\}) = P(A = \{0\}|B = \{0\})P(B = \{0\}) + P(A = \{0\}|B = \{1\})P(B = \{1\})$$

$$= (1 - 0.2)0.4 + (1 - 0.7)0.6 = 0.5.$$

Hence, $P(A = \{1\}) = 1 - P(A = \{0\}) = 0.5$.

b. By the definition of conditional probability,

$$P(B = \{1\}|A = \{1\}) = \frac{P(B = \{1\})P(A = \{1\}|B = \{1\})}{P(A = \{1\})} = \frac{0.6 \cdot 0.7}{0.5} = 0.84.$$

14. a. Let N_i be the number of different bytes emitted by the source in state S_i. Then,

$$N_i = \sum_{k=0}^{i} \binom{8}{k}.$$

Notice that the bytes emitted in states S_1 and S_2 are also emitted in state S_3. Therefore,

$$n = N_3 = \sum_{k=0}^{3} \binom{8}{k} = 93.$$

b. $P(B = 11000000 | S = S_2) = \frac{1}{N_2} = \frac{1}{37} \approx .027.$

c. By Total Probability Theorem:

$$P(B) = \sum_{i=1}^{3} P(B|S = S_i)P(S = S_i) = \frac{1}{3}\left(0 + \frac{1}{N_2} + \frac{1}{N_3}\right) = \frac{1}{3}\left(\frac{1}{37} + \frac{1}{93}\right) \approx 0.013.$$

d. By Bayes' Theorem:

$$P(S = S_2|B) = \frac{P(S = S_2)P(B|S = S_2)}{P(B)} = \frac{\frac{1}{3} \cdot \frac{1}{37}}{\frac{1}{3}\left(\frac{1}{37} + \frac{1}{93}\right)} = \frac{1}{1 + \frac{37}{93}} \approx 0.715.$$

23. a. By the Total Probability Theorem:

$$P(Y = 1) = \sum_{j=0}^{\infty} P(Y = 1|X = j)P(X = j) = \frac{e^{-1}}{1!} \cdot e^{-\frac{1}{2}} + \frac{e^{-1}}{0!} \cdot e^{-\frac{1}{2}} \cdot \frac{1}{2} = \frac{3}{2}e^{-\frac{3}{2}}.$$

b. By Bayes' Theorem:

$$P(X = k|Y = 1) = \frac{P(X = k)P(Y = 1|X = k)}{P(Y = 1)} = \begin{cases} \dfrac{e^{-1/2}e^{-1}}{\frac{3}{2}e^{-3/2}} = \dfrac{2}{3} & \text{if } k = 0 \\ \dfrac{1}{3} & \text{if } k = 1 \\ 0 & \text{if } k > 1 \end{cases}.$$

c. Since $P(X = 0|Y = 1) = \frac{2}{3} > P(X = 1|Y = 1) = \frac{1}{3}$, $\hat{X}_{MAP} = 0.$

28. Binary erasure channel (BEC) has inputs $X \in \{0, 1\}$ and outputs $Y \in \{0, 1, E\}$ with $P(X = 0) = .35,$

$$P(Y = 0|X = 0) = .7, \quad P(Y = E|X = 0) = .3, \quad P(Y = 0|X = 1) = 0,$$

$$P(Y = E|X = 1) = .2.$$

a. Hence, by Total Probability,

$$P(Y = E) = P(Y = E|X = 0)P(X = 0) + P(Y = E|X = 1)P(X = 1)$$

$$= .3(.35) + .2(1 - .35) = .235.$$

b. By Bayes,

$$P(X = 0|Y = E) = \frac{P(Y = E|X = 0)P(X = 0)}{P(Y = E)} = \frac{.3(.35)}{.235} = \frac{21}{47}.$$

c. What is the most probable X given that $Y = E$?

$$P(X = 0 | Y = E) = \frac{21}{47} \Rightarrow P(X = 1 | Y = E) = 1 - \frac{21}{47} = \frac{26}{47}.$$

Therefore, $X = 1$ is the most probable channel input given that $Y = E$.

CHAPTER 12

1.

$$f_X(x) = \int_{-\infty}^{\infty} f_{X,Y}(x, y) \, dy = \int_0^1 4xy \, dy = 2x.$$

Therefore, by the definition of bivariate conditional pdf:

$$f_{Y|X}(y|x) = \frac{f_{X,Y}(x, y)}{f_X(x)} = \frac{4xy}{2x} = 2y.$$

4. $X \sim N(0, 1)$ can be written as $f_X(x) = \frac{1}{\sqrt{2\pi}} e^{-\frac{x^2}{2}}$. Then,

$$f_{X,Y}(x, y) = f_{Y|X}(y|x) f_X(x)$$

and

$$f_Y(y) = \int_{-\infty}^{\infty} f_{X,Y}(x, y) \, dx = \int_{-\infty}^{\infty} f_{Y|X}(y|x) f_X(x) \, dx = \int_{-\infty}^{\infty} \frac{1}{\sqrt{2\pi}} e^{-\frac{x^2}{2}} \frac{1}{2} e^{-|y-x|} \, dx.$$

By definition of bivariate conditional pdf:

$$f_{X|Y}(x|y) = \frac{f_{X,Y}(x, y)}{f_Y(y)} = \frac{\frac{1}{2\sqrt{2\pi}} e^{-\frac{x^2}{2} - |y-x|}}{\int_{-\infty}^{\infty} \frac{1}{2\sqrt{2\pi}} e^{-\frac{x^2}{2} - |y-x|} \, dx}.$$

5. a. The system S is such that the output, Y, and input, X, are independent standard normals. S is described by $f_{Y|X}$, which in this case is simply

$$f_Y(y) = \frac{1}{\sqrt{2\pi}} e^{-\frac{y^2}{2}}.$$

b.

$$f_Y(y) = \int_{-\infty}^{\infty} f_{X,Y}(x, y) \, dx = \int_{-\infty}^{\infty} \frac{1}{2\pi} e^{-\frac{1}{2}(x^2 + y^2)} \, dx = \frac{e^{-\frac{y^2}{2}}}{\sqrt{2\pi}} \int_{-\infty}^{\infty} \frac{1}{\sqrt{2\pi}} e^{-\frac{x^2}{2}} \, dx = \frac{e^{-\frac{y^2}{2}}}{\sqrt{2\pi}}.$$

Thus, $Y \sim N(0, 1)$.

c.

$$f_{X|Y}(x|y) = \frac{f_{X,Y}(x,y)}{f_Y(y)} = \frac{\frac{1}{2\pi}e^{-\frac{1}{2}(x^2+y^2)}}{\frac{e^{-\frac{y^2}{2}}}{\sqrt{2\pi}}} = \frac{e^{-\frac{x^2}{2}}}{\sqrt{2\pi}}.$$

Having observed $Y = y$, we know about X no more than we knew before, $X \sim N(0, 1)$.

14. a.

$$P(S = s|V = 1/2) = \frac{P(S = s)f_{V|S}(1/2|s)}{f_V(1/2)}$$

where $f_V(1/2) = \sum_{i=1}^2 P(S = i)f_{V|S}(1/2|i)$. Hence,

$$P(S = 1|V = 1/2) = \frac{\frac{1}{3} \cdot 2 \cdot \frac{1}{2}}{\left(\frac{1}{3} \cdot 2 \cdot \frac{1}{2}\right) + \left(\frac{2}{3} \cdot 3 \cdot \left(\frac{1}{2}\right)^2\right)} = \frac{2}{5}$$

and

$$P(S = 2|V = 1/2) = 1 - P(S = 1|V = 1/2) = \frac{3}{5}.$$

So, as $P(S = 2|V = 1/2) > P(S = 1|V = 1/2)$, implies given $V = 1/2$, $\hat{S}_{MAP} = 2$.

b. This estimate is correct if $\hat{S}_{MAP} = S$. Therefore, it is correct with probability $P(\hat{S}_{MAP} = S = 2|V = 1/2) = \frac{3}{5}$.

21. a. Assuming that the number of received photons is distributed by Poisson, $\mathcal{P}(x)$:

$$P(K = k|X = x) = e^{-x}\frac{x^k}{k!}.$$

b. By the Total Probability Theorem:

$$P(K = k) = \int_0^\infty P(K = k|X = x)f_X(x)\,dx = \int_0^\infty e^{-x}\frac{x^k}{k!}e^{-x}\,dx.$$

Then using substitution, $y = 2x$ and $dy = 2\,dx$, we get

$$P(K = k) = \int_0^\infty e^{-y}\frac{(y/2)^k}{k!}\frac{1}{2}\,dy = \frac{1}{2}\left(\frac{1}{2}\right)^k\frac{1}{k!}\int_0^\infty e^{-y}y^k dy = \frac{1}{2}\left(\frac{1}{2}\right)^k\frac{k!}{k!} = \frac{1}{2}\left(\frac{1}{2}\right)^k.$$

Thus, K is geometrically distributed with $\beta = \frac{1}{2}$.

c. We know the conditional pdf on X:

$$f_{X|K}(x|K = 2) = \frac{f_X(x)P(K = 2|X = x)}{P(K = 2)} = \frac{e^{-x} \cdot e^{-x} \cdot \frac{x^2}{2!}}{\frac{1}{2}\left(\frac{1}{2}\right)^2} = 4x^2e^{-2x}.$$

25. a. Given $f_{Y|X}$ and f_X, what can you say about f_Y?

$$f_Y(y) = \int_{-\infty}^{\infty} f_{Y|X}(y|x)f_X(x)\,dx,$$

so f_Y is completely determined.

b. Given $f_{Y|X}$ and f_Y, what can we say about the input distribution f_X? This is a more complicated question with a variety of answers. It is still the case that

$$f_Y(y) = \int_{-\infty}^{\infty} f_{Y|X}(y|x)f_X(x)\,dx.$$

This is an integral equation for the unknown function $f_X(x)$ in terms of the known functions $f_Y(y)$ and the so-called kernel function $f_{Y|X}(y|x)$. If, say, $f_{Y|X}$ is in fact just f_Y, as can happen, then we know nothing about f_X. The integral equation reduces to the tautology that $f_Y = f_Y$. However, if, say,

$$f_{Y|X}(y|x) = \frac{1}{\sigma\sqrt{2\pi}}e^{-\frac{(y-x)^2}{2\sigma^2}},$$

then f_X is recoverable by Fourier transformation on y of both sides of the integral equation with the right-hand side now being a familiar convolution.

CHAPTER 13

1. The model for received X and transmitted M is

$$f_{X|M}(X|m_1) = e^{-x}U(x),\ f_{X|M}(x|m_2) = 2e^{-2x}U(x),$$
$$P(M=m_1) = .4, P(M=m_2) = 1-.4 = .6.$$

a. We determine the minimum error probability receiver $\hat{m}(x)$ by the methods of Section 13.3. We calculate the likelihood ratio

$$\Lambda(x) = \frac{f_{X|M}(x|m_1)}{f_{X|M}(x|m_2)} = \frac{1}{2}e^x,$$

and the threshold

$$\tau = \frac{P(M=m_2)}{P(M=m_1)} = \frac{3}{2}.$$

Hence,

$$\hat{m}(x) = \begin{cases} m_1 & \text{if } \Lambda > \tau \\ m_2 & \text{otherwise} \end{cases}.$$

More specifically, after minor simplification,

$$\hat{m}(x) = \begin{cases} m_1 & \text{if } x > \log 3 \\ m_2 & \text{otherwise} \end{cases}.$$

b. We evaluate the error probability P_e of $\hat{m}$ from

$$P_e = P(\hat{m} = m_1, M = m_2) + P(\hat{m} = m_2, M = m_1)$$

$$= P(X \geq \log 3 | M = m_2)P(M = m_2) + P(X < \log 3 | M = m_1)P(M = m_1)$$

$$= .6 \int_{\log 3}^{\infty} 2e^{-2x} \, dx + .4 \int_0^{\log 3} e^{-x} \, dx = .6\frac{1}{9} + .4\left(1 - \frac{1}{3}\right) = \frac{1}{3}.$$

Note that if we just guessed that $M = m_2$, then we would have an error probability of $P(M = m_1) = .4$. Hence, we are doing a little better than this simple decision rule.

10. Our calculation follows Section 13.4. Let H_0 represent absent reflected signal and H_1 present reflected signal.

$$f_{Y|H}(y|H_0) = \frac{1}{\sqrt{2\pi}}e^{-\frac{y^2}{2}}, \ f_{Y|H}(y|H_1) = \frac{1}{2\sqrt{\pi}}e^{-\frac{y^2}{4}},$$

$$\Lambda(y) = \frac{f_{Y|H}(y|H_1)}{f_{Y|H}(y|H_0)} = \frac{1}{\sqrt{2}}e^{\frac{y^2}{4}},$$

$$\phi(y) = \begin{cases} H_1 & \text{if } \Lambda > \tau \\ H_0 & \text{otherwise} \end{cases}.$$

a. The threshold τ will be converted into another nonnegative constant c after we simplify this expression.

$$P_D = P(\phi = H_1 | H = H_1) = P(|y| > c | H = H_1)$$

$$= \int_{-\infty}^{-c} \frac{1}{2\sqrt{\pi}}e^{-\frac{y^2}{4}} \, dy + \int_c^{\infty} \frac{1}{2\sqrt{\pi}}e^{-\frac{y^2}{4}} \, dy = \int_{-\infty}^{-c} \frac{1}{\sqrt{\pi}}e^{-\frac{y^2}{4}} \, dy = 2\Phi\left(-\frac{c}{\sqrt{2}}\right),$$

where Φ is the $\mathcal{N}(0, 1)$ cdf. Similarly,

$$P_{FA} = P(\phi = H_1 | H = H_0) = 2\int_{-\infty}^{-c} \frac{1}{\sqrt{2\pi}}e^{-\frac{y^2}{2}} \, dy = 2\Phi(-c).$$

The receiver operating characteristic (ROC) displays P_D as a function of P_{FA}. A parameterized representation is available to us through $P_D(c)$ vs. $P_{FA}(c)$ as c varies from 0 (where $P_D(0) = P_{FA}(0) = 1$) through infinity (where $P_D(\infty) = P_{FA}(\infty) = 0$).

b. We are now told that $P(H = H_1) = .3$ and that we are interested in the error probability P_e when we use a Neyman–Pearson detector designed for $P_{FA} = .1$.

$$P_e = P(\phi = H_1 | H = H_0)P(H = H_0) + P(\phi = H_0 | H = H_1)P(H = H_1)$$

$$= P_{FA}(1 - .3) + (1 - P_D).3 = .37 - .3P_D.$$

From

$$P_{FA}(c) = .1 = 2\Phi(-c)$$

we can use Matlab or a table to identify $c = 1.18$ and then to determine $P_D(c) = .238$. Hence, $P_e = .29$. Note that if we just guessed that $H = H_0$, then our error probability would be .3 and not that different.

16. We observe $X = S + N$ arising from a signal $S \sim \mathcal{P}(\lambda_0)$ and additive, independent noise $N \sim \mathcal{P}(\lambda_1)$, with $\lambda_0 = \frac{3}{2}\lambda_1$.

   a. Specify $p_{X|S}(x|s)$.

   $$p_{X|S}(x|s) = p_N(x - s) = e^{-\lambda_1}\frac{\lambda_1^{x-s}}{(x-s)!}U(x - s).$$

   b. Provide an expression for $p_{S|X}(s|x)$. Use Bayes:

   $$p_{S|X}(s|x) = \frac{p_{X|S}(x|s)p_S(s)}{\sum_s p_{X|S}(x|s)p_S(s)}.$$

   We first compute $p_X(x)$:

   $$\begin{aligned}
   p_X(x) &= \sum_{s=-\infty}^{\infty} p_{X,S}(x, s) = \sum_{s=-\infty}^{\infty} p_{X|S}(x|s)p_S(s) \\
   &= \sum_{s=0}^{\infty} e^{-\lambda_1}\frac{\lambda_1^{x-s}}{(x-s)!}U(x - s)\frac{\lambda_0^s}{s!}e^{-\lambda_0}U(s) \\
   &= e^{-(\lambda_0+\lambda_1)}\sum_{s=0}^{x}\frac{\lambda_1^{x-s}}{(x-s)!}\frac{\lambda_0^s}{s!} = \frac{e^{-(\lambda_0+\lambda_1)}}{x!}\sum_{s=0}^{x}\frac{x!}{(x-s)!s!}\lambda_1^{x-s}\lambda_0^s \\
   &= e^{-(\lambda_0+\lambda_1)}\sum_{s=0}^{x}\binom{x}{s}\lambda_1^{x-s}\lambda_0^s \\
   &= e^{-(\lambda_0+\lambda_1)}\frac{(\lambda_0 + \lambda_1)^x}{x!},
   \end{aligned}$$

   for $x \geq 0$, and $p_X(x) = 0$ for $x < 0$. So

   $$\begin{aligned}
   p_{S|X}(s|x) &= \frac{p_{X,S}(x, s)}{p_X(x)} = \frac{e^{-\lambda_1}\frac{\lambda_1^{x-s}}{(x-s)!}U(x - s)\frac{\lambda_0^s}{s!}e^{-\lambda_0}U(s)}{e^{-(\lambda_0+\lambda_1)}\frac{(\lambda_0+\lambda_1)^x}{x!}} \\
   &= \binom{x}{s}\frac{\lambda_1^{x-s}\lambda_0^s}{(\lambda_0 + \lambda_1)^x}U(x - s)U(s).
   \end{aligned}$$

   c. Determine the MAP rule $\hat{S}(2)$ for estimating S from X when $X = 2$. (Hint: just try the different possible values of S.)

   $$\begin{aligned}
   p_{S|X}(s|2) &= \binom{2}{s}\frac{\lambda_1^{2-s}\lambda_0^s}{(\lambda_0 + \lambda_1)^2}U(2 - s)U(s) = \frac{2!\lambda_1^2}{(\lambda_0 + \lambda_1)^2}\frac{1}{(2 - s)!s!}\left(\frac{\lambda_0}{\lambda_1}\right)^s \\
   &= c\frac{1}{(2 - s)!s!}\left(\frac{3}{2}\right)^s,
   \end{aligned}$$

where c does not depend on s. We have:

$$p_{S|X}(0|2) = c\frac{1}{2}, \quad p_{S|X}(1|2) = c\frac{3}{2}, \quad p_{S|X}(2|2) = c\frac{1}{2} \cdot \frac{9}{4} = \frac{9}{8},$$

$$p_{S|X}(s|2) = 0 \quad \text{for } s > 2.$$

Therefore,

$$\hat{S}(2) = 1.$$

d. Evaluate the error probability P_e incurred by using X itself to estimate S.

$$P_e = P(S \neq X) = P(N > 0) = 1 - P(N = 0) = 1 - e^{-\lambda_1}.$$

22. a. The MAP rule $\hat{S}(x)$ is the value of $S = s$ that maximizes the posterior probability $f_{S|X}(s|x)$. Hence, from the given information,

$$f_{S|X}(s|x) = \frac{f_{X|S}(x|s)f_S(s)}{\int_s f_{X|S}f_S \, ds} \propto f_{X|S}f_S = \frac{1}{2}e^{-s-|x-s|}.$$

$$\hat{S}(x) = \text{argmax} f_{S|X}(s|x) = \underset{s \geq 0}{\text{argmin}}[s + |x - s|],$$

where we use the fact that

$$\underset{x}{\text{argmax}}[f(x)g(y)] = \underset{x}{\text{argmax}}[f(x)].$$

Consider the case $x \leq 0$ to determine

$$\underset{s \geq 0}{\text{argmin}}[s + |x - s|] = \underset{s \geq 0}{\text{argmin}}[s + s - x] = 0.$$

If $x \geq 0$, then

$$\underset{s \geq 0}{\text{argmin}}[s + |x - s|] = x.$$

Hence, $\hat{S}(x) = xU(x)$.

b. The MLE rule $\tilde{S}$ is given by

$$\tilde{S}(x) = \text{argmax} f_{X|S}(x|s),$$

and does not require knowing f_S. In this case

$$\tilde{S}(x) = \underset{s \geq 0}{\text{argmax}} \frac{1}{2}e^{-|x-s|} = \underset{s \geq 0}{\text{argmin}} |x - s|,$$

where we assume that from the "physics" of the exercise we still believe that $S \geq 0$. We find the same solution $\tilde{S}(x) = xU(x) = \hat{S}(x)$.

c. The mean square error performance is given by

$$E(\hat{S} - S)^2 = E(\tilde{S} - S)^2 = E(XU(X) - S)^2$$

$$= \int_0^\infty e^{-s}\,ds \int_{-\infty}^\infty (xU(x) - s)^2 \frac{1}{2} e^{-|x-s|}\,dx$$

$$= \frac{1}{2}\int_0^\infty e^{-s}\,ds \left[\int_{-\infty}^0 e^{x-s}(0 - s)^2\,dx + \int_0^s (x - s)^2 e^{x-s}\,dx \right.$$

$$\left. + \int_s^\infty (x - s)^2 e^{s-x}\,dx \right].$$

The first integral in the square brackets is trivially seen to be $s^2 e^{-s}$, once one factors this out of the integrand. A change of variable reduces the second integral in the square brackets to

$$\int_0^s y^2 e^{-y}\,dy = 2 - (s^2 + 2s + 2)e^{-s},$$

where we have used the simple integration facts, derivable by integration by parts,

$$\int_a^b ye^{-y}\,dy = (a + 1)e^{-a} - (b + 1)e^{-b}, \quad \int_a^b y^2 e^{-y}\,dy$$

$$= (a^2 + 2a + 2)e^{-a} - (b^2 + 2b + 2)e^{-b}.$$

The third integral in the square brackets is easily shown to be 2 by a simple change of variables. Hence,

$$E(\hat{S} - S)^2 = E(\tilde{S} - S)^2 = \frac{1}{2}\int_0^\infty e^{-s}[s^2 e^{-s} + 4 - (s^2 + 2s + 2)e^{-s}]\,ds = \frac{5}{4},$$

after further evaluation as above. Note that $\frac{5}{4} > 1 = \text{VAR}(S)$. Hence, in a mean square sense the MAP and MLE estimators perform worse than just estimating S by its mean of one.

CHAPTER 14

1. $P(A \cap B) = p_2, P(A) = p_1 + p_2 + p_4, P(B) = p_2 + p_3 + p_5, A \perp\!\!\!\perp B \Rightarrow p_2 = (p_1 + p_2 + p_4)$
   $(p_2 + p_3 + p_5)$ (*). Consider $p_1 = p_3 = p_4 = p_5 = \frac{1}{9}$, $p_2 = \frac{1}{9}$ and $p_6 = \frac{4}{9}$. This satisfies
   equation (*) and $\sum_{i=1}^6 p_i = 1$, so it is a pmf.

3. $F = (A_1 \cup A_2) \cap A_3^c$, $P(F) = P((A_1 \cup A_2) \cap A_3^c) = P(A_1 \cup A_2)P(A_3^c)$, since $\perp\!\!\!\perp_i A_i$ implies
   that A_3^c is independent of $A_1 \cup A_2$. But $P(A_1 \cup A_2) = P((A_1^c \cap A_2^c)^c) = 1 - P(A_1^c \cap A_2^c) =$
   $1 - P(A_1^c)P(A_2^c)$, where the first equality follows from de Morgan's law and the last equality
   follows from the independence between A_1^c and A_2^c. So, $P(F) = [1 - (0.6)(0.65)](0.9) =$
   0.549.

6. Since X_i are independent and A, B, and C depend on nonoverlapping bits, they are also independent. And we have

$$P(A) = p^2; \quad P(B) = (1-p); \quad P(C) = p(1-p).$$

a. $P_a = P(A \cap B \cap C) = P(A)P(B)P(C) = p^3(1-p)^2.$
b. $P_b = P(A^c \cap B^c \cap C^c) = (1-p^2)p(1-p(1-p)).$
c. $P_c = 1 - P_b = 1 - (1-p^2)p(1-p(1-p)).$
d. $P_d = P(A \cap B \cap C^c) = P(A)P(B)P(C^c) = p^2(1-p)(1-p(1-p)).$

17.

$$F_{Z_n}(z) = P(Z_n \leq z) = P\left(\frac{2}{n}\max_i X_i \leq z\right)$$

$$= P\left(\max_i X_i \leq \frac{nz}{2}\right) = P\left(\cap_{i=1}^n X_i \leq \frac{nz}{2}\right)$$

$$= \prod_{i=1}^n P\left(X_i \leq \frac{nz}{2}\right) \qquad \text{(by independence)}$$

$$= \left[F_X\left(\frac{nz}{2}\right)\right]^n = \begin{cases} 0 & \text{if } z \leq 0 \\ \left(\frac{nz}{2\theta}\right)^n & \text{if } 0 < \frac{nz}{2} \leq \theta \\ 1 & \text{if } \frac{nz}{2} > \theta \end{cases}.$$

Differentiating,

$$f_{Z_n} = \begin{cases} \dfrac{n^{n+1}}{(2\theta)^n} z^{n-1} & \text{if } 0 < \frac{nz}{2} \leq \theta. \\ 0 & \text{otherwise} \end{cases}$$

18.

$$P(X > Y) = \int_{-\infty}^{\infty} P(X > Y | Y = y) f_Y(y)\, dy$$

$$= \int_{-\infty}^{\infty} P(X > y | Y = y) f_Y(y)\, dy$$

$$= \int_{-\infty}^{\infty} P(X > y) f_Y(y)\, dy \qquad \text{(by independence)}$$

$$= \int_{-\infty}^{\infty} (1 - F_X(y)) f_Y(y)\, dy = \int_0^1 e^{-y}\, dy = 1 - e^{-1}.$$

27.

$$P(Z = z) = P(X_1 + X_2 = z) = \sum_{k=0}^{z} P(X_1 + X_2 = z, X_2 = k)$$

$$= \sum_{k=0}^{z} P(X_1 + X_2 = z | X_2 = k) P(X_2 = k)$$

$$= \sum_{k=0}^{z} P(X_1 = z - k | X_2 = k) P(X_2 = k)$$

$$= \sum_{k=0}^{z} P(X_1 = z - k) P(X_2 = k) \qquad \text{(by independence)}$$

$$= \sum_{k=0}^{z} \frac{e^{-\lambda_1}(\lambda_1)^{z-k}}{(z-k)!} \cdot \frac{e^{-\lambda_2}(\lambda_2)^{k}}{k!}$$

$$= e^{-(\lambda_1+\lambda_2)} \frac{1}{z!} \sum_{k=0}^{z} (\lambda_1)^{z-k}(\lambda_2)^{k} \frac{z!}{(z-k)!k!} \qquad \text{(after dividing/multiplying by } z!)$$

$$= \frac{e^{-(\lambda_1+\lambda_2)}(\lambda_1 + \lambda_2)^{z}}{z!} \qquad \text{(by Binomial Theorem),}$$

$$p_{Z|X_1}(z|x_1) = P(X_1 + X_2 = z | X_1 = x_1) = P(X_2 = z - x_1 | X_1 = x_1)$$

$$= P(X_2 = z - x_1) \qquad \text{(by independence)}$$

$$= \frac{e^{-\lambda_2}(\lambda_2)^{z-x_1}}{(z-x_1)!},$$

$$p_{X_1|Z}(x_1|z) = \frac{p_{Z|X_1}(z|x_1) \cdot p_{X_1}(x_1)}{p_Z(z)} = \frac{\dfrac{e^{-\lambda_2}(\lambda_2)^{z-x_1}}{(z-x_1)!} \cdot \dfrac{e^{-\lambda_1}(\lambda_1)^{x_1}}{x_1!}}{\dfrac{e^{-(\lambda_1+\lambda_2)}(\lambda_1 + \lambda_2)^{z}}{z!}}$$

$$= \frac{(\lambda_1)^{x_1}(\lambda_2)^{z-x_1}}{(\lambda_1 + \lambda_2)^{z}} \binom{z}{x_1}.$$

32. For this exercise, we will use the following trigonometric relations:
    1. $\sin^2 a + \cos^2 a = 1$
    2. $\cos(a+b) = \cos a \cos b - \sin a \sin b \Rightarrow \cos(a)\cos(b) = \frac{1}{2}[\cos(a+b) + \cos(a-b)]$
    3. $\sin(a+b) = \sin a \cos b + \sin b \cos a$

a.
$$E X(t) = E(A\cos(\omega_0 t + \theta)) = E A E \cos(\omega_0 t + \theta) \quad \text{(by independence)},$$

$$E \cos(\omega_0 t + \theta) = \int_{-\pi}^{\pi} \cos(\omega_0 t + \theta) \cdot \frac{1}{2\pi} \, d\theta = \frac{1}{2\pi} \sin(\omega_0 t + \theta)\big|_{-\pi}^{\pi} = 0.$$

So, $E X(t) = 0$.

Then,
$$\text{COV}(X(t), X(s)) = EX(t)X(s) - EX(t)\,EX(s)$$
$$= EX(t)X(s)$$
$$= E(A\cos(\omega_0 t + \theta)A\cos(\omega_0 s + \theta))$$
$$= E A^2 E(\cos(\omega_0 t + \theta)\cos(\omega_0 s + \theta)),$$

where

$$E A^2 = \text{VAR}(A) + (E A)^2 = \frac{1}{\alpha^2} + \left(\frac{1}{\alpha}\right)^2 = \frac{2}{\alpha^2},$$

and

$$2E\big(\cos(\omega_0 t + \theta)\cos(\omega_0 s + \theta)\big) = E \cos(\omega_0(t - s)) + E \cos(\omega_0(t + s) + 2\theta),$$

by using (2). However, $E \cos(\omega_0(t - s)) = \cos(\omega_0(t - s))$ and $E \cos(\omega_0(t + s) + 2\theta) = \frac{1}{2\pi} \int_{-\pi}^{\pi} \cos(\omega_0(t + s) + 2\theta) \, d\theta = 0$.

Hence,

$$E\big(\cos(\omega_0 t + \theta)\cos(\omega_0 s + \theta)\big) = \frac{1}{2} \cos(\omega_0 t - \omega_0 s).$$

Therefore,

$$\text{COV}(X(t), X(s)) = \frac{1}{\alpha^2} \cos(\omega_0(t - s))$$

$$\text{VAR}(X(t)) = \text{COV}(X(t), X(t)) = \frac{1}{\alpha^2}.$$

b. From Section 9.12.1,

$$\hat{X}(t) = \frac{\text{COV}(X(t), X(s))}{\text{VAR}(X(s))} X(s) + E X(t) - \frac{\text{COV}(X(t), X(s))}{\text{VAR}(X(s))} E X(s).$$

Therefore, $\hat{X}(t) = \cos(\omega_0(t - s))X(s)$.

37.

$$f_{X,Y,Z}(x,y,z) = f_X(x)f_Y(y)f_Z(z)$$

$$= \frac{1}{\sqrt{2\pi(1/2)}}e^{-\frac{x^2}{2(1/2)}} \cdot \frac{1}{\sqrt{2\pi(1/2)}}e^{-\frac{y^2}{2(1/2)}} \cdot \frac{1}{\sqrt{2\pi(1/2)}}e^{-\frac{z^2}{2(1/2)}}$$

$$= \frac{1}{\pi^{3/2}}e^{-(x^2+y^2+z^2)}.$$

40. a.

$$F_Z(z) = P(Z \le z) = P(\max_{i \le n} X_i \le z) = P(\cap_{i \le n} X_i \le z)$$

$$= \prod_{i=1}^{n} P(X_i \le z) = \begin{cases} z^n & \text{if } 0 \le z \le 1 \\ 1 & \text{if } z > 1 \\ 0 & \text{otherwise} \end{cases},$$

$$f_Z(z) = \begin{cases} nz^{n-1} & \text{if } 0 \le z \le 1 \\ 0 & \text{otherwise} \end{cases}.$$

$$EZ = \int_0^1 z f_Z(z)\,dz = \int_0^1 nz^n\,dz = \frac{n}{n+1}.$$

b.

$$F_Y(y) = P(Y \le y) = P(\min_{i \le n} X_i \le y) = P(\cup_{i \le n} X_i \le y)$$

$$= P\big((\cap_{i \le n} X_i > y)^c\big) = 1 - P(\cap_{i \le n} X_i > y)$$

$$= 1 - \prod_{i=1}^{n} P(X_i > y) = \begin{cases} 1-(1-y)^n & \text{if } 0 \le y \le 1 \\ 1 & \text{if } y > 1 \\ 0 & \text{if } y < 0 \end{cases},$$

$$f_Y(y) = \begin{cases} n(1-y)^{n-1} & \text{if } 0 \le y \le 1 \\ 0 & \text{otherwise} \end{cases},$$

$$EY = \int_0^1 y f_Y(y)\,dy = \int_0^1 ny(1-y)^{n-1} = \int_0^1 n(1-x)x^{n-1}\,dx$$

(after substituting $x = 1 - y$, $dx = -dy$)

$$= n\left[\frac{x^n}{n} - \frac{x^{n+1}}{n+1}\right]\Bigg|_0^1 = 1 - \frac{n}{n+1} = \frac{1}{n+1}.$$

CHAPTER 15

1. a.

$$f_X(x) = \int_{-\infty}^{\infty} f_{X,Y}(x,y)\,dy = e^{-x}U(x).$$

So,

$$f_{Y|X}(y|x) = \frac{f_{X,Y}(x,y)}{f_X(x)} = e^{-y}U(y).$$

Then,

$$E(Y|X) = \int_0^{\infty} y e^{-y}\,dy = 1.$$

b.

$$f_X(x) = \int_{-\infty}^{\infty} f_{X,Y}(x,y)\,dy = \int_0^{\infty} \frac{(x+y)}{2} e^{-(x+y)} U(x)\,dy.$$

Let $z = x + y$ and $dz = dy$. Then, using this substitution and then integration by parts:

$$f_X(x) = U(x)\int_x^{\infty} \frac{z}{2} e^{-z}\,dz = U(x)\left[\frac{-z}{2}e^{-z}\Big|_x^{\infty} + \int_x^{\infty} \frac{1}{2}e^{-z}\,dz\right] = \frac{e^{-x}}{2}(1+x)U(x).$$

So,

$$f_{Y|X}(y|x) = \frac{f_{X,Y}(x,y)}{f_X(x)} = \frac{(x+y)}{(1+x)}e^{-y}U(y).$$

Then,

$$E(Y|X) = \int_0^{\infty} y\frac{(x+y)}{(1+x)}e^{-y}\,dy = \frac{1}{(1+x)}\int_0^{\infty}(xy+y^2)e^{-y}\,dy = \frac{(x+2)}{(1+x)}.$$

4. a. Using the subconditioning property,

$$E(V_1 V_2) = E(E(V_1 V_2|V_1)) = E(V_1 E(V_2|V_1)) = E(V_1 \cdot V_1^3) = E(V_1^4)$$

$$= \int_{-1}^{1} v_1^4 \frac{1}{2}\,dv_1 = \frac{1}{5}.$$

b. Not independent, since $E(V_2|V_1)$ depends on V_1.

5. a. From Section 15.2,

$$E(Z) = E(E(Z|X)) = E(X^2) = \int_0^1 x^2 \, dx = \frac{x^3}{3}\bigg|_0^1 = \frac{1}{3}.$$

   b.

$$E(XZ) = E(E(ZX|X)) = E(XE(Z|X)) = E(X^3) = \int_0^1 x^3 \, dx = \frac{1}{4}.$$

9. a. First, notice that $\sigma_1^2 = \sigma_2^2 = \frac{1}{2}$ and $\rho = \frac{1}{2}$. Then, using Example 12.5,

$$f_{V_2|V_1}(v_2|v_1) = \frac{f_{V_1,V_2}(v_1, v_2)}{f_{V_1}(v_1)}$$

$$= \frac{1}{\sqrt{2\pi\sigma_2^2(1-\rho^2)}} e^{-\frac{(v_2 - \frac{\rho\sigma_2}{\sigma_1}v_1)^2}{2\sigma_2^2(1-\rho^2)}}$$

$$= \frac{1}{\sqrt{2\pi(3/8)}} e^{-\frac{(v_2 - \frac{1}{2}v_1)^2}{2(3/8)}}.$$

   b. From Section 15.4.1,

$$\hat{V}_2(V_1) = E(V_2|V_1) = \int_{-\infty}^{\infty} v_2 f_{V_2|V_1}(v_2|v_1) \, dv_2 = \frac{1}{2}v_1.$$

15. a. From Section 15.4.1,

$$\hat{Y}(X) = E(Y|X) = \int_{-\infty}^{\infty} y f_{Y|X}(y, x) \, dy = \frac{1}{2}\int_{-\infty}^{\infty} y e^{-|y-\cos x|} \, dy.$$

   Substituting $z = y - \cos x$ and $dz = dy$,

$$= \frac{1}{2}\int_{-\infty}^{\infty} (z + \cos x) e^{-|z|} \, dz$$

$$= \frac{1}{2}\underbrace{\int_{-\infty}^{\infty} z e^{-|z|} \, dz}_{0} + \frac{1}{2}\cos x \underbrace{\int_{-\infty}^{\infty} e^{-|z|} \, dz}_{2}$$

$$= \cos x.$$

b.

$$E(Y - \hat{Y}(X))^2 = \int_{x=-\infty}^{\infty} \int_{y=-\infty}^{\infty} (y - \cos x)^2 f_{X,Y}(x, y) \, dx \, dy$$

$$= \int_{x=-\infty}^{\infty} \int_{y=-\infty}^{\infty} (y - \cos x)^2 f_{Y|X}(y|x) f_X(x) \, dx \, dy$$

$$= \int_{x=0}^{\infty} \alpha e^{-\alpha x} \left(\int_{y=-\infty}^{\infty} (y - \cos x)^2 \frac{1}{2} e^{-|y - \cos x|} \, dy \right) dx.$$

Substituting $z = y - \cos x$ and $dz = dy$,

$$= \int_{x=0}^{\infty} \alpha e^{-\alpha x} \underbrace{\left(\int_{z=-\infty}^{\infty} (z)^2 \frac{1}{2} e^{-|z|} \, dz \right)}_{=2 \text{ (variance of Laplacian r.v.)}} dx$$

$$= \int_{x=0}^{\infty} \alpha e^{-\alpha x} \, 2 \, dx = 2.$$

17. a. From the solution of Exercise 14.27 we obtain

$$P(Y = y) = e^{-(\lambda_0 + \lambda_1)} \frac{(\lambda_0 + \lambda_1)^y}{y!}.$$

b. We know the conditional probability of $X = x$ given $Y = y$. Again, using the solution of Exercise 14.27:

$$p_{X|Y}(x|y) = \begin{cases} \dfrac{(\lambda_1)^x (\lambda_0)^{y-x}}{(\lambda_1 + \lambda_0)^y} \dbinom{y}{x} & \text{if } x \leq y \\ 0 & \text{otherwise} \end{cases}.$$

c.

$$\hat{X}(Y) = E(X|Y) = \sum_{x=0}^{y} x p_{X|Y}(x|y) = 0 + \sum_{x=1}^{y} x \frac{(\lambda_1)^x (\lambda_0)^{y-x}}{(\lambda_1 + \lambda_0)^y} \binom{y}{x}$$

$$= \sum_{x=1}^{y} x \frac{(\lambda_1)^x (\lambda_0)^{y-x}}{(\lambda_1 + \lambda_0)^y} \frac{y!}{x!(y-x)!}$$

$$= \frac{1}{(\lambda_1 + \lambda_0)^y} \sum_{x=1}^{y} (\lambda_1)^x (\lambda_0)^{y-x} \frac{y(y-1)!}{(x-1)!(y-x)!},$$

after changing index to be $z = x - 1$,

$$= \frac{y}{(\lambda_1 + \lambda_0)^y} \sum_{z=0}^{y-1} (\lambda_1)^{z+1} (\lambda_0)^{y-z-1} \frac{(y-1)!}{z!(y-z-1)!}$$

$$= \frac{y\lambda_1}{(\lambda_1 + \lambda_0)^y} \sum_{z=0}^{y-1} (\lambda_1)^z (\lambda_0)^{(y-1)-z} \binom{y-1}{z} \quad \text{(by the Binomial Theorem)}$$

$$= \frac{y\lambda_1}{(\lambda_1 + \lambda_0)^y} (\lambda_1 + \lambda_0)^{y-1} = \frac{y\lambda_1}{\lambda_1 + \lambda_0}.$$

Although X must be an integer, its least mean square estimator $\hat{X}$ need not be. In the next chapter we consider estimators that respond to this issue.

$$\hat{X}(Y) = \frac{Y\lambda_1}{\lambda_1 + \lambda_0}.$$

d. The mean square error performance of $\hat{X}$ is given by

$$E(\hat{X}(Y) - X)^2 = \sum_{y=0}^{\infty} \sum_{x=0}^{y} \left(\frac{y\lambda_1}{\lambda_1 + \lambda_0} - x \right)^2 P(Y = y) p_{X|Y}(x|y)$$

$$= \sum_{y=0}^{\infty} \sum_{x=0}^{y} \left(\frac{y\lambda_1}{\lambda_1 + \lambda_0} - x \right)^2 \frac{e^{-(\lambda_0 + \lambda_1)}}{y!} (\lambda_1)^x (\lambda_0)^{y-x} \binom{y}{x}.$$

CHAPTER 16

1. a. The Bayes rule $d^*(x)$ assigns the action $a^* \in \mathcal{A}$ that minimizes the conditional risk $EL(\Theta, a)|X = x)$. In this exercise we are told that

$$EL(\Theta, a)|X = x) = |x - a| + x^2.$$

Hence, for each x we need only find the minimizing a. In this case $\mathcal{X} = \mathbb{R}$ but $\mathcal{A}$ is restricted to the nonnegative integers. Thus,

$$d^*(x) = \underset{a \in \mathcal{A}}{\mathrm{argmin}}[|x - a| + x^2] = \underset{a \in \mathcal{A}}{\mathrm{argmin}}[|x - a|] = \begin{cases} 0 & \text{if } x \le .5 \\ k \in \mathcal{A} & \text{if } |x - k| < 0.5, x > .5 \end{cases}.$$

b. If $X \sim \mathcal{U}(-.5, .5)$ then for all such values of X, $d^*(X) = 0$. Hence, the Bayes risk

$$r^* = EL(\Theta, d^*) = E\left[|X - 0| + X^2\right] = \int_{-.5}^{.5} [|x| + x^2] \, dx = \frac{1}{3}.$$

4. We need to first evaluate, but only for $x > 1$,

$$EL(\Theta, a)|X = x) = \int_{\theta} [(\theta - a)^2 + a + \theta] f_{\Theta|X}(\theta|x) \, d\theta.$$

$$f_{\Theta|X}(\theta|x) = \frac{f_{X|\Theta}(x|\theta)f_\Theta(\theta)}{f_X(x)} = \frac{\frac{1}{2}e^{-|x-\theta|}\frac{1}{2}}{\int_{-1}^{1}\frac{1}{4}e^{-|x-\theta|}\,d\theta}$$

$$= \frac{e^{-|x-\theta|}}{e^{-x}(e-e^{-1})},$$

when $x > 1$. Hence, for $x > 1$

$$f_{\Theta|X}(\theta|x) = \frac{e^{x-|x-\theta|}}{e-e^{-1}}.$$

As $\Theta \sim \mathcal{U}(-1, 1)$, this simplifies further to

$$f_{\Theta|X}(\theta|x) = \frac{e^\theta}{e-e^{-1}}U(1-|\theta|).$$

Hence, under the assumptions we have made,

$$E[L(\Theta, a)|X = x] = \int_{-1}^{1} \frac{e^\theta}{e-e^{-1}}[(\theta-a)^2 + a + \theta]\,d\theta$$

$$= a^2 + a + \int_{-1}^{1} \frac{e^\theta}{e-e^{-1}}[\theta^2 + (1-2a)\theta]\,d\theta$$

$$= a^2 + a + \frac{e - 5e^{-1} + (1-2a)2e^{-1}}{e-e^{-1}},$$

after some tedious integration. We now need to identify the minimizing value of $a \in \mathcal{A} = \mathbb{R}$. The expression for the conditional expected loss is simply a quadratic in a, and we determine that

$$d^*(x) = -\frac{1}{2} + \frac{2e^{-1}}{e+e^{-1}}.$$

Curiously, $d^*(x)$ does not depend upon x as a consequence of our assumptions that $x > 1$ and $|\Theta| \le 1$.

7. Let H_1 denote that the light source is present and H_0 denote that it is absent. Then,

$$P(X = k|H = H_1) = e^{-2}\frac{2^k}{k!}, \quad P(X = k|H = H_0) = e^{-1}\frac{1}{k!}, \quad P(H = H_0) = \frac{2}{3},$$

$$P(H = H_1) = \frac{1}{3}.$$

The loss $L(H, a)$ is 5 if $a \ne H$ and 3 if $a = H$. The analyses of Section 16.5.3 inform us that the optimal decision rule is again of the likelihood-ratio-threshold form with

$$\Lambda(k) = \frac{P(X = k|H = H_1)}{P(X = k|H = H_0)} = e^{-1}2^k, \quad \tau = \frac{P(H = H_0)}{P(H = H_1)}\frac{L_{01} - L_{00}}{L_{10} - L_{11}},$$

$$d^*(k) = \begin{cases} H_1 & \text{if } \Lambda(k) > \tau \\ H_0 & \text{otherwise} \end{cases} = \begin{cases} H_1 & \text{if } e^{-1}2^k > 2 \\ H_0 & \text{otherwise} \end{cases} = \begin{cases} H_1 & \text{if } k \ge 3 \\ H_0 & \text{otherwise} \end{cases}.$$

14. a. From the solution in Exercise 16.13 for the Bayes decision rule, given the absolute value loss function, as the median of the conditional cdf $F_{\Theta|X}$, we first need to determine

$$F_{\Theta|X}(\theta|x) = \int_{-\infty}^{x} f_{\Theta|X}(\theta|x)\,d\theta$$

and then solve for a such that

$$F_{\Theta|X}(a|x) = \frac{1}{2}.$$

$$f_{\Theta|X}(\theta|x) = \frac{f_{X|\Theta}f_{\Theta}}{f_X} = \frac{\theta e^{-\theta(x+1)}U(\theta)U(x)}{\int_0^{\infty}\theta e^{-\theta(x+1)}\,d\theta} = (1+x)^2\theta e^{-\theta(x+1)}U(\theta)U(x),$$

where we omit details of elementary integration. Hence, the desired

$$F_{\Theta|X}(\theta|x) = \int_0^{\theta}(1+x)^2 z e^{-(1+x)z}\,dz = [1-(2+x)e^{-(1+x)\theta}]U(\theta)U(x).$$

Setting

$$(2+x)e^{-(1+x)a} = \frac{1}{2}$$

yields for the conditional median

$$a = \frac{1}{1+x}\log[2(2+x)] = d^*(x) \in \mathbb{R}, \text{ for } x \geq 0,$$

the desired Bayes rule for this loss function and given probability information.

b. The MMSE

$$\hat{\Theta}(x) = E(\Theta|X=x) = \int_{-\infty}^{\infty}\theta f_{\Theta|X}(\theta|x)\,d\theta = \int_0^{\infty}\theta^2(1+x)^2 e^{-(1+x)\theta}\,d\theta = \frac{2}{1+x}.$$

c. Clearly, the two estimators do not agree. It matters to the end result what is the choice of loss function. Different loss functions can be expected to yield different estimators.

CHAPTER 17

2. a.

$$\phi_X(u) = \int_{-\infty}^{\infty}e^{iux}f_X(x)\,dx = \int_{-a}^{a}e^{iux}\frac{1}{2a}\,dx = \frac{1}{2aiu}[e^{iua}-e^{-iua}] = \frac{\sin ua}{ua}.$$

b.

$$\phi_X(u) = \int_{-\infty}^{0} e^{iux} \frac{\alpha}{2} e^{\alpha x} \, dx + \int_{0}^{\infty} e^{iux} \frac{\alpha}{2} e^{-\alpha x} \, dx$$

$$= \frac{\alpha}{2(iu + \alpha)} e^{x(iu+\alpha)} \Big|_{-\infty}^{0} + \frac{\alpha}{2(iu - \alpha)} e^{x(iu-\alpha)} \Big|_{0}^{\infty}$$

$$= \frac{\alpha}{2(iu + \alpha)} - \frac{\alpha}{2(iu - \alpha)} = \frac{\alpha^2}{u^2 + \alpha^2}.$$

3. a. Using definition of the Binomial pmf and of characteristic function,

$$\phi_X(u) = \sum_{k=0}^{n} P(X = k) e^{iuk} = \sum_{k=0}^{n} \binom{n}{k} \left(\frac{1}{2}\right)^n e^{iuk} = \left(\frac{1}{2}\right)^n \sum_{k=0}^{n} \binom{n}{k} 1^{n-k} e^{iuk}$$

$$= 2^{-n}(1 + e^{iu})^n.$$

5. Differentiating $\phi_X(u)$, we get

$$\phi_X'(u) = \frac{-2u}{(1 + u^2)^2},$$

and differentiating once more,

$$\phi_X''(u) = \frac{-2(1 + u^2)^2 + (2u)(2)(2u)(1 + u^2)}{(1 + u^2)^4} = \frac{6u^2 - 2}{(1 + u^2)^3}.$$

Then, we have

$$E(X) = -i\phi_X'(0) = 0$$

and

$$\text{VAR}(X) = E(X^2) - (E(X))^2 = E(X^2) = -\phi_X''(0) = 2.$$

16. a. By definition of characteristic function,

$$\phi_{S_n}(u) = E(e^{iuS_n}) = E(e^{iu \frac{1}{\sqrt{n}} \sum_{k=1}^{n} X_k}) = E\left(\prod_{k=1}^{n} e^{\frac{iuX_k}{\sqrt{n}}}\right),$$

and by independence and the definition of the characteristic function,

$$\phi_{S_n}(u) = \prod_{k=1}^{n} E(e^{\frac{iuX_k}{\sqrt{n}}}) = \prod_{k=1}^{n} \phi_{X_k}\left(\frac{u}{\sqrt{n}}\right) = \left(\phi_X\left(\frac{u}{\sqrt{n}}\right)\right)^n.$$

b.

$$\phi_{S_n - E(S_n)}(u) = E(e^{iu(S_n - E(S_n))}) = e^{-iuE(S_n)} E(e^{iuS_n}) = e^{-iuE(S_n)} \phi_{S_n}(u).$$

17. b. Since X and N are independent,

$$\phi_Y(u) = \phi_X(u)\phi_N(u).$$

As N follows Laplacian distribution with parameter 1,

$$\phi_N(u) = \frac{1}{1+u^2}.$$

So, as $\phi_N(u) > 0$,

$$\phi_X(u) = \frac{\phi_Y(u)}{\phi_N(u)} = \frac{\frac{e^{-u^2}}{(1+u^2)}}{\frac{1}{(1+u^2)}} = e^{-u^2}.$$

And this is the characteristic function of a normal random variable with mean 0 and variance 2, so,

$$f_X(x) = \frac{1}{\sqrt{4\pi}} e^{\frac{-x^2}{4}}.$$

37. a. By definition of characteristic function,

$$\phi_X(u) = Ee^{iuX} = \frac{1}{3} + \frac{2}{3}e^{iu}.$$

From the table in Section 17.2,

$$\phi_N(u) = e^{\lambda T(e^{iu}-1)}.$$

If we denote $S = \sum_{j=1}^{N} X_j$, then by Section 17.10,

$$\phi_S(u) = \phi_N(-i\log\phi_X(u)) = \exp\left(\lambda T(e^{\log(\frac{1}{3}+\frac{2}{3}e^{iu})} - 1)\right) = \exp\left(\frac{2}{3}\lambda T(e^{iu}-1)\right).$$

b.

$$E(I) = E\left(\frac{1}{T}\sum_{i=1}^{N}X_i\right) = \frac{1}{T}E\left(\sum_{i=1}^{N}X_i\right) = \frac{1}{T}E(S)$$

where, by Section 17.10,

$$E(S) = E(X)E(N) = \frac{2}{3}\lambda T.$$

Thus,

$$E(I) = \frac{1}{T}E(S) = \frac{2}{3}\lambda.$$

39. From the table in Section 17.2, we know:

$$\phi_N(u) = e^{2T(e^{iu}-1)}.$$

Let $B = \sum_{i=1}^{N} L_i$, and from Section 17.10,

$$\phi_B(u) = \phi_N(-i \log \phi_L(u)) = e^{2T(\phi_L(u)-1)} = \exp\left(2T\left(\frac{1}{n}\frac{e^{iun}-1}{e^{iu}-1} - 1\right)\right).$$

CHAPTER 18

2. By the basic quadratic Chebychev,

$$P\left(|r_A(n) - P(A)| \geq \epsilon n^{-\frac{1}{2}}\right) \leq \frac{nE(r_A(n) - P(A))^2}{\epsilon^2}.$$

$$E(r_A(n) - P(A))^2 = E\left(\frac{1}{n}\sum_{i=1}^{n}(I_A(X_i) - EI_A(X_i))\right)^2 = \frac{\mathrm{VAR}(I_A(X_1))}{n},$$

by the summands being $i.i.d.$ Note that

$$\mathrm{VAR}(I_A) = P(A)(1 - P(A)) = F_X(1)(1 - F_X(1)).$$

Hence,

$$P(|F_X(1) - r_A(n)| \geq \epsilon n^{-\frac{1}{2}}) \leq \frac{F_X(1)(1 - F_X(1))}{\epsilon^2}.$$

The factors of n cancelled.

5. Let

$$X_i = \begin{cases} 1 & \text{if the } i\text{th symbol is in error} \\ 0 & \text{otherwise} \end{cases}.$$

The $\{X_i\}$ are $i.i.d.$ We wish to bound $P\left(\sum_{i=1}^{100} X_i \geq 20\right)$. Since $P(X_i = 1) = .1$, we have that

$$EX_i = .1, \quad \mathrm{VAR}(X_i) = .1(1 - .1) = .09.$$

We use the basic quadratic Chebychev bound to write

$$P\left(\sum_{i=1}^{100} X_i \geq 20\right) = P\left(\sum_{i=1}^{100}(X_i - EX_i) \geq 10\right) \leq \frac{E\left(\sum_{i=1}^{100}(X_i - EX_i)\right)^2}{10^2} = \frac{100\,\mathrm{VAR}(X_1)}{100}$$

$$= \mathrm{VAR}(X_1) = .09.$$

We now turn to an exponential upper bound of the form

$$P\left(\sum_{i=1}^{100}(X_i - EX_i) \geq 10\right) \leq e^{-10\lambda}Ee^{\lambda\left(\sum_{i=1}^{100}(X_i-EX_i)\right)}.$$

By *i.i.d.*,

$$Ee^{\lambda\left(\sum_{i=1}^{100}(X_i-EX_i)\right)} = \left(Ee^{\lambda(X_1-.1)}\right)^{100},$$

$$Ee^{\lambda(X_1-.1)} = .1e^{.9\lambda} + .9e^{-.1\lambda}.$$

We now have an exponential upper bound

$$P\left(\sum_{i=1}^{100}(X_i - EX_i) \geq 10\right) \leq \left(.1e^{.9\lambda} + .9e^{-.1\lambda}\right)^{100}e^{-10\lambda}.$$

There remains the choice of $\lambda > 0$. Let B denote the upper bound and consider

$$\log B = -10\lambda + 100\log(.1e^{.9\lambda} + .9e^{-.1\lambda}).$$

To optimize the choice of λ,

$$\frac{d\log B}{d\lambda} = 0 = -10 + \frac{100}{.1e^{.9\lambda} + .9e^{-.1\lambda}}[.09e^{.9\lambda} - .09e^{-.1\lambda}].$$

Simplifying we find that

$$e^\lambda = \frac{9}{4}.$$

Hence,

$$P\left(\sum_{i=1}^{100}X_i \geq 20\right) \leq \left(\frac{9}{4}\right)^{-10}\left(.1\left(\frac{9}{4}\right)^{.9} + .9\left(\frac{9}{4}\right)^{-.1}\right)^{100} = .0118.$$

The optimized exponential bound gives us a lower, and therefore better, upper bound than does the quadratic Chebychev.

8. We have that $X \sim \Gamma(\alpha, \lambda)$ and are interested in an exponential upper bound to $P(X \geq \tau)$. For this we need the characteristic function

$$\phi_X(u) = \left(\frac{\lambda}{\lambda - iu}\right)^\alpha,$$

which can be thought of as being the characteristic function of α *i.i.d.* $\mathcal{E}(\lambda)$ random variables (even when α is not an integer). Hence, from the Chebychev exponential bound

$$P(X \geq \tau) \leq e^{-\theta\tau}\phi_X(-i\theta) = e^{-\theta\tau}\left(\frac{\lambda}{\lambda - \theta}\right)^\alpha.$$

Note that this requires that $\theta < \lambda$.

To optimize the choice of $0 < \theta < \lambda$, let B denote the upper bound and consider

$$\log B = -\theta\tau + \alpha \log \lambda - \alpha \log(\lambda - \theta),$$

For an optimum choice of θ

$$\frac{d \log B}{d\theta} = -\tau + \frac{\alpha}{\lambda - \theta} = 0.$$

Hence,

$$\theta = \lambda - \frac{\alpha}{\tau},$$

and

$$P(X \geq \tau) \leq e^{-(\lambda\tau - \alpha)} \left(\frac{\lambda\tau}{\alpha}\right)^{\alpha}.$$

17. We are given $\{X_1, \ldots, X_n\}$ $i.i.d.$ as $\beta(3, 4)$ and will determine a Hoeffding upper bound to

$$P\left(\left|\sum_{i=1}^{n}(X_i - EX_i)\right| \geq \epsilon\right) \leq 2e^{-\frac{2\epsilon^2}{\sum_1^n (b_i - a_i)^2}}.$$

The constants b_i and a_i are determined by

$$P(a_i \leq X_i - EX_i \leq b_i) = 1.$$

For the case of $X \sim \beta(3, 4)$ we would need to evaluate EX to determine a and b. However, if you examine the Hoeffding bound you see that it depends upon a and b only through their difference $b - a$. This difference is determined by the range of variation of X and not at all by EX. Hence, for $X \sim \beta$, $b - a = 1 - 0 = 1$. The Hoeffding bound is

$$P\left(\left|\sum_{i=1}^{n}(X_i - EX_i)\right| \geq \epsilon\right) \leq 2e^{-\frac{2\epsilon^2}{n}}.$$

Clearly, this bound is only of interest if it is (substantially) less than one. Hence, it is only of interest if ϵ^2/n is large compared to one.

CHAPTER 19

1. Given integers $m < q$ and $p = m/q$, we construct a $\{0, 1\}$-valued binary sequence $\{X_i\}$ satisfying

$$r_n(1) = \frac{1}{n}\sum_{i=1}^{n}X_i, \quad \lim_{n\to\infty} r_n(1) = p.$$

Choose, say,

$$X_i = \begin{cases} 1 & \text{if } (\exists k \geq 0) \; kq \leq i < kq + m \\ 0 & \text{otherwise} \end{cases}.$$

(Alternatively, we could have chosen at random m locations for ones in each of the consecutive blocks of length q, but we do not do so.) We now verify that this choice works. Say, $n = jq + r$ with the remainder $0 \leq r < q$. Then

$$jm \leq \sum_{i=1}^{n} X_i \leq jm + m = (j + 1)m.$$

Therefore, an upper bound is

$$\frac{1}{n} \sum_{i=1}^{n} X_i \leq \frac{(j + 1)m}{jq + 0} = \frac{j + 1}{j} \frac{m}{q}.$$

Hence, as $j \to \infty$ as $n \to \infty$,

$$\limsup_{n \to \infty} \frac{1}{n} \sum_{i=1}^{n} X_i \leq \frac{m}{q} = p.$$

Correspondingly,

$$\frac{1}{n} \sum_{i=1}^{n} X_i \geq \frac{jm}{(j + 1)q},$$

$$\liminf_{n \to \infty} \frac{1}{n} \sum_{i=1}^{n} X_i \geq \frac{m}{q} = p.$$

We conclude that

$$\lim_{n \to \infty} \frac{1}{n} \sum_{i=1}^{n} X_i$$

exists and equals p, as desired. We conclude that any rational number in the unit interval can be a limiting relative frequency for a binary-valued sequence. This is generalized in Exercise 19.2.

4. a. We are given that $X_n = a_n Z$ for Z an almost surely (with probability one) finite random variable. If Z is 0 with probability one, then X_n trivially converges almost surely to 0 with probability one. More generally, we assert that if $\{a_n\}$ converges to a finite value a, then $X_n \to aZ$ almost surely. In this case,

$$\{\omega : X_n(\omega) = a_n Z(\omega) \to X(\omega) = aZ(\omega)\} = \{\omega : Z(\omega) \text{ finite }\},$$

and convergence is with probability one or almost surely since Z is almost surely finite.

b. For convergence in mean square, we use the mutual convergence condition

$$E[a_n Z - a_m Z]^2 = (a_n - a_m)^2 EZ^2 \to 0,$$

irrespective of the choices of $\{n\}$ and $\{m\}$, each tending to infinity. In a trivial case of $P(Z = 0) = 1$ then $EZ^2 = 0$, and we have convergence to 0 no matter what the choice for the sequence $\{a_n\}$. Convergence in mean square of X_n to $X = aZ$ is assured if a_n converges to some finite a and EZ^2 is finite. If EZ^2 is infinite, then there is no mean square convergence for any choice of sequence $\{a_n\}$.

9. We are given that

$$X_n(\omega) = \begin{cases} \sqrt{n} & \text{if } n \le Z(\omega) \le n + \sqrt{n} \\ Z(\omega) & \text{otherwise} \end{cases}, \quad \Omega = \mathbb{R}, \ f_Z(z) = \frac{1}{2} e^{-|z|}.$$

a. Following the hint, to show that $X_n(\omega) \to Z(\omega)$ almost surely we need to use the first half (finitely often case) of the Borel–Cantelli Lemma that is the subject of Section 20.3. This part of the lemma could have been presented as early as Chapter 3. We will show that X_n is unequal to the candidate limit Z only finitely often with probability one. Let

$$A_n = \{\omega : X_n(\omega) \ne Z(\omega)\} = \{\omega : X_n(\omega) = \sqrt{n}\} = \{\omega : n \le Z(\omega) \le n + \sqrt{n}\}.$$

$$P(A_n) = \int_n^{n+\sqrt{n}} \frac{1}{2} e^{-|z|} \, dz = \frac{1}{2}[e^{-n} - e^{-n-\sqrt{n}}] < e^{-n}.$$

Therefore,

$$\sum_{n=1}^{\infty} P(A_n) < \sum_{n=1}^{\infty} e^{-n} = \frac{e^{-1}}{1 - e^{-1}} < \infty.$$

Invoking the Borel–Cantelli Lemma, we see that we have shown that $\{A_n\}$ occurs only finitely often with probability one or that $X_n \ne Z$ occurs only finitely often with probability one. Hence, $X_n \to Z$ almost surely.

b. To examine convergence in mean square, with X_n defined in terms of Z, we evaluate

$$E(X_n - Z)^2 = \int_{-\infty}^{\infty} \frac{1}{2} e^{-|z|} (X_n^{(z)} - z)^2 \, dz = \frac{1}{2} \int_n^{n+\sqrt{n}} e^{-|z|} (\sqrt{n} - z)^2 \, dz.$$

Using the integration facts that

$$\int_a^b y e^{-y} \, dy = (a+1)e^{-a} - (b+1)e^{-b} \ \text{and} \ \int_a^b y^2 e^{-y} \, dy$$

$$= (a^2 + 2a + 2)e^{-a} - (b^2 + 2b + 2)e^{-b}$$

yields

$$E(X_n - Z)^2 = \frac{1}{2}e^{-\sqrt{n}}\left((n - \sqrt{n})^2 + 2(n - \sqrt{n}) + 2\right)e^{-n+\sqrt{n}}$$

$$- \frac{1}{2}e^{-\sqrt{n}}\left(n^2 + 2n + 2\right)e^{-n} = O(e^{-n}).$$

Observe that this expression decays as order of e^{-n} to conclude that

$$E(X_n - Z)^2 \to 0,$$

as desired to establish $X_n \xrightarrow{ms} Z$.

c. For convergence in probability $X_n \xrightarrow{p} Z$ we need to establish that for any $\epsilon > 0$,

$$\lim_{n\to\infty} P(|X_n - Z| > \epsilon) = 0.$$

In our case, we can establish something even stronger,

$$\lim_{n\to\infty} P(X_n \neq Z) = 0.$$

$$P(X_n \neq Z) = P(n \leq Z \leq n + \sqrt{n}) = \int_n^{n+\sqrt{n}} \frac{1}{2}e^{-|z|}\,dz$$

$$= \frac{1}{2}[e^{-n} - e^{-n-\sqrt{n}}] < e^{-n} \to 0.$$

Hence, $X_n \xrightarrow{p} Z$, as desired.

d. To establish convergence in distribution of X_n to Z, $X_n \xrightarrow{D} Z$, we must show that at all points of continuity of the cdfs,

$$\lim_{n\to\infty} F_{X_n}(z) = F_Z(z).$$

We note first that

$$\{\omega : X_n(\omega) \leq z\} \subset \{\omega : Z(\omega) \leq z\} \cup \{\omega : X_n(\omega) \neq Z(\omega)\}.$$

Hence, by the union bound,

$$P(X_n \leq z) = F_{X_n}(z) \leq F_Z(z) + P(X_n \neq Z).$$

Interchanging X_n and Z yields

$$|F_{X_n}(z) - F_Z(z)| \leq P(X_n \neq Z).$$

However, we already know from **(c)** that $P(X_n \neq Z) \to 0$. Hence, we have shown that $F_{X_n}(z) - F_Z(z) \to 0$ and $X_n \xrightarrow{D} Z$, as desired.

CHAPTER 20

2. In parts (a)–(d), we are asked to evaluate the probability of an event A that is defined in terms of infinitely many $i.i.d.$ random variables $X_0, X_1, \ldots$. Our strategy is to approximate such an event by ones $B^{(n)}$ defined in terms of only n random variables, with these new events so chosen that, by Kolmogorov's monotone continuity axiom K4, $P(B^{(n)}) \to P(A)$. In this exercise, $P(X_t = 1) = p = 1 - P(X_t = 0)$.

   a. In this case we are interested in $P(A)$ for $A = \cap_{t=0}^{\infty}(X_t = 0)$. Let $B^{(n)} = \cap_{t=0}^{n}(X_t = 0)$. By $i.i.d.$,

$$P(B^{(n)}) = (1 - p)^n.$$

Hence,

$$P(A) = \lim_{n \to \infty} P(B^{(n)}) = 0.$$

   b. Let A_k be the event that no more than k ones will appear in the infinite sequence of random variables. Let B_j be the event that exactly j ones will appear in the infinite sequence,

$$A_k = \cup_{j=0}^{k} B_j, \ \ P(A_k) = \sum_{j=0}^{k} P(B_j).$$

Consider the approximation $B_j^{(n)}$ to B_j that is the event that exactly j ones occur in the first n random variables $X_0, \ldots, X_{n-1}$. From the binomial,

$$P(B_j^{(n)}) = \binom{n}{j} p^j (1 - p)^{n-j}.$$

By axiom K4,

$$\lim_{n \to \infty} P(B_j^{(n)}) = P(B_j).$$

We verify that this limit is zero by considering the upper bound

$$P(B_j^{(n)}) \le \frac{n^j}{j!} \frac{p^j}{(1 - p)^j} (1 - p)^n.$$

The value j is fixed and

$$\lim_{n \to \infty} n^j (1 - p)^n = 0;$$

the factor n^j grows as a polynomial but the factor $(1 - p)^n$ decays to zero exponentially fast. Assembling the pieces, we see that each $P(B_j) = 0$ and therefore that $P(A_k) = 0$.

c. Fix some t_0 and let A_{t_0} be the event that there exists a $t > t_0$ such that $X_t = 1$,

$$A_{t_0} = \cup_{t > t_0}(X_t = 1).$$

It will be easier given independence to consider the complementary event

$$A_{t_0}^c = \cap_{t > t_0}(X_t = 0).$$

Approximate this complementary event by

$$B_{t_0}^{(n)} = \cap_{n \geq t > t_0}(X_t = 0).$$

By independence, $P(B_{t_0}^{(n)}) = (1 - p)^{n - t_0}$. By axiom K4,

$$P(A_{t_0}^c) = \lim_{n \to \infty} P(B_{t_0}^{(n)}) = 0.$$

Hence, we conclude that $P(A_{t_0}) = 1$.

d. Let A be the probability that we will observe infinitely many ones. A^c is the event that we observe finitely many ones and can be represented as

$$A^c = \cup_{k=0}^{\infty} B_k,$$

where, as in part (b), B_k is the event of exactly k ones in the infinite sequence. Furthermore, as we saw in part (b),

$$P(B_k) = 0.$$

Hence,

$$P(A^c) = \sum_{k=0}^{\infty} P(B_k) = 0 \Rightarrow P(A) = 1.$$

e. From the Chebychev inequality for these $i.i.d.$ random variables with $EX_t = p$,

$$P\left(\left|\frac{1}{n}\sum_{t=0}^{n-1} X_t - p\right| \geq \epsilon\right) \leq \frac{\text{VAR}(X_0)}{n\epsilon^2}.$$

The limit as $n \to \infty$ of the upper bound is zero, hence

$$\lim_{n \to \infty} P\left(\left|\frac{1}{n}\sum_{t=0}^{n-1} X_t - p\right| < \epsilon\right) = 1.$$

5. c. X_t is given as a GRP on $T = \mathbb{R}$ with mean $EX_t = t$ and stationary covariance function. As the mean is time varying, the GRP is nonstationary and therefore not ergodic (time averages need not equal ensemble or statistical averages). If, instead, $EX_t = m$, constant,

then the given GRP is stationary, and by the theorem of Section 20.11.1 it will be ergodic if the covariance function is absolutely integrable,

$$\int_{-\infty}^{\infty} e^{-|\tau|} \, d\tau = 2 \int_{0}^{\infty} e^{-\tau} \, d\tau = 2 < \infty.$$

Hence, in this modified case, X_t is ergodic.

6. a. We are given that X_t is stationary and ergodic for $T = [0, \infty)$ and wish to estimate EX_5^2 from the observed sample function X_t. By stationarity, $EX_5^2 = EX_t^2$. An estimator is

$$\frac{1}{a} \int_{0}^{a} X_t^2 \, dt, \ E\left(\frac{1}{a} \int_{0}^{a} X_t^2 \, dt\right) = EX_0^2 = EX_5^2,$$

and is such that

$$\lim_{a \to \infty} \frac{1}{a} \int_{0}^{a} X_t^2 \, dt = EX_5^2 \ a.s.$$

7. For a Markov chain, the probability of a transition in one step from a state i to a state j is given by p_{ij}. In this exercise, $p_{00} = .9$ and $p_{11} = .05$.
   a. The one-step transition matrix (in which the row sums are one)

$$\mathbb{P} = \begin{pmatrix} .9 & .1 \\ .95 & .05 \end{pmatrix}$$

   b. The vector $\pi^{(t)}$ specifies the probabilities of the chain being in each of the states at time t. The limiting distribution exists when the Markov chain is homogeneous, irreducible, and aperiodic, which this chain is. In this case

$$\lim_{t \to \infty} \pi^{(t)} = \pi,$$

and π satisfies

$$\pi \mathbb{P} = \pi = \begin{pmatrix} \pi_0 & \pi_1 \end{pmatrix} \begin{pmatrix} .9 & .1 \\ .95 & .05 \end{pmatrix}.$$

We have the equations

$$\pi_0 + \pi_1 = 1$$

(since π is a probability distribution over the states) and

$$\pi_0 = .9\pi_0 + .95\pi_1.$$

The solution is

$$\lim_{t \to \infty} \pi^{(t)} = \begin{pmatrix} \dfrac{19}{21} & \dfrac{2}{21} \end{pmatrix}.$$

c. From Section 20.6.5, the mean waiting time $T_{j,j}$ to return to state j is given by the reciprocal of the asymptotic probability of being in that state, $1/\pi_j$. Hence, the expected waiting time between successive ones is

$$T_{1,1} = \frac{1}{\pi_1} = \frac{21}{2}.$$

This answer is consistent with the low probability of .05 of returning to one in one step.

d. Given that we start in state zero, the probability that the first one will not appear in the next k symbols (transitions) is simply that we make k successive transitions from state zero to state zero, and is given by

$$(p_{0,0})^k = .9^k.$$

8. a. If in the preceding exercise the initial state distribution is

$$\pi^{(1)} = \begin{pmatrix} .7 & .3 \end{pmatrix},$$

then

$$\pi^{(2)} = \pi^{(1)}\mathbb{P} = \begin{pmatrix} .7 & .3 \end{pmatrix} \begin{pmatrix} .9 & .1 \\ .95 & .05 \end{pmatrix} = \begin{pmatrix} .915 & .085 \end{pmatrix}.$$

Calculation of

$$\pi^{(5)} = \pi^{(2)}\mathbb{P}^3$$

is best done with computational resources such as those provided by Matlab or Mathematica.

14. $\{N_t, t \geq 0\}$ is a Poisson process (one having independent increments) with a mean rate of jumps of $\lambda = 5$ per unit time. We wish to evaluate the trivariate probability $P(N_{t_3} = n_3, N_{t_2} = n_2, N_{t_1} = n_1)$ for $t_3 > t_2 > t_1$. As a Poisson process has nondecreasing sample functions, this probability is zero unless $n_3 \geq n_2 \geq n_1$. We now assume this to be the case. We express this probability in terms of conditional probabilities as

$$P(N_{t_3} = n_3, N_{t_2} = n_2, N_{t_1} = n_1)$$
$$= P(N_{t_3} = n_3 | N_{t_2} = n_2, N_{t_1} = n_1)P(N_{t_2} = n_2 | N_{t_1} = n_1)P(N_{t_1} = n_1).$$

By the Poisson-distributed, independent increments property of the Poisson process and the Poisson distribution of increments,

$$P(N_{t_3} = n_3 | N_{t_2} = n_2, N_{t_1} = n_1) = P(N_{t_3} - N_{t_2} = n_3 - n_2) = e^{-\lambda(t_3-t_2)}\frac{(\lambda(t_3 - t_2))^{n_3-n_2}}{(n_3 - n_2)!},$$

$$P(N_{t_2} = n_2 | N_{t_1} = n_1) = P(N_{t_2} - N_{t_1} = n_2 - n_1) = e^{-\lambda(t_2-t_1)}\frac{(\lambda(t_2 - t_1))^{n_2-n_1}}{(n_2 - n_1)!},$$

$$P(N_{t_1} = n_1) = e^{-\lambda t_1}\frac{(\lambda t_1)^{n_1}}{n_1!}.$$

Taking a product of these results yields the desired evaluation

$$P(N_{t_3} = n_3, N_{t_2} = n_2, N_{t_1} = n_1) = e^{-\lambda t_3} \frac{\lambda^{n_3} (t_3 - t_2)^{n_3 - n_2} (t_2 - t_1)^{n_2 - n_1} t_1^{n_1}}{(n_3 - n_2)!(n_2 - n_1)!n_1!}.$$

18. $\{X_n, n \in N\}$ is a martingale. For $m > k \geq j$,

$$E\left(X_j (X_m - X_k)\right) = E\left[E\left(X_j (X_m - X_k)|X_0, \ldots, X_k\right)\right],$$

$$E\left(X_j (X_m - X_k)|X_0, \ldots, X_k\right) = X_j (E(X_m|X_0, \ldots, X_k) - X_k) = X_j (X_k - X_k) = 0,$$

by the martingale property. Hence, $E\left(X_j (X_m - X_k)\right) = 0$.

26. X_t is a GRP with $EX_t = 0$ and $R_X(t, s) = \sigma^2 \delta(t - s)$. Hence, X_t is a $w.s.s.$ white or thermal noise GRP. We define

$$Y_t = \int_0^t X_s \, ds$$

and note that it is also a GRP whose complete description requires EY_t and $R_Y(t, s)$.

a.

$$EY_t = E \int_0^t X_s \, ds = \int_0^t EX_s \, ds = 0.$$

$$R_Y(t, s) = EY_t Y_s = E \int_0^t X_u \, du \int_0^s X_v \, dv = \int_0^t du \int_0^s dv R_X(u, v)$$

$$= \sigma^2 \int_0^t du \int_0^s dv \delta(u - v).$$

In order to evaluate this last double integral we break up the range (assume $t \geq s$)

$$\int_0^t du \int_0^s dv \delta(u - v) = \int_0^s du \int_0^s dv \delta(u - v) + \int_s^t du \int_0^s dv \delta(u - v).$$

In the second integral there is no overlap in the values u and v. Hence,

$$\int_s^t du \int_0^s dv \delta(u - v) = 0.$$

In the first integral

$$\int_0^s du \int_0^s dv \delta(u - v) = \int_0^s du = s.$$

We conclude that

$$R_Y(t, s) = \sigma^2 \min(t, s).$$

b. For a Brownian motion we have a GRP with $EY_t = 0$ and $C_Y(t, s) = \min(t, s)$. As the mean is 0, the correlation and covariance are the same. We see that these are indeed the properties of the process Y_t defined above, and it is a Brownian motion.

28. We are given that X_t is a $w.s.s.$ process with autocorrelation $R_X(\tau)$ and power spectral density $S_X(\omega)$ related to the autocorrelation through

$$S_X(\omega) = \int_{-\infty}^{\infty} R_X(\tau)e^{-i\omega\tau} \, d\tau.$$

Assume that $EX_t = m$.

a. Define $Y_t = aX_t + b$.

$$R_Y(t, s) = EY_tY_s = a^2EX_tX_s + abEX_t + abEX_s + b^2 = a^2R_X(t - s) + 2abm + b^2,$$

or

$$R_Y(\tau) = a^2R_X(\tau) + c.$$

Correspondingly,

$$S_Y(\omega) = \int_{-\infty}^{\infty} [a^2R_X(\tau) + c]e^{-i\omega\tau} \, d\tau = a^2S_X(\omega) + 2\pi c\delta(\omega).$$

b. For fixed t_0, define $Y_t = X_{t-t_0}$.

$$R_Y(t, s) = EX_{t-t_0}X_{s-t_0} = R_X(t - s),$$

or $R_Y(\tau) = R_X(\tau)$. Hence, also $S_Y(\omega) = S_X(\omega)$.

c.

$$Y_t = \frac{dX_t}{dt} \Rightarrow R_Y(t, s) = E\frac{dX_t}{dt}\frac{dX_s}{ds} = \frac{\partial^2 R_X(t - s)}{\partial t \partial s}.$$

Hence,

$$R_Y(\tau) = -R_X''(\tau).$$

As differentiation with respect to τ corresponds to multiplication by $i\omega$ in the frequency domain,

$$S_Y(\omega) = |i\omega|^2 S_X(\omega) = \omega^2 S_X(\omega).$$

Note that the key properties of a psd are preserved, as they should be.

d. Now we have

$$Y_t = X_t \cos(\omega_c t + \Theta), \quad \Theta \sim \mathcal{U}(-\pi, \pi),$$

and the random phase angle Θ is independent of the modulating X_t process.

$$R_Y(t, s) = E[X_t \cos(\omega_c t + \Theta)X_s \cos(\omega_c s + \Theta)]$$

$$= R_X(t, s)E[\cos(\omega_c t + \Theta) \cos(\omega_c s + \Theta)],$$

by the independence of X_t and Θ. Recall the trigonometric identity

$$2\cos(A)\cos(B) = \cos(A+B) + \cos(A-B),$$

to write

$$2E[\cos(\omega_c t + \Theta)\cos(\omega_c s + \Theta)] = E\cos(\omega_c(t+s) + 2\Theta) + E\cos(\omega_c(t-s)).$$

The second term is simply $\cos(\omega_c(t-s))$. That the first term is zero is seen from

$$E\cos(\omega_c(t+s) + 2\Theta) = \int_{-\pi}^{\pi} \cos(\omega_c(t+s) + 2\theta)\, d\theta = 0.$$

Hence,

$$R_Y(\tau) = \frac{1}{2}R_X(\tau)\cos(\omega_c\tau).$$

Using the complex exponential representation for cosine, we have

$$S_Y(\omega) = \frac{1}{4}\int_{-\infty}^{\infty} R_X(\tau)[e^{i\omega_c\tau} + e^{-i\omega_c\tau}]e^{-i\omega\tau}\, d\tau.$$

Hence,

$$S_Y(\omega) = \frac{1}{4}[S_X(\omega - \omega_c) + S_X(\omega + \omega_c)].$$

38. We are given that $X_t, t \geq 0$ is a GRP with $EX_t = \sin(t)$ and $C_X(t,s) = \min(t,s)$. We are asked to define the process

$$Y_t = \int_0^t X_s\, ds.$$

As Y_t is an integral of a GRP, it is also a GRP. Hence, a complete description requires only EY_t and $R_Y(t,s)$ or $C_Y(t,s)$.

$$EY_t = E\int_0^t X_s\, ds = \int_0^t EX_s\, ds = \int_0^t \sin(s)\, ds = 1 - \cos(t).$$

$$C_Y(t,s) = E\left[\int_0^t (X_u - EX_u)\, du \int_0^s (X_v - EX_v)\, dv\right]$$

$$= \int_0^t du \int_0^s dv\, C_X(u,v) = \int_0^t du \int_0^s dv\, \min(u,v).$$

We break up this integral into two parts, assuming $t \geq s$,

$$\int_0^t du \int_0^s dv\, \min(u,v) = \int_0^s du \int_0^s dv\, \min(u,v) + \int_s^t du \int_0^s \min(u,v)\, dv.$$

The first part, by symmetry, evaluates as

$$\int_0^s du \int_0^s dv \min(u, v) = 2 \int_0^s du \int_0^u \min(u, v)\, dv = 2 \int_0^s du \int_0^u v\, dv = \int_0^s u^2\, du = \frac{1}{3}s^3.$$

The second part evaluates as

$$\int_s^t du \int_0^s \min(u, v)\, dv = \int_s^t du \int_0^s v\, dv = \int_s^t \frac{1}{2}s^2\, du = \frac{1}{2}(t - s)s^2.$$

Assembling results yields, for $t \geq s$,

$$C_Y(t, s) = \frac{1}{3}s^3 + \frac{1}{2}(t - s)s^2.$$

Index